# Engineering Plasticity and Its Applications

From Nanoscale to Macroscale

# Engineering Plasticity and Its Applications

## From Nanoscale to Macroscale

Proceedings of the 9th AEPA 2008
Daejeon, Korea     20 – 24 October 2008

*editors*

**Hoon Huh**
*Korea Advanced Institute of Science & Technology, Korea*

**C G Park**
*Pohang University of Science & Technology, Korea*

**C S Lee**
*Pohang University of Science & Technology, Korea*

**Y T Keum**
*Hanyang University, Korea*

NEW JERSEY · LONDON · SINGAPORE · BEIJING · SHANGHAI · HONG KONG · TAIPEI · CHENNAI

*Published by*

World Scientific Publishing Co. Pte. Ltd.

5 Toh Tuck Link, Singapore 596224

*USA office:* 27 Warren Street, Suite 401-402, Hackensack, NJ 07601

*UK office:* 57 Shelton Street, Covent Garden, London WC2H 9HE

**British Library Cataloguing-in-Publication Data**
A catalogue record for this book is available from the British Library.

**ENGINEERING PLASTICITY AND ITS APPLICATIONS NANOSCALE TO MACROSCALE**
**(With CD-ROM)**
**Proceedings of the 9th AEPA2008**

ISBN-13 978-981-4261-56-2
ISBN-10 981-4261-56-4

Printed by FuIsland Offset Printing (S) Pte Ltd, Singapore

# PREFACE

This Book Volume includes selected full papers for the Proceedings of the Ninth Asia-Pacific Conference on Engineering Plasticity and Its Applications, AEPA2008, which was held in Daejeon, Korea on the 20th to 24th of October, 2008. The Asia-Pacific Conference on Engineering Plasticity and Its Applications has been very successful with its exotic and extensive research topics as an open forum to exchange ideas and introduce the latest research findings in the broad disciplines of plasticity including solid mechanics, material science, metal forming, structural analysis as well as physics and chemistry.

During the Ninth Asia-Pacific Conference on Engineering Plasticity and Its Applications, AEPA2008, following the previous successful conferences in Hong Kong (1992), Beijing (1994), Hiroshima (1996), Seoul (1998), Hong Kong (2000), Sydney (2002), and Shanghai (2004), Nagoya (2006), 149 full papers were presented in oral and poster sessions. The papers have been selected from submission of over 200 abstracts and then from 186 full papers for a broad range of engineering plasticity such as Constitutive modeling, Damage, fracture, fatigue and failure, Dynamic loading and crash dynamics, Experimental and numerical techniques, Molecular dynamics, Nano, meso, micro and crystal plasticity, Phase transformations, Plasticity in advanced materials, Plasticity in materials processing technology. Among the papers presented, 138 full papers have been selected as a special issue of the International Journal of Modern Physics B after peer review by two experts in the international scientific committee and the organizing committee for each paper.

The organizing committee wishes to express their hearty appreciation to authors, reviewers and participants as well as plenary lecturers and keynote lecturers for their continual support and contribution for the success of the AEPA2008. Sincere thanks should also go to all members of the international steering committee, the international scientific committee and the organizing committee for their devoted effort to make the AEPA2008 a great success. Many thanks should go to all supporters of Daejeon Convention Center, Sejong Convention Services and students at KAIST for their elaborated devotion. On behalf of the organizing committee, I would like to thank all the generous sponsors for their financial support which enabled the organizing committee to prepare and organize the conference and the publication of the proceedings and the special issue. Special thanks are due to the editors at World Scientific Publishing Co. Pte. Ltd. for their kind support to this special issue and proceedings volume. Thanks to all of them, the AEPA could make significant contribution to the progress and innovation of engineering plasticity and related fields.

Hoon Huh

Organizing Chairman, the AEPA2008, Daejeon, Korea

# AEPA2008 CONFERENCE ORGANIZERS

**Conference Chair**

Prof. Hoon Huh (Korea Advanced Institute for Science and Technology, Korea)

**Conference Co-Chair**

Prof. Chan Gyung Park (Pohang University of Science and Technology, Korea)
Prof. Young Tak Keum (Hanyang University, Korea)

**General Secretary**

Prof. Hoon Young Kim (Kangwon National University, Korea)

**International Steering Committee**

Prof. N. Ohno (Nagoya University, Japan)
Prof. T. Abe (Tsuyama National College of Technology, Japan)
Prof. D. N. Lee (Seoul National University, Pohang University of Science and Technology, Korea)
Prof. W. B. Lee (The Hong Kong Polytechnic University, Hong Kong)
Prof. S. I. Oh (Seoul National University, Korea)
Prof. M. Tokuda (Mie University, Japan)
Prof. Y. Tomita (Kobe University, Japan)
Prof. B. Y. Xu (Tsinghua University, China)
Prof. J. Q. Xu (Shanghai Jiaotong University, China)
Prof. W. H. Yang (The University of Michigan, USA)
Prof. T. X. Yu (The Hong Kong University of Science and Technology, Hong Kong)
Prof. L. C. Zhang (The University of Sydney, Australia)
Prof. H. Huh (Korea Advanced Institute for Science and Technology, Korea)
Prof. C. G. Park (Pohang University of Science and Technology, Korea)
Prof. Y. T. Keum (Hanyang University, Korea)

**Program Committee**

Prof. J. H. Lee (Chair, Korea Institute of Materials Science, Korea)
Prof. H. S. Kim (Pohang University of Science and Technology, Korea)
Prof. B. M. Kim (Pusan National University, Korea)
Prof. B. S. Kang (Pusan National University, Korea)
Prof. M. S. Joun (Gyeongsang National University, Korea)
Prof. H. T. Jeong (Kangnung National University, Korea)
Dr. M. E. Kim (Daerim MTI, Korea)

Dr. M. C. Kang (Korea Magnesium Technology Research Association, Korea)
Dr. B. Y. Kim (Ki Sung Highest, Korea)
Dr. S. W. Choi (Korea Institute of Industrial Technology, Korea)
Dr. H. J. Park (Korea Institute of Industrial Technology, Korea)
Dr. Y. S. Lee (Korea Institute of Materials Science, Korea)
Dr. Y. N. Kwon (Korea Institute of Materials Science, Korea)

## Scientific Committee

Prof. C. S. Lee (Chair, Pohang University of Science and Technology, Korea)
Prof. Y. S. Lee (Kookmin University, Korea)
Prof. I. S. Kim (Kumoh National Institute of Technology, Korea)
Prof. H. S. Kim (Pohang University of Science and Technology, Korea)
Prof. W. J. Nam (Kookmin University, Korea)
Prof. W. J. Chung (Seoul National University of Technology, Korea)
Dr. M. H. Seo (POSCO, Korea)
Prof. K. A. Lee (Andong National University, Korea)
Dr. M. G. Lee (Korea Institute of Materials Science, Korea)
Dr. Y. S. Lee (Korea Institute of Materials Science, Korea)
Prof. T. K. Ha (Kangnung National University, Korea)
Dr. S. S. Han (Kumoh National Institute of Technology, Korea)
Prof. H. N. Han (Seoul National University, Korea)
Prof. B. B. Hwang (Inha University, Korea)

## Financial Committee

Prof. W. J. Chung (Chair, Seoul National University of Technology, Korea)
Prof. I. S. Kim (Kumoh National Institute of Technology, Korea)

## Organizing Committee

Dr. J. J. Yi (Research Institute of Industrial Science & Technology, Korea)
Prof. J. J. Park (Hongik University, Korea)
Prof. Y. H. Kim (Chungnam National University, Korea)
Prof. Y. S. Kim (Kookmin University, Korea)
Prof. W. J. Nam (Kookmin University, Korea)
Prof. S. I. Kang (Yonsei University, Korea)
Dr. S. J. Kim (Korea Institute of Materials Science, Korea)
Prof. Y. S. Kim (Kyungpook National University, Korea)
Prof. Y. H. Moon (Pusan National University, Korea)
Dr. M. Y. Lee (Sung Woo Hitech, Korea)
Dr. S. H. Choi (Kwang Ho Precision Co., Korea)
Dr. S. T. Choi (Korea)
Dr. Y. D. Hahn (Korea Institute of Machinery and Materials, Korea)

Dr. K. H. Hwang (Taeyang Metal Industrial Co., Ltd., Korea)
Dr. B. C. Koh (Samsung Electro-Mechanics, Korea)
Dr. J. T. Kim (Doosan Heavy Industries & Construction Co., Ltd., Korea)
Dr. S. M. Na (HYUNDAI HYSCO, Korea)
Dr. J. S. Park (GM Daewoo Auto & Technology, Korea)
Dr. J. C. Bae (Korea Institute of Industrial Technology, Korea)
Dr. H. Y. Chung (ENERGREEN Inc., Korea)
Dr. S. S. Hong (Agency for Defense Development, Korea)
Dr. J. S. Hwang (Hyundai-Motor Company, Korea)

## International Scientific Committee

Prof. T. Abe (Tsuyama National College of Technology, Japan)
Prof. H. Altenbach (Martin-Luther University, Germany)
Prof. D. Banabic (Technical University of Cluj-Napoca, Romania)
Dr. S. Berveiller (Ecole Nationale Supérieure d'Arts et Métiers, France)
Prof. O. T. Bruhns (Ruhr-Universität Bochum, Germany)
Prof. E. P. Busso (Ecole des Mines de Paris, France) Prof. J. Cao (Northwestern University, USA)
Dr. J. L. Chaboche (Onera (Office National d'Etudes et Recherches Aérospatiales), France)
Prof. Y. W. Chang (Pohang University of Science and Technology, Korea)
Prof. C. L. Chow (University of Michigan, USA)
Prof. I. F. Collins (University of Auckland, New Zealand)
Prof. R. Du (Chinese University of Hong Kong, Hong Kong)
Prof. F. Ellyin (University of Alberta, Canada)
Prof. J. Fan (Alfred University, USA)
Prof. F. D. Fischer (Montanuniversität Leoben, Austria)
Prof. N. A. Fleck (Cambridge University, UK)
Prof. Q. Gao (Southwest Jiaotong University, China)
Prof. S. Ghosh (The Ohio State University, USA)
Dr. N. K. Gupta (Indian Institute of Technology, Bombay, India)
Prof. K. Higashi (Osaka Prefercture University, Japan)
Prof. Z. P. Huang (Peking University, China)
Prof. K. C. Hwang (Tsinghua University, China)
Prof. C. G. Kang (Pusan National University, Korea)
Prof. D. N. Lee (Seoul National University, Korea)
Prof. W.B. Lee (The Hong Kong Polytechnic University, Hong Kong)
Prof. G. Lu (Swinburne University of Technology, Australia)
Dr. S. L. Mannan (IGCAR (Indira Gandhi Centre for Atomic Research), India)
Dr. D. L. McDowell (Georgia Institute of Technology, USA)
Prof. K. W. Neale (Université de Sherbrooke, Canada)
Prof. A. Needleman (Brown University, USA)

Prof. N. Ohno (Nagoya University, Japan)
Prof. S. R. Reid (The University of Manchester, UK)
Prof. D. W. Shu (Nanyang Technological University, Singapore)
Prof. P. F. Thomson (Monash University, Australia)
Prof. M. Tokuda (Mie University, Japan)
Prof. Y. Tomita (Kobe University, Japan)
Prof. V. Tvergaard (Technical University of Denmark, Denmark)
Prof. B. Y. Xu (Tsinghua University, China)
Prof. J. Q. Xu (Shanghai Jiaotong University, China)
Prof. G. T. Yang (Taiyuan University of technology, China)
Prof. W. Yang (Tsinghua University, China)
Prof. F. Yoshida (Hiroshima University, Japan)
Prof. T. X. Yu (The Hongkong University of Science and Technology, Hong Kong)
Prof. L. C. Zhang (The University of Sydney, Australia)
Prof. H. M. Zbib (Washington State University, USA)

# AEPA2008 Sponsoring Companies

 POSCO

 KISWIRE

 Marketing LAB

 Samsung Electro-Mechanics Co. Ltd.

 Sungwoo Hitech Co. Ltd.

 HWASHIN Co. Ltd

ESI GROUP Hankook ESI Co. Ltd.

# AEPA2008 Sponsoring Organizations

 Korea Research Foundation

 Guwon Scholarship Foundation

 Daejeon Convention Bureau

 Korea Institute of Material Science

 BK21 KAIST Valufacture Institute of ME

Photograph of delegates in AEPA 2008 conference, Daejeon, Korea.

# CONTENTS

## Part A   Constitutive Modeling

# Part B    Damage, Fracture, Fatigue and Failure

## Part C  Dynamic Loading and Crash Dynamics

## Part D   Engineering Applications and Case Studies

## Part E    Experimental and Numerical Techniques

## Part F    Molecular Dynamics

## Part G    Nano, Meso, Micro and Crystal Plasticity

## Part H    Phase Transformations

## Part N    Structural Plasticity

## Part O    Superplasticity

## Part P    Time-Dependent Deformation

# Part A
# Constitutive Modeling

# STUDY ON SOFTENING CONSTITUTIVE MODEL OF SOFT ROCK USING STRAIN SPACE BASED UNIFIED STRENGTH THEORY

LI SONG[1], CHONGDU CHO[2][†], SHENG LU[3]

*Department of Mechanical Engineering, Inha University,*
*253 Yong-hyun Dong, Nam Ku, Incheon 402-751, KOREA,*
*songli0125@hotmail.com, cdcho@inha.ac.kr, lxlusheng@hotmail.com*

HONGJIAN LIAO[4]
*Department of Civil Engineering, Xi'an Jiaotong University,*
*Xi'an 710049, P. R. CHINA,*
*hjliao@mail.xjtu.edu.cn*

Received 15 June 2008
Revised 23 June 2008

This study attempts to modify the unified strength theory by considering compression as a positive load in geotechnical engineering. It also aims to establish a unified elastoplastic strain softening constitutive model which can accurately describe the strain softening behavior of one kind of soft rocks distributed in Japan. The hardening function parameters of the unified elastoplastic strain softening constitutive model are determined from experiments. In addition, numerical simulations of this model are performed to compare the pre-peak, post-peak and the residual strengths of soft rock predicted by this study and experimental results. Simulation results demonstrated that the proposed constitutive equations in strain space can well describe the softening behavior and accurately predict the peak and residual strengths of soft rock. While the proposed equation is applicative for normally consolidated state and overconsolidated state according to the simulation results.

*Keywords*: Soft rock; strain softening; strain space; unified elasto-plastic; constitutive model.

## 1. Introduction

Soft rock[1] is often found in the construction of dams[2], tunnels[3], mines[4], nuclear power stations, and bridges. It is a geotechnical material featuring notable plastic deformation. As a weather sensitive nonlinear material in geotechnical engineering[5], soft rock has attracted many researches, with emphasis placed on investigation of constitutive models[6] that can accurately predict the soft rock deformation and failure[7,8], softening, and peak and residual strengths[9].

Although most studies on deformation[4], failure[5] and constitutive models[6] of soft rock are performed in stress space, they are inadequate in terms of fully describing the behavior of unstable materials such as geotechnical materials in stress space[10], because the rupture and failure of geotechnical materials are functions of strain, which can be

[†]Corresponding Author.

determined experimentally. Therefore, a constitutive model involving strain space criteria can be established with respect to the three principal strain axes. While some models[8] consider strain space, they do not consider the difference in tensile and compressive strengths, and the effect of intermediate principal stress of soft rock. In practice, however, it is important to establish a constitutive model of soft rock in strain space, which can describe both the difference of tensile and compressive strengths and the effect of intermediate principal stress.

This study proposes a softening model of a diatom soft rock in strain space, where compression is taken as a positive load by modifying the unified strength theory for accommodating practical convenience[11,12]. A modified strain space yield function of unified strength theory is derived, and a triaxial elasto-plastic constitutive relationship of soft rock is established. Finally, numerical simulations are conducted to verify the validity of this relationship.

## 2.　Theoretical Background

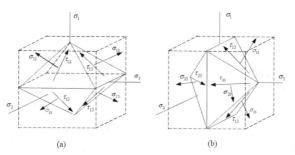

Fig. 1. Stress components on twin-shear stress element model.

An orthogonal octahedral element was introduced in the unified strength theory[12] shown in Fig.1. For accommodating practical convenience in geotechnical engineering, this study modifies all equations of the unified strength theory by taking compression as a positive load. The modified parameters $\beta$ and $C$ of the unified strength theory can be determined as $\beta = (\alpha - 1)/(\alpha + 1)$ and $C = \sigma_c(1+b)\alpha/(\alpha+1)$ by using test results, taking compression as a positive load as $\sigma_1 = \sigma_c$, $\sigma_2 = \sigma_3 = 0$ and $\sigma_1 = \sigma_2 = 0$, $\sigma_3 = -\sigma_t$. Substituting the modified $\beta$ and $C$ back into the unified strength theory, the modified equations are then expressed as follows respectively for different stress Lode angles:

$$F = \alpha\sigma_1 - \frac{1}{1+b}(b\sigma_2 + \sigma_3) = \alpha\sigma_c, \qquad \text{when } \sigma_2 \leq \frac{\alpha\sigma_1 + \sigma_3}{\alpha + 1}, \qquad (1a)$$

$$F' = \frac{\alpha}{1+b}(\sigma_1 + b\sigma_2) - \sigma_3 = \alpha\sigma_c, \qquad \text{when } \sigma_2 \geq \frac{\alpha\sigma_1 + \sigma_3}{\alpha + 1}. \qquad (1b)$$

where $\alpha$ is the ratio of tension and compression strength of the material, $b$ is the parameter of intermediate principal stress effect.

## 3.   The Unified Elasto-Plastic Softening Constitutive Model

The modified unified strength theory in the form of stress invariants can be obtained by substituting the principal stresses in the form of the stress invariants into the modified unified strength theory in Eqs. (1a) and (1b) as:

$$F = (\alpha - 1)\frac{I_1}{3} + \frac{1-b}{1+b}\sqrt{J_2}\sin\theta + \sqrt{\frac{J_2}{3}}\cos\theta = \alpha\sigma_c , \quad \text{when } \theta \le \tan^{-1}\frac{\sqrt{3}\alpha}{\alpha+2}, \tag{2a}$$

$$F' = (\alpha-1)\frac{I_1}{3} + \sqrt{\frac{J_2}{3}}\left(\frac{2-b}{1+b}\alpha+1\right)\cos\theta + \sqrt{J_2}\left(\frac{b}{1+b}\alpha+1\right)\sin\theta = \alpha\sigma_c , \quad \text{when } \theta \ge \tan^{-1}\frac{\sqrt{3}\alpha}{\alpha+2}, \tag{2b}$$

where $I_1$ and $J_2$ are the first invariant of the stress tensor and the second invariant of deviatoric stress tensor, respectively[10,12] and $\theta$ is Lode angle.

The yield function of the unified strength theory in stress space for a softening geotechnical material is expressed as:

$$\Phi = F(\sigma_{ij}) - \kappa = 0 . \tag{3}$$

As for triaxial consolidated undrained condition[10], $\varepsilon_v = 0$, $I_1 = 3KI_1' = 3K \cdot \varepsilon_v = 0$, and $\theta = 0 \le \tan^{-1}\sqrt{3}\alpha/(\alpha+2)$, where $\varepsilon_v$ is the volume strain.

In addition, $\sigma_1 > \sigma_2 = \sigma_3$; then $J_2 = \frac{1}{3}(\sigma_1 - \sigma_3)^2 = \frac{1}{3}q^2 .$

Eq. (2a) can be reduced to $F = \sqrt{\dfrac{J_2}{3}} = \sigma_t ,$

and Eq. (3) to $\psi = \sqrt{\dfrac{J_2}{3}} - \sigma_t - \kappa = 0$, i.e., $\psi = \frac{1}{3}q - \sigma_t - \kappa = 0$, or

$$\kappa = \frac{1}{3}q - \sigma_t . \tag{4}$$

If the hardening function is chosen as the function of the plastic shear strain, $\kappa = \kappa(\varepsilon_s^p)$, then

$$q = 3G\varepsilon_s^e = 3G(\varepsilon_s - \varepsilon_s^p) . \tag{5}$$

The consistency condition is introduced as:

$$d\psi = \frac{\partial\psi}{\partial\varepsilon_s}d\varepsilon_s + \frac{\partial\psi}{\partial\varepsilon_s^p}d\varepsilon_s^p + \frac{\partial\psi}{\partial\kappa}d\kappa = 0 , \tag{6}$$

where $\varepsilon_s$ is the shear strain and $\varepsilon_s^p$ is the plastic shear strain.

Since $d\kappa = \dfrac{\partial\kappa}{\partial\varepsilon_s^p}d\varepsilon_s^p$, then Eq. (6) can be written as:

$$d\psi = \frac{\partial\psi}{\partial\varepsilon_s}d\varepsilon_s + \left(\frac{\partial\psi}{\partial\varepsilon_s^p} + \frac{\partial\psi}{\partial\kappa}\cdot\frac{\partial\kappa}{\partial\varepsilon_s^p}\right)d\varepsilon_s^p = 0 , \tag{7}$$

and

$$d\varepsilon_s^p = d\lambda\left[\left(\frac{\partial Q}{\partial q}\right)^2 + \left(\frac{1}{q}\frac{\partial Q}{\partial\theta_\sigma}\right)^2\right]^{\frac{1}{2}} , \tag{8}$$

where $\theta_\sigma$ is the stress Lode angle and $Q$ is the potential function.

Substituting Eq. (8) into Eq. (7), $d\lambda$ can be obtained and incremental form of Eq. (5) can be written as

$$dq = 3G \cdot d\varepsilon_s^e = 3G(d\varepsilon_s - d\varepsilon_s^p) = 3G \cdot \left(1 - \dfrac{\dfrac{\partial \psi}{\partial \varepsilon_s}}{\dfrac{\partial \psi}{\partial \varepsilon_s} - \dfrac{\partial \psi}{\partial \kappa} \cdot \dfrac{\partial \kappa}{\partial \varepsilon_s^p}}\right) d\varepsilon_s = 3G \cdot \left(1 - \dfrac{\dfrac{\partial \psi}{\partial \varepsilon_s}}{\dfrac{\partial \psi}{\partial \varepsilon_s} + \dfrac{\partial \kappa}{\partial \varepsilon_s^p}}\right) d\varepsilon_s$$

According to the conventional triaxial compression tests[10], $\partial \psi / \partial \varepsilon_s = G$, hence,

$$dq = 3G\left(1 - \dfrac{G}{G + \dfrac{\partial \kappa}{\partial \varepsilon_s^p}}\right) d\varepsilon_s \qquad (9)$$

Thus, the elasto-plastic constitutive relationship under triaxial stress is presented by using the elastic plastic theory in strain space and the unified strength theory, taking compression as a positive load, as established in this paper.

## 4.    Numerical Simulation and Discussion

Experimental data[8] are compared with the results derived in this paper. The specimen is a diatom soft rock in Japan. Specimens were made with a diameter of $0.05m$ and a height of $0.1m$. Consolidated undrained triaxial experiments controlled by stress and/or strain were conducted under different confining pressures and different loading rates. The specimens with high void ratios were fully saturated. The preconsolidation pressure was $1,500\ kPa$.

The hardening function in Eq. (9) is chosen as given in reference 13, shown in Fig. 2.

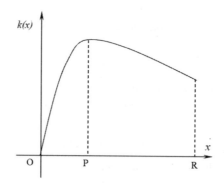

Fig. 2. Hardening function.

$$\kappa(x) = \dfrac{H}{A} e^{Ax}\left(1 + \dfrac{1}{PA} - \dfrac{x}{P}\right) - \dfrac{H}{A}\left(1 + \dfrac{1}{PA}\right)\ , \qquad (10)$$

where $P, H$, and $A$ are hardening function parameters determined from experiments. $P$ is the corresponding value of $x$ when the stress reaches its peak value. Here $x$ is plastic shear strain. $H$ and $A$ can be determined by using Eq. (10) with the peak and residual strengths of the soft rock, respectively, listed in Table 1.

The stress-strain curves of the proposed model can be drawn using the numerical integration results obtained with Eqs. (9) and (10) shown in Figs.3 and 4 (preconsolidation pressure 1500kPa and strain rate $\dot{\varepsilon}_a = 0.175\%/min$ ). The strain softening behavior as well as the peak and residual strengths are described fairly well which are significant for practical engineering applications. The error between simulation and experimental results lies within 10% for normally consolidated state and 5% for overconsolidated state, respectively. So the established model in this paper is found to be effective and applicable.

Table 1.   Hardening function parameter values of the unified elasto-plastic softening constitutive model.

| Parameter | 100 kPa | 500 kPa | 1000 kPa | 1500 kPa | 2000 kPa | 2500 kPa | 3000 kPa | 3500 kPa |
|---|---|---|---|---|---|---|---|---|
| P | 0.0127 | 0.0161 | 0.0225 | 0.0318 | 0.0285 | 0.0302 | 0.0283 | 0.0312 |
| A | -114.597 | -92.898 | -65.503 | -48.663 | -58.876 | -56.356 | -61.222 | -62.251 |
| H（e4） | 9.8887 | 8.7834 | 7.1651 | 5.6624 | 6.6302 | 7.0142 | 8.3980 | 9.8378 |

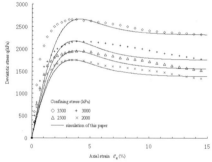
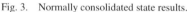

Fig. 3.   Normally consolidated state results.

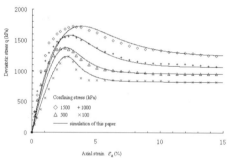

Fig. 4.   Overconsolidated state results.

## 5.   Conclusion

From the study in this paper, by taking compression as a positive load, the unified strength theory equation is modified to accommodate practical convenience in geotechnical engineering, where compression is customarily taken as positive. Also, elasto-plastic constitutive models in strain space are proven to be applicable to the softening behavior of soft rock. Finally, a triaxial unified elasto-plastic softening constitutive equation of soft rock is derived in strain space, the associated flow rule, and the unified strength theory. The equation is used to simulate the strain softening behavior of a type of soft rock with a proper hardening function. The simulation results show that this model can well describe softening behavior and can predict the peak and residual strengths of soft rock. The proposed softening constitutive model can describe the residual and peak strengths of overconsolidated soft rock better than the ones of normally consolidated rocks.

## Acknowledgment

This work was supported by Inha University.

## References

1.　X. X. Miao and Z. C. Chen, *Mechanics of soft rock* (China University of Mining and Technology Press, China, 1995).
2.　H. Basarir, *Turkey. Eng. Geol.* **86,** 225 (2006).
3.　N. Yoshida, M. Nishi, M. Kitamura and T. Adachi, *Int. J. Rock. Mech. Min. Sci.* **34**, 353.e1 (1997).
4.　C. Wang, Y. Wang and S. Lu, *Int. J. Rock. Mech. Min. Sci.* **37**, 937 (2000).
5.　D. A. Sun, H. Matsuoka, D. Muramatsu, T. Hara, A. Kudo, Z. Yoshida and S. Takezawa, *Int. J. Rock. Mech. Min. Sci.* **41,** 87 (2004).
6.　R. Yoshinaka, T. V. Tran and M. Osada, *Int. J. Rock. Mech. Min. Sci.* **35**, 941 (1998).
7.　H. J. Liao, W. C. Pu, J. H. Yin, M. Akaishi and A. Tonosaki, *Int. J. Rock. Mech. Min. Sci.* **41**, 423 (2004).
8.　H. J. Liao, W. C. Pu and W. C. Qing, *Int. J. Rock. Mech. Min. Sci.* **27**, 1861 (2006).
9.　G. R. Krishan, X. L. Zhao, M. Zaman and J. C. Roegiers, *Int. J. Rock. Mech. Min. Sci.* **35**, 695 (1998).
10.　Y. R. Zheng, Z. J. Shen and X. N. Gong, *The principles of geotechnical plasticity mechanics.* (China Architecture and Building Press, Beijing, 2002).
11.　L. Song, Ph.D. Thesis, Xi'an Jiaotong Univ., China, 2006.
12.　M.H. Yu, *Generalizes Plasticity* (Springer, U.S.A. 2005).
13.　L. Y. Zhang, Ph.D. Thesis, Logistical Engineering Univ. of PLA, China 2004.

# DYNAMIC CRUSHING SIMULATION OF METALLIC FOAMS WITH RESPECT TO DEFORMATION AND ENERGY ABSORPTION

Y. F. ZHANG[1†], L. M. ZHAO[2]

*Taiyuan University of Technology, Taiyuan, P.R.CHINA,*
*zhangyifen16@126.com, zhaolm123@126.com*

Received 15 June 2008
Revised 23 June 2008

In this paper, a two-dimensional nonlinear elastic-plastic mass-spring-damper-rod element model is employed to simulate the crush behavior of metallic foams. The density heterogeneity and pore fluid of metallic foams are considered. The metallic foams are compressed by applying a planar pulse loading or by giving a deformation rate. Several numerical results show the deformation patterns and the energy absorption regime of metallic foams under crush loading. The influence of heterogeneous density, cell fluid, loading intensity, deformation rate on the deformation and the energy absorption of metallic foams is assessed.

*Keywords*: Metallic foams; the crushing simulation; deformation and energy absorption.

## 1. Introduction

When vehicles such as aircrafts collide, their kinetic energy is dissipated by the plastic deformation and collapse of their structures. Thus, the ability of absorbing energy plays an important role in overall safety performance of structures. Metallic foams have been commonly considered as good security-guard materials which resist blast shock or high speed crushing due to their superior energy absorption capability. For this reason, investigations into their responses to dynamic crush have been a subject of great topical interest. Even though some experimental observations have been reported,[1,2] numerical modeling and theoretical analysis are still very limited. In the current work, a two-dimensional nonlinear elastic-plastic mass-spring-damper-rod model is developed. The model is used to simulate the deformation and energy absorption of metallic foams under crush loading. The effects of the pore fluid and the heterogeneous density distribution to the deformation and energy absorption of the foams are considered. High deformation rate compression behaviors of metallic foams are discussed.

[†]Corresponding Author.

## 2.  Model

We use a simple two-dimensional model to simulate the crush behaviors of a foam block of length $L$, width $W$ and thickness $H$, as depicted in Fig.1. The model consists of $N \times M$ discrete lump masses connected by elastic-plastic nonlinear springs and dampers in parallel in loading direction and constrained by hinged extensible rods transverse to the loading direction. The dampers stand for the contribution of the cell fluid to the strength of foam block. The extensible rods satisfy the deformation compatibility of transversal foam material. The crush loading ( pressure pulse or deformation rate ) is applied on $N$th row lump mass. The first row springs are fixed on a rigid platform.

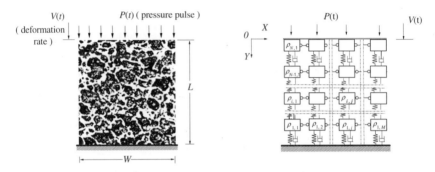

Fig. 1. An aluminum foam block under crush loading (left) and sketch of two dimensional mass-spring-damper-rod simulation model used in the present study (right).

### 2.1.  *Dynamical equilibrium equations*

When a foam block experiences a planar crushing, the lump mass with the density $\rho_{i,j}$ ($i =1,\cdots, N$ and $j = 1,\cdots, M$) displaces an amount, defining two components of the displacement as $x_{i,j}$ and $y_{i,j}$. The dynamical equilibrium equations are described by

$$\rho_{i,j}\, \ddot{y}_{i,j} = (\sigma_{i+1,j} - \sigma_{i,j} + \sigma'_{i+1,j} - \sigma'_{i,j})\frac{N}{L} + \left( \sigma''_{i,j} \sin\theta_{i,j} - \sigma''_{i,j-1} \sin\theta_{i,j-1} \right)\frac{M}{W} \qquad (1)$$

$$\rho_{i,j}\, \ddot{x}_{i,j} = (\sigma''_{i,j}\cos\theta_{i,j} - \sigma''_{i,j-1}\cos\theta_{i,j-1})\frac{M}{W} \qquad (2)$$

where $\sigma_{i,j}$, $\sigma'_{i,j}$ and $\sigma''_{i,j}$ are the compressive stress of the nonlinear spring, the contribution of the damper to the strength and the tensile or compressive stress of the elastic bar with a rotation angle $\theta_{i,j}$, respectively. The change of cross section area of the foam block is neglected in this analysis. The relation between the compressive strain $\varepsilon_{i,j}$ and lump mass displacement of a nonlinear spring is given by

$$\varepsilon_{i,j} = \frac{N}{L}(y_{i,j} - y_{i-1,j}). \qquad (3)$$

The relation between the strain $\varepsilon''_{i,j}$ of elastic rod and the displacements of lump mass is

$$\varepsilon''_{i,j} = \frac{M}{W}\sqrt{(y_{i,j+1} - y_{i,j})^2 + (\frac{W}{M} + x_{i,j+1} - x_{i,j})^2} - 1.0. \tag{4}$$

## 2.2. The constitutive relationships and energy absorption

The meso-scopic heterogeneity of foams block can be represented by density $\rho_{i,j}$ of lump masses. In terms of compressive behavior of metallic foams, the relative modulus (the nonlinear spring modulus, $E_{i,j}$, divided by the solid modulus, $E_s$), the plastic collapse stress $(\sigma_{i,j})_{pl}$ of the nonlinear spring relative to the yield strength $\sigma_{ys}$ of the solid cell edge materials, and "densification strain" $(\varepsilon_{i,j})_d$ of the non-linear spring are defined as functions of the relative density (the ratio of the density $\rho_{i,j}$ of the foam lump mass to that of the solid $\rho_s$), which can be found at Refs. 3 and 4. The constitutive relations are adopted for the nonlinear elastic - plastic spring, i.e.

$$\sigma_{i,j} = E_{i,j}\varepsilon_{i,j} \quad \text{for} \ \ 0 \le \varepsilon_{i,j} \le (\varepsilon_{i,j})_{pl} \tag{5}$$

$$\sigma_{i,j} = (\sigma_{i,j})_{pl} \quad \text{for} \ \ (\varepsilon_{i,j})_{pl} < \varepsilon_{i,j} \le (\varepsilon_{i,j})_d \tag{6}$$

$$\sigma_{i,j} = E_s\left(\varepsilon_{i,j} - (\varepsilon_{i,j})_d\right) + (\sigma_{i,j})_{pl} \quad \text{for} \ \ \varepsilon_{i,j} > (\varepsilon_{i,j})_d \tag{7}$$

where $(\varepsilon_{i,j})_{pl}$ is the plastic collapse strain ( $(\varepsilon_{i,j})_{pl} = (\sigma_{i,j})_{pl}/E_{i,j}$ ). The contribution of cell fluid to the strength is obtained by the following equations:

$$\sigma'_{i,j} = \frac{\mu}{4K_{i,j}}\left(\frac{W}{M}\right)^2 \frac{N}{L}(\dot{y}_{i,j} - \dot{y}_{i,j-1}) \qquad \text{(Open-cell)} \tag{8}$$

$$\sigma'_{i,j} = \frac{p_0\varepsilon_{i,j}\left(1 - 2v^*\right)}{1 - \varepsilon_{i,j}\left(1 - 2v^*\right) - \rho_{i,j}/\rho_{i,j}} \qquad \text{(Closed-cell)} \tag{9}$$

where $\mu$ is the dynamic viscosity of the fluid, $K_{i,j}$ is the permeability,[3] $p_0$ is the initial gas pressure within the cell and $v^*$ is Poisson's ratio, taking $v^* = 0.3$.
The stress-strain relation of the extensible rod satisfies that :

$$\sigma''_{i,j} = \frac{2\,E_{i,j}E_{i,j+1}}{(E_{i,j} + E_{i,j+1})}\,\varepsilon''_{i,j}. \tag{10}$$

The constitutive equation for nonlinear springs includes loading and unloading laws. Loading complies with the stress-strain relation described in Eqs. (5)~(7). Unloading is elastic to the tension cutoff and continuing unloading follows a horizontal plateau. Subsequent reloading follows the unloading curve.
The energy absorption per unit volume $W_{i,j}$ of the element subscripted as $i, j$ can be found by integrating over strain

$$W_{i,j} = \int_0^{\varepsilon_{i,j}} \left(\sigma_{i,j} + \sigma'_{i,j}\right)d\varepsilon_{i,j}. \tag{11}$$

The maximum useful energy absorption per unit volume ends at the densification strain $(\varepsilon_{i,j})_d$. The total energy absorption per unit volume of the foam block, $W$, is then

$$W = \sum_{i=1}^{n}\sum_{j=1}^{m} W_{i,j} \,. \tag{12}$$

The nonlinear dynamic equations (1) and (2) are numerically integrated by the explicit algorithm. The numerical algorithm is set up in FORTRAN program.

## 3.  Results and Discussion

The foam block is crushed by applying a planar rectangular pressure pulse $P(t)$ of duration $T$ or by giving a deformation rate $V(t)$ which starts at $V_0$ but decreases linearly to 0 within a time interval of $T$. Defining non-dimensional variable $\tau = t/L/\sqrt{E_s/\rho_s}$ and two non-dimensional numbers $\tau_d = T/L/\sqrt{E_s/\rho_s}$ and $p = P/\sigma_{ys}$, the dimensionless pulse loading $p(\tau)$ demands that

$$p(\tau) = p \quad \text{for } 0 \le \tau \le \tau_d \text{ and } p(\tau) = 0 \text{ for } \quad \tau > \tau_d. \tag{13}$$

Defining non-dimensional initial deformation rate $v_0 = V_0/\sqrt{E_s/\rho_s}$, the dimensionless deformation rate $v(\tau)$ is given by

$$v(\tau) = v_0 \left(1 - \tau/\tau_d\right) \quad \text{for } 0 \le \tau \le \tau_d \text{ and } v(\tau) = 0 \text{ for } \quad \tau > \tau_d. \tag{14}$$

Taking $\rho_s = 2.7 Mg/m^3$, $E_s = 69 GN/m^2$ and $\sigma_{ys} = 195 MPa$ for a typical aluminum foam, some examples are conducted for $N = 10$ and $M = 10$ to demonstrate the deformation and energy absorption of the block of metallic foams in crushing events, as follows.

### 3.1.  *The deformation analysis of aluminum foams*

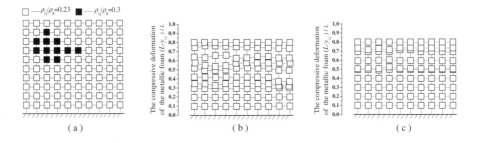

Fig. 2. The deformation patterns of the heterogeneous foam deformed to 20% at different initial deformation rates: (a) Undeformed model. (b) The deformed foam at $v_0 = 0.1$. (c) The deformed foam at $v_0 = 0.5$.

The deformation patterns of closed-cell aluminum foam with heterogeneous density distribution are simulated with the model shown in Fig. 2(a). Model consists of one hundred lump masses. Black color lump masses represent heterogeneous density region. Their relative density is marked in Fig. 2(a). Fig. 2(b) shows the deformed patterns of the foam block compressed by 20% engineering strain for $v_0 = 0.1$ and Fig. 2(c) for $v_0 = 0.5$, where engineering strain $\varepsilon$ is defined by the end displacement of the foam block, corresponding to the displacement $y_{N,j}$ of lump mass in $N$th row, divided by the initial length of the foam block. The overall strain is not homogeneous and localized deformation becomes obvious. For the block of foam with homogeneous density distribution, the dynamic compression characteristic is evident from a number of the simulation results. Fig. 3 shows three of those deformation patterns, where the blocks of the foam with relative density of 0.1 were compressed dynamically by 20% engineering strain. It can be seen that most of deformation is concentrated in two bands. Examination of large number of foams simulated at high displacement rates revealed that the locations of the crush bands depend on initial deformation rate.[1]

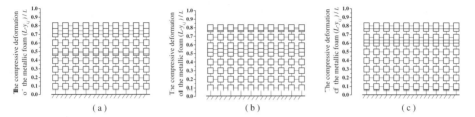

Fig. 3.  The deformation patterns for homogeneous foam compressed by 20% engineering strain at different initial deformation rates. (a) $v_0 = 0.1$. (b) $v_0 = 0.5$. (c) $v_0 = 1.0$.

## 3.2. *The energy absorption of aluminum foams*

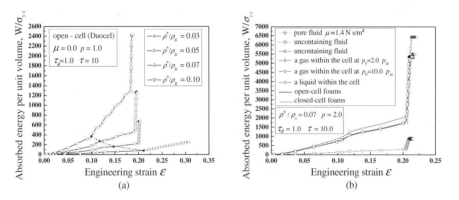

Fig. 4.   The energy absorption diagram of aluminum foam block. (a) Influence of the relative density on the energy absorption of aluminum foams. (b) Influence of pore fluid on the energy absorption of aluminum foam.

Fig. 4(a) shows the energy absorption diagram of the open-cell aluminum foam with the relative density ranging from 0.03 to 0.1. The applied rectangular pulse is characterized by $p = 1.0$ and $\tau_d = 1.0$, which represents the pulse pressure being equal to the yield strength of the cell edge materials and lasting a duration of elastic wave propagation through foams block. Very little energy is absorbed during a loading process (meaning that $\tau \leq 1.0$). The loading portions of the curves end at an engineering strain which depends on the relative density of foams. This is plotted at a dotted line on Fig. 4(a).

After foam block is unloaded $(\tau > 1.0)$, micro-inertia of foam tends to suppress the more compliant cell collapse or "densification", so it increases the crushing stress and diffuses the crushing wave front, which allows large energy absorption up to a constant value. So the curves are followed by steep rise and then a nearly vertical rise with the foam block unloading. Because the higher relative density foam is sensitive to the inertia effect, they are superior in the energy absorption capacity respects to lower relative density foams. Fig. 4(b) illustrates that the pore fluid has a less influence on the energy absorption of aluminum foam and the closed cell foam is especially good at the energy absorption capacity as compared with the open cell aluminum foam.

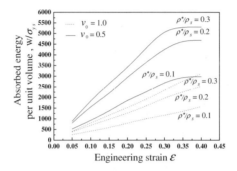

Fig. 5.   The energy absorption of closed-cell aluminum foam deformed to the different engineering strain.

The absorption energy per unit volume of closed-cell foams deformed to a series of engineering strain, 5%, 10%, etc., is calculated and the results are plotted as a family of curves on Fig.5. When foam block is compressed to the engineering strain $\varepsilon_0$ at a deformation rate given by Eq.(14), the time interval $\tau_d$ can be determined by $2\varepsilon_0 / v_0$, which means the interval $\tau_d$ being inversely proportional to the initial deformation rate $v_0$ and a crushing process lasting longer time under lower initial deformation rate. So, from Fig.5, we can see that the absorption energy per unit volume of foam block decreases with the initial deformation rate rise.

## 4.  Conclusions

(1)  Under a constant engineering strain, the energy absorption increases with the relative density and decrease with the deformation rate. Closed-cell foams are superior in the energy absorption capacity.

(2)  The foams with a heterogeneous density show the localized deformation mode. The deformation of homogeneous foams is concentrated in two bands, the location of which depends on initial deformation rate.

(3)  The pore fluid has less influence on the deformation and energy absorption of aluminum foam.

## Acknowledgments

The authors would like thank The National Natural Science Foundation of China (No. 10572100, 90716005) and Youth Science and Technology Research Foundation of Shanxi (No. 2007021005) for supporting this work.

## References

1. V.S Deshpande and N.A Fleck, *Int. J. Impact Eng.* 24(3), pp.277-298 (2000).
2. K.A Dannenmann and J. Lankford Jr, *Mater. Sci. Eng.* A293, pp. 157-164 (2000).
3. L.J. Gibson and M..F. Ashby, *Cellular Solids: Structure and Properties* (Cambridge, UK, Cambridge Univ. Press, 2$^{nd}$ ed, 1997).
4. L.J. Gibson, Annu. *Rev. Mater. Sci.* 30, pp.191-227 (2000).

# COMPARATIVE STUDY OF SINGLE CRYSTAL CONSTITUTIVE EQUATIONS FOR CRYSTAL PLASTICITY FINITE ELEMENT ANALYSIS

MYOUNG GYU LEE[1†]

*Ferrous Alloys Research Group,Korea Institute of Materials Science,531 Changwondaero,
Changwon, Gyeongnam 641-010,KOREA,
mang92@kims.re.kr*

ROBERT H. WAGONER[2]

*Department of Materials Sci.&Eng.,Ohio State University,241 College Rd.,
Columbus,OH 43210, USA
wagoner@matsceng.ohio-state.edu*

SUNG-JOON KIM[3]

*Advanced Metallic Materials Department, Korea Institute of Materials Science,531 Changwondaero,
Changwon, Gyeongnam 641-010,KOREA,
sjkim@kims.re.kr*

Received 15 June 2008
Revised 23 June 2008

Two sets of single crystal constitutive equations used for the crystal plasticity finite element analysis are comparatively investigated by simulating simple deformation of oriented single crystals. The first of these consists of conventional constitutive equations, which have been adopted for the prediction of deformation texture and their parameters are generally obtained by back-fitting polycrystalline stress-strain response. The other set uses interactions between moving dislocations on the primary slip system and the corresponding forest dislocations. The idealized Orowan hardening mechanism is adopted for the calculation of the critical force, and constitutive parameters are determined by the geometry of dislocations, thus less fitting procedure is involved. The stress-strain curves of copper single crystal are used to demonstrate how the two models work for the orientation dependent stress-strain responses.

*Keywords*: Crystal plasticity; constitutive equation; single crystal; hardening.

## 1. Introduction

The characterization procedure by back-fitting of polycrystalline behavior in the crystal plasticity finite element method has successfully predicted the texture evolution of polycrystals[1-4]. However, these constitutive models have not well represented the real

---

†Corresponding Author.

single crystal behavior such as orientation dependent stress-strain response[5-6]. The single crystal constitutive equation is phenomenological model based on strength of polycrystals, which does not consider the crystallographic interactions of dislocations[7]. Therefore, the evolution equations for the hardening might not be able to capture the detailed behavior of single crystals.

As a second approach, the evolution of flow stress is expressed by the dislocation densities as internal state variables[8]. These models have captured orientation dependent stress-strain behavior of cubic single crystals. Arsenlis and Parks[6] developed continuum based model considering dislocation density and applied to the orientation dependent tension of single crystal aluminum. The results could capture the stress-strain responses with different orientations although many fitting parameters make the application of the constitutive equations difficult for complicated boundary value problems.

In the present work, comparative studies on the single crystal constitutive equations between the phenomenological model and dislocation interaction model were presented. The parameters for the phenomenological model are fitted from the simple tension or compression of polycrystal, while no arbitrary fittings are involved in the dislocation density based model.

## 2.  Single Crystal Constitutive Equations for Crystal Plasticity

The basic kinematics used for the crystal plasticity is generally based on the multiplicative decomposition[9] of the total deformation gradient, $\mathbf{F} = \mathbf{F}^e \mathbf{F}^p$. Here, $\mathbf{F}^e$ is elastic lattice distortion, while $\mathbf{F}^p$ defines the slip by the dislocation in the unrotated configuration[8]. The plastic velocity gradient is assumed as $\overline{\mathbf{L}}^p = \sum_{\alpha=1}^{n} \dot{\gamma}^\alpha \mathbf{s}_0^\alpha \otimes \mathbf{n}_0^\alpha$ by the work by Rice[10]. Here $\mathbf{s}_0^\alpha$ and $\mathbf{n}_0^\alpha$ are slip direction and slip plane normal, respectively. For a rate-dependent crystal plasticity model, the plastic shear rate of each slip system $\alpha$ is expressed as a power law function of the resolved shear stress $\tau^\alpha$ as

$$\dot{\gamma}^\alpha = \dot{\gamma}_0 \left( \frac{\tau^\alpha}{g^\alpha} \right)^{\frac{1}{m}} \text{sign}(\tau^\alpha) \tag{1}$$

where $g^\alpha$ is the slip resistance and $m$ is the rate sensitivity exponent[11].

Existing texture calculations utilize phenomenological models for the evolution of flow stress,

$$\dot{g}^\alpha = \sum_\beta h_{\alpha\beta} \dot{\gamma}^\beta \tag{2}$$

where $h_{\alpha\beta}$ are hardening coefficients. Many previous models for the texture analysis adopted the following simplest form for the hardening coefficients, $h_{\alpha\beta}$ as

$$h_{\alpha\beta} = h_\beta \left( q + (1-q)\delta_{\alpha\beta} \right) \tag{3}$$

where $\delta_{\alpha\beta}$ is Kronecker delta and $q$ determines the ratio between self and latent hardening. For cubic metals $1 \le q \le 1.6$ is acceptable range of this parameter for the simulation of texture development[12]. The form of $h_\beta$ in Eq.(3) has been proposed by several works[2,6,8,11,13], and here the specific form proposed by Brown et al.[13] is adopted as

$$h_\beta = h_0 \left(1 - \frac{g^\beta}{g_s}\right)^a \tag{4}$$

where $h_0$ and $g_s$ are the initial hardening rate and the saturated flow stress, respectively, and a is hardening exponent.

For the dislocation density based model, the same power-law type is adopted to relate the rate of shear deformation to the applied shear stress on slip system $\alpha$ as shown in Eq. (1). The slip hardening is expressed by the interaction of primary dislocations with corresponding forest dislocations which act as obstacles to motion of mobile dislocations. To derive the hardening equation by the dislocation interactions, some assumptions should be made; (1) the idealized interaction between moving dislocation and corresponding forest dislocation. (2) perfect obstacles for forest dislocations. As Orowan pointed out[14], the applied stress should be sufficiently high to bow the dislocations in a semicircular form between two obstacles. The critical stress necessary to bend a dislocation to a radius r is calculated by the equilibrium with line tension of dislocation,

$$g^\alpha \cdot b = \frac{T}{r} \tag{5}$$

where T is a dislocation line tension and b is it's Burger's vector. The line tension equal to its self-energy per unit length and approximated for the radius of curvature r as $T \approx \frac{1}{2}\mu b^2$. Then [15],

$$g^\alpha = \frac{\mu b}{2r} = \frac{\mu b}{l} \tag{6}$$

If the applied stress is raised to a value greater than that given in Eq. (6), the dislocation line results in the dislocation loops around each obstacle, which is well known Orowan hardening mechanism. Since the obstacle distance $l$ depends on the density of dislocation passing through the primary slip plane, Eq.(6) can be written as

$$g^\alpha = \mu b \sqrt{\rho_f} \tag{7}$$

where $\rho_f$ is density of forest dislocation corresponding to the dislocation on the slip system $\alpha$. The flow stress derived in Eq. (7) is valid only if the slip plane normal direction is perfectly parallel to the line direction of its forest dislocation. When the angle of both directions is $\theta$, the effective forest dislocation density $\overline{\rho}_f$ can be simply defined.

$$\overline{\rho}_f = \rho_f \cos\theta = \rho_f \cdot \left(\mathbf{n}^\alpha \cdot \xi_f\right) \tag{8}$$

where $\mathbf{n}^\alpha$ and $\xi_f$ are the slip plane normal of mobile dislocation and line direction of corresponding forest dislocation, respectively. The flow stress considering these effective quantities is

$$g^\alpha = \mu b \sqrt{\rho_f \cdot \left(\mathbf{n}^\alpha \cdot \xi_f\right)} \tag{9}$$

Because there are many slip systems in crystals, the relation Eq. (9) should be generalized by considering the geometry of dislocations. Franciosi and Zaoui[16] observed in their experiments that the obstacle density in slip system $\alpha$ could be obtained by linear combination of interactions with dislocations among all other slip systems. If there are n different slip systems, the Eq. (9) becomes

$$g^{\alpha} = \mu b \sqrt{\sum_{\beta=1}^{n} h'_{\alpha\beta}\rho_{\beta}} \qquad (10)$$

with $h'_{\alpha\beta} = \mathbf{n}^{\alpha} \cdot \boldsymbol{\xi}^{\beta}$.

For the dislocation density evolution, commonly used evolution equation of the edge dislocation density originally proposed by Essmann and Mughrabi[17] is used in this study.

$$\dot{\rho}^{\alpha} = \frac{1}{b}\left(\frac{\sqrt{\sum_{\beta}^{n}\rho^{\beta}}}{k_a} - k_b\rho^{\alpha}\right)\cdot\left|\dot{\gamma}^{\alpha}\right| \qquad (11)$$

where $k_a$ and $k_b$ are material parameters.

## 3. Results and Discussion

The stress-strain responses in different orientations are calculated and compared with those of experimental observations. As for the experimental data, stress-strain responses of single copper crystal are reproduced from the previous published data[18]. Three different orientations, <100>, <111> and <123> orientations aligned with tensile axes, are selectively considered.

### 3.1. *Stress-strain response by the phenomenological model*

For copper crystal considered in the present work, crystallographic slip is assumed to occur on the 12 {111}<110> slip systems. The anisotropic elasticity tensor for the copper single crystal is obtained as: $C_{11}=170$ GPa, $C_{12}= 124$ GPa and $C_{44}=75$ GPa. Because the data for Eqs. (1) and (3) are unavailable, the reported data in literature are used[2]; a reference shearing rate $\dot{\gamma}_0 =0.001$/sec, m=0.012, q=1.4.

By using FE analysis, the stress-strain curve oriented along <100> is fitted. A set of fitting parameters $h_0=200$ MPa, $g_s=150$ MPa, $g_0=2.5$ and a=2.25 gives reasonable result by the correspondence between the simulated stress-strain response and the experimental data as shown in Fig. 1(a). The other two oriented stress-strain curves are calculated as shown in Fig. 1(b)-(c). The predicted stress-strain curve for the <111> orientation reasonably agrees with measured one, while the simulated stress-strain curve for the <123> orientation far deviate from the experimental data. The results show that the fitting parameters by the multi-slip stress-strain curve do not well represent the stress-strain curves with having less favored slip systems, especially for the single slip case. This is due to the commonly assumed self/latent hardening ratio q=1.4 in Eq. (3), which results in significant self hardening. However, from the experimental data, the hardening rate of the <123> oriented stress-strain curve is very low, which reflects negligible self hardening.

### 3.2. *Stress-strain response by dislocation density based model*

Simulation results by adopting the dislocation density based single crystal constitutive equations are also adopted. The same procedure is iterated to obtain the material parameters for the Eq. (10) and Eq. (11). The stress-strain curve oriented along <100> is fitted for the copper single crystal. The burger's vector is 0.25nm and the shear modulus for the copper is 48 GPa. The fitting parameters $k_a=20$, $k_b=25b$ and initial dislocation density $\rho_0 =10e3$ /mm$^2$ for all slip systems give reasonable result as shown in Fig. 2(a). Note that the components of interaction matrix in Eq. (10) are geometrically determined. The predicted stress-strain behavior for the other orientations is shown in Fig.2(b)-(c) for the copper single crystal.

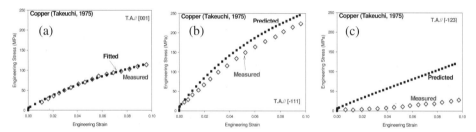

Fig. 1. Comparison of stress-strain curves between prediction by the phenomenological model and experiment: (a) fitted curve by the <100> orientation, (b) <111> prediction, (c) <123> prediction.

Figs. 1 and 2 show that the prediction capability of dislocation single crystal constitutive equation for the stress-strain response of single crystal is much better than that by the phenomenological single crystal constitutive model. The results generally agree well with those from measurements although there exist some deviations for the <123> orientation. On the contrary to the phenomenological model, dislocation based single crystal constitutive equation gives better agreement with the stress-strain response of single slip case: <123> orientation. The good agreement between prediction and measurement for the single slip cases for suggests that the contribution by the self hardening might be negligibly small compared with that by latent hardening.

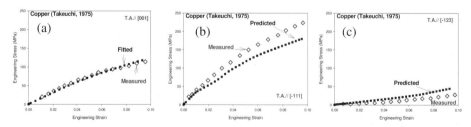

Fig. 2. Comparison of stress-strain curves between prediction by the dislocation density based model and experiment: (a) fitted curve by the <100> orientation, (b) <111> prediction, (c) <123> prediction.

## 4. Summary

Two single crystal constitutive models used for the crystal plasticity finite element analysis were comparatively investigated by simulating simple deformation of oriented single crystals. For the new model based on dislocation density, the idealized Orowan hardening mechanism was adopted for the calculation of the critical force, and constitutive parameters are determined by the geometry of dislocations. As results of comparative studies, the following conclusions were made: (1) Single crystal constitutive parameters could be obtained by fitting orientation dependent stress-strain curves of copper single crystal. The phenomenological constitutive model with commonly suggested self/latent hardening ratio did not represent true single crystal behavior. (2) On the other hand, the dislocation density based model gave better predictions for the orientation dependent stress-strain response. More detailed analysis with a self hardening by dislocations on the same slip plane and applications to the polycrystal has recently been reported in other work[19].

## Acknowledgments

This work was sponsored through contract No. FA9550-05-0068 with AFOSR, which is greatly appreciated. SJK appreciates the support by the Center for Advanced Materials Processing (CAMP) of the 21st Century Frontier R&D Program.

## References

1. K.K. Mathur and P.R. Dawson, *Int. J. Plasticity* **5** (1989) 67.
2. C.A. Bronkhorst, S.R. Kalidindi, L. Anand, *Philos. Trans. Roy. Soc. London A* **341** (1992) 443.
3. S. Nemat-Nesser, L.Q. Ni, T. Okinaka, *Mech. Mater.* **30** (1998) 325.
4. Kumar, P.R. Dawson, *Comput. Methods Appl. Mech. Eng.* **153** (1998) 259.
5. R. Becker, S. Panchanadeeswaran, *Acta Metall Mater.* **43** (1995) 2701.
6. Arsenlis, D.M. Parks, *J. Mech. Phys. Solids* **50** (2002) 1979.
7. J.L.Bassani, T.Y. Wu, *Proc. Roy. Soc. London A* **435** (1991) 21.
8. A.M. Cuitino, M. Ortiz, *Modelling Simul. Mater. Sci. Eng.* **1** (1992) 225.
9. E.H. Lee, *J. Appl. Mech.* **36** (1969) 1.
10. J.R. Rice, *J. Mech. Phys. Solids* **19** (1971) 433.
11. D. Peirce, R.J. Asaro, A. Needleman, *Acta Metall.* **30** (1982) 1087.
12. U.F. Kocks, *Metall. Trans.* **1** (1972) 1121.
13. S.B. Brown, K.H. Kim, L. Anand, *Int. J. Plasticity* **5** (1989) 95.
14. E. Orowan, Philos. Trans. Roy. Soc. London A **53** (1940) 8.
15. J. Weertman and J.R. Weertman, J.R., *Elementary dislocation theory* (Oxford Univ. Press, 1992).
16. P. Franciosi and A. Zaoui, *Acta Metall.* **30** (1982) 1627.
17. U. Essmann and H. Mughrabi, *Phil. Mag. A* **40** (1979) 731.
18. T. Takeuchi, *Trans. Jpn. Inst. Metals* **16** (1975), 629.
19. M.G.Lee, H. Lim, B.L.Adams, J.P.Hirth, R.H.Wagoner, *in preparation.*

# PLASTIC DEFORMATION BEHAVIOR OF HIGH STRENGTH STEEL SHEET UNDER NON-PROPORTIONAL LOADING AND ITS MODELING

TAKESHI UEMORI[1†]

*Department of Mechanical Engineering, School of Engineering, Kindai University,*
*1, Takayaumenobe, Higashi-Hiroshima, 739-2116, JAPAN,*
*uemori@hiro.kindai.ac.jp*


YUJI MITO[2], SATOSHI SUMIKAWA[3], RYUTARO HINO[4], FUSAHITO YOSHIDA[5]

*Department of Mechanical System Engineering, Faculty of Engineering, Hiroshima University,*

*1-4-1, Kagamiyama, Higashi-Hiroshima, 739-8527, JAPAN,*

*m080069@hiroshima-u.ac.jp, rhino@hiroshima-u.ac.jp, fyoshida@hiroshima-u.ac.jp*

TETSUO NAKA[6]

*Yuge National College of Maritime Technology,*

*1000, Yuge, Kamijima-Cho, Ochi-Gun, Ehime, 794-2593, JAPAN,*

*naka@ship.yuge.ac.jp*

Received 15 June 2008
Revised 23 June 2008

This paper deals with plastic deformations of a high tensile strength steel sheet (HTSS sheet) under biaxial stress condition including strain path. Using a cruciform specimen of a HTSS sheet of 780MPa-TS, experiments under proportional and non-proportional loadings were investigated. Numerical simulations of stress-strain responses for several strain paths after biaxial stretching were conducted using a large-strain cyclic plasticity model (Yoshida-Uemori model). The results of numerical simulation agrees well the corresponding experimental results, which is attributed to the accurate modeling of the backstress evolution of the anisotropic yield function.

*Keywords*: Kinematic hardening model; high tensile strength steel sheet; non-proportional loading; constitutive model; yield surface; yield function.

## 1. Introduction

In a last decade, with great advances of the computational technologies, finite element simulations of sheet metal stamping, became very popular in press-forming industries, however, still accuracy of numerical simulation is not always guaranteed, especially in

---

[†]Corresponding Author.

springback analysis. One of the reasons for that is the problem of constitutive models used in the simulation. In most of cases of stamping simulations, the isotropic hardening model is employed, however, it causes the errors of stress analysis since it cannot describe the Bauschinger effect of materials[1-5].

In order to solve the above mentioned problem, two of the present authors (Yoshida and Uemori[2,3]) have recently proposed an accurate constitutive model of large-strain cyclic plasticity which describes well cyclic plasticity characteristics, such as the Bauschinger effect and cyclic hardening, as well as the sheet plastic anisotropy (hereafter, called 'Yoshida and Uemori model'). High capability of this model in simulating cyclic behavior has been already verified by comparing the numerical simulations with the corresponding experimental data of uniaxial cyclic tension-compression on several types of sheet metals. However, for multi-axial stress conditions, the verification of this model has never been reported, although the constitutive modeling for multi-axial stress state is one of the most important issues to be investigated.

In the present research, in order to examine multi axial plastic deformations of HTSS sheet, experiments of uniaxial tension, as well as plane strain stretching test, after a bi-axial stretching were performed using a cruciform specimen of a 780MPa HTSS sheet. The performance of Yoshida-Uemori model in simulating stress-strain responses under biaxial stretching with strain path changes is demonstrated by comparing the numerical simulation with the experimental results.

## 2. Experimental Procedure

### 2.1. *Uniaxial tension tests*

Uniaxial tension tests were conducted for specimens cut from a HTSS sheet of 780MPa-TS in three directions of $0°$ (the rolling direction), $45°$ and $90°$. The Lankford values for the sheet are listed in Table 1. From these results, it is found that the anisotropy of this sheet is rather weak.

Table 1. Lankford values of HTSS sheet of 780 MPa-TS for three directions of $0°$ (rolling direction), $45°$ and $90°$.

| $r_0$ | $r_{45}$ | $r_{90}$ |
| --- | --- | --- |
| 0.71 | 0.98 | 0.90 |

### 2.2. *Bi-axial tension experiments*

A cruciform specimen (see Fig. 1) was subjected to stress-controlled biaxial stretching. For strain measurement, strain gauges were used. As schematically illustrated in Fig. 2, the following two types of experiments were performed:

*(1) Proportional loading experiment:*

The specimens were radially loaded in several directions, where $(\sigma_x, \sigma_y) = (1:0), (2:1), (1:1), (1:2),$ and $(0:1)$.

*(2) Non-proportional loading experiment:*

First, equi-balanced bi-axial stretching ($\sigma_y/\sigma_x=1$) had been imposed on the specimen, and then fully unloaded (O→A→O in Fig. 2). Next, it was stretched under uniaxial (O→B) or plane stress condition (O→C, $\sigma_y/\sigma_x =1/2$).

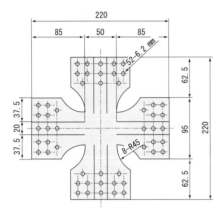

Fig. 1.  Cruciform specimen for biaxial tension experiments ( in mm).

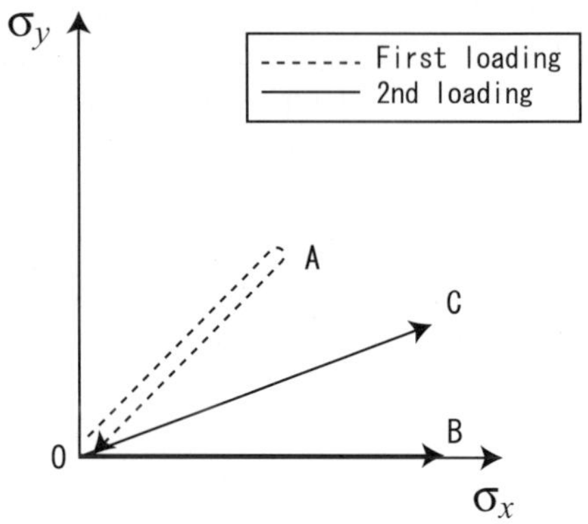

Fig. 2. Schematic illustration of stress paths in bi-axial stretching.

## 3.  Yoshida-Uemori Model

The present constitutive model of plasticity (Yoshida-Uemori model) has been constructed within the framework of two-surface modeling, wherein the yield surface moves kinematically within a bounding surface. Since the experimental stress-strain responses under reverse deformation show that the re-yielding starts at a very early stage of stress reversal, we assume only the kinematic hardening for the yield surface, while for the bounding surface mixed isotropic-kinematic hardening.

When the yield function at the initial (non-deformed) state, $f_o$, has a general form:

$$f_o = \phi(\sigma_{ij}) - Y = 0, \tag{1}$$

where $\phi$ denotes a function of the Cauchy stress $\sigma_{ij}$, and $Y$ is the initial yield strength, the subsequent yield function $f$ is given by the equation:

$$f = \phi(\sigma_{ij} - \alpha_{ij}) - Y = 0, \tag{2}$$

where $\alpha_{ij}$ denote the backstress. For our HSS, Gotoh's bi-quadratic yield function[6] was employed. The associated flow rule is written as

$$d\varepsilon_{ij}^p = \frac{\partial f}{\partial \sigma_{ij}} d\lambda \tag{3}$$

The bounding surface $F$ is expressed by the equation:

$$F = \phi(\sigma_{ij} - \beta_{ij}) - (B + R) = 0, \tag{4}$$

where $\beta_{ij}$ denotes the center of the bounding surface, and $B$ and $R$ are its initial size and isotropic hardening (IH) component.

The kinematic hardening of the yield surface describes the transient Bauschinger deformation characterized by early re-yielding and the subsequent rapid change of workhardening rate, which is mainly due to the motion of less stable dislocations, such as piled-up dislocations. The relative kinematic motion of the yield surface with respect to the bounding surface is expressed by

$$\alpha_{ij}^* = \alpha_{ij} - \beta_{ij}$$

(5)

where $\alpha_{ij}^*$ indicates the kinematic hardening of the bounding surface. The kinematic hardening rules for the yield surface and the bounding surface are given by

$$\overset{\circ}{\alpha}_{ij} = C\left\{\left(\frac{a}{Y}\right)(\sigma_{ij} - \alpha_{ij}) - \sqrt{\frac{a}{\alpha}}\alpha_{ij}^*\right\}\dot{p},$$

(6)

$$a = B + R - Y$$

(7)

$$\overset{\circ}{\beta}_{ij} = m\left[\left(\frac{b}{B+R}\right)(\sigma_{ij} - \beta_{ij}) - \beta_{ij}\right]\dot{p},$$

(8)

$$\dot{p} = \sqrt{\frac{2}{3}\dot{\varepsilon}_{ij}^r \dot{\varepsilon}_{ij}^r}$$

(9)

where $C$, $b$ and $m$ are material constants. One of the strong features of this model is that it involves limited numbers of material parameters (for cyclic plasticity, seven parameters) which can be easily determined from experiments (automatic identification of material parameters is possible, e.g., refer to Yoshida et al[7].).

## 4. Results and Discussion

Figure 3 shows the cyclic stress-strain response of 780 MPa HTSS, together with the corresponding calculated results by Yoshida-Uemori model. The experimental data clearly shows the Bauschinger effect of the material, not only at the small strain range, but also at large strain range. The calculated stress strain response shows an excellent agreement with the experimental data. From this result, the seven material parameters in the model were determined.

Figure 4 shows the experimentally determined surfaces of equi-plastic works under proportional loadings, together with the calculated results by the present model (Yoshida-Uemori model using Gotoh's yield function). The calculated results have a good agreement with the experimental observations. As far as proportional loadings without stress reversals are concerned, the surfaces of equi-plastic work calculated by simple isotropic hardening model would be always almost the same as those predicted by the present model, since during a monotonic loading the yield surface of Yoshida-Uemori model approaches the isotropically hardened bounding surface.

In contrast, under non-proportional loading, it is hardly possible for the isotropic hardening model to accurately simulate the stress-strain behavior. Figures 5 (a) and (b) show the stress-strain responses under the subsequent uniaxial stretching (O→A→O→B in Fig. 2) and plane strain stretching (O→A→O →C) after preloading of equi-balanced biaxial stretching, respectively. The calculated results by Yoshida-Uemori model and also by the isotropic hardening model are depicted in these figures. For both cases of strain

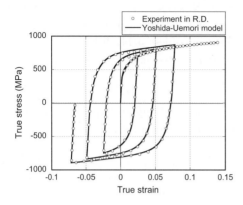

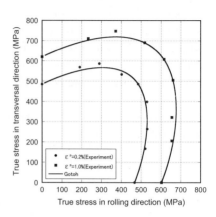

Fig. 3. Comparison of experimental data of cyclic tension-compression with the calculated result by Yoshida-Uemori model.

Fig. 4. Experimental results of equi-plastic work surfaces under proportional loadings, together with the calculated results by Gotoh's yield function.

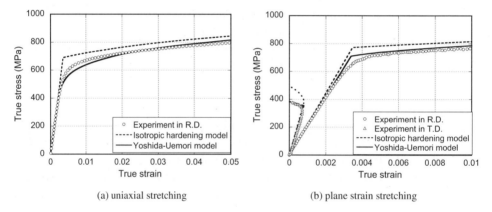

(a) uniaxial stretching

(b) plane strain stretching

Fig. 5. Stress-strain responses of HTSS under (a) uniaxial stretching(O→A→O→B in Fig.2) ; (b) plane strain stretching(O→A→O→C) after preloading of equi-balanced biaxial stretching. Broken line and solid line are the results of calculation by the isotropic hardening model and Yoshida Uemori model.

path changes, the experimentally observed flow stresses are apparently lower than those predicted by the isotropic hardening model. In contrast, Yoshida-Uemori kinematic hardening model captures this behavior very well.

## 5. Conclusions

Bi-axial tension experiments under proportional and non-proportional loadings were conducted on HTSS sheet of 780 MPa-TS. Particularly for cases of strain-path changes, the flow stresses are apparently lower than those predicted by the classical model of the isotropic hardening. The accuracy of the present authors' constitutive model of plasticity (Yoshida-Uemori model) in simulating such bi-axial stress-strain responses under non-proportional loadings has been confirmed by comparing the calculated results with the corresponding experimental observations.

## References

1.  T. Uemori, T. Okada, and F. Yoshida : Metals and Materials, Vol.4 (2000), p.311-313.
2.  F. Yoshida, and T. Uemori : Int. J. Plasticity, Vol. 18, (2002), p.661-686.
3.  F. Yoshida, and T. Uemori : Int. J. Mechanical Sciences, Vol. 45 (2003), p.1687-1702.
4.  F. Yoshida, T. Uemori, and K. Fujiwara : Int. J. Plasticity, Vol. 18, (2002), p. 633-659.
5.  T. Uemori, T. Okada, and F. Yoshida : Key Engineering Materials, Vols.233-236, (1998), p.177-180.
6.  M. Gotoh : Int. J. Mechanical Sciences, Vol. 19 (1979), p.505-512.
7.  F. Yoshida, M. Urabe, and V.V. Toropov : Int. J. Mechanical Sciences, Vol. 40 (1998), p.237-249.

# A FULL CONSTITUTIVE RELATION OF SILICON–STEEL SHEET WITH ANISOTROPY CONSIDERED

S.M. BYON[1]

*Department of Mechanical Engineering, Dong-A University, Busan, 604-714, KOREA,*
*smbyon@dau.ac.kr*

U.K. YOO[2], Y. LEE[3†]

*Department of Mechanical Engineering, Chung-Ang University, Seoul, 156-756, KOREA,*
*dain1999@hanmail.net, ysl@cau.ac.kr*

Received 15 June 2008
Revised 23 June 2008

We present a full constitutive relation of silicon steel which can describe the anisotropy effect as well. Using a pilot rolling machine, initial silicon strip with thickness of 2.5mm is rolled into sheet with several thicknesses as reduction ratio increases from 10% to 90%. To examine the effect of anisotropy on the stress-strain behavior, the specimen was cut out from the sheet so that the direction of specimen and sheet is $0^0$, $30^0$, $45^0$, $60^0$ and $90^0$, respectively. A series of tensile test are then performed with the specimens. The stress-strain curves computed from the proposed constitutive relation are compared with the experimental data. Results show that the predicted curves are in overall in a good agreement with measured ones. The work hardening and unstable softening behaviors of silicon steel during rolling are predicted by the proposed full constitutive relation.

*Keywords*: Constitutive relation; silicon steel; cold rolling; anisotropy.

## 1. Introduction

Silicon-steel is special steel containing a lot of silicon. Since high-efficient magnetic field is formed around it when electrical current flows, it is commonly used as a core material of transformer and motor. But the production is restricted owing to its poor workability. Brittle property of silicon causes breakage of strip during rolling. The crack is initiated at the first rolling pass and propagated from edge to center during rolling.

Recently, high-silicon steel is in great demand and mill engineers together with researchers in silicon steel rolling process are making efforts to improve the productivity. But they have a big obstacle to achieve high-productivity since the possibility of the strip breakage sharply increases as the content of silicon increases. To minimize the occurrence of the strip breakage, we must first understand stress-strain behavior of high-silicon steel and set up a constitutive relation of it.

Researchers[1-2] proposed an equation[3] which calculates yield stresses in a given reduction ratio, which is called 'equation of deformation resistance (EDR)'. This is

---

[†]Corresponding Author.

illustrated as Fig. 1(a). To analyze the breakage of strip, one might need a full constitutive relation (FCR) which can depict the ultimate tensile stress and failure stress in a given reduction ratio at the same time. Figure 1(b) shows the FCR experimentally measured in a wide range of reduction ratios.

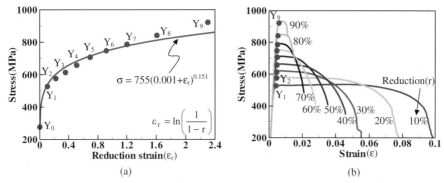

Fig. 1.  Stress-strain relation of the silicon-sheet (a) Equation of deformation resistance[3] (EDR) for the pass schedule design (b) Measured stress-strain relation in a wide range of reduction ratios of the silicon-sheet.

In this paper, a series of tensile test are performed using specimens machined from silicon-steel sheet. Through a pilot cold rolling, sheets with different thicknesses are produced while reduction ratio increases from 10% to 90%. Specimens are cut out from those cold rolled sheet. Anisotropic properties at each rolled sheet are examined by testing the specimens obliquely cut relative to rolling direction. The oblique orientations are $0^0$, $30^0$, $60^0$ and $90^0$.

We present a full constitutive relation (FCR) of silicon-steel sheet which covers from elastic region up to fracture state. To characterize the stress-strain behavior subjected to work hardening, we adopt Ramberg-Osgood model[4] as one part of FCR. We propose a new type of model to describe softening region of stress-strain curve. The proposed full constitutive relation is verified by comparing it with experimental data.

## 2.  Experiments

### 2.1.  *Pilot cold rolling*

To measure the stress-strain curve of a silicon-steel with reduction ratio variation, we should make sheet through pilot cold rolling with raw strip (i.e. hot rolled strip). Figure 2 shows the pilot cold rolling mill with a single stand driven by a 75kW constant torque DC motor. Ductile casting iron (DCI) rolls are used, with a maximum diameter of 310 mm and a face width of 320 mm. Some part of hot rolled strip is slit into many pieces with rectangle shape appropriate for the pilot cold rolling. It is important to take the pieces with keeping up the rolling direction of hot rolled coil. Figure 3 shows the cold rolled sheets. Lengths are different since the cold rolled sheets were elongated with different reduction ratios (10% ~ 90%).

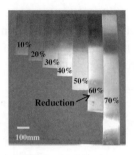

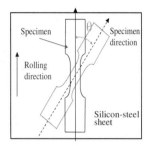

Fig. 2. A pilot cold rolling mill with two-high rolls.

Fig. 3. Cold rolled sheets at different reduction ratios.

Fig. 4. Tensile specimen cut out from rolled sheet.

## 2.2.  Tensile test

The stress-strain curves of cold rolled silicon-steel sheets are obtained by performing tensile test with Zwick Z100 machine. Tensile specimens are cut out from cold rolled sheets at each reduction ratio. They are machined to the strict dimension according to ASTM A370 for tensile test. Gauge length of the specimen is 50 mm. To investigate the effect of anisotropy, the specimens were cut obliquely relative to rolling direction of sheet. The orientations are $0^0$, $30^0$, $60^0$ and $90^0$. Figure 4 shows the orientation ($\theta$) of tensile specimens from a cold rolled sheet. For example, $0^0$ and $90^0$ represent the rolling direction and transverse one, respectively.

## 3.  Constitutive Relation

The stress-strain behavior of steel is divided into two regions which are a hardening region up to ultimate strength and a softening region until fracture of specimen, as illustrated in Fig. 5.

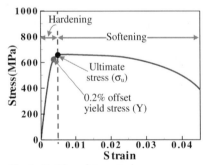

Fig. 5. Partition of stress-strain curve.

Table 1. Material parameters in the hardening region.

| Parameters | Symbols | Values |
|---|---|---|
| Elastic modulus | E (GPa) | 210 |
| Yield stress | $Y^*$ (MPa) | 527, 576, 613, 657, 706, 747, 786, 842, 922 |
| Hardening exponent | $n$ | 20 |

*Yield stress at each reduction ratio (10%~90%).

## 3.1.  Hardening region

We adopt Ramberg-Osgood model[4] to depict the hardening region. Equation (1) shows the Ramberg-Osgood's equation:

$$\varepsilon = \frac{\sigma}{E} + 0.002 \left( \frac{\sigma}{Y} \right)^n \quad (\varepsilon \le \varepsilon_u), \tag{1}$$

$n$ denotes an optimized hardening parameter to fit all experimental data to curves of E and Y. $\varepsilon_u$ represents the ultimate strain corresponding to the ultimate stress ($\sigma_u$). Since this model was suggested on the basis of many tests of sheet material[4, 5], modifications of the model are not needed. But elastic modulus (E) and 0.2% offset yield strength (Y) are required to complete the model. These material parameters are summarized in Table 1.

### 3.2. *Softening region*

To give a picture of the softening behavior, we propose the new type of equation based on Ramberg-Osgood's model. Since the decrement of stress from ultimate stress ($\sigma_u$) is proportional to the increment of strain from ultimate strain ($\varepsilon_u$), the unstable softening is mathematically described as:

$$\varepsilon = \varepsilon_u + f_1 \left( \frac{\sigma_u - \sigma}{\sigma_u} \right)^{f_2} \quad (\varepsilon_u \le \varepsilon), \tag{2}$$

where $f_1$ and $f_2$ are material parameters dependent on the reduction ratio (i.e. r = ($h_0$-h)/$h_0$ $\times$100, $h_0$ and h denote the thicknesses of raw strip before cold rolling and rolled sheet at a given pass). Details on material parameters in Eq. (2) are summarized in Table 2.

Table 2. Material parameters of silicon-steel strip in the softening region.

| Parameters | Symbols | Functions of reduction ratio ($r$) and specimen orientation ($\theta$) |
|---|---|---|
| Conversion coefficient | $f_1$ | $f_1 = \alpha((105\text{-}r)/100)^{1.6}$ <br> where, $\alpha = 5.5 \times 10^{-8}\theta^4 - 1.025 \times 10^{-5}\theta^3 + 5.71 \times 10^{-4}\theta^2 + 8.722 \times 10^{-3}\theta + 0.11$ |
| Softening exponent | $f_2$ | $f_2 = -\beta(r + 5)(r - 90) + 0.1$ <br> where, $\beta = 1.3 \times 10^{-6}\theta + 0.0001$ |
| Ultimate stress | $\sigma_u$ (MPa) | $\sigma_u = \lambda r + \delta$ <br> where, $\lambda = 0.302 \times 10^{-3}\theta^2 + 0.221 \times 10^{-1}\theta + 4.566$ <br> $\delta = 0.397 \times 10^{-4}\theta^4 - 0.688 \times 10^{-2}\theta^3 + 0.365\theta^2 + 5.49\theta + 491.13$ |

## 4.  Results and Discussion

Figure 6 shows measured and calculated stress-strain curves of silicon-steel while the reduction ratio varies from 10% to 90% in the rolling direction (i.e. $\theta = 0^0$). Solid lines represent the computed ones and dashed lines do the experimentally measured data. In overall, a good accord between experiments and calculations is noted, whereas partial disagreements are also observed at the vicinity of fracture point. This may be attributable to the fluctuation of tensile test at the unstable region.

Figure 7 illustrates the differences between the predicted constitutive relations and the measured values when specimen orientation ($\theta$) is $45^0$. In comparison with the case of $\theta = 0^0$ (Fig. 6), it is shown that the predicted curves are biased to the left with respect to measured ones. This is due to that the assumption made in modeling, i.e., the ultimate stress ($\sigma_u$) is nearly equals to the yield stress (Y) owing to brittleness of silicon-steel sheet (See Fig. 5). The assumption is mainly adequate for zero degree orientation, but somewhat differences in relevance are observed with the specimen orientation change. To predict its behavior more accurately, the variation of the difference between $\sigma_u$ and Y with the orientation change should be modeled appropriately and included in the Eq. (1).

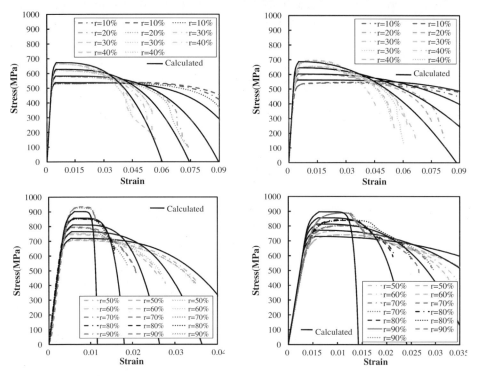

Fig. 6. Measured and calculated stress-strain curves for different reduction ratios when strip orientation ($\theta$) is $0^0$.

Fig. 7. Constitutive relations for different reduction ratios when strip orientation ($\theta$) is $45^0$.

Those effects caused by the assumption are also shown in Fig. 8, which is the case that specimen orientation ($\theta$) is $60^0$. The amount biased to the left in this orientation is less than that in $45^0$. It is also observed that the maximum elongation of specimen with orientation of $60^0$ is larger than that of $45^0$, indicating that silicon-steel sheet is deformed with direction during cold rolling.

In the case that the specimen orientation is $90^0$, the comparison between the predicted and the measured constitutive relations is made in Fig. 9. Among all specimen orientations, this case reveals the maximum difference in the magnitude of stress between the predicted and the measured one. This is attributed to the modeling declination of $\sigma_u$ in Table 2, which represents the underestimation at the small orientations (under $45^0$) whereas the overestimation at the larger orientations (over $45^0$).

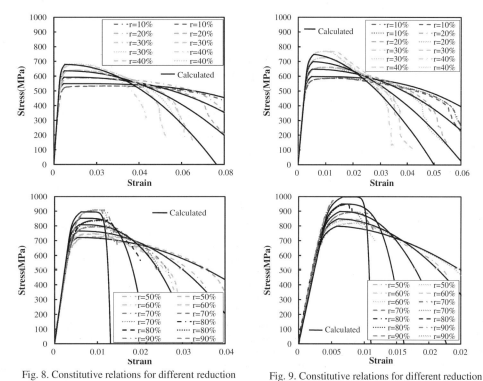

Fig. 8. Constitutive relations for different reduction ratios when strip orientation ($\theta$) is $60^0$.

Fig. 9. Constitutive relations for different reduction ratios when strip orientation ($\theta$) is $90^0$.

## 5. Concluding Remarks

In this paper, following a series of pilot cold rolling for making specimens and tensile tests, a full constitutive relation for silicon-steel sheet has been presented. In the proposed relation, two models are needed to describe the work hardening region and the unstable softening region, respectively. The effect of reduction ratio on the stress-strain behavior was a monotonically increasing of yield and ultimate tensile stress whereas the effect of specimen orientation with respect to rolling direction of the sheet, i.e., anisotropy, on the stress-strain relation revealed somewhat a periodic behavior in elongation.

### Acknowledgments

This study was supported by research funds from Dong-A University.

### References

1. W. L. Roberts, *Cold Rolling of Steel* (Marcel Dekker, New York and Basel, 1978).
2. V. B. Ginzburg, *Steel-Rolling Technology* (Marcel Dekker, New York and Basel, 1989).
3. S. M. Byon, S. I. Kim and Y. Lee, *J. Mater. Proc. Tech.* **201**, 106 (2008).
4. W. Ramberg, W. R. Osgood, *NACA Technical Note No. 902* (National Advisory Committee for Aeronautics, Washington DC, 1943).
5. K. J. R. Rasmussen, *J. Constr. Steel Res.* **59**, 47 (2003).

# SOME REMARKS ON RATE-SENSITIVITY OF NITI SHAPE MEMORY ALLOYS

OTTO T. BRUHNS[1†]

*Institute of Mechanics Ruhr University Bochum,*
*44801 Bochum, GERMANY,*
*bruhns@tm.bi.rub.de*

Received 15 June 2008
Revised 23 June 2008

Uniaxial and bi-axial tests on tubular specimens are presented. The specimens consist of NiTi alloy exhibiting pseudoelastic behavior at room temperature. A novel device to measure and control axial and torsional strains within a well defined gage section of the specimen was developed. This device, based on inductive transducers, uses the same reference points for both twist and elongation measurements. The viscous and rate dependent behavior of binary, pseudoelastic NiTi is investigated. The main focus is on the decoupling of thermal and viscous effects on the transformation stress as the specimen material is subject to heating and cooling due to latent heat generation and absorption during phase transition. On this account, an active temperature control is proposed to account for swift temperature variations. In addition to uniaxial testing of the sample, two-dimensional tension/torsion experiments are conducted in order to generalize the uniaxial findings. It is concluded that the material under consideration is independent of the rate of deformation for the applied temperature range.

*Keywords*: NiTi shape memory alloys; pseudoelasticity; isothermal testing; bi-axial loading.

## 1.  Introduction

During the last years, more and more experimental data on shape memory alloys and especially NiTi have been accumulated. In this context, dog bone shaped tubular specimens are frequently used when complex tension/torsion tests are considered. Extensive studies have been performed [1-3] in connection to material modeling.

Detailed examinations have been performed with reference to the connection between specimen temperature and transformation stress. Moreover, as a large amount of mechanical experiments is performed under strain control, multiple works have been dedicated to the characterization and determination of the strain rate effect. Ref. 4 have been among the first to report a direct connection between higher strain rates and higher transformation stress levels. Similar findings have been made by Ref. 5. Furthermore, experimental data for different ambient media are presented, resulting in different specimen temperatures and different stress-strain curves due to the temperature dependence of the transformation stress [6]. Quite recently Ref. 7 examined the rate dependence of NiTi shape-memory alloy at low and high strain rates and at room

---

[†]Corresponding Author.

temperature. Although the main focus in this work is on high strain rates up to $10^3$ s$^{-1}$, they also observed a moderate rate sensitivity for small rates from $10^{-4}$ to 1 s$^{-1}$.

It is well accepted that the transformation process results in production of latent heat leading to an exchange of heat between the specimen and the environment. Hence, the specimen temperature is changed. Thus, different specimen geometries and different surrounding media may lead to different experimental results, which renders the data from distinct experimental setups incomparable. This is why it is highly important to realize an experimental setup, allowing for a decoupled examination of strain rate and temperature effects on the material behavior.

On this account, temperature controlled experiments are imperative. First experiments of that kind were conducted on 0.1mm wire [8-10]. Consequently, only uniaxial tension tests were possible and the actual strain is deduced just from the movement of the clamping. The specimen temperature is measured using a single thermocouple. Based on the results of these tests, it is reported that the material behavior is independent of strain rate. By contrast, Ref. 11 arrives at an approach utilizing a Perzyna-type inelastic multiplier, thus, realizing a rate-dependent theory. In order to generalize these findings even for bulk material, isothermal experiments are to be carried out on three-dimensional specimens.

## 2. Material and Experimental Procedures

The material used is a 50.7 at. % NiTi exhibiting a pseudoelastic behavior at room temperature. The specimen has an hourglass-like tubular shape and is machined from tubular stock of a single charge of material. Incorporating a radius-to-wall-thickness ratio of 5:1 resulting from an outer, inner diameter of 9.6 mm, 7.92 mm, respectively, the specimen can be considered as thin-walled [12]. After the fabrication of the specimen, a special heat treatment which consists of a 1 hour solution annealing at 850°C, followed by a 0.5 hour precipitation annealing at 350°C is performed. After each annealing step the specimen is quenched in water.

For the measurement and control of strain a novel device based on inductive displacement transducers is used. This device allows for accurate measurements of twist and elongation at the same measuring points. Two thin aluminum discs are clamped within the gage section onto the specimen so that the rotation axes of the discs and the specimen are identical. Inductive transducers are used for both, measuring the axial displacement of the discs by a lever system and the rotational movement transferred by two steel cords. The levers allow for relative rotational movement of the discs which is realized by using gliding feeler heads. Two steel cords with a diameter of 0.45 mm are fixed to each of the discs. The cords are guided around the discs within a guideway groove. Two weights of $m = 40$ g are used to guarantee a constant tension force of the cords.

To realize isothermal control paths, an active temperature control is used. For heating a Joule heating is used. As current flows through an electrical conductor, part of the energy is converted into thermal energy. Consequently, the conductor gets warmer. Thus, knowing the effective resistivity of the conductor, the heating power can be controlled by changing the driven current. In this manner, the heat is generated throughout the whole cross-sectional area. In contrast to most other heating mechanisms the Joule heating produces heat homogeneously throughout the specimen material.

Since cooling of a specimen can only take place at surfaces or through cross-sectional areas, forced convection is the method of choice for cooling. Here, the volume flow is set via controlling the dynamic pressure of the cooling fluid, nitrogen, by means of a pressure control system. The desired fluid temperature is adjusted within a heat exchanger system. Thereupon, the cooling fluid is blown through a nozzle system onto the gage section of the specimen. Additionally, some part of the cooling fluid is blown through an inward fluid feed into the interior of the specimen. Seven thermocouples are applied onto the outer and one additional thermocouple onto the inner surface of the specimen. For the inner thermocouple, small-diameter thermo wires are used.

## 3.   Experimental Results and Discussion

Two different types of experiments are discussed, simple uniaxial and two-dimensional tests. For each experiment type, isothermal and non-isothermal material behavior are compared. The given data pertain mechanical and thermal data so that, for each stress-strain diagram, the respective temperature diagram is shown. First, uniaxial tension tests are presented. These experiments can be directly linked to the aforementioned uniaxial wire experiments[8,9]. Subsequently, two-dimensional box experiments are treated.

### 3.1.   *Simple tension*

Three different strain rates are applied ranging from $\dot{\varepsilon} = 10^{-5}$ s$^{-1}$ to $10^{-3}$ s$^{-1}$. The maximum strain is chosen to be $\varepsilon_{max}= 3.5\%$, which is well below the end of the pseudoelastic hysteresis. The material response under non-isothermal conditions is illustrated by Fig. 1. Here, Fig. 1(a) comprehends the stress-strain curves for the different strain rates while Fig. 1(b) represents the respective temperature-strain diagrams. Furthermore, as a measure of the local heterogeneity of the specimen temperature field, the temperature difference between highest and lowest temperature reading of the thermocouples is given.

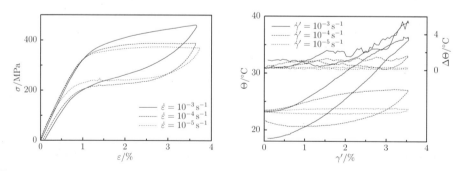

Fig. 1. Uniaxial tension tests for different strain rates, starting at room temperature: (a) Stress-strain diagram. (b) Temperature-strain diagram.

It is evident from Fig. 1(a) that the slope of the stress "plateau" steepens for higher strain rates. Simultaneously, the reverse transformation stress values are shifted to lower values with respect to the forward transformation stress curve, thus, yielding a larger hysteresis for larger strain rates. Naturally, a higher rate of latent heat is generated for higher strain rates. Since the amount of heat transfer between the specimen and the environment is finite, the specimen gets significantly warmer for the higher strain rates, leading to a maximum heating of 13K for $\dot{\varepsilon} = 10^{-3}$ s$^{-1}$ and only a slight increase of approximately 1K for the lowest strain rate.

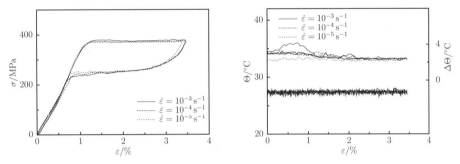

Fig. 2. Uniaxial isothermal tension tests for different strain rates: (a) Stress-strain diagram. (b) Temperature-strain diagram.

In Fig. 2, the respective diagrams for the isothermal test conditions are given. It is obvious that the mean specimen temperature is kept constant at 27.5 °C. It is of further interest to inspect the $\Delta\Theta$-curves in order to estimate the quality of the temperature control. Different from the respective curves for the non-isothermal conditions, the $\Delta\Theta$-curves in Fig. 2(b) are almost identical for the different strain rates.

Fig. 2(a) shows that the stress-strain curves are identical for the different strain rates. The plateaus exhibit the same slope and the hystereses coincide. Consequently, under the applied isothermal conditions the material behavior is independent of strain rate for the given strain rate range. Moreover, it is straightforward to claim that the achieved, macroscopically isothermal conditions mean effectively and sufficiently isothermal conditions for the transformation process. Consistently, the quality of the temperature control is approved.

## 3.2. *Combined box tests in the first axial/torsional strain-strain quadrant*

A box-shaped loading path in the first axial/torsional strain-strain quadrant is chosen as two-dimensional experiment. Here, von Mises equivalent quantities are used, although from the comparison of simple tension and torsion experiments it can be concluded that due to the strong tension/torsion asymmetry of the material the theoretical background of von Mises equivalence has to be revised thoroughly [13]. The loading path is implemented so that the maximum torsional and axial equivalent strains are equal, i.e. $\varepsilon_{max} = \gamma'_{max} = 2.0$ % resulting in a square in the strain-strain space. First, the specimen is loaded in axial direction until the maximum strain value is reached. Subsequently, the axial strain is held constant while the specimen is distorted up to the maximum shear strain. The unloading is realized accordingly. Loading velocities are applied with $\dot{\varepsilon} = 10^{-5} \ldots 10^{-3} \, s^{-1}$.

Figs. 3 and 4 present the mechanical and thermal behavior of the specimen under non-isothermal conditions. While Fig. 3(a) illustrates the strain-strain progression, Fig. 3(b) shows the respective stress-stress response. Similar curves for non-isothermal conditions have been reported [1]. In the first part of the box test, loading corresponds to the uniaxial case of simple tension exhibiting the behavior discussed in Sec. 3.1. Thereafter, when the specimen is loaded in torsional direction, the axial stress continuously decreases, which motivates the notion that the forward transformation may be ascribed to an equivalent stress measure. At the beginning of the axial unloading process, a shear stress dip is obvious. Different from the findings in Ref. 1, the renewed increase of the shear stress does not correspond to a change of the axial stress direction. However, during axial unloading the normal stress indeed migrates into the compression domain. A continuous increase of the shear stress after the dip is evident until the shear strain is finally reduced, resulting in a concurrent decrease of the shear and the compressive stress.

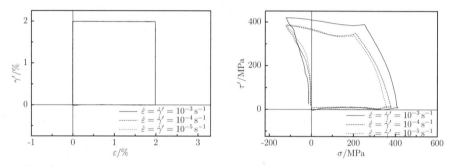

Fig. 3. Combined box test for different strain rates, starting at room temperature: (a) Strain-strain diagram. (b) Stress-stress diagram.

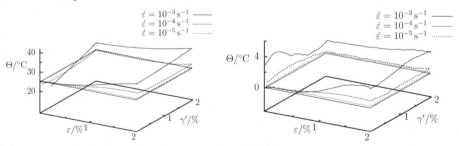

Fig. 4. Temperature response for the combined box test for different strain rates, starting at room temperature: (a) Temperature-strain diagram. (b) Temperature heterogeneity-strain diagram.

These diagrams show that the conclusions drawn in the case of uniaxial experiments can be analogously applied to the two-dimensional test. Again, the higher the strain rates, the larger is the specimen temperature rise, see Fig. 4(a). Consequently, a transformation stress shift takes place so that hystereses get larger for higher strain rates even though this behavior is not as pronounced as in the uniaxial cases.

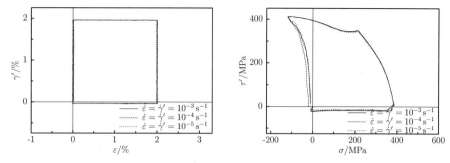

Fig. 5. Combined isothermal box test for different strain rates: (a) Strain-strain diagram. (b) Stress-stress diagram.

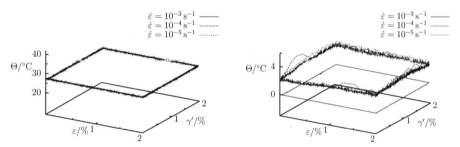

Fig. 6. Temperature response for the isothermal combined box test for different strain rates: (a) Temperature-strain diagram. (b) Temperature heterogeneity-strain diagram.

The material behavior of the respective isothermal experiments is depicted in Figs. 5 and 6. Clearly, the stress response is identical for the different strain rates. Analogous to the uniaxial case, this is ascribed to the isothermal conditions with the macroscopic mean temperature being constantly kept at 27.5 °C. Irrespective of two outliers, the local temperature heterogeneity is constant for all three experiments featuring an order of 2K.

## 4. Conclusions

Contrary to what has been stated in several articles [1,7] for the present NiTi alloy it can be accounted that the material shows no strain rate dependence within the regime of quasi-static processes at room temperature. Consequently, the stress-strain curves for various strain rates coincide under isothermal testing conditions. This does not only hold true for the uniaxial case but also for complex loading paths and complex specimen geometries.

Although the stress-strain curves in Fig. 1 (and similar curves in the above references) may suggest to conjecture a rate dependence of the behavior, it has to be considered that as a consequence of the non-isothermal conditions the increase in the slopes of the stress-strain curves may also be attributed to the latent heat of transformation and the heat of deformation[5,7,10]. A comparison of isothermal and non-isothermal tests with varying strain rates underlines our observation that this increase is mainly a temperature effect rather than a strain rate effect.

## References

1.  D. Helm and P. Haupt, *Int. J. Solids Struct.* **40**, 827 (2003).
2.  V. Imbeni *et al.*, in *Proc. Shape Memory and Superelastic Technol. Conf. 2003 (SMST-2003),* eds. A. Pelton and T.W. Duerig. (SMST Society, Menlo Park, CA, 2003), pp. 267-276.
3.  J.M. McNaney *et al.*, *Mech. Mater.* **35**, 969 (2003).
4.  K. Mukherjee, S. Sircar and N.B. Dahotre, *Mater. Sci. Eng.* **74**, 75 (1985).
5.  J.A. Shaw and S. Kyriakides, *J. Mech. Phys. Solids* **43**, 1243 (1995).
6.  J. Ortin and A. Planes, *Acta metall.* **37**, 1433 (1989).
7.  S. Nemat-Nasser, *et al.*, *J. Eng. Mater. Technol.* **127**, 83(2005).
8.  P. Lin, *et al.*, *JSME Int. J. Ser. A - Mech. Mater. Eng.* **39**, 108 (1996).
9.  H. Tobushi, *et al.*, *Mech. Mater.* **30**, 141 (1998).
10. H. Tobushi *et al.*, *Proc. Inst. Mech. Eng. Part L J. Mater.: Design Appl.* **213**, 93 (1999).
11. D. Helm, *J. Mech. Mater. Struct.* **2**, 87 (2007).
12. T.J. Lim and D.L. McDowell, *J. Eng. Mater. Technol.* **121**, 9 (1999).
13. C. Grabe and O.T. Bruhns, *Int. J. Solids Struct.* **45**, 1876 (2008).

# CREEP ANALYSIS FOR A WIDE STRESS RANGE
# BASED ON STRESS RELAXATION EXPERIMENTS

HOLM ALTENBACH[1†], KONSTANTIN NAUMENKO[2]

*Chair of Engineering Mechanics, Martin-Luther-University Halle-Wittenberg,*
*Kurt-Mothes-Straße 1, Halle (Saale), 06120, GERMANY,*
*holm.altenbach@iw.uni-halle.de, konstantin.naumenko@iw.uni-halle.de*


YEVGEN GORASH[3]

*Chair of Dynamics and Strength of Machines, National Technical University "KhPI",*
*Frunze street 21, Kharkiv, 61002, UKRAINE,*
*yevgen.gorash@gmail.com*

Received 15 June 2008
Revised 23 June 2008

Many materials exhibit a stress range dependent creep behavior. The power-law creep observed for a certain stress range changes to the viscous type creep if the stress value decreases. Recently published experimental data for advanced heat resistant steels indicates that the high creep exponent (in the range 5-12 for the power-law behavior) may decrease to the low value of approximately 1 within the stress range relevant for engineering structures. The aim of this paper is to confirm the stress range dependence of creep behavior based on the experimental data of stress relaxation. An extended constitutive model for the minimum creep rate is introduced to consider both the linear and the power law creep ranges. To take into account the primary creep behavior a strain hardening function is introduced. The material constants are identified for published experimental data of creep and relaxation tests for a 12%Cr steel bolting material at 500°C. The data for the minimum creep rate are well-defined only for moderate and high stress levels. To reconstruct creep rates for the low stress range the data of the stress relaxation test are applied. The results show a gradual decrease of the creep exponent with the decreasing stress level. Furthermore, they illustrate that the proposed constitutive model well describes the creep rates for a wide stress range.

*Keywords*: Creep; stress relaxation; advanced heat-resistant steel.

## 1. Introduction

Many components of power generation equipment and chemical refineries are subjected to high temperature environments and complex loading over a long time period. For such conditions the structural behavior is governed by various time-dependent processes including creep deformation, stress relaxation, stress redistribution as well as damage evolution in the form of micro-cracks, micro-voids, and other defects. The aim of "creep

---

[†]Corresponding Author.

41

mechanics" is the development of methods to predict time-dependent changes of stress and strain states in engineering structures up to the critical stage of creep rupture, see e.g. Refs. 1 and 2. To this end various constitutive models which reflect time-dependent creep deformations and processes accompanying creep like hardening (recovery) and damage have been recently developed. One feature of the creep constitutive modeling is the response function of the applied stress which is usually calibrated against the experimental data for the minimum (secondary) creep rate. An example is the Norton-Bailey law (power law) which is often applied because of the easier identification of material constants, the mathematical convenience in solving structural mechanics problems and the possibility to analyze extreme cases of linear creep or perfect plasticity by setting the creep exponent to unity or to infinity, respectively. Therefore, the majority of available solutions within the creep structural mechanics are based on the power law creep assumption, e.g. Refs. 1 and 3–5. On the other hand, it is known from the materials science that the "power law creep mechanism" operates only for a specific stress range and may change to the linear, e.g. diffusion type mechanism, with a decrease of the stress level (see Ref. 6). As the recently published experimental data (Refs. 7–10) show, advanced heat resistant steels exhibit the transition from the power law to the linear creep at the stress levels relevant for engineering applications. To establish creep behavior for low and moderate stress levels special experimental techniques were employed. Furthermore, experimental analysis of creep under low stress values requires expensive long-time tests. Although, the results presented in Refs. 7–10 indicate that the power law may essentially underestimate the creep rate, experimental data for many other materials are not available, and the power law stress function is usually preferred.

On the other hand, the ranges of "low" and "moderate" stresses are specific for many engineering structures under in-service loading conditions (see Ref. 2). The reference stress state in a structure may significantly change during the creep process. Stresses may slowly relax down during the service time, and thereby the application of the power law stress response function might be questionable.

In this paper we analyze the creep behavior of a 12%Cr steel based on the experimental data on creep and relaxation from Refs. 11–13. The data for the minimum creep rate are only available for the "moderate" and "high" stress levels. To recover the creep rates in the "low" stress range we apply the relaxation test data.

## 2. Creep Modeling for a Wide Stress Range

Creep behavior of modern heat-resistant steels can be analyzed basing on the creep and the stress relaxation data. Experimental data for the 12Cr-1Mo-1W-0.25V steel bolting material at 500°C presented in Refs. 12 and 13 has been selected for the formulation of the constitutive creep model for a wide stress range.

Experimental relaxation curves from Ref. 12 can be transformed into creep strain vs. time dependence by the use of the following equation

$$\varepsilon^{cr}(t) = \left[\sigma_0 - \sigma(t)\right]/E \,, \tag{1}$$

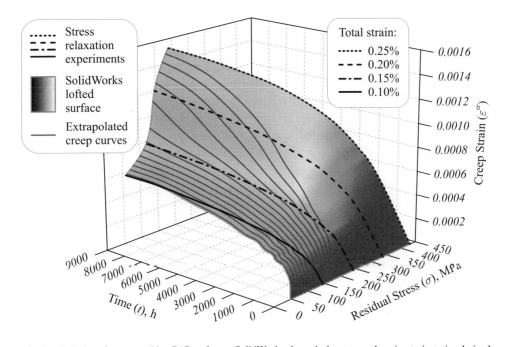

Fig. 1.   Lofted surface created by CAD-software SolidWorks through the stress relaxation trajectories derived from the experimental data (Ref. 12) for the 12Cr-1Mo-1W-0.25V steel bolting material at 500°C.

where $\varepsilon^{cr}$ denotes the creep strain, $\sigma_0$ is the initial stress, $\sigma$ is the residual stress, $t$ is the time and $E$ is the Young's modulus, respectively. The experimental values for $\sigma_0$ and $\sigma(t)$ are given in Ref. 12 and the value $E = 164.8$ GPa – in Ref. 14.

The experimental dependencies presenting residual stress vs. time and creep strain vs. time can be combined into a general three-dimensional plot with the time $t$, the residual stress $\sigma$ and the creep strain $\varepsilon^{cr}$ as orthogonal axes containing relaxation trajectories corresponding to the defined values of the constant total strain $\varepsilon = 0.25\%, 0.20\%, 0.15\%, 0.10\%$ (Fig. 1). These trajectories are formed by the interpolation of measurement points with $\varepsilon^{cr}(t)$, $\sigma(t)$ and $t$ as coordinates into the solid lines applying the CAD-software SolidWorks. The relaxation experimental data in the form of the trajectories are fitted by a surface created applying loft operation through the trajectories.

The creep curves shown in Fig. 2 were obtained from the lofted surface as cross-sections perpendicular to the residual stress axis. They exhibit both primary and secondary stages and demonstrate the necessity to apply a combined primary and secondary creep constitutive model. To describe the minimum creep rate for a wide stress range the following constitutive equation is applied (see Refs. 15 and 16 for the detailed description of the proposed creep constitutive model)

$$\dot{\varepsilon}^{cr} = a_1(\sigma H) + a_2(\sigma H)^n = A\sigma H\left[1 + \left(\frac{\sigma H}{\sigma_0}\right)^{n-1}\right]. \tag{2}$$

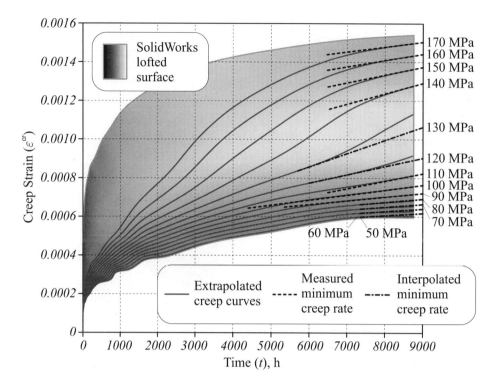

Fig. 2.   Creep curves derived from the lofted surface as cross-sections orthogonal to the residual stress axis.

Equation (2) includes the strain hardening function $H(\varepsilon^{cr})$ from Ref. 17

$$H(\varepsilon^{cr}) = 1 + \alpha e^{-\beta \varepsilon^{cr}} . \tag{3}$$

The constants in (2) are identified based on the available data of the minimum creep strain rate for the moderate stress range as follows: $a_1 = A = 2.4 \cdot 10^{-10}$ MPa$^{-1}$/h, $a_2 = 5.0 \cdot 10^{-20}$ MPa$^{-5}$/h, $n = 5$ and $\sigma_0 = 263$ MPa. The results of fitting are presented in Fig. 3. To identify the primary creep material parameters $\alpha$ and $\beta$ in (3) the stress relaxation data are applied. In this case the total strain $\varepsilon$ is kept constant while the rate of change of the residual stress is related to the creep rate as follows

$$\frac{d\sigma}{dt} = -E \frac{d\varepsilon^{cr}}{dt} , \tag{4}$$

and with Eqs. (1)–(4) the following form of the relaxation problem is obtained

$$\frac{d\sigma}{dt} = -EA\sigma \left(1 + \alpha e^{-\beta \varepsilon^{cr}}\right) \left[1 + \left(\frac{\sigma \left(1 + \alpha e^{-\beta \varepsilon^{cr}}\right)}{\sigma_0}\right)^{n-1}\right] . \tag{5}$$

The differential equation (5) is solved numerically using the MathCAD software for various values of initial conditions, i.e. the initial stress values $\sigma_0 = 145$ MPa, 240 MPa,

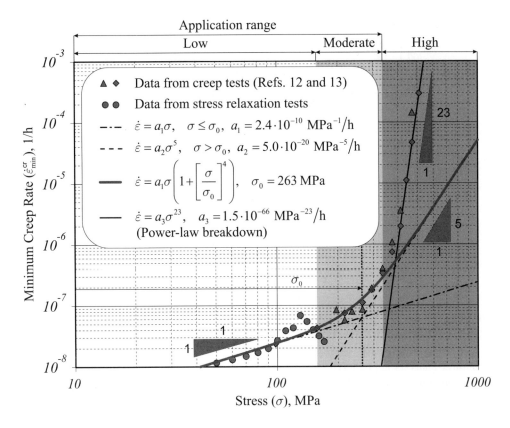

Fig. 3.    Modeling of minimum creep strain rate for the 12Cr-1Mo-1W-0.25V steel bolting material at 500°C fitting the data derived from creep and relaxation experiments (Refs. 12 and 13).

336 MPa corresponding to the total strain $\varepsilon = 0.10\%, 0.15\%, 0.20\%$ in the stress relaxation tests from Ref. 12. As a result the creep material constants in (3) are identified as $\alpha = 6$ and $\beta = 4500$. Figure 4 presents the results, where a good agreement of the numerical solution of (5) with the data in Ref. 12 can be observed. The results confirm the applicability of the constitutive model (2) to the description of the creep behavior for advanced heat resistant steels in stress application area. Furthermore, they allow to reconstruct the minimum creep rates for the low stress levels from the results of the relaxation testing. The obtained minimum creep rates are presented in Fig. 3 by circles.

## 3.  Conclusions and outlook

The aim of this paper was to analyze the creep behavior for a wide stress range. For both the moderate and the high stress levels the minimum creep rates were taken from creep tests, while for the low stress levels they were reconstructed from the stress relaxation data. The obtained results illustrate that the creep exponent gradually decreases with the decrease of the stress level. This behavior must be considered in creep modeling for

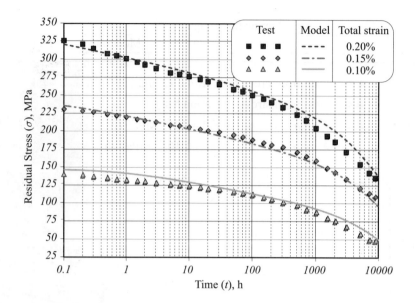

Fig. 4. Comparison of the numerical solution of the uniaxial relaxation problem by the creep constitutive model (2) with experimental data (Ref. 12) for the 12Cr-1Mo-1W-0.25V steel bolting material at 500°C.

structural analysis applications. Future work should be related to the formulation of a stress-range-dependent multi-axial creep constitutive equations and to analyze dominant softening/damage mechanisms that control the tertiary creep stage.

## References

1.    J. Betten, *Creep Mechanics* (Springer-Verlag, Berlin, 2005).
2.    K. Naumenko and H. Altenbach, *Modeling of Creep for Structural Analysis* (Springe-Verlag, Berlin, 2007).
3.    J. T. Boyle and J. Spence, *Stress Analysis for Creep* (Butterworth, London, 1983).
4.    J. A. Hult, *Creep in Engineering Structures* (Blaisdell Publishing Co., Waltham, 1966).
5.    R. K. Penny and D. L. Mariott, *Design for Creep* (Chapman & Hall, London, 1995).
6.    H. J. Frost and M. F. Ashby, *Deformation-Mechanism Maps* (Pergamon, Oxford, 1982).
7.    L. Kloc and V. Sklenička, *Mater. Sci. Eng.*, **A387-A389**, 633 (2004).
8.    L. Kloc, V. Sklenička and J. Ventruba, *Mater. Sci. Eng.*, **A319-A321**, 774 (2001).
9.    M. Rieth, *J. Nucl. Mater.*, **367-370(2)**, 915 (2007).
10.   M. Rieth *et al.*, "Creep of the Austenitic Steel AISI 316 L(N). Experiments and Models", Scientific Report: FZKA-7065 (Forschungszentrum Karlsruhe GmbH, Karlsruhe, 2004).
11.   F. V. Ellis and S. Tordonato, *Trans. ASME. J. Pres. Ves. Tech.*, **122(1)**, 66 (2000).
12.   NRIM Creep Data Sheet No. 44, National Research Institute for Metals, Tokyo (1997).
13.   NRIM Creep Data Sheet No. 10B, National Research Institute for Metals, Tokyo (1998).
14.   C. Tanaka and T. Ohba, *Trans. NRIM*, **26(1)**, 1 (1984).
15.   H. Altenbach, Y. Gorash and K. Naumenko, *Acta Mech.*, **195(1-4)**, 263 (2008).
16.   K. Naumenko, H. Altenbach and Y. Gorash, *Arch. Appl. Mech.*, in press.
17.   Y. Kostenko *et al.*, in *Proc. IMECE06* (ASME, Chicago, 2006), pp. 1–10.

# DAMAGE-COUPLED CONSTITUTIVE MODEL FOR UNIAXIAL RATCHETING AND FATIGUE FAILURE OF 304 STAINLESS STEEL

GUOZHENG KANG[1†], JUN DING[2], YUJIE LIU[3]

*Department of Applied Mechanics and Engineering, Southwest Jiaotong University,*
*Chengdu, Sichuan 610031, P. R. CHINA,*
*guozhengkang@yahoo.com..cn, djcxl@163.com, yjliu6@163.com*

Received 15 June 2008
Revised 23 June 2008

Based on the existed experimental results of 304 stainless steel, the evolution of fatigue damage during the stress-controlled cyclic loading was discussed first. Then, a damage-coupled visco-plastic cyclic constitutive model was proposed in the framework of unified visco-plasticity and continuum damage mechanics to simulate the whole-life ratcheting and predict the fatigue failure life of the material presented during the uniaxial stress-controlled cyclic loading with non-zero mean stress. In the proposed model, the whole life ratcheting was described by employing a non-linear kinematic hardening rule, i.e., the Armstrong-Frederick model combined with the Ohno-Wang model I, and considering the effect of fatigue damage. The damage threshold was employed to determine the failure life of the material. The simulated whole-life ratcheting and predicted failure lives are in a fairly good agreement with the experimental ones of 304 stainless steel.

*Keywords*: Ratcheting; low cycle fatigue; visco-plastic constitutive model; damage.

## 1. Introduction

Ratcheting is one of key issues, which should be addressed in the safety assessment and fatigue life estimation of the materials and structure components subjected to an asymmetrical stress-controlled cyclic loading. Since early 1990's, the ratcheting has been extensively studied as reviewed by Refs. 1-3. However, the existed literatures focused mainly on the ratcheting deformation and its constitutive modeling. The whole-life ratcheting and fatigue failure life were not addressed, since the prescribed number of cycles was relatively small there. The models did not consider the effect of fatigue damage on the ratcheting, and failed to simulate the whole-life ratcheting observed by Refs. 4 and 5. Therefore, it is necessary to discuss the ratcheting-fatigue interaction and simulate the whole-life ratcheting and fatigue failure life of the materials due to the significance in the design and assessment of structure components. The ratcheting-fatigue interaction was investigated recently by some researchers, and some failure models[6-9] were established. However, such failure models are constructed by using the linear or

---

[†]Corresponding Author.

nonlinear fitting of the experimental data, and cannot provide any information about the cyclic stress-strain response of the material presented during the cyclic loading. Therefore, it is necessary to consider the ratcheting-fatigue interaction in the framework of cyclic constitutive model. A damage-coupled cyclic constitutive model is a good candidate for describing the whole-life ratcheting and predicting the failure life of the materials.

Therefore, in this work, based on the previous experimental observation of the uniaxial ratcheting-fatigue interaction for 304 stainless steel[4], the evolution of fatigue damage during the stress cycling was first discussed. Then, a damage-coupled cyclic constitutive model was proposed in the framework of unified visco-plasticity and continuum damage mechanics[10] to simulate the whole-life ratcheting and predict the fatigue failure life of the material. Finally, the capability of the proposed model is verified by comparing the simulated results with the experimental ones.

## 2.  Evolution of Damage

To address the ratcheting-fatigue interaction, the evolution of damage during the stress cycling is discussed from the existed experimental results. In this work, it is assumed that the damage is isotropic and can be represented by the decrease of unloading elastic modulus after each cycle. It implies that the damage variable $D$ is defined as follow:

$$D = 1 - E / E_0. \tag{1}$$

Where, $E$ is the current unloading elastic modulus after each cycle, and $E_0$ is the initial unloading elastic modulus in the first reverse, which is the same as the initial tensile elastic modulus of the material.

From the definition of $D$ and corresponding experimental data, the evolution curves of damage variable $D$ with respect to the accumulated plastic strain $p$ are derived and the results are shown in Fig. 1 for the stress cycling with peak stress of 325MPa and various stress ratios.

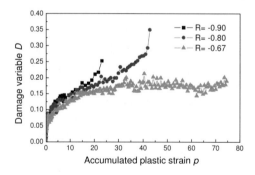

Fig. 1. Evolution of damage variable $D$ with respect to accumulated plastic strain $p$ (unit: mm/mm) obtained from the tests with peak stress of 325MPa and various stress ratios $R$.

Based on the work done by Ref. 10, the evolution law of damage variable $D$ is proposed as follow:

$$\dot{D} = \left( \frac{\sigma_{eq}^{2}}{2E_0 S(1-D)^2} \right) \dot{p} . \tag{2}$$

where, $S$ is a material constant and can be obtained directly from the experimental relation of damage variable $D$ vs accumulated plastic strain $p$ as shown in Fig. 1.

## 3. Damage-coupled Visco-plastic Constitutive Model

A damage-coupled constitutive model is constructed in the framework of unified visco-plasticity and continuum damage mechanics. Based on the strain equivalent principle[10], the isotropic damage is coupled with the elasticity and visco-plastic flow in the damage-coupled unified visco-plastic constitutive model by using the effective stress. Thus, the main equations of the damage-coupled constitutive model are outlined as follows:

$$\varepsilon_{ij} = \varepsilon_{ij}^{e} + \varepsilon_{ij}^{vp} \tag{3a}$$

$$\varepsilon_{ij}^{e} = \frac{1+\nu}{E_0} \left( \frac{\sigma_{ij}}{1-D} \right) - \frac{\nu}{E_0} \left( \frac{\sigma_{kk} \delta_{ij}}{1-D} \right) \tag{3b}$$

$$\dot{\varepsilon}_{ij}^{vp} = \frac{3}{2} \frac{S_{ij}/(1-D) - \alpha_{ij}}{(S_{ij}/(1-D) - \alpha_{ij})_{eq}} \left( \frac{\dot{\lambda}}{1-D} \right) \tag{3c}$$

$$F_y = (S_{ij}/(1-D) - \alpha_{ij})_{eq} - Q \tag{3d}$$

and

$$\frac{\dot{\lambda}}{1-D} = \dot{p} = \left\langle \frac{F_y}{K} \right\rangle^n \tag{3e}$$

where, $E_0$, $\nu$, $K$ and $n$ are material constants, which can be determined from the experimental data.

The nonlinear kinematic hardening rule is adopted as

$$\alpha_{ij} = \sum_{k=1}^{M} \alpha_{ij}^{(k)} \tag{4a}$$

$$\dot{\alpha}_{ij}^{(k)} = \xi^{(k)} \left[ \frac{2}{3} r^{(k)} (1-D) \dot{\varepsilon}_{ij}^{vp} - \mu \alpha_{ij}^{(k)} \dot{\lambda} - H(f^{(k)})(1-\mu) \alpha_{ij}^{(k)} \dot{\lambda} \right] \tag{4b}$$

and its non-damage version is similar to the Abdel-Karim-Ohno model[11].

The evolution law of isotropic hardening becomes

$$\dot{Q} = \gamma(Q_\infty - Q) \dot{\lambda} \tag{5}$$

If the saturated isotropic deformation resistance $Q_\infty$ is set to be higher than the initial one $Q_0$, Eq. (5) can describe the cyclic hardening feature of 304 stainless steel with suitable choice of material parameter $\gamma$. It should be noted that although the evolution laws of kinematic and isotropic hardening are implicitly influenced by the damage, all the material constants used in the evolution laws are obtained from the experimental data without the damage considered for simplicity.

## 4.  Simulations and Discussion

It should be noted that only the uniaxial whole-life ratcheting and failure life of 304 stainless steel are simulated and predicted in this work, even if the damage-coupled vicso-plastic cyclic model is deduced in a three-dimensional form.

### 4.1. *Determination of material parameters and failure criterion*

The material parameter relative to the evolution of damage can be determined from the experimental results by the method mentioned in the Section 2. For simplicity, other material parameters are determined from the experiments without any fatigue damage considered. The $K$ and $n$ can be determined from the monotonic tensile curves at different strain rates, and the $\zeta^{(k)}$ and $r^{(k)}$ are obtained from one monotonic tensile stress-plastic strain curve at certain strain rate after the effect of cyclic hardening is extracted by the method similar to that used by Ref. 12. The $Q_\infty$ and $\gamma$ are set from the stress response curve obtained in the strain-controlled cyclic test with moderate strain amplitude (here, 0.6%). The ratcheting parameter $\mu$ is obtained from the experimental ratcheting in the initial stage of cyclic loading by trials-and-errors. The values of all material parameters are listed in Table 1.

Table 1.  The values of material constants used in the proposed model.

| |
|---|
| $M=12$, $Q_0=120.0$MPa, $K=90$, $n=13$, $E=190$GPa, $v=0.3$, $\mu=0.05$, $m=6$, $\gamma=0.2$ |
| $\zeta^{(1)}= 6024$, $\zeta^{(2)}= 2000$, $\zeta^{(3)}= 724$, $\zeta^{(4)}= 320$, $\zeta^{(5)}= 188$, $\zeta^{(6)}= 86$, $\zeta^{(7)} =45$, $\zeta^{(8)}= 27$, $\zeta^{(9)} = 17$, $\zeta^{(10)}= 11$, $\zeta^{(11)} =6.8$, $\zeta^{(12)} = 4.0$ |
| $r^{(1)}= 34.7$, $r^{(2)} =18.3$, $r^{(3)} =11.0$, $r^{(4)} =7.0$, $r^{(5)} =8.0$, $r^{(6)} =13.5$, $r^{(7)} =13.9$, $r^{(8)} =13.0$, $r^{(9)} =9.5$, $r^{(10)} =25.5$, $r^{(11)} =16.2$, $r^{(12)} =139.7$MPa |
| $\gamma=0.25$, $Q_{sa}=200.0$MPa, $S=25$MPa, $D_c=0.3$ |

   To predict the failure life of the material by the proposed damage-coupled constitutive model, a suitable failure criterion is necessary. It has been observed from the experiments that most of specimens fail when the damage variable $D$ reaches a critical value ranged from 0.2 to 0.4. Therefore, the critical damage variable $D_c$ is set as 0.3 to obtain reasonable prediction for the fatigue lives of the material presented in the prescribed load cases. It means that if the damage $D$ reaches the $D_c$, and the material fails.

### 4.2. *Simulations to uniaxial whole-life ratcheting and failure lives*

The uniaxial whole-life ratcheting of 304 stainless steel is simulated by the proposed damage-coupled constitutive model. Figs. 2 show the experimental and simulated results obtained by the damage-coupled model and the model without damage, respectively. It is seen that it is necessary to introduce the effect of damage into the constitutive model to obtain a reasonable simulation of the whole-life ratcheting, especially for the accelerated evolution of ratcheting strain at the end of cycling. Here, the ratcheting strain is defined as $\varepsilon_r = 1/2(\varepsilon_{max} + \varepsilon_{min})$, where the $\varepsilon_{max}$ and $\varepsilon_{min}$ are the maximum and minimum axial strain in each cycle, respectively.

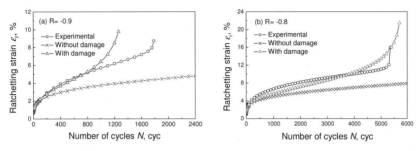

Figs. 2. Simulated and experimental results of whole-life ratcheting for the tests with peak stress of 325MPa and various stress ratios: (a) R= -0.9; (b) R= -0.8.

Based on the failure criterion proposed in Section 4.2, the failure life of the material is predicted by the proposed damage-coupled constitutive model in the stress-controlled cyclic loading with various stress levels. All the predicted and correspondent experimental lives as well as their relations are shown in Fig. 3. It is seen from the figure that the proposed model provides a good prediction for the failure life of the material presented in the cyclic loading with the ratcheting concerned. All the data are located within the twice-error band.

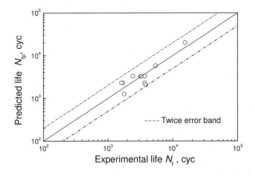

Fig. 3. Relation of predicted failure life $N_{fP}$ and experimental fatigue life $N_f$ of the material under the stress-controlled cyclic loading.

## 4.3. *Discussion*

It should be noted that only the uniaxial ratcheting-fatigue interaction of the material at room temperature is discussed and simulated in this work. The experimental observation and constitutive modeling of multiaxial ratcheting-fatigue interaction are now in progress, and will be discussed in further work. Moreover, the tests show that the final ratcheting strains for the material are larger than 10% in many loading cases. The assumption of small deformation adopted in this work is not very suitable to represent the stress-strain response of the material up to such high ratcheting strain. However, the main features of whole-life ratcheting and the damage evolution of the material are captured by the proposed damage-coupled model in the framework of small deformation theory. The adoption of finite deformation theory only makes the simulated values of ratcheting strains closer to the experimental ones, while does not change the simulated evolution feature of the whole-life ratcheting.

## 5.  Conclusions

(1) A damage-coupled cyclic constitutive model is proposed. In the model, the effects of cyclic hardening feature and fatigue damage on the ratcheting are addressed by employing the nonlinear kinematic hardening and isotropic hardening rules containing the damage variable $D$, respectively. The model simulates the whole-life ratcheting behavior of the material reasonably.

(2) A failure criterion of critical damage $D_c$ (which is set as 0.3 in this work) is proposed to predict the failure life of the material. It is shown that the proposed damage-coupled constitutive model provides a reasonable prediction to the failure life with the ratcheting considered.

## Acknowledgments

Financial supports by National Natural Science Foundation of China (10402037), the project of "973" with contract number of 2007CB714704 and the project of "863" with contract number of 2006AA04Z406 are gratefully acknowledged.

## References

1. N. Ohno, *Appl. Mech. Rev.* **43**, 283 (1990).
2. N. Ohno, *Mater. Sci. Res. Int.* **3**, 1 (1997).
3. G. Z. Kang, *Int. J. Fatigue*, (2008) doi:10.1016/j.ijfatigue.2007.10.002.
4. G. Z. Kang, Y. J. Liu, Z. Li, *Mater. Sci. Eng. A* **435-436**, 396 (2006).
5. G. Z. Kang and Y. J. Liu, *Mater. Sci. Eng. A* **472**, 258 (2008).
6. S. Kwofie and H. D. Chandler, *Int. J. Fatigue* **29**, 2117 (2007).
7. R. J. Rider, S. J. Harvey, H. D. Chandler, *Int. J. Fatigue* **17**, 507 (1995).
8. Z. Xia, D. Kujawski, F. Ellyin, *Int. J. Fatigue*, **18**, 335 (1996).
9. Y. J. Liu, G. Z. Kang, Q. Gao, *Int. J. Fatigue*, **30**, 1065 (2008).
10. J. Lemaitre and R. Desmorat, *Engineering damage mechanics: ductile, creep, fatigue and brittle* failures (Springer-Verlag, Heidelberg, 2005).
11. M. Abdel-Karim and N. Ohno, *Int. J. Plasticity*, **16**, 225 (2000).
12. G. Z. Kang, Q. Gao, X. J. Yang, *Mech. Mater.*, **34**, 521 (2002).

# AN IMPROVED ANALYTICAL CONSTITUTIVE RELATION FOR NORMAL WEIGHT HIGH-STRENGTH CONCRETE

Z. H. LU[1†], Y. G. ZHAO[2]

*Department of Architecture, Nagoya Institute of Technology, Nagoya, JAPAN,*
*lzh2076@hotmail.com, zhao@nitech.ac.jp*

Received 15 June 2008
Revised 23 June 2008

In this paper, some available analytical models for the complete stress-strain curve for high-strength concrete (HSC) under uniaxial compression are examined and compared with experimental curves published in the literature. Based on these findings, a new analytical constitutive relation is proposed to generate the complete stress-strain curves for normal weight concrete subjected to uniaxial compression with a strength range of 60-120MPa. To demonstrate the validity of the model, comparisons are made with published experimental data for uniaxial compressive tests on high-strength concrete specimens. The present model is shown to give a quite good representation of mean behavior of the actual stress-strain response.

*Keywords*: Constitutive relation; High strength concrete; Elastic modulus; Peak strain.

## 1. Introduction

In recent years, the demand for high-strength concrete (HSC) has been growing at an ever-increasing rate, and many new structures have been built using concrete with a compressive strength as high as 120MPa. With the increasing use of HSC as a structural material, more information on its mechanical properties is needed. For rational design of concrete structures, the complete stress-strain curve is essential. Due to its importance, a number of empirical expressions for the stress-strain relationship of HSC are proposed in the literature[1-11]. These models usually compare well with some experiments, especially with the results used for the model development, but provide poor prediction in many other cases. Therefore, it is difficult to choose an appropriate model for the non-linear analysis of structural members.

In this paper, some of these formulations are examined and compared with the experimental stress-strain curves of HSC subjected to uniaxial compression published in the literature. Based on these findings, a new proposal for the complete stress-strain model for HSC with a strength range of 60-120MPa is developed based on the published

---

[†]Corresponding Author.

experimental data. As the parameters of the model, new formulas for predicting the initial tangent modulus of elasticity and the strain at peak stress are also proposed. Comparisons of stress-strain curves generated by the new proposal with the published experimental data show that the present model provides a quite good representation of mean behavior of the actual stress-strain response.

## 2. Some of Existing Analytical Models for the Complete Stress-Strain Curves

A set of experimental stress-strain curves has been used by Wee et al.[10] to assess the analytical models proposed by Hongnestad[12], Wang et al.[3], Carreira and Chu[5], and CEB-FIP[7]. It was concluded that: (1) For HSC with compressive strength higher than 60 MPa, the equation proposed by Hongnestad[12] is no longer applicable even if the experimental values $\varepsilon_0$ for each individual specimen instead of a fixed value of 0.002 as suggested is used; (2) Because the coefficients in the Wang et al.'s model[3] were determined directly from the characteristics of each individual curve, it can give good predictions for the experimental curves. However, generalization of the coefficients would include normal variations in the shape of the stress-strain curve, and hence a similar variation would be, expected in the predicted response. Also, the computation involved in generating the stress-strain curve is rather tedious except with the aid of a computer; (3) CEB-FIP

Table 1. Three stress-strain models for HSC.

| Researchers | Equations |
|---|---|
| Hsu and Hsu[8] | $0 \le \varepsilon \le \varepsilon_d$, $f_c = n f_c' \beta (\varepsilon/\varepsilon_0)/[n\beta - 1 + (\varepsilon/\varepsilon_0)^{n\beta}]$; $\beta = (f_c'/65.23)^3 + 2.59$ <br> For $0 \le \varepsilon \le \varepsilon_0$, $n = 1$; <br> For $\varepsilon_0 \le \varepsilon \le \varepsilon_d$, $n = 1$ if $0 < f_c' < 62$ MPa; $n = 2$ if $62 \le f_c' < 76$ MPa; <br> $n = 3$ if $76 \le f_c' < 90$ MPa; and $n = 5$ if $f_c' \ge 90$ MPa. <br> $\varepsilon > \varepsilon_d$, $f_c = 0.3 f_c' \exp[-0.8(\varepsilon/\varepsilon_0 - \varepsilon_d/\varepsilon_0)^{0.5}]$ <br> $E_{it} = 1.2431 \times 10^2 f_c' + 2.26371 \times 10^4$, $\varepsilon_0 = 1.29 \times 10^{-5} f_c' + 2.114 \times 10^{-3}$ |
| Wee et al.[10] | $0 \le \varepsilon \le \varepsilon_0$, $f_c = f_c' \beta (\varepsilon/\varepsilon_0)/[\beta - 1 + (\varepsilon/\varepsilon_0)^\beta]$; <br> $\varepsilon > \varepsilon_0$, $f_c = k_1 f_c' \beta (\varepsilon/\varepsilon_0)/[k_2 \beta - 1 + (\varepsilon/\varepsilon_0)^{k_2 \beta}]$; <br> $\varepsilon_0 = 780(f_c')^{1/4} \times 10^{-6}$; $E_{it} = 10200(f_c')^{1/3}$; <br> $\beta = 1/[1 - f_c'/(\varepsilon_0 E_{it})]$; $k_1 = (50/f_c')^3$ and $k_2 = (50/f_c')^{1.3}$ |
| Van Gysel and Taerwe[9] | $0 \le \varepsilon \le \varepsilon_0$, $f_c = f_c'[(E_{it}/E_0)(\varepsilon/\varepsilon_0) - (\varepsilon/\varepsilon_0)^2]/[1 + (E_{it}/E_0 - 2)(\varepsilon/\varepsilon_0)]$ <br> $\varepsilon > \varepsilon_0$, $f_c = f_c'/\{1 + [(\varepsilon/\varepsilon_0 - 1)/(\varepsilon_{max}/\varepsilon_0 - 1)]^2\}$ <br> $\varepsilon_{max} = \varepsilon_0 \{[E_{it}/(2E_0) + 1]/2 + [E_{it}/(2E_0) + 1]^2/4 - 1/2]^{1/2}\}$ <br> $E_{it} = 21500 \alpha_E (f_c'/10)^{1/3}$; $\varepsilon_0 = 700(f_c')^{0.31} \times 10^{-6}$ |

Note: $f_c$ = stress of concrete; $e$ = strain of concrete; $E_{it}$ = the initial tangent modulus of elasticity, MPa; $\varepsilon_0$ = the strain at peak stress (mm/mm); $f_c'$ = $150 \times 300$ mm cylinder compressive strength of concrete, MPa; $E_0$ = the secant modulus at peak stress ($E_0 = f_c'/\varepsilon_0$); $\varepsilon_{max}$ is the concrete strain when concrete stress is equal to $0.5 f_c'$ on the descending part of the stress-strain curve; $\varepsilon_d$ = the strain corresponds to a stress value of $0.3 f_c'$ in the descending part of the stress-strain curve; $n$, $\beta$ = the material parameters; $\alpha_E$ = coarse aggregate coefficient; and $k_1$, $k_2$ = correction factors.

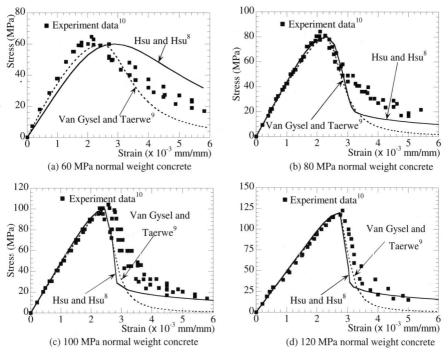

Fig. 1. Comparisons of stress-strain curves generated by Hsu and Hsu[8] and Van Gysel and Taerwe[9] with experimental data.[10]

model[7] predicts too steep a drop in the postpeak region of the curve for HSC. Moreover, the model gives hardly residual strength at high stains, which is contrary to the experimental data; (4) Although Carreira and Chu's model[5] gives a fairly good prediction for the ascending portion of the stress-strain curves, the model does not adequately represent the descending portion of the stress-strain curve of HSC for a wide range of concrete strength.

Therefore, the analytical models mentioned above will not discussed herein. In the present study, three different analytical expressions for the stress-strain curve of HSC under axial compression are briefly reviewed and shown in Table 1.

The complete stress-strain curves of HSC with compressive strength $f_c' =$ 60, 80, 100, and 120 MPa generated by Hsu and Hsu[8] and Van Gysel and Taerwe[9] are depicted in Fig. 1, compared with the experimental curves obtained by Wee et al.[10]. Fig. 1 reveals:
(1) There is a great divergence between the predictions of both the analytical Models and the experimental curves due to lower initial tangent modulus and larger peak strain compared with the corresponding test results;
(2) The residual strength of Van Gysel and Taerwe's model[9] at high strains tends to zero, which is contrary to the experimental observations;
(3) The discontinuity of the tangent at the point $\varepsilon_d$, which seems to be a shortcoming for the Hsu and Hsu's model.

Based on the same test data, Wee et al.[10] proposed a modification of Carreira and Chu's equation[5]. A comparison of the predictions of this improved model with test data generally shows good agreement. However, there is a discontinuity of the tangent at the peak stress.

In view of the foregoing, it is necessary to propose a simple, but more appropriate analytical model to represent the complete stress-strain curves for high-strength concrete.

## 3.  New Proposal for the Complete Stress-Strain Curve of HSC

### 3.1.  *Formulations for the complete stress-strain curve*

The new proposal for the complete stress-strain curve of HSC is based on the experimental results reported by Wee et al.[10] and the following conditions are considered in proposing equations to represent the stress-strain relationship of HSC.

(1) The equations should compare as favorably with experimental data as possible.

(2) Ascending and descending branches of the stress-strain curve should be implied, and the equations should represent both ascending and descending branches of the curve.

(3) The mathematical form should be as simple as possible and easily usable in any analysis.

(4) The equations should be based on physically significant parameters that can be determined experimentally. At the point of origin, $f_c = 0$ and $d(f_c)/d\varepsilon = E_{it}$. At the point of maximum stress, $f_c = f_c'$ and $d(f_c)/d\varepsilon = 0$..

(5) $\varepsilon \to \infty, f_c \to 0$.

Assimilating all this conditions, an improved stress-strain relationship for HSC ($60 \leq f_c' \leq 120$ MPa) is proposed as follows:

$$f_c = f_c' \frac{(E_{it}/E_0)(\varepsilon/\varepsilon_0) - (\varepsilon/\varepsilon_0)^2}{1 + (E_{it}/E_0 - 2)(\varepsilon/\varepsilon_0)}; \ \ 0 \leq \varepsilon \leq \varepsilon_0 \tag{1a}$$

$$f_c = \frac{f_c'}{1 + k_1(\varepsilon/\varepsilon_0 - 1)^{k_2}}; \ \ \varepsilon > \varepsilon_0 \tag{1b}$$

where $E_0 = f_c'/\varepsilon_0$ and $k_1, k_2$ = correction factors. $k_1$ and $k_2$ are given by

$$k_1 = 0.075 f_c' - 3.5, \text{ and } k_2 = 1.65 - 0.1 k_1 \tag{2}$$

### 3.2.  *Initial tangent modulus of elasticity and strain at peak compressive stress*

From Eq. (1) and Eq. (2), one can clearly see that the most important parameters with physical significance used to define the stress-strain relationship include the concrete strength $f_c'$, the initial tangent modulus of elasticity $E_{it}$, and the strain at peak stress $\varepsilon_0$, and needless to say, all of these parameters can be obtained experimentally. In this paper, the peak strain and the initial tangent modulus of elasticity are obtained using the regression analysis of experimental data provided by Wee et al.[10] and Mansur et al.[13], as shown in Fig. 2. The empirical expressions are given as follows:

$$\varepsilon_{0,reg} = 430(f_c')^{0.38} \times 10^{-6} \tag{3}$$

$$E_{it,reg} = 13500(f_c')^{0.27} \tag{4}$$

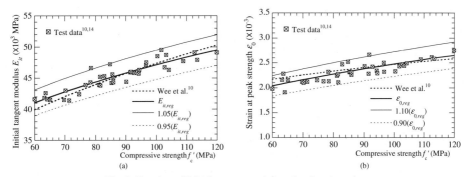

Fig. 2. Test data of initial tangent modulus of and peak strain.

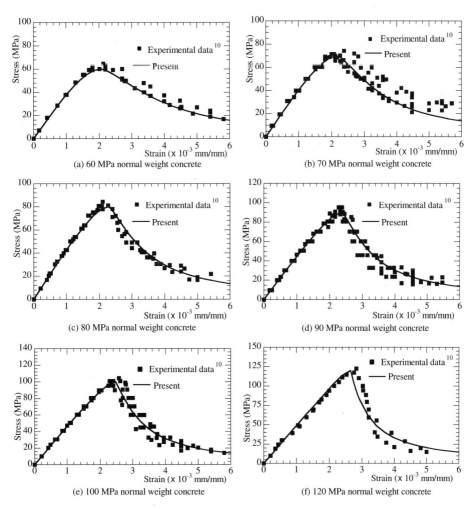

Fig. 3. Comparison of proposed model with experimental stress-strain curves for normal weight high strength concrete.

### 3.3.  *Verification of the new proposal*

In order to verify the present model, comparisons are made with the experimental data of the complete stress-strain curves of HSC in compression provided by Wee et al.[10]. The test data read directly from the publication is depicted in Fig. 3 for different grades of concrete. One can clearly see that in general, the present model fits the experimental stress-strain curves quite well.

### 4.  Conclusions

From the investigation of the present study, the following conclusions can be made for normal weight concrete with compressive strength ranging from 60 MPa to 120 MPa.

(1) New proposals for predicting the initial tangent modulus of elasticity and the peak stain of high strength concrete are proposed.
(2) Although the proposed analytical model for generating the complete stress-strain curves of high strength concrete is simple, it gives a quite good representation of mean behavior of the actual stress-strain response.

### Acknowledgments

This study is partially supported by a "Grant-in-Aid for Scientific Research (*Tokubetsu Kenkyuin Shorei-hi*)" from Japan Society for the Promotion of Science (JSPS) (No: 19·07399). The support is gratefully acknowledged.

### References

1.  M. Sargin, S. K. Ghosh, and V. K. Handa, *Mag. Concrete Res.*, 23(75-76), 99(1971).
2.  S. Popovics, *Cement & Concrete Res.*, 3(5), 583 (1973).
3.  P. T. Wang, S. P. Shah, and A. E. Naaman, *ACI J.*, 75(11), 603(1978).
4.  A. Tomaszewicz, *FCB/SINTEF rapport STF 65 A84605.* (1984).
5.  D. J. Carreira and K. H. Chu, *ACI J.*, 82(6), 797(1985).
6.  K. K. B. Dahl, *Denmarks Tekniske Hojskole, Afdelingen for Baerende Konstruktioner*, Ser. R, N 282. (1992).
7.  Comité Euro-International du Béton-Fédération Internationale de la Précontrainte (CEB-FIP). (Thomas Telford, London, 1993).
8.  L. S. Hsu and C. T. T. Hsu, *Mag. Concr. Res.*, 46(169), 301(1994).
9.  A.Van Gysel and L. Taerwe, *Mater. Struct.*, 29,529(1996).
10.  T. H. Wee, M. S. Chin, and M. A. Mansur, *J. Mat. Civ. Eng.*, 8(2), 70(1996).
11.  P. Kumar, *Mater. Struct.*, 37,585 (2004).
12.  E. Hognestad, *Bull. Ser. No. 399*, (Univ. Illinois Engrg. Experimental Station, Champaign, Ill. 1951).
13.  M. A. Mansur, M. S. Chin, and T. H. Wee, *J. Mat. Civ. Eng.*, 11(1), 21(1999).

# THERMOMECHANICAL RESPONSE OF THE ROTARY FORGED WHA OVER A WIDE RANGE OF STRAIN RATES AND TEMPERATURES

W. G. GUO[1†], C. QU[2], F. L. LIU[3]

*School of Aeronautics, Northwestern Polytechnical University, Xi'an 710072, P. R. CHINA,*
*weiguo@nwpu.edu.cn, hahaha5001@yahoo.com.cn, lfl_zt@sina.com*

Received 15 June 2008
Revised 23 June 2008

This paper is to understand and model the thermomechanical response of the rotary forged WHA, uniaxial compression and tension tests are performed on cylindrical samples, using a material testing machines and the split Hopkinson bar technique. True strains exceeding 40% are achieved in these tests over the range of strain rates from 0.001/s to about 7,000/s, and at initial temperatures from 77K to 1,073K. The results show: 1) the WHA displays a pronounced changing orientation due to mechanical processing, that is, the material is inhomogeneous along the section; 2) the dynamic strain aging occurs at temperatures over 700K and in a strain rate of $10^{-3}$ 1/s; 3) failure strains decrease with increasing strain rate under uniaxial tension, it is about 1.2% at a strain rate of 1,000 1/s; and 4) flow stress of WHA strongly depends on temperatures and strain rates. Finally, based on the mechanism of dislocation motion, the parameters of a physically-based model are estimated by the experimental results. A good agreement between the modeling prediction and experiments was obtained.

*Keywords*: Rotary forged WHA; Strain rate; Temperature; Modeling.

## 1. Introduction

Tungsten heavy alloys (WHA) mainly consist of 90–95% tungsten (W) and the rest nickel/iron(Ni-Fe). W-phase particles (bcc) are embedded in Fe–Ni-phase (fcc) matrix. The bcc structure of this refractory metal does not have phase transformations until melting point 3695K[1]. Since bcc metals exhibit a much greater temperature- and strain-rate sensitivity, the WHA has a useful combination of properties that makes it attractive to a number of military and commercial applications. In the past decades, most emphasises have mainly been placed on shearband formation, high-strain-rate dynamic deformation of the WHA[1-3]. In the recent years, the more technical processes have been attempted to enhance dynamic performance of WHA. In the present paper, main

---

[†]Corresponding Author.

objective is to evaluate mechanical characteristic of the high-speed rotary forged WHA, therefore uniaxial compression and tension tests are performed on cylindrical samples over the range of strain rates from 0.001/s to about 7,000/s, and a temperature range from 77K to 1,073K. These results are also analyzed to describe characteristics of this rotary forged WHA. To predict flow stresses of WHA, using these experimental results, and based on the mechanism of dislocation motion, a physically-based model is established and the results are compared with experimental data.

## 2. Experimental Procedure and Results

### 2.1. *Material and samples*

In this work, tungsten heavy alloy bar is fabricated by liquid-phase sintering of the blended elemental powders and then completely forged by high speed rotary forged machine. Fig.1 is the energy dispersive analysis of WHA, Major chemical composition is given in Table 1. All testing samples are taken from the rotary forged WHA bar by EDM. These samples have a 4 mm nominal diameter and 4 mm height.

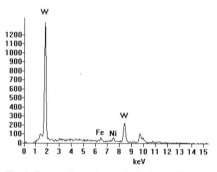

Fig. 1. Energy dispersive analysis sheet of X-ray.

Table 1. Major chemical composition of the rotary forged WHA.

| Element | % (wt.) | % (At.) |
|---------|---------|---------|
| Fe | 1.8010 | 5.1985 |
| Ni | 4.6554 | 2.7822 |
| W | 93.5437 | 82.0193 |
| other | Bal. | Bal. |

First to check mechanical orientation of the rotary forged working process, samples are taken along two directions of the bar with a diameter 30mm. Then they are tested at a $10^{-3}$ 1/s and room temperature. Figs. 2 and 3 show the comparable results.

In Fig. 2, samples are along radial direction of the bar, and in Fig. 3 the samples is along axial direction. Both figures clearly show, 1) the WHA bar has an inhomogeneous response along the section; 2) the stress or strength in center part of the bar is lower than in outer part; 3) the stresses in outer part of the WHA bar are almost the same (along zero and 90 degree radial directions). Therefore, in the work, all samples are only taken in outer part of the bar.

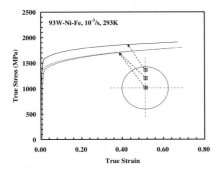

Fig. 2. True stress-true strain curves along 90 degree radial direction.

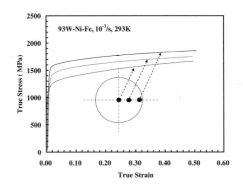

Fig. 3. True stress-true strain curves along 0 degree radial direction.

### 2.2. *Low and high strain-rate experiments*

Compression tests at strain rates of $10^{-3}$ 1/s and $10^{-1}$ 1/s are preformed over the temperature range from 77K to 1,000K, using a material testing machines, with true strains exceeding about 45%. Dynamic tests at a strain rate of 800 1/s upward are performed using a split Hopkinson bar equipment at a temperature range from 77 to about 1,000K, with strains exceeding 40%.

### 2.3. *Experimental results and discussion*

Figs 4 and 5 show the strong dependence of the flow stress on strain rates. When strain rates are below $10^{-1}$ 1/s, the flow stresses are not sensitive on strain rates. Figs 5 and 6 also show that flow stress of WHA strongly depend on the temperature. When the temperature is over about 900K (Fig. 6), flow stress of WHA seems to trend to a stable value, that is, it is not sensitive to temperature or strain rate. This non-sensitivity value is sometimes thought as athermal stress component $\tau_a$.

As also seen in Figs 5 and 6, when the temperature is increased from 77K (liquid nitrogen) to about 600K, the flow stress decreases quickly, but at temperatures between 600K and 800K, it is basically constant. Actually, an temperature non-sensitive range can be seen in Fig. 6 from 700K to 1100K, where the flow stress may even increase with the increasing temperature (around 900K), possibly due to dynamic strain aging[4,5].

### 2.4. *Failure strain evaluation of the rotary forged WHA*

In engineering use and numerical simulation, researchers and engineers often need to know failure strains of WHA. To obtain the reference values, several tensile tests are carried out at room temperature 296K and the results are shown in Fig. 7. Failure strain for the WHA at lower strain rates only is about 4%. It can also seen that failure strain of the WHA decreases with increasing strain rates. It is about 1.5% at a strain rate of 3,000/s.

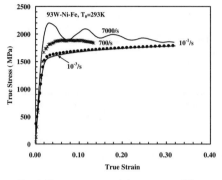

Fig. 4. True stress-true strain curves at different strain rates and a room temperature 273K.

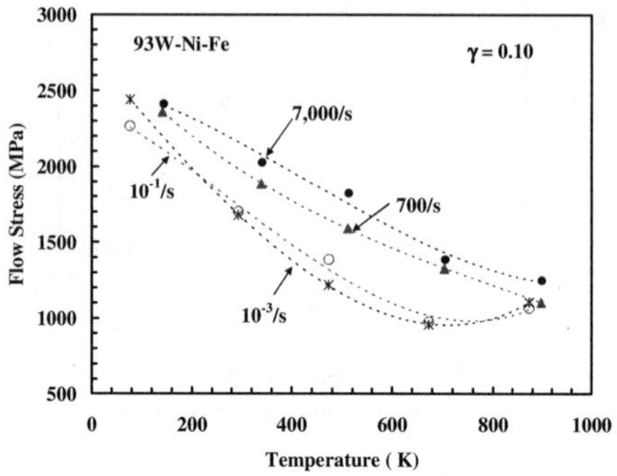

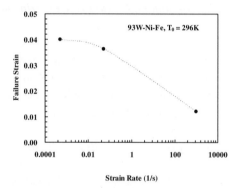

Fig. 5. Flow stress as a function of temperature for indicated strain rates and at 10% strain.

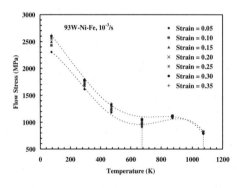

Fig. 6. Flow stress as a function of temperature for indicated strains and $10^{-3}$ 1/s strain rate.

Fig. 7.  Failure strains of the rotary forged WHA at different strain rates.

## 3.  A Physically Based Constitutive Model

Over the years, although various theories have been put forth to account for dislocation movement during plastic flow under an applied stress and in the presence of obstacles such as interstitial atoms, a common feature of these theories is that the flow stress can be thought as the sum of two components[6,7], that is,

$$\tau = \tau_a + \tau^*. \tag{1}$$

The athermal part, $\tau_a$, of the flow stress, $\tau$, is independent of the strain rate, $\dot{\gamma}$. The temperature effect on $\tau_a$ is only through the temperature dependence of the elastic modulus, especially the shear modulus, $\mu(T)$ [8]. $\tau_a$ mainly depends on the microstructure of the material, e.g., the dislocation density, grain sizes, point defects, and various solute atoms. $\tau_a$ would be proportional to $\mu(T)$. Hence, set[6,7],

$$\tau_a = f(\rho, d_G, ...)\mu(T)/\mu_0 \qquad (2)$$

where $\rho$ is the average dislocation density, $d_G$ is the average grain size, the dots stand for parameters associated with other impurities, and $\mu_0$ is a reference value of the shear modulus. In a general loading, the strain $\gamma$ represents the effective plastic strain which is a monotonically increasing quantity in plastic deformation. Therefore, as a first approximation, we may use a simple power-law representation, and choose an average value for $\mu_0$ so that $\mu(T)/\mu_0 \approx 1$. Then, $\tau_a$ may be written as,

$$\tau_a \approx a_0 + a_1 \gamma^n + ... \qquad (3)$$

where $a_0$, $a_1$ and $n$ are free parameters which must be fixed experimentally.

The thermally-activated part of the flow stress, $\tau^*$, represents the resistance to the motion of dislocations by the short-range barriers. The quantity $\tau^*$, in general, is a function of temperature, $T$, strain rate, $\dot{\gamma}$. To obtain a relation between $\dot{\gamma}$, $T$, and $\tau^*$, let $\Delta G$ be the energy that a dislocation must overcome its short-range barrier by its thermal activation. Kocks et al.[9] suggest the following relation between $\Delta G$ and $\tau^*$, representing a typical barrier encountered by a dislocation:

$$\Delta G = G_0 \left[ 1 - \left( \frac{\tau^*}{\hat{\tau}} \right)^p \right]^q, \quad G_0 = \tau b \lambda \ell = \tau V_0 \qquad (4)$$

where $0 < p \leq 1$ and $1 \leq q \leq 2$ define the profile of the short-range barrier, $\hat{\tau}$ is the shear stress above which the barrier is crossed by a dislocation without any assistance from thermal activation, and $G_0$ is the energy required for a dislocation to overcome the barrier solely by its thermal activation; $b$ is the magnitude of the Burgers vector; $\lambda$ and $\ell$ are the average effective barrier width and spacing, respectively; and $V_0$ is the activation volume. The plastic strain rate is defined,

$$\dot{\gamma} = \dot{\gamma}_r \exp(-\frac{\Delta G}{\kappa T}), \qquad (5)$$

After combine Eqs (4) and (5),

$$\tau^*(\dot{\gamma}, \gamma, T) = \hat{\tau}[1 - (-\frac{kT}{G_0} \ln \frac{\dot{\gamma}}{\dot{\gamma}_r})^{1/q}]^{1/p} \qquad , \qquad (6)$$

Ono[10] and Kocks et al.[9] suggest that $p = 2/3$ and $q = 2$ are suitable values for these parameters for many metals.

Here, for WHA, making a combination of Eqs (4) and (5), then following the parameter estimation method established by Nemat-Nasser and co-authors[6,7], the final constitutive relation for this material becomes, for $T \leq T_c$,

$$\tau = 1500\gamma^{0.1} + 2800\{1 - [-5.1 \times 10^{-5} \cdot T \cdot \ln(\frac{\dot{\gamma}}{1.21 \times 10^{11}})]^{1/2}\}^{3/2}, \qquad (7)$$

$$T = T_0 + \Delta T, \quad \Delta T = \frac{\eta}{\rho' C_V} \int_0^\gamma \tau d\gamma = 0.42 \int_0^\gamma \tau d\gamma,$$

where, $T_c = \left( -5.1 \times 10^{-5} \ln \frac{\dot{\gamma}}{1.21 \times 10^{11}} \right),$

Fig. 8 compares the experimental results with the model predictions at strain rates of $10^{-3}$ /s to 7,000/s. As is seen, a good correlation between these data and the model predictions is obtained. But this model does not include the dynamic strain aging effects, which occur in the temperature about over 600K and at a low strain rates of $10^{-3}$ /s.

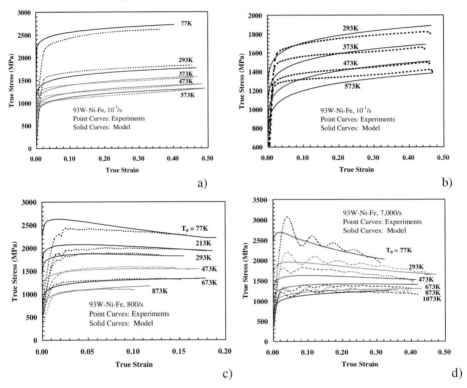

Fig. 8. Comparison of model predictions with experimental results for a Rotary Forged WHA at indicated strain rates and temperatures; a- $10^{-3}$ $^1$/s; b- $10^{-1}$ $^1$/s; c- 800 1/s; d- 7000 1/s.

## 4. Conclusions

1. The rotary forged WHA displays a pronounced changing orientation by mechanical processing, that is, the material is inhomogeneous along the section;
2. The dynamic strain aging occurs at a temperature over 700K and at $10^{-3}$ 1/s;
3. The failure strains of the rotary forged WHA decrease with increasing strain rate under uniaxial tension, it is about 1.2% at a strain rate of 1,000 1/s;
4. Flow stress of the rotary forged WHA strongly depends on temperatures and strain rates;
5. Based on the experimental results, a physically-based model is established. In the absence of dynamic strain aging, the model predictions are in good agreement with the experimental results over a wide range of temperatures for low strain rates.

**Acknowledgments**

This work has been supported by the National Defense Pre-Research Foundation under contract A2720060277 with the northwestern polytechnic university.

**References**

1.  D.S. Kim, S. Nemat-Nasser, et al., *Mech. Mater.* **28**, 227 (1998).
2.  W.-S. Lee, G.-L. Xiea, C.-F. Lin, *Mat. Sci. Eng. A.* **257**, 256 (1998).
3.  J.B. Stevens, R.C. Batra, *Inter. J. Plasticity.* **14**, 841 (1998).
4.  W.-G. Guo, *Key Eng. Mater.* **340-341**, 823 (2007).
5.  W.-G. Guo, S. Nemat-Nasser,. *Mech. Mater.* **38**, 109 (2006).
6.  S. Nemat-Nasser, W.-G. Guo, M.-Q.Liu, *Scripta Mat.* **40**, 859 (1999).
7.  S. Nemat-Nasser, W.-G. Guo, D.P.Kihl, *J. Mech. Phys. Solids.* **49**, 1823 (2001).
8.  H. Conrad. *Mat. Sci. Eng.* **6**, 260 (1970).
9.  U.F. Kocks, A.S. Argon, M.F. Ashby, *Progress in Mater. Sci.* **19**, 221 (1975).
10. K. Ono, *J. Appl. Phys.* **39**, 1803 (1968).

# CREEP CONSTITUTIVE RELATIONSHIPS AND CYCLIC BEHAVIORS OF Sn96.5Ag3Cu0.5 UNDER HIGH TEMPERATURES

JI-HONG LIU[1]

*Environmental Technology Laboratory, Daikin Industries, Ltd.*
*1304 Kanaoka-cho, Kita-ku, Sakai, Osaka, 591-8511, JAPAN,*
*jihong.liu@daikin.co.jp*

XIANG-QI MENG[2], JIN-QUAN XU[3†]

*Department of Engineering Mechanics, Shanghai Jiaotong University*
*800 Dongchuan Road, Minhang, Shanghai, 200240, P. R. CHINA,*
*jqxu@sjtu.edu.cn*

Received 15 June 2008
Revised 23 June 2008

As a lead-free solder, Sn96.5Ag3Cu0.5 has a wide application in electronic packaging. Since the solder materials usually work under cyclic temperature surroundings, creep constitutive relationships and cyclic behaviors are necessary to carry out the thermal stress analysis of a package with such a solder for its strength and life evaluations. This paper has investigated the creep constitutive relationships by constant (non-cyclic) loadings firstly, based on the creep test results at various stress and temperature levels. The complete form of the constitutive relationship containing both the linear viscous and hyperbola-sine creeps is proposed. Secondly, through the tests under cyclic stress loadings, the cyclic stress-strain relationships have been illustrated.

*Keywords*: Creep; Solder; Constitutive Relationship; Cyclic Behavior.

## 1. Introduction

Solder joints have been widely applied in the modern electronic devices. The quality of solder joints generally dominates the strength and life of electronic devices. Traditionally, solder materials of Sn-Pb series are the main joint materials due to their excellent strength, thermal properties and cheap costs. However, with the increasing needs of health care for human's surroundings, lead-free solders have been developed rapidly in the last decade. Moreover, due to the increase of packaging density, solder joints now appear in various combinations with different sizes, shapes and materials. Quantitative evaluation method of strength and life of solder joints is thereby strongly expected and has called extensive interests. To develop such an evaluation method, the mechanical properties of solder material especially its creep behaviors have to be clarified firstly. Many studies[1-5] have been reported on the time-dependent constitutive relationships both for Pb-contained and

---

†Corresponding Author.

66

Pb-free solders. However, the constitutive relationships obtained by different researchers are usually different. The creep behavior of solders generally can be distinguished into 3 stages: transient, steady and accelerating creeps. The constitutive relationships are usually prepared only for the steady creep stage by fitting the experimental stress-strain rate results with a pre-determined form. Many kinds of the constitutive relationship form have been proposed. The most famous form [1-3,6-8] is

$$\frac{d\varepsilon}{dt} = C\sigma^n e^{-\frac{Q}{RT}} \tag{1}$$

where $n$ is the power law creep index, $Q$ is the creep active energy which means the critical energy for an atom to escape from its original balance, and its dimension is $J/mol$, $R$ is the atmosphere constant $8.31\,J/(K\bullet mol)$, $T$ is the absolute temperature and $C$ is a test constant. Zhang [9] has proposed the constitutive relationship as follows:

$$\frac{d\varepsilon}{dt} = \frac{C}{T}\sigma^n e^{-\frac{Q}{RT}} \tag{2}$$

Compared with Eq.(1), the effect of temperature is different here. To take the exponent creep at high stress levels into account, the following hyperbola-sine constitutive relationship has also been proposed [10,11].

$$\frac{d\varepsilon}{dt} = A(\sinh B\sigma)^n e^{-\frac{Q}{RT}} \tag{3}$$

Here $A,B$ are also test constants. To fit experimental results well, double power law creep relationships can also be found in literature [6,7].

$$\frac{d\varepsilon}{dt} = A_1\left(\frac{\sigma}{\sigma_N}\right)^{n_1} e^{-\frac{Q_1}{RT}} + A_2\left(\frac{\sigma}{\sigma_N}\right)^{n_2} e^{-\frac{Q_2}{RT}} \tag{4}$$

Here $\sigma_N$ is a reference stress, $A_1, A_2$, $n_1, n_2$ and $Q_1, Q_2$ are test constants. From the discussion above, it can be seen that there are various constitutive relationship forms to describe the creep property of a solder material. For engineering applications, besides the difficulty of selecting a proper constitutive relationship for a specified solder material, the constants appeared in the constitutive relationship still have to be determined by experiments. One shall also be aware that all above constitutive relationships cannot deal with the creep at low stress level since it will behave linear viscous. In this study, the creep tests of Sn96.5Ag3Cu0.5 have been carried out. The complete form of constitutive relationships has been proposed. The cyclic deformation tests have been carried out too, and some cyclic deformation behaviors have been clarified.

## 2. Creep Experiments and Results

All creep tests are carried out by a Shimadzu test machine (capacity of 100kN) with a temperature box as shown in Fig.1. The specimen is cut out from a Sn96.5Ag3Cu0.5 plate, and the test part is polished by a fine sand paper. The specimen geometry is shown in Fig.2, with high temperature strain gages sticking to its two surfaces. Before the creep tests, static tests are carried out under different temperatures with loading head speed of 3mm/min to obtain the static mechanical properties. Table 1 shows the test numbers and conditions in this study. However, the stress levels shown in the table are only for reference, since the exact stress values will be recorded again by a force sensor during the test. Fig.3 shows one example of the static stress-strain curves. The data plots in the figure are the average of the strain values measured by the strain gages at the two surfaces. From the curves for different temperatures, the yield strengths corresponding to 0.2% residual strain and the effective elastic modulus corresponding to the slope of the linear part can be determined. By the polynomial approximation, it is found they can be expressed by the function of temperature as:

$$E = 43.71 - 0.1034(T - T_R) + 1.458 \times 10^{-4}(T - T_R)^2 \quad GPa \qquad (5)$$

$$\sigma_Y = 46.48 - 1.071(T - T_R) + 0.0125(T - T_R)^2 - 5.03 \times 10^{-5}(T - T_R)^3 \quad MPa \qquad (6)$$

Here $T_R = 273K$ corresponds to $0\,^\circ C$, $E$ is the effective elastic modulus, and $\sigma_Y$ is the yield strength. It can be found that they strongly depend on the temperature, and both of them decease with the increase of temperature. Creep tests are then carried out under different temperatures and stress levels referred to the yield stress as shown in Table 1. The loading stress is applied by the quick loading head speed (3mm/min) to the expected stress level, and then is kept constant by stress controlling process of the machine. Fig.4 and 5 show examples with all transient, steady and accelerating creep regions. The strain rates for steady creep can be determined by the slope of the linear part in the curves. It shall be noted that absolutely linear part of the creep curve is impossible in experimental results. Therefore the creep strain rate shall be determined carefully by the linear fitting technique to cover as more data as possible within the pre-determined accuracy.

Fig. 1. Test configurations.

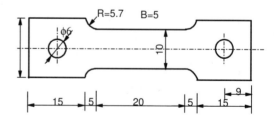

Fig. 2. Specimen geometry.

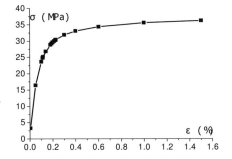

Table 1. Test conditions and numbers.

| Temperature °C | | 20 | 60 | 90 | 120 |
|---|---|---|---|---|---|
| Static tests | | 2 | 2 | 2 | 2 |
| Creep Tests | $0.5\sigma_{YT}$ | 2 | 2 | 2 | 2 |
| | $0.6\sigma_{YT}$ | 2 | 2 | 2 | 2 |
| | $0.7\sigma_{YT}$ | 2 | 2 | 2 | 2 |
| | $0.8\sigma_{YT}$ | 2 | 2 | 2 | 2 |
| | $0.9\sigma_{YT}$ | 2 | 2 | 2 | 2 |
| | $1.0\sigma_{YT}$ | 2 | 2 | 2 | 2 |

Fig. 3. Strain-stress curve at room temperature.

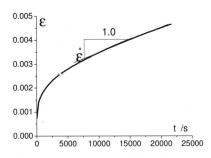

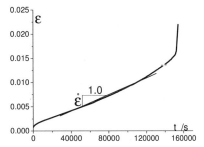

Fig. 4. Creep curve at $90\,°C$, 75% $\sigma_Y$.     Fig. 5. Creep curve at $120\,°C$, 95% $\sigma_Y$.

## 3. Constitutive Relationships

Fig.6 shows the experiment results in the double logarithmic scales. It can be seen that the creep behaviors for the steady stage must be distinguished at least into two regions. The creep at the low stress level is linear viscous, and its slope of stress-strain rate curve is about 1.0. The creep at high stress levels contains power law and exponent creeps, and the slope of the curve at the beginning is about 5.85, which means the power-law creep index, agrees well with that in Ref.[12]. It shall be noted that low or high stress levels shall be referred to the yield strength. The power law creep will appear at the relatively small stress values if the temperature is high. The linear viscous creep limit $\sigma_V$, which is the boundary of the linear viscous and power law creeps, can be obtained from Fig.6 as:

$$\sigma_V = 17.36 - 0.1219\,(T - T_R) + 2.456 \times 10^{-4} (T - T_R)^2 \quad MPa \tag{7}$$

It can be seen that linear viscous creep limit is strongly dependent of temperature. To cover both power law and exponent creeps, we use Eq.(3) to fit the experiment results at high stress levels, and use the linear viscous creep relationship to fit that at low stress. Finally, we obtain:

$$\frac{d\varepsilon}{dt} = \begin{cases} A_0 \sigma e^{-\frac{Q}{RT}} & \text{if } \sigma \leq \sigma_V \\ A(\sinh B\sigma)^n e^{-\frac{Q}{RT}} & \text{if } \sigma > \sigma_V \end{cases} \tag{8}$$

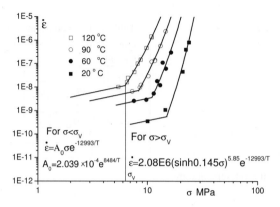

Fig. 6. Creep strain rates at different stress levels and temperatures.

where $\sigma_V$ is shown in Eq.(7), $Q/R = 12993K$, $n = 5.85$, $A = 2.08 \times 10^6\ s^{-1}$, $B = 0.145\ MPa^{-1}$, $A_0 = 2.039 \times 10^{-4} e^{8484/T}\ MPa^{-1}s^{-1}$. All the lines shown in Fig.6 are plotted according to Eq.(8) with coefficients shown above. It can be seen that the constitutive relationships of Eq.(8) are accurate enough (the maximum relative error is smaller than 6%) to describe the experimental creep behaviors. It is interesting that the coefficient $A$ in the hyperbola-sine relationship is independent of temperature, but the coefficient $A_0$ in the linear viscous case seems dependent of temperature if the active energy is kept the same as that at the high stress level. However, if we consider that the active energy at low stress levels is different with that at high stress levels, then the coefficient $A_0$ can also be regarded as independent of temperature by taking the active energy $Q/R = 12993 - 8484 = 4509\ K$.

## 4.  Cyclic Deformation Behaviors

To investigate the cyclic behavior of the solder materials, cyclic tests at various cyclic stresses are carried out by the sine wave stress with stress ratio of -1 at low frequencies. The specimen is the same with that shown in Fig.2. Cyclic tests are carried out by MTS809 fatigue machine with a temperature box, and the stress and strain are recorded automatically. Fig.7 shows the cyclic curves for the maximum stress of 15 MPa with 1 Hz frequency under the room temperature. It can be seen that the curves will be saturated after only about ten cycles. The maximum strain in each cycle increases with the cycles, and finally saturates to a constant. It is valuable to note that though the maximum stress is only 60% of the yield stress in this case, there is also obvious cyclic saturation behavior. The curves for the first several cycles are not symmetrical due to the creep and relaxation effects, but it becomes symmetrical when the deformation is saturated. Fig.8 shows the saturated cyclic curves for the maximum stress of 10 MPa with 0.2 Hz frequency under $60°C$, Fig.9 shows that for the maximum stress of 12 MPa with 0.2 Hz frequency under $90°C$. Fig.10 shows the saturated cyclic stress-strain relationships which can be obtained from the saturated maximum stress and strain. It can be seen the relationships can be approximated by linear functions except the case of 13MPa under room temperature (since at this case, the creep is linear viscous), and the saturated strain decreases with the frequency increase. The saturated stress-strain relationships can be expressed as:

$$\sigma = 6.7 + 2500\varepsilon \text{ MPa} \quad \text{for } f = 1Hz$$
$$\sigma = 4.2 + 2650\varepsilon \text{ MPa} \quad \text{for } f = 0.2Hz$$

(9)

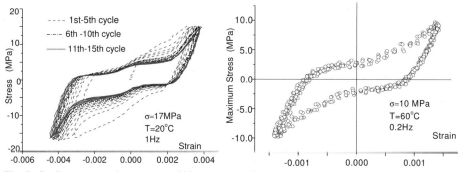

Fig. 7. Cyclic stress-strain curves at $20°C$.　　　Fig. 8. Saturated stress-strain curves at $60°C$.

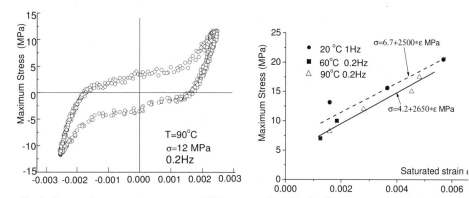

Fig. 9. Saturated stress-strain curves at $90°C$.　　　Fig. 10. Cyclic stress-strain relationships.

## 5. Conclusions

Creep and cyclic deformation tests of the lead-free solder Sn96.5Ag3Cu0.5 have been carried out. The main results obtained can be concluded as:

1) The yield stress and effective elastic modulus of the solder material are dependent of temperature, and their empirical formulas have been determined.
2) Creep deformation appears even if the stress is smaller than the yield stress.
3) The complete form of the constitutive relationship has been determined.
4) The boundary stress between the linear viscous and hyperbola-sine creep is also dependent of temperature.

5) The active energy should be considered different at low and high stress levels.
6) Cyclic deformation behavior appears even if the maximum stress is smaller than the yield stress, but it will be saturated after several cycles soon.
7) The saturated cyclic stress-strain relationship is dependent of loading frequency.

## Acknowledgment

This study is supported by the 5th SJTU-Daikin co-project.

## References

1. R. Mahmudi, A. R. Geranmayeh, *Material Science and Engineering,* **A448**, 287(2007).
2. F. Gao, T. Takemoto, *Materials Letters,* **60**, 2315(2006).
3. I. Shohji, T. Yoshida, T. Takahashi, *Materials Science and Engineering*, **A366**, 50(2004).
4. I. Dutta, D. Pan, R. A. Marks, *Materials Science and Engineering,* **A410–411**, 48(2005).
5. H. Lu, H. Shi, M. Zhou, *Microelectronics Reliability*, **46**, 1148(2006).
6. S. Wiese, S. Rzepka, *Microelectronics Reliability*, **44**, 1893(2004).
7. S. Wiese, F. Feustel, E. Meusel, *Sensors and Actuators*, **A99**, 188(2002).
8. S. Wiese, E. Meusel, *Journal of Electronic Packaging*, **125**, 531(2003).
9. X. P. Zhang, C. B. Yu, Y. P. Zhang, *Journal of Materials Processing Technology*, **35**, 842(2007).
10. A. Schubert, In *Proc. of IEEE 53rd electronic components and technology conference* (New Orleans, Louisiana, USA, 2003), pp. 603-610.
11. Q. Zhang, In *Proc. of IEEE 53rd electronic components and technology conference* (New Orleans, Louisiana, USA, 2003), pp. 1862-1868.
12. J. W. Kim, D. G. Kim, S. B. Jung, *Microelectronics and reliability*, **46**, 535(2006).

# Part B
## Damage, Fracture, Fatigue and Failure

# PLASTIC DEFORMATION OF POLYMER INTERLAYERS DURING POST-BREAKAGE BEHAVIOR OF LAMINATED GLASS - PARTIM 2: EXPERIMENTAL VALIDATION

D. DELINCÉ[1], D. CALLEWAERT[3], W. VANLAERE[4], J. BELIS[5†]

*Laboratory for Research on Structural Models, Ghent University, Ghent, BELGIUM,*
*didier.delince@UGent.be, dieter.callewaert@UGent.be, wesley.vanlaere@UGent.be, jan.belis@UGent.be*

J. DEPAUW[2]

*Jan De Nul n.v. , Hofstade-Aalst, BELGIUM,*
*jeffreydepauw@yahoo.com*

Received 15 June 2008
Revised 23 June 2008

Transparent polymer interlayer foils are widely used to increase the safety of glass applications, mainly in construction and automotive industry. In case one or more glass sheets in a laminate fracture, the remaining structural capacity (i.e. stiffness and strength) highly depends on the solicitation and mechanical properties of the interlayer. In such a post-breakage state, the interlayer is subjected to a complex combination of phenomena such as partial delamination, large strain deformation, strain hardening, rupture, etc. In part 1, an analytical model was presented to describe the mechanical behavior of glass laminates in a post-breakage state in which both glass plates are broken. In addition, a comparison of the theoretical model with experimental results is presented in part 2 below.

*Keywords*: Laminated glass; post-breakage; plastic deformation; experiments.

## 1. Introduction

In Partim 1 of this paper[1], an analytical approach was presented to describe the residual strength and stiffness of a laminate composed of two broken annealed float glass plates and one adhesive interlayer foil.

This paper focuses on the main results observed for stage III, more specifically when cracks have appeared in the two glass plates. Four-points bending tests were carried out on Glass/SGP laminates to study the different post-breakage stages as described by Kott[2] for Glass/PVB laminates. The analytical model mentioned above[1] is used to process the test results in order to evaluate the compatibility of analytical and experimental methods.

---

[†] Corresponding Author.

## 2. Materials

Twenty-seven Laminated glass samples composed of annealed float glass have been tested (cf. Table 1, below). The nominal thickness of the samples was 8 mm float glass – 1.52 mm SentryGlas® Plus (SGP) interlayer[3] – 8 mm float glass. Four different sample series were tested, each with the same length $L_t$ (1100 mm) but with a different width $b$ (120, 150, 200 and 300 mm). Compared to the widely used polyvinyl butyral (PVB), the SGP interlayer is a relatively stiff thermoplastic polymer with a glass transition temperature $T_g$ of about 55° C to 60°C, and thus in a glassy state at room temperature.

## 3. Methods and Test Description

Four series of four-points bending tests were performed on laminated glass plates about their weak axis. Fig. 1 schematically depicts the testing configuration.

The tests were started on two series of unbroken samples (stage I: $b$ = 120 mm and 200 mm), and on two series of pre-cracked samples (stage II: $b$ = 150 mm and 300 mm).

For the latter, the initial crack was processed in the central section of the upper plate previously to the test.

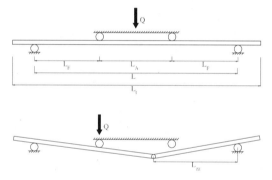

Fig. 1. Four-point bending test: configuration in stage I/II (upper part) and in stage III (lower part).

The bending load $Q(t)$ was applied with a roughly constant rate between 0.01 mm/s and 0.02 mm/s. The transition between the initial stages (I or II) to the post-breakage stage III is characterized by brittle fracture and the crack propagation in the glass plates. The crack initiation points are randomly distributed (occasionally even outside the central loading zone). Subsequently, the cracks develop mainly in transversal direction, resulting in a cracked zone of limited length. Unlike Glass/PVB laminates in similar configurations, the Glass/SGP laminates showed a significant residual stiffness to the imposed deformation. As illustrated in Fig. 2, a "yield-section" configuration appears at that point for increasing deflections. Finally, failure occurs, generally due to tearing of the interlayer. Fig. 3 depicts a typical curve of the applied force $Q$ in function of the time $t$ and the different stages.

Fig. 2. Four-point bending test in stage III: local behavior in the yield-section. Left: test WTC21A. Right: yield-section model.

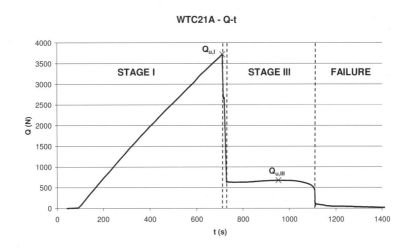

Fig. 3. Typical loading curve (example shown for specimen WTC21A).

## 4. Results

An overview of the test results are tabled in Table 1 and depicted in Fig. 4.

Table 1. Overview of the samples and test results.

| Test nr. | a1,eff [mm] | d,eff [mm] | a2,eff [mm] | b,eff [mm] | Qu,I [N] | Qu,III [N] | Lzz [mm] | mIII,Q | mIII,G [Nmm/mm] | mIII,E |
|---|---|---|---|---|---|---|---|---|---|---|
| **120mm** | | | | | | | | | | |
| WTC2A | 7.85 | 1.76 | 7.83 | 119.81 | 1801.9 | 371.3 | 350 | 349.5 | 35.8 | 385.3 |
| WTC2A* | 7.86 | 1.76 | 7.84 | 119.92 | 1987.0 | 297.5 | 450 | 359.8 | 37.8 | 397.5 |
| WTC2B | 7.85 | 1.71 | 7.84 | 120.33 | 1183.8 | 545.1 | 280 | 408.7 | 32.1 | 440.8 |
| WTC2B* | 7.85 | 1.77 | 7.84 | 120.02 | 2842.5 | 407.8 | 340 | 372.2 | 35.4 | 407.6 |
| WTC2C | 7.82 | 1.82 | 7.84 | 120.49 | 2008.5 | 331.0 | 430 | 380.6 | 37.6 | 418.2 |
| WTC2D* | 7.94 | 1.77 | 7.79 | 120.59 | 1900.7 | 637.7 | 233 | 397.0 | 28.6 | 425.6 |
| average | | | | | 1954.1 | 431.7 | | | | |
| stdev | | | | | 531.5 | 132.4 | | | | |
| **200mm** | | | | | | | | | | |
| WTC21A | 8.03 | 1.69 | 8.04 | 200.53 | 3719.6 | 543.5 | 410 | 358.1 | 37.9 | 396.0 |
| WTC21A* | 7.94 | 1.69 | 8.03 | 200.44 | 2869.3 | 763.5 | 338 | 414.8 | 35.6 | 450.4 |
| WTC21B | 7.93 | 1.80 | 7.97 | 200.32 | 3504.7 | 477.4 | 450 | 345.6 | 38.1 | 383.7 |
| WTC21B* | 7.93 | 1.75 | 8.02 | 200.54 | 2876.1 | 464.2 | 450 | 335.6 | 37.7 | 373.3 |
| WTC21C | 7.96 | 1.76 | 8.05 | 199.98 | 3398.1 | 641.1 | 370 | 382.2 | 36.6 | 418.8 |
| WTC21C* | 7.96 | 1.76 | 8.05 | 200.06 | 2672.7 | 572.5 | 410 | 378.1 | 37.9 | 416.0 |
| WTC21D* | 7.90 | 1.83 | 7.91 | 200.00 | 2817.1 | 547.7 | 435 | 383.8 | 38.2 | 422.0 |
| average | | | | | 3122.5 | 572.8 | | | | |
| stdev | | | | | 408.0 | 102.8 | | | | |

| Test nr. | a1,eff [mm] | d,eff [mm] | a2,eff [mm] | b,eff [mm] | Qu,II [N] | Qu,III [N] | Lzz [mm] | mIII,Q | mIII,G [Nmm/mm] | mIII,E |
|---|---|---|---|---|---|---|---|---|---|---|
| **150mm** | | | | | | | | | | |
| WTC10A* | 7.92 | 1.69 | 7.96 | 150.78 | 2221.5 | 507.6 | 360 | 390.5 | 37.0 | 427.5 |
| WTC10B | 7.90 | 1.79 | 7.92 | 150.12 | 1841.6 | 603.1 | 305 | 394.8 | 34.2 | 429.1 |
| WTC10B* | 7.93 | 1.87 | 7.92 | 150.17 | 2078.9 | 517.5 | 370 | 410.8 | 37.0 | 447.8 |
| WTC10C | 7.81 | 1.84 | 7.85 | 150.36 | 1716.0 | 404.8 | 435 | 377.4 | 38.3 | 415.7 |
| WTC10C* | 7.90 | 1.76 | 7.85 | 150.53 | 1416.0 | 459.5 | 430 | 422.9 | 38.4 | 461.4 |
| WTC10D | 7.93 | 1.73 | 7.96 | 150.53 | 1955.0 | 545.4 | 360 | 420.3 | 36.9 | 457.2 |
| WTC10D* | 7.93 | 1.73 | 7.97 | 150.61 | 1897.7 | 469.3 | 418 | 419.7 | 37.8 | 457.5 |
| average | | | | | 1875.2 | 501.0 | | | | |
| stdev | | | | | 260.0 | 64.2 | | | | |
| **300mm** | | | | | | | | | | |
| WTC32A | 7.92 | 1.75 | 7.92 | 300.84 | 1993.7 | 884.1 | 405 | 383.5 | 37.7 | 421.2 |
| WTC32A* | 7.93 | 1.80 | 7.94 | 300.85 | 2271.7 | 794.4 | 408 | 347.1 | 37.8 | 385.0 |
| WTC32B | 7.88 | 1.77 | 7.94 | 300.89 | 2835.8 | 876.6 | 430 | 403.7 | 38.0 | 441.7 |
| WTC32B* | 7.94 | 1.79 | 7.87 | 300.86 | 3717.2 | 718.1 | 450 | 346.1 | 38.0 | 384.1 |
| WTC32C | 7.89 | 1.76 | 7.88 | 300.89 | 2698.0 | 1000.8 | 362 | 388.0 | 36.4 | 424.4 |
| WTC32D | 7.93 | 1.80 | 7.86 | 300.89 | 3519.7 | 833.5 | 438 | 391.0 | 38.0 | 428.9 |
| WTC32D* | 7.93 | 1.84 | 7.86 | 300.85 | 3265.2 | 728.2 | 445 | 347.1 | 38.0 | 385.1 |
| average | | | | | 2900.2 | 833.7 | | | | |
| stdev | | | | | 638.5 | 98.6 | | | | |

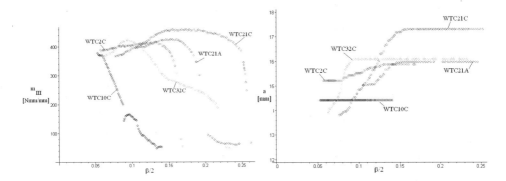

Fig. 4. Test results in stage III in the yield-section. Left: moment – deformation. Right: calculated delaminated length *a*.

## 5. Analysis and Discussion

The test results were processed for stage III using the analytical model presented in partim 1[1], considering the following values[3,2] of the interfacial properties established for Glass/PVB laminates to calculate the delaminated length *a*:

$$\Gamma_0 = 50 \; J/m^2 \;\; and \;\; \tau = 3 \; N/mm^2 \tag{13}$$

The used analytical model shows some shortcomings regarding the experiments:

(i) there was no single yield-section, but usually a broken zone of limited length (cf. fig. 3) due to the applied test sequence. However, a "yield-section" deformation does appear in most cases;

(ii) the model simplifies the damage occurring in the compression zone of the upper glass plate, while some small splitting pieces of glass were observed during the test. We consider this as one of the weakest assumptions of the analytical model in this case.

In spite of those shortcomings, the processing of the test results allows to put some interesting facts in evidence:

- despite the random breakage patterns in the glass plates developing during stage III, variation of the residual strength is limited (see values of $Q_{u,III}$ in Table 1);
- the calculated delaminated length *a*, even if depending on the assumed values for the mechanical properties at the interface Glass/SGP, has a credible order of magnitude (about 1 cm to 1.8 cm). In this case, the delaminated length can be interpreted as being spread across the various cracks of the broken zone instead of near a single cracked yield-section.
- comparing left and right graphs in fig. 4, it appears clearly that a greater resisting moment corresponds to a greater delaminated length;

- the calculated stresses in the interlayer in the yield-section, which are depending on the assumed values for the mechanical properties at the Glass/SGP interface, are in the order of magnitude of the yield stress of the interlayer resulting from uniaxial tensile tests[5].

## 6.  Conclusions

The post-breakage behavior of a Glass/SGP laminate during a four-points bending test presenting a "yield-section" configuration has been analyzed using a simple model, which takes into account delamination effects. As a result, the role of the interlayer stiffness and the adhesion level between glass and interlayer is clarified. From this it can be concluded that the post-breakage behavior of laminated glass cannot be characterized by the mechanical properties of the interlayer only, but that also the adhesion level and delamination ability between the interlayer and the glass plates are of major importance. Consequently, a higher level of adhesion between the glass and the interlayer does not always provide a higher global residual strength to the laminate.

The next step in this research will be to take explicitly into account the plastic properties of the interlayer in the model. It is expected that this will have a more important factor on the post-breakage behavior of Glass/SGP laminates than of Glass/PVB laminates.

## Acknowledgments

This work was supported by the Fund for Scientific Research-Flanders (FWO-Vlaanderen, grant nr. 3G018407). Additionally, the authors wish to acknowledge the support of the Laboratory for Research on Structural Models at Ghent University and Lerobel (Groep Leroi, Belgium).

## References

1.  J. Belis, D. Delincé, J. Depauw, D. Callewaert and R. Van Impe, *Mod. Phys. Lett. B*, partim 1, in press.
2.  A. Kott, Ph.D. Thesis, ETH Zürich, Zürich, 2006.
3.  DuPont de Nemours and Company, Inc., DuPont SentryGlas® Plus Specifying and Technical Data, Doc. Ref. SGP030718_1, v.1, (2003).
4.  M. Seshadri, S.J. Bennison, A. Jagota and S. Saigal, *Acta Mater.*, **50**, 4477 (2002).
5.  J. Belis, J. Depauw, D. Delincé, D. Callewaert and R. Van Impe, *Eng. Fail. Anal.*, (expected 2009).

# STUDY ON THE FRACTURE BEHAVIOR OF W-NI-FE HEAVY ALLOYS

WEIDONG SONG[1†], JIANGUO NING[2]

*State Key Laboratory of Explosion Science and Technology, Beijing Institute of Technology,*
*Beijing 100081, P. R. CHINA,*
*swdgh@bit.edu.cn, jgning@bit.edu.cn*


HAIYAN LIU[3]

*School of Science, Beijing Institute of Technology,Beijing 100081, P. R. CHINA,*
*liuhy58@yahoo.com.cn*

Received 15 June 2008
Revised 23 June 2008

The fracture behaviors of tungsten alloys 91W-6.3Ni-2.7Fe were investigated by tensile tests and numerical simulations. Firstly, tensile tests were conducted on the S-570 SEM with an in-situ tensile stage. With this system, the process of deformation, damage and evolution in micro-area can be tracked and recorded, and at the same time, the load-strain curve can be drawn. Secondly, the 2D finite element model of a unit cell for the tungsten alloys was established by using finite element program. By copying the unit cell model, the macro-model of the alloys was given. Dozen of cases were performed to simulate the fracture behaviors of tungsten alloys. Thirdly, the random model of the alloys was established. The fracture patterns of the alloys were investigated by the model. The interface between the tungsten particle and the matrix was explored in details. The effect of interface strength on the fracture patterns of the alloys was taken into account. A good agreement was achieved between the experimental results and the numerical predictions.

*Keywords*: Tungsten alloys; Fracture behavior; Micro-crack; Numerical simulation.

## 1. Introduction

Tungsten alloys are particulate reinforced composites which contain a very large percentage of nearly spherical tungsten particles embedded in a ductile matrix of nickel-iron. These alloys are characterized by high strength, moderate ductility and high toughness resulting from their special microstructure features. The good combination of strength and ductility makes the composites an attractive candidate for many military and civil applications[1-3]. Rabin and German[4] conducted tensile tests at room temperature and found that the flow stress increased and the fracture strain decreased with increasing tungsten content and crack initiation occurred on the sample surfaces at the tungsten-tungsten particle boundaries. Churn and German[5] found that crack propagation occurred

---

[†] Corresponding Author.

by cleavage of tungsten particles ahead of the crack tip in bend tests. In contrast, several authors have reported crack initiation at tungsten-tungsten particle boundaries[6].

The objective of this paper was to investigate the fracture behavior of 91W-6.3Ni-2.7Fe by using tensile tests and numerical simulations. The two-dimensional finite element model and random distribution model of tungsten alloys were established and dozens of cases were performed to investigate the fracture modes of tungsten alloy target under tensile loading.

## 2.  Tensile Tests

### 2.1.  *Experimental set-up and procedure*

The tungsten alloy used in present study is 91W-6.3Ni-2.7Fe, which was made by the refined alloy powder through solid-phase sintering. In Fig.1, a rectangle notch (1mm × 0.1mm) was fabricated in the middle of the specimen to easily control the initiation point of the crack. The tensile tests were conducted on the S-570 SEM with an in-situ tensile stage. With this system, the process of deformation, damage and evolution in micro-area can be tracked and recorded, and at the same time, the load-strain curve can be drawn.

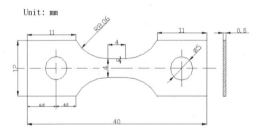

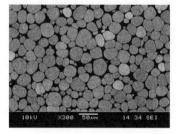

Fig. 1.  Schematic of the specimen.          Fig. 2.  Microstructure of tungsten alloys.

### 2.2.  *Experimental results*

Fig.2 shows the original microstructure of the tungsten alloys used in the tests.  It can be seen from Fig.2 that the average tungsten particle size is about 30~40μm. The particles are oblong with different diameters and surrounded by Ni-Fe matrix and the tungsten particles appear well bonded to the Ni-Fe matrix.

Fig.3 illustrated the detail transgranular process of the primary crack. Fig.3 (a) demonstrated the crack would perforate a particle. Fig.3 (b) presented that the primary crack perforated the matrix and extended to the next particle. Fig.3 (c) depicted the obvious slippage phenomenon on the surface of the particle which would be perforated by the primary crack. In Fig.3 (d), there were some microvoids formed in the particle.

Fig.3 (e) and Fig.3 (f) showed that the microvoids tore and connected with the primary crack. The crack propagation is of mixed character, containing both intergranular and transgranular features.

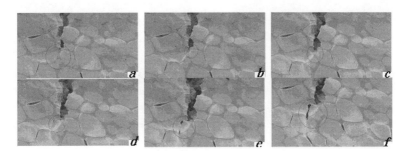

Fig. 3. Initiation and propagation of micro-cracks.

## 3. Numerical Simulations

### 3.1. *Finite element model*

The finite element program used for the study is Ansys and the element is 2-D SOLID42. The material model for both the tungsten particle and the matrix is plastic kinematic model. The bond strength between tungsten particle and matrix is set by CONSTRAINED_TIE-BREAK with the failure strain of 0.015. The parameters used in the model are listed in Table 1.

Table 1. Parameters used in numerical simulations.

|  | $\rho(g/cm^3)$ | $E(GPa)$ | $\mu$ | $\sigma_y(MPa)$ | $\sigma_t(MPa)$ |
|---|---|---|---|---|---|
| W particle | 19 | 410 | 0.27 | 800 | 500 |
| Matrix | 15 | 200 | 0.31 | 300 | 150 |

### 3.2. *Numerical simulations on the fracture patterns*

#### 3.2.1. *Ductile fracture of the matrix*

Fig.4 illustrated the fracture process of the interface between tungsten particle and the matrix of tungsten alloys with the particle size of 35μm under a tensile loading. From it, it can be seen that the stress of high value concentrates on tungsten particles. The matrix near the notch demonstrates ductile fracture firstly accompanying with the tungsten-matrix interface debonding under the tensile loading. The cracks propagate gradually in the tensile process leading to the macro-failure of the material.

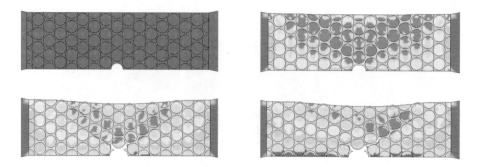

Fig. 4.  Fracture process of W-M interface and matrix.

### 3.2.2.  *Interface debonding and particle fracture*

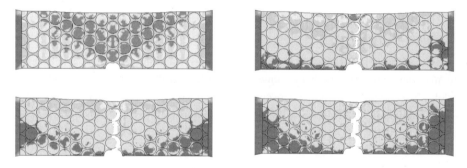

Fig. 5.  Fracture process of tungsten particle and matrix.

When the interface and the matrix possess high bond strength, tungsten particles (particle size: 35μm) present high bearing capacity (see Fig.5). With the increasing of the tensile loading, lots micro-cracks occur firstly in tungsten particles and there is a high amount of particle fracture in the material. The matrix shows ductile fracture during the crack propagation, which means that the matrix plays a role in the resistance of the crack propagation and in the strengthening of ductility of tungsten alloys. With the increasing of the tensile loading, the cracks propagated and connected with each other leading to the total failure of the material.

### 3.2.3.  *Random model of tungsten alloy*

In order to simulate the response of the material with real micro-structure, a multi-particle model was established based on the random distribution of tungsten alloy. The random distribution program was developed by using Matlab. Then, a command stream file was created for the APDL in Ansys. The command stream file was read into Ansys. The multi-particle model was then generated and adopted to simulate the mechanical properties and the fracture behavior of tungsten alloys by using Ls-Dyna.

In mesh generation, solid162 element, a quadrilateral element with four modes, was chosen. The material models were the same with the above simulation. Due to the random distribution of the particles, free meshing was adopted in the mesh generation. ANSYS/LS-DYNA MAT_ADD_EROSION supply many failure criteria. Because of the brittle feature of tungsten particle, maximum principle stress failure criteria was adopted here, while strain failure criteria was chosen for the matrix due to its ductility.

Fig.6 showed the random distribution model of tungsten alloys, a multi-particle model. The diameter of the particles ranges from 8μm to 40μm. Fig.7 demonstrated the experimental results and the numerical predictions of the force-strain. It can be seen that both of them are in good agreement, which means that the finite element model can be used to predict the fracture behavior of tungsten alloys.

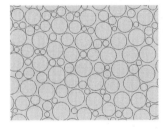

Fig. 6. Random distribution model of tungsten alloys.

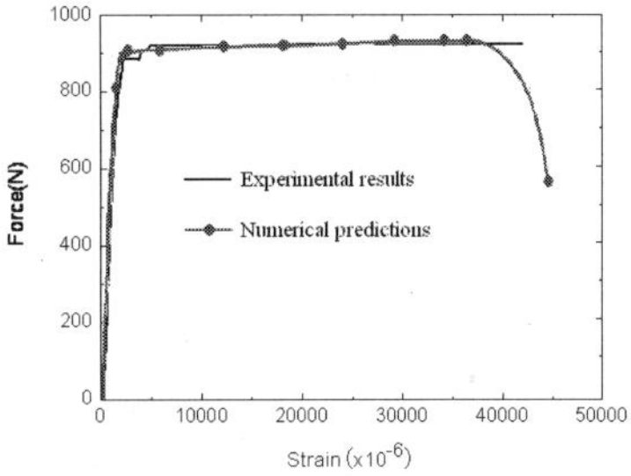

Fig. 7. Comparison of experimental results with numerical predictions.

In this section, several cases were performed to explore the four possible fracture modes of tungsten alloys: matrix failure, tungsten cleavage, tungsten-tungsten intergranular failure, and tungsten-matrix interfacial separation.

From Fig. 8, the stress in the particle has a higher value, acting as reinforcement in the matrix. The micro-crack initiated in matrix firstly in $45^0$ direction and then the stress began to decrease. With increasing the tensile loading, there were more and more micro-cracks showing in the above direction. Then, the micro-cracks propagated and connected with each other leading to a main crack. During the damage process, the composite under tensile loading demonstrated ductile fracture of the matrix. There was few particle fracture found in the composite from initial deformation to final failure, which illustrated that the tungsten particles appeared to be well bonded to the Ni-Fe-W matrix.

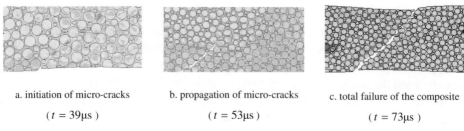

| a. initiation of micro-cracks | b. propagation of micro-cracks | c. total failure of the composite |
| :---: | :---: | :---: |
| ($t = 39\mu s$) | ($t = 53\mu s$) | ($t = 73\mu s$) |

Fig. 8.  Fracture behavior of tungsten alloys.

## 4.  Conclusions

In-situ experiments were carried out to investigate the fracture behavior of 91W-6.3Ni-2.7Fe tungsten alloys. Different finite element (FE) modes were established and series of cases were performed to explore the typical fracture patterns of tungsten alloys under tensile loading. In the simulations, three of the four possible fracture patterns for the tungsten alloy were observed: matrix failure, tungsten cleavage and tungsten-matrix interfacial debonding. From numerical simulations, it could be found that initially micro-cracks started preferentially at matrix or tungsten-matrix boundaries. Comparison of the experimental results and the numerical predictions shows a good agreement between them, which verifying the rationality of the FE models.

## Acknowledgments

The work was supported by the National Natural Science Foundation of China (No. 10625208, 10602008).

## References

1.  K.T. Ramesh, R.S. Coates, *Metall. Trans. A*. **23A**, 2625 (1992).
2.  D. Rittel, G. Weisbrod, *Int. J. of Fract.* **212**, 87(2001).
3.  S.J. Cimpoeru, R.L. Woodward, *J Mater Sci Lett.* **9**, 187(1990).
4.  B.H. Rabin, R.M. German, *Metall. Trans. A*. **19A**. 1523 (1998).
5.  K.S. Churn, R.M. German, *Metall. Trans. A*. **B35**, 8723 (1987).
6.  Islam S. Humail, F. Akhtar, S.J. Askari, et al. *Int. J. Refractory Metals and Hard Materials.* **25**, 380(2007).

# ANALYSIS OF COLLISION BETWEEN DRILLSTRING AND WELL SIDEWALL

G.H. ZHAO[1†], Z. LIANG[2]

*Southwest Petroleum University, Chengdu, P.R. CHINA,*
*wy_zgh@yahoo.com.cn, liangz_2242@126.com*

Received 15 June 2008
Revised 23 June 2008

Drillstring is the most important tool in petroleum drilling engineering. Alternating stress has been found to be responsible for the premature failure of drillstring. Propagation of stress wave, induced by collision between tool-joints of drillstring and borehole wall, is studied in this paper. The condition that all the tool-joints of drill pipes (DPs) strike borehole wall at the same time has been considered. Because of symmetry, the middle cross section of the DP is simplified as fixed end, and mechanical model is established as the beam with both ends fixed. Propagation of lateral displacement wave and stress wave in the DP is investigated by means of Eigen-frequency method and the Finite Element Analysis software ANSYS. The theoretic results coincide with those obtained from numerical modeling very well and also explain the drillstring accidents in gas fields.

*Keywords*: Drillstring; tool-joint; collision; stress wave; fracture.

## 1. Introduction

Air drilling is a new drilling method, whose drilling rate is almost four times of that for traditional mud drilling. Damaging of drillstring in air drilling is much severer than that in mud drilling, and this has been a significant issue which needs to be solved urgently. By analyzing the actual data of Puguang gas field in China, we found that most of fracture failure had happened in bottom hole assemblies (BHA) and in the DPs, which were near BHA. For DPs, fracture failure always happened at the section next to tool-joint and the average length of broken DP was 0.59m. International Association of Drilling Contractors and American Petroleum Institute[1] had analyzed 1785 failing DPs, among which 8% failures happened at the first DP above drill collar (DC), and almost 23% happened within the five DPs above DC. These drillstrings are always in a longitudinally compressive state, and incline to parametric resonance under axial excitation. Therefore, collision between these drillstrings and borehole wall is strengthened greatly.

During collision impact force varies rapidly. Thus lateral stress wave will come into being and propagate in drillstring. By using nonlinear theory of finite elements, M.S. Li[2]

---

[†] Corresponding Author.

studied collision between single DP and borehole wall, and presented the lateral shift and the first principal stress of DP. J.B. Liu[3] developed dynamic gap element method, they analyzed the nonlinear transient dynamics of impact and contact between drillstring and borehole wall, and also designed the eccentric BHA. Y.C. Kuang[4] studied collision between bit and Member Sha-2 sandstone experimentally and determined restoration coefficient. In this paper, lateral displacement wave and stress wave, which are induced by collision between DP and borehole wall, are investigated. The key factors of fracture for drillstring and sections in risk of failure are presented in this study.

## 2. Mechanical Analysis

Influenced by centrifugal force, DPs always touch borehole wall during drilling and the borehole wall will thus exert pressure and friction force to the DPs. As a result, the DPs will precess backward,[5] and the drillstring that is precessing backward at a high speed will strike borehole wall. Since the joint of the DP is wider than DP's body, collision always happens between joints and borehole.

To simplify the analysis model, it is assumed that only joint contacts with the borehole wall, and that both joints of the DP collide with the borehole at the same time.

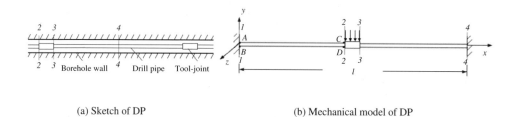

(a) Sketch of DP                                 (b) Mechanical model of DP

Fig. 1. Sketch of collision between drillstring and borehole.

A typical example of which the DP is simplified as a beam with both ends fixed (shown in Fig.1) is analyzed herein. The length of the DP is $l$=10 $m$, outer and inner diameter of the DP are $D$=127 mm and $d$=108 mm respectively. The mass of a single DP is $M$=292 kg, and the elastic modulus is $E$=206×10⁹ Pa. Joint's outer diameter and length are $d_w$=165.1 mm and 0.4 m, the inner diameter of the borehole is $D_w$=314 mm and the speed of rotation (SOR) is $n$=65rpm. Due to symmetry of geometry and loading, the middle cross section of the DP (section *1-1* and *4-4*) doesn't change direction and can be treated as a fixed end, thus the DP could be simplified as a beam with two fixed ends. Segment *1-2* and *3-4* are two half parts of the DP that are connected with each other by joint *2-3*. The collision effects between joint and borehole is simplified to be the lateral impact force exerted on the outer surface of joint. Based on Bernoulli-Euler beam theory, the governing equation of lateral forced vibration is written as:

$$EI \frac{\partial^4 y}{\partial x^4} + \rho A \frac{\partial^2 y}{\partial t^2} = F(x,t) , \tag{1}$$

with the boundary conditions:

$$y(0,t) = y(l,t) = \frac{\partial y(0,t)}{\partial x} = \frac{\partial y(l,t)}{\partial x} = 0 , \tag{2}$$

and initial conditions:

$$y(x,0) = \partial y / \partial t = 0 , \tag{3}$$

in which $\rho$ is the density of the DP, $I$ is the moment of inertia of the DP. $F(x, t)$ is impact force exerted on unit longitudinal length of joint, and it can be approximated as half sine function of time and assumed to distribute uniformly along space:

$$F(x,t) = F_1(x) \cdot F_2(t) , \tag{4}$$

where,

$$F_1(x) = \begin{cases} -1 & x \in [4.8,5.2] \\ 0 & x \in [0,4.8) \cup (5.2,10] \end{cases} , \qquad F_2(t) = \begin{cases} F_0 \sin \dfrac{\pi t}{T} & t \in [0,T] \\ 0 & t > T \end{cases} ,$$

and $T$ is the time for collision. For corresponding homogeneous equation expressed by Eq.(1), natural frequencies $\omega_n$ and mode functions $X_n(x)$ that meet the fixed boundary conditions could be determined by separation of variables:

$$\cos k_n l \cdot \cosh k_n l = 1 , \qquad k_n^2 = \omega_n / \sqrt{EI/(\rho A)} , \tag{5}$$

$$X_n(x) = \frac{\sin k_n x - \sinh k_n x}{\sin k_n l - \sinh k_n l} - \frac{\cos k_n x - \cosh k_n x}{\cos k_n l - \cosh k_n l} . \tag{6}$$

Under zero initial conditions expressed by Eq. (3), the forced vibration equation (1) could be solved by means of eigen function method:

$$y(x,t) = \sum_{n=1}^{\infty} X_n(x) q_n(t) . \tag{7}$$

Substituting Eq.(7) into Eq.(1) and employing the orthogonality of mode functions, the solution of Eq.(1) can be obtained:

$$y(x,t) = \sum_{n=1}^{\infty} \frac{X_n(x)}{\omega_n \cdot \int_0^l X_n^2(x)\mathrm{d}x} \int_0^l X_n(x) \int_0^t F(x,\tau) \sin \omega_n(t-\tau) /(\rho A) \mathrm{d}\tau \mathrm{d}x . \tag{8}$$

To determine the impact force, we consider the following instance. For drillstring rotating at angular velocity of $\omega_r = 2n\pi/60$ , the corresponding angular rate of precessing backward can be given by $\omega_p = \omega_r \cdot d_w /(D_w - d_w)$ . Single-sided annular gap between joint and borehole is $R = (D_w - d_w)/2$ , and the line velocity of DP's axis is $v = \omega_p \cdot R$ . Here the borehole wall is made up of Member Sha-2 sandstone, and the restoration coefficient is 0.1.[4] So the average impact force $\overline{F}$ and collision time $T$ could be derived as:

$$\overline{F} = \frac{120}{40} \cdot Mg \cdot v, \quad T = \frac{M\Delta v}{\overline{F}}. \tag{9}$$

The amplitude of impact force could be obtained by means of equivalent momentum:

$$F_0 = \frac{\overline{F} \cdot T}{\int_0^T \sin \frac{\pi t}{T} dt}. \tag{10}$$

Due to the constraints from borehole wall, the DP will strike the borehole at the line speed of precessing. By numerical integration, displacement wave and stress wave that propagate in the DP could be obtained. The lateral deflection varying with time and space is shown in Fig.2 (a). The maximum amplitude of the deflection occurs in the middle cross section of the mechanical model where the joint is located, the peak value of 32.04mm occurs at $t=51.9$ms. The longitudinal normal stress varying with time and space is presented in Fig.2 (b). The maximum stress amplitude occurs at the fixed ends of the mechanical model, i.e. the middle cross section of DP, and the peak value of 116.715MPa occurs at $t=48.3$ms, i.e. 10.9ms after the collision. In addition, the secondary peak of the longitudinal normal stress of 71.959MPa occurs in the mid section of mechanical model at $t=55.1$ms. Therefore the cross section next to the joint is also endangered by fracture.

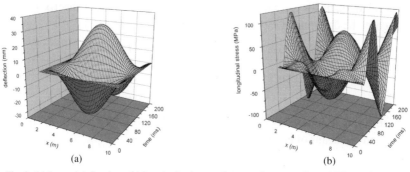

Fig. 2. (a) Lateral deflection, (b) longitudinal normal stress of outer surface of DP, vs. time and space.

When SOR is altered, the strength of stress wave will change accordingly. Through Eq.(8), (9), (10), the relation between the maximum amplitude of longitudinal normal stress and SOR is found to be almost linear (shown in Fig.3). The amplitudes of the longitudinal normal stresses, which are caused by backward precession of the DP's joints, backward precession of the DC and axial vibration, are also illustrated in Fig.3.[6] It can be seen that the stress wave caused by collision is much stronger than that caused by longitudinal vibration and backward precession. Therefore, collision may be one of the main causes of fracture of drillstring. The following two field cases agree with this result well. Firstly, in air drilling the mud cake, which always exists in mud drilling, doesn't exist between drillstring and borehole wall. So the failure rate of drillstring in air drilling is much higher than that in mud drilling. Secondly, for the BHA, which is compressed longitudinally under drilling condition, parametric resonance may happen in the case of axial excitation, thus the BHA will strike borehole wall severely and be prone to fracture. In another word, fracture usually occurs where a severe collision exists.

That both tool-joints of the DP strike borehole wall together are considered in this study since this kind of collision is regarded as the most dangerous case. Actually, joints collide with borehole wall non-symchronously, causing complicated propagation, reflection and superimposition of stress wave. Thereby the extreme stress may appear at different sections of the DP.

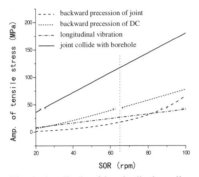

Fig. 3. Amplitude of longitudinal tensile stress vs. speed of rotation.

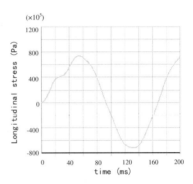

Fig. 4. Longitudinal normal stress vs. time for point *D*.

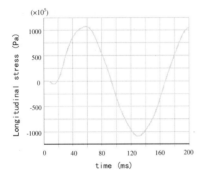

Fig. 5. Longitudinal normal stress vs. time for point *A*.

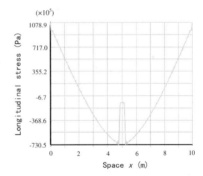

Fig. 6. Longitudinal normal stress varying with space at the upper surface of DP at *t*=57.5ms.

## 3. Numerical Simulation

Transient dynamic analysis is carried out using the Finite Element Analysis software ANSYS to analyze the variation of the stress with time at the sections along the beam risking of fracture.

Fig.4 shows the longitudinal normal stress varying with time at point *D*, which is on the body of the DP and next to the joint. The maximum of normal stress of 74.28MPa occurs at *t*=54ms. Fig.5 presents the longitudinal normal stress varying with time at point

*A*, which is located in the middle cross section of the DP (corresponding to the fixed end in Fig.1). It reaches the maximum value of 107.89MPa at $t$=57.46ms, which is almost 20ms after the collision. The longitudinal normal stress varying with space at $t$=57.5ms is also shown in Fig.6. It can be seen that propagation of stress wave influences stress field of the DP significantly. These results obtained from finite element analysis are consistent with those obtained from the theoretic analysis very well. In addition, Y.L. Zhang[5] had ever measured collision stress of drillstring by means of model experiment, and the extreme stress that occurred near the tool-joint was 50~80MPa. The results obtained in this investigation agree with the experimental data well.

## 4.  Conclusion

The case that both joints of DP strike borehole wall together is studied herein. Due to symmetry of the geometry of the structure and loading, a simplified model of a beam with both ends fixed is chosen for the analysis of bending stress wave in the DP using eigen function method and finite element modeling technique. The following conclusions could be drawn:

(1) Due to reflection and superimposition of stress wave, the maximum amplitude of the longitudinal normal stress of 108MPa occurs in the middle cross section of the DP at almost 20ms after the collision.

(2) The extreme stress also appears in the section that is next to the joint, and this agrees well with the field data that fracture failure of the DP always happens at this position.

(3) The maximum amplitude of stress wave caused by collision is almost linear to SOR.

(4) The stress wave caused by collision is much stronger than that caused by longitudinal vibration and backward precession. Therefore collision between drillstring and borehole wall may be one of the main causes of fracture of drillstring.

Therefore, choosing proper drilling parameters to prevent BHA from striking borehole wall severely may reduce failure of the BHA. Furthermore, welding protective belt in the outer surface of the joints is also an effective method to weaken collision and protect the DPs from damage.

## Acknowledgments

This work is supported by Key Laboratory of Oil and Gas Equipment from Ministry of Education (2006STS04) and Science Foundation of Southwest Petroleum University.

## References

1.  IADC, API, in: *Drillstring Failures are Scrutinized*, 1990(12)/1991(1).
2.  M.S. Li, X.Z. Yang and D.L. Gao, *China Petroleum Machinery*, **34**, 15 (2006).
3.  J.B. Liu, H.J. Ding and X.H. Zhang, *Chinese J. Computational Mechanics*, **19**, 456 (2002).
4.  Y.C. Kuang, *Natural Gas Industry*, **22**, 55 (2002).
5.  Y.L. Zhang, in: *Kinematics and Dynamics of Drillstring*, (Petroleum Industry Press, Beijing, 2001).
6.  G.H. Zhao and Z. Liang, in *proc. 9th Int. Conference on Engineering Structural Integrity Assessment*, ed. S.J. WU *et al.* (China Machine Press, Beijing, 2007), pp. 411-413.

# FE SIMULATION OF EDGE CRACK INITIATION AND PROPAGATION OF CONVENTIONAL GRAIN ORIENTATION ELECTRICAL STEEL

D. H. NA[1], Y. LEE[2†]

*Department of Mechanical Engineering, Chung-Ang University, Seoul, KOREA,*
*dhna@wm.cau.ac.kr, ysl@cau.ac.kr*

Received 15 June 2008
Revised 23 June 2008

Three-dimensional finite element simulation has been carried out to understand better the crack initiation and growth at the edge side of silicon steel sheet during cold rolling, which is attributable to elastic deformation of work roll, i.e., roll bending. Strain-controlled failure model was coupled with finite element method and a series of FE simulation has been carried out while three different roll bending modes are considered. FE simulation shows that the negative roll bending mode during rolling affects significantly the crack initiation behavior. When the strain for failure was reduced by 20%, number of elements removed was increased by about 305%. If an initial crack with 2.5mm in length was assumed on the strip, the initial edge crack propagated toward inner region of strip and the propagated length is about 10times of the initial edge crack length.

*Keywords*: FE simulation of edge crack initiation; Roll bending; CGO silicon steel; Cold rolling.

## 1. Introduction

Electrical steel, called silicon steel as well, is a kind of ferrite steel which has excellent electrical quality and guarantees a magnetic property. Hence it has been employed for manufacturing an equipment or apparatus which produces magnetism and electricity. Recently, it has been in great demand in the area of energy-saving related industry since the silicon steel has excellent electric character required in the energy-saving equipment.

In comparison with general cold rolled steel, the silicon steel includes a lot of silicon component and subsequently its deformation behavior is of a limited ductility at even room temperature. The ratio of width over thickness of silicon steel strip is very large like the general cold steel strip. In most cases, the width and thickness of the silicon steel is 1000mm and 2.3mm, respectively. Hence the length of work roll is more than 1000mm, say 1200mm. Rolling with such a wide width of strip usually causes uneven thickness variation across the width of the strip or to wavy edges on both sides of the strip. It usually causes elastic bending of work roll due to high deformation resistance of silicon steel. This leads to creation of tensile stress at the edge of the strip and generation of compressive stress at its center during rolling. Therefore, cracks might initiate at the edge region of the strip during rolling.

[†]Corresponding Author.

To prevent the crack initiation at the edge region, operators in the mill yard heat up the strip up to 200°C and roll it. The amount of crack growth at the edge region is then reduced but the crack still initiates. Other possibilities for the edge crack initiation might be attributable to inferior control of the roll speed for winding strip and that for releasing strip, and eccentricity of work rolls. However, there is very low probability for the crack initiate due to these reasons.

Grimwade[1] described the defects which possibly occur at the surface of strip during cold rolling and proposed a scheme which prevents the generation of defects. However, he didn't specify the scheme in detail. Ghosh et. al[2] predicted the occurrence of edge crack of aluminum alloys during cold rolling by taking the effective stress, equivalent plastic strain and void volume fraction as a damage parameter. They tried to predict the occurrence of edge crack but didn't simulate the failure of crack. Ogawa et. al[3] reported that flatness of the work roll leads to much less probability of the creation of edge crack. They also suggested that intermediate roll shifting, adjustment of back-up roll crown and intermediate roll bending force might be effective method to draw the work roll flat. Investigating the previous technical reports and/or articles revealed that no study which predicted crack initiation and simulated propagation been reported up to date.

This study presents 3-D finite element simulation of crack initiation and propagation of silicon steel strip during rolling. For this purpose, we first examine the effect of the deformation mode of work roll (no bending, negative bending and positive bending) on the behavior of edge crack of strip when the strip with no edge crack is rolled. Second, we simulate the crack propagation when the steep strip with an initial edge crack is rolled. The initial crack tip angle was taken as a parameter to examine how it depends on crack propagation sideward of the strip under a given rolling condition.

## 2.  Roll Bending During Cold Rolling

Work roll under the action of the rolling force is bent elastically and this roll bending leads to uneven thickness across the width of the strip or to wavy edges on both sides of the strip. Figure 1 illustrates schematic of wavy edges due to roll bending. If excessive tension is applied to the strip with wavy edge, the edge crack might propagate sideward of the strip as shown in Fig. 2.

Fig. 1.   Wavy edges due to roll bending.          Fig. 2.   Growth of edge crack sideward of strip.

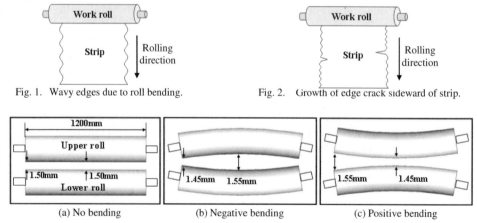

(a) No bending          (b) Negative bending          (c) Positive bending

Fig. 3.   Possible elastic deformation mode of work rolls during cold rolling.

Even a four-high rolling mill where small diameter work rolls are backed by larger rolls to prevent the deformation of work rolls, bending of work rolls is inevitable during rolling. Figure 3 illustrates three possible deformation modes of work rolls which might occur during cold rolling. Figure 3(a) shows an ideal case that the upper and lower work rolls are not bent at all during rolling. It is referred 'no bending' in this study and it is considered as one of roll bending modes as well. Figure 3(b) illustrates a negative bending mode which occurs commonly due to higher deformation resistance at the center area of strip. 'Edge rolling' is then applied to the work rolls to reduce the negative bending. But excessive application of edge rolling sometimes leads to positive bending (Fig. 3(c)). In actual rolling, the state of no bending doesn't exist. Negative bending and positive bending occur consecutively during rolling.

## 3. Finite Element Analysis

Figure 4(a) illustrates a schematic of an actual cold reversing rolling process. Strip is uncoiled in the Tension Reel 2 and is wind up by the Tension Reel 1 during rolling.[4] Initial strip with thickness of 2.3mm is rolled in general cold rolling process. Following five passes, its thickness becomes 0.292mm. Diameter of the work roll is usually 230mm and its width is about 1200mm. Width of the high silicon-contained electrical steel strip is generally 1000mm.

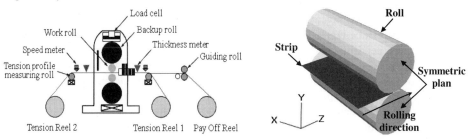

Fig. 4. (a) Schematic of cold reversing rolling mill.[4]     Fig. 4. (b) FE modeling for work rolls and strip.

### 3.1. *FE modeling and mesh*

To reduce computational time, FE model focuses on the work rolls and the length of steel strip is shortened by 200mm. Three-dimensional FE analysis was conducted using ABAQUS.(Ver. 6.7-1) The explicit time integration method was employed and element type used for the strip is C3D8R(8-node linear brick, reduced integration with hourglass control). Friction coefficient, 0.1 was used and the work roll speed was set as 3.467 m/s. The FE simulation was conducted for the first pass only and measured tensile stress in z-direction acting the strip is applied. Figure 4(b) shows FE modeling. An half of the strip was analyzed due to its geometrical symmetry. The number of elements used is 64,000 and size of all elements is equal to 2.5×2.5×0.575mm. Number of element layers toward the thickness direction of the strip is four.

### 3.2. *Failure simulation*

In FE simulation, material failure implies a complete loss of load carrying capacity of an element which stems from progressive degradation of the stiffness of the element. In case of ductile failure, stiffness degradation initiates when the equivalent plastic strain, for example, reaches a specified strain for failure, $\varepsilon_f$ of material. In FE analysis with

ABAQUS, the equivalent plastic strain at an element is supposed to depend on a ratio, $\eta$ (pressure stress over von Mises stress) because deformation and failure of the silicon steel are assumed to be sensitive to stress triaxiality.[5] Note that in this study the ratio, $\eta$ was fixed as 0.25 to keep consistency of FE simulation and the strain for failure was determined following a tensile test of the silicon steel. Once stiffness degradation initiates, softening of the yield stress starts based on a scalar damage approach. Softening response is described by a stress-displacement relation. To minimize the effect of mesh size on damage evolution, a characteristic length is computed on the basis on element geometry. In this study, an element under severe local deformation condition is assumed to fail fully when the equivalent plastic displacement of the element is 0.025.

## 4.  Result and Discussion

### 4.1.  *Crack initiation and propagation when no initial edge crack is on the strip*

Figure 5 shows the edge crack initiation and growth for three roll bending modes. Initial strip has no crack on the edge at all. Failure strain, $\varepsilon_f$ in FE simulation is specified as 0.23 since the ultimate tensile stress of the material used in this study is 0.23. In other words, whenever an element reaches $\varepsilon_f$, the element begins to experience damage and subsequently loses load carrying capacity step by step until stiffness of the element is degraded fully. The edge crack initiation and growth are represented by the elements removed. As expected, positive bending yields smallest number of elements removed at the edge of strip since equivalent strain at the edge is smaller than that under any other cases. Negative bending creates more edge cracks since strain to z-direction, i.e., rolling direction, at the edge is much larger than that at the center, as shown in Fig. 5(b). Before the strip rolls in the first pass, its surface is washed by acid. Hence strip temperature drops to almost room temperature when the strip is about to be rolled in the first pass. As temperature of strip decreases, the strain for failure for a given stress decreases as well. If the strain for failure, $\varepsilon_f$ is reduced, much more crack initiation and growth might be expected. Therefore, we may need to simulate those when $\varepsilon_f$ is about 80 percent of 0.23.

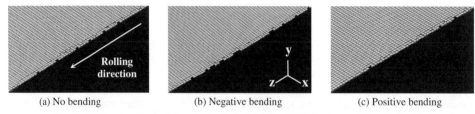

(a) No bending                  (b) Negative bending                  (c) Positive bending

Fig. 5.   Different pattern of edge crack occurs when $\varepsilon_f = 0.23$ for three roll bending mode.

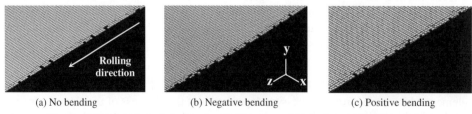

(a) No bending                  (b) Negative bending                  (c) Positive bending

Fig. 6.   Different pattern of edge crack occurs when $\varepsilon_f = 0.18$ for three roll bending mode.

Figure 5 and 6 illustrate this phenomenon. In all cases, we can see the cracks propagated from the very edge region to the center region.

### 4.2. *Crack initiation and propagation when initial an edge crack is on the strip*

To examine the effect of roll bending mode on an initial crack, an initial edge crack is assumed to exist at the side before rolling. Figure 7(a) illustrates SEM picture of the edge crack shape of aluminum alloy strip after cold rolling.[2] With reference to this, an initial edge crack on the silicon steel is modeled as shown in Fig. 7(b). $\alpha$ represents for the angle of the initial crack.

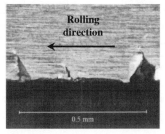

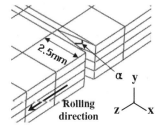

Fig. 7. (a) Edge cracks on the side of aluminum alloy strip.[2]     Fig. 7. (b) FE model of the edge crack.

Figure 8 shows the propagated edge crack length for three modes of roll bending viewed from y-direction (see Fig. 7(b)) when the initial crack angle is $20^0$. For all cases, the initial edge crack propagated into inner region of the strip. Its schematic description is shown in Fig. 2. This indicates that roll bending mode influences the crack propagation behavior significantly. The crack length propagated for positive bending is the smallest than any other bending modes. This result is consistent with that obtained when no initial edge crack is assumed on the strip. It is interesting that the propagated crack length for no bending mode is almost 90% for negative bending mode. This may be explained by the failure strain used in FE simulation.

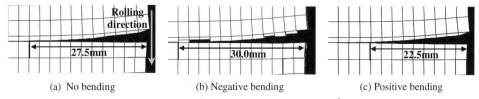

(a) No bending            (b) Negative bending            (c) Positive bending

Fig. 8.  Edge crack propagation and length propagated when angle is $20^0$. Failure strain, $\varepsilon_f$ is 0.23.

Figure 9 shows the crack propagation behavior when the angle of an initial edge crack is increased to $30^0$. In overall, the propagated edge crack length was reduced as the angle was increased, except positive bending case. This is attributable to the drop of stress concentration at the ahead of the initial edge crack. The amount of the drop, however, is not proportional to the change of angle. Fig. 9(b) illustrates that some elements located to sideward of the initial edge crack were removed. It indicates the energy required for crack initiation is concentrated at a point, i.e., the tip of the initial edge crack. Hence the crack length propagated for the case of negative bending is almost equal to that for positive bending. Regardless the angle, the length of the initial edge crack was enlarged to about 25mm, which is approximately 10 times of the initial crack length. Thus, besides minimizing the occurrence of negative bending mode, trimming the initial edge crack before rolling is a method.

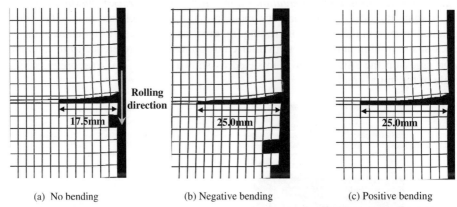

(a)  No bending                 (b) Negative bending                (c) Positive bending

Fig. 9.    Edge crack propagation and length propagated when angle is $30^0$. Failure strain, $\varepsilon_f$ is 0.23.

## 5.  Conclusions

We conducted FE simulation of edge crack initiation and propagation of silicon steel during cold rolling. Three types of deformation mode of work roll, i.e., no bending, negative bending and positive bending were considered. The conclusions are summarized as followings: When an initial crack doesn't exist on the strip, negative bending mode of work roll causes more crack creations than other deformation modes. This is because the strain at edge region is larger than other regions due to deformed work roll shape. Therefore minimizing the negative bending during rolling is crucial to inhibit the edge crack creations. When the strain for failure is reduced by 20%, number of elements removed is increased by about 305%. When an initial crack is assumed on the strip, negative bending mode leaded to the noteworthy crack propagation into inner region of the strip. More elements at side of the initial edge crack were removed due to the strain component acting to the thickness direction as the angle increases from $20\,^\circ$ to $30\,^\circ$.

## Acknowledgments

This research was supported by Chung-Ang University Research Grants in 2008.

## References

1.   M. F. Grimwade, *Gold Technology*, **36**, (2002).
2.   S. Ghosh, M. Li and  D. Gardiner, *Journal of Manufacturing Science and Engineering*, **126**, 74 (2004).
3.   S. Ogawa, S. Hamauzu, H. Matsumoto and T. Kawanami, *ISIJ International*, **31**, 599 (1991).
4.   S.M. Byon, S.I. Kim and Y. Lee, *Journal of Materials and Processing Technology*, **201,** 106 (2008).
5.   C. Yan, L. Ye and Y.-W. Mai, *Materials Letters*, **58**, 3219 (2004).

# PREDICTION OF THE DELAMINATION IN THE PEARLITIC STEEL FILAMENTS BY 3 DIMENAIONAL ATOM PROBE TOMOGRAPHY

Y. S. YANG[1], C. G. PARK[2†]

*Department of Materials Science and Engineering, Pohang University of Science and Technology (POSTECH), Pohang 790-784, KOREA,*
*polaris@postech.ac.kr, cgpark@postech.ac.kr*

J. G. BAE[3], D. Y. BAN[4]

*R&D center KISWIRE, Pohang 790-841, KOREA,*
*jgbae@kiswire.com, dyban@kiswire.com*

Received 15 June 2008
Revised 23 June 2008

We have tried to find out the critical factor governing the delamination in the pearlitic steel filaments. Steel filaments were fabricated depending on the carbon content from 0.72 to 1.02 wt.%. The delamination was identified by a torsion tester specially designed for thin-sized wires and scanning electron microscopy. The results showed that as the carbon content increased, the number of twists to fracture decreased, and the delamination only occurred in the filament with 1.02 wt.% C. In order to elucidate this behavior, the microstructure of the filaments was observed using advanced analysis techniques such as 3 dimensional atom probes tomography (3-DAPT).

*Keywords*: Delamination; cementite dissolution; 3 dimensional atom probe tomography.

## 1. Introduction

The pearlitic steel cords (twisted filaments) are widely used for the reinforcement parts, such as belt and carcass, of automotive tires due to their outstanding strength as well as acceptable ductility.[1-3] The steel cords are generally produced via several steps of cold drawing from wires containing 0.6 ~ 0.9 wt.% C, patenting to produce a fine pearlite microstructure and stranding to make the final cords. In response to the market trend toward lighter tires, the strength required for steel cords has been increased. Over the past years, many efforts have been made to improve the strength of the filaments. The strength exceeding 5400 MPa has been reported.[1] To increase the strength, it is necessary to decrease the lamellar spacing and to increase the drawing strain in accordance with the Embury-Fisher equation.[2] The torsional ductility is important when filaments are twisted during stranding. It has been known that the delamination, longitudinal splitting into several layers, is one of the factors predicting the degree of the torsional ductility. There

---

[†]Corresponding Author.

have been a number of works on the delamination in conjunction with the cylindrical texture,[3] lamellar spacing[4] and strength.[5] However, the critical parameter affecting the delamination is not clearly understood yet.

The purpose of the present study is, thus, to find out the critical factor governing the delamination in the pearlitic steel filaments. The number of twist and delamination, which depends carbon content from 0.72 to 1.02 wt.%, has been investigated. In addition, the changes of microstructural parameters were also identified. The relation between microstructure and delamination phenomena observed in heavily cold drawn steel filaments has been discussed.

## 2.   Experimental Procedure

Pearlitic steel filaments containing 0.72, 0.82, 0.92 and 1.02 wt.% C, respectively, were fabricated by KISWIRE. The diameter of final wire was approximately 0.175 mmϕ. The torsional ductility of the filaments was measured by using a torsion-torque tester (ATM TO-15) at a rotation speed of 30 rpm. Back tension was applied to the specimen under a weight of 1 % of the maximum tensile strength.

Microstructure parameters, such as the lamellar spacing, cementite thickness and volume fraction of cementite, were identified by TEM (JEOL 2010F). In order to identify cementite dissolution, AP analyses were carried out by using a laser assisted 3D-APT (CAMECA LAWATAPT™). The samples were prepared using a two-stage electro-polishing procedure. Steel filaments were first thinned in a solution of 25 % perchloric acid in glacial acetic acid at 20 V d.c. and then micro-polished in 2 % perchloric acid in 2-butoxy ethanol at 8 V d.c. The AP analyses were carried out in the ultra-high vacuum ($\sim 1 \times 10^{-10}$ mbar) at tip temperature of 50 K.

## 3.   Results

### 3.1.   *Microstructure*

Fig. 1. exhibits TEM micrographs showing the longitudinal microstructure of the filaments with the changes of carbon content. All the microstructure was a pearlite composed of ferrite ($\alpha$) and cementite ($\theta$) lamellae, regardless of the carbon content. Due to the heavy cold drawing, the pearlite lamella was aligned straight along the wire axis. Many dislocations generated and tangled were also observed mainly in ferrite lamella. The microstructural parameters, such as lamellar spacing ($\lambda_P$), cementite thickness ($t_C$) and cementite volume fraction ($V_C$), were measured, as listed in Table 1. The parameters were strongly influenced by carbon content. That is, as carbon content increased from 0.72 to 1.02 wt.%, the lamellar spacing and cementite thickness decreased from 16.4 to 9.4 nm and 1.8 to 1.3 nm, respectively. The volume fraction of cementite increased from 10.8 to 15.3 and then decreased down to 13.3 with the increase in carbon content. It is

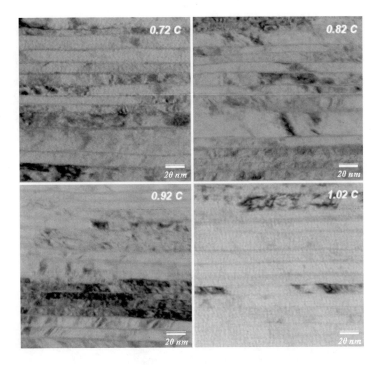

Fig. 1. TEM micrographs showing the longitudinal microstructure of the steel filaments depending on carbon content.

Table 1. The variations of microstructural parameters and mechanical properties of the steel filaments depending on carbon content.

| | Microstructural parameters | | | Mechanical properties | | |
|---|---|---|---|---|---|---|
| C (wt.%) | $\lambda_P$ (nm) | $t_C$ (nm) | $V_C$ (%) | $\sigma_{TS}$ (MPa) | No. of twist (N) | Delamination |
| 0.72 | 16.4 ± 1.0 | 1.8 ± 0.6 | 10.8 | 3206 | 26.7 | × |
| 0.82 | 13.5 ± 1.2 | 1.7 ± 0.9 | 12.1 | 3480 | 25.9 | × |
| 0.92 | 11.6 ± 1.5 | 1.5 ± 0.8 | 15.3 | 3744 | 23.4 | × |
| 1.02 | 9.41 ± 0.5 | 1.3 ± 0.5 | 13.3 | 3908 | 16.2 | O |

worth to note the decreasing volume fraction of cementite in the filament with 1.02 wt.% C, since it is widely reported that increase of carbon content increased the volume fraction of cementite.[2,4]

In order to find out the reasons why the volume fraction of cementite decreased, APT tests were carried out. The 3D tomographic images and carbon concentration profiles are shown in Fig. 2. The analysis volume was approximately $5 \times 5 \times 15$ nm$^3$. Each red dot represents the location at which a carbon was detected. And the density of dots in the analyzed volume is proportional to the density of carbon atoms. That is, the carbon-enriched regions correspond to the cementite, and some carbon atoms appeared to be present even in the ferrite. One or two cementite lamellae were observed in the analysis

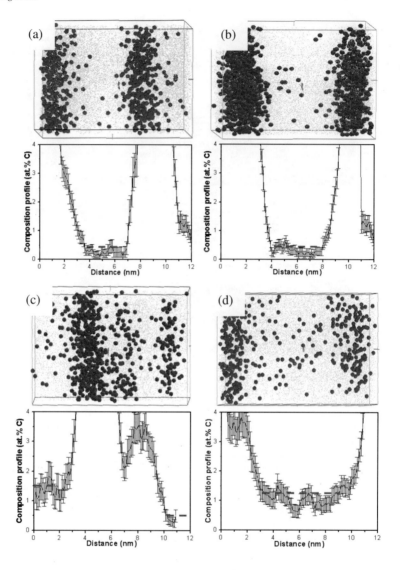

Fig. 2. 3D elemental map and composition profile showing the distribution of carbon in cementite and ferrite; (a) 0.72 wt.% C, (b) 0.82 wt.%, (c) 0.92 wt.%, (d) 1.02 wt.%.

direction. From comparison between composition profiles, it was found that amount of dissolved carbon was strongly influenced by the carbon content. That is, in case of steel filaments containing carbon of 0.72, 0.82 and 0.92 wt.%, the carbon concentration in ferrite was approximately 0.5 at. %. However, its value in the steel filaments with 1.02 wt.% C increased to 1.5 at.% C. It means that the carbon in the cementite lamellae actively diffused into ferrite in the case of steel filaments with 1.02 wt.% C.

## 3.2. *Tensile strength and torsional ductility*

The change of the tensile strength values depending on the carbon content are shown in Table 1. The increase in the carbon content improved the tensile strength ($\sigma_{TS}$). That is, as the carbon content increased from 0.72 to 1.02, the tensile strength of the steel filaments increased linearly from 3206 to 3908 MPa.

Fig. 3 (a). shows the torsional behavior of the steel filaments with increasing carbon content. The number of twists to fracture decreased from approximately 26 to 16 with the increase in thecarbon content from 0.72 to 1.02 wt.%. However, it is noted that torque in the filament with 1.02 wt.% C was abruptly dropped from 3.7 to 2.8 after a short period of plastic deformation. The SEM micrographs showing the fracture topography of the steel filaments after torsion test are shown in Fig. 3 (b). From the observation of fracture surface, spiral cracks along the drawing axis, named as delamination, were only detected in the steel filament with 1.02 wt.% C, while the fracture surface corresponding to a normal torsion fracture mode was observed in the steel filament with 0.72 to 0.92 wt.% C.

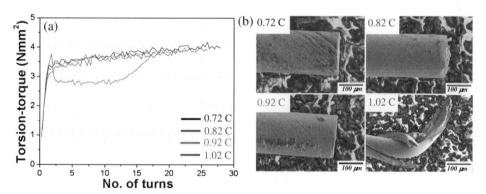

Fig. 3. (a) torsion-torque curve obtained from the torsion tester and (b) fracture topography of the steel filaments depending on carbon content.

## 4. Discussion

The lamellar spacing of the filaments with the carbon contents decreased. The tensile strength was, thus, gradually improved with the decreasing lamellar spacing. Reduced lamellar spacing provides an increased number of cementites, which can act as effective barriers against the dislocation motion. This result is well coincident with Embury-Fisher relationship that tensile strength is inversely proportional to lamellar spacing.[2]

A number of twists to fracture decreased with the increase in carbon content as shown in Fig.3 (a). This result can be well explained by the decrease in lamellar spacing. That is, the decrease in lamellar spacing increases the interfacial area between the cementite and ferrite which is one of the strong crack initiation sites. Since the voids and cracks are formed along the interface, the decreased lamellar spacing causes an easy failure. Since the delamination occurred during the torsion in the filament with 1.02 wt.% C, however, the maximum number of twist to failure reduced drastically to 2. From the microstructural observations as already shown, the source of the delamination was identified as the amount of remaining carbon induced during heavy cold drawing. As mentioned above, the maximum carbon concentration in ferrite of the filament with 1.02 wt.% C exceeded 1.5 at.% And this value was comparatively higher than that of the filaments with 0.72, 0.82 and 0.92 wt.% C. That is, when remaining carbon in ferrite is higher than 1.5 at.%, the delamination occurred. It is believed that the increase in dissolved carbon remained in ferrite improved the strength of ferrite, resulting in the decrease in ductility. When the applied stress concentrated in that sites, sub nanometer size cracks or voids are, thus, easily formed, resulting in the delamination. Therefore, the behavior of cementite dissolution is suspected to play an important role in the occurrence of the delamination.

## 5.  Summary

In order to find out the critical factor affecting the delamination, newly designed 3D-APT, which provides the position and composition profiles of alloy elements, has been used. The increase in the carbon content decreased the lamellar spacing, resulting in the increase in tensile strength and decrease in torsional ductility. However, delamination was not influenced by lamellar spacing, but the amount of carbon in ferrite induced during heavy drawing. That is, when the carbon concentration in ferrite exceeded 1.5 at.%, the delamination occurred. Therefore, the amount of carbon remaining in ferrite can be used as a critical parameter governing the delamination.

## Acknowledgments

The authors would like to express thanks to KISWIRE R&D Center, NCNT, NCRC and BK21 for supplying steel filaments and supporting this work financially.

## References

1.  T. Tarui, J. Takahashi,H. Tashrio, N. Maruyama, S. Nishida, *Nippon steel technical reports*. **91**, 56 (2005).
2.  J.D. Embury, R.M. Fisher, *Acta. Metall.* **14**, 147 (1966).
3.  K. Shimizu, N. Kawabe, *Wire J. Inter.,* **15** 88 (2002).
4.  G. Langford, *Metall. Trans. A*, **8** 861 (1977).
5.  D.B.Park, E.G.Kang, W.J.Nam, *Journal of materials processing technology*, **187-188,** 178 (2007).

# LOW CYCLE FATIGUE BEHAVIOR OF Zn-22Al ALLOY IN SUPERPLASTIC REGION AND NON-SUPERPLASTIC REGION

ATSUMICHI KUSHIBE [1][†]

*Research and Development Institute, Takenaka Corporation, JAPAN,*
*kushibe.atsumichi@takenaka.co.jp*

TSUTOMU TANAKA [2], YORINOBU TAKIGAWA [3], KENJI HIGASHI [4]

*Department of Materials Science, Osaka Prefecture University, JAPAN,*
*t_tanaka@tri.pref.osaka.jp, takigawa@mtr.osakafu-u.ac.jp, higashi@mtr.osakafu-u.ac.jp*

Received 15 June 2008
Revised 23 June 2008

The crack propagation properties for ultrafine-grained Zn–22wt%Al alloy during low cycle fatigue (LCF) in the superplastic region and the non-superplastic region were investigated and compared with the corresponding results for several other materials. With the Zn– 22wt%Al alloy, it was possible to conduct LCF tests even at high strain amplitudes of more than ±5%, and the alloy appeared to exhibit a longer LCF lifetime than the other materials examined. The fatigue life is higher in the superplastic region than in the non-superplastic region. The rate of fatigue crack propagation in the superplastic region is lower than that in the other materials in the high $J$-integral range. In addition, the formation of cavities and crack branching were observed around a crack tip in the supereplastic region. We therefore conclude that the formation of cavities and secondary cracks as a result of the relaxation of stress concentration around the crack tip results in a reduction in the rate of fatigue crack propagation and results in a longer fatigue lifetime.

*Keywords*: Low cycle fatigue; superplasticity; $J$ Integral; crack propagation rate.

## 1. Introduction

Isolators and seismic dampers have been receiving great attention as a means of ensuring a secure and safe habitable space during earthquakes[1-2]. Seismic dampers absorb the energy of an earthquake by undergoing plastic deformation before the main structure of a construction undergoes such deformation. Deformation data for buildings measured during large earthquakes shows that a seismic damping material is exposed to severe deformation with a strain amplitude of at least ±6 % and a strain rate of the order of > $1.0 \times 10^{-3}$ s$^{-1}$. Low yield-point steel has been widely used as a damping material because of its low yield stress and good energy absorption properties. However, it is necessary to carry out maintenance work on seismic dampers after every earthquake because work hardening leads to a degradation of the seismic response[3]. This situation led us to an interest in certain characteristics of the superplastic phenomenon: low deformation stress, no work hardening and high ductility. Recently, we have reported that Zn-22wt.%Al

[†]Corresponding Author.

alloy, produced with very fine grains through an equal-channel-angular extrusion (ECAE) process, exhibits superplastic behavior even at room temperature and at a high-strain-rate of $10^{-2}$ s$^{-1}$, suggesting potential for application to seismic dampers [4-8].

Meanwhile, seismic dampers are liable to fail under low cycle fatigue (LCF) due to repeated plastic deformation. Therefore, the LCF properties of the Zn-22wt.%Al alloy need to be understood. However, there have been no reports on the LCF behavior of this alloy. In a previous report [9], we did report on the fatigue life and fracture appearance of Zn-22wt.%Al alloy tested under typical service conditions encountered by seismic dampers. However, a more thorough investigation is required in order to enhance understanding of the LCF damage tolerance characteristics of this alloy. Analysis results of superplastic region have recently been reported [25]. However, fatigue behavior in non-superplastic region is still not clarified. Thus, the objective of this work is to look into the fatigue mechanisms of Zn-22wt.%Al alloy, including crack initiation and crack growth behavior, in both the superplastic region and the non-superplastic region.

## 2. Experimental Procedures

The material used in this study is an extruded Zn-22wt.%Al alloy. It is well known that reducing the grain size increases the superplastic strain rate and/or decreases the superplastic temperature[10-11]. In this study, the as-received alloy was subjected to the ECAE process, which is capable of producing considerable grain refinement[12]. The ECAE die incorporated a 90°angled channel and so-called processing route Bc was used, in which the specimen was removed from the die and then rotated by +90°between each pass.

First, a rod with a diameter of 19.5 mm and a length of 90 mm was solution-treated in air at 623 K for a day and water-quenched. Subsequently, eight passes of ECAE were carried out at 333 K using a hydraulic press operating at 0.2 mm/s on the first pass and 1 mm/s thereafter. Typical scanning electron micrographs of (a) the as-received alloy and (b) the ECAE-treated alloy are shown in Fig. 1. The resulting alloy consisted of eutectoid phases of Zn (bright) and Al (dark) and had a homogeneous equiaxed microstructure. The average grain sizes were approximately 1.2 μm and 0.5 μm for the as-received and ECAE-treated alloys, respectively.

LCF tests were implemented using an automatically controlled tensile testing machine using push-pull cycles at a constant strain amplitude. The specimens used in the LCF tests were cylindrical and each had a diameter of 6 mm and a length of 6 mm in the gage section, in common with our previous work[9]. To avoid local deformation and stress concentration at the contact points between extensometer probes and the specimen surface, a non-contact displacement measurement system was used in these strain-controlled fatigue tests. The strain amplitude was measured from the length between the shoulders of the specimen by using a LED/CCD(light-emitting diode/charge-coupled

Fig. 1.  Microstructures of  (a) as- received alloy and
(b) ECAE-treated alloy.

device) optical micrometer (LS-7030 , Keyence Corp.).
As part of this study, it was necessary to separate LCF testing into the superplastic and non-superplastic regions. To this end, a deformation mechanism map for the Zn-22wt.%Al alloy is shown in Fig. 2 at 303 K. As this figure indicates, the cyclic deformation condition of $\dot{\varepsilon} = 1.0 \times 10^{-1}$ s$^{-1}$ in the as-received alloy corresponds to the dislocation creep region (that is, the non-superplastic region), while that of $\dot{\varepsilon} = 1.0 \times 10^{-3}$ s$^{-1}$ in the ECAE-treated alloy corresponds to the superplastic region[25]. In the study, we carried out fatigue tests at these two levels of strain rate. Fatigue life was defined as the number of cycles corresponding to a 25% reduction in maximum tensile load. The depth of surface cracks was observed by using a color laser microscope (VK 9500, Keyence Corp.).

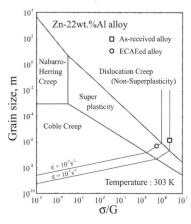

Fig. 2. Deformation mechanism map for Zn-22wt.%Al.

## 3. Results and Discussion

It is well known that the relationship between plastic strain range and the number of cycles to failure follows the Coffin-Manson equation[13-14].

$$\Delta \varepsilon_p N_f^{\alpha} = \theta \qquad (1)$$

where $\Delta \varepsilon_p$ is the plastic strain range, $N_f$ is the fatigue life, $\alpha$ is the fatigue ductility exponent and $\theta$ is the fatigue ductility coefficient. Coffin-Manson relationships derived using the apparent plastic strain range obtained from the width of the hysteresis loop at zero stress in superplastic region[25] and the non-superplastic region are shown in Fig. 3. For comparison, data for carbon steel[15], high-strength steel[15], and Ti alloy[15] are also included in this plot. The plastic strain range obtained in this study is larger than the values for the other materials because the Zn-22wt.%Al alloy is intended to use as a seismic damper. The figure

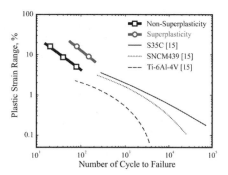

Fig. 3. Relationship between plastic strainrange and number of cycles to failure.

makes clear that the Zn-22wt.%Al alloy tested in the superplastic region exhibited a much longer fatigue life than that in the non-superplastic region; it is also longer than for all the other materials. That is, the fatigue life of Zn-22wt.%Al alloy appears to be greatly improved by the occurrence of superplasticity. When exhibiting superplasticity, this alloy is an optimum material for use in seismic dampers.
In a previous report[9] it was shown that specimens fracture through propagation of surface cracks formed during cyclic deformation. This means that the initiation and growth process of surface cracks would occupy a large part of the fatigue life. For this reason,

crack initiation and propagation behavior in the superplastic region and the non-superplastic region were investigated. The growth rate of a surface small crack, $da/dn$, was analyzed in terms of the cyclic $J$-integral range, $\Delta J$. In this study, $\Delta J$ was estimated by reference to the values given by Raju-Newman[16], Dowling[17] and Hatanaka et.al.[18-20]. For uniaxial loading of smooth specimens with small semi-circular surface cracks, $J$ is given by Dowling[17] as

$$\Delta J = 2\pi a \{M_k/\phi(\lambda)\}^2 \{\Delta W_e + f(n)\Delta W_p\} \quad (2)$$

where $a$ is the crack length, $M_k$ is the modification coefficient given by Raju-Newman[16], $\phi(\lambda)$ is the complete elliptic integral of the second kind, $\lambda$ is the aspect ratio of the semi-circular cracks, $\Delta W_e$ is the elastic strain energy density[19], $\Delta W_p$ is the plastic strain energy density[19] and $f(n)$ is a function of the work hardening exponent $n$ that is given by Hatanaka[20] as

$$f(n) = (n+1)\{3.85(1-n)/\sqrt{n} + \pi n\}/2\pi \quad (3)$$

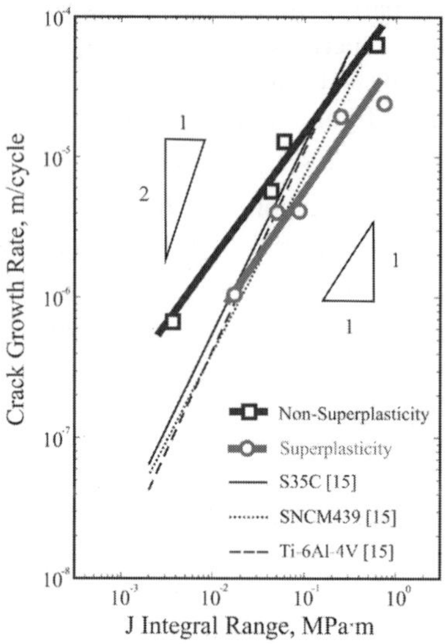

Figure 4 shows the relationship between the crack propagation rate and $\Delta J$. For comparison, data for

Fig. 4. Relationship between crack propagation rate and $\Delta J$.

carbon steel[15], high-strength steel[15] and Ti alloy[15] are also included in this plot. The fatigue crack growth rate in the superplastic region is significantly lower than that in the non-superplastic region. Moreover, the exponent of the fatigue crack growth plot for Zn-22wt.%Al alloy is the lowest of all the alloys, so Zn-22wt.%Al alloy in the superplastic region exhibits the lowest fatigue crack growth rate in the high J-integral range above 0.3. This result suggests that Zn-22wt/%Al alloy is suitable as a seismic damping material because it exhibits a lower fatigue crack growth rate when the strain amplitude is large; that is, in the high $J$-integral range. The crack growth curve of the Zn-22wt.%Al alloy can be represented by the following equations

$$da/dN = 1.8 \times 10^{-4} (\Delta J)^{1.0} \quad (4) \quad \text{(Non-superplastic region)}$$

$$da/dN = 5.5 \times 10^{-5} (\Delta J)^{1.0} \quad (5) \quad \text{(Superplastic region)}$$

These equations can be used in calculating the fatigue life spent for crack propagation.
The crack initiation behavior was analyzed in terms of the crack density on the surface as observed by a color laser microscope. Figure 5 shows the relationship between the density of cracks on the surface observed in 1 mm$^2$ and the number of cycles at a total strain of 10% for both cyclic deformation conditions. The density of small cracks increases linearly with increasing number of cycles, subsequently reaching saturation at a certain crack density. It is supposed that the enough large crack is formed at the saturation point of the crack number and propagates with the deformation. Moreover, the crack density to failure in the superplastic region is higher than that in the non-superplastic region. This suggests that the superplastic deformation mechanism contributes to retarding propagation of the main crack. In addition, Figure 6 shows the situation adjacent to the crack tip in the case of the specimen tested in the superplastic region.

Many small cavities (~0.2 μm) are visible in the vicinity of the crack. It is possible that crack propagation occurs through the joining up of cavities nucleated predominantly at the triple-point junction, with the specimen exhibiting the intergranular type of fracture typical of superplastic deformation. On the other hand, the formation of cavities was not observed in the vicinity of the crack in the case of the specimen tested in the non-superplastic region.

Figures 3-6 suggest that crack initiation and propagation behavior probably depend on the deformation mechanism of the material. It is well known that grain boundary sliding (GBS) is the predominant mode of deformation during superplastic flow[21-24]. Further, GBS generally causes a stress concentration at the triple point. Therefore, it is inferred that the significant superplastic elongation is due to the accommodation of the stress concentration by diffusional flow and dislocation annihilation[21-24]. In cyclic deformation, likewise, it is considered that the process of accommodating superplasticity and the cavity formation observed in Fig. 6 would relax and/or disperse the stress concentration around the crack tip, resulting in the retardation of fatigue crack propagation and the corresponding evolution of surface crack populations.

A schematic summary of this crack initiation and propagation behavior in the superplastic region is shown in Fig. 7. It is clear that efforts to develop a superplastic seismic damper with even higher damping performance should focus on refining the microstructure of the Zn-22wt.%Al alloy.

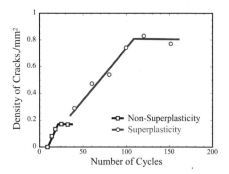

Fig 5  Relationship between density of cracks and number of cycles to failure.

Fig. 6. Micrograph adjacent to crack tip of specimen tested in the superplastic region.

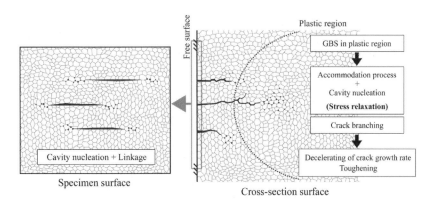

Fig. 7. Schematic representation of crack initiation and propagation behavior in Zn-22wt.%Al alloy tested in superplastic region.

## 4. Summary

The mechanisms of crack initiation and propagation in the superplastic and non-superplastic regions are investigated for Zn-22wt.%Al alloy and compared with the results for various other materials. Fatigue life of the alloy is greater in the superplastic region than in the non-superplastic region and also exceeds that of the other materials. Further, the propagation of cracks in the superplastic region can be attributed to the linking up of small cavities, while the crack propagation rate is lower than that in the non-superplastic region. At large strain amplitudes, especially, the alloy has excellent resistance to crack growth rate in comparison with the results for the other materials. Multiple cracks form on the surface of an alloy specimen tested in the superplastic region, which is indicative of a decrease in the fatigue crack propagation rate. These results are considered attributable to the relaxation of stress concentration around the crack tip due to an accommodation process related to superplasticity and cavity nucleation.

## Acknowledgment

The authors are grateful to H. Takahashi (Graduate student, Osaka Prefecture University) for his helpful cyclic testing results.

## References

1. C.S. Tsai, K.C. Tsai, J. Eng. Mech. ASCE 121 (1995) 1075.
2. N.P. Plakhtienko, Int. Appl. Mech. 37 (2001) 414.
3. A. Kushibe, K. Makii ,L.F.Chang, T. Tanaka, M.Kohzu,K. Higashi, Mater.Sci.Forum. 475-479 (2005) 3055.
4. T. Tanaka, K. Makii, A. Kushibe, M. Kohzu, K. Higashi, Scripta Mater. 49 (2003) 361.
5. T. Tanaka, H. Watanabe, K. Higashi, Mater. Trans. 44 (2003) 1891.
6. T. Tanaka, K. Higashi, Mater. Trans. 45 (2004) 1261.
7. T. Tanaka, S.W. Chung, L.F. Chaing, K. Makii, A. Kushibe, M. Kohzu, K. Higashi, Mater. Trans. 45 (2004) 2542.
8. T. Tanaka, K. Higashi, Mater. Trans. 45 (2004) 2547.
9. T. Tanaka, M. Kohzu, Y. Takigawa, K. Higashi, Scripta Mater. 52 (2005) 231.
10. Sherby, J. Wadsworth, Prog. Mater. Sci. 33 (1989) 169.
11. J. Pilling, N. Ridley, Superplasticity in crystalline solids, The Institute of Metal, London, 1989.
12. K. Nakashima, Z. Horita, M. Nemoto, T.G. Langdon, Acta mater. 46 (1998) 1589.
13. L.F. Coffin, Trans. ASME 76 (1954) 931.
14. S.S. Manson, Heat Transfer Symposium (1953) 9.
15. K. Hatanaka, T. Fujimitsu, S. Nishida, Trans. Jpn. Soc. Mech. Eng. A 57 (1991) 244.
16. I.S. Raju, J.C. Newman Jr, Eng. Fract. Mech. 11 (1979) 817.
17. N.E. Dowing, ASTM STP 637 (1977) 97.
18. K. Hatanaka, T. Fujimitsu, H. Watanabe, Trans. Jpn. Soc. Mech. Eng. 50 (1984) 737.
19. K. Hatanaka, T. Fujimitsu, H. Watanabe, Trans. Jpn. Soc. Mech. Eng. 51 (1985) 790.
20. K. Hatanaka, T. Fujimitsu, ASTM STP 942 (1988) 257.
21. A. Ball, M.M. Hutchison, Metal. Sci. J. 3 (1969) 1.
22. T.G. Langdon, Philo. Mag. 22A (1970) 689.
23. A.K. Mukhajee, Mater. Sci. Eng. 8 (1971) 83.
24. M.F. Ashby, R.A. Verrall, Acta Metall. 21 (1973) 149.
25. T. Tanaka, A.Kushbe,M. Kohzu, Y. Takigawa, K. Higashi, Scripta Mater. 59 (2008) 215.

# TEMPERATURE DEPENDENT FRACTURE MODEL AND ITS APPLICATION TO ULTRA HEAVY THICK STEEL PLATE USED FOR SHIPBUILDING

YUN CHAN JANG[1], YOUNGSEOG LEE[6†]

*Department of Mechanical Engineering, Chung-Ang University, KOREA,*
*ycjang@wm.cau.ac.kr, ysl@cau.ac.kr*

GYU BAEK AN[2], JOON SIK PARK[3], JONG BONG LEE[4]

*POSCO Technical Research Laboratories, Pohang, KOREA,*
*gyubaekan@posco.com, poolside@posco.com, jongblee@posco.co.kr*

SUNG IL KIM[§]

*POSCO Technical Research Laboratories, Kwangyang, KOREA,*
*ksimetal@posco.com*

Received 15 June 2008
Revised 23 June 2008

In this study, experimental and numerical studies were performed to examine the effects of thickness of steel plate on the arrest fracture toughness. The ESSO tests were performed with the steel plates having temperature gradient along the crack propagation direction. A temperature dependent crack initiation criterion was proposed as well. A series of three-dimensional FEA was then carried out to simulate the ESSO test while the thickness of the steel plate varies. Results reveal that a temperature dependent brittle criterion proposed in this study can describe the fracture behavior properly.

*Keywords*: Arrest toughness (arrestability); ultra heavy thick steel plate; ESSO test; failure simulation.

## 1. Introduction

Recently, shipbuilding companies have tried to use steel plates with thickness of 80 mm in building super large-sized container vessels. Inoue et al.[1] ,however, reported that brittle crack arrestability at weld and even base metal is suspicious if the thickness of steel plate is more than 65mm.

The Study of brittle crack arrest was started in 1953 by Robertson[2] as he introduced the concept of crack arrest temperature. The crack arrest can be the second line of safety defense of structures which experience temperature gradients. A crack might initiate in a cold region of the structure and be arrested when it encounters a material with higher temperature. In the 1970s, research results of brittle crack arrest were applied to design of pressure vessels, storing tanks and maritime structures which requires higher safety.

---

†Corresponding Author.

Since 1980, Japanese researchers have studied of brittle crack arrest fracture toughness in connection with development of steel plate for shipbuilding.[3]

Evaluation method of brittle crack arrestability (arrest fracture toughness) largely is divided into a large-scale experiment and a small-scale experiment. Note it is difficult to analyze the crack arrestability character of a small-sized specimen if we perform the small-scale experiment. This character can be evaluated by the large-scale experiment but it requires high cost for test. One of methods to overcome this problem is employing numerical approach such as finite element method and simulates the crack propagation and crack arrest. Key point of FE simulation of material failure is how to set up a model for crack initiation.

Ritchie et. al[4] proposed a temperature independent crack initiation criterion under mode-I loading condition: It was assumed the crack initiates when the maximum principal stress at the ahead of crack tip with a certain distance reaches a critical value. Using a local stress criterion, Susumu et al.[5] simulated brittle crack propagation of steel-A and E used for shipbuilding. They reported that the dynamic fracture toughness was dependent on the crack propagation speed and the temperature change along crack growth path. The studies cited so far, however, didn't investigate the effect of thickness on the crack arrestability. For this reason, the variation of arrest fracture toughness of steel plate with thickness varied is highly desirable.

In this study we conducted ESSO test and finite element analysis to examine the arrest fracture toughness of steel plate with thickness of 50mm and 80mm. The material used in this study is EH36. Temperature gradient (−80~40°C) was applied to the steel plate along the crack propagation direction before test. We propose a temperature dependent cleavage crack initiation model and coupled it with FE analysis. Crack propagation is simulated using an element removing scheme. Static and thermal analysis was carried out by implicit integration and explicit integration technique was used to simulate the crack initiation and propagation under impact loading.

## 2.  Experiment

### 2.1.  *Esso test*

We performed ESSO experiment using a heavy (3000ton) tensile testing machine (Fig. 1). Temperature gradient (−80 to 40 °C) was assigned in the specimen with a notch along the crack propagation direction(Y-direction). The specimen is under various side load (nominal stress) before the crack initiates.

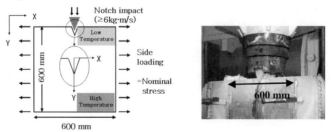

Fig. 1.  Schematic of ESSO test and specimen. The range of temperature gradient is min. −80 ~ max. 40 ℃.

When an impact load with a hammer is applied to the specimen, the crack starts initiating and propagates.

Table 1. ESSO test conditions.

| Specimen name and number | Thickness [mm] | Nominal stress [MPa] | Temperature gradient [°C/mm] |
|---|---|---|---|
| EH36-50 (No. 1) | 50 | 163 | 18 |
| EH36-50 (No. 2) | 50 | 107 | 23 |
| EH36-80 (No. 4) | 80 | 188 | 22 |
| EH36-80 (No. 5) | 80 | 148 | 22 |
| EH36-80 (No. 6) | 80 | 93 | 22 |

As shown in Table 1 we carried out the ESSO test three times for each specimen (50mm and 80mm thickness). We measured the crack arrest length after test. One of them was not successful. Numbers, '50' and '80' of EH36-50 and EH36-80 indicate the specimen thickness, respectively. The chemical compositions of EH36 steel is as follows; C: 0.057, Si: 0.14, Mn: 1.5, P: 0.14, S: 0.0017 (weight percentage). Its yield stress and ultimate tensile stress is 375MPa and 610MPa, respectively.

### 2.2. *Variations of yield stress dependent on temperature and strain rate*

Variation of yield stress with temperature change was measured through a uniaxial axis tensile test and results are shown in Fig. 2. Even though the variation of yield stress was not linear with the temperature change, we assumed a linear relationship to predict the yield stress in terms of temperature. Equation for this relationship appears in Fig. 2. $\sigma_o$ and $T_o$ represents yield stress at room temperature and room temperature, respectively. 'A' is a material constant and A=1.8 is used in this study.

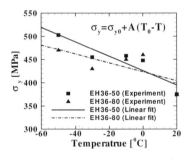 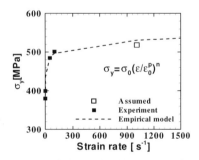

Fig. 2. Dependence of yield stress on temperature.    Fig. 3. Dependence of yield stress on strain rate.

We also conducted a high strain rate tensile test in the ranges of 0.01~ 100s$^{-1}$ using a test equipment, VHS 8800$^®$. Figure 3 shows test results. Since we couldn't obtain the yield stress at higher strain rates, say 500 s$^{-1}$, we assumed a yield stress at the strain rate of 1000s$^{-1}$ and set up an empirical-based constitutive equation as shown in Fig. 3 to predict the variation of yield stresses in a wide range of strain rates. The yield stress in Fig. 3 is based on the yield stresses obtained from high strain rate test of many steels.[5] $\sigma_o$

implies a static yield stress at room temperature. $\dot{\bar{\varepsilon}}^P, \dot{\bar{\varepsilon}}_0^{\ P}$ are equivalent plastic strain rate and reference equivalent plastic strain rate, respectively. 'n' is a material constant. $\sigma_o = 375$ [MPa], $n = 0.028$, $\dot{\bar{\varepsilon}}_0^{\ P} = 0.01$ are used in this study.

## 3. Finite Element Simulation of Crack Initiation and Propagation

Uniform meshes (6mm×3mm×5mm) are generated along the crack propagation direction (see Fig. 1). Maximum size of element at the corner region of specimen is 20mm. The number of elements used is 57,000 and element type for the specimens is C3D8R (8-node linear brick, reduced integration with hourglass control).

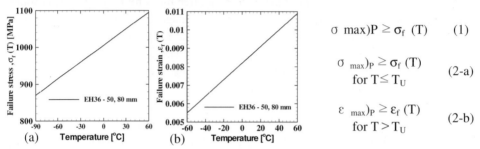

$$\sigma_{max})P \geq \sigma_f \ (T) \qquad (1)$$

$$\sigma_{max})_P \geq \sigma_f \ (T) \\ \text{for } T \leq T_U \qquad (2\text{-}a)$$

$$\varepsilon_{max})_P \geq \varepsilon_f \ (T) \\ \text{for } T > T_U \qquad (2\text{-}b)$$

Fig. 4.   Temperature dependencies of (a) stress-controlled failure
model and  (b) strain-controlled failure model.

We first use the stress-controlled failure model which depends on temperature, i.e., RKR-model, (see Eq. (1)). We then propose a mixed-controlled failure model that consists of the stress-controlled and strain-controlled failure criterion together. (see Eq. (2-a, b)) $\sigma_f(T)$ and $\varepsilon_f(T)$ represent temperature dependent failure stress and failure strain. $T_U$ indicates upper shelf energy temperature. Selection of crack initiation model in Eq. (2) is absolutely dependent on the temperature, $T_U$. In this study $T_U$ is set up as $-60\,^\circ$C. Figure 4 shows temperature dependent failure stress and strain for crack initiation. Crack tip stress is calculated first using measured fracture toughness at a temperature. A series of FE analysis with assumed failure stress or strain was then performed. Computed crack arrest lengths were compared with measured ones and finally the temperature dependent failure stress and strain was determined. For different temperatures and specimen thicknesses, this kind of FE analysis was performed and those stresses and strains was decided and plotted in Fig. 4 We use the element removing method to simulate crack propagation. Since we simulate mode-I crack propagation, we remove elements whenever their maximum principle stress or strain at the ahead of crack tip reaches a failure stress/strain.

## 4. Results and Discussion

### 4.1.   *Crack arrest length and crack propagation speed*
Table 2 summaries the measured crack arrest lengths and the ones computed by FEA. The mean crack speed is calculated by either the stress-controlled failure model (Eq.1) or the mixed-controlled failure model (Eq. 2-a, b). We can observe the mean crack speed

computed from the mixed-controlled model is smaller than that by stress controlled model. This is attributable to ductile failure when the mixed-controlled model is activated at lower temperature than −60 °C. As nominal stress was reduced, calculated crack arrest length and mean crack speed decreased. But in case of the specimen with 50mm thickness mean crack speed increased. This is because temperature gradient of specimen No. 2 is higher than that of specimen No. 1 (see Table. 1). The crack arrest lengths calculated by mixed-controlled model is much closer to the measured ones, in comparison with stress-controlled model. This indicates mixed-controlled failure model might describe the crack initiation of the ESSO test appropriately. The same specimen thickness doesn't necessary imply that one can get the same crack arrest length because different crack arrest lengths measured depend on different nominal loading conditions and temperature gradients.

Table 2. Crack arrest length and mean crack speed - measured *vs* computed.

| Specimen name and number | Measured Crack arrest length [mm] | Stress-controlled failure Crack arrest length [mm] | Stress-controlled failure Mean crack speed [m/s] | Mixed-controlled failure Crack arrest length [mm] | Mixed-controlled failure Mean crack speed [m/s] |
|---|---|---|---|---|---|
| EH36-50 (No 1) | 450 | 444 | 1394 | 438 | 1352 |
| EH36-50 (No 2) | 390 | 402 | 1637 | 354 | 1602 |
| EH36-80 (No 4) | 390 | 594 | 2112 | 426 | 1671 |
| EH36-80 (No 5) | 340 | 582 | 1841 | 312 | 1332 |
| EH36-80 (No 6) | 281 | 276 | 1334 | 228 | 1371 |

Figure 5(a), (b) and (c) shows the crack length propagated inside specimen is much longer than the one at near its surface.

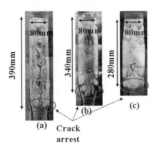 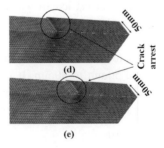

Fig. 5. Experimentally measured and calculated fractured surface (a) EH36-80, No. 4 (b) EH36-80, No. 5 (c) EH36-80, No. 6 (d) Crack arrest by stress model (No.2) and (e) Crack arrest by mixed model (No. 2).

This is confirmed from FE simulation (Fig. 5(e)) in which stress-controlled failure model was used for crack initiation. This is attributable to higher tri-axial stresses at the inside material than at surface. If mixed-controlled failure model is used, the crack front is of a linear line (Fig. 5(d)). This is because the magnitude of maximum principle strain is very even at the inside and near surface of specimen.

### 4.2.  *Arrest fracture toughness*

To estimate the effect of specimen thickness on the arrest fracture toughness, $K_{ca}$ quantitatively, we evaluated $K_{ca}$ using an equation which appears in Fig. 6. Figure 6 shows a relationship between temperature and arrest fracture toughness. $C_a$ represents crack arrest length, $\sigma_g$ side loading and B width of specimen. (see Fig. 1). Note that the crack arrest lengths computed from the mixed-controlled failure model is similar with those measured. Therefore we computed $K_{ca}$ based on the crack arrest length computed by the mixed-controlled model. Table 3 shows that crack the computed crack arrest toughness of 50mm thickness specimen is larger than that of 80mm thickness specimen. The arrest fracture toughness of specimen is reduced as its thickness increases.

Table 3.  Measured vs computed of arrest fracture toughness.

| Specimen name and number | $K_{ca}$ [MPa√m] | |
|---|---|---|
| | Measured | Computed |
| EH36-50 (No 1) | 277 | 265 |
| EH36-50 (No 2) | 150 | 135 |
| EH36-80 (No 4) | 263 | 294 |
| EH36-80 (No 5) | 180 | 167 |
| EH36-80 (No 6) | 97 | 83 |

\* 'Measured' implies that $K_{ca}$ is calculated using the crack arrest length measured.

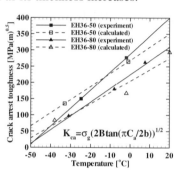

Fig.  6.  Relation between $K_{ca}$ and temperature.

### 5.   Concluding Remarks

Crack arrest lengths obtained from the ESSO test illustrate that the temperature dependent mixed-controlled failure model is appropriate for numerical simulation of ESSO test. Crack arrest length was largely dependent on nominal stress applied to the specimen before the crack initiates, not on the specimen thickness. FE simulation shows that the crack propagation speed is strongly dependent on temperature gradient. The arrest fracture toughness was decreased by about 60% while the specimen thickness increases from 50mm to 80mm. It is attributable to that the plastic deformation at the moving crack tip is suppressed by high local tri-axial stress and the mixed-controlled fracture mechanism acting on the crack front as the specimen thickness increases.

### Acknowledgments

This research was supported by the Chung-Ang University Research Grants in 2008 and Research Grants of POSCO Tech. Res. Lab.

### References

1.  T. Inoue, T. Ishikawa, S. Imai, T. koseki, K. Hirota, M. Tada, H. Kitada, Y. Yamaguchi and H. Yajima, *Proc. the 16th Int. offshore and polar engineering conference*, 132 (SF, USA, 2006).
2.  T. S. Robertson, *J. Iron Steel Inst London.*, **175,** 361 (1953).
3.  Y. Nakano and M. Tanaka, *ISIJ, 22,* 147 (1982).
4.  R.O. Ritchie, J.F. Knott, and J.R. Rice, *J. Mech. Phys. Solids,* **21,** 395 (1973).
5.  S. Machida, H. Yoshinari and S. Aihara, *Fatigue and fracture mechanics*, 617, (1997).

# INVESTIGATION OF FRETTING FATIGUE BEHAVIOR OF TI811 ALLOY AT ELEVATED TEMPERATURE

XIAO-HUA ZHANG[1†]

*Aviation School, Northwestern Polytechnical Universit,*
*Box 398 127 You Yi Xi Road, Xi'an, ShaanXi, P. R. CHINA,*
*yhzhangxh@nwpu.edu.cn*

DAO-XIN LIU[2]

*Aviation School, Northwestern Polytechnical University*
*Box 398 127 You Yi Xi Road, Xi'an, ShaanXi, P.R. CHINA,*
*liudaox@nwpu.edu.cn*

Received 15 June 2008
Revised 23 June 2008

The fretting fatigue behavior of the Ti811titanium alloy, as influenced by temperature, slip amplitude, and contact pressure, was investigated using a high-frequency fatigue machine and a home-made high-temperature apparatus. The fretting fatigue failure mechanisms were studied by observing the fretting surface morphology features. The results show that the sensitivity to fretting fatigue is high at both 350°C and 500°C. The higher the temperature is, the more sensitive the alloy is to fretting fatigue failure. Creep is an important factor that influences the fretting fatigue failure process at elevated temperature. The fretting fatigue life of the Ti811 alloy does not change in a monotonic way as the slip amplitude and contact pressure increase. This is due to the fact that the slip amplitude affects the action of fatigue and wear in the fretting process, and the nominal contact pressure affects the distribution and concentration of the stress and the amplitude of fretting slip at the contact surface, and thus further influences the crack initiation probability and the driving force for propagation.

*Keywords*: Fretting fatigue; elevated temperature; titanium alloy; wear; creep.

## 1. Introduction

Fretting fatigue (FF) occurs whenever a small-amplitude oscillatory motion between the two contacting bodies is combined with an applied cyclic axial load. Damage caused by FF leads to premature crack nucleation and results in a reduction of fatigue life as compared with plain fatigue. FF exists widely in many industries, such as the aviation, space, traffic, and nuclear industries. FF damage is universal and prevalent in the aviation industry [1-3].

[†]Corresponding Author.

The titanium alloy Ti811 has many advantages, including low density, high Young's modulus, excellent vibration damping capacity, good thermal stability, and good welding and molding performance; particularly, its ratio of tensile strength to density is the highest among industrial titanium alloys. As a result, this alloy has become one of the important materials selected for rotating components of the high-temperature parts of advanced aircraft engine compressors[4]. However, titanium alloys are also characterized by poor tribological properties, such as a high and unstable friction coefficient, severe adhesive wear, susceptibility to fretting, and poor FF resistance[5]. As a result, the reliability and service life of titanium alloy blade/disk attachments in gas turbines are affected by FF damage[6,7]. Research has been carried out on the thermal stability, oxidation resistance, thermohaline stress corrosion resistance, and creep properties of Ti811 alloy at elevated temperatures[8, 9]. However, very little attention has been paid to the FF behavior of Ti811 alloy at elevated temperature. This paper presents a study of the effects of temperature, fretting slip amplitude, and contact pressure on the FF behavior of Ti811 alloy at elevated temperature.

## 2.  Experiment Procedures

FF specimens and fretting pads were obtained from Ti811 titanium alloy bars (Ø 16 mm). Ti811 alloy is almost all $\alpha$-phase and contains 7.9% Al, 1.0% Mo, 0.99% V, 0.05% Fe, 0.1% C, 0.01% N, 0.001% H, 0.06% O, and balance, Ti. The material was treated by double annealing (910 °C for 1 h, cooled in air, 580 °C for 8 h, cooled in air). The resulting microstructure is an equiaxial $\alpha$-phase and intergranular $\beta$-phase. The mechanical properties of the alloy are: $\sigma_b$=931 MPa, $\sigma_{0.2}$=890 MPa, $\delta$=23% and $\Psi$=46%.

A PLG-100C high-frequency fatigue machine was used to conduct FF tests. The load was set in pull-pull. The contact state between the pad and specimen was flat to flat, with a rectangular contact area of 2 mm$\times$ 6 mm. Relative slip between the specimen and pad was introduced by the difference in elastic deformation between them. The relative slip amplitude could be changed by adjusting the fretting pad length. The contact pressure of the pads to the specimen was controlled using the stressing ring. The FF susceptibility of the titanium alloy to the temperature was evaluated by the S-N curves. The cycle load was in a sinusoidal form at 110 Hz with a stress ratio of 0.1. Test temperatures of 350 and 500 °C were adopted to simulate the working conditions of aircraft engine compressors.

## 3.  Results and Discussion

### 3.1.  *Effect of temperature on fretting fatigue life*

The working temperature of compressor components increases along the sequence of compressors in an aircraft engine. The maximum temperature is about 500°C. Fig. 1 shows the FF S-N curves of the Ti811 titanium alloy at room temperature, at 350°C, and at 500°C. The results show that the FF limit decreases and the fretting fatigue life (FFL) is reduced at the same cyclic stress when the temperature is increased. Especially at high stress levels, the FFL is reduced more significantly.

Two effects were introduced when the temperature increased. One was that the effect of creep was augmented, and so the holistic fatigue capability of the material decreased; the other was that the oxidation of the material surface was accelerated. Three major factors that affected the creep behavior of the material were the stress level, the temperature, and time. The results indicated that the effect of the temperature was more significant at high stress levels, mostly owing to the role of creep. Previous research has indicated [8] that the diffusion mechanism was dominant in the creep mechanism when the temperature was below 425°C. The creep sensitivity of the alloy increased at temperatures above 425°C owing to an increased number of slip systems and the initiation of intergranular sliding. Though the FF specimen fractured in the fretting contact area under a test condition of 530 MPa (maximal cyclic stress) and 500°C, most of the cracks that were produced by plain fatigue existed outside the fretting area of the specimen. The results further illuminated the fact that creep produced by the synergism of the stress and temperature severely affected the FF at high temperature and stress.

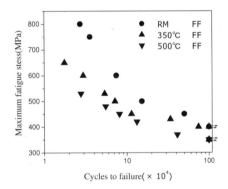

Fig. 1.    S-N curve of fretting fatigue of Ti811 alloy at different temperature.

### 3.2.  *Effect of slip on ffl*

Fig. 2 shows the effect of slip on the FFL of the Ti811 titanium alloy. The maximal stress in the cycle was 530 MPa and the contact pressure was 85 MPa. The results indicated that the FFL changed nonmonotonically with increasing slip amplitude. A minimum FFL was observed in a certain range of relative slip. The fact that the contact condition in the fretting area was affected by the slip amplitude was the primary reason. In the range of large relative slip, the contact condition was gross slip. The area and degree of wear on the surface increased when gross slip occurred. Not only was the probability of forming propagating cracks reduced, but also the cracks formed could be removed. The position of the stress concentration was at the edge of the contact area in the case of partial slip, but this position changed to the central section of the contact area in the case of gross slip

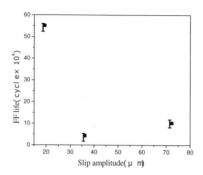

Fig. 2.   Slip amplitude on FF life of Ti811 alloy.

(Fig. 3(a)). Simultaneously, the surface energy of the contact area was altered by the oxidized debris. Hence the degree of stress concentration in the material surface could be reduced, and the fatigue effect of the partial contact could be reduced. Therefore the FFL increased when the relative slip was large.

In the range of small relative slip, the contact condition was partial slip. The sticking region (the region where the pads and the material surface were immobile relative to each other) was enlarged and the slip region decreased (Fig. 3(c)). Elastic deformation occurred in tiny peaks on the material surface in the sticking region; this was caused by the tangential displacement. The area and degree of wear on the surface were small in the case of partial slip. The probability of forming propagating cracks was reduced, and so the FFL increased. A mixed state existed when the range of relative slip was between those of gross slip and partial slip. The sticking region was surrounded by a slip region (Fig. 3(b)). The tangential friction force varied wildly during the slip. The rather large tangential friction force acted iteratively on the contact area, which was the boundary between the slip region and the sticking region. A very large partially compressive stress appeared in front of the slip region, and the maximal tensile stress occurred behind the slip region. Intensive plastic deformation and partial wear occurred at the surface of the material. Cracks were then easily initiated at the boundary between the slip region and the sticking region and rapidly propagated, and so the minimum FFL was observed in this range of relative slip.

(a) Slip amplitude 72 μ m          (b) Slip amplitude 36μm          (c) Slip amplitude 18μm

Fig. 3.   Slip amplitude on FF surface scar of Ti811 alloy.

### 3.3. *Effect of contact pressure on ffl*

Fig. 4 shows the effect of the nominal contact pressure on the FFL of the Ti811 titanium alloy. The maximal cycle stress was 530 MPa and the length of the fretting pad was 15 mm. The results indicated that the FFL changed nonmonotonically with increasing contact pressure. Because the sticking region was enlarged and the plastic deformation of the contacted microasperities increased at the surface of the material when

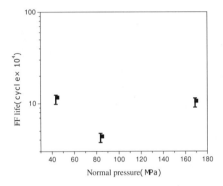

Fig. 4. Surface nominal pressure on FF life of Ti811 alloy.

the contact pressure was increased, the true contacted area increased and the true contact pressure was reduced, and so the probability of forming propagating cracks was reduced and the FFL increased. Gross slip and intensive wear occurred in the contact area when the contact pressure was low. Under these conditions, the nucleation pit of the microcracks might be removed and the probability of forming propagating cracks reduced. The debris which collected at the interface might reduce the direct action between the surface and the tangential force by becoming compacted into powder beds. The non-fretting area of the contact surface was narrow and the contact pressure was concentrated at the boundary between the slip region and the non-slip region when the contact pressure was in a certain range, and so the number of cracks initiated in this area was low, and the FFL was low. In addition, the relation between the changes of the contact pressure and the slip amplitude was one of coupling adjustment. The true slip amplitude was smaller when the contact pressure was larger under the condition of the same fretting-pad span and cyclic stress. In contrast, the results in the previous paragraph indicate that the FFL increased when the slip amplitude was small.

When the contact pressure was large, the gross wear was light and the pressure was the dominant factor among all factors that influenced the FF damage. Severe plastic-deformation flow occurred near the fracture (Fig.5). When the contact pressure was low, the gross wear was severe. The mechanism of wear was a mixed mechanism of delamination and abrasion. The number of propagating cracks was reduced, and the fracture position transferred from the edge of the contact area to the middle of the contact area (Fig.6 ). When the contact pressure was in a certain range between the preceding two conditions, the synergistic action of pressure and partial wear resulted in the FF cracks initiating and propagating, and so the FFL decreased.

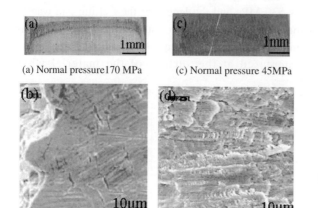

(a) Normal pressure170 MPa        (c) Normal pressure 45MPa

(b) Normal pressure170MPa        (d) Normal pressure 45MPa

Fig. 5.  FF surface feature of Ti811 under different pressure.

Fig. 6.  FF fracture feature under 45 MPa.

## 4.  Conclusions

(1). The Ti811 titanium alloy studied here was susceptible to FF damage at both 350°C and 500°C. The sensitivity to FF increased when the temperature increased. Creep was the dominant factor in the FF damage at the elevated temperature.

(2). The FF life changed nonmonotonically with increasing slip amplitude because the slip amplitude affected the action of fatigue and wear in the FF process

(3). The FF life changed nonmonotonically with increasing contact pressure because the change of the nominal contact pressure altered the stress distribution, the stress concentration, the fretting slip amplitude, the degree of wear, and the wear mechanism in the contact area, and affected the probability of initiating cracks and the driving force for propagation.

## References

1.  S. Chakravarty, A. K. Koul, *J. of metals,* **47**, 31(1995).
2.  T.C. Lindley. *Int. J. Fatigue,* **19**, 39(1997).
3.  B. P. Conner, A. L. Hutson, L, *Wear,* **255**, 259(2003).
4.  R. Q. Zhao, B. N. Liu, *Rare Metal Mater Eng.,* **23**, 59(1994).
5.  D. X. Liu, B. Tang, J. W. He, *Chin. J. Nonferrous Metals,* **11**, 454(2001).
6.  S. K. Bhaumik, M. Rangaraju, M. A. Venkataswamy, *Eng. Failure Anal,* **9**, 255 (2002).
7.  L. L. WU, B. C. Holloway, *Surf. Coat. Technol,* **130**, 207 (2000).
8.  R. Q. Zhao, K. Y. Zhu. Z. C. Li. *Dev. Appl. Mater,* **10**, 16 (1995).
9.  W. F. Zhang, X. L. Liu, W. G. Zhao. *Trans. Mater. Heat Treat.,* **24**, 55(2003).

# EFFECT OF VARIOUS HEAT TREATMENT PROCESSES ON FATIGUE BEHAVIOR OF TOOL STEEL FOR COLD FORGING DIE

S. U. JIN[1], S. S. KIM[5]

*School of Materials Science and Engineering, Gyeongsang National Univ., Chinju, KOREA,*

*sanguk@gnu.ac.kr, sang@ gsnu.ac.kr*

Y. S. LEE[2][†], Y. N. KWON[3], J. H. LEE[4]

*Department of Materials Processing Technology, KIMS, Changwon 641-010, KOREA,*

*lys1668@kims.re.kr, kyn1740@ kims.re.kr, ljh1239@ kims.re.kr*

Received 15 June 2008
Revised 23 June 2008

Effects of various heat treatment processes, including "Q/T (quenching and tempering)", "Q/CT/T (Quenching, cryogenic treatment and tempering)", "Q/T (quenching and tempering) + Ti-nitriding" and "Q/CT/T (Cryogenic treatment and tempering) + Ti-nitriding", on S-N fatigue behavior of AISI D2 tool steel were investigated. The optical micrographs and Vicker's hardness values at near surface and core area were examined for each specimen. Uniaxial fatigue tests were performed by using an electro-magnetic resonance fatigue testing machine at a frequency of 80 Hz and an R ratio of -1. The overall resistance to fatigue tends to decrease significantly with Ti-nitriding treatment compared to those for the general Q/T and Q/CT/T specimens. The reduced resistance to fatigue with Ti-nitriding is discussed based on the microstructural and fractographic analyses.

*Keywords*: S-N Fatigue; Cold Forging; AISI D2 Steel; Cryogenic Treatment; Ti-nitriding.

## 1. Introduction

Cold forging has become one of the most common processes in the mass production of components, and the dies for cold forging have been subjected to more severe environment with broadly applied high-strength materials for the automobile and electronics industries [1,2,3]. Depending on the process conditions and the characteristics of material and surface conditions, various modes of tool failure can be encountered [4]. Wear and sudden fracture due to fatigue are the most common forms of failure [5]. Since most fatigue cracks originate at the surface, the surface condition has a substantial influence on fatigue behavior. Thus the improvement of surface conditions can have a beneficial effect on wear resistance and fatigue behavior of materials. Nitriding is one of the most widely used thermo-chemical surface modification methods, and produces a strong and shallow case with high compressive residual stress on the

[†]Corresponding Author.

surface [6].  The carbides in tool steel also serve as fatigue initiation sites and may affect the fatigue behavior of tool steel for cold forging die significantly [7,8,9].  The cryogenic treatment at -170°C for 15 minutes in between quenching and tempering is known to encourage the fine carbide distribution [ref].

Controversy exists on the effect of nitriding on fatigue behavior of tool steel, such that nitriding may either improve the resistance to fatigue with increased residual stress [6] or decrease the resistance with easy fatigue crack initiation at surface [10].  In this work, various heat treatment processes, including "Q/T (quenching and tempering)", "Q/CT/T (Quenching, cryogenic treatment and tempering), "Q/T (quenching and tempering) + Ti-nitriding" and "Q/CT/T (Cryogenic treatment and tempering) + Ti-nitriding" were applied for AISI D2 tool steel to understand the fatigue behavior.  The microstructural observation and SEM fractographic analysis were conducted to identify the effect of each heat treatment on the resistance to S-N type fatigue.

## 2.  Experimental Procedures

The material used in this study was AISI D2 steel with the chemical composition shown in Table 1.  The various heat treatments, including Q/T, Q/CT/T, Q/T + Ti-nitriding and Q/CT/T + Ti-nitriding were applied for the specimen.  Each heat treatment process is schematically illustrated in Figure 1.  For Ti-nitriding, approximately 2 to 20 g of electrolyzed metallic titanium is added in salt bath.  For Ti-nitriding specimens, the tempering prior to Ti-nitriding was conducted at low temperature of 180°C, since high temperature tempering effect is subdued with nitriding at 500°C for 120 minutes.

For fatigue test, a magnetic resonance fatigue testing machine was employed operating at 80 Hz at an R ratio of -1.  Figure 2 shows (a) the magnetic resonance fatigue testing machine and (b) the fatigue specimen, respectively, used in this study.  The microstructure of cross-sectioned specimen was documented by using an optical microscope after chemical-etching using a natal solution of 10 ml $HNO_3$ + 90 ml ethanol. Detailed fractographic analysis for the fatigued specimen was conducted by using a scanning electron microscope (SEM) to identify the characteristics of fatigue crack initiation.

Table 1.  Chemical composition (wt%).

|  | C | Si | Mn | Cr | Mo | V | Fe |
|---|---|---|---|---|---|---|---|
| AISI D2 (SKD11) | 1.50 | 0.60 | 0.50 | 12.0 | 0.80 | 0.9 | Bal. |

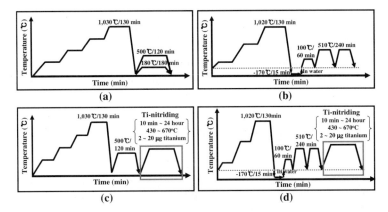

Fig. 1. Schematic illustration of various heat treatment processes: (a) Q/T (quenching and tempering), (b) Q/CT/T (Quenching, cryogenic treatment and tempering), (c) Q/T (quenching and tempering) + Ti-nitriding and (d) Q/CT/T (Cryogenic treatment and tempering) + Ti-nitriding.

Fig. 2. Photographs showing (a) magnetic resonance fatigue testing machine and (b) fatigue specimen used in this study.

## 3. Results and Discussion

Figure 3 shows the optical micrographs of the AISI D2 specimens with different heat treatment conditions of (a) Q/T, (b) Q/CT/T, (c) Q/T + Ti-nitriding and (d) Q/CT/T + Ti-nitriding, respectively. Equiaxed prior austenite grain boundaries with approximately 10 μm in diameter and carbides are observed for each specimen. For Ti-nitrided specimens, approximately 30 μm deep diffusion layer is observed on the surface area. The average prior austenite grain size and the average volume fraction of carbides are summarized in Table 2. Any notable change in prior austenite grain size for the AISI D2 specimen was not observed with different heat treatment processes used in this study. The average volume fraction of carbide appears to be marginally increased with Ti-nitriding process. Considerable shape change in carbides with different heat treatment processes was not observed.

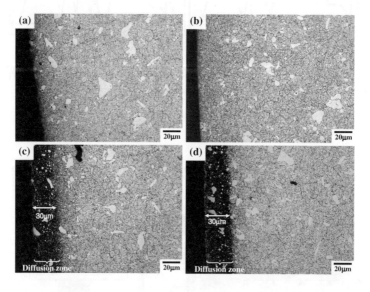

Fig. 3. Optical micrographs of AISI D2 specimens with different heat treatment conditions of (a) Q/T, (b) Q/CT/T, (c) Q/T + Ti-nitriding and (d) Q/CT/T + Ti-nitriding, respectively.

Table 2. Average prior austenite grain size and volume fraction of carbides for AISI D2 steel with different heat treatment processes.

| Heat treatment | Q/T | Q/CT/T | Q/T + Ti-nitriding | Q/CT/T + Ti-nitriding |
|---|---|---|---|---|
| Average prior austenite grain size (μm) | 10.5 ± 1.6 | 10.1 ± 0.9 | 9.7 ± 1.0 | 10.4 ± 1.1 |
| Average volume fraction of carbide (%) | 10.3 ± 1.0 | 10.9 ± 1.6 | 11.4 ± 0.2 | 11.6 ± 1.3 |

Table 3 shows Vicker's hardness values for each specimen measured at near surface and core area, respectively. Approximately over 200 Hv increase in surface hardness with Ti-nitriding treatment was observed, regardless of prior heat treatment condition. It was also observed that the overall hardness values for the Q/CT/T specimens, with and without Ti-nitriding, are slightly higher than that for Q/T counterparts. Such a change in hardness with different heat treatment processes would affect the fatigue behavior of tool steel.

Table 3. Vicker's hardness values for AISI D2 steel with different heat treatment processes and measured at near surface and core area, respectively.

| Heat treatment | Q/T | Q/CT/T | Q/T + Ti-nitriding | Q/CT/T + Ti-nitriding |
|---|---|---|---|---|
| near surface | 637 | 650 | 830 | 879 |
| core | 634 | 650 | 695 | 737 |

Table 4 shows the number of cycles to failure for each specimen tested at various applied stresses and an R ratio of -1. The overall resistance to fatigue tends to be either similar or slightly higher for the Q/CT/T specimens compared to that for the Q/T specimen. With Ti nitriding, the resistance to fatigue decreased significantly for both Q/T and Q/CT/T specimens. For example, approximately one-fold decrease in the number of cycles to failure ($N_f$) was observed for the Q/T specimens at the same applied stresses. Further decrease in $N_f$ was noted for the Q/CT/T specimens.

Table 4. Number of cycles to failure for AISI D2 specimens with different heat treatment conditions, fatigue tested at various applied stresses and an R ratio of -1.

| Maximum stress (MPa) | Number of cycles to failure ($N_f$) | | | |
|---|---|---|---|---|
| | Q/T | Q/CT/T | Q/T + Ti-nitriding | Q/CT/T + Ti-nitriding |
| 1200 | | $8.1 \times 10^3$ | $3.8 \times 10^2$ | $7.2 \times 10$ |
| 1100 | $1.5 \times 10^4$ | $1.6 \times 10^4$ | $9.8 \times 10^3$ | |
| 900 | $1.5 \times 10^6$ | $1.6 \times 10^6$ | $1.2 \times 10^6$ | $4.7 \times 10^2$ |
| 800 | $10^7$ * | | | |
| 750 | | $10^7$ * | | |
| 600 | | | $10^7$ * | $8.8 \times 10^6$ |

* run-out

In order to understand the reason for the reduced resistance to fatigue for the Ti-nitrided specimens, the fracture surface of fatigue failed specimens was observed by using an SEM. Figure 4 shows the SEM fractographs of fatigued (a) Q/T, (b) Q/CT/T, (c) Q/T + Ti-nitriding and (d) Q/CT/T + Ti-nitriding specimen, respectively. For the specimens without Ti-nitriding treatment, failed carbide particle near surface area tend to serve as a single fatigue crack initiation site, as shown in Figures 4(a) and 4(b). For the

Ti-nitrided specimens, on the other hand, fatigue crack initiation occurs over broad surface area rather than a single location, as shown in Figures 4(c) and 4(d). At relatively low applied stresses of 800 and 900 MPa for the Q/T and the Q/CT/T specimens, internal crack initiation was also observed at carbide clusters, as shown in Figure 5. Figure 5 shows (a) low and (b) high magnification SEM fractographs for the fatigued Q/T specimen at an applied stress of 900 MPa. Figure 6 shows the SEM micrographs of the specimen with (a) Q/T, (b) Q/CT/T, (c) Q/T + Ti-nitriding and (d) Q/CT/T + Ti-nitriding, respectively. This figure clearly shows that Ti-nitriding introduces porous surface structure, which would encourage easy crack initiation, reducing the resistance to fatigue. Moreover, relatively larger pores are often observed the Q/CT/T + Ti-nitriding specimen compared to the Q/T + Ti-nitriding specimen.

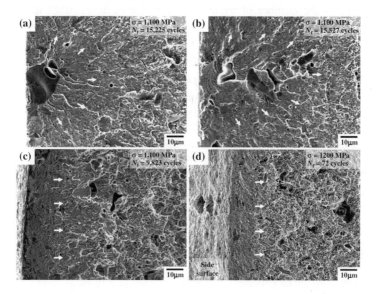

Fig. 4.   SEM fractographs of fatigued (a) Q/T, (b) Q/CT/T, (c) Q/T + Ti-nitriding and (d) Q/CT/T + Ti-nitriding specimen, respectively. The fractographs are documented in the vicinity of crack initiation sites.

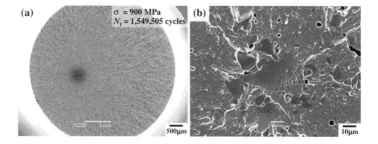

Fig. 5. SEM fractographs at (a) low and (b) high magnification, respectively, for the fatigued Q/T specimen at an applied stress of 900 MPa, showing internal fatigue crack initiation.

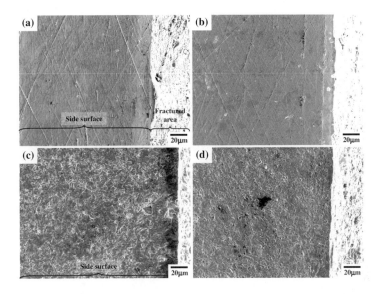

Fig. 6.   SEM micrographs of the (a) Q/T, (b) Q/CT/T, (c) Q/T + Ti-nitriding and (d) Q/CT/T + Ti-nitriding specimen, respectively, showing the surface condition of each specimen.

In this study, cryogenic treatment prior to tempering improves the hardness of AISI D2 steel compared to general Q/T counterpart, as demonstrated in Table 3. Ti-nitriding process on both Q/T and Q/CT/T specimens can further increase the surface hardness significantly. Such an improvement in hardness would be beneficial for wear resistance [10]. Despite the increase in the hardness of AISI D2 specimen with Ti- nitriding and/or cryogenic treatment, the resistance to S-N fatigue either does not improve with cryogenic treatment or significantly decreases with Ti-nitriding treatment. From the fractographic and micrographic analyses in Figures 4, 5 and 6, it was observed that Ti-nitriding produces porous surface structure, which encourages easy fatigue crack initiation on the surface area. It is well known that approximately 40 to 60% of total cycles to failure is used for the initiation of crack [11]. With the formation of porous surface structure, which would effectively decrease the initiation component of fatigue life, the resistance to fatigue for the Ti-nitrided ASIS D2 specimen decreases significantly. The main purpose of cryogenic treatment for tool steel prior to tempering is to obtain the fine distribution of carbides. In this study, the expected carbide refinement with cryogenic treatment was not observed, as demonstrated in Figure 3. Slight improvement in hardness with cryogenic treatment did not play an important role on determining fatigue life of AISI D2 steel. Unexpectedly larger pores observed on the surface area of the Ti-nitrided ASIS D2 specimen further decrease the resistance to fatigue.

## 4.  Conclusions

Effects of various heat treatment processes, including Q/T, Q/CT/T, Q/T + Ti-nitriding and Q/CT/T + Ti-nitriding, on S-N fatigue behavior of AISI D2 tool steel were examined and the following conclusions are drawn.

(1) Despite the increase in the hardness of AISI D2 specimen with Ti- nitriding and/or cryogenic treatment, the resistance to S-N fatigue either does not improve with cryogenic treatment or significantly decreases with Ti-nitriding treatment.

(2) The formation of porous surface structure with Ti-nitriding, which would effectively decrease the initiation component of fatigue life, appears to be responsible for the significant decrease in the resistance to fatigue for the Ti-nitrided AISI D2 specimens.

## Acknowledgments

This work has been supported by the Automotive Foundation Technology program, which is funded by the Ministry of Knowledge Economy, R.O. Korea.

## References

1. F. A. Kivk, *Met. Technol.* **vol. 9**, 198 (1982).
2. S. Inoue, *J. Jpn. Inst. Iron Steel.* **vol. 73**, 1461 (1987).
3. T. Arakawa and M. Suzuki, *Bull. Jpn. Inst. Met.* **vol. 5**, 269 (1966).
4. G. H. Farrahi, H. Ghadbeigi, *Journal of Materials Processing Technology,* **vol. 174**, 318 (2006)
5. ASM Metals Handbook, vol. 11, failure Analysis and Prevention (American Society for Metals, Metals Park, OH, 1986).
6. ASM Metals Handbook, vol. 1, Properties and Selection: Irons, Steels, and High Performance Alloys (American Society for Metals, Metals Park, OH, 1978).
7. H. Berns, J. Leng, W. Trojahn, R. Wahling and H. Wisell, *Powder Metall. Int.* **vol. 19**, 22 (1987).
8. J. Yoshida, M. Katsumata and Y. Yamazaki, *J. Jpn. Inst. Iron Steel*, **vol. 84**, 79 (1988).
9. Y. Natsume, K. Murakami and T. Miyamoto, *Proc. Jpn. Soc. Mech. Eng.,* **vol. 900,** 323 (1990).
10. *ASM Metals Handbook, vol. 4, Heat Treating* (American Society for Metals, Metals Park, OH, 1991).
11. George E. Dieter, *Mechanical Metallurgy* (McGraw-Hill, New York, 1988).

# FATIGUE LIFE PREDICTION OF ROLLED AZ31 MAGNESIUM ALLOY USING AN ENERGY-BASED MODEL

SUNG HYUK PARK[1], SEONG-GU HONG[2], BYOUNG HO LEE[3], CHONG SOO LEE[4†]

*Center for New and Renewable Energy Measurement,*
*Division of Industrial Metrology Korea Research, Institute of Standards and Science,*
*209 Gajeong-Ro, Yuseong-Gu, Daejeon 305–340, South Korea*
*shpark@postech.ac.kr, sghong@postech.ac.kr, popoml@posco.com, cslee@postech.ac.kr*

Received 15 June 2008
Revised 23 June 2008

Fatigue behavior of rolled AZ31 magnesium alloy, which shows an anisotropic deformation behavior due to the direction dependent formation of deformation twins, was investigated by carrying out stress and strain controlled fatigue tests. The anisotropy in deformation behavior introduced asymmetric stress-strain hysteresis hoops, which make it difficult to use common fatigue life prediction models, such as stress and strain-based models, and induced mean stress and/or strain even under fully-reversed conditions; the tensile mean stress and strain were found to have a harmful effect on the fatigue resistance. An energy-based model was used to describe the fatigue life behavior as strain energy density was stabilized at the early stage of fatigue life and nearly invariant through entire life. To account for the mean stress and strain effects, an elastic energy related to the mean stress and a plastic strain energy consumed by the mean strain were appropriately considered in the model. The results showed that there is good agreement between the prediction and the experimental data.

*Keywords*: AZ31 Mg alloy; Deformation twin; Fatigue life prediction; Mean stress/strain.

## 1. Introduction

Magnesium (Mg) and its alloy have recently received considerable attention due to their excellent properties such as low density, high strength-to-weight ratio, high specific stiffness, and good damping characteristics. These advantages make Mg alloys very attractive as structural materials in a wide variety of applications, in particular, as components for automotive and electronic industries. In these applications, their manufacturing process is mainly high-pressure die casting.[1] However, casting defects such as microscopic shrink holes, micropores, and inclusions, as well as the rather low ductility of many cast alloys, restrict the wide use of common cast Mg alloy components. Since wrought Mg alloys are known to have superior mechanical properties compared to cast Mg alloys[2], applications of Mg alloys processed by extrusion, forging, rolling or strip-casting[3] are recently receiving wide interest.

---

†Corresponding Author.

131

Understanding the fatigue characteristics of Mg alloys is essential when they are used as structural members. There have been many studies on the fatigue properties of Mg alloys. However, it is still lacking of studies relating the fatigue properties to the microstructural characteristics. It is well known that wrought Mg alloys are strongly textured and this texture, characterized by grain elongation and/or orientation pole figure, can lead to a pronounced asymmetry of mechanical properties, mainly caused by the direction dependant activation of twins.[4] This asymmetric feature of deformation behavior and the deformation twins developed may influence the fatigue characteristic of the material because the mean stress (or strain) induced by the deformation asymmetry and the deformation twins can affect the crack initiation and propagation. Moreover, the asymmetric feature of stress-strain relation makes it difficult to use common fatigue life prediction models (i.e. stress and strain based models) by giving rise to the difficulty in defining a plastic strain amplitude in the stress-strain hysteresis loop.

In this study, it was attempted to correlate the fatigue resistance of rolled AZ31 Mg alloy with its microstructural characteristic by introducing an energy-based fatigue damage parameter, where mean stress and strain effects were appropriately considered.

## 2.  Material and Experimental Procedures

The Mg alloy used in this study was a hot-rolled AZ31 Mg plate (thickness = 50mm) with a chemical composition of 3.6 Al- 1.0 Zn- 0.5 Mn and the balance Mg (in wt pct), which was homogenized by heat-treating at 400 °C for 4hrs. The average linear intercept grain size was about 50 μm. Observation on the initial texture revealed that (0002) basal planes perpendicular to the normal direction of rolling plane were strongly developed.

Fatigue test specimens with a gauge length of 10 mm and a gauge diameter of 5 mm were machined from the homogenized plate and mechanically polished to remove the surface defects. The strain and stress controlled fatigue tests were performed using INSTRON 8801 testing machine at 1 Hz in laboratory air at room temperature. Stress and strain signals were measured and controlled by a load cell and an extensometer attached to the sample. The stress controlled tests were conducted using a sine wave form with the stress ratio, $R_\sigma = \sigma_{min}/\sigma_{max}$, of 0. Strain controlled tests were carried out with a triangular wave form with the strain ratios, $R_\varepsilon = \varepsilon_{min}/\varepsilon_{max}$, of -1 and 0.5.

## 3.  Results and Discussion

### 3.1.  *Cyclic stress-strain behavior*

#### 3.1.1.  *Strain-controlled fatigue tests under $R_\varepsilon = -1$*

Stress-strain hysteresis loops at the half-life under the total strain amplitude ranging from 0.4 to 1.2% are presented in Fig. 1(a). Mechanical anisotropy occurred in the stress-strain curves due to plastic deformation by twinning, and this caused asymmetric hysteresis

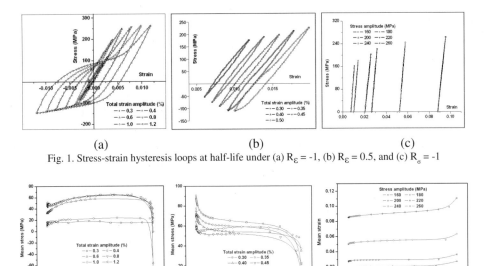

(a)  (b)  (c)

Fig. 1. Stress-strain hysteresis loops at half-life under (a) $R_\varepsilon = -1$, (b) $R_\varepsilon = 0.5$, and (c) $R_\sigma = -1$

(a)  (b)  (c)

Fig. 2. Mean stress/strain evolutions with life fraction under (a) $R_\varepsilon = -1$, (b) $R_\varepsilon = 0.5$, and (c) $R_\sigma = -1$.

loops distorted at the compressive region. These low compressive yield strength and strain hardening plateau are typically observed in materials which deform by twinning.

The evolution of mean stress is plotted as a function of life fraction in Fig. 2(a). During fully reversed strain controlled fatigue tests, tensile mean stresses occur due to the anisotropic behavior between tension and compression. The tensile mean stress values are nearly same in the total strain amplitude range of 0.8~1.2 % because tensile peak stress values in this range are similar owing to the concave feature of stress-strain curve (i.e. the low strain hardening rate).

### 3.1.2. *Strain-controlled fatigue tests under $R_\varepsilon = 0.5$*

Fig. 1(b) shows hysteresis loops at the half-life, obtained from the tests under $R_\varepsilon = 0.5$. When the strain amplitude is greater than 0.45%, hysteresis loops become asymmetric because compressive stress, which is enough to activate twinning, takes place during cyclic loading. Besides, hysteresis loop in compression is distorted due to the deformation by twinning.

As shown in Fig. 2(b), mean stress dramatically decreases in the early stage of life due to stress relaxation and it keeps constant up to the final stage corresponding to the crack opening. For the case of $\Delta\varepsilon_t/2 = \pm 0.5\%$ with distorted loop by twinning, however, mean stress gradually increases with increasing cycle after mean stress relaxation at the initial life. As twinning mechanism reduces cyclic hardening, the extent of cyclic hardening in compression becomes smaller compared to that in tension with increasing cycle. Accordingly, mean stress increases up to the crack opening point.

### 3.1.3. *Stress-controlled fatigue tests under $R_\sigma = 0$*

Hysteresis loops obtained under $R_\sigma = 0$ are depicted in Fig. 1(c). During the testing, the load range and mean value were adjusted for each 1% accumulation in axial strain in order to keep the true stress amplitude and the true mean stress constant. For $R_\sigma = 0$ tests, since only tensile stress is applied to the specimen, most of basal planes are under the state of the c-axis compression. Hence, it is hard to form twinning and hysteresis loop becomes symmetric.

Fig. 2(c) shows the evolution of mean strain as a function of life fraction. The mean strain was fully developed during just a few cycles and its value maintained constant until the final fracture. As the applied stress amplitude increased, the developed mean strain slightly increased with the number of cycles, but it was too small to have a ratcheting effect.

### 3.2. *Fatigue life prediction*

Total strain amplitude vs. fatigue life data for three testing conditions are presented in Fig. 3(a), where for the convenience of comparison the stress amplitude vs. fatigue life data obtained under $R_\sigma = 0$ were converted to the total strain amplitude vs. fatigue life data. The

results show that the fatigue resistance of rolled Mg alloy decreased in the following order of test conditions: $R_\varepsilon = -1$, $R_\varepsilon = 0.5$, and $R_\sigma = 0$. This reduction of fatigue resistance depending on the test conditions was most likely attributed to the mean stress and/or strain, induced by the deformation anisotropy due to the twinning mechanism or imposed in the test. Namely, increased mean stress and imposed mean strain reduced the fatigue resistance.

Fatigue damage is in general represented by the parameters based on stress, strain, and energy.[5,6] All of the fatigue life prediction models using these parameters are based on the assumption that the behavior of materials undergoing cyclic loading is stabilized at the early stage of life and the damage accumulated per each cycle is identical through the entire life.[7] However, continuous change and the asymmetric feature of hysteresis loop made it difficult to use stress and strain based parameters as fatigue damage. On the contrary, an energy based damage parameter (i.e. plastic strain energy density) composed of both stress and strain was stabilized at the early stage of life and this stabilized state was maintained through the entire life. Therefore, it is considered to be a suitable fatigue parameter for the assessment and prediction of fatigue life of rolled Mg alloys. By incorporating the elastic and plastic strain energy terms, where the elastic strain energy term is used to account for the mean stress effect, a strain energy-based fatigue life prediction model can be represented as[8]

$$\Delta W_t = \Delta W_p + \Delta W_{e^+} = k_u N_f^\alpha + C_u \tag{1}$$

where $\Delta W_p$ is the plastic strain energy density per cycle and $\Delta W_{e^+}$ is the tensile elastic strain energy density per cycle affected by the mean stress. $\alpha$, $k_u$ and $C_u$ are material constants, which can be obtained from a fully reversed fatigue test. $C_u$ is the energy value

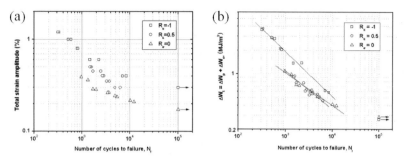

Fig. 3. Fatigue life behavior: (a) total strain amplitude vs. fatigue life data and (b) life predictions with the energy-based model considering mean stress, where solid lines represent the predictions by Eq. (1).

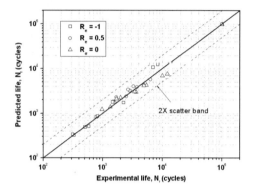

Fig. 4. Comparison of fatigue lives between the prediction, Eq. (3), and experiment.

corresponding to the fatigue limit of the material.

Fatigue life curves based on the total strain energy damage parameter are shown in Fig. 3(b), where solid lines represent the predictions by Eq. (1) and the values for the plastic strain energy density and tensile elastic strain energy density were calculated from the hysteresis loop at half-life. It can be noted that although the tensile elastic strain energy density was considered in the life prediction model, the fatigue data under $R_\varepsilon$=0.5 and $R_\sigma$=0 were deviated from those of $R_\varepsilon$=-1.

It should be noted that Eq. (1) is just considering the mean stress effect. However, considerable damage is accumulated by the large plastic deformation developed in tensile loading of the first cycle (i.e. mean strain) under the test conditions of $R_\varepsilon$=0.5 and $R_\sigma$=0. To account for this mean strain effect, the plastic strain energy consumed by the mean strain is added to Eq. (1). Accordingly, the fatigue life prediction model considering both the mean stress and mean strain effects can be expressed as

$$\Delta W_t = \Delta W_p + \Delta W_{e^+} + \left\{ \frac{f(\varepsilon_m)}{N_f} \right\}^{\gamma} = k_u N_f^\alpha + C_u \qquad (2)$$

where $f(\varepsilon_m)$ is the plastic strain energy associated with mean strain and $\gamma$ is the material constant. $f(\varepsilon_m)$ represents the tensile plastic strain energy of hysteresis loop at the first cycle and can be calculated from positive region of the hysteresis loop at first cycle. Its values increase with increasing stress or strain amplitude. Material constants of $k_u$ and $C_u$ can be acquired from $R_\varepsilon = -1$ test data without mean strain. The value of $\gamma$ can be calculated with $f(\varepsilon_m)$ values and fatigue lives obtained from $R_\varepsilon = 0.5$ and $R_\sigma = 0$ tests with positive mean strains. Substituting the calculated values to Eq. (2) gives the final form of the energy-based fatigue life prediction model of rolled AZ31 Mg:

$$\Delta W_t = \Delta W_p + \Delta W_{e^+} = 178.4 N_f^{-0.685} + 0.3 - \left\{ \frac{f(\varepsilon_m)}{N_f} \right\}^{0.13} \tag{3}$$

A comparison between the experimental data and the prediction by the proposed model, Eq. (3), is presented in Fig. 4. The result shows that the suggested model provides a good correlation on the fatigue life of rolled Mg alloy undergoing the mean stress/strain evolution during fatigue deformation; all experimental data could be included within a 2X scatter band of the model prediction.

## 4. Conclusions

The fatigue characteristic of rolled Mg alloy, which has a strong texture with the basal planes parallel to the rolling direction, was investigated by carrying out the stress and strain controlled low-cycle fatigue tests. The direction dependent activation of twins induced the deformation anisotropy, which led to the asymmetric hysteresis loops, and changed the cyclic hardening characteristics in tension and compression. These facts made it difficult to use common fatigue damage parameters such as a stress amplitude and plastic strain amplitude. In addition, the anisotropic characteristic of stress-strain relation induced the mean stress (or strain) during fatigue deformation even under a fully reversed condition. It was found that an energy based parameter shows the desirable features as a fatigue parameter; it was stabilized at the early stage of life and nearly invariant through the entire life. An energy-based model was developed to describe the fatigue life behavior, where the mean stress and strain effects were taken into account by introducing an elastic strain energy related to the mean stress and a plastic strain energy consumed by the mean strain during the first cycle. The results showed that the prediction by the suggested model is correlated well with the experimental data within a factor of 2.

## References

1.  C. D. Lee, *Met. Mater. Int.* 12 (2006) 377.
2.  Z. B. Sajuri, Y. Miyashita, Y. Hosokai, and Y. Mutoh, *Int. J. Mech. Sci.* 48 (2006) 198.
3.  B. H. Lee, W. Bang, S. Ahn, and C. S. Lee, *Mater. Trans. A* 39 (2008) 1426.
4.  X.Y. Lou, M. Li, R.K. Boger, S.R. Agnew, and R.H. Wagoner, *Int. J. Plasticity* 23 (2007) 44.
5.  J. T. Yeom, C. S. Lee, J. H. Kim, D. G. Lee, and N. K. Park, *Key Eng. Mat.* 240 (2007) 235.
6.  J. S. Park. S. J. Kim, K. H. Kim, S. H. Park, and C. S. Lee, *Int. J. Fatigue* 27 (2005) 1115.
7.  S. G. Hong, S. B. Lee, and T. S. Byun, *Mater. Sci. Eng. A* 4457(2007) 139.
8.  K. Golos and F. Ellyin, *ASME J. Press. Vessel Technol.* 110 (1998) 36.

# PLASTIC DEFORMATION OF POLYMER INTERLAYERS DURING POST-BREAKAGE BEHAVIOR OF LAMINATED GLASS - PARTIM 1: ANALYTICAL APPROACH

J. BELIS[1†], D. DELINCÉ[2], D. CALLEWAERT[4], R. VAN IMPE[5]

*Laboratory for Research on Structural Models, Ghent University, Ghent, BELGIUM,*
*jan.belis@UGent.be, didier.delince@UGent.be, dieter.callewaert@UGent.be, rudy.vanimpe@UGent.be*

J. DEPAUW[3]

*Jan De Nul n.v. , Hofstade-Aalst, BELGIUM,*
*jeffreydepauw@yahoo.com*

Received 15 June 2008
Revised 23 June 2008

Transparent polymer interlayer foils are widely used to increase the safety of glass applications, mainly in construction and automotive industry. In case one or more glass sheets in a laminate fracture, the remaining structural capacity (i.e. stiffness and strength) highly depends on the solicitation and mechanical properties of the interlayer. In such a post-breakage state, the interlayer is subjected to a complex combination of phenomena such as partial delamination, large strain deformation, strain hardening, rupture, etc. In partim 1, an analytical model is presented in order to describe the mechanical behavior of glass laminates in post-breakage state for a simple configuration. In addition, the theoretical model is compared to experimental results in part 2, which is published together with part 1.

*Keywords*: Laminated glass; post-breakage; plastic deformation; delamination.

## 1. Introduction

Trends in construction and automotive industry show an increasing use of glazed surfaces in designs. Since these glass components are increasingly used for structural applications, such as load-bearing beams supporting glass floors or windshields rigidifying car bodies, structural safety and post-breakage behavior have become major glass design issues.

However, only little information is available on the post-breakage behavior of laminated glass plates. Theoretical and experimental work has been conducted by Kott[1] on laminates composed of two glass layers and one layer of polyvinyl butyral (PVB), the most-used interlayer in industry. This author distinguishes three different stages during the failure process of such laminates, as illustrated in Fig. 1.

---

[†] Corresponding Author.

137

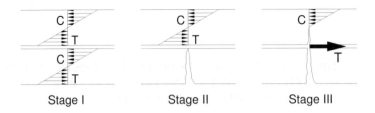

Fig. 1. Schematic representation of the three stages of behavior in flexure of laminated glass with related stress distribution according to Kott[1] and Belis et al.[2]. C and T represent Compressive and Tensile stress respectively.

Stage I applies as long as the flexural strength of the glass is not reached. Each glass panel is exposed to simple bending stresses presumed that the interlayer shows considerable shear deformation, as is the case with PVB under static loading conditions.

In stage II the critical tensile stress of the glass is reached, and at least one glass panel has been broken. The high adhesion of glass to the interlayer prevents the broken pieces of glass falling down. The static function of the interlayer is reduced to the carrying of the bond stresses.

Stage III is the final stage, in which all glass panes are broken but the whole element has not collapsed yet. Equilibrium is possible when the flexural internal forces are equilibrated by compressive stresses in the partly broken glass layer on the one hand, and by tensile stresses in the interlayer on the other. Obviously, the interlayer is of major importance for the residual stiffness and strength of the laminate in this stage.

The model of Kott intends to describe the post-breakage behavior of laminates for various breakage patterns of the glass plates and different supporting conditions, but its experimental research was limited to laminates with a PVB interlayer. However, PVB has a relatively low stiffness and strength compared to more recently developed interlayer films for special applications (e.g. hurricane resistance), such as SentryGlas® Plus (SGP)[3]. Therefore, a more elaborated alternative expression is presented for the specific case of laminates composed of annealed glass and a SGP interlayer.

## 2.  Methods

This paper focuses on the modeling of the post-breakage behavior (stage III). The analytical model proposed by Kott[1] to describe the post-breakage behavior of the Glass/PVB laminate is adapted and further developed to describe the particular deformation scheme observed for Glass/SGP laminates in a four-point bending configuration, where the development of a "yield-section" due to the fracture scheme is observed (see fig. 2). A comparison of the developed theory to experimental results is presented in "partim 2: Experimental validation[4]".

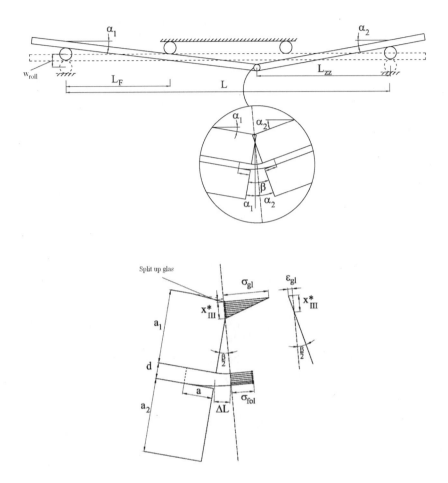

**Fig. 2.** Schematic representation of four-points bending test in Stage III with a "yield-section" configuration (above) and detail of the modeling of the yield-section with indication of compressive stresses in the upper glass plate and tensile stresses in the interlayer (below).

## 3. Analytical Model

Kott[1] proposed a simple expression to calculate the ultimate resisting moment of the laminated glass plate in stage III, $M_{u,III}$:

$$M_{u,III} = f_{y,t} . d . b . \left( \frac{d}{2} + a_1 - \frac{1}{3} . x_{III} \right) \tag{1}$$

$$\text{with } x_{III} = 2 . n_{pl} . d \; ; \; n_{pl} = \frac{f_{y,t}}{f_{cu,g}} \tag{2}$$

where $f_{y,t}$ is the yield stress of the interlayer, $f_{cu,g}$ the compression strength between glass pieces, $b$ the width of the sample, $a_1$ and $d$ the thickness of the upper glass plate and of the interlayer respectively, and $x_{III}$ the distance between the upper flange of the sample and the neutral axis (cf. fig. 2).

This model is adapted and extended to describe the equilibrium between external and internal forces for the "yield-section" configuration. First, the acting moment in the "yield-section" configuration cannot be calculated anymore according to the usual theory for four-points bending tests:

$$m_{III,E}(t) = \frac{M_{III,E}(t)}{b} = m_{III,Q}(t) + m_{III,G} = \frac{M_{III,Q}(t) + M_{III,G}}{b} \tag{3}$$

$$m_{III,Q}(t) = \frac{Q(t).L_F.L_{zz}}{L.b} \tag{4}$$

$$m_{III,G} = \frac{\gamma_{gl}.(a_1 + a_2)}{2}.\left[ L_t.L_{zz} - (L_t - L).\left( L_{zz} + \frac{L_t - L}{4} \right) - L_{zz}^2 \right] \tag{5}$$

with $m_{III,Q}$ and $m_{III,G}$ the acting moment per unit of width respectively due to the load $Q$ and to the own weight of the glass plate ($\gamma_{gl} = 25$ kN/m³ is the volumetric weight of the glass; the weight of the interlayer is neglected), and $L_{zz}$, $L_F$, $L$ and $L_t$ are defined in Fig. 2.

Secondly, the opening of the yield-section $\beta$ can be deduced:

$$\beta = \alpha_1 + \alpha_2 \tag{6}$$

$$\tan(\alpha_1) = \frac{w_{roll}}{L_F} \tag{7}$$

$$\tan(\alpha_2) = \tan(\alpha_1).\frac{L - L_{zz}}{L_{zz}} = \frac{w_{roll}}{L_F}.\frac{L - L_{zz}}{L_{zz}} \tag{8}$$

Finally, the equilibrium of the internal forces in the yield-section can be written considering a plane stress state:

$$\sigma_{int}.d = \frac{\sigma_{gl}.x_{III}^*}{2} \tag{9}$$

with $\sigma_{int}$, $\varepsilon_{int}$ respectively the constant tension stress and strain in the interlayer, and $\sigma_{gl}$ and $\varepsilon_{gl}$ respectively the compression stress and strain in the upper flange of the laminate, in the yield-section.

Being a linear elastic material, the glass stresses can be determined as follows:

$$\sigma_{gl} = E_g.\varepsilon_{gl} \tag{10}$$

with $E_g = 70.000$ MPa the elastic modulus of glass. The strain in the upper flange can be related to the opening $\beta$:

$$\varepsilon_{gl} = \tan\left(\frac{\beta}{2}\right).x_{III}^*$$ (11)

A finite strain in the interlayer is only possible if the interlayer delaminates from the glass layer near the crack. Considering a linear elastic material for the interlayer, the model proposed by Seshadri[5] is used:

$$\sigma_{int} = \frac{\tau.a + \sqrt{2.\Gamma_0.E_{int}.d}}{d}$$ (12)

with

$$\sigma_{int} = \varepsilon_{int}.E_{int}$$ (13)

Consequently, the strain in the interlayer is then a function of the delaminated length $a$:

$$\varepsilon_{int} = \frac{\Delta L}{a} = \frac{\left(a_1 + \frac{d}{2} - x_{III}^*\right).\frac{\beta}{2}}{a}$$ (14)

with $\tau$ the frictional force per unit area between the delaminated interlayer and the glass on the length $a$, and $\Gamma_0$ the interfacial adhesion (equaling the fracture energy of the interface), supposed to be independent of the strain rate.

The resisting moment in the yield-section can be determined:

$$m_{III,R} = \frac{M_{III,R}}{b} = \sigma_{int}.d.\left(\frac{d}{2} + a_1 - \frac{1}{3}.x_{III}^*\right) = \frac{\sigma_{gl}.x_{III}^*}{2}.\left(\frac{d}{2} + a_1 - \frac{1}{3}.x_{III}^*\right)$$ (15)

Substituting (10) and (11) in (15):

$$m_{III,R} = \frac{E_g.\tan\left(\frac{\beta}{2}\right).\left(x_{III}^*\right)^2}{2}.\left(\frac{d}{2} + a_1 - \frac{1}{3}.x_{III}^*\right)$$ (16)

Considering a quasi-static equilibrium between the acting moment given by Eq. (3) and the resisting moment given by Eq. (16), the parameter $x_{III}^*$ can be calculated at each time step $t$. Then, the delaminated length $a$ can be calculated solving the following equation, resulting from the combination of (12), (13) and (14):

$$a = \frac{\sigma_{int}.d - \sqrt{2.\Gamma_0.\frac{\sigma_{int}.a}{\Delta L}.d}}{\tau}$$ (17)

The strain in the interlayer can then be calculated using Eq. (14).

The model is very useful to illustrate the importance of the different parameters during the post-breakage stage (Stage III).

## 4.    Conclusions

A simple theoretical model has been proposed to describe the post-breakage behavior (in stage III) of laminated glass composed of two layers of glass and one interlayer in a particular configuration, namely a four-point bending test with a "yield-section" configuration.

The main feature of this model is to include the effect of local delamination of the interlayer in a comprehensive analytical way. This model is compared to experimental results of bending tests performed on Glass/SGP laminates in Partim 2 of this paper.

## Acknowledgments

This work was supported by the Fund for Scientific Research-Flanders (FWO-Vlaanderen, grant nr. 3G018407). Additionally, the authors wish to acknowledge the support of Lerobel (Groep Leroi, Belgium).

## References

1.  A. Kott, Ph.D. Thesis, ETH Zürich, Zürich, 2006.
2.  J. Belis, R. Van Impe, W. Vanlaere, G. Lagae, P. Buffel and M. De Beule, *Key Eng. Mat.*, **274-276**, 975 (2004).
3.  DuPont de Nemours and Company, Inc., DuPont SentryGlas® Plus Specifying and Technical Data, Doc. Ref. SGP030718_1, v.1, (2003).
4.  D. Delincé, J. Depauw, D. Callewaert, W. Vanlaere and J. Belis,, *Mod. Phys. Lett. B*, partim 2, in press.
5.  M. Seshadri, S.J. Bennison, A. Jagota and S. Saigal, *Acta Mater.*, **50**, 4477 (2002).

# LOAD CARRYING CAPACITY OF CORRODED REINFORCED CONCRETE BEAMS SUBJECTED TO REPEATED LOAD

CONGQI FANG[1†], MEIYING YI[2], SHUJIAN CHENG[3]

*Department of Civil Engineering, Shanghai Jiaotong University,*
*Shanghai,200240, P. R. CHINA,*
*cqfang@sjtu.edu.cn, yimeiying1983@yahoo.com.cn, csj02@sjtu.edu.cn*

Received 15 June 2008
Revised 23 June 2008

Load carry capacity of corroded reinforced concrete beams under repeated load was investigated experimentally. A total of fifteen test samples, including three non-corrosion and twelve corroded reinforced concrete (RC) beams, were experimentally evaluated for midspan deflection, cracking of concrete, and failing modes under repeated loading. Beams of different corrosion percentage were tested for the effect of steel corrosion and repeated loading. For corroded RC beams width and spacing of cracks increased significantly with steel corrosion. Before the yielding of the reinforcement steel, midspan deflection for beams of lower corrosion percentage was less than that of non-corrosion beams. However, as the reinforcement steel yielded, the midspan deflection of corroded beams was significantly larger than those of non-corrosion beams, and this tended to increase with corrosion percentage. As the reinforcing steel was severely corroded, the beam showed a mechanical characteristic similar to a camber, where arch action was more obvious.

*Keywords*: Load carrying capacity; reinforcement corrosion; repeated load.

## 1. Introduction

Durability aspects of concrete structures has initiated much more concern in recent years than ever before and a lot of effort has been devoted to understanding the mechanisms and causes of steel corrosion.[1-3] However, relatively little concern has been devoted to assessment of the residual capacity of corroded structures subjected to repeated loading. The appropriate assessment of the actual existing structure demands evaluation of the structural behavior in terms of such as ultimate load carrying capacity and failing mode of the corroded RC beams on monotonic loading and repeated loading as well.

The present study experimentally investigated the influence of corrosion of reinforcement steel on load carrying capacity of RC beams subjected to repeated loading. The load carrying capacity was evaluated, in terms of load-deflection curve, cracking modes of concrete, and failing modes of the beam.

---

[†]Corresponding Author.

## 2.  A General Description of Load Carrying Capacity

### 2.1.  *The effect of corrosion*

Load carrying capacity of a RC beam is usually affected by concrete strength, reinforcement distribution, steel strength, effective depth of cross section, etc. For existing RC structures, the influence of corrosion is of particularly concern. Corrosion is considered one of the main causes for the limited durability of a RC structure.[4] The volume increase due to corrosion leads to weakening of the bond, which directly affects the serviceability and ultimate load carrying capacity of RC members.[5]

### 2.2.  *Effect of repeated loads*

Substantial degradation in the bond capacity took place during repeated loading, and thus reduced ultimate load carrying capacity of the beam. The strain in the concrete when the beam is subjected to repeated loading includes two parts, as showed in Eq. (1):

$$\varepsilon = \varepsilon_u + \varepsilon_{res}.\tag{1}$$

where $\varepsilon_u$ is strain caused by stress. $\varepsilon_{res}$ is residual strain, as showed in Eq. (2).

$$\varepsilon_{res} = \sum_{i=1}^{N} \varepsilon_{res}(i).\tag{2}$$

where N is the total number of loading step cycles, $i$ the $i$-th loading. The accumulation of the strains of every repeating unloading in each cycles results in the total residual strain.

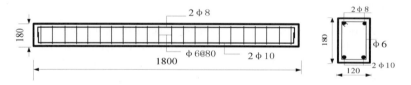

Fig. 1.  Geometry and reinforcement of the experiment beam. All dimension in mm.

## 3.  Experimental Program

### 3.1.  *Specimens*

The geometry and reinforcement of specimens are indicated in Fig. 1. A total of fifteen sample beams were cast in two batches of different concrete strength, 28-33 MPa and 38-52 MPa, respectively..The reinforcement steel was corroded to different corrosion percentage, using accelerated corrosion method in laboratory. [6-7]

### 3.2.  *Loading*

The load for each step was chosen according to the computed cracking load and the yielding strength of reinforcement. As showed in Fig. 2, deflection of the beam was measured by dial indicators. Data measured in the strain gauges were imputed to strain measuring instruments and transferred to a computer. The load steps, load value, crack widths, reading from dial indicators, as well as the failing modes were recorded manually.

If reinforcement yielded, deflection exceeded 1/50 of the beam span, or concrete in compressive region crushed, the previous step load was defined as the maximum limit state load, or the ultimate load carrying capacity of the beam.

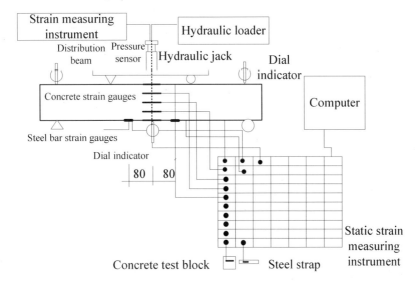

Fig. 2. Test schematic diagram.

## 4.  Experimental results and discussion

A total of 15 beams were tested, as shown in Table 1. Strains and deflections at the midspan were recorded during loading, using strain measuring instruments and computer. Cracking, maximum crack widths and average crack spacing were recorded manually.

### 4.1.  *Cracking of concrete*

Fig. 3 is a typical figure showing cracking of the corroded RC beams during loading. Cracking modes were generally very similar with each other for low corrosion beams. Cracks began to occur near the midspan and the loading points as load was increased to 20-35% of the failing load, with the maximum crack width and length being about 0.1 mm wide and 15-40 mm long. As the load was increased thereafter, new cracks occurred until the load reached to 50% of the maximum load, when the existing cracks continued to open wider but new cracks no longer to occur. For all corroded beams, crack width was rapidly increased to penetrate cross the beam as the reinforcement yielded, with concrete in compressive region crushed, and the beam failed.

Table 1. Test results.

| Beam No. | Corrosion percentage (%) | Crack load (kN) | Yield load (kN) | Failing load (kN) |
|---|---|---|---|---|
| 1 | 0 | | 21 | 24 |
| 2 | 0 | 7 | 19 | 26 |
| 3 | 0 | 7 | 18 | 25 |
| 4 | 1.81 | 6 | 16 | 24 |
| 5 | 2.16 | 5 | 15 | 21 |
| 6 | 2.29 | 6 | 14 | 21 |
| 7 | 3.94 | 8 | 17 | 23 |
| 8 | 3.53 | 8 | 18 | 23 |
| 9 | 4.27 | 8 | 16 | 23 |
| 10 | 1.51 | 6 | 15 | 23 |
| 11 | 1.59 | 6 | 14 | 25 |
| 12 | 2.09 | 6 | 16 | 27 |
| 13 | 1.67 | 7 | 18 | 25 |
| 14 | 1.58 | 6 | 16 | 23 |
| 15 | 1.65 | 6 | 15 | 23 |

### 4.2. *Load-deflection curve*

A typical load-deflection curve is showed in Fig. 4. Results from the test demonstrated that: (1) Deflection increased linearly and no cracks were observed when the applied loads were low. The deflection returned to zero when totally unloading; (2) From point A to B, small residual deflection existed. However, the residual deflection in this stage was relatively small, e.g., less than 1 mm, which was only about 3% of the maximum deflection at midspan that occurred when the beam failed. (3) Load increased from point B, Crack width increased as the load continued to increase. The neutral axis was accordingly shifted up in the cross section of the beam. Plastic deformation occurred in the compressive region. (4) As the applied load increased to point C, concrete in the compressive region crushed, and the turning point in the load-deflection curve was observed. The deflection was increased rapidly thereafter.

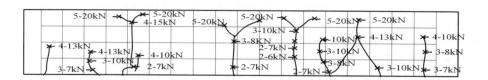

Fig. 3. A typical cracking mode of the beam.

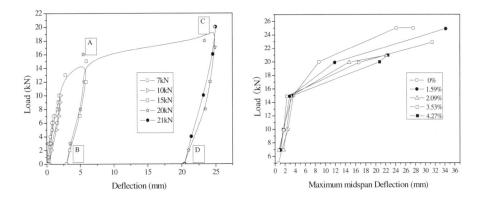

Fig. 4. Load-deflection curve.               Fig. 5. Typical maximum load-deflection curve.

The load-deflection curves for beams of different corrosion percentage were showed see Fig. 5. Before the yielding of reinforcement steel, deflection in beams of lower corrosion percentage were less than those of non-corrosion beams. The explanation for this may be the corrosion products expanded and bond strength increased accordingly, but cracking not yet occurred. As the reinforcement steel yielded, the deflection at midspan of corroded beams was larger than those of non-corrosion beams, and this tended to increased with corrosion percentage. The explanation for this may be that, as steel reinforcement yielded, bond between reinforcement steel and concrete failed, and the deflection increased with higher corrosion percentage.

Fig. 6. Failing of a corroded RC beam subjected to repeated loads.

## 4.3. *Failing modes*

The failing modes of corroded RC beams subjected to repeated load were different from those of normal RC beam. Fig. 6 shows a typical failing of the corroded beams. For low corrosion percentage, the failing modes were similar to those of non-corrosion RC beams.

For a higher corrosion percentage, cracks due to corrosion occurred. A few stress cracks occurred in large crack spacing, and uneven distribution of cracks was observed. As corrosion percentage was increased to a higher level, the strength and plastic capacity of reinforcement steel was also deteriorated. The failing mode of the beam tended to change to brittle failure from plastic failure. Take No. 9 beam which had a corrosion percentage of 4.27, for example, failed as the tension reinforcement broke suddenly.

## 5.  Conclusions

(1) The resistance to bending of corroded RC beams under repeated loading rapidly decreased in the cracking stage, and eventually decreased again in the failing stage. (2) Width and spacing of cracks of corroded RC beams increased with steel corrosion percentage under repeated loading. (3) When reinforcement steel yielded the residual deflection at the midspan of the beams was about 3% of the failing load. (4) Deflection of low corrosion beams were less than those of non-corrosion beams until the reinforcement steel yielded. As steel yielded, the deflection at midspan of corroded beams was larger than those of non-corrosion beams.

## References

1.   G.J. Al-Sulaimani, M. Kaleemullah, I.A. Basunbul, Rasheeduzzafar, Influence of corrosion and cracking on bond behaviour and strength of reinforced concrete member, *ACI Struct. J.* 87 (2) (1990) , pp220-231.
2.   Y. Auyeung, Bond properties of corroded reinforcement with and without confinement, PhD thesis, New Brunswick Rutgers, The State University of New Jersey, May, 2001.
3.   C. Fang, K. Lundgren, L. Chen, C. Zhu, Effect of corrosion on bond in reinforced concrete, *Cem. Concr. Res.* 34(11) (2004), pp2159-2167.
4.   M. Maslehuddin, I. M. Allam, G. J. Al-Sulaimani, A.I. Al-Mana, and S.N. Abduljauwad, Effect of Rusting of Reinforcing Steel on its Mechanical Properties and Bond with Concrete, *ACI Mater. J.* 87 (5) (1990) 496-502.
5.   D. Coronelli, M.G.Mulas, ngth of reinforced concrete member, *ACI Struct. J.* 87 (2) (1990) , pp220-231.
6.   M. Y. Yi, Load bearing capacity of corroded reinforced concrete beams under repeated loading. M.Eng. Thesis, Shanghai Jiaotong Univ. 2008.
7.   C. Fang, K. Gylltoft, K. Lundgren, M. Plos, Effect of corrrosion on bond in reinforced concrete under cyclic loading. *Cement and Concrete Research,* Vol. 36, Issue 3 (2006), pp 548-555.

# A MODEL FOR GALLING BEHAVIOR IN FORMING OF HIGH STRENGTH STEEL UNDER TENSION-BENDING

Y.K. HOU[1]

*Auto Body Manufacturing Technology Center, Shanghai JiaoTong University, Shanghai 200240, CHINA,*
*houyingke@sjtu.edu.cn*

W.G. ZHANG[2†]

*School of Naval Architecture and Ocean Engineering, Shanghai JiaoTong University, Shanghai200240, CHINA,*
*wgzhang@sjtu.edu.cn*

Z.Q. YU[3], S.H. LI[4]

*School of Mechanical Engineering, Shanghai JiaoTong University,Shanghai 200240, CHINA,*
*yuzhq@sjtu.edu.cn, lishuhui@sjtu.edu.cn*

Galling is a known failure mechanism in sheet metal forming (SMF), especially in the forming of high strength steels. It results in increased cost of die maintenance and scrap rate of products. In this study, U-channel forming tests were conducted to study the galling behavior under the condition of tension-bending. The effects of tool surface characteristics (hardness and surface roughness) and process parameters (blank holder force and sliding distance of the contact surfaces) on galling were investigated. Experimental results indicate that galling tendency can be reduced through hardening and polishing of the forming tool and galling becomes more severe with increasing blank holder force and sliding distance in SMF. A numerical model between the galling and these factors was established through non-linear curve fitting of the large amount of experimental data.

*Keywords*: Galling; Sheet metal forming; high strength steel; tension-bending.

## 1. Introduction

Sheet metal forming (SMF) is an important technology in automotive industry, covering a broad range of processes such as bending, stretching and deep drawing. One of the major causes for tool failure and rejection of products in SMF is transfer and accumulation of adhered sheet material to the tool surfaces, generally referred to as galling. Problems with galling are of major interest for high strength steels, since forming pressures are high, as well as shear stresses in the forming operations.

Galling is the consequence of complex interactions between large number of parameters including forming tool, workpiece, lubricant and production conditions[1]. By proper polishing of the tool surface, to remove irregularities and asperities that can act as initiation sites for material transfer, the galling tendency can be greatly reduced and the tools life can be considerably prolonged in both lubricated and unlubricated systems[2-3].

---

†Corresponding Author.

Furthermore, galling can be successfully hindered or delayed by modification of the tools surface by PVD, CVD or DLC coatings[4-6]. However, the majority of the forming tools are still uncoated in automotive industry due to the larger size and the complex shape of the most forming tools. Special designed tool steels, introducing specific carbides and nitrides effectively increase the potential to resist work material transfer[7-9]. The galling performance of coating or surface treatment greatly depends on the type of material to be formed[10]. At the same time, tool hardness influences the frictional behavior between tool and workpiece and affects the galling behavior further in SMF.

Various test methods were also developed to evaluate galling and die wear in some literatures. For example, the ball scratching and pin-on-disk tests are extensively used to evaluate galling of sheet materials[11-13]. In these tests, the plastic deformation of sheet specimen that occurs in SMF is ignored. Moreover, in pin-on-disk test tool material is in contact with the same disc surface during the entire testing duration. Similarly, the twist-compression test (TCT) is also based on the repeated contact surface[14]. U bending and draw-bead tests are also extensively used to evaluate galling by many researchers[10], which are more representative of the actual stamping conditions .

The aim of the present work is to investigate the effects of tool surface characteristics (hardness and surface roughness) and process parameters (blank holder force and sliding distance) on galling and to establish a numerical model for the galling evaluation aimed at the forming process of high strength steel under tension-bending.

## 2. Experiment

### 2.1. Experimental Procedures

The deep drawing process is a typical method in SMF, which include two contact conditions between tool and workpiece. One is the sliding condition under compression and another is the sliding condition under tension-bending. The forming tests of U-channel were conducted in this study. The draw-beads were used to limit the material flow and maintain a constant strain distribution in a part. The contact condition between tool and workpiece in the tests is the sliding condition under tension-bending. The cross-section view of the forming tool is shown in Fig.1. The shape of the draw-bead was semicircular and the radius was 6mm. The forming tests were carried out with a single-action hydraulic press. All tests were performed with only rust-preventative oil as lubrication. The draw depth was 50mm. Fig.2 is the formed U-channel workpieces.

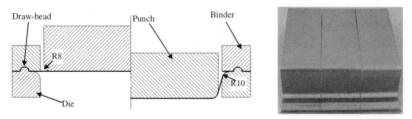

Fig. 1. The cross-section view of the test tool.          Fig. 2. The formed U-channel workpieces.

The sheet material used in this study was 0.7mm cold rolled high strength steel B170P1. Table 1 shows the mechanical properties of the sheet material. 60mm×300mm sheet specimens were used in the U-channel forming tests. The tool material was Mo-Cr cast iron which was usually used as tool material in automotive industry. The tool material was hardened and tempered to get three different grades of hardness 35HRC, 45HRC and 52HRC. Finally, the forming tools were ground and polished resulting in an average surface roughness $R_a$ of about 0.1μm, 0.21μm, 0.56μm, 0.95μm and 1.2μm. Table 2 shows the forming tools used in this study and their surface characteristics and symbols.

Table 1. Mechanical properties of the sheet material.

| Material | Yield Strength (MPa) | Tensile Strength (MPa) | Elongation (%) | n | r |
|---|---|---|---|---|---|
| B170P1 | 245 | 376 | 37 | 0.23 | 1.98 |

Table 2. Tool material specification and symbol for the tests.

| Tool material | Hardness (HRC) | Roughness $R_a$(μm) | Symbol |
|---|---|---|---|
| Mo-Cr cast iron | 35 | 0.21 | Tool-A |
| Mo-Cr cast iron | 45 | 0.21 | Tool-B |
| Mo-Cr cast iron | 52[a] | 0.21 | Tool-C |
| Mo-Cr cast iron | 35 | 0.56 | Tool-D |
| Mo-Cr cast iron | 45 | 0.56 | Tool-E |
| Mo-Cr cast iron | 52[a] | 0.56 | Tool-F |
| Mo-Cr cast iron | 45 | 0.1 | Tool-G |
| Mo-Cr cast iron | 45 | 0.95 | Tool-H |
| Mo-Cr cast iron | 45 | 1.2 | Tool-I |

[a] The maximum attainable hardness using conventional heat treatment process.

The main characteristic of galling is the scratches left behind on the products. In automotive industry, the depth of scratches left on parts is one of the main criterions for surface qualities of products. In case a scratch on a part is deep enough that the scratch can not be masked with lacquer, the part will be rejected. Therefore, the roughness parameter $R_y$ was used to evaluate galling in the study. The $R_y$ value was measured by means of a stylus profilometer. Five times of measurements were performed and the mean value was used. The measuring position was the outer sidewall of the U-channel. The galling areas were also analyzed by scanning electron microscopy (SEM).

## 2.2. Experimental results

Fig.3 shows the variations of the $R_y$ value with the number of forming with different forming tools. The blank holder force used in the tests was 2.5T. It can be seen from Fig.3 that tool surface characteristics (hardness and surface roughness) have crucial effects on the galling behavior in SMF. As shown in Fig.3a, galling tendency can be reduced by increasing the hardness of the forming tool. In this case, the forming tool with a hardness of 52HRC shows the best galling property. Polishing of the tool surface also improves the anti-galling properties of the forming tools, see Fig.3b. The roughest tool results in the severest galling. The SEM micrographs of the galling areas on the U-channel outer surfaces with Tool-A are shown in Fig.4. Fig.4 reveals that galling severity increases with the number of forming tests and deep scratches appear at the 300th forming test. The typical characteristics (furrow, scratched surface) of galling are observed in Fig.4c and Fig.4d.

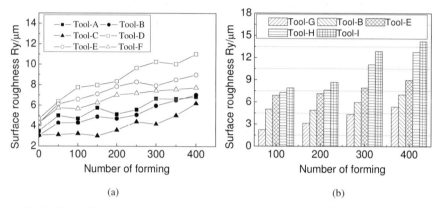

(a)                                      (b)

Fig. 3. Effect of tool surface characteristics on galling: (a) tool hardness; (b) tool surface roughness.

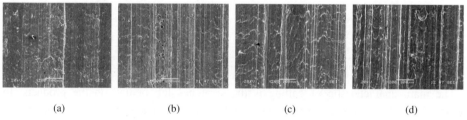

(a)                  (b)                  (c)                  (d)

Fig. 4. SEM micrographs of the galling surfaces with number of forming:
(a) the 100th; (b) the 200th; (c) the 300th; (d) the 400th.

The investigation on the effect of blank holder force on galling behavior included five different grades of blank holder force, 1.5, 2, 2.5, 3 and 3.5T, in this study. The forming tools used in the investigation were Tool-C and Tool-F. After each experiment of one blank holder force, the forming tool was reground resulting in the average surface roughness $R_a$ of 0.21μm or 0.56μm. Fig.5 shows the variations of part surface roughness of $R_y$ value in the forming process under different blank holder force. It is noted that blank holder force has negative effect on galling performance in SMF. For Tool-C, see Fig.5a, heavy galling appears at the 400th forming tests when the blank holder force is 3.5T. Fig.5b clearly indicates that galling is very severe at the 400th forming tests when the blank holder force is more than 3T.

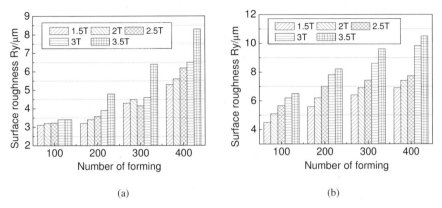

Fig. 5. Effect of blank holder force on galling: (a) Tool-C; (b) Tool-F.

## 3. The Model for Galling Behavior

The factors investigated in this study include tool hardness (H), tool surface roughness (μ), blank holder force (P) and the sliding distance. After the forming tests and the measurements of the $R_y$ values, an equation for the galling evaluation in SMF was established through non-linear curve fitting of the large amount of experimental data using the software of Origin 7.0, see Eq. (1) and Eq. (2). In Eq. (1), 'n' corresponds to the number of forming which stands for the accumulative value of the sliding distance between the forming tool and the sheet materials. Eq. (2) is the expression of the function $f(n)$.

$$R_y = \frac{P^{1.8079}\mu^{0.39759}}{H^{0.59516}} f(n) \tag{1}$$

$$f(n) = 11.82062 + 0.03972n - (1e-4)n^2 + (1.7967e-7)n^3 \tag{2}$$

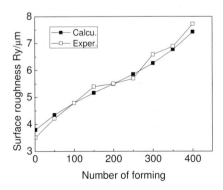

Fig. 6. The comparison between the calculated and experimental values.

The reliability of the equation was validated using another forming experiment of the U-channel. The parameters, H, $\mu$, P, n were 48HRC, 0.8$\mu$m, 2T and 400 respectively. The draw depth was also 50mm. The calculated values of the surface roughness $R_y$ along with the experimental values are shown as a function of the number of forming, see Fig.6. It is evident form this figure that the calculated values are in close proximity with the experimental ones. The fact suggests reasonably good reliability of the equation to predict the galling behavior in SMF within the selected experimental conditions.

## 4.  Conclusions

In this study, galling behavior in the forming of high strength steels under the condition of tension-bending was investigated through the U-channel forming tests. The effects of tool hardness, tool surface roughness and blank holder force on galling were studied in the forming of high strength steels. Experimental results indicate galling tendency can be reduced through hardening and polishing of the forming tool and galling becomes more severe with increasing blank holder force and sliding distance in SMF. A numerical equation for galling evaluation aimed at the sliding condition under tension-bending was established through non-linear curve fitting of the large amount of experimental data. The reliability of the equation was validated.

## Acknowledgments

This work was supported by the National Natural Science Foundation of China under grant No. 50605043. The authors also thank the Dr. Xinping Chen and Dr. Haomin Jiang (Baosteel technical center) for their support and advice throughout this study.

## References

1.    Hyunok Kim, *J. Mater. Process. Tech.*, doi:10.1016/j.jmatprotec.2007.11.281 (2008).
2.    B. Podgornik and S. Hogmark, *J. Mater. Process. Tech.*, 174, 334 (2006).
3.    B. Podgornik, J.Vizintin and S. Hogmark, *Surface Engineering*, 22, 235 (2006).
4.    V. Imbeni, C. Martini and E. Lanzoni, *Wear*, 251, 997 (2001).
5.    P.Q. Wu, H. Chen, M. Van Stappen, and L. Stals, *surf. Coat. Technol.*, 127, 114 (2000).
6.    Teisuke Sato, Tatsuo Besshi and Isao Tsutsui, *J. Mater. Process. Tech.*, 104, 21 (2000).
7.    A. gaard, P.V. Krakhmalev, J. Bergstrom and N. Hallback, *Tribology Letters*, 26, 67 (2007).
8.    Shrinidhi Chandrasekharan, Hariharasudhan Palaniswamy, Nitin Jain, Gracious Ngaile and Taylan Altan, *International Journal of Machine Tools & Manufacture*, 45, 379 (2005).
9.    C.M. SUH, G.W. CHOI, K.R. KIM and M.S. HAN, *Int. J. Mod. Phys. B*, 20, 27 (2006).
10.   T. Skare and F. Krantz, *Wear*, 255, 1471 (2003).
11.   K. Gurumoorthy, M. Kamaraj and K. Prasad Rao, *Materials and Design*, 28, 987 (2007).
12.   E.van der Heide and D.J. Schipper, *Wear*, 254, 1127 (2003).
13.   S.C. SHARMA and M. KRISHNA, *Int. J. Mod. Phys. B*, 20, 27 (2006).
14.   M. Tajdari and M. Javadi, *J. Mater. Process. Tech.*, 177, 247 (2006).

# DYNAMIC FAILURE OF A SPOT WELD IN LAP–SHEAR TESTS UNDER COMBINED LOADING CONDITIONS

J. H. SONG[1], J. W. HA[2], H. HUH[3†]

*School of Mechanical, Aerospace & System Engineering, KAIST,335,Gwahangno, Deadoek Science Town, Daejeon, 305-701 Korea,*
*jhsong_me@kaist.ac.kr, hajiwoong@kaist.ac.kr, hhuh@kaist.ac.kr*

J. H. LIM[4], S. H. PARK[5]

*Automotive Steel Research Center, POSCO,699, Gumho-dong, Gwangyang-si, Jeonnam, 545-090 Korea,*
*jiholim@posco.com, sunghopark@posco.com*

Received 15 June 2008
Revised 23 June 2008

This paper is concerned with the evaluation of the dynamic failure load in the lap-shear tests of a spot weld. Dynamic lap-shear tests of a spot weld in SPRC340R were conducted with different tensile speeds ranging from $5 \times 10^{-5}$ m/sec to 5.0 m/sec. Dynamic effects on the failure load of a spot weld are examined based on the experimental data. Experimental results indicate that failure strength increases with increasing loading rates. Finite element analyses of dynamic lap-shear tests were also performed considering the failure of a spot weld. A spot weld is modeled with a beam element and dynamic failure model is utilized in order to describe the failure of a spot weld in the simulation. The failure loads obtained from the analyses are compared to those from the lap-shear tests. The comparison shows that the failure loads obtained from the analyses are close in consistence with those obtained from the experiments.

*Keywords*: Resistance spot weld; lap-shear test; dynamic failure load.

## 1. Introduction

The electric resistance spot welding process is an indispensable assembling process of steel auto-panels in the automotive industries since its introduction in 1950's. As a modern auto-body contains several thousands of spot welds, the strength of spot welds under impact loading conditions becomes extremely important in the evaluation of the crashworthiness of auto-body members.[1]

Usually, the failure of a spot weld is assumed to be independent of the strain rate in the early design stage so that the failure behavior of a spot weld is obtained using static failure tests.[2-4] However, under rapid collapse in a crash situation, the failure behavior of

---

†Corresponding Author.

a spot weld can be quite different to the statically loaded case. In order to evaluate the dynamic failure load of spot welds, dynamic failure tests have been conducted using the coach-peel specimen and the cross-tension specimen. [5-6]

With the advancement of computer simulation technology, it is attempted to develop the accurate failure model of a spot weld and implement the failure model into the crash analysis in order to describe the failure of spot welds in vehicle crash simulations. Currently, a common method for modeling a spot weld in the automotive industry is to use a rigid link or a beam element. The onset of failure of a spot weld is determined with the failure model as a function of the forces acting on the spot weld. Song and Huh[7] proposed a dynamic failure model with analyzing experimental results under combined loading conditions since spot welds in auto-body components are subjected to many different types of combined loads when they deform under impact loading conditions.

In this paper, quasi-static and dynamic lap-shear tests were conducted with different tensile speeds ranging from $5 \times 10^{-5}$ m/sec to 5.0 m/sec. Dynamic effects on the failure load of a spot weld, which is critical for structural crashworthiness, are examined based on the experimental data. FE analyses of lap-shear tests were also performed considering the failure of a spot weld. Dynamic failure model of a spot-weld is utilized to describe the failure of a spot weld in the simulations. The failure loads obtained from the numerical simulation are compared to those from the lap-shear test.

## 2.  Dynamic Lap-Shear Tests of a Spot Weld

### 2.1.  *Experimental conditions*

Dynamic failure loads of a spot weld in the lap-shear specimens are evaluated in this paper. The spot-welded material was SPRC340R with a thickness of 1.2 mm. Spot welding of the steel sheet was performed using a static spot/projection welding machine. The welding current of 7.6 kA was imposed during the time of 15cycles at 60 Hz with the holding force of 3.0 kN. The diameter of the welded nugget is about 6.5 mm.

In the lap-shear test, a specimen begins to bend and the nugget rotates after the shear load is applied on the spot weld. Such an out-of-plane deformation is induced to eliminate the bending moment in the non-clamping region of the specimen as shown in Fig. 1. Due to the rotation of the nugget, the combined axial and shear loads are applied onto the spot weld in the specimen. As the applied load increases, failure caused by localized necking occurs at one of the localized necking points.

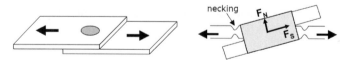

Fig. 1. Failure mechanism of a spot weld in the lap-shear test.

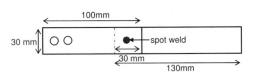

Fig. 2. Dimension of a specimen for dynamic lap-shear tests.

Fig. 3. High speed material testing machine.

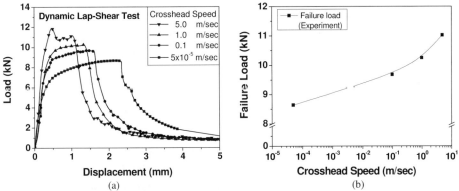

(a)                                                              (b)

Fig. 4. Experimental results for dynamic failure tests of a spot weld: (a) load–displacement curves at various crosshead speeds: (b) effect of the loading rate on the failure load of a spot weld.

Fig. 2 shows schematic description of a specimen used in the dynamic lap-shear test. The specimen is mounted on a high speed material testing machine as shown in Fig. 3 for the dynamic lap-shear test of a spot weld. The high speed material testing machine is servo-hydraulic equipment that has the maximum stroke velocity of 7800 mm/sec and the maximum load of 30 kN.[8] The load is acquired by a piezoelectric-type load cell, Kistler 9051A, and the displacement is obtained by a LDT from Sentech company. In order to achieve the constant velocity during tests, a special jig is used to move to some distance under no loading and then seize a specimen instantly. Using the high speed material testing machine, dynamic lap-shear tests were conducted at the crosshead speeds of 0.1 m/sec, 1.0 m/sec and 5.0 m/sec. Quasi-static test was also performed using the INSTRON 4206 at the crosshead speed of $5 \times 10^{-5}$ m/sec.

## 2.2. Experimental results

Load–displacement curves of the lap-shear specimens are depicted in Fig. 4(a) at various crosshead speeds. The failure loads of a spot weld are also plotted in the logarithmic scale of the loading rate, as shown in Fig. 4(b). Figures indicate that the failure load increases with increasing crosshead speed while the associated displacement at failure decreases. When the crosshead speed increases from $5 \times 10^{-5}$ m/sec to 5.0 m/sec, the failure load increases from 8.69 kN to 11.06 kN while the displacement at failure

(a)                                    (b)

Fig. 5. Deformed shape of the lap-shear specimen: (a) onset of failure: (a) after failure.

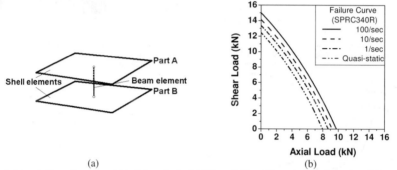

(a)                                    (b)

Fig. 6. Description of a spot weld in FE analysis: (a) FE modeling; (b) strain-rate dependent failure contour[7].

decreases from 2.48 mm to 1.09 mm. Fig. 5(a) represents the deformed shapes of the specimen when localized necking occurs. Rotation of the nugget as explained in Fig. 1 is observed from the specimen; thus, the specimen fails under combined axial and shear loads as shown in Fig. 5(b).

## 3. FE Analysis of Lap-Shear Tests Using the Dynamic Failure Model

Finite element analyses of the lap-shear specimen were also performed considering the failure of a spot weld. The spot weld in the specimen is modeled with a single beam element as shown in Fig. 6(a), which connects a pair of shell elements by constraining all of the degrees of freedom. The mesh size of the shell elements was set to 8 mm, which is widely used in the modeling guide of the auto-body.[9] In order to describe the failure of a spot weld in the simulation, the strain-rate dependent failure contours shown in Fig. 6(b) are considered using the dynamic failure model[7] of a spot weld. The failure model expressed in Eq. (1) was implemented into LS-DYNA3D[10] via a user-subroutine.

$$\left(\frac{f_n}{F_N}\right)^2 + 1.35\left(\frac{f_n}{F_N}\right)\left(\frac{f_s}{F_S}\right) + \left(\frac{f_s}{F_S}\right)^2 = 1 \tag{1a}$$

$$F_N = 7.96\left(1 + 0.00714\left(\ln\frac{\dot{\varepsilon}}{\dot{\varepsilon}_{ref}}\right)^{1.49172}\right) \text{ kN} \tag{1b}$$

$$F_S = 12.25\left(1 + 0.00714\left(\ln\frac{\dot{\varepsilon}}{\dot{\varepsilon}_{ref}}\right)^{1.49172}\right) \text{ kN} \tag{1c}$$

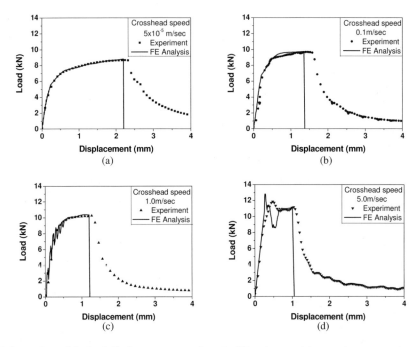

Fig. 7. Comparison of the load–displacement curves from the FE analyses and the experiments at various crosshead speeds: (a) 5x10$^{-5}$ m/sec; (b) 0.1 m/sec; (c) 1.0 m/sec; (d) 5.0 m/sec.

Table 1. Comparison of the failure load obtained from the FE analyses and the experiments.

| Crosshead speed (m/sec) | Failure load (kN) | |
|---|---|---|
| | FE Analysis | Experiment |
| 5x10$^{-5}$ | 8.80 | 8.69 |
| 0.1 | 9.74 | 9.62 |
| 1.0 | 10.35 | 10.23 |
| 5.0 | 10.92 | 11.06 |

Fig. 7 compares load–displacement curves obtained from the analyses to those from the experiments. The failure loads obtained from the analyses and the experiments are summarized in Table 1. The comparison represents that the failure loads of the spot weld obtained from the analyses are close in coincidence with those obtained from the experiments. In the FE analysis, the strain rate of the spot-welded region should be calculated to identify the failure contour of a spot weld. In this paper, the strain rate of the spot weld is calculated from the relative velocity of two nodes in a beam element as shown in Fig. 8(a). Using the calculated strain rates, failure contours at different crosshead speeds are determined from the failure model in Eq. (1). The axial and shear loads acting on the spot weld are also plotted in Fig. 8(b). The graph explains that the shear load is initially applied on the spot weld and that it decreases while the axial load increases due to the rotation of the nugget as the specimen deforms. When the

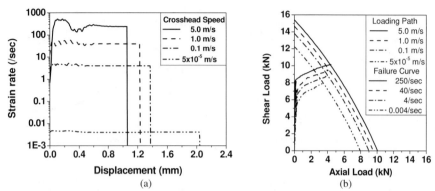

Fig. 8. Identification of the failure of a spot weld in FE analyses at various crosshead speeds: (a) calculation of the strain rate; (b) failure contours and loading path of a spot weld.

combination of the forces meets the failure model at each strain rate as shown in Fig. 8(b), the constraint condition imposed on the beam element is released to simulate the failure of the spot weld.

## 4. Conclusion

Dynamic failure tests of spot-welded lap-shear specimens are conducted at the different tensile speeds ranging from $5 \times 10^{-5}$ m/sec to 5.0 m/sec. The failure loads and failure behavior of the specimen are investigated using the experimental results. The results indicate that failure of a spot weld under impact loading condition is quite different to the statically loaded case. In the dynamic loading case, the failure load increases with increasing crosshead speed while the displacement at failure decreases. Finite element analyses of the lap-shear specimen are also performed considering the failure of a spot weld. The dynamic failure model is utilized to describe the failure of a spot weld during the simulation. With the aid of the dynamic failure model, the failure loads obtained from the analyses are close in coincidence with those obtained from the experiments. It can be concluded that dynamic failure model is appropriate to describe the failure behavior of spot welds in vehicle simulations.

## References

1. Y. J. Chao, *J. Eng. Mater.- T. ASME.* **125** (2003) 125.
2. B. Langrand and A. Combescure, *Int. J. Solids Struct.* **41** (2004) 6631.
3. S.-H. Lin, J. Pan, S.-R. Wu, T. Tyan and P. Wung, *Int. J. Solids Struct.* **39** (2002) 19.
4. J. H. Song, H. Huh, J. H. Lim and S. H. Park, *Int. J. Mod. Phys.* **22** (2008) 1469.
5. F. Schneider and N. Jones, *Int. J. Mech, Sci.* **45** (2003) 2061.
6. X. Sun and M. A. Khaleel, *Int. J. Impact Eng.* **34** (2007) 1668.
7. J. H. Song and H. Huh, *Int. J. Solids Struct.* Submitted for publication
8. H. Huh, S. B. Kim, J. H. Song and J. H. Lim, *Int. J. Mech, Sci.* **50** (2008) 918.
9. J. H. Song, H. Huh, H. G. Kim and S. H. Park, *Int. J. Automot. Techn.* **7** (2006) 329.
10. J. O. Hallquist, LS-DYNA3D Keyword User's Manual-Vers. 9.7 (LSTC, Livermore, 2003).

# EVALUATION OF DAMAGE IN STEELS SUBJECTED TO EXPLOITATION LOADING - DESTRUCTIVE AND NON-DESTRUCTIVE METHODS

ZBIGNIEW L. KOWALEWSKI[1†], SŁAWOMIR MACKIEWICZ[2], JACEK SZELĄŻEK[3]

*Institute of Fundamental Technological Research, Świętokrzyska 21, Warsaw, 00-049, POLAND,*
*zkowalew@ippt.gov.pl, smackiew@ippt.gov.pl, jszela@ippt.gov.pl*


KRYSTYNA PIETRZAK[4]

*Motor Transport Institute, Jagiellońska 80, Warsaw, 00-049, POLAND,*
*krystyna.pietrzak@its.waw.pl*


BOLESŁAW AUGUSTYNIAK[5]

*Gdańsk University of Technology, Narutowicza 11/12, Gdańsk, 80-952, POLAND,*
*augustyniak@mag-lab.pl*

Received 15 June 2008
Revised 23 June 2008

Damage due to creep and plastic flow is assessed using destructive and non-destructive methods in steels (40HNMA and P91). In the destructive methods the standard tension tests were carried out after prestraining and variations of the selected tension parameters were taken into account for damage identification. In order to assess a damage development during the creep and plastic deformation the tests for both steels were interrupted for a range of the selected strain magnitudes. The ultrasonic and magnetic techniques were used as the non-destructive methods for damage evaluation. The last step of the experimental programme contained microscopic observations. A very promising correlation between parameters of methods for damage development evaluation was achieved. It is well proved for the ultimate tensile stress and birefringence coefficient.

*Keywords*: Creep; plastic strain; damage; ultrasonic and magnetic techniques.

## 1. Introduction

There are many testing techniques commonly used for damage assessments. Among them we can generally distinguish destructive and non-destructive methods. Having the parameters of destructive and non-destructive methods for damage development evaluation it is worth to analyze courses of their variation in order to find a possible

---

†Corresponding Author.

correlation. This is because of the fact that typical destructive investigations like creep or standard tension tests give the macroscopic parameters characterizing the lifetime, strain rate, yield point, ultimate tensile stress, ductility, etc. without any information concerning a microstructural damage development and material microstructure variation. On the other hand, the non-destructive methods enable to provide a knowledge concerning damage development in a particular time of the entire working period of an element, however, without sufficient information dealing with the microstructure and time of its further secure exploitation. Therefore, it seems to be reasonable to plane future damage development investigations in the form of interdisciplinary tests connecting results achieved using destructive and non-destructive methods with microscopic observations in order to find mutual correlation between their parameters. This paper and accompanying papers[1-3] can be treated as an attempt in this direction.

## 2.   Experimental Details

### 2.1.  *Destructive technique*

Two kinds of steel commonly used in the selected elements of polish power plants were investigated: 40HNMA and P91.

The experimental programme comprised tests for the materials in the as-received state and for the same materials subjected to a range of selected magnitudes of prior deformation due to creep at elevated temperatures, and due to plastic flow at room temperature. Uniaxial tension creep tests were carried out on the 40HNMA and P91 steels using plane specimens. For each steel all tests were conducted in the same conditions. In the case of 40HNMA the stress level was equal to 250 MPa, and temperature 773 K, whereas for the P91 – 290 MPa, and 773 K, respectively. Details of the destructive tests programme as well as its main results are presented by Kowalewski *et al.*[1]. In order to assess a damage development during the process of creep the tests for the 40HNMA steel were interrupted for a range of the selected time periods: 100h, 241h, 360h, 452h, 550h, 792h, 929h and 988h, which correspond to the increasing amounts of creep strain equal to 0.34%, 0.8%, 1%, 1.1%, 1.2%, 2.3%, 4.0%, and 6.5%, respectively. In the case of P91 the tests were interrupted after 40h (0.85%), 180h (1.85%), 310h (3.15%), 390h (4.6%), 425h (5.9%), 440h (7.9%) and 445h (9.3%). After each prestraining test a damage of specimen was assessed using the non-destructive methods (magnetic and ultrasonic) and in the next step the same specimens were stretched until the failure was achieved.

### 2.1.1.  *Magnetic  techniques*

Magnetic properties were measured using standard laboratory method of magnetisation, where hysteresis loop with the HBE (Barkhausen effect) and also the MAE (magneto-acoustic emission effect) can be tested[4]. Two sensors were used: (a) the pickup coil (PC), and (b) the acoustic emission transducer (AET). A voltage signal induced at PC was used for the magnetic hysteresis loop $B(H)$ evaluation (low frequency component) as well as for HBE analysis (high frequency component). An intensity of the HBE is given by the rms (root mean square) voltage $Ub$ envelopes. We compared here the maximal values (*Ubpp*) of $Ub$ for one period of magnetisation. The analogue analysis was performed for the MAE voltage signal from the AET. In this case we compared the maximal values (*Uapp*) of *Ua* voltage envelopes. The magnetic coercivity *Hc*, evaluated from the *B(H)* hysteresis loop plots, was also considered.

### 2.1.2. *Ultrasonic technique*

In order to evaluate a damage progress in specimens made of the 40HNMA and P91 steels the acoustic birefringence *B* was measured[5, 6]. The acoustic birefringence *B* is a measure of material acoustic anisotropy. It is based on the velocity difference of two shear waves polarized in the perpendicular directions. In specimens subjected to creep the shear waves were propagated in the specimen thickness direction and were polarized along its axis and in the perpendicular direction.

## 3.  Experimental Results and Their Discussion

### 3.1. *Evaluation of damage development using destructive tests*

The tension characteristics for both materials after prestraining are presented in Fig. 1 for the 40HNMA steel and in Fig. 2 for the P91 steel. In all of these diagrams the characteristics for the prestrained steels are compared with tension curves of the materials in the as-received state. Taking into account the results presented for the 40HNMA steel, it is easy to note that this material due to prestraining by creep exhibits significant softening effect, expressed by a big decrease of the ultimate tensile stress. An opposite effect can be observed for this material prestrained due to plastic deformation at room temperature. In this case the prior deformation leads to the hardening effect. Similar results were obtained for the P91 steel. On the basis of tension characteristics, Figs. 1 and 2, variations of the basic mechanical properties of both steels, due to deformation achieved by prior creep or plastic flow were determined. It was observed for both materials that the Young's modulus is almost not sensitive on the magnitude of creep and plastic deformations. Contrary to the Young's modulus the other considered tension test parameters, especially the yield point and ultimate tensile stress, exhibit clear dependence on the level of prestraining. An example of the comparative results concerning variations of the yield point for both steels tested after two different kinds of prestraining is presented in Table1.

Table 1. Variations of the yield point due to creep ($\varepsilon_c$) or plastic ($\varepsilon_p$) deformations.

**40HNMA**

| $\varepsilon_c$ [%] | 0 | 0.34 | 0.8 | 1.1 | 1.2 | 2.3 | 4.0 | 6.5 | $\varepsilon_p$ | 0.5 | 1.5 | 6.5 | 10.5 |
|---|---|---|---|---|---|---|---|---|---|---|---|---|---|
| $R_{0.2}$ [MPa] | 1040 | 1010 | 995 | 949 | 944 | 895 | 857 | 763 | $R_{0.2}$ | 1056 | 1071 | 1084 | 1144 |

**P91**

| $\varepsilon_c$ [%] | 0 | 0.85 | 1.85 | 3.15 | 4.6 | 5.9 | 7.9 | 9.3 | $\varepsilon_p$ | 3.0 | 4.5 | 7.5 | 10.5 |
|---|---|---|---|---|---|---|---|---|---|---|---|---|---|
| $R_{0.2}$ [MPa] | 508 | 534 | 529 | 526 | 514 | 513 | 506 | 520 | $R_{0.2}$ | 628 | 664 | 718 | 720 |

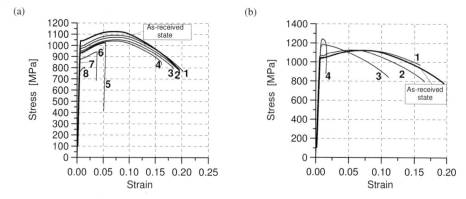

Fig. 1. Tension characteristics of 40HNMA steel: (a)  material after prior deformation due to creep interrupted in different phases of damage development, (b) material after prior deformation due to plastic flow interrupted in different phases of the process, (numbers corresponding to the increasing values of prior creep deformation are described in the Section 2.1, while for plastic deformation: 1 (0.5%); 2 (1.5%); 3 (6.5%); 4 (10.5%)).

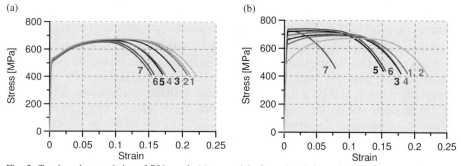

Fig. 2. Tension characteristics of P91 steel: (a)  material after prior deformation due to creep interrupted in different phases of damage development, (b) material after prior deformation due to plastic flow interrupted in different phases of the process, (numbers corresponding to the increasing values of prior creep deformation are described in the Section 2.1, while for plastic deformation: 1 (2%); 2 (3%); 3 (4.5%); 4 (5.5%); 5 (7.5%); 6 (9%); 7 (10.5%)).

### 3.2.  Evaluation of damage development using magnetic technique

The results of investigations using the magnetic techniques are presented in Figs. 3 and 4. The materials are labelled as follows: 1) P91pd – P91 steel after plastic deformation due to tension at room temperature, 2) P91cd – P91 steel after creep deformation, 3) 40HNMcd – 40HNMA steel after creep deformation. Figure 3 shows a monotonous increase of the coercivity force. The highest increase (of order + 65 %) is observed for 'P91pd'. There is evident a much lower increase of the $Hc$ for P91 steel subjected to creep. The HBE intensity modifications as a function of prestraining are more complex. Plastic deformation at room temperature of the P91 steel leads to a monotonous decrease of the $Ub$ envelopes maximum (at least for $\varepsilon_p > 2\%$) while creep deformation leads at first to a very high increase of the $Ub$ envelopes maximum (with decrease of the HBE envelope maximum width). The HBE intensity maximum of this steel decreases then monotonously down to the level of not creep damaged specimen. The results show that creep of the 40HNMA steel leads to relatively very small modification of the HBE intensity: the *Ubpp* parameter is nearly constant. Because of small decreases of the width of the *Ub* envelope one can find that the integral of the HBE intensity decreases of about − 10%. Three plots of the *Uapp* parameter in Fig. 4 show properties of the MAE. Plastic deformation of the P91 steel at room temperature leads to the evident broadening of the MAE envelope (two maxima appears) and to an abrupt decrease of the *Uapp* parameter value, as the plot 1 shows in Fig. 4. Creep process does not increase, but decreases the width of the MAE maximum and also leads to a decrease of the *Uapp* parameter value, as the plot 2 shows in Fig. 4.

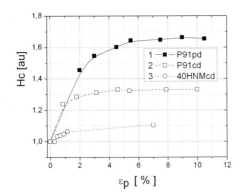

Fig. 3. An influence of prestraining on the magnetic hysteresis force.

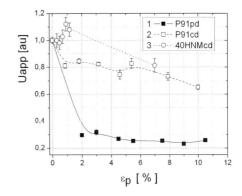

Fig. 4. An influence of prestraining on the MAE intensity.

### 3.3. *Evaluation of damage development using ultrasonic technique*

Figure 5 presents mean values of the acoustic birefringence measured in specimens after creep or plastic deformation. The birefringence was measured in the fixtures, where a texture of material was assumed to be unchanged during creep test, and in the working part of the specimen. In the deformed part of specimen a value of the birefringence depends on the deformation amount.

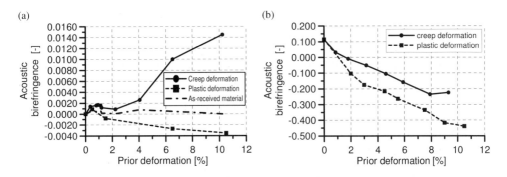

Fig. 5. Acoustic birefringence B variations due to prior deformation for: (a) 40HNMA steel, (b) P91 steel.

It can be noticed that the birefringence variations due to creep are significantly higher than the birefringence scatter for the non-deformed material. The acoustic birefringence was measured in several points along the working part of each specimen, thus enabling to find its maximum.

The plots presented in Fig. 5 indicate that the acoustic birefringence is sensitive on the amount of prior deformation. Another advantage of this parameter is also well represented in Fig. 5. Namely, it is very sensitive on the form of prior deformation. This feature is especially well revealed in the case of the birefringence determined for the 40HNMA steel, Fig. 5a. For specimens prestrained due to creep the increase of this parameter is observed with the increase of prior deformation. An opposite effect was achieved for specimens prestrained due to plastic deformation at room temperature, i.e. with the increase of prior deformation a decrease of the birefringence was obtained. Also, the results for P91 steel allow to distinguish a type of prior deformation, however, the effect is not as clear as for the 40HNMA steel observed.

## 4. Conclusions

The paper presents the results of tests for damage assessments in steels commonly used for pipelines at power stations. The results show that the selected ultrasonic and magnetic parameters can be good indicators of material degradation and can help to localize the regions where material properties are changed due to prestraining. Moreover, the correlation achieved between parameters of the damage evaluation methods applied indicates that investigations concerning creep problems should go in the direction of interdisciplinary tests connecting parameters assessed by the classical macroscopic destructive investigations with parameters coming from the non-destructive tests.

## Acknowledgment

The support from the State Committee for Scientific Research (Poland) under grant 4 T07A 018 26 is greatly appreciated.

## References

1. Z.L. Kowalewski, B. Augustyniak, J. Szelążek, S. Mackiewicz, in *Proc. of the XIV International Symposium on Plasticity and Its Current Applications*, ed. A.S. Khan (Neat Press, Fulton, Maryland, 2008).
2. Z.L. Kowalewski, T. Skibiński, J. Szelążek, S. Mackiewicz, in *Proc. of the XII International Symposium on Plasticity and Its Current Applications*, ed. A.S. Khan (Neat Press, Fulton, Maryland, 2006), pp. 619-621.
3. K. Pietrzak, Z.L. Kowalewski, D. Rudnik, A. Wojciechowski, in *Proc. of the XIV International Symposium on Plasticity and Its Current Applications*, ed. A.S. Khan (Neat Press, Fulton, Maryland, 2008).
4. B. Augustyniak, M. Chmielewski, M.J. Sablik, *IEEE- Trans. on Magnetics*, **36**, no. 5, 3624-3626 (2000).
5. Z.L. Kowalewski, S. Mackiewicz, J. Szelążek, J. Deputat, in *Proc. of XXI Symp. on Experimental Mechanics of Solid Body*, ed. J. Stupnicki (Warsaw University of Technology, 2004).
6. S. Mackiewicz, Z.L. Kowalewski, J. Szelążek, J. Deputat, *Mechanical Review*, 7/8, 15-24 (2005).

# FATIGUE PROPERTIES OF AUTOMOBILE HIGH-STRENGTH BOLTS

CONGLING ZHOU [1†]

*College of Mechanical Engineering, Tianjin University of Science and Technology, 1038, Dagu Nanlu, Hexi Dist., Tianjin, 300222, P.R.CHINA,*
*zhoucongling@yahoo.com*

SHIN-ICHI NISHIDA [2], NOBUSUKE HATTORI [3]

*Faculty of Science and Engineering, Saga University, 1, Honjo-machi, Saga, 840-8502, JAPAN,*
*nishida@saga-u.ac.jp, hattori@saga-u.ac.jp*

Received 15 June 2008
Revised 23 June 2008

This study is focused on the fatigue properties of automobile high-strength bolts, including the effect of mean stress level, pre-processing schedule and the residual stresses. And the mean stress levels are 0.3, 0.5 and 0.7 times to the tensile strength ($\sigma_B$) of the material respectively. The main results obtained are as follows: 1) the fatigue strength increases under the mean stress loading, but the differences between the loading levels are not so evident; 2) most of the cases in this study are broken from the bottom of the screw thread, and the crack initiated from the impurities.

*Keywords*: High-strength bolt; Automobile; Mean stress; Residual stress.

## 1. Introduction

As it is known that high-strength bolts are widely used in modern machines, and they are especially important in the modern transport tools, such as cars, trains, airplanes and so on. And it is reported that more than 1000 bolts and 1,500,000 bolts are used in a car and an airplane, respectively[1]. Moreover, there are many transport accidents caused by the bolts fatigue failure. On the other hand, automobile is becoming more and more close to our daily life. Therefore, it is very necessary to study the fatigue properties of high-strength bolts used in automobiles.

Pre-loading level is a very important factor on the fatigue properties of the high-strength bolts[2-5]. In this study, it is focused on the effects of different pre-loading levels, that is, mean stress levels, on the fatigue properties of high-strength bolts, as well as how the different pre-processing procedure and methods affect the fatigue properties of automobile high-strength bolts.

[†]Corresponding Author.

## 2.  Materials and Experimental Procedure

Materials used in this study are automobile high-strength bolts. Table 1 and Table 2 list the chemical composition, mechanical properties and the pre-processing conditions of the test materials, respectively. Eight types of bolts are used in this study, denoted by type 1 to type 8 respectively. All of the bolts are jointed with nuts made under the same condition, which is listed in Table 3.

Table 1.  Chemical compositions, mass%×100.

| | Material | ND | C | Si | Mn | P | S | Cu | Ni | Cr | Mo | B |
|---|---|---|---|---|---|---|---|---|---|---|---|---|
| Type1 | SWRCHB323M | | 20 | 21 | 81 | 1.5 | 2.8 | 1 | 1 | 17 | − | 0.21 |
| Type2 | SWRCHB323M | | 23 | 18 | 93 | 1.5 | 0.7 | 1 | 1 | 16 | − | 0.14 |
| Type3 | SWRCHB334M | M12 | 34 | 24 | 88 | 2.0 | 1.3 | 2 | 2 | 14 | − | 0.15 |
| Type4 | SCM435 | | 41 | 19 | 75 | 1.6 | 1.5 | 1 | 2 | 101 | 16 | − |
| Type5 | SCM440 | | 40 | 19 | 64 | 0.8 | 0.4 | 1 | 2 | 93 | 22 | − |
| Type6 | SWRCHB323M | | 24 | 19 | 90 | 1.9 | 0.7 | 2 | 2 | 15 | − | 0.13 |
| Type7 | SWRCHB325M | M10 | 27 | 16 | 110 | 2.2 | 0.8 | − | − | 15 | − | 0.15 |
| Type8 | SCM440 | | 41 | 22 | 66 | 1.0 | 0.7 | 1 | − | 101 | 21 | − |

Table 2.  Mechanical property and pre-process conditions.

| Material | ND | $\sigma_B$, MPa | Process | Quenching | Tempering |
|---|---|---|---|---|---|
| Type 1 | | 800 | Thread-rolling →Heat treatment | 870°C, ≥5min | 470°C, ≥60min |
| Type 2 | | 900 | Thread-rolling →Heat treatment | 870°C, ≥5min | 445°C, ≥60min |
| Type 3 | M12 | 1000 | Thread-rolling →Heat treatment | 870°C, ≥5min | 445°C, ≥60min |
| Type 4 | | 1200 | Thread-rolling →Heat treatment | 870°C, ≥5min | 450°C, ≥60min |
| Type 5 | | 1200 | Heat treatment→ Thread-rolling | 870°C, ≥5min | 460°C, ≥60min |
| Type 6 | | 800 | Thread-rolling →Heat treatment | 870°C, ≥5min | 470°C, ≥60min |
| Type 7 | M10 | 1000 | Thread-rolling →Heat treatment | 870°C, ≥5min | 470°C, ≥60min |
| Type 8 | | 1200 | Thread-rolling →Heat treatment | 870°C, ≥5min | 460°C, ≥60min |

Table 3.  Detail of nuts.

| Type of bolt | ND | Material | Pitch, mm | Hardness, HRC |
|---|---|---|---|---|
| Type 1-5 | M12 | S45C | 1.25 | 34 |
| Type 6-7 | M10 | (Refining steel) | 1.25 | 33 |
| Type 8 | M10 | | 1.5 | 36 |

The minimum diameter of bolt is measured before the tensile-compressive fatigue test, in order to get the mean stress value loading on the bolts. The fatigue test is done on a Shimazu-EHF10 machine (±15t, ±25mm) and the mean stress loadings are 0.3, 0.5 and 0.7 times to the tensile strength ($\sigma_B$) of the test specimen. In the test, all of the nuts are loaded to the same station under the same load (15Nm) to make sure the fatigue failure happened at the same position.

A scanning electrical microscope (SEM) was employed to investigate the initiation and propagation behavior of the fatigue crack by observing the fracture surface of the specimen. Hardness is tested with a Vickers hardness tester under the load of 2.94N. The residual stress is measured by X-ray diffraction method (SHIMADZU X-ray stress measurement equipment DX-10).

## 3. Experimental Results and Discussion

### 3.1. *Fatigue strength*

Fig. 1 to Fig. 3 shows the S-N curves of the test specimens under the mean stress level of 0.3, 0.5 and 0.7 times to their tensile strength ($\sigma_B$), respectively. It can be shown that under the mean stress of $0.3\sigma_B$, the fatigue limits of type 1 to 4 and type 7 are almost the

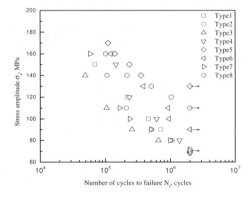

Fig. 1.  S-N curves under $\sigma_m=0.3\sigma_B$.

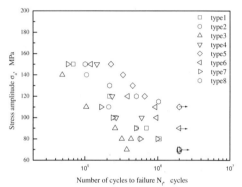

Fig. 2.  S-N curves under $\sigma_m=0.5\sigma_B$.

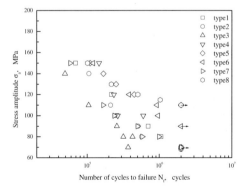

Fig. 3.  S-N curves under $\sigma_m=0.7\sigma_B$.

same, while the fatigue limit for type 5, type 6 and type 8 increases 85%, 28% and 57% to the former 5 types, which is 60MPa, 40 MPa and 20MPa, respectively. To the case of $0.5\sigma_B$ and $0.7\sigma_B$, there are almost no difference between the fatigue limits of type 1, 2, 3, 4 and 7, while the fatigue limits of the other three types are higher than the fatigue limit of the former 5 types, which is 20MPa for type 6 and 40MPa higher for type 5 and 8, that is, 28% and 57% higher, respectively. Moreover, there is almost no difference between the results under the mean stress $0.5\sigma_B$ and $0.7\sigma_B$.

On the other hand, for the different type specimen, under different mean stress value, $0.3\sigma_B$, $0.5\sigma_B$ and $0.7\sigma_B$, the fatigue limit presents different results. For specimens except type 5, the mean stress value has hardly effect on their fatigue limits; and for the case of type 5, the fatigue limit under $0.3\sigma_B$ is 20MPa higher than those under mean stress of $0.5\sigma_B$ and $0.7\sigma_B$. Fig. 4 shows the effect of mean stress on the fatigue limits of different type specimens. It is shown from this figure that the mean stress value affects the fatigue limits differently for different type of high strength bolt. And, based on Table 1, it is shown that the mean stress affects the fatigue limit easier for the specimen of processing procedure" Heat treatment → Thread-rolling" than that of the opposite procedure.

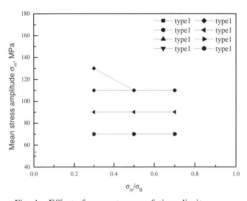

Fig. 4.   Effect of mean stress on fatigue limit.

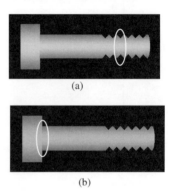

Fig. 5.   Failure position of specimens.

### 3.2.  *Micro-structure behaviors*

The representative microstructure of the fractured cross section are observed by a electric micro-scope and a SEM. Firstly, the broken position on the bolt are different for different pre-processing procedure, which is shown in Fig. 5, (a) is the microstructure of the specimens except type 5, and (b) is that of type 5. Fig.6 shows the representative microstructure of the fracture surface for these two cases in Fig. 6(a) and (b), respectively. From the observation, it is shown that the fatigue initiated from the impurities no matter where the fracture position is. This appearance may be explained from the microstructure of the screw thread root, which is shown in Fig. 7, where (a) is the microstructure for the specimen under the "Thread-rolling →Heat treatment" procedure, and (b) is the microstructure for the opposite processing procedure (type 5).

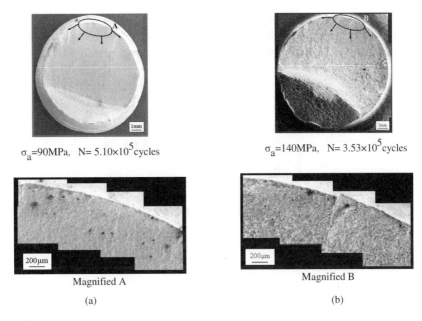

$\sigma_a$=90MPa,   N= 5.10×10$^5$cycles          $\sigma_a$=140MPa,   N= 3.53×10$^5$cycles

Magnified A                                         Magnified B

(a)                                                (b)

Fig. 6.  Representative microstructure of fracture surface.

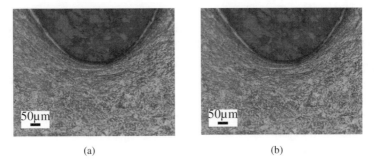

(a)                                                (b)

Fig. 7.  Representive microstructure of screw thread root of specimen.

It is shown that, there are thread flow at the screw thread root of type 5, which means that structure miniaturization happened under the thread-rolling position, while for other specimens, this miniaturization have been removed by the heat treatment after the thread-rolling. Therefore, the fracture position for type 5 is different to other type specimens. Moreover, in the test specimen, there are some impurities in the material. No matter the fracture position is, the impurities can cause stress concentration in the specimen. This may be a main reason why all the fatigue cracks initiated from the impurities.

On the other hand, the hardness at the screw thread root is checked, which is shown in Fig. 8 and Table 4. It is clear that the surface hardness of type 5 is much higher than the inner hardness, while for other types of specimen, the difference is not so clear to contribute for the fatigue strength. Also the residual stress at the screw thread root is measured by the X-ray method, which presented that the residual stress for type 5 is much higher than the other types of specimens. These may be some other reasons for the special fracture position of type 5.

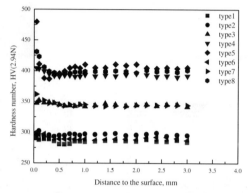

Fig. 8.  Hardness of screw thread root.

## 4.  Conclusions

This study is focused on the fatigue properties of some kinds of automobile high-strength bolts, including the effect of mean stress levels, procedure of pre-processing and the hardness of the fracture surface. The residual stress at the screw thread root is also checked. The main results obtained in this study are:

(1)  The mean stress affects the fatigue limits more easily for specimens thread rolling after the heat treatment, and the effect to specimens under the opposite procedure is not evident.

(2)  Most of the automobile high-strength bolts are fractured at the screw thread root, and the fatigue crack initiates from the impurities in the bolt.

## Acknowledgments

This study is supported by the Talent Fetching Funding of Tianjin University of Science and Technology (20070401).

## References

1.  NAKAHARA K. Master Thesis, Saga University, Japan, 2006.
2.  S. NISHIDA, *Failure Analysis in Engineering Applications* (Butterworth Heinemann Ltd., Oxford 1992).
3.  YANG Qiang, in Journal of Tianjin University of Technology and Education, (China, 2007), pp.29-32.
4.  LI Yingguo. in Mechanics, (China, 1993), pp.20-24, 39.
5.  YAN Fei, in Journal of Wuhan University of Technology, (China, 2007), pp.72-74.

# RESEARCH ON THE EFFECT OF LOADING CONDITIONS ON THE STRENGTH AND DEFORMATION BEHAVIORS OF ROCKS

PENG-ZHI PAN[1,†], XIA-TING FENG[2], HUI ZHOU[3]

*State Key Laboratory of Geomechanics and Geotechnical Engineering, Institute of Rock and Soil Mechanics, Chinese Academy of Sciences, Wuhan, 430071, P. R. CHINA, pzpan@whrsm.ac.cn, xtfeng@whrsm.ac.cn, hzhou@whrsm.ac.cn*

Received 15 June 2008
Revised 23 June 2008

The paper aims at a numerical study of strength and deformation behaviors of rocks under different loading conditions in uniaxial compression with an elasto-plastic cellular automaton (EPCA$^{2D}$) code. Two loading conditions, i.e. with and without considering frictions between loading platens and rock specimen's ends, are used in the failure processes of heterogeneous rocks. Rock specimens are assumed to be the same heterogeneity, i.e. the specimens with different sizes have the same probability of containing flaws. Under this condition, it is concluded that the strength and deformation behaviors of rocks are influenced by the heterogeneity a little. The mismatch of elastic properties of the platen and the rock in influencing the stress distribution at the ends of the specimen is the dominant cause for the so-called strength and deformation size effects, which in turn affects the final failure patterns of rocks.

*Keywords*: Elasto-plastic cellular automaton; rock failure processes; heterogeneity; scale effect.

## 1. Introduction

In experimental investigations people found that the experimental results are very sensitive to the loading conditions. Any small change in the contact condition between sample and loading platens may result in different strength and deformation behaviors as well as failure modes of the sample. For example, size and shape effects in the failure process in experiments are said to be caused by different loading conditions. And this phenomenon has been greatly concerned by many researchers in the last decades. Most researchers believed that a scale effect does exist in the failure process of rocks[1-6], but some found that there is no scale effect of rocks under uniaxial compression[7, 8]. The mechanism of scale effect is still ambiguous.

In present work, elasto-plastic cellular automaton (EPCA$^{2D}$) which is a self-developed numerical code, is used to investigate the mechanism of strength and deformation size effects that happen in the failure processes of rocks under uniaxial compression. The elasto-plastic cellular automaton model can expediently incorporate the heterogeneity of materials, and has advantage of accommodating localization and parallelization etc. By

---

[†]Corresponding Author.

173

constructing some local simple rules and adopting a simple failure criterion, the model can aptly simulate the self-organizational process and damage localization in the failure processes of heterogeneous rocks.

## 2. Brief Introduction to EPCA$^{2D}$

The Elasto-Plastic Cellular Automaton (EPCA) approach[9] is used to study the influence of loading conditions on the strength and deformation behavior. With this approach, the rock specimen is discretized into a system composed of cell elements firstly. Then a heterogeneous material behavior is adopted with homogeneous index *m* and the elemental seed parameter *s* for the heterogeneous mechanical properties of rock, such as Young's modulus, Poisson's ratio and cohesive strength etc. The mechanical parameters of rocks are assumed to conform to Weibull's distribution.

According to the initial and boundary conditions, a certain loading control method such as constant strain rate or linear combination of stress and strain[10] etc. is adopted to simulate the loading process of rock specimen in order to obtain the complete stress-strain curves of rock failure process. In each loading step, the cell state is updated by the cellular automaton updating rule[9] to get the displacement field and stress field.

The calculated stresses are substitute into the modified Mohr-Coulomb criterion to check whether or not cell element yield occurs. If the strength criterion is not satisfied, the external force is increased further. Otherwise, the cell element yields and the corresponding plastic strain produces according to the elasto-brittle-plastic constitutive theory.

The stress and deformation distributions throughout the domain are adjusted instantaneously after each rupture to reach the equilibrium state. In the process of stress adjustment, the stress of some cell elements may satisfy the critical value and further ruptures are caused. The process is repeated until no yield cell elements are presented. On the basis of equilibrium of stress adjustment, further load will be subjected to the system until the macro failure of rock specimen occurs. Thus the EPCA model links the mesoscopic mechanical model to macrostructure failure, which has been regarded as one of the most challenging tasks in the area of brittle failure.

## 3. Geometry, Loading Condition and Material Properties

In order to determine the influence of specimen geometry on the complete stress-strain curves for intact rock, specimens with the same diameter (D=50mm) and different lengths (L=15mm, 30mm, 50mm, 75mm, 100mm, 125mm and 150mm) are numerically tested in uniaxial compression using the EPCA$^{2D}$ code. Mohr-Coulomb criterion and associated plastic flow rule are adopted. The mechanical properties of cell elements are assumed to conform to the Weibull's distribution, in which the homogeneous index *m=6.0* and seed parameter *s=10*, respectively. Young's modulus, Poisson's ratio, cohesion and friction angle are 4.8GPa, 0.3, 15MPa and 49 degrees, respectively. Softening coefficient and residual coefficient are 1.0 and 0.0667, respectively.

Fig. 1a and b illustrate the loading conditions with and without considering loading platens. Therefore, Fig. 1a is the case under pure uniaxial compression. Fig. 1b is the case considering the platen-rock interaction because of the elastic properties mismatch of the platen and the rock. In the present modeling, the Young's modulus and Poisson's ratio of steel platen are 120GPa and 0.25, respectively. Platen length is 10mm.

In the simulation, the upper end of the platen is subjected to a stepwise strain increment (1e-5); and the bottom end of the platen is fixed. The complete stress-strain curve is obtained from the average value of stress and strain in axial direction (y) of every cell element of the rock specimen, the platen is not included. The cellular automaton convergence precision is 1e-11; and the convergence of non-linear iteration is 1%. Each cell element conforms to the elasto-brittle-plastic constitutive relation (Fig. 1c). In the properties assignment, the same homogeneous index (6.0) is used for each specimen so that the specimens with different sizes have almost (not exact) the same probability of containing flaws. In this case, for example, the specimens with the same diameter 50mm but different lengths have properties assigned via the same homogeneous index 6.0. From Fig. 1d, it can be seen that different lengths have almost the same distribution.

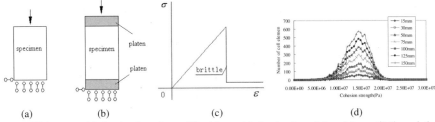

(a)    (b)    (c)    (d)

Fig. 1. (a) Without considering loading platen; (b) with considering loading platen; (c) constitutive relation of cell element and (d) cohesive strength distribution in the rock specimens with diameter 50mm and different lengths.

## 4. Results Analysis

### 4.1. *Strength and deformation behavior*

The results of complete stress-strain curves for rock specimens with different sizes by subjecting them to uniaxial compression with and without considering loading platens are presented. Numerical results show that, for the conditions with and without loading platen effect, the deformational behaviors are significantly different. Fig. 2 presents the results with and without considering platen effect. If there is no friction between loading platens and specimen's ends, even though the specimen is heterogeneous, the strength and deformation behaviors are only a little difference (Fig. 2a).

However, when considering the platen effect, the strength and deformation behaviors are significantly different. It can be seen from Fig. 2b that, before the peak stress, the influence of the specimen length (with the same diameter) on the deformational behavior is small. However, with the same specimen diameter, the deformational behavior in the post-peak region is greatly influenced by different specimen lengths, i.e. the shape effect. The stress-strain curves for specimens with larger lengths have a steeper slope in the post-peak region, i.e. the stress-strain curve drops more quickly – while, for specimens with a smaller length, the stress-strain curve in the post peak region drops more gently. This simulated deformation behaviors of specimens with the same diameter but different lengths are consistent with the experimental results[1]. The strength behaviors of the specimens are also influenced by the sizes. From Fig. 2b, it is shown that the strength of the specimen with same diameter but smaller length is higher than that of the specimens with larger lengths. From the results, it is found that, the effect of platen on the stress-strain curves is reduced when the length of specimen is long enough. For example, when the L:D ratio is larger than 1.5, especially larger than 2, the difference of strength behavior is not significant. Therefore, if we want to reduce the shape effect, specimens with L:D ratios larger than 2 are preferred.

Fig. 3a is the EPCA$^{2D}$ simulated result of the uniaxial strength with respect to length/diameter ratio of rock specimen with different sizes. It is evident that the friction between loading platens and specimen's ends is the important reason to produce shape effect. However, when the length/diameter ratio is large enough especially larger than 2, the effect of loading platen can be ignored. This is in good agreement with experimental result shown in Fig. 3b[11].

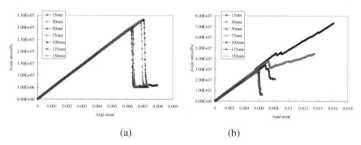

(a)                                                    (b)

Fig. 2. Complete stress-strain curves of rocks with the same diameter 50mm but different lengths using different loading conditions. (a) Without loading platen and (b) with loading platen.

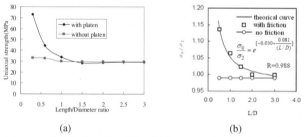

(a)                                                    (b)

Fig. 3. Relationship between uniaxial strength and length/diameter ratio of rock specimen with different sizes, with and without considering loading platen effect. (a) EPCA$^{2D}$ simulated result and (b) experimental result[11].

## 4.2. *Failure patterns of rocks*

With the same diameter, but different lengths, the final failure patterns of the simulated rock specimens, which consider the loading platen effect, are shown in Fig. 4. The existence of friction between loading platen and specimen's ends has a great influence on the stress distribution in the rock specimen. Near the specimen's ends, a triaxial compressive stress zone is formed. For rock specimen with larger length, the centre part of the specimen is less influenced by this compressive stress state. Therefore, a shear plane will be formed near the centre part of the specimen with larger length. The failure of shorter specimen is complex because the triaxial compressive stress zone may have influence on the entire specimen. As a result, spalling at the surface is the main failure pattern of rock specimens with shorter length. The bearing capacity falls down more slowly because the inner part of the specimen is hard to fail. This is the reason why shorter specimen has a higher uniaxial strength. Fig. 5 shows the failure pattern of rock specimens with different sizes, not considering the effect of loading platens. It also can be seen that complex failure is found for the specimens with shorter lengths (noting that this is pure compression, i.e. without the confining effect of the steel platen in the numerical tests). Many split surfaces are found along the axial direction. For specimens with a longer length, shear failure is the main failure pattern. However, depending on the heterogeneity of the rock specimens, a variety of failure patterns is possible.

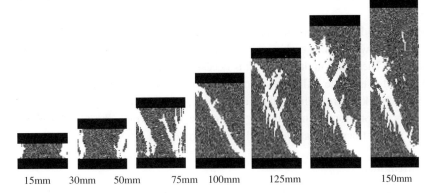

15mm    30mm    50mm    75mm    100mm    125mm    150mm

Fig. 4. Final failure patterns of rocks with the same diameter 50mm but different lengths with considering loading platens.

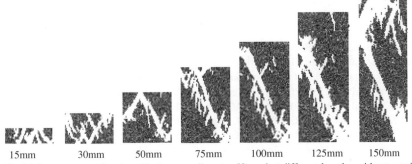

15mm        30mm        50mm        75mm        100mm        125mm        150mm

Fig. 5. Final failure patterns of rocks with the same diameter 50mm but different lengths, without considering loading platens.

## 5. Conclusions

Self-developed numerical code $EPCA^{2D}$ is used to study the influence of loading conditions on the strength and deformation behaviors of heterogeneous rocks. By using this code, the perfect loading conditions can be subjected on the specimen.

Numerical results indicate that, for rock specimens with the same homogeneity, the size and shape effects do not seem to be dominated by the heterogeneity of the rock specimens. With the same heterogeneity, if there is no friction between the specimen end and platen, no discernible size effect will occur. The existence of friction leads to the rock specimen being in a state of triaxial stress during the uniaxial compressive failure processes. The same heterogeneity implies that the properties of the rock specimens with different lengths conform to the same stochastic random distribution, which means that the rock specimens contain flaws with the same probability. This case of rock specimens with the same homogeneity does exist in experiment as the specimens are always drilled out from the same rock mass especially from the rock mass without naked flaws. Another case that the rock specimens have different homogeneity, i.e. larger specimens have more probability of containing flaws, is not considered in present paper and may be studied in future.

Present work on numerical simulation of rock failure processes is conducted in 2D framework. However, most physical problems in geotechnical engineering are three dimensional. Therefore, in the future, it is necessary to conduct failure processes of rocks in 3D space to study the influence of loading conditions on the deformation and strength behaviors of rocks more realistically. Furthermore, parallelism is one of the advantages of the cellular automaton. Hence, in the future, we will develop the parallel algorism in EPCA to realize finer simulation of rock failure processes.

## Acknowledgments

This work is supported financially by the National Natural Science Foundation of China (Grant No. 50709036) and State Key Laboratory of Geomechanics and Geotechnical Engineering Foundation (Grant No. O710121Z01)

## References

1. J. A. Hudson, E. T. Brown, C. Fairhurst. Proceedings of the 13<sup>th</sup> Symposium on Rock Mechanics, University of Illinois, Urbana, ASCE., 773 (1971).
2. P. C. Papanastasiou and I. G. Vardoulakis. *Int. J. Num. Ana. Meth. GeoMech.* **13**, 183 (1989).
3. Berthelot, J. M., and Robert, J. L. *J. Engng. Mech. ASCE.* **116**(3), 587 (1990).
4. Otsuka, K. and Date, H. *Engng. Frac. Mech.* **65**, 111 (2000).
5. Guinea, G. V., Elices, M., and Planas, J. *Engng. Frac. Mech.* **65**, 189 (2000).
6. Z. Bažant, Er-Ping Chen. *Appl. Mech. Rev.* **50**(10), 593 (1997).
7. Hodgson K, Cook N G W. In: 2nd Cong. Inter. Soc. Rock Mech., Begrade, **2**, 3 (1970).
8. Obert L, Windes S L, Duvall W I. *US Bureau of Mines Report of Investigations*, 3891 (1946).
9. X.T. Feng, P.Z. Pan and H. Zhou. *Int. J. Rock Mech. and Min. Sci.*, **43**(7), 1091 (2006).
10. P.Z. Pan, X.T. Feng and J.A. Hudson. *Int. J. Rock Mech. and Min. Sci.*, **43**(7), 1109 (2006).
11. S.Q. Yang, C.D Su, W.Y. Xu. *Engng. Mech.* **22**(4), 112 (2005).

# Part C
## Dynamic Loading and Crash Dynamics

# NUMERICAL SIMULATION OF DYNAMIC FAILURE FOR ALUMINA CERAMIC

REN HUILAN[1†], NING JIANGUO[2]

*State key laboratory of Explosion Science and Technology, Beijing Institute of Technology,*
*Beijing, 100081, P. R. CHANA,*
*huilanren@bit.edu.cn, jgning@bit.edu.cn*

Received 15 June 2008
Revised 23 June 2008

The plate impact experiments have been conducted to investigate the dynamic behavior of alumina. Based on the experimental observations, the three-dimensional finite element models of flyer and alumina target are established by adopting ANSYS/LS-DYNA, several cases were performed to investigate the fracture behavior of alumina target under impact loading. By analyzing the fracture mechanism and damage process of the alumina target, it is concluded that the nucleation and growth of great number of radial and axial cracks and circumferential cracks play a dominant role in the fracture behavior of alumina target. The stress histories of alumina target are simulated. By the comparison of experimental results with the numerical predictions, a good correlation is obtained.

*Keywords*: Alumina; dynamic fracture; numerical simulation.

## 1. Introduction

In the past three decades ceramics have been used armor applications. Compared to the other materials such as metal, they tend to be light weight, high compression strength, high Hugoniot elastic limit and excellent ballistic performance against projectile. These characteristics make ceramics well suited for armor applications. Research groups all over the world have investigated the high strain rate and ballistic properties of ceramics. The reason for this interest in alumina is because it offers moderate ballistic performance for relative small cost. Sintered alumina has been preferred, again for reason of cost. Because of the characteristic, impact velocities, impact tilted angle or the thickness of target and projectile, it is difficult and complicated to study the response of impact loading by the experimental methods under all of the loading conditions. In addition, under high velocity impact the target will be destroyed and comminuted so that they can't be recycled to be studied furthermore. Then numerical simulation are widely developed to study the ballistic impact response for alumina target[1-3].Numerical simulation provide insight into the response of alumina material and instruct the experimental plan to be designed.

---

†Corresponding Author.

In objective of this paper is to investigate the mechanical properties and fracture behavior of alumina by using plate impact experiments and numerical simulations. The three-dimensional finite element models of projectile and alumina target are established and some cases were performed to investigate the fracture behavior of alumina target under impact loading, which can not be observed directly from experiments.

## 2.  Plate Impact Experiment

The mechanical behaviors of the AD90 alumina were investigated by performing plate impact experiment using a 100mm diameter single-stage gas-gun. A schematic illustration of the impact configuration is shown in Fig.1 The high pressure is released suddenly to drive the projectile to move quickly along the gun pipe. The flyer plate mounted at the head of an aluminum projectile is used to impact the alumina target. The target is composed of five circular disks of alumina specimen, in which three manganin sensors are placed to record the stress histories at different locations of target. The specimens are 60mm in diameter and 3mm in thickness.

The experimental setup is designed so that a planar-parallel compressive shock-wave propagates through the target and flyer; the compression wave-profiles are measured with little interference from radial reflected waves. Compressive wave make the target reach the Hugoniot state. When the left-propagated compressive wave arrive at the left free surface of flyer, then reflected (release) wave propagates the target to make the specimen unloading from the Hugoniot state, as shown in Fig.2.

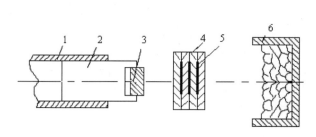

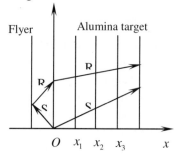

1-projectile support    2-light chamber 3-flyer

4-alumina target  5-manganin transducers 6-recycle bin

Fig. 1. Apparatus of one-stage light gas gun.

Fig. 2. Drawing of shock wave and release wave propagation.

## 3.  Numerical Model

### 3.1. *Material models and equations of state for flyer*

Constitutive model describe the mechanical behaviors of materials against the loading. The Johnson-Cook models are used widely under large strain, high strain rate and high temperature for the metal materials.

The yield criterion is described as given[4]

$$\sigma_y = (A + B\bar{\varepsilon}^{p^n})(1 + C\ln\dot{\varepsilon}^*)(1 - T^{*m})$$  (1)

where $\bar{\varepsilon}^p$ is equivalent plastic strain, $\dot{\varepsilon}^*$ is normal strain rate, $\dot{\varepsilon}^* = \dot{\varepsilon}/\dot{\varepsilon}_0, \dot{\varepsilon}_0 = 1.0s^{-1}$, $T^*$ is normal temperature, $T = \dfrac{T^* - T_{room}}{T_{melt} - T_{room}}$, $T$ is temperature, $T_{room}$ is room temperature, $T_{melt}$ is temperature of melted materials $A, B, C, n, m$ are the parameters of material.

The failure criterion for a grid is expressed as

$$D_f = \Sigma\frac{\Delta\varepsilon}{\varepsilon^f}$$  (2)

Where $\Delta\varepsilon$ is equivalent plastic strain during a cycle of integration, $\varepsilon^f$ is equivalent strain to fracture under current strain rate, temperature, pressure, equivalent stress.

The Mie-Gruneisen equation of state (EOS) is used in the simulation. It is of the form

$$P - P_{ref} = \frac{\Gamma}{V}(E - E_{ref})$$  (3)

Where $P$ is pressure, $E$ is specific energy, $V$ is specific volume;

For the reference state the shock Hugoniot is chosen, so $P_{ref}$ and $E_{ref}$ become

$$P_{ref} = P_H = \frac{\rho_0 C_0^2 \eta}{(1 - S\eta^2)}$$  (4a)

$$E_{ref} = E_H = \frac{P_H \eta}{2\rho_0}$$  (4b)

Using $K = C_0^2\rho_0$ and $\beta = \rho/\rho_0 - 1$, Eq.(4a) can be expressed as

$$P_H = K_1\beta + K_2\beta^2 + K_3\beta^3$$  (5)

Where $\eta$ is the compressibility, $\eta = 1 - \rho_0/\rho$, $C_0$ is the bulk sound speed, Eq.(6a) is derived assuming a linear relationship between shock speed $U$ and particle velocity $v$, given by $U = C_0 + Sv$. The parameters used in the model are listed by Table1.

Table 1. Parameters of copper flyer.

| Density $\rho/(kg \cdot m^{-3})$ | Elastic moduli $E/GPa$ | Shear moduli $G/GPa$ | Poisson ratio $v$ | Yield strength $Y_d/GPa$ | Wave velocity $C_0/(km \cdot s^{-1})$ |
|---|---|---|---|---|---|
| 8930 | 81.8 | 47.7 | 0.3 | 0.34 | 3.94 |

### 3.2. Material models and equations of state for alumina target

MAT_JOHNSON_HOLMQUIST_CERAMICS model is adopted to describe the dynamic response for alumina target. MAT_JOHNSON_HOLMQUIST_CERAMICS model is a plastic failure model for modeling ceramics, glass and other brittle material. The equivalent stress is given by

$$\sigma^* = \sigma_i^* - D_t\left(\sigma_i^* - \sigma_f^*\right)$$  (6)

Where

$$\sigma_i^* = a\left(p^* + t^*\right)^n \left(1 + c\ln\dot{\varepsilon}\right) \tag{7}$$

$$D_t = \sum \Delta\varepsilon^p / \varepsilon_f^p \tag{8}$$

$$\varepsilon_f^p = d_1\left(p^* + t^*\right)^{d_2} \tag{9}$$

$$\sigma_f^* = b\left(p^*\right)^m\left(1 + c\ln\dot{\varepsilon}\right) \le sf\,\max \tag{10}$$

$\sigma_i^*$ represents the intact, undamaged behavior, $\sigma_f^*$ represents the damaged behavior, and $D_t$ represent the accumulated damaged based upon the increase in plastic strain per computational cycle. Fracture occurs when the damage parameter $D_t$ reaches the value of 1. $\varepsilon_f^p$ represents the plastic strain. The parameters used in the model are listed by Table2.

Table 2.  Parameters  of alumina target .

| density $(kg \cdot m^{-3})$ | shear modulus G $(GPa)$ | Maximum tensile strength T $(GPa)$ | HEL $(GPa)$ | PHEL $(GPa)$ |
|---|---|---|---|---|
| 3625 | 90.16 | 0.2 | 5.3 | 3.6 |

## 4. Simulation Results

### 4.1. *Fracture characteristic*

Micro-cracks nucleate either at in homogeneities such as inclusions, grain interfaces, glass phase and reinforcements or at defects such as micro-cracks, and pores within the sintered ceramics under dynamic loading[5]. When tensional stress wave released from the free surface propagate the target, once tensional failure criterion are satisfied, the tensional damage will accumulate and fracture the target.

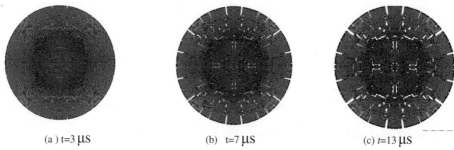

(a) t=3 μs          (b)  t=7 μs          (c) t=13 μs

Fig. 3.   Characteristic of fracture of the third alumina sample.

Fig.3 shows the fracture behaviors of alumina from numerical simulations. They are summarized as (1) Lots of radial cracks from the center of target nucleate and grow. (2) Circumferential cracks around center are occurred, as shown in Fig.3. Lots of interconnected cracks result in the fragments of alumina target. (3) Between the last piece of alumina sample and the others the distinguished difference is that the protruded figure is emerged because of fewer confined pressure on the side of last one and declination of shock wave.

## 4.2. *Comparison between the simulation and experiments*

Comparison is presented between the experimental results and the numerical results for stress histories in the alumina target at different locations, as shown in Fig.4. The good agreement is obtained at the peak stress and trend of shock wave declination. From these figures, it can be observed that the stress-time curves are nonlinear, which is related with the internal damage in the material. The maximum value of the stress and the dynamic yield strength increased as the strain rates were increased, which indicated that the alumina was a rate dependent material.

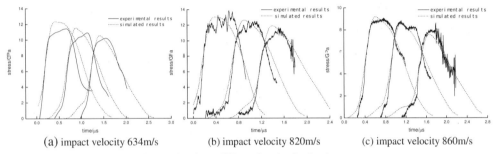

      (a) impact velocity 634m/s        (b) impact velocity 820m/s        (c) impact velocity 860m/s

Fig. 4. Stress histories at different location in alumina target.

Stress histories at different position with different distances from the center of target are shown to describe the influence of side (release) wave on stress wave propagation. From results of numerical simulations it is shown that, side (release) wave has more influence on stress wave propagation away form the center of target (24mm,26mm from the center of target),where stress peak is significantly lower than the peak value of experiment because of unloading of side (release) wave; and less influence near the center of target, where stress doesn't decline until (release) wave form free surface of flyer arrive, so that peak stress is comparable with experimental result. Stress distribution basically tends to be evenly at position within 20.7mm from the center of target and agree well with experimental result, as shown in Fig.5.

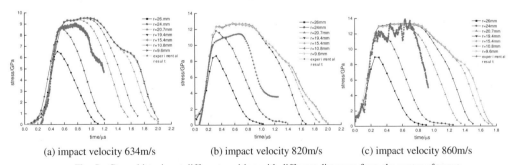

      (a) impact velocity 634m/s        (b) impact velocity 820m/s        (c) impact velocity 860m/s

Fig. 5. Stress histories at different position with different distances from the center of target.

## 5. Conclusions

Plate impact experiments were performed to investigate the dynamic behaviors of alumina. The stress-time curves were obtained under different strain rates. Based on experimental results, finite element model for flyer and alumina target were established by adopting LS-DYNA. Several cases were performed to investigate the fracture behavior of the alumina target under different impact velocities. It is concluded that the nucleation and growth of cracks play an important role in the macroscopic fracture of alumina target. A good agreement between numerical predictions and experimental results was obtained, which suggests that the finite element model is efficient and credible to simulate the mechanical properties of alumina.

## Acknowledgments

Supported by the National Natural Science Foundation of China (10772027, 10625208).

## References

1.   J. H. Timothy and W. T. Douglas, Constitutive modeling of aluminum nitride for large strain, *Inter. J. Impact Eng.* **25**, 211 (2001).
2.   A. M. Rajendran and D. J. Grove, *CMES,* **3(3)**, 367 (2002).
3.   Li Ping. Ph.D. Thesis, Beijing Institute of Technology, 2002.
4.   D. J. Steinberg, *Journal of Physics*, C3 suppl. 837 (1991).
5.   H. D. Epinosa and G. Raiser, *J. Hard Mater.* **3-4**, 285 (1992).

# ANALYSIS OF DYNAMIC STRESS-STRAIN RELATIONSHIP OF LOESS

HONGJIAN LIAO[1]

*Department of Civil Eng., Xi'an Jiaotong University, Xianning West Road 28, Xi'an, Shannxi,710049, CHANA,*
*Key Lab. for the Exploitation of Southwestern Resources & the Environmental Hazard Control Engineering of*
*Chinese Ministry of Education, Chongqing University, Chongqing, 400044, CHINA,*
*hjliao@mail.xjtu.edu.cn*

ZHIGANG ZHANG[2]

*Department of Civil Eng., Xi'an Jiaotong University, Xianning West Road 28, Xi'an , Shannxi,710049,CHANA,*
*Department of Engineering, Air force Engineering University, Xi'an, Shannxi, 710038, CHINA,*
*uxzzx1@163.com*

CHUNMING NING[3][†]

*Dept. of Capital Construction, Xi'an Jiaotong Univ., Xianning West Road 28, Xi'an, Shannxi 710049, CHANA,*
*cmning@xjtu.edu.cn*

JIAN LIU[4]

*Department of Civil Eng., Xi'an Jiaotong University, Xianning West Road 28, Xi'an, Shannxi, 710049,CHANA,*
*liujian0430@163.com*

LI SONG[5]

*Department of Mechanical Eng., Inha Univ., 253 Yong-hyun Dong, Nam Ku, Incheon 402-751, KOREA,*
*songli0125@hotmail.com*

Received 15 June 2008
Revised 23 June 2008

This paper aims to study dynamic properties of loess. This study is helpful to the subject on how to avoid or decrease the seismic disasters on loess ground. Dynamic triaxial tests are carried out with saturated remoulded soil samples taken form loess sites in Xi'an, China. Dynamic stress and strain relationship as well as the rule of the accumulated residual strain are obtained from the test results. Linear relationship between accumulated residual strain and vibration circle under constant amplitude circular loading is presented. A hypothesis about the accumulated residual strain is proposed. 1D dynamic constitutive relationship model which can well describe the real relationship between dynamic stress and strain under irregular dynamic loading is established. Numerical program with this model is developed and an example is tested. Numerical results of hysteresis loop, accumulated residual strain, amplitude of dynamic stress and damping ratio show good agreement with test results. It is indicated that the hypothesis of accumulated residual strain and the 1D dynamic constitutive relationship model can accurately simulate the dynamic triaxial tests of saturated remoulded loess.

*Keywords*: Loess; Constitutive relationship; Dynamic properties; Numerical analysis; Triaxial tests.

[†]Corresponding Author.

## 1. Introduction

Dynamic constitutive relationship of soil is usually regarded as the relationship between dynamic stress and strain under dynamic loading[1, 2]. Dynamic stress and strain is caused by circulate load of the external environment. An important aspect of constitutive relationship research is the study on mathematical model of constitutive relationship. Presently there are mainly two types of dynamic constitutive relationship models of soil. One is based on elastic-plastic mechanics and the other is established through the analysis of 1D dynamic charateristics of soil. It is easier to apply the latter in dynamic response analysis in the engineering practice. Some basic concepts of the latter are as follows[1, 4, 5].Under constant amplitude circular loading, dynamic stress-strain curve of soil can be obtained shown in Fig.1. Unloading curve ABC and reloading curve CDE are usully called as hysteresis-curve and the loop ABCDE is named as hysteresis loop. With different loading amplitudes, different hysteresis loops can be obtained and the curve linking the summits of these loops is called as backbone-curve which is shown in Fig.2.

1D dynamic constitutive relationship model can be established by assembling a backbone-curve model and a hysteresis-curve model. In this type of constitutive model, the relationship of dynamic stress and strain agrees with the backbone-curve in the initial loading process, and when unloading or reloading appears, the relationship agrees with the hysteresis-curve. Much research of this type of models has been carried out[6-11]. However, the accumulated residual strain is not considered in the research. Therefore, this paper focus on the subject of the effect of the accumulated residual strain on constitutive model. Dynamic triaxial tests of saturated remoulded loess are carried out and a hypothesis of the accumulated residual strain is proposed. Based on the hypothesis of the accumulated residual strain, Hardin-Drnevich backbone-curve model[4, 5] and a model of hysteresis-curve[6-11], 1D dynamic constitutive relationship model is established in this paper. Numerical program for the established model is developed and verified by an practical example.

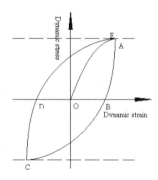

Fig. 1. Hysteresis-curve.

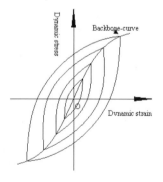

Fig. 2. Backbone-curve.

## 2. Backbone-Curve Model and Hysteresis-Curve Model

Backbone-curve equation is written as follows:

$$\sigma_d = \frac{\varepsilon_d}{A + B|\varepsilon_d|} \tag{1}$$

Where $\sigma_d$ is axial dynamic stress, $\varepsilon_d$ is axial dynamic strain, $1/A$ is the initial modulus $E_0$ and $1/B$ is the dynamic stress ultimate value $\sigma_{ult}$. $E_0$ and $\sigma_{ult}$ can be obtained by experimental dynamic triaxial tests.

The hypothesis based on which hysteresis-curve is constructed is as follows: in the unloading or reloading process, hysteresis-curve passes through the maximum-stress point in the loading history, and hysteresis-curve is closer to the dynamic stress ultimate value when dynamic strain increases. According to the hypothesis, hysteresis-curve equation is obtained as follows:

$$\sigma_d = \frac{\varepsilon_d - \varepsilon_0}{A + nB|\varepsilon_d - \varepsilon_0|} + \sigma_0 \quad \text{or} \quad \varepsilon_d = \frac{A(\sigma_d - \sigma_0)}{1 - nB|\sigma_d - \sigma_0|} + \varepsilon_0 \tag{2}$$

$$n = -\frac{A}{B|state \cdot \varepsilon_{d,\max} - \varepsilon_0|} + \frac{1}{B|state \cdot \sigma_{d,\max} - \sigma_0|} \tag{3}$$

Where $\sigma_d$ is axial dynamic stress, $\varepsilon_d$ is axial dynamic strain, $\sigma_0$ is dynamic stress of the transforming point of loading state, $\varepsilon_0$ is dynamic strain of the transforming point of loading state, n is a parameter which ensure hypsteresis-curve satisfying the hypothesis and needs to be recounted when the loading state changes, state is a parameter to record the loading state(reloading 1, unloading -1), $\sigma_{d,max}$ is the maximum stress in the loading history, and $\varepsilon_{d,max}$ is the maximum strain in the loading history.

## 3. Dynamic Triaxial Tests

Loess samples are taken from loess sites in Xi'an, China. And it is Malan loess. Natural density of the soil is 1.986 $g \cdot cm^{-3}$ and dry density is 1.665 $g \cdot cm^{-3}$. Moisture content is 19.3%. Plastic limit is 20.81% and liquid limit is 30.14%. Cylinder samples with diameter 39.1mm and height 80.0mm are made. Constant amplitude harmonic loading is applied axially and its frequency is 1 Hz.

Relationship between accumulated residual strain and vibration circle is obtained shown in Fig.3. It is obvious that accumulated residual strain is linearly related to vibration circle. Based on this viewpoint, a hypothesis of accumulated residual strain is proposed in the following part of this paper.

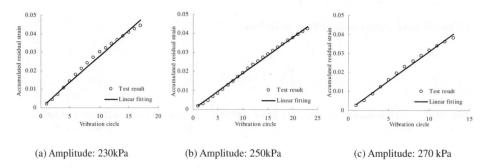

(a) Amplitude: 230kPa    (b) Amplitude: 250kPa    (c) Amplitude: 270 kPa

Fig. 3. Relationship between accumulated residual strain and vibration circles for different amplitudes of dynamic stress.

## 4. Hypothesis of Accumulated Residual Strain

A hypothesis of accumulated residual strain is proposed in this paper. It is as follows: in the dynamic process, when an increment of dynamic compressive strain occurs, a part of the increment is unrecoverable and can not cause stress increment; in other words, this part decreases the tangent modulus of soil; the proportion coefficient of this part in the strain increment is named as coefficient of residual strain.

The hypothesis can be realized through mobile reference axis. The concrete method is as follows:

1. Firstly, it is assumed that compressive stress is positive stress and compressive strain is positive strain; $\Delta\varepsilon_i=\varepsilon_i-\varepsilon_{i-1}$ is the increment of dynamic strain; $\Delta\sigma_i=\sigma_i-\sigma_{i-1}$ is the increment of dynamic stress; $p$ is the coefficient of residual strain.

2. When we use test dynamic strain to calculate dynamic stress, the following process is adopted. When $\Delta\varepsilon_i>0$, $\Delta\varepsilon_i$ consists of two parts: one part is residual strain $p\Delta\varepsilon_i$ which causes an movement of reference axis and the other part is $(1-p)\Delta\varepsilon_i$ which is recoverable and causes a dynamic stress increment $\Delta\sigma_i$ which is calculated according to the equations of constitutive model. When $\Delta\varepsilon_i<0$, there is not residual strain, and $\Delta\sigma_i$ is computed according constitutive equations with $\Delta\varepsilon_i$.

3. When we use test dynamic stress to calculate dynamic strain, the following process is adopted. When $\Delta\sigma_i >0$, a calculating strain $\Delta\varepsilon'_i$ can be gained and the real increment of dynamic strain is $\Delta\varepsilon_i =(1+p)\Delta\varepsilon'_i$; and a part of the increment is the residual strain $p\Delta\varepsilon'_i$ which causes an movement of reference axis. When $\Delta\sigma_i <0$, there is not residual strain, and $\Delta\varepsilon'_i$ is just the real increment of dynamic strain $\Delta\varepsilon_i$.

4. When the 1D dynamic constitutive model is applied in the numerical analysis of dynamic response, the hypothesis is realized by changing the tangent modulus: When current step is positive loading or reloading, tangent modulus $G_t$ changes to $G_t/(1+q)$; otherwise, $G_t$ remains constant.

It can be found that when the hypothesis proposed in this paper is used in the dynamic constitutive relationship model, equations of the constitutive model do not need to be changed and what needs to be changed is only the reference axis. Therefore, it is convenient to consider accumulated residual strain in the constitutive model with this hypothesis.

## 5. Analysis of Example

Backbone-curve model, hysteresis-curve model and hypothesis of accumulated residual strain constitute a total 1D dynamic constitutive model. Numerical program is developed to realize the constitutive model. In the program, dynamic stress can be obtained by inputing test dynamic strain, in addition dynamic strain can be obtained by inputing test dynamic stress.

An example is carried out. Comparison of hysteresis loops is shown in Fig. 4. It can be found that test hysteresis loops and computed hysteresis loops are similar in Fig. 4. Residual strain, amplitude of dynamic stress and damping ratio are compared in Fig. 5. Likewise, test results and computed results of accumulated residual strain, amplitude of dynamic stress and damping ratio are coincident.

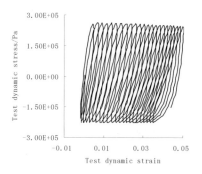

(a) Test result of dynamic stress-strain

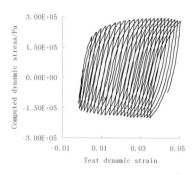

(b) Result of computed dynamic stress with test strain

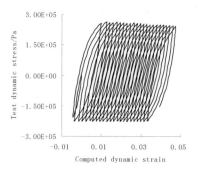

(c) Result of test dynamic stress with computed strain

Fig. 4. Comparison of hysteresis loops.

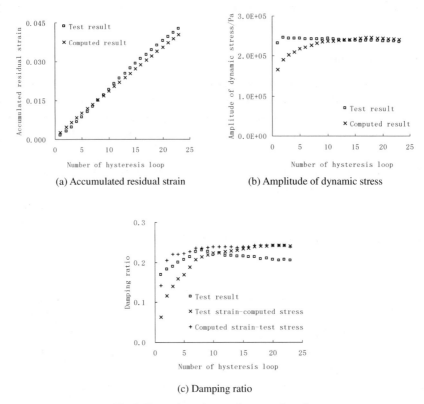

(a) Accumulated residual strain          (b) Amplitude of dynamic stress

(c) Damping ratio

Fig. 5. Comparison of test and computed results.

## 4. Conclusions

In this paper, through dynamic triaxial tests of saturated remoulded loess, it is found that accumulated residual strain is linearly related to vibration circle. Based on this viewpoint, a hypothesis of accumulated residual strain is proposed. 1D dynamic constitutive model of loess is established and it consists of Hardin-Drnevich backbone-curve model, a hysteresis-curve model and hypothesis of accumulated residual strain. Constitutive model is introduced to the developed numerical program. Numerical results of hysteresis loop, accumulated residual strain, amplitude of dynamic stress and damping ratio show good agreement with test results. It is indicated that the hypothesis of accumulated residual strain and the 1D dynamic constitutive model can accurately simulate the dynamic triaxial tests of saturated remoulded loess.

## Acknowledgments

This paper are funded by the national natural science foundation of China (50379043) and key laboratory for the exploitation of southwestern resources & the environmental hazard control engineering (Chongqing university) of Chinese ministry of education.

# References

1.  Xie Dingyi. *Soil Dynamics* (Xi'an Jiaotong University Press, Xi'an, 1988).
2.  Zheng Yingren, Shen Zhujiang, Gong Xiaonan. *Fundamental Principles of Geotechnical Plastic Mechanics* (China building industry press, Beijing. 2002), p.209.
3.  Kong Liang, Wang Yanchang, Zheng Yingren. *Development in Soil Dynamic Constitutive Model. J. Ningxia university, Natural science ed.* **22(1)**, 17 (2001).
4.  Hardin BO, Drnevich VP. *Shear modulus and damping in soils: measurement and paramenter effects, J. soil mech. Found. Eng. Div. ASCE*, **98(6)**, 603 (1972).
5.  Hardin BO, Drnevich VP. *Shear modulus and damping in soils: design equations and curves. J. soil mech. Found. Eng. Div. ASCE*, **98(7)**, 667 (1972).
6.  Wang Lanmin. Loess Dynamics (Earthquake engineering press, Beijing. 2003), pp.1~60.
7.  Zhang Kexu, Li Mingzai, Wang Zhikun. *Dynamic Elactic-Plastic Models of Soils Based on Non-mansing's Rule, Earthquake eng. and eng. Vib.* **17(2)**, 74 (1997).
8.  Wang Zhiliang, Han Qingyu. *Analysis of Wave Propagation for the Site Seismic Response, Using the Visco-Elastoplastic Model, Earthquake eng. and eng. Vib.* **1(1)**, 117 (1981).
9.  Li Xiaojun. *An Simple Function Expression of Dynamic Constitutive Relationship of Soil, Yantu Gongcheng Xuebao*, **14(5)**, 90 (1992).
10. Ruan Maotian. *Ramberg-Osgood Constitutive Model with Variable Parameters for Dynamic Nonlinear Analysis of Soils, Earthquake eng. and eng. Vib.* **12(2)**, 69 (1992).
11. Li Xiaojun, Liao Zhenpeng, Zhang Kexu. *A Functional Formula of Dynamic Skeleton Curve Taking Accout of Damping Effect, Earthquake eng. and eng. Vib.* **14(1)**, 30 (1994).

# STUDY OF CUTTING AERIAL PMMA WITH MINIATURE DETONATION CORD

Z.Q. LI[1], L.M. ZHAO[2†], Y.G. ZHAO[3]

*Institute of Applied Mechanics, Taiyuan University of Technology, Taiyuan, 030024, P. R. CHINA,*
*lizhiqiang@tyut.edu.cn, zhaolm@tyut.edu.cn, ygzhao@126.com*

Received 15 June 2008
Revised 23 June 2008

In order to shorten the time of through-the-canopy-ejection, and to ensure pilot safely escape and survive. The application of linear cutting technique using miniature detonation cord( MDC) in through-the-canopy-ejection-system is proposed. A series of different kinds of MDC are designed. Firstly experimental study on the cutting process of the PMMA plate with MDC is carried out. Material of metal cover explosive types and the range of charge quantities are determined. Consequently the phenomena of spallation is observed, and the relationship between the cutting depth and charge quantities is obtained. For the comparison, the process of explosion cutting PMMA plate is simulated by means of nonlinear dynamic analysis code LS-DYNA. Spallation phenomena which occurs in the experiment, is also observed in the simulation. Simulation results present the relationship of cutting depth of PMMA plate versus charge linear density, which well agree with experimental ones.

*Keywords:* Explosion cutting; MDC; finite element analysis; spallation.

## 1. Introduction

Explosion cutting is an important approach developed in recent year, which is widely used in civil and military engineer area. Explosion cutting technic is applicable to special work situation, where normal mechanic industry methods such as saw cutting, grinding, gas cutting, turning can not be executed. All kinds of linear shaped charge, called cutting explosives, are manufactured to enhance cutting quality. It is well known that high-speed piece shaped jet due to the metal cover under blasting load can cut steel plate. In practice, diversities of cutting explosive can satisfy the requirements of every special work condition [1]. The effect of cutting directly depends on the types of explosives and charge quantities. Based on the detonation principle and uncompress fluid related theory, more high detonation velocity produces more high jet velocity and better cutting effect. In the case of invariant explosive and charge, its cutting depth is determined by material of metal cover, vertex angle, explosive height, wall thickness of metal cover, charge density, with cover and without cover etc. Special application requirement, which is the cutting depth must be exactly controlled and spallation section occurs at the opposite side of MDC, has been provided in the engineering field. It is a case that PMMA is cut in the

---

†Corresponding Author.

aerial escape system[2].Aerial PMMA is a kind of high molecular material, which strength is far lower than metal and rock. It has the feature of spallation after directional process. MDC with appropriate charge is blasted to generate detonation waves acting on the surface of PMMA. When the detonation waves arriving to free surface are reflected to be tension stress waves, whose value reaches failure strength of PMMA, spallation damage occurs and accompanies exactly controlled cutting depth at the opposite of MDC. Therefore, spallation is a principal failure form of cutting directional PMMA with MDC. It is also material internal structure failure under strong impact loading.

Study on the explosion cutting PMMA has been hardly reported to date at home and oversea. It is very complicate nonlinear dynamics problems being microsecond order of magnitude, so it is very difficult to establish a theory model to analysis the process of cutting. With the improvement of finite element method and the rapid development of the computer, large nonlinear dynamic analysis codes such as LS-DYAN, DYTRAN are developed to fulfill the computer simulation of cutting process based on the experiment, which save development time and reduce cost.

## 2. Experiment

### 2.1. *The description of experiment*

The main goal of the experiment is to obtain the relationship of PMMA cutting depth with charge linear density by using apparatus shown in Fig.1, which consists of semicircular MDC, PMMA and fixture. Semicircular MDC consists of explosive, metal cover and protective, shown in Fig.2. Geometry of Metal cover is shown in Fig.3, where $L = 2 - 2.2mm$, $H = 0.9 - 1.2mm$, $d_1 = 0.2mm$, $d_2 = 0.1 - 0.3mm$. The main effects on linear cutting PMMA plate with MDC depend on the type of explosive, the charge quantities, the material of metal cover, the igniting point of explosive and boundary support condition of PMMA. Therefore, three types of explosives as RDX and HMX and HNS , two kinds of metal cover as Plumbum and Aluminum, and various charge quantities from 0.6g/m to 1.4g/m, are choosed to find proper explosive, metal cover and the range of charge quantities in the experiment. Forty MDC are manufactured through different combination. No.3 directional PMMA often used as canopy of warcraft is selected, and its size is 180mm long, 140mm wide, 7mm thick. Protective is employed to protect MDC, and it is made of XM-23 sealing gum. MDC, protective and PMMA are adhered with XM-23 adhesive. Sulfurized MDC is symmetrically displaced at the center of PMMA. According to different charge linear density, exploratory experiments are performed forty-five times by using end detonation at room temperature at a Chinese Research Institute, certified to qualify to make blast test. Explosion cutting process is recorded by high-speed pickup camera made in Japan , which is able to capture 100 thousand images every second.

### 2.2. *Experiment results*

The spallation of PMMA at the opposite side of explosive is shown in Fig. 4. Here, two photos from high speed pickup camera and digital camera, respectively, indicate that spallation phenomena with some fragments occurrs at the free surface of PMMA, and PMMA is cut. The cutting depth of PMMA is measured after experiment and listed in Table 1 through statistical process. Detonation velocity relates with charge quantities and the diameter of MDC. In the experiment, the value of detonation velocity is approximate from 7000m/s to 8000m/s provided for numerical simulation by means of electro probe method.

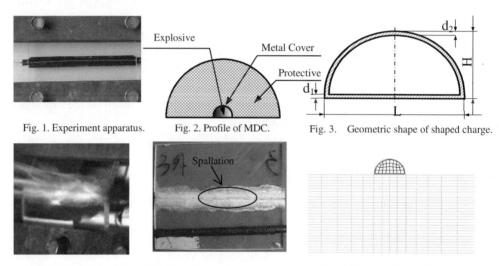

Fig. 1. Experiment apparatus.    Fig. 2. Profile of MDC.    Fig. 3.    Geometric shape of shaped charge.

Fig. 4. The Spallation of No.3 specimen.    Fig. 5. FE model (front view).

Table 1. The relationship of cutting depth(mm) with linear charge density(g/m).

| Dens. Expl. Dep. | 0.60 | 0.70 | 0.83 | 0.90 | 1.01 | 1.07 | 1.22 | 1.39 |
|---|---|---|---|---|---|---|---|---|
| HNS(Pb) | 6.00 | 6.21 | 6.50 | 7.00 | | penetration | penetration | |
| HNS(Al) | | | | 3.90 | | 3.90 | 5.98 | 7.00 |
| HMX(Pb) | | | 7.00 | 7.00 | 7.00 | | | |
| RDX(Pb) | | | 4.60 | 5.40 | 5.39 | | 7.00 | |

In spite of experimental scatter, it can also be seen that HNS with Pb cover is more great penetration capability than the other two explosives. Besides, HNS is a kind of heat resisting explosive with the feature of low mechanic sensitiveness, small electrostatic sensitiveness and good radioresistance and low toxicity. Therefore, it is very suitable to be used as MDCof escape system. It is found that various charge linear density from 0.6g/m to 0.9g/m is the best range. The conclusion is verified in the computer simulation below.

## 3. Numerical Simulation

The numerical model consists of three parts, Fig.5, and is solved by using the nonlinear, explicit FE code LS-DYNA(version ls-970). No.3 directional PMMA and the HNS explosive are modeled by the brick element. Pb cover is modeled by the default shell element using five section points. Hourglass control is used for shells and bricks. The element size is 0.25mm for HNS and uniform through the structure. The element size of PMMA, some 0.35mm, must be enough small to successfully simulate the spallation phenomena. The total numbers of elements and nodes are 150,000 and 155,000, respectively. XM-23 sealing gum is not considered in the FE model. Therefore, it is approximate agreement between model and experiments. The Pb cover and PMMA are represented by Lagrangian element formulation, while the HNS is represented by Arbitary Lagrangian Eulerian (ALE) element formulation. Automatic-surface-to-surface contact options is generally employed. Only slide contact option is used to complete fluid–structure interaction between the HNS and PMMA plate. The two sides of PMMA parallel to HNS length direction are fixed.

### 3.1. *Material model*

According to the mechanics properties of PMMA presented in the experiment, it is modeled by material type 105 of LS-DYNA (*MAT_DAMAGE_2). This is an elastic vi-scoplastic material model combined with continuum damage mechanics (CDM), seeing Berstad, Hopperstad, et al.[4]. For PMMA, material parameters are used: $\rho = 1.19 g/cm^3$, $E = 3.6 GPa$, $v = 0.4$, $\sigma_s = 69.10 MPa$ and discrete stress-strain relationship shown in Table 2.

HNS is modeled by material model 8 (MAT_HIGH_EXPLOSIVE_BURN), which requires JWL equation of state to define the pressure-volume relationship. The equation of state defines pressure as a function of relative volume, $V$ and internal energy per initial volume, $E$

$$p = A\left(1 - \frac{\omega}{R_1 V}\right)e^{-R_1 V} + B\left(1 - \frac{\omega}{R_2 V}\right)e^{-R_2 V} + \frac{\omega E}{V}. \tag{1}$$

Where $\omega$, $A$, $B$, $R_1$ and $R_2$ are user defined input parameter given by Dobratz [5] for a variety of high explosive materials. Here, input parameters $A = 5.409$, $B = 0.094$, $R_1 = 4.6$, $R_2 = 1.35$, $\omega = 0.35$, $E = 0.1$, $V = 1.0$ are used in the simulation. For HNS, detonation velocity, mass density and Chapman-Jouget pressure ($P_{CJ}$) corresponding to different charge linear density are used. Pb cover is modeled by material model 3 ( *MAT_PLASTIC_KINEMATIC) with failure.

### 3.2. *Numerical results*

The computer simulation is performed in HPJ6750 2CPU workstation. Four different HNS charge linear density from 0.6g/m to 0.9g/m are simulated using the model above.

The propagation process of detonation wave along explosive length direction from start point to end point needs some 27us, so the physical termination time is set as 30us. Many results are obtained from the simulation. Here, our interesting results are only given below.

Fig.6 shows the sequence of deformation behavior of PMMA corresponding to different linear density after explosion cutting terminated. It can be clearly seen spallation phenomena occurs at the free surface of PMMA, as is similar to the experiment observation. By numbering the spallation layers from free surface to the location of

Table 2. Stress-strain relationship.

| Strain ($\varepsilon$) | 0.020 | 0.023 | 0.026 | 0.030 | 0.033 | 0.038 | 0.058 |
|---|---|---|---|---|---|---|---|
| Stress (MPa) | 69.10 | 70.90 | 71.70 | 72.60 | 73.91 | 74.30 | 74.78 |

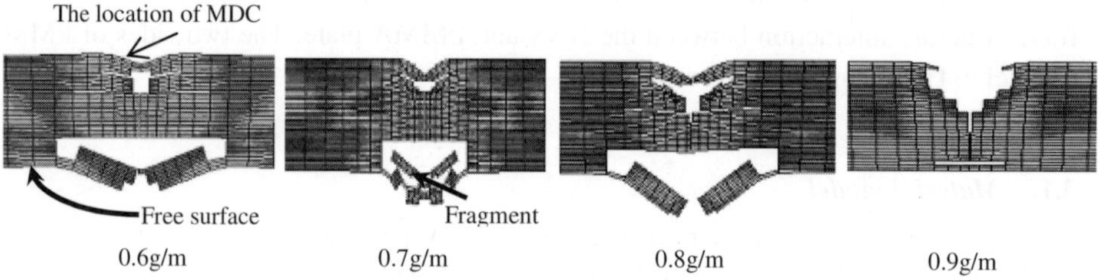

Fig. 6. Sequence from PMMA plate based on different linear density.

MDC, the cutting depth of PMMA plate corresponding to linear density 0.6g/m, 0.7g/m, 0.8g/m and 0.9g/m is 6.0mm, 6.2mm, 6.4mm and 7mm, respectively. The simulation results are good agreement with experimental ones.

## 4.   Conclusions

The study has shown the potential of using the finite element method in testifying spallation phenomena in explosion cutting experiments. From a lot of simulation results, the relationship of cutting depth with charge linear density is obtained, which provides accurate control over cutting depth in use of charge quantities for engineering technologist. In addition, the simulation with boundary condition and without boundary condition is made respectively. As a result, the effect of boundary conditions can be neglected for transient problem being microsecond order of magnitude. The work also provides strong references for the cutting experiment on canopy of aircraft in the future.

## Acknowledgments

Author is very grateful to experiment participants. The work is sponsored by National Natural Science Foundation of China (10672112), Shanxi Province Youth Science Foundation of China (200703005).

# References

1. Q.C. Song, S.H.Jin, W.Li, Chinese Journal of Explosion and Shock Waves, **17**, 382 (1997).
2. W. Wang, Aircraft Design, **3**, 28 (1985).
3. J.O. Hallquist, LS-DYNA theory manual-Version 970 (LSTC, CA, 2003).
4. T. Berstad, O.S.Hopperstad, O.G.Lademo and K.A.Malo, second European LS-DYNA Conference, (Gothenburg, Sweden, 1999).
5. B.M.Dobratz, LLNL explosives handbook, Lawrence Livermore national laboratory, Rept. UCRL-529997 (1981).

# A STUDY ON THE BEHAVIOR OF A THIN STS 304 SHEET WITH A FREE BOUNDARY CONDITIONS SUBJECTED TO IMPACT LOADING

D. G. AHN[1†], G. J. MOON[2]

*Department of Mechanical Engineering, Chosun University, 375 Seosuk-dong, Gwang-ju, 501-759, KOREA,*
*smart@mail.chosun.ac.kr,breakgun@hanmail.net*

C. G. JUNG[3], D. Y. YANG[4]

*Department of Mechanical Engineering, KAIST, 335 Science Street, Yoosung-gu, Deajeon, 305-701, KOREA,*
*cgj@kaist.ac.kr, dyyang@kaist.ac.kr*

Received 15 June 2008
Revised 23 June 2008

The objective of this paper is to investigate into impact behaviors of a STS 304 sheet with a thickness of 0.7 mm in a free boundary condition subjected to impact loading by a hemispherical impact head using drop impact tests and the three-dimensional FE analyses. The drop impact tests and the FE analyses were conducted with different impact energy ranging from 37.0 J to 45.7 J. From the results of the impact tests, the influence of the impact energy on the force-deflection curve, the absorption mechanism of the impact energy and deformation behaviors of specimen were examined quantitatively. Through the FE analyses, the variation of stress-strain distributions and characteristics of the local deformation during the impact of the specimen were investigated.

*Keywords*: Impact behavior; STS 304 sheet; free boundary condition; impact energy absorption.

## 1. Introduction

Because STS 304 stainless steel has several advantageous characteristics, such as high corrosion resistance, excellent fire resistance and a good weldability, the STS 304 stainless steel has been widely used as a structure material in chemical plants, steel plants, nuclear power plants, and ocean structures.[1] Recently, STS 304 stainless steel has been utilized in the components of transport vehicles, including automobiles, railway vehicles and ships, due to its good formability, the excellent ductility and high work hardening characteristics.[2] Moreover, the importance of the crashworthiness of transport vehicles has been emphasized in an effort to improve the safety of passengers.[3] Hence, various studies of the impact and high strain rate characteristics of structure materials of the transport vehicles have been actively undertaken in effort to improve the dynamic structural integrity of these vehicles.[4-5] Lepareux et al. investigated the dynamic response of stainless steel plates with a range of thickness of 4 - 10 mm using ballistic impact tests and FE analyses.[6] Radford et al. studied the response of clamped stainless steel and

---

†Corresponding Author.

sandwich beams through ballistic impact tests.[7] Previous research works related to the impact characteristics of structural materials have been focused on the impact behaviors of materials subject to fixed or clamped boundary conditions to examine the dynamic fracture characteristics of the materials. However, the absorption mechanism of the crash energy of the transport vehicles is almost close to the free boundary crash phenomenon with a successive wrinkling of the materials.

The objective of this paper is to investigate into impact behaviors of STS 304 stainless sheet with a thickness of 0.7 mm in a free boundary condition subjected to impact loading using by a hemispherical impact head through drop impact experiments and three-dimensional FE analyses. From the results of the impact tests, the influence of the impact energy on the force-deflection curve, the absorption mechanism of the impact energy and the deformation behaviors of the specimen in a free boundary condition were examined quantitatively. Through the FE analyses, the variation of the stress-strain distributions and characteristics of the local deformation during the impact of the specimen were investigated.

## 2. Experiments and Finite Element Analysis

### 2.1. *Impact tests*

In order to examine the impact response of STS 304 stainless steel sheet subjected to a free boundary condition, several impact tests are carried out using a drop impact testing machine with micro-processor based signal processing system. Fig. 1 shows the empirical set-up of the drop impact tests.

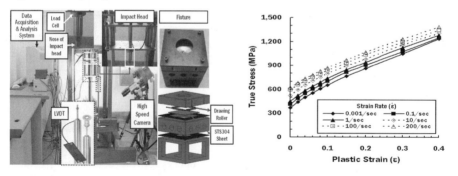

Fig. 1. Experimental set-up and fixture of impact tests.  Fig. 2. Stress-strain relationships with strain rate effects.

The contact force between the impact head and the specimen is measured by the load cell that is attached to the top side of the impact head. The deflection of the specimen is measured via the LVDT. Micro-processor based data acquisition devices can acquire $1 \times 10^4$ ea/sec of the force and the deflection data simultaneously. An impact head with a hemispherical nose shape is used in the impact tests. The diameter of the impact head is 20 mm. A drawing-type fixture with drawing rollers is designed to generate a free boundary condition on the specimen, as shown Fig. 1. The drawing roller draws the specimen in the hole of the fixture without rebounding by the specimen. The dimension of the specimen is 120 mm (W) × 120 mm (L) × 0.7 mm (t). The weights of the specimen and the impact head are 78.3 g and 11.1 kg, respectively. The impact energy and the impact velocity of the tests range 37.2-45.7 J and 2.5-2.9 m/sec, respectively. The absorbed impact energy by the specimen is estimated by a direct integration of the force-deflection curve.

### 2.2. *Finite element analyses*

The stress distributions, the strain distributions and the local deformation behaviors of STS 304 stainless steel sheet according to the impact energy are simulated by a three-dimensional finite element analysis using the explicit commercial code ABAQUS V6.5. In the FE analysis, the impact head and the fixture are assumed as analytical rigid surfaces. Rectangular shell elements with four nodes are used as the mesh of the specimen. The impact head, the fixture and the specimen are represented by 4,986 elements and 5,164 nodes. In the impact analysis, the nose diameter of impact head and the friction coefficient are set to be 20 mm and 0.12, respectively. Young's modulus, Poisson's ratio and the yield strength of the STS 304 stainless steel are measured by tensile tests. *Table 1.* shows chemical compositions and mechanical properties of the specimen. The strain-stress relationships of the STS 304 with the effects of the strain rate are measured through high-speed tensile tests, as shown in Fig. 2. The piecewise linear plasticity model is adopted to directly apply the measured strain-stress relationships with the effects of strain rate to finite element analysis.

Table 1. Chemical compositions (wt %) and mechanical properties of the STS 304 stainless steel sheet.

| Fe (wt %) | C (wt %) | Si (wt %) | Mn (wt %) | P (wt %) | S (wt %) | Ni (wt %) | Cr (wt %) | Young's Modulus (GPa) | Poisson's ratio | Yield Strength (MPa) |
|---|---|---|---|---|---|---|---|---|---|---|
| 68.6 | 0.08 | 1.0 | 2.0 | 0.045 | 0.03 | 9.2 | 19.0 | 180.0 | 0.29 | 283.0 |

## 3.  Results and Discussions

### 3.1.  *Results of impact tests*

Figs 4 and 5 show the influence of the impact energy on the deformation behaviors, force-deflection curves and the absorbed impact energy-deflection curves of the STS 304 stainless steel sheet subjected to a free boundary condition, respectively. Fig. 4 shows that the dominant failure mode of the specimen subjected to the free boundary condition is the mixed mode with a depression of the impact area and a one-sided wrinkle in the center line. From this result, it was noted that the compressive stress is induced in the wrinkled region and the tensile stress is formed on the opposite side of the wrinkled region. The Fig. 4 also shows that the edge width of the wrinkle decreases as the impact energy increases. The force-deflection curves show that several small peak forces take place after the first peak force occurred, as shown Fig. 5. This may be ascribed to the successive wrinkling of the specimen and the continuous work hardening of the wrinkled area during the collision of the impact head with the specimen. In addition, the Fig. 5 shows that the maximum load and the maximum deflection of the specimen increase simultaneously as the impact energy increases. In the Fig. 5, it was observed that that the absorption rate of the impact energy decreases when wrinkling of the specimen initiates. This is due to the fact that the absorption mechanism of the impact energy changes from the drawing and stretching of the specimen to the drawing and wrinkling of the specimen. In addition, it was shown that absorbed impact energy and the absorption rate of impact energy of the specimen lie in the ranges of 29.3-38.2 J and 79.2-84.5 %, respectively.

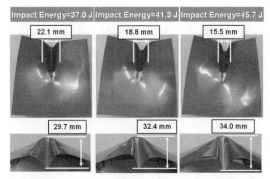

Fig. 4. Variation of the deformed shape according to different impact energies.

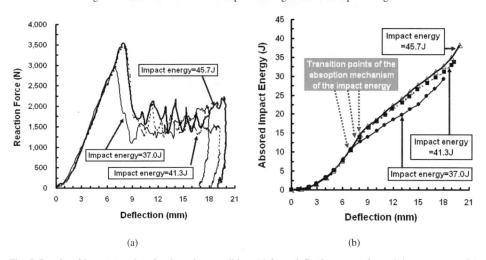

(a)　　　　　　　　　　　　　　　(b)

Fig. 5. Results of impact tests in a free boundary condition: (a) force-deflection curves for each impact energy: (b) absorbed impact energy-deflection relationships for each impact energy.

## 3.2.  *Results of FE analyses*

Figs. 6 and 7 show the results of FE analyses. The results of the FE analyses were compared to those of the impact tests to obtain a proper FE model, as shown in Fig. 6. From this comparison, it was found that the FE model can appropriately simulate the behaviors of STS 304 stainless steel sheet subjected to an impact load in a free boundary condition.

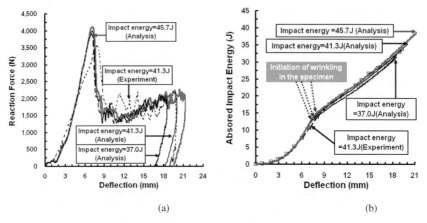

(a)                                                                     (b)

Fig. 6. Results of FE analysis and comparison of the results of FE analyses with those of impact tests: (a) Force-deflection curves: (b) absorbed impact energy-deflection relationships.

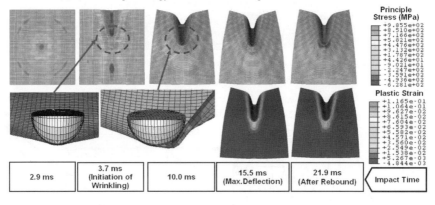

Fig. 7. Deformed shapes, principle stress and plastic strain distributions of the specimen (Impact energy= 45.7J).

Fig. 7 shows the variation of deformed shapes, principle stress and plastic strain distributions of the specimen according to the impact time. In Fig. 7, it was seen that the specimen draws symmetrically in the hole of the fixture before the initiation of the wrinkle in the specimen, but on the other hand, asymmetric deformation of the specimen with compressive stresses in the wrinkled area and tensile stresses opposite to the wrinkled area takes place after the occurrence of the wrinkling. Through a comparison of Fig. 6 with Fig. 7, it was found that the rapid decrease of the reaction force after the first peak of the force in the force-deflection curves appears due to the reduction of the contact area between the impact head and the specimen induced by the wrinkling of the specimen. In addition, it was observed that the fluctuation of the force in the force-deflection curves is caused by the successive contact and detachment of impact head and the specimen due to the growth of the wrinkle. From the Fig. 7, it was noted that the increase of the reaction force in the vicinity of the maximum deflection, as shown in

Fig. 6, results from a secondary plastic deformation of the specimen in the contact area of the impact head with the wrinkle. The results of the FE analysis show that the maximum principle stress and the maximum plastic strain of the specimen lie in ranges of 958.4-985.5 MPa and 0.094-0.117, respectively. Through a comparison of the maximum stress and plastic strain with the strain-stress relationship, as shown in Fig. 2, it was shown that the specimen does not fractured in the test range. From the results of the FE analyses, it was confirmed that the absorption mechanism of the impact energy in the free boundary condition changes from drawing and stretching of the specimen to an asymmetrical drawing of the specimen after the wrinkle in the specimen initiates.

## 4. Conclusions

This study examines the impact behaviors of STS 304 stainless steel sheet with a thickness of 0.7 mm in a free boundary condition subjected to impact loading by a hemispherical impact head through drop impact experiments and three-dimensional FE analyses. The results of the impact tests showed that the dominant failure mode of the STS 304 stainless steel sheet in the free boundary condition is a mixed mode with a depression in the impact area and a one-sided wrinkle on the center line of the specimen. From the results of the impact tests, force-deflection and the absorbed impact energy-deflection relationships of the STS 304 stainless steel sheet with for different impact energy were obtained. The results of the FE analysis showed that the behavior of the specimen changes from symmetric drawing and stretching to asymmetric drawing with compressive stresses in the wrinkled area and tensile stresses opposite to the wrinkled area after the wrinkle appears. In addition, it was shown that fluctuation of the force-deflection curves is induced by the successive wrinkling of the specimen and the transition of the absorption mechanism of the impact energy takes place when the wrinkle of the STS 304 stainless steel sheet initiates.

## Acknowledgments

This study was supported by research funds from Chosun University, 2008.

## References

1. D. Kaczorowski, P. Combrade, J. P. Vernot, A. Beaudouin and C. Crenn, Tribol. int., **39**, 1503 (2006).
2. R. Anderson, E. Schedin, C. Magnusson, J. Ocklund and A. Persson, SAE transactions, **111**, 1918 (2002).
3. S. M. Sohn, B. J. Kim, K. S. Park and Y. H. Moon, J. Mater. Process. Technol., **187**, 286 (2007).
4. E. Ratte, S. Leonhardt, W. Bleck, M. Franzen and P. Urban, Steel Research International, **77**, 692 (2006).
5. Z. Yuan, Q. Dai, X. Cheng, K. Chen and W. Xu, Mater. Sci. eng., A **475**, 202 (2008).
6. M. Lepareux, Ph. Jamet, Ph. Matheron, J. L. Lieutenant, J. Couilleaux and D. Duboelle, Nucl. eng. des., **115**, 105 (1989)
7. D. D. Radford, N. A. Fleck and V. S. Deshpande, Int. j. impact eng., **32**, 968 (2006).

# DESIGN OF THE CROSS SECTION SHAPE OF AN ALUMINUM CRASH BOX FOR CRASHWORTHINESS ENHANCEMENT OF A CAR

S. B. KIM[1], H. HUH[2†]

*School of Mechanical, Aerospace & System Engineering, KAIST*
*335, Gwahangno, Deadoek Science Town, 305-701, KOREA,*
*ksb79@kaist.ac.kr, hhuh@kaist.ac.kr*


G. H. LEE[3]

*Electrotechnology R&D Center, LS Industrial Systems, Songjeong-dong,*
*Heungdeok-gu, Cheongju-si, Chungcheongbuk-do, 361-720, KOREA,*
*ghlee@lsis.biz*


J. S. YOO[4], M. Y. LEE[5]

*Technology Institute, Sungwoo Hitech, Gijang-Gun, Busan, 619-961, KOREA,*
*ryusjs@swhitech.co.kr, mylee@swhitech.co.kr*

Received 15 June 2008
Revised 23 June 2008

This paper deals with the crashworthiness of an aluminum crash box for an auto-body with the various shapes of cross section such as a rectangle, a hexagon and an octagon. First, crash boxes with various cross sections were tested with numerical simulation to obtain the energy absorption capacity and the mean load. In case of the simple axial crush, the octagon shape shows higher mean load and energy absorption than the other two shapes. Secondly, the crash boxes were assembled to a simplified auto-body model for the overall crashworthiness. The model consists of a bumper, crash boxes, front side members and a sub-frame representing the behavior of a full car at the low speed impact. The analysis result shows that the rectangular cross section shows the best performance as a crash box which deforms prior to the front side member. The hexagonal and octagonal cross sections undergo torsion and local buckling as the width of cross section decreases while the rectangular cross section does not. The simulation result of the rectangular crash box was verified with the experimental result. The simulation result shows close tendency in the deformed shape and the load–displacement curve to the experimental result.

*Keywords*: Crash box; crashworthiness; aluminum; cross section shape.

## 1. Introduction

Automotive company has recently developed a crash box which absorbs the crash energy at the low speed to decrease the damage and to increase crashworthiness of an auto-body.

---

†Corresponding Author.

The low speed crash test is carried out to evaluate the damageability and repairability of an auto-body with RCAR (Research Council Automobile Repairs) regulation.[1] RCAR is an international organization that endeavors to reduce insurance costs by improving automotive damageability, repairability, safety and security. The RCAR test is carried out by impacting a car into a 40% offset barrier at the speed of 15+1km/h. The damageability is evaluated using the measured deformation of an auto-body.

In order to maintain the high performance of an auto-body and to satisfy various regulations, it is necessary to develop a light weight auto-body system with light materials. Generally, the aluminum auto-body could achieve to reduce the weight by 30% compared to the steel auto-body at the same body stiffness.[2,3] In addition, an aluminum body enhances the strength, the endurance and noise-vibration characteristics.[4-6] Under these circumstances, automotive companies have applied extruded aluminum parts as structural components.

There are two analysis methods for the design of the crash box: One is the single part crash analysis considering the crashworthiness of a single crash box; and the other is the full car crash analysis considering the total vehicle including crash boxes. The single part crash analysis can be carried out with less computing effort than the full car analysis, but the full car simulation needs a large computation memory and computing time to consider the correlation between the crash box and other parts. In order to reduce the computing effort, there should be an efficient analysis method without losing the accuracy. One way is to adopt a simplified auto-body model to realize the current design analysis.

This paper proposes an efficient design method of crash boxes by introducing an aluminum crash box with the low speed crash analysis. The simplified auto-body model was suggested for an effective and reliable design method of a crash box comparing the performance of a full car model and a single crash box model. The selected shapes of cross section for a crash box are rectangle, hexagon and octagon.

## 2.  Crashworthiness of the Crash Box using the Single Part Model

### 2.1.  *Dynamic Tensile Characteristics of Extruded Aluminum*

The mechanical properties of an extruded aluminum, AA7003-T7(2.9t) were obtained from the static and dynamic tensile test. AA7003-T7 is widely used for the structural materials for aircraft and vehicles for its high elongation due to the aging process at the high temperature. Quasi-static tensile tests were carried out at the strain rate of 0.003/sec using the Instron 5583. The yield stress, UTS and elongation of AA7003-T7 are 247 MPa, 291 MPa and 17%, respectively. Dynamic tensile tests were carried out using a high speed material testing machine.[7,8] Fig. 1 describes true stress–true strain curves obtained from experiments at strain rates ranging from 0.003 to 200/sec. These flow curves are applied to the finite element analysis for the dynamic crash.

Table 1.  Dimension of cross section.

| Shape | Model | w [mm] | h [mm] | h/w |
|---|---|---|---|---|
| | R1 | 80 | 80 | 1.00 |
| Rectangle | R2 | 60 | 100 | 1.67 |
| | R3 | 40 | 120 | 3.00 |
| | H1 | 92.37 | 106.66 | 1.16 |
| Hexagon | H2 | 66.18 | 121.79 | 1.84 |
| | H3 | 40 | 136.9 | 3.42 |
| | O1 | 96.57 | 96.57 | 1.00 |
| Octagon | O2 | 68.27 | 115.15 | 1.69 |
| | O3 | 40 | 133.72 | 3.34 |

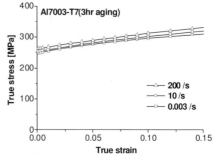

Fig. 1. True stress–true strain curves of AA7003-T7.

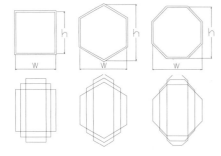

Fig. 2. Cross section shapes of crash box.

Fig. 3. Deformed shape of crash boxes.

## 2.2.  *Crashworthiness of a Crash Boxes*

The various cross-sectional shapes of a crash box such as a rectangle, a hexagon and an octagon were selected to evaluate the performance of a crash box. A crash box at each cross section has the same height of 170 mm, the same thickness of 3 mm and the same mass of 440.64 g. The cross section varies with 3 different shapes from an equilateral polygon by reducing the sides while maintaining the same area of 970 mm$^2$ as shown in Table 1 and Fig. 2. The vehicle crashes on the rigid wall inclined by 10° with the velocity of 16 km/h based on the RCAR regulation. The total vehicle weight is assumed to be 1600 kg and the vehicle absorbs the crash energy of 16000 J approximately. Thus the bumper and crash box system is assumed to sustain the crash energy of 8000 J. Considering the crash test to be performed, the mass and the initial velocity of rigid wall are selected to 250 kg and 8 m/sec, respectively for the crash energy of 8000 J. A commercial explicit finite element code, LS-DYNA3D, was used in simulation.

Fig. 3 shows the deformed shape of crash boxes. In the rectangular cross sections of R1 and R2, the deformation initiates at the middle region of the tube with an inextensional mode while the tube wall of H1, H2, O1 and O2 is folded with an extensional mode at the upper region of the tube. Fig. 4 and 5 show the absorbed energy and the mean load of crash sections. The octagonal and hexagonal shapes show higher mean load and energy absorption than rectangular shapes. In case of R3, H3 and O3 cross sections, the mean load and absorbed energy are the same due to the similar dimension of cross sections.

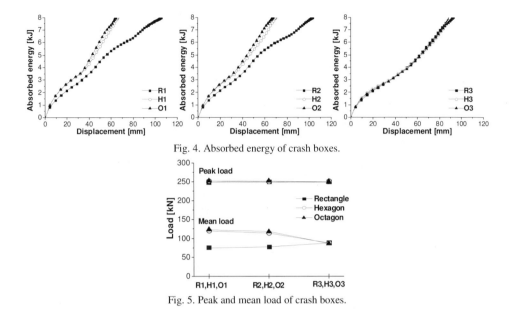

Fig. 4. Absorbed energy of crash boxes.

Fig. 5. Peak and mean load of crash boxes.

## 3. Crashworthiness of the Crash Box using the Simplified Auto-body Model

### 3.1. *Simplified Auto-body Model*

Fig. 6 shows the simplified auto-body model based on ULSAB-AVC model.[9] This model is composed of a bumper, crash boxes, front side members and a sub-frame. The suspension, engine and a body are replaced by mass points and connected to front side members and a sub-frame with rigid bar elements. The mass and the inertia moment are controlled to represent the behavior of the full car. The cross section of R1 is assembled to the simplified auto-body model and the full car model. Two models impact into a 40% offset barrier with the velocity of 16 km/h based on the RCAR regulation. The bumper and crash boxes of each model show similar absorbed energy as shown in Fig. 7. The total computing time (IBM p690, 4CPU) for the simplified auto-body model was 5hr 56min compared to the computing time of 39hr 13min for the full car model.

### 3.2. *Crashworthiness of a Crash Boxes*

The rectangular, hexagonal and octagonal cross sections are assembled to the simplified auto-body model. The rectangular shape deforms progressively, on the other hands, hexagonal and octagonal shapes undergo torsion and buckling as the width of the cross section decreases as shown in Fig. 8. The cross sections of R1, R2 and R3 absorb the energy of 52, 49.3 and 55.4% of the total energy, respectively. This result represents that the rectangular cross section absorbs the energy equivalently with the bumper. On the other hands, hexagonal and octagonal cross sections such as H1, H2, H3, O1, O2 and O3 absorb the energy of 25, 28.4, 50.7, 26.1, 40.1 and 53.5% of the total energy, respectively.

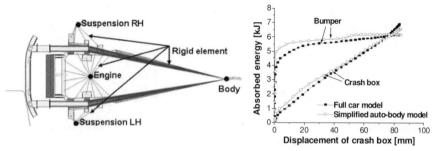

Fig. 6. Simplified auto-body model (top view).     Fig. 7. Absorbed energy of bumper and crash box.

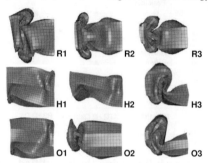

Fig. 8. Deformed shape of crash boxes assembled to the simplified auto-body model (top view).

As the width of a cross section decreases, absorbed energy of the hexagonal and octagonal cross sections increases but the torsion and the buckling effect also increase gradually. Thus the rectangular cross section shows the best performance as a crash box for the simplified auto-body model contrast to the octagonal cross section for the single part model. The simplified auto-body model considers the correlation between the crash box and other parts while the single part model does not. These results demonstrate that the simplified auto-body model is an effective and reliable design model for the crash box because this model reduces the computing effort compared to the full car model and is still accurate compared to the single part model.

## 4.  Verification of the Numerical Simulation

The simulation result of the rectangular crash box was verified with experimental result. The shape of cross section is rectangle with 72 mm in width and 115 mm in height (h/w=1.6). A crash box has the length of 177 mm and the thickness of 3 mm. The crash test of the crash box was carried out using the high speed crash testing machine. The carrier of 250 kg moves to the crash box with the velocity of 8 m/sec. Deformed shape of the experiment and the simulation is compared to each other in Fig. 9. The deformed shape of the simulation result is nearly the same as the experimental result. The middle region undergoes severe deformation with folding. The reaction forces are plotted with respect to the crushing distance in Fig. 10. The simulation result shows close tendency in the load–displacement curve and the initial peak load to the experimental result.

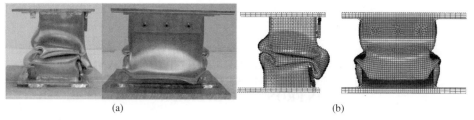

Fig. 9. Deformed shapes of rectangular crash box: (a) experimental result; (b) finite element analysis result.

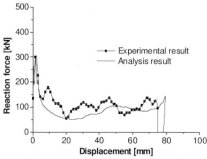

Fig. 10. Comparison of load-displacement curves obtained from the experiment and the analysis.

## 5. Conclusions

This paper has investigated the crashworthiness of aluminum crash boxes with rectangular, hexagonal and octagonal cross sections. In the numerical simulation of the single part model, the octagonal cross section shows higher energy absorption and mean load than the other two cross sections. The simplified auto-body model based on the full car was proposed for effective and reliable design of crash boxes. The simulation result shows that the rectangular cross section shows the best performance as a crash box. The hexagonal and octagonal cross sections undergo torsion and buckling as the width of cross section decreases while the rectangular cross section does not. The simulation result of the rectangular crash box was verified by the experimental result. Comparison demonstrates that the simulated deformed shape and load–displacement curve are nearly the same as the experimental result.

## References

1. Research Council for Automobile Repairs, www.RCAR.org, (RCAR, 1999).
2. V. Lakshminrayan, H. Wang, W. J. Willams and Y. Harajli, *SAE 951080* (1995).
3. B. J. Kim and S.-J. Heo, *Int. J. Auto. Tech.*, **4**, 141 (2003).
4. M. L. Bieber, J. S. Werner, K. Knoblauch, J. Neitz, M. Neitz, S. Santosa and T. Wierzbicki *Computers and Structures*, **68**, 343 (1998).
5. M. Langseth and O. S. Hopperstad, *Int. J. Impact Eng.*, **22**, 829 (1999).
6. J. S. Qiao, J. H. Chen and H. Y. Che, *Thin-Walled Structures*, **44**, 692 (2006).
7. J. S. Kim, H. Huh, K. W. Lee, D. Y. Ha, T. J. Yeo and S. J. Park, *Int. J. Auto. Tech.*, **7**, 571 (2006).
8. H. Huh, S. B. Kim, J. H. Song and J. H. Lim, *Int. J. Mech. Sci.*, **50**, 918 (2008).
9. Porche Engineering Services, ULSAB-AVC Engineering Report (2001).

# REGRESSION MODEL FOR LIGHT WEIGHT AND CRASHWORTHINESS ENHANCEMENT DESIGN OF AUTOMOTIVE PARTS IN FRONTAL CAR CRASH

GIHYUN BAE[1], HOON HUH[2†]

*School of Mechanical, Aerospace & System Engineering, KAIST, 335, Gwahangno, Deadoek Science Town, Daejeon, 305-701 Korea*
*baegh102@kaist.ac.kr, hhuh@kaist.ac.kr*

SUNGHO PARK[3]

*Automotive Steel Research Center, POSCO, Gwangyang, Jeonnam, 545-090, Korea*
*sunghopark@posco.co.kr*

Received 15 June 2008
Revised 23 June 2008

This paper deals with a regression model for light weight and crashworthiness enhancement design of automotive parts in frontal car crash. The ULSAB-AVC model is employed for the crash analysis and effective parts are selected based on the amount of energy absorption during the crash behavior. Finite element analyses are carried out for designated design cases in order to investigate the crashworthiness and weight according to the material and thickness of main energy absorption parts. Based on simulations results, a regression analysis is performed to construct a regression model utilized for light weight and crashworthiness enhancement design of automotive parts. An example for weight reduction of main energy absorption parts demonstrates the validity of a regression model constructed.

*Keywords*: Crashworthiness; Light weight design; Regression model; ULSAB-AVC model; Frontal car crash.

## 1. Introduction

The recent megatrend of both automobile users and industries is to demand better vehicle safety and crashworthiness although the light-weight design for the reduction of fuel consumption becomes a very challenging issue. Confronting these contradictory requirements, recent researches are mostly focused on replacing conventional low strength steels by advanced high strength steels (AHSS) such as DP and TRIP steels. One of the remarkable attempts is the ULSAB-AVC (Ultra Light Steel Auto Body - Advanced Vehicle Concepts) project[1]. The light-weight vehicle using AHSSs should accompany the quantitative evaluation of the crashworthiness since it should always meet the requirement of the structural safety performance.

---

†Corresponding Author.

Various researches were carried out to design an auto-body structure for weight reduction and the improved crashworthiness. Marklund and Nillson[2] performed the shape optimization of the center-pillar model with time dependent boundary condition. Bae et al.[3] replaced conventional steels with AHSSs and carried out the optimum design of center-pillar parts with a simplified side impact analysis. Zhang et al.[4] carried out the robust optimization of automotive front side rails for the light weight design. Oh et al.[5] conducted the size optimization for the thicknesses of the engine room member using response surface. However, most of researchers have dealt with the optimum design of automotive parts and did not concern how to employ higher strength steels systematically.[6-8]

This paper deals with a regression model for light weight and crashworthiness enhancement design of automotive parts in frontal car crash. The ULSAB-AVC model is employed for the crash analysis and main parts are selected based on the amount of energy absorption during the crash behavior. Finite element analyses are carried out for designated design cases in order to investigate the crashworthiness and weight according to the material and the thickness. Based on simulations results, a regression model is constructed for light weight and crashworthiness enhancement design of automotive parts. An example for weight reduction of main energy absorption parts demonstrates the validity of the regression model constructed.

## 2. Frontal Impact Analysis

### 2.1. *Frontal Impact Condition and Analysis Model*

The US-NCAP (U.S. New Car Assessment Program) for the frontal crash test is based on the FMVSS 208 of the NHTSA (National Highway Traffic Safety Administration). The US-NCAP test regulation is introduced in Table 1. The ULSAB-AVC model which is composed with the AHSS more than 85 % of the structure is employed for the frontal car crash analysis. The finite element analysis is performed with the commercial software LS-DYNA3D v971[9].

### 2.2. *Analysis Result*

The analysis result of the ULSAB-AVC model in the frontal crash condition is represented in Fig. 1. Automotive parts near the engine room deform to absorb the crash energy transmitted from the bumper as shown in Fig. 1(b). The maximum overall deformation is 642.69 mm and it satisfies the safety target established in ULSAB-AVC project. The deceleration is filtered by the CFC60 filter as presented in ULSAB-AVC engineering report and the maximum deceleration among peaks is 37.81g. Main parts in

Table 1. US-NCAP test regulation.

| Impact regulation | US-NCAP |
|---|---|
| Impact condition | Impact velocity: 35 mph (56 km/h) |
|  | Full face rigid barrier, Zero degree impact |
| Crashworthiness assessment | Overall deformation (OD), Deceleration |

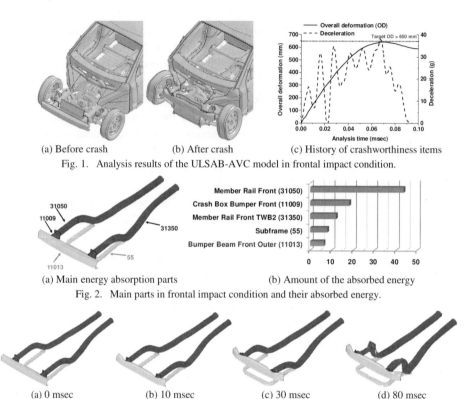

(a) Before crash          (b) After crash          (c) History of crashworthiness items
Fig. 1.   Analysis results of the ULSAB-AVC model in frontal impact condition.

(a) Main energy absorption parts          (b) Amount of the absorbed energy
Fig. 2.   Main parts in frontal impact condition and their absorbed energy.

(a) 0 msec          (b) 10 msec          (c) 30 msec          (d) 80 msec
Fig. 3. Sequential deformation mode of main energy absorption parts with respect to the analysis time.

Table 2.   Reference of main energy absorption parts of the ULSAB-AVC model in frontal car crash.

| Part name | Part number | Mass (kg) | Thickness (mm) | Material |
|---|---|---|---|---|
| Member Rail Front | 31050 | 3.68 | 1.5 | DP500/800 |
| Assy Crash Box Bumper Front | 11009 | 1.37 | 1.1 | DP400/700 |
| Member Rail Front TWB2 | 31350 | 13.14 | 1.3 | DP500/800 |
| Subframe | 55 | 6.27 | 2.0 | HSLA350/450 |
| Bumper Beam Front Outer | 11013 | 2.56 | 1.0 | MART1250/1520 |

the frontal crash condition are selected based on the amount of the energy absorption during the car crash behavior as shown in Fig. 2. During the car crash, main parts deform sequentially to absorb the crash energy as shown in Fig. 3. Table 2 introduces the reference of main energy absorption parts of the ULSAB-AVC model in frontal car crash.

## 3.   Regression Model for Light Weight and Crashworthiness Enhancement Design

### 3.1.   *Parameterization of the Part Strength*

The crashworthiness of an auto-body depends directly on the strength of automotive parts. The part strength is determined by the used material and the part volume. Therefore, the part strength $S$ can be defined as follows:

$$S = M \times V \tag{1}$$

Table 3. Parameterization result of dynamic material properties of the representative AHSS.

| Material name | $\sigma_y$ (MPa) | $\sigma_t$ (MPa) | R | M |
|---|---|---|---|---|
| DP350/600 | 350 | 600 | 1.107 | 595.0 |
| DP400/700 | 400 | 700 | 1.143 | 714.4 |
| DP500/800 | 500 | 800 | 1.076 | 780.1 |
| DP700/1000 | 700 | 1000 | 1.054 | 975.0 |

where $M$ and $V$ denote the parameterized dynamic material properties and the part volume, respectively. Then, the parameterized dynamic material properties are calculated as follows:

$$M = (0.25\sigma_y + 0.75\sigma_t) \times R \tag{2}$$

where $\sigma_y$ and $\sigma_t$ denotes the yield stress and the tensile stress. And $R$ is the strain rate hardening ratio between static (0.001/s) and dynamic (100/s) flow stress. The static material properties are parameterized simply using weight factors as shown in Equation (1). Finally, the strain rate hardening ratio is multiplied to the parameterized static material properties to describe the dynamic material properties. Table 3 show the parameterization result of the dynamic material properties of the representative AHSS to be utilized for the crashworthiness assessment according to the strength change.

The total strength change $\Delta S_{total}$ is calculated by the sum of the part strength change considering the crashworthiness sensitivity in frontal impact as follows:

$$\begin{aligned} \Delta S_{total} &= \sum_{i=1}^{n} \{\alpha \times \Delta S\}_i \\ &= \sum_{i=1}^{n} \left\{ \frac{E_0 m_0}{S_0} \times (S - S_0) \right\}_i \\ &= \sum_{i=1}^{n} \left\{ \frac{E_0 m_0}{M_0 t_0} \times (Mt - M_0 t_0) \right\}_i \end{aligned} \tag{3}$$

where $\alpha$ is the coefficient of the crashworthiness sensitivity of an automotive part; $E$ is the amount of the energy absorption during the frontal crash behavior; $m$ is the part mass; and $t$ is the part thickness, respectively. And index $0$ denotes the initial design of an auto-body. The coefficient of the crashworthiness sensitivity is determined by the absorbed energy, the part mass and the part strength of the initial auto-body design. And $n$ is the number of automotive parts to be designed for light weight and crashworthiness enhancement. The total strength change is selected as a design parameter to construct a regression model for light weight and crashworthiness enhancement design of automotive parts in the frontal car crash condition.

## 3.2. Regression Analysis

The crashworthiness according to the strength change is investigated to construct a regression model for light weight and crashworthiness enhancement design of automotive parts. A quantitative evaluation of the crashworthiness according to the strength change is

Table 4.  Design table to investigate crashworthiness and measured results.

| Design case | Part # | Material | Thickness (mm) | $\Delta S_{total}$ | OD (mm) | Deceleration (g) | Remark |
|---|---|---|---|---|---|---|---|
| 1 | 31050 | DP700/1000 | 1.7 | 20.30 | 597.64 | 42.70 | Increase[a] |
| 2 | 31050 | DP400/700 | 1.3 | -13.29 | 687.89 | 39.83 | Decrease |
| 3 | 11009 | DP500/800 | 1.3 | 8.67 | 629.51 | 35.93 | Increase |
| 4 | 11009 | DP350/600 | 0.9 | -4.04 | 656.70 | 38.35 | Decrease |
| 5 | 31350 | DP700/1000 | 1.5 | 5.93 | 594.03 | 41.73 | Increase |
| 6 | 31350 | DP400/700 | 1.1 | -3.87 | 669.04 | 34.76 | Decrease |

[a] Increase: (Thickness) + 0.2 mm, (Material) + 1 grade, Decrease: (Thickness) - 0.2 mm, (Material) - 1 grade

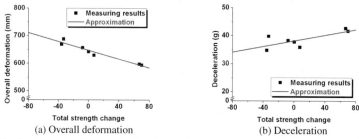

(a) Overall deformation         (b) Deceleration

Fig. 4.   Approximation results of crashworthiness assessment items in frontal car crash.

performed using main energy absorption parts of three tube-shaped parts. Table 4 shows measured results of the overall deformation and the deceleration which are important items to assess the vehicle safety in the frontal car crash condition. The material properties and the part thickness change simultaneously in each design case.

Based on simulation results, regression analysis is carried out to approximate measured results with a linear regression equation. The result of the regression analysis is presented in Fig. 4. Linear regression equations are as follows:

$$Overall\ Deformation(mm) = 647.25 - 0.79 \times \Delta S_{total} \qquad (4)$$

$$Deceleration(g) = 38.23 - 0.05 \times \Delta S_{total} \qquad (5)$$

The overall deformation decreases according to the increase of the total strength of an auto-body. On the contrary, the deceleration increases according to the increase of the total strength of an auto-body. Regression models can be utilized to light weight and crashworthiness enhancement design of automotive parts by predicting the crashworthiness of an auto-body without the numerical simulation.

## 4.  Verification with Light Weight Design

An example for weight reduction of main energy absorption parts is used to demonstrate the validity of regression models constructed. Three tube-shaped parts are selected as a target for weight reduction since they are main energy absorption parts. The amount of weight reduction is 4.14 kg (22.8 %) when the part thickness decreases equally 0.3 mm. The initial material is replaced by the higher strength steel correspondingly to compensate the thickness reduction. Table 5 shows the modified design and the strength

Table 5. An example for weight reduction of main energy absorption parts.

| Design | P1: 31050 | | P2: 11009 | | P3: 31350 | | $\Delta S_{total}$ |
|---|---|---|---|---|---|---|---|
| | Thickness (mm) | Material | Thickness (mm) | Material | Thickness (mm) | Material | |
| Initial | 1.5 | DP500/800 | 1.1 | DP400/700 | 1.3 | DP500/800 | - |
| Modified | 1.2 | DP700/1000 | 0.8 | DP500/800 | 1.0 | DP700/1000 | -11.2 |

Table 6. Verification of the regression model utilized for light weight design of an auto-body.

| Design | Mass (kg) | OD (mm) | Deceleration (g) |
|---|---|---|---|
| Initial | 18.19 | 642.69 | 37.81 |
| Prediction | 14.05 | 656.17 | 37.67 |
| Simulation | 14.05 | 650.85 | 38.09 |
| Error (%) | - | - 0.8 | 1.1 |

change ($\Delta S_{total}$) calculated by Equation (3). The crashworthiness of the modified design is predicted by regression equations and also measured from the simulation. Table 6 represents the quantitative comparison between the prediction and the simulation. It demonstrates that the regression model validates for reduction of the weight while the crashworthiness of automotive parts remains the same.

## 5. Conclusion

A regression model for light weight and crashworthiness enhancement design of automotive parts is constructed after the regression analysis. Main parts for the crashworthiness in the frontal car crash condition are selected based on the amount of absorbed energy during the crash behavior. A design parameter which indicates the amount of the total strength change is newly proposed for the strength distribution of the regression analysis. Based on simulation results of designated design cases generated by main energy absorption parts, regression equations for predicting the overall deformation and the deceleration are constructed in the form of the linear equation as a result of the regression analysis. An example for weight reduction of three tube-shaped parts demonstrates that the regression model constructed validates for light weight and crashworthiness enhancement design of automotive parts in frontal car crash.

## References

1. IISI, *ULSAB-AVC Engineering Report* (Porsche Engineering Services, 2001).
2. P. O. Marklund and L. Nilsson, *Struct. Multidisc. Optim.* **21**, 383 (2001).
3. G. H. Bae, J. H. Song, H. Huh and S. H. Kim, *Trans. KSAE* **13**, 84 (2005).
4. Y. Zhang, P. Zhu and G. Chen, *Thin-walled Struct.* **45**, 670 (2007).
5. S. Oh, B.-W. Ye and H.-C. Sin, *Int. J. Automotive Tech.* **8**, 93 (2007).
6. A. M. Elmarakbi and J. W. Zu, *Int. J. Automotive Tech.* **6**, 491 (2005).
7. J. H. Yoon, H. Huh, S. H. Kim, H. K. Kim and S. H. Park, in *Proc. IPC-13*, (KSAE, Gyeongju, 2005), pp. 742-747.
8. H. Huh, J. H. Lim, J. H. Song, K.-S. Lee, Y.-W. Lee and S. S. Han, *Int. J. Automotive Tech.* **4**, 149 (2003).
9. J. O. Hallquist, *LS-DYNA Keywork User Manual* (LSTC, 2006).

# AN EXPERIMENTAL STUDY ON THE IMPACT DEFORMATION AND THE STRAIN RATE SENSITIVITY IN SOME STRUCTURAL ADHESIVES

TOSHIMASA NAGAI[1†], TAKESHI IWAMOTO[2], TOSHIYUKI SAWA[3], YASUHISA SEKIGUCHI[4]

*Graduate School of Engineering, Hiroshima University,*
*1-4-1 Kagamiyama, Higashi-Hiroshima, 739-8527 JAPAN,*
*m071450@hiroshima-u.ac.jp, iwamoto@mec.hiroshima-u.ac.jp, sawa@mec.hiroshima-u.ac.jp,*
*seki@mec.hiroshima-u.ac.jp*

HIDEAKI KURAMOTO[5], NORIO UESUGI[6]

*Hiroshima City industrial Promotion Center,*
*3-8-24 Senda-machi, Naka, Hiroshima, 730-0052 JAPAN,*
*hide22@itc.city.hiroshima.jp, uesugi-n@itc.city.hiroshima.jp*

Received 15 June 2008
Revised 23 June 2008

The impact deformation behavior and the strain sensitivity of structural adhesives are experimentally investigated by using INSTRON-type universal testing machine and split Hopkinson pressure bar apparatus. The experimental results show some fundamental features of the typical compressive stress-strain behavior of polymers with linear elastic and nonlinear inelastic deformation stages. In the inelastic deformation, the peak stress, and the strain-softening stage after the peak can be observed at the entire range of strain-rate from $10^{-4}$ to $10^3$ /s. In addition, it can be found that the relationship between the peak stress at the strain-softening stage and strain-rate for a semi-logarithm curve is linear in a range of low strain rate, however, that becomes nonlinear at high strain rate. Finally, some constitutive models try to be applied for to describe the stress-strain behavior of structural adhesives.

*Keywords*: Structural adhesive; split Hopkinson pressure bar; strain rate sensitivity.

## 1. Introduction

Because of many advantages such as its high energy absorption characteristic, flexibility on joining different materials and possibility to lighten the weight of mechanical structures, adhesive bonding is increasingly applied for in automobile and aircraft structures subjected to not only static loading but also dynamic or impulsive loading, Many research works about the strength of adhesive joints subjected to static loading can be referred[1-2] and static joint strength has been improved vastly.

It is desired for adhesive joints in principal parts of mechanical structures to have enough strength against high loading rates and large dynamic deformation. A few

---

†Corresponding Author.

218

research works about adhesive joints subjected to dynamic loading or the impact deformation of structural adhesives themselves can be also referred[3-6], however, they are limited in case that a few kinds of structural adhesives are used. Therefore, it is greatly important to collect more detail information about many kinds of structural adhesives to estimate the strength of adhesive joints under dynamic loading.

The split Hopkinson pressure bar (SHPB) technique, originally developed by Kolsky[9], has been modified to determine the mechanical properties of a variety of engineering materials under such high-speed deformation. Instead of the single load-time curve with unknown strain rate obtained by the Charpy impact test, the SHPB can provide dynamic stress-strain curves as a function of strain rates. Recently many experimental research works employ the pulse shaping techniques for the SHPB method to measure impact deformation behavior of polymers[7-8] more precisely. On the other hand, it is well known that the mechanical properties of polymers, including structural adhesives, are sensitive in the deformation rate. Since the strain rate sensitivity of the adhesives is unclear, the utility of the adhesive can be increased by the experimental results with a wide range of strain rate from quasi-static to impact deformation.

In this study, quasi-static to impact compressive tests are carried out by using INSTRON-type universal testing machine and SHPB apparatus to reveal the impact deformation behavior and the strain rate sensitivity of some kinds of structural adhesives. The stress-strain curves are measured more accurately at a wide range of strain rate by control of the loading pulse with a copper pulse shaper and stress bars made of extra-super duralmin. Then, the peak stress is plotted against the strain-rate in the semi-logarithm scale. The obtained data are compared with the prediction based on the some constitutive equations and the past-published experimental data[13].

## 2. Experimental Method

### 2.1. *Materials and specimens*

The commercial structural adhesives used in this work are a thermosetting epoxy adhesive (Scotch-Weld 1838, Sumitomo 3M Limited, Japan) and a carboxyl-terminated acrylonitrile butadiene (CTBN) modified epoxy adhesive (DP-460, Sumitomo 3M Limited, Japan).

The aspect ratios of the cylindrical specimens are typically 0.5-1.0 in order to minimize the inertial[9] and frictional effects at the ends. Cylindrical specimens with 14.0 mm in diameter and 7.0 mm in length were manufactured for compressive experiments. Figure 1 show the cylindrical adhesive specimen for compressive experiments. All specimens were made by an injection of the adhesives into a pipe of polyvinyl chloride and machined to the above-mentioned geometry. Then,

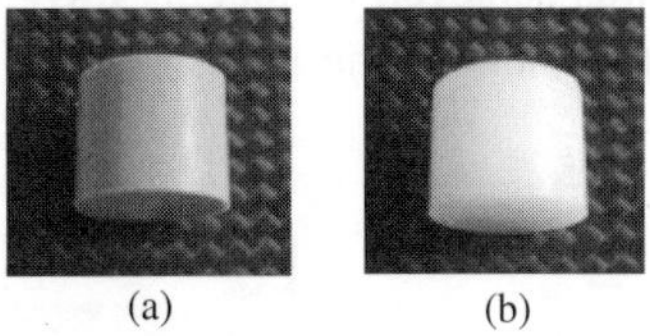

(a)  (b)

Fig. 1. Photographs of cylindrical adhesive specimens made of adhesives (a) bi-component thermosetting epoxy adhesive (b) CTBN modified epoxy adhesive.

specimens were cured in a furnace at the temperature for the constant time after curing at room temperature for 24 hours. As the curing temperature and time in the furnace depend on material, the curing time and temperature for thermosetting epoxy adhesive specimens are chosen to 2 hours and 65° C, and those for CTBN modified epoxy adhesive specimens are 3 hours and 49° C. Finally, the surfaces on the both side of edge of specimens are ground by abrasive paper (#2000) to reduce the friction during the testing. Furthermore, the interface between the die and the surface on the end of the specimen is lubricated with graphite grease.

## 2.2. Impact compressive experiments using split Hopkinson pressure bar method

The INSTRON-type universal testing machine (Autograph AG-1 250kN, Shimadzu Corporation, Japan) is employed to conduct the compressive experiments at the quasi-static to dynamic strain rate from $10^{-4}$ /s to $10^0$ /s archived by a constant crosshead speed.

Figure 2 shows the schematic of experimental setup for impact compressive experiments at strain rate over $10^2$ /s using SHPB. A calculation based on the one-dimensional elastic wave propagation theory by Kolsky[10] shows the nominal strain rate $\dot{\varepsilon}(t)$ and nominal stress $\sigma(t)$ in the specimen are to be

$$\dot{\varepsilon}(t) = -\frac{2C_0}{L}\varepsilon_r(t), \sigma(t) = \frac{A_0}{A_s}E\varepsilon_t(t),\qquad(1)$$

where $L$ and $A_s$ are the original sample length and cross-sectional area; $c_0$, $E$ and $A_0$ are the elastic bar-wave velocity, Young's modulus, and cross-sectional area of bars; $\varepsilon_r(t)$ and $\varepsilon_t(t)$ are measured reflected and transmitted strain signals on the bar surfaces.

When the material of specimens has low strength and mechanical impedance such as polymeric materials, the two following limitations of the SHPB technique must be modified before valid data can be obtained[7]. First, to capture the weak strain signals transmitted from the low-impedance specimens for testing of adhesive specimens by using the SHPB apparatus, an extra super duralumin is chosen for the stress bars. As schematically shown in Fig.2, the lengths of 7075-T6 aluminum alloy bars with the common diameter of 16 mm were 1250, 1250 and 500 mm for the incident, transmission and striker bars, respectively.

In addition, the pulse shaping technique[11] was used. The pulse shaping techniques has recently been developed to control the loading pulse. The control of the rising time was achieved by placing a C11000 copper disk with 5.0 mm in diameter and 0.5 mm in thickness at the impact end of the incident bar. The plastic deformation of the copper

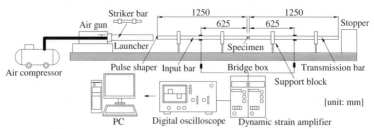

Fig. 2. Schematic of the split Hopkinson pressure bar apparatus for impact compressive experiment.

pulse shaper upon impact effectively increases the rise time of the incident pulse in the bar.

Two semiconductor strain gages with a gage length of 2 mm (KSP-2-120-E4, Kyowa Electronic Instruments Co. Ltd., Japan) are attached to the surfaces of the opposite sides in the direction of diameter for the incident and transmission bars at 625 mm from the interface between the bar and specimen to record strain signals.

## 3. Experimental Results and Discussion

Assuming that the wave propagation in the bars is approximately one-dimensional, the force histories at the incident bar end (front end) of the specimen, $P_1(t)$, and at the transmitted bar end (back end) of the specimen, $P_2(t)$, can be calculated as

$$P_1(t) = \{\varepsilon_i(t) + \varepsilon_r(t)\}EA_0, P_2(t) = \varepsilon_t(t)EA_0, \qquad (2)$$

Figure 3 shows the force histories of front and back end on the interfaces between bars and specimen calculated by Eq. (2) during the impact compressive test with the pulse shaper. From this result, since the oscillations in force histories are negligible in amplitude, time history of front end force history nearly agrees with the back end force history. Therefore, the specimen is in the approximately stress-equilibrium state during the experiment.

Figure 4 shows the compressive stress-strain behavior of two structural adhesives over a strain rate range of $1.0 \times 10^{-4}$ s$^{-1}$ to $1.0 \times 10^3$ s$^{-1}$. From this figure, some fundamental features of the typical compressive stress-strain behavior of polymers with linear elastic

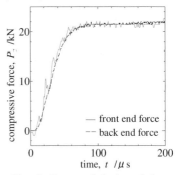

Fig. 3. Front and back end force histories calculated from Eq. (2) during an impact compressive test for thermosetting epoxy adhesive using pulse shaping techniques.

and nonlinear inelastic deformation stages. In the inelastic deformation stage, the curves have a peak stress and a strain-softening stage after the peak can be observed.

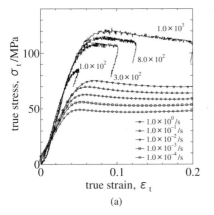

(a)

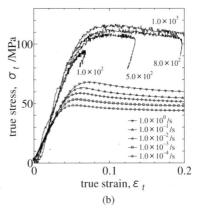

(b)

Fig. 4. Compressive stress-strain behavior of structural adhesives over a strain rate range from $1.0 \times 10^{-4}$ s$^{-1}$ to $1.0 \times 10^3$ s$^{-1}$ (a) thermosetting epoxy adhesive (b) CTBN modified epoxy adhesive.

In this research work, some constitutive models are used for description of the plastic flow characteristic of structural adhesives. For the constitutive model of the structural adhesive, Cowper-Symonds factor[12] depending on the strain rate is used. The equivalent Mises stress $\overline{\sigma}$ is defined as

$$\overline{\sigma} = \left[ A + B\left(\overline{\varepsilon}^{p}\right)^{n} \right] \cdot \left[ 1 + \left(\frac{\dot{\varepsilon}}{C}\right)^{\frac{1}{p}} \right], \tag{3}$$

where $A$ is the quasi-static yield stress, $B$ is a hardening constant, $n$ is the hardening exponent, $C$ and $p$ are the Cowper-Symonds strain rate parameters. Using the results obtained from quasi-static to impact compressive experiments, constants $A$, $B$, $n$, $C$, and $p$ are determined to be 30.012 MPa, 21.946 MPa, 0.0894, 104.592, and 6.729, respectively.

Figure 5 (a) shows the equivalent stress-equivalent plastic strain curves of thermosetting epoxy adhesive, as compared to Cowper-Symonds model over a range of strain rate from $1.0 \times 10^{-4}$ s$^{-1}$ to $1.0 \times 10^{3}$ s$^{-1}$. From this figure, it can be found that it is possible to describe the plastic flow characteristic using Cowper-Symonds model except for the strain-softening region in quasi-static deformation.

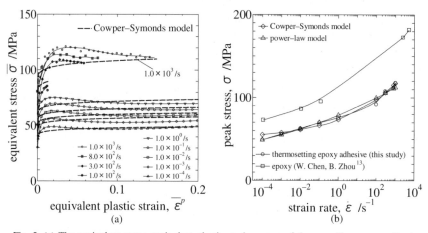

Fig. 5. (a) The equivalent stress-equivalent plastic strain curves of thermosetting epoxy adhesive, as compared to Cowper-Symonds model over a strain-rate range of $1.0 \times 10^{-4}$ s$^{-1}$ to $1.0 \times 10^{3}$ s$^{-1}$. (b) Variation of peak stress of thermosetting epoxy adhesive and an epoxy[13] with strain-rate.

Figure 5 (b) shows the strain rate dependence of the stress-strain behavior in thermosetting epoxy adhesive of peak stress, which means the maximum value of the equivalent stress just before the strain-softening region, versus strain-rate in the semi-logarithm scale. The strain rate dependence of peak stress versus strain-rate curve in an epoxy resin obtained by Chen and Zhou[13] is plotted in this figure. From this figure, a linear behavior can be found in the relationship between peak stress and strain-rate of thermosetting epoxy adhesive at low strain rate, and nonlinear behavior can be observed at the high strain region. These tendencies are corresponding to the results by Chen and Zhou[13] and can be also found in deformation behavior of the other polymeric materials such as polyvinyl chloride[14] or polyurethane[15]. In addition, the descriptions for the peak stress-strain-rate relationship based on Cowper-Symonds and power-law models are shown in this figure. From this figure, a fairly good agreement can be found between experimental results and prediction based on the constitutive models. Chen and Zhou[13] formulated the strain-rate effect of an epoxy resin using a complicated hyperbolic tangent function, which made it capable to capture the strain-rate effect of epoxy resin until the stress archives its peak value. However, this result is telling us that simple constitutive model for strain rate effect is sufficient.

## 4. Concluding Remarks

Here, compressive experiments from quasi-static to impact deformation were carried out using universal testing machine and SHPB apparatus to reveal the strain rate sensitivity in some kinds of structural adhesives. The pulse shaping technique and aluminum alloy bars are applied to conventional SHPB test to obtain stress-equilibrium state in the specimen during experiments. In low strain-rate regime, it is possible to describe the stress-strain behavior, except for the strain-softening behavior, of structural adhesives using the Cowper-Symonds constitutive model.

## Acknowledgments

Authors wish to thank Professor Chiaki SATO of Tokyo Institute of Technology in Japan for providing his valuable experimental data and discussions.
Financial support by the Suzuki foundation is gratefully acknowledged.

## References

1. Lucas F.M. da Silva, R.D. Adams, *International Journal of Adhesion and Adhesives*, **27**, 362 (2007).
2. T. Sawa, K. Temma, T. Nishigaya, H. Ishikawa, *Journal of Adhesion Science and Technology*, **9**, 215 (1995).
3. R.D. Adams, J. A. Harris, International Journal of Adhesion and Adhesives, **16**, 61 (1996).
4. T. Yokoyama, *J. Strain Anal.*, **38**, 233 (2003).
5. C. Sato, K. Ikegami, International Journal of Adhesion and Adhesives, **20**, 17 (2000).
6. L. Goglio, M. Peroni, L. Peroni, M. Rossetto, *International Journal of Adhesion and Adhesives*, **28**, 329 (2008).
7. W. Chen, F. Lu, D. J. Frew, M. J. Forrestal, *J. Applied Mechanics, T. ASME.*, **86**, 214 (2002).
8. O.S. Lee, M.S. Kim, *Nuclear Engineering and Design*, **226**, 119 (2003).
9. A. M. S. Hamouda, Journal of Materials Processing Technology, **124**, 209 (2002).
10. H. Kolsky, *Proc. Phys. Soc.*, **B 62**, 676 (1949).
11. S. Netmat-Nassar, J. B. Isaacs, J. E. Starrrett, *Proc. Roy. Soc. Lond.*, **A 435**, 371 (1991).
12. G. R. Cowper, P. S. Symonds, *Brown Univ. Applied Mathematics Report*, 28 (1958).
13. W. Chen, B. Zhou, Mechanics of Time-Dependent Materials, **2**, 103 (1998).
14. S. Y. Soong, R. E. Cohen, M. C. Boyce, W. Chen, Polymer, **49**, 1440 (2008).
15. J. Yi, M. C. Boyce, G. F. Lee, E. Balizer, *Polymer*, **47**, 319 (2006).

# EFFECTS OF IMPACT VELOCITY AND SLENDERNESS RATIO ON DYNAMIC BUCKLING LOAD FOR LONG COLUMNS

K. MIMURA[1†], T. UMEDA[2]

*Division of Mechanical Engineering, Graduate School of Engineering, Osaka Prefecture University*
*1-1 Gakuen-cho, Naka-ku, Sakai-city, Osaka 599-8531, JAPAN,*
*mimura@me.osakafu-u.ac.jp, umeda@me.osakafu-u.ac.jp*


M. YU[3], Y. UCHIDA[4]

*Students of Graduate School of Engineering, Osaka Prefecture University*
*1-1 Gakuen-cho, Naka-ku, Sakai-city, Osaka 599-8531, JAPAN,*
*m07yu @ me.osakafu-u.ac.jp, m09 yuuk @me.osakafu-u.ac.jp*


H. YAKA[5]

*Mechanical Engineering Research Laboratory, Hitachi, Ltd*
*832-2 Horiguchi, Hitachinaka-city, Ibaraki 312-0034, JAPAN,*

Received 15 June 2008
Revised 23 June 2008

In this research, the buckling behavior of long columns under dynamic load was investigated both experimentally and numerically, and an effective buckling criterion for dynamic load was derived from the results in terms of the impact velocity and the slenderness ratio. In the experiments, a free fall drop-weight type impact testing machine was employed. The dynamic buckling loads were measured by the load sensing block, and the displacements were measured by a high speed magnetic-resistance device. In the numerical analyses, dynamic FEM code 'MSC-Dytran' was used to simulate the typical experimental results, and the validity and the accuracy of the simulations were checked. The dynamic buckling loads at various impact velocities were then systematically investigated. From both experimental and simulated results, it was found that the dynamic to static buckling load ratios can be successfully described as a square function of the slenderness ratio of the columns, while they can be also described by a power law of the applied impact velocity.

*Keywords*: Dynamic buckling; Axial impact load; Long columns; Slenderness ratio; Impact velocity.

## 1. Introduction

Recently, a growing need arises for precise understanding and modeling of dynamic buckling phenomena of long columns since they are essential and fundamental elements of machine and architectural structures, and there may be many occasions that they are subjected to impact load due to collision, seismic oscillation and so on. In general, dynamic buckling loads of long columns show strong positive impact-velocity dependence which may come from the restriction on the lateral deflection of a column due to the inertia effect. However, its precise

---

†Corresponding Author.

mechanism has been still unknown. In addition, the occurrence of bending deflection with high mode numbers at higher impact velocities leads to a complicated size (length and/or diameter) dependence of the buckling load which is considerably different from that in the quasistatic loading condition. Clarification of generalized relations between the dynamic buckling load and the impact velocity, the length of a rod, the dimensions of a cross-sectional area is, therefore, very important. An early experimental study on the dynamic buckling of rods subjected to axial impact has been done by Abrahamson et al.[1] , and a series of experimental and analytical works soon followed by Hayashi and Sano.[2-4] The latter dealt with three different impact cases in the experiments, that is, the low velocity impact due to low frequency vibration, the intermediate impact loading that causes 'dynamic elastic buckling' and very high velocity impact that causes plastic compressive deformation preceding the dynamic buckling. They also performed the numerical analyses and pointed out that the propagation of high mode number deflection from the impact end to the other end was the main reason for the increase of the buckling load with impact velocity. As for theoretical approaches, Karagiozova and Jones[5] proposed a discrete model for elastic-plastic buckling of an axially loaded rod, and they successfully described the post-buckling shapes of the rods under large mass impact although the applied impact velocity is very high enough to cause plastic deformation before buckling. Recently, Hong and Shijie et al.[6, 7] calculated elastic buckling loads for long rods subjected to considerably low velocity impacts caused by fluid-solid slamming. They assumed a sinusoidal load-time history for the dynamic buckling, and pointed out that as a time to attain the maximum load (=buckling load) decreased, a ratio of the dynamic buckling load to static buckling load increased. They also concluded though their calculations that in the case of dynamic buckling, this ratio became small for a large slenderness ratio's rod.

In this study, the dynamic buckling behavior of long columns was investigated systematically both by experiments and by numerical simulations. In the experiments, the newly developed free fall type drop-weight impact testing device was employed, and aluminum rods with various diameters and lengths were axially loaded at the impact velocity from 1m/s to 3m/s. For the intermediate velocity range from 0.01m/s to 1m/s, numerical simulations by using dynamic FEM code 'MSC-Dytran' were performed. From both experimental and simulated results, we show the fact that the ratio of the dynamic buckling load to the static one was successfully described by power laws of impact velocity and slenderness ratio.

## 2. Experimental Procedure and Specimens

In the present research, slender cylindrical rod specimens having diameters of 6mm~10mm at intervals of 1mm and also having lengths of 600mm~1500mm at intervals of 300mm were dynamically loaded by using the drop-mass impact loading apparatus shown in Fig.1. The specimen is made of A2017 aluminum alloy the strain rate sensitivity of which is considerably small so that the contribution of plastic flow to the post-buckling behavior can be supposed not to be affected by an impact velocity. In the experiments, three different 'intermediate' impact velocities of 1000mm/s, 1450mm/s and 3000mm/s as well as quasi-static velocity (approximately 0.02mm/s by hydraulic jack) were employed. Drop masses of 10kg~40kg were used according to the applied impact velocities so that the kinematic energy of the mass at each impact velocity was approximately equal. Supporting conditions of the upper and the

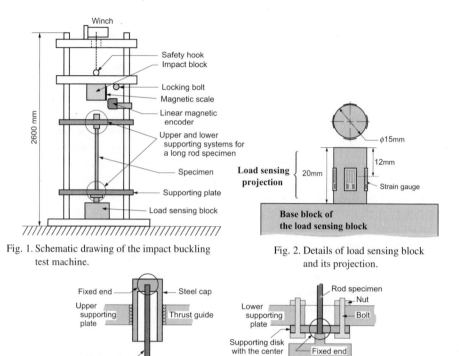

Fig. 1. Schematic drawing of the impact buckling test machine.

Fig. 2. Details of load sensing block and its projection.

(a) Upper supporting system

(b) Lower supporting system

Fig. 3. Details of the supporting systems.

lower ends of the specimen were both 'clamped' (or 'fixed'). This condition was accomplished by using a steel cap with a deep hole (for the upper end, see Fig.3(a)) and also a steel disc with a through-hole (for the lower end, see Fig.3(b)) the diameters of which were exactly equal to that of the tested rod. The impact load generated at the contact surface between the impact mass and the specimen was travelling downward, and then detected at the lower end of the specimen by means of the load sensing block method.[8-10] The details of the load sensing block and its sensing projection are shown in Fig.2. When a stress wave comes into the sensing projection, its stress state becomes almost stable after several wave reflections inside the projection, while a stress wave transmitted from the projection to the base block is scattered and weakened so that any stress disturbance comes back to the projection. From this reason, we can measure an impact load of any time duration precisely using the very compact device. In the present experiment, an axial displacement of the specimen was supposed to be identical to that of the impact mass, and was detected by a pair of the magnetic scale on the lateral surface of the mass and the linear magnetic encoder attached to the right supporting rod of the impact loading apparatus. These load-time and displacement-time signals were amplified through high speed strain amplifiers whose effective frequency was of up to 500MHz (CDV-700A, Kyowa Electric Instruments Co. LTD), and then recorded onto high speed digital memories (DS-8617, Iwatsu Co. Ltd.).

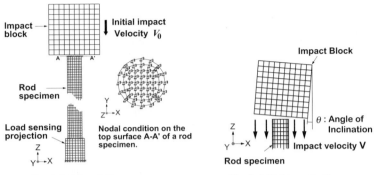

Fig. 4. FEM modeling.          Fig. 5. Initial imperfection.

## 3.   Numerical Modeling and Simulations Based on Dynamic Finite Element Method

As mentioned in the previous section, only the quasistatic velocity and the intermediate impact velocities of order of $10^3$mm/s were employed in the experiment to examine the buckling behavior of long columns; these test velocities could not cover enough range in impact velocity to construct an effective criterion for dynamic buckling. In order to overcome the situation, numerical analyses covering a wide range of impact velocities were performed by Dynamic FEM Code 'MSC-Dytran'. In the calculations, only the impact block, the specimen and the load sensing projection were modeled to reduce computation time. A typical finite element discretization for the standard computational model is shown in Fig.4. As shown in the figure, 8-node hexahedral isoparametric elements were used in the discretization. The number of elements in the cross-section of the specimen was 48, the mesh size in Z direction along the specimen was 2.5mm. Time step for iteration was so chosen as to satisfy the Courant condition for dynamic problems. The total number of nodes for the rod specimen with the diameter of 10mm and the length of 1200mm was, for example, 27417 and the number of elements was 23040.

In the present study, a small inclination of the impact block against the vertical axis of the specimen was introduced as an initial imperfection in order to initiate and develop dynamic instability, such as that shown in Fig.5. Here, through comparison between simulations and experiments, this inclination angle was determined as 1.0 degree. On the other hand, to simulate the clamped conditions at both ends of the specimen in the experiment, all nodal displacements in X- and Y-directions at the upper and the lower end of the specimen were assumed to be zero, while the nodal displacements in Z-direction at both ends were allowed so that both surfaces could maintain shear free contact with the impact block (at the upper end) or the sensing projection (at the lower end). It should be noted that since Z-direction displacements at the upper end of the specimen is not forced to be the same, and since the impact block is inclined to introduce the initial imperfection, there is a possibility that the upper end surface may rotate, and the supporting condition for the upper end is, therefore, strictly speaking, 'pseudo-clamped'. From this reason, as mentioned later, simulated dynamic buckling loads are, in many cases, slightly smaller than those in the experiments. Finally, for the input condition of dynamic loading, only an initial impact velocity of $V_0$ was given to the impact block.

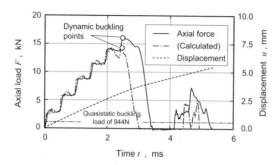

Fig. 6. Typical experimental force-time curve at the impact velocity of 1450mm/s.

## 4.    Results and Discussion

Fig. 6 shows typical load- and displacement-time diagrams for a specimen with 10mm diameter and 1200mm length at the impact velocity of 1450mm/s in the experiment. In the figure, the simulated load-time curve based on the dynamic FEM and the experimentally obtained quasistatic buckling load (almost equal to the value predicted by Euler's equation) are also shown by the single- and double- chain line, respectively. Note that the load-time curves like stairs are due to the wave reflections inside the rod specimen. In this study, an abrupt load drop without accompanying the decrease in displacement such as that shown in the figure is considered as a 'dynamic buckling point'. According to the above definition, the dynamic buckling load in this case is 14~15 times as large as the quasistatic buckling load. Namely, dynamic buckling load shows the strong velocity dependence. The simulated load-time curve shows fairly good agreement with the experimental one and this implies the validity of the present simulation method, although the predicted buckling load is slightly smaller than the actual buckling load because of the 'pseudo-clamped' supporting condition at the upper end of the specimen, as mentioned previously. In order to evaluate the effects of cross-sectional size and shape on the buckling load, let us introduce slenderness ratio $\lambda$ ($= l/k$, $k$: radius of gyration) of the specimen and dynamic to static buckling load ratio $\alpha$ ($=$ the ratio of the dynamic buckling load to the quasistatic one) here. Fig.7 and Fig.8 show the 'experimentally obtained' and 'numerically obtained' $\alpha - \lambda$ relations at the impact velocity of 1450mm/s, respectively. In the figures, it is clearly shown that the value of $\alpha$ increases with increasing $\lambda$, and this result is in strong contrast to the experimental results obtained by Hong et al.[6] as discussed previously in Section 1. Furthermore, independently of the differences in diameter and length of the specimens, it seems that there is a unified relationship between $\alpha$ and $\lambda$ for all range of experimental data. Taking account of the condition that the value of the dynamic to static buckling load ratio $\alpha$ should be approaching to 1.0 when the slenderness ratio $\lambda$ is gradually decreasing to zero (in this case, the buckling load or buckling stress given by Euler's equation becomes infinitely large), a unified relation between $\alpha$ and $\lambda$ may be written as:

$$\alpha = a \cdot \lambda^n + 1, \tag{1}$$

where $n$ denotes an exponent, a parameter $a$ may be a function that represents the velocity dependence of $\alpha$. By using the least square method, the exponent $n$ was found to take the value of 2.0 for both experimental and simulated cases independently of applied impact velocities. The approximation curves based on Eq.(1) and $n=2$ are shown in Figs.7 and 8 by the thick chain lines, and they are in good agreement with the plotted results. While, values of Parameter $a$ are dependent on the applied impact velocity. For example, in experimental cases, the values of $a$ were given as $5.3 \times 10^{-5}$ for $V_0=1000$mm/s, $6.57 \times 10^{-5}$ for $V_0=1450$mm/s and $10.7 \times 10^{-5}$ for $V_0=3000$mm/s as shown in Fig.9. Approximation curves depicted in the figure by the solid lines well express the experimental plots. Fig.10 shows the relationship between the parameter $a$ and the impact velocity. From the figure, we assume that the parameter $a$ can be described by a power law of the initial impact velocity $V_0$ as:

$$a = \beta \cdot V_0^m \,, \tag{2}$$

where $\beta$ and $m$ are constants. By using the least square method again, $\beta$ and $m$ are identified as $m=0.68$, $\beta = 4.7 \times 10^{-7}$ for the experiments and $\beta = 3.4 \times 10^{-7}$ for the simulations. You can see that the calculated curves based on Eq.(2) well express the experimental and simulated plots. Although the identified values of $\beta$ for the simulations were about 25% smaller than that for the experiments, this difference may be caused by the insufficient clamped condition for the upper end of the specimen in the simulations as mentioned previously, and the proposed

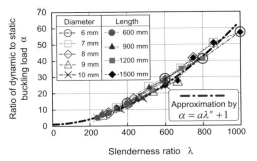

Fig. 7. Experimentally obtained relations between dynamic to static buckling load ratio $\alpha$ and slenderness ratio $\lambda$ at the impact velocity of 1450mm/s.

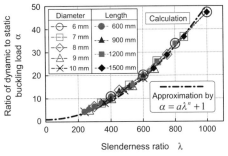

Fig. 8. Simulated relations between dynamic to static buckling load ratio $\alpha$ and slenderness ratio $\lambda$ at the impact velocity of 1450mm/s.

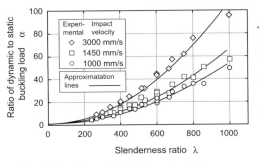

Fig. 9. Dependence of initial impact velocity on parameter $a$; value of $a$ is $5.3 \times 10^{-5}$ for $V_0=1000$mm/s, $6.57 \times 10^{-5}$ for $V_0=1450$mm/s and $10.7 \times 10^{-5}$ for $V_0=3000$mm/s if the exponent $n$ takes the value of 2.0.

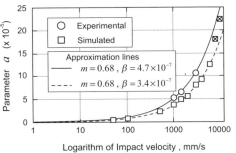

Fig. 10. Relation between the parameter $a$ and the initial impact velocity $V_0$. $a$ is able to be described as $a = \beta V_0^m$, where $m$ commonly takes 0.68, $\beta$ takes $4.7 \times 10^{-7}$ for experimental data and $3.4 \times 10^{-7}$ for simulated data.

criterion for dynamic buckling loads is considered to be still effective for various shapes of slender rods.

References

1.  G.R. Abrahamsonand J.N. Goodier, *J. Appl. Mech.* **33**, 241 (1966).
2.  T. Hayashi and Y. Sano, *Bulletin of the JSME*, **38-306**, 277 (1972).
3.  T. Hayashi and Y. Sano, *Bulletin of the JSME,* **38-306**, 286-293 (1972).
4.  T. Hayashi, Y. Sano and S. Asai, *Bulletin of the JSME*, **38-307**, 474 (1972).
5.  D. Karagiozova and N. Jones, *Int. J. of Impact Engineering*, **18**, No.7-8, 919 (1996).
6.  H. Hong, K.C. Hee and C. Shijie, *Int. J. of Solids and Structures*, **37**, 5297 (2000).
7.  C. Shijie, H. Hong and K.C. Hee, *Int. J. of Mechanical Sciences*, **44**, 687 (2002).
8.  K.Mimura,  S.Hirada,   Y.Chuma and S.Tanimura, *Trans. JSME*, **62-603**, 2609 (1996).
9.  Y.Chuman, K.Mimura, K.Kaizu and S.Tanimura, *Int. J. Impact Eng.* **19-2**, 165 (1997).
10. T. Umeda, K. Mimura and S. Tanimura, *Materials Science Research International* (*JSMS International,* Special Technical Publication-2, 429 (2001).

# CRASHWORTHINESS DESIGN OF THE SHEAR BOLTS FOR LIGHT COLLISION SAFETY DEVICES

JIN SUNG KIM[1], HOON HUH[2†]

*KAIST, Daejeon, KOREA,*
*soitgoes@kaist.ac.kr, hhuh@kaist.ac.kr*

TAE SOO KWON[3]

*Korea Railroad Research Institute, Uiwang-City, Gyonggi-Do, KOREA,*
*tskwon@krri.re.kr*

Received 15 June 2008
Revised 23 June 2008

This paper introduces the jig set for the crash test and the crash test results of shear bolts which are designed to fail at train crash conditions. The tension and shear bolts are attached to Light Collision Safety Devices(LCSD) as a mechanical fuse when tension and shear bolts reach their failure load designed. The kinetic energy due to the crash is absorbed by the secondary energy absorbing device after LCSD are detached from the main body by the fracture of shear bolts. A single shear bolt was designed to fail at the load of 250 kN. The jig set designed to convert a compressive loading to a shear loading was installed to the high speed crash tester for dynamic shear tests. Two strain gauges were attached at the parallel section of the jig set to measure the load responses acting on the shear bolts. Crash tests were performed with a carrier whose mass was 250 kg and the initial speed of the carrier was 9 m/sec. From the quasi-static and dynamic experiments as well as the numerical analysis, the capacity of the shear bolts were accurately predicted for the crashworthiness design.

*Keywords*: Shear bolt;Light collision safety devices;Crashworthiness;Dynamic shear test.

## 1. Introduction

The crashworthiness of trains is now a major concern since a crash accident of a train leads to a fatal disaster accompanying loss of lives and properties although the train accident is less frequently reported than car accidents. In order to design a reliable LCSD satisfying the standard for the train crashworthiness to minimize passenger injuries and fatalities, a more systematic approach is required based on improved energy management concepts and design involving new structural arrangements of higher absorbing capacity in a controlled manner. LCSD is employed as an energy absorber in low speed collision. Repairing charges can remarkably decrease while passenger safety is also secured by absorbing most crash energy in LCSD. The energy absorbing mechanism of LCSD is operated sequentially in each energy absorber by corresponding levels of load as shown

---

†Corresponding Author.

231

in Fig. 1. The coupler is the first energy absorber whose load-carrying capacity is 1,000 kN and the second energy absorber is an expansion tube whose driving force is over 1,500 kN. The tension bolts are installed between the first and second energy absorber and make the energy absorbing mechanism to be sequential to the levels of load.[3] After the energy absorption of the expansion tube, LCSD should be detached from the train when carrying load is over 2,000 kN and eight shear bolts are broken. A single shear bolt sustains the shear load of 250 kN respectively since eight shear bolts are designed to carry the load of 2,000 kN. The maximum load of the designed shear bolt should be verified in dynamic loading conditions experimentally. The load responses are measured with strain gauges attached to the jig set and calibrated by the reference load cell. Since the designed shear bolt undergoes dynamic shear deformation, the dynamic material properties of the base material, SCM440H, should be provided in order to take account of crashworthiness design of the shear bolts. This paper demonstrates that the maximum load of shear bolts in the quasi-static test is distinguishably different from that in the dynamic shear test.

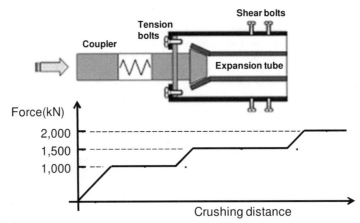

Fig. 1. Energy absorbing mechanism of Light Collision Safety Devices.

## 2.  Design of Shear Bolts and Jig System

### 2.1.  *Shear bolts*

Basic shapes of shear bolts are commercial standard bolt except they have a narrow groove as shown in Fig. 2. The flat region right below the bolt head is longer than that of conventional bolts since shear bolts need a narrow groove where shear deformation takes place. The basic dimensions of the designed shear bolt are based on the M30 standard bolt. The outer diameter of the shear bolt is 30 mm. Two kinds of shear bolts were prepared by the size of a groove whose diameter, D, is 20 and 22 mm respectively. The gap of a groove is 4 mm for all shear bolts. The material for shear bolts is SCM440H

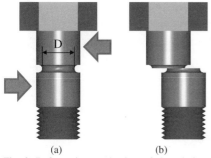

|  |  |
| --- | --- |
| (a) | (b) |

Fig. 2. Deformation mechanism of shear bolts: (a) applied forces before fracture; (b) after fracture.

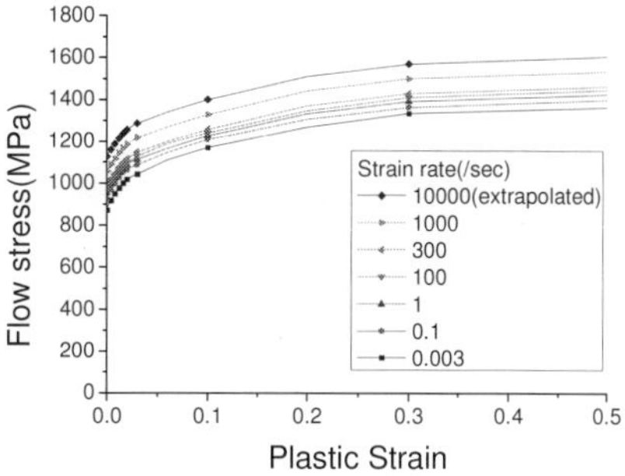

Fig. 3. Strain rate dependent stress-strain curves of SCM440H.

after heat treatment. The material is heated up to 850℃ and held for 3 hours. After the heat treatment, the material is tempered three times at 600℃ for 5 hours. The stress-strain curves of the high strength steel of SCM440H are shown in Fig. 3, which has lower strain rate sensitivity than the other conventional mild steels. The stress-strain relations were obtained by high speed material tests up to the strain rate of 1,000/sec and estimated by modified Johnson-Cook model[4] up to the strain rate of 10,000/sec. The maximum strain rate locates at the center of a groove and is around 3,000/sec at finite element analyses. For the reason, the upper bound of strain rates in piecewise linear data was determined to be 10,000/sec which can cover the maximum strain rate during high speed shear deformation.

## 2.2. *Shear-off jig system for crash tests*

The shear bolts are applied to install LCSD to the front head of a train. The shear bolt fails when the crush load exceeds the designated load, 250 kN. Design of the shear-off jig system is obviously simple for quasi-static shear-off tests, but becomes very complicated for dynamic crash shear-off tests since a crash test needs sufficient loading speed, crash energy and reliable measurement system. Therefore, the shear-off jig system should be carefully designed and verified for a corresponding crash condition. The shear-off jig system in Fig. 4 converts compressive loading to shear loading. A carrier of the crash tester impacts the end of the shear-off jig and the polyurethane pad stops the moving jig after the fracture of the shear bolt. A half bridge of strain gages is devised for a load measurement since the load measurement using load cells sustaining the full jig system has a severe load ringing problem. The strain gages are attached both sides of a lower jig which is fixed on the left side as shown in Fig. 4 (b). Load calibration of an output signal from strain gauges is performed in a quasi-static UTM(Universal Test Machine) by comparing an output signal from strain gauges with the load signal from UTM as shown in Fig 5. Two signals are in proportion and the scale factor is obtained by dividing the load signal by the output signal from strain gauges. The shear-off jig system can perform both quasi-static and dynamic shear-off tests using the same load measurement method.

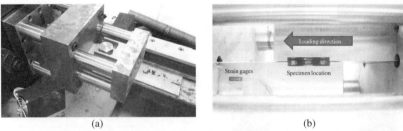

(a)                                                    (b)

Fig. 4. Shear-off jig system: (a) fixed on the wall; (b) shear loading mechanism.

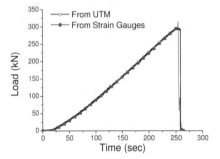

Fig. 5. Strain gage output voltage versus load signal from UTM.

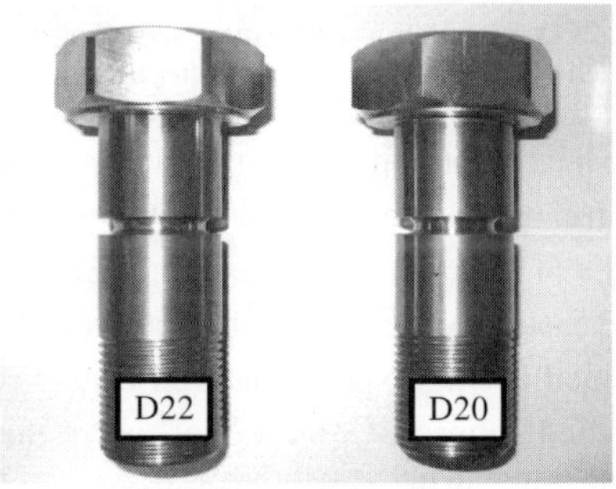

Fig. 6. Prepared shear bolts.

## 3.  Experiment

The shear bolts are prepared by a groove as shown in Fig. 6. Quasi-static shear tests are performed with a static UTM whose maximum capacity is 50 tonf. The shear-off jig system is installed upon the bed of UTM. A data acquisition board on PC captures the signals for the load and displacement from a strain conditioning amplifier and a linear displacement transducer. Sequential deformed shapes during shear deformation are shown in Fig. 7 (a). Deformed shapes show that the specimen is tilted during shear deformation by 7 degree which is measured by image processing of the third picture of Fig. 7 (a). The load responses during quasi-static shear tests are shown in Fig. 8 (a). The maximum loads of shear bolts are 259.6 kN for D22 specimens and 216.1 kN for D20 specimens. The total stroke until failure is about 3.8 mm for D22 specimens and 3.5 mm for D20 specimens. Deformed shapes after quasi-static and dynamic shear tests are shown in Fig. 9. The specimen, D20–2, shows abnormal fracture at quasi-static deformation. The abnormal fracture seems to be affected by initial defects of an original specimen. D22 shear bolts seem to satisfy targeting shear-off load while D20 shear bolts cannot satisfy targeting shear-off load in quasi-static shear tests. But dynamic shear tests are still needed to evaluate the crashworthiness of the shear bolts since the targeting shear load, 250 kN, should be evaluated in crash conditions. Dynamic shear tests were performed in High Speed Crash Tester which was horizontal-type. The shear-off jig system is fixed on the wall horizontally. The mass of a moving carrier is 250 kg and crash speed is 9 m/sec. The deformed shapes are continuously taken by a high speed camera at

7,000 frames/sec. Load responses are obtained from the strain conditioning amplifier at the sampling rate of 500 kHz as shown in Fig. 8 (b). The load responses at dynamic shear tests are highly reliable after 0.2 msec since the load oscillation caused by load ringing phenomena decreases after 0.2 msec. The maximum shear-off loads in dynamic shear tests are 312.6 kN for D22 specimens and 261.1 kN for D20 specimens. Both D22 and D20 specimens show clear fracture surfaces after quasi-static and dynamic shear tests as shown in Fig. 10. Duration of the dynamic shear tests is ranged from 0.43 to 0.49 msec and the total stroke is ranged from 3.9 mm to 4.5 mm. Consequently, D20 shear bolts satisfy design criteria, the maximum shear load and clear fracture surface, in the crash speed of 9 m/sec.

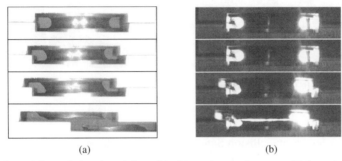

(a)                                        (b)

Fig. 7. Deformed shapes during shear deformation: (a) quasi-static shear tests; (b) dynamic shear tests.

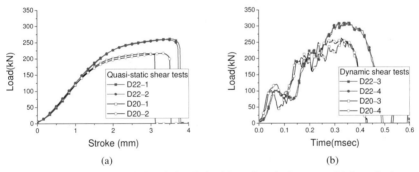

(a)                                        (b)

Fig. 8. Load responses of the designed shear bolts: (a) quasi-static shear tests; (b) dynamic shear tests.

Table 1. Maximum shear load in quasi-static and dynamic shear tests.

| Test method | Min. diameter (mm) | Specimen # | Max. shear load (kN) |
| --- | --- | --- | --- |
| Quasi-static shear tests | 22 | D22–1 | 260.7 |
| | | D22–2 | 258.4 |
| | 20 | D20–1 | 215.7 |
| | | D20–2 | 216.5 |
| Dynamic shear tests | 22 | D22–3 | 312.1 |
| | | D22–4 | 313.2 |
| | 20 | D20–3 | 259.2 |
| | | D20–4 | 262.9 |

Fig. 9. Deformed specimens after quasi-static and crash tests.

(a)                                                    (b)

Fig. 10. Typical fracture surface after dynamic shear tests: (a) D22; (b) D20.

## 4. Conclusion

The shear bolts for Light Collision Safety Devices were designed and evaluated by quasi-static and dynamic shear tests. The designed shear bolts, D20 specimens, showed the maximum shear load of 216.1 kN for quasi-static shear tests and 261.1 kN for dynamic shear tests while D22 specimens showed the maximum shear load of 259.6 kN for quasi-static shear tests and 312.6 kN for dynamic shear tests. The maximum shear load of D20 specimens is 261.1 kN which is 4.4% larger than targeting shear-off load. Accordingly, D20 shear bolts with a material of SCM440H operate sufficient function in Light Collision Safety Devices.

## References

1.  SAFETRAIN, BRITE/EURAM Project n.BE-3092, *Dynamic tests*, SAFETRAIN Technical Report T8.2-F, Deutsche Bann, Berlin, Germany (2001).
2.  J. H. Lewis, *Proc. of WCRR '94* (Paris, 1994), pp. 893–900.
3.  J. S. Kim, H. Huh, W. M. Choi and T. S. Kwon, *Proc. of The Korean Soc. of Automot. Engineers spring conf.* (Changwon, 2007), **3**, pp. 2037–2041.
4.  H. Huh, W. J. Kang and S. S. Han, *Exp. Mech.* **42**, 1, 8 (2002).
5.  J. –S. Koo and Y. H. Youn, *Int. J. Automot. Techn.* **5**, 3, 173 (2004).
6.  J. S. Kim, H. Huh, K. W. Lee, D. Y. Ha, T. J. Yeo and S. J. Park, *Int. J. Automot. Techn.* **7**, 5, 571 (2006).
7.  H. Huh, S. B. Kim, J. H. Song and J. H. Lim, *Int. J. Mech. Sci.* **50**, 918 (2008).

# COMPUTATIONAL PERFORMANCE EVALUATION OF A SIDE STRUCTURE CONSIDERING STAMPING EFFECTS

SE-HO KIM[1†]

*School of Automotive, Industrial and Mechanical Engineering, Daegu University*
*15, Naeri, Jillyang, Gyeongsan, Gyeongbuk, 712-714 KOREA,*
*mvksh@daegu.ac.kr*

KEE-POONG KIM[2]

*Automotive Components Service Center, KITECH*
*971-35, Wolchul-dong, Buk-gu, Gwangju, 500-460 KOREA,*
*keepkim@kitech.re.kr*

Received 15 June 2008
Revised 23 June 2008

In this paper, the influence of the stamping effect is investigated in the performance analysis of a side structure. The analysis covers the performance evaluation such as crashworthiness and NVH. Stamping analyses are carried out for a center pillar, and then, numerical simulations are carried out in order to identify the stamping effect on the crashworthiness and the natural frequency. The result shows that the analysis considering the forming history leads to a different result from that without considering the stamping effect, which demonstrates that the design of auto-body should be carried out considering the stamping history for accurate assessment of various performances.

*Keywords*: Performance evaluation; side structure; stamping effects.

## 1. Introduction

An analysis of crashworthiness, NVH, durability, and other performance measures has been conducted as a means of predicting major performances of an automobile to reduce the development cost and period in automobile design. In today's performance analyses, finite element models must be more precise as computer technology develops. In this regard, active research has been carried out by carmakers and researchers to achieve the accuracy which guarantees accurate prediction of material properties, part characteristics, and so on. The thin-walled members of an auto-body have stamping effects caused by the stamping process including initial residual stress, work hardening, and irregular distribution of thickness. Taking into consideration stamping effects, the behaviors of thin-walled members show considerably different characteristics. A performance analysis

---

[†]Corresponding Author.

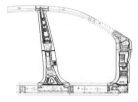

Fig. 1. CAD data of the side structure assembly.

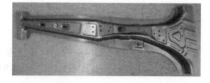

Fig. 2. Shape of the center pillar member.

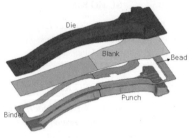

Fig. 3. Finite element model for the stamping analysis of the center-pillar member.

should consider any change in mechanical properties that occur in the production process of members, and the related effects should be identified from the preceding stage[1-3].

In this paper, a stamping analysis is conducted for the major member constituting a side structure of an auto-body; based on the results, analyses for crashworthiness and NVH are performed on the side structure. From the results of these analyses, the deforming characteristics of the side structure, change in energy-absorbing capabilities, and change in natural frequency are identified in order to select variables to be reflected in future performance analyses of a full car.

## 2.  Stamping Analysis of the Center Pillar

This study analyzes the side structure of the auto-body illustrated in Fig. 1. In particular, a stamping analysis is performed on a center pillar which is the most important load-carrying and energy-absorbing members in the case of side impact. The shape of center pillar used for the stamping analysis is described in Fig. 2.

The stamping analysis of the center pillar is performed only for the first stage, the draw-forming process (OP10), where deformation mostly takes place. The material of the blank is SPRC440E, which is a tailor welded blank (TWB) with a thickness of 1.4 mm and 0.8 mm in the upper and lower parts, respectively. The stress-strain curve of the plate is expressed in $\bar{\sigma} = 1498(\bar{\varepsilon}^p + 0.000919)^{0.0437}$ MPa. A commercial code LS-DYNA3D[4] is used for the analysis. Fig. 3 portrays the finite element model of the tools and the blank used for analysis. A blank holding force of 100 kN is imposed; assuming non-lubricating conditions, the Coulomb friction coefficient is set as 0.15.

In Fig. 4, the deformed shape obtained from the analysis is compared to that of an actual product; overall, they show fairly good agreement with each other. Fig. 5 compares the distribution of the thickness reduction ratio obtained from the experiment and analysis

(a) analysis  (b) experiment

Fig. 4. Comparison of deformed shapes of the blank between the analysis and the experiment.

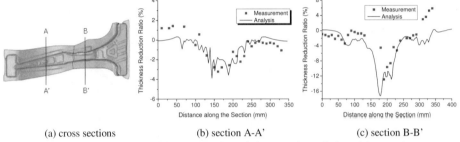

(a) cross sections  (b) section A-A'  (c) section B-B'

Fig. 5. Comparison of the thickness reduction ratio between the analysis and the experiment.

(a) thickness reduction ratio(%)  (b) effective plastic strain

Fig. 6. Distribution of the thickness reduction ratio and the effective plastic strain after the stamping analysis.

at cross sections, which illustrates similar distribution. As seen in the distribution in Fig. 6, the side wall of the member is reduced by a maximum of 25% in thickness, and the maximum effective plastic strain is 0.3, which means that mechanical properties are changed dramatically by the manufacturing process.

## 3. Performance Analysis with Consideration of Stamping Effects

Using the results of the stamping analysis described in section 2, a side impact analysis and an NVH analysis of the side structure are conducted to identify stamping effects. The performance analysis considers the side structure having 12 members as described in Fig. 1: LS-DYNA3D is used in the performance analysis.

### 3.1. Conditions for performance analysis

A side impact analysis is conducted in accordance with test procedure of US-SINCAP[5]. In this impact test method, a moving deformable barrier (MDB) with a slope of $27^\circ$ crashes into the side of a stopped vehicle at a speed of 38.5 mph. In a side impact, most of the collision energy is absorbed by the center pillar, with the remainder being hardly

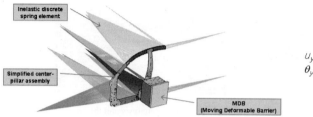
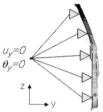

Fig. 7. Finite element model of a simplified side structure assembly with inelastic spring elements and a simplified moving deformable barrier.

Fig. 8. Finite element model of a simplified side structure assembly with the simplified boundary condition for the NVH analysis.

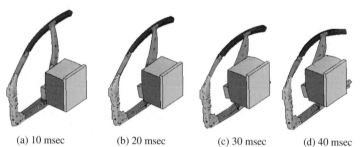

(a) 10 msec      (b) 20 msec      (c) 30 msec      (d) 40 msec

Fig. 9. Deformed shapes of the side structure assembly during the side impact analysis.

Table 1. Difference of the intrusion amount of the center pillar at 30 msec.

| Location | w/o stamping effect (mm) | w/ stamping effect (mm) | Difference (mm) |
|---|---|---|---|
| Pelvis(z=400mm) | 332.87 | 324.93 | -7.94 |
| Rib(z=700mm) | 230.80 | 221.74 | -9.06 |
| Head(z=1000mm) | 76.53 | 82.69 | +6.34 |

deformed and undergoing rigid body motion. Thus, using the method suggested by Bae *et al.*[6], a simplified center pillar model is constructed with the modified MDB in order to reduce the analysis time as shown in Fig. 7. The initial speed condition of MDB is set at 15.33 m/s toward the cabin room direction. A comparison is made for deformation amount, deformation mode, amount of energy absorption, and impact force with MDB.

For the NVH evaluation of the side structure, a simplified model is constructed to conduct the modal analysis, and changes in the natural frequency are compared considering of stamping effects. The innermost nodal points toward the width(y) direction, as in Fig. 8, are restrained with respect to translational and rotational degrees of freedom toward the y direction in order to impose the symmetric condition.

## 3.2. Results of performance analysis

Based on the analysis conditions explained in Section 3.1, a side impact analysis and an NVH analysis are performed. Fig. 9 shows deformed shapes of the side structure and the MDB upon collision. Table 1 presents the tendency of the intrusion amount of the center pillar toward the cabin room at regions of the dummy, and Fig. 10 illustrates the shape of

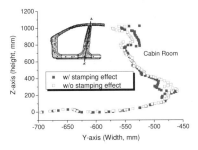
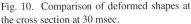

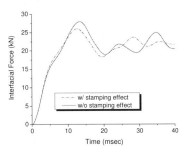

Fig. 10.  Comparison of deformed shapes at the cross section at 30 msec.

Fig. 11.  Comparison of impact forces between the side structure assembly and the MDB.

Table 2. Difference of energy absorbed in the side structure assembly.

| Time (msec) | w/o stamping effect (J) | w/ stamping effect (J) | Difference (%) |
|---|---|---|---|
| 10 | 1513.20 | 1577.64 | 4.26 |
| 20 | 4148.14 | 4231.92 | 1.24 |
| 30 | 6284.99 | 6625.40 | 5.42 |
| 40 | 7911.19 | 8294.88 | 4.85 |

deformation toward the width direction in a cross section of the center pillar. In the upper region of the center pillar, the intrusion amount is greater when stamping effects are considered than when not; opposite results are shown in the lower region of the center pillar. In the stamping process of the center pillar, the thickness at the side wall is decreased significantly while the plastic strain is increased at that area. Taking these contradicting effects into consideration together, the strength of the members is determined. In the upper region, small amount of bending deformation occurs without contact with MDB, and thus effects of thickness reduction become dominant. Therefore, a reduction in thickness leads to decreased strength, and larger deformation occurs when the forming history is taken into account. In the lower region, considerable plastic deformation results from collision with the MDB. An increase in the effective plastic strain makes a greater contribution to enhanced strength than does thickness reduction, and consequently smaller deformation takes place when the forming history is considered than when not taken into account.

In Fig. 11, a comparison is made between impact forces calculated from the contact force between the MDB and the side structure. Table 2 compares the amounts of energy absorption of the side structure. The impact force is smaller when stamping effect is taken into consideration, which means that, upon initial deformation, the effects of collapse load reduction by thickness reduction are greater than those of increased yield strength by prehistoric strain, for deformation by bending, not axial collapse, is dominant in side impacts. The amount of energy absorption is greater when stamping effects are considered than otherwise. This implies that as collision-induced deformation proceeds, effects of work hardening caused by forming increase the energy absorption ratio. When

Table 3. Comparison of the natural frequency of the side structure assembly with respect to the stamping effect.

| Mode | w/o stamping effect (Hz) | w/ stamping effect (Hz) | Error (%) |
|------|------|------|------|
| 1st | 77.9 | 77.5 | -0.51 |
| 2nd | 122.3 | 122.0 | -0.25 |
| 3rd | 159.7 | 159.5 | -0.13 |
| 4th | 177.7 | 177.1 | -0.34 |

stamping effects are considered based on the above results, it is noted that deformation modes of side structure change dramatically by height, and that the amount of energy absorption grows larger.

In Table 3, a comparison is made between natural frequencies that are shown when the forming history is considered and when it is not taken into account; in both cases, the frequency differs with each other to below 1%. The mass of members is not changed significantly by forming while stiffness of the structure is decreased slightly due to thickness reduction in some regions. Changes of the mass and the stiffness are not significant compared to the effect of shape change, and thus it is concluded that no significant difference is shown in the natural frequencies.

## 4.  Conclusions

This study has performed a crash analysis and an NVH analysis of the side structure, in consideration of stamping effects induced by the sheet metal forming process, in order to enhance the accuracy of automobile performance analyses. A stamping analysis has been carried out on the center pillar constituting the side structure. Results from the succeeding performance analysis have been compared between when stamping effects are taken into consideration and when not taken into account. It is noted that the different deformation mode is obtained by the combination of contradicting effects of thickness reduction and work hardening in the impact analysis. By comparing natural frequencies in the NVH analysis, it has been clarified the consideration of forming history has little effect on the analysis results. It has been demonstrated that in an automobile analysis when the large plastic deformation takes place, it is essential to consider effects of reduced thickness owing to the forming of members as well as increased strength induced by work hardening in designing auto parts.

## References

1. H. Huh, K. P. Kim, S. H. Kim, J. H. Song, H. S. Kim and S. K. Hong, *Int. J. Mech. Sci.* **45**, 1645 (2003).
2. T. Dutton, R. Sturt, P. Richardson and A. Knight, *Trans. SAE 2001-01-3050* (2001).
3. S. Simunovic, J. Shaw and G. A. Aramayo, *Trans. SAE 2001-01-1056* (2001).
4. J. O. Hallquist, *LS-DYNA3D Keyword User's Manual-Vers. 9.7* (LSTC, CA, 2003)
5. NHTSA, *Laboratory test procedure for new car assessment program side impact testing* (2002).
6. G. H. Bae, J. H. Song, H. Huh and S. H. Kim, *Int. J. Impact Eng.* submitted.

# A NEW METHOD FOR VIBRATION RESPONSE OF BEAM ON FOUNDATION UNDER MOVING LOAD

YOUZHE YANG[1†], XIURUN GE[2]

*School of Naval Architecture, Ocean and Civil Engineering Shanghai Jiao tong University, Room 2302 Haoran*

*High Technique Building, 1954 HushanRoad, Shanghai, 200030, P. R. CHINA,*

*yangyzh@sjtu.edu.cn,xiurunge@public.wh.hb.cn*

Received 15 June 2008
Revised 23 June 2008

In this research, an approach combining the precise time integration method (PTIM) and mode decomposition method is proposed to compute the response of beam structures resting on viscoelastic foundation. The PTIM has high precision, high efficiency, but it still suffers from the problem of large-size matrices when directly computing structure. And this problem can be overcome by mode superposition. Thus, the present paper integrates the PTIM and mode decomposition method together which holds the explicit recurrence form of precise algorithm. Comparing with the other numerical methods, it is found that the presented method is much more precisely and time-saving. The effect of the speed of moving load, the foundation stiffness and the length of the beam on the response of beam have also been studied. These numerical computation results show that the present method is effective and feasible.

*Keywords*: Precise time integration; mode superposition; moving load; viscoelastic foundation.

## 1. Introduction

The dynamic analysis of beam resting on foundation as a dynamic soil-structure interaction problem is of great importance in structural and foundation engineering. It has been studied by many investigators (See Refs. 1–4) who adopted the numerical methods such as the central difference, Newmark-$\beta$, and $Wilson-\theta$ method. Undoubtedly, these mainstream methods have played important role, but they still have some limitations[5]. To improve the accuracy of time step integration, Zhong and Williams[6] proposed the Precise Time Step Integration Method—PTSIM with arbitrary order of accuracy. This paper applies PTSIM to study the dynamic responses of beam on foundation. For a beam structure, without an elastic foundation, the authors (See Refs. 7–8) have applied the integration method to calculate the dynamic response of Euler-Bernoulli beam under moving load.

In this research work, a procedure using the mode decomposition method together with the precise time integration method (MDPIM) is proposed to investigate the

---

[†]Corresponding Author.

dynamic behavior of Euler-Bernoulli beam resting on viscoelastic foundation subjected to moving load. The mode decomposition method serves to transform the dynamic equations from geometric coordinates to normal coordinates and makes reduction of the degrees, and then the precise integration method serves to solve the transformed equations whose order has been reduced. Numerical examples are presented for moving load across the beam and the results show that this new coupling method can improve the computation efficiency significantly.

## 2. Equations of Motion

Let y(x, t) denote the vertical deformation of the beam with x being the traveling direction of the moving load and t being time. The well-known governing equation of a Bernoulli-Euler beam on a foundation [9] is:

$$EI\frac{\partial^4 y}{\partial^4 x} + ky + c\frac{\partial y}{\partial t} + m\frac{\partial^2 y}{\partial^2 t} = F(x,t) \tag{1}$$

Where $EI$ is the rigidity of the beam; m is the unit mass of the beam, $k$ is the modulus of sub-grade reaction, c is a dashpot of viscosity, and $F(x, t)$ is applied external load.
The deflection $y(x, t)$ at position x can be expresses by

$$y(x,t) = \sum_{n=1}^{\infty} A_n(t)\Phi_n(x) \tag{2}$$

Where $A_n(t)$ the amplitude or the generalized coordinate; $\Phi_n(x)$ is the mode shape for the nth mode, respectively.

Applying Lagrange's equation and mode orthogonality, the normal equation for the nth mode could be obtained and given in the standard form as:

$$\ddot{A}_n + 2\xi_n\omega_n \dot{A}_n(t) + \omega_n^2 A_n(t) = F_n(t)/M_n \tag{3}$$

Where $\omega_n$ and $\xi_n$ =natural frequency and the damping ratio; and $M_n$ and $F_n(t)$ = generalized mass and force for the nth mode, respectively.

## 3. The High-Precision Integration Scheme

Eq. (3) can be rewritten in matrix form:

$$\ddot{A} + 2\xi\Omega \dot{A} + \Omega^2 A = f(t) \tag{4}$$

Where $2\xi\Omega = (\Phi^T M\Phi)^{-1}(\Phi^T C\Phi)$, $\Omega^2 = (\Phi^T M\Phi)^{-1}(\Phi^T K\Phi)$, $f(t) = (\Phi^T M\Phi)^{-1}(\Phi^T F)$

M is a diagonal matrix of the generalized modal masses $M_n$, K is a diagonal matrix of the generalized modal stiffness $K_n$, C is a diagonal matrix of coefficients $C_{nm}$, and $f(t)$ is a column vector of the generalized modal forces $F_n(t)$.
According to the precision time-step integration method [6], the equation of the motion of the beam on foundation in Eq. (4) can be written as

$$u = Hu + P \tag{5}$$

Where u is the response vector of size $2n \times 1$, H is a $2n \times 2n$ matrix, and P is the force vector of size $2n \times 1$, with:

$$u = \left\{ \begin{array}{c} A \\ \vdots \\ A \end{array} \right\}, H = \left[ \begin{array}{cc} 0 & I \\ -\Omega^2 & -2\xi\Omega \end{array} \right], P = \left\{ \begin{array}{c} 0 \\ f(t) \end{array} \right\} \tag{6}$$

Eq. (5) can be written into discrete equations using the exponential matrix representation. Integrating Eq. (5) and then can be expressed in the following discrete form:

$$u((j+1)h) = e^{Hh}u(jh) + \int_{jh}^{(j+1)h} e^{H((j+1)h-\tau)} P(\tau) d\tau \tag{7}$$

$$= e^{Hh}u(jh) + H^{-1}[e^{Hh} - I]P(jh) \quad (j=0,1,2\ldots)$$

Where h is the time step of integration. The force $P(\tau)$ is assumed to be constant within the time interval from j h to (j+1) h. Thus, the basic recursive formula of MDPIM is :

$$\left\{ \begin{array}{c} y \\ \vdots \\ y \end{array} \right\}_{j+1} = \left[ \begin{array}{cc} \Phi & 0 \\ 0 & \Phi \end{array} \right] \left( T \left\{ \begin{array}{c} A \\ \vdots \\ A \end{array} \right\}_j + H^{-1}[T-I]P_j \right) \tag{8}$$

The precision of integration depends on the accuracy of exp (Hh). The 2N algorithm is presented [6] to compute $T = \exp(Hh)$

According to Eq. (8), for a structure with N DOF, if only the first p modes are considered, the order of matrix T will be reduced from $2n \times 2n$ to $2p \times 2p$. Moreover, Eq. (8) indicates that the present method holds the explicit recurrence form of precise algorithm; therefore, the present method is highly efficient.

## 4. Simulation and Results

To examine the performance of the present method, the following computation examples are presented.

### 4.1. *A beam on Winkler foundation subjected to a concentrated moving load*

In this numerical analysis the beam is assumed to be a typical rail, which is subjected to a concentrated external force at the middle of the beam with the following properties [10].

Table 1. The parameters for the beam on foundation.

| L(m) | m(kg/m) | E( $Nm^{-2}$ ) | I( $m^4$ ) | k( $Nm^{-2}$ ) | v( $ms^{-2}$ ) |
|------|---------|---------------|-----------|---------------|---------------|
| 50 | 43650.86 | 3.303e10 | 18.638 | 1.14e7 | 90 |

A load traveling with a velocity v takes $t_d = L/v$ to cross the beam. The theoretical formulation of this problem is [11] :

When $t \leq t_d$, $y(x,t) = \dfrac{2F_0}{mL} \sum_{j=1}^{\infty} \dfrac{1}{\omega_j^2 - (j\pi v/L)^2} (\sin\dfrac{j\pi vt}{L} - \dfrac{j\pi v}{\omega_j L} \sin\omega_j t) \sin(\dfrac{j\pi x}{L})$

When   $t > t_d$ ,   $y(x,t) = \dfrac{2F_0}{mL} \displaystyle\sum_{j=1}^{\infty} \dfrac{1}{\omega_j^2 - (j\pi v/L)^2} \dfrac{\pi v t}{\omega_j L} *[(-1)^j *(\sin\omega_j(t-t_d) - \sin\omega_j t)]\sin(\dfrac{j\pi x}{L})$        (9)

Where,  $\omega_j = a(\pi/L)^2 \sqrt{j^4 + \lambda^4}, j = 1, 2, ...,$        $a^2 = EI/m, \lambda^4 = (L/\pi)^4 k/EI$

$F_0$ is  the amplitude of the applied load, which is equal to one here.

The vertical response of the beam is calculated by the central difference, Houbolt, Wilson-$\theta$, where $\theta$ is 1.4 and Newmark-$\beta$ ,where $\alpha$ is 0.5, $\delta$ is 0.25 for the time step $\Delta t = 0.01$. The comparison of the solutions of MDPIM and other numerical methods are shown in Fig.1, Table 2. In Fig.1 the current method is consistent with the theoretical solutions, better than other numerical methods. The example shows that MDPIM has a higher precise than other numerical methods and three times faster than Newmark-$\beta$ .

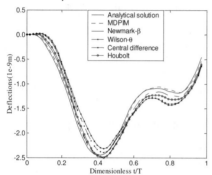

Fig. 1. Comparison of the results of the proposed method with other several numerical methods.

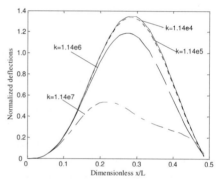

Fig. 2.  The effect of foundation stiffness on normalized deflections in mid-span.

Table 2. Computation time for time analysis for example 4.3.

| Method | Proposed method | Newmark-$\beta$ |
|---|---|---|
| Computation time(s) | 4.06 | 11.49 |

### 4.2. *A beam on a viscoelastic foundation under moving load*

The basic parameters are also the same as in the first example.

#### 4.2.1. *The effect of foundation stiffness*

The effect of the stiffness $k$ variation of the viscoelastic foundation is examined for the same length and damping $\xi = 0.1$. The normalized deflections under moving load are shown in Fig.2. As expected, the deflection under moving decreases with increased stiffness of the foundation and decreases significantly especially when the stiffness increases beyond the values of 1.14e5 $Nm^{-2}$ . In general, the practical values of  $k$ are greater than 1.14e5 $Nm^{-2}$ . So these results indicate that when a concentrated force with high-speed travels along a beam, the dynamic effects are greatly influenced by the foundation stiffness.

### 4.2.2. *The effect of traveling speed*

The same beam as above is also considered. But now let the beam resting on viscoelastic foundation with a stiffness $k$=1.14e7 $N/m^2$ and damping $\xi = 0.1$. The velocity of the moving load varies from 30m/s to 100m/s. The results for the dynamic normalized deflections in mid-span are shown in Fig.3 (a). It is interesting to note that the dynamic normalized deflections are reasonably constant for traveling speeds which are within the range of 60m/s, and only increase slightly as the traveling speed increasing for the velocity is beyond the value of 60m/s. When the stiffness reaches to 1.14e8 $N/m^2$, the peak of normalized deflections are constant. However, when the value of the foundation stiffness becomes smaller, this is not the case, shown in Fig.3 (a). When the foundation stiffness is lower than 1.14e7 $N/m^2$, the peak of normalized deflections increase distinctly with the traveling speed increasing, but remain constant for different foundation stiffness. Hence, when the foundation stiffness is relatively large, as in the case of foundation of railway tracks, the influence of the traveling speed, especially, within the range 60m/s, is negligible as the results in Fig.3 (a) demonstrates.

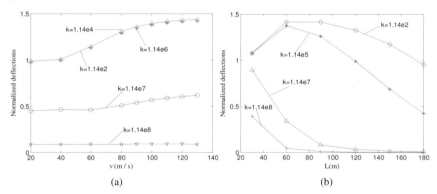

(a)                                    (b)

Fig. 3. Variation of peak values of dynamic normalized deflections for different stiffness:   (a) with traveling speed; (b) with span length.

### 4.2.3. *The effect of the span length of the beam*

In order to study the influence of the length of beam on its dynamic response, a simply supported beam with a constant foundation $k$=1.14e7 $N/m^2$, damp ratio $\xi = 0.1$ and the load velocity of 90m/s is considered. The length of the beam varies from 30m to 200m and the dynamic normalized deflections on the mid-span are shown in Fig.3 (b). It can be seen that the normalized deflections decrease as the lengths increase. And the decreasing is significant when span length is lower than 120m. When the span length is beyond the value of 120 m, the normalized deflections almost remain constant. So does that of $k$=1.14e8 $N/m^2$. However, at lower values of $k$, it is not the case, as shown in Fig.3 (b). When the foundation stiffness $k$ is reduced to 1.14e2 $N/m^2$ and 1.14e5 $N/m^2$, the peak normalized deflections, which initially increase as the span length increases, then decrease with the span beyond the values of 60m. Practical values of k are greater than $1.14e5 \ N/m^2$ , so the effect of length for $L \geq 120m$ is not significant in this case.

## 5. Conclusions

The present paper, based on the mode decomposition and precise time integration method, analyzed the dynamic analysis of beam on a viscoelastic foundation subjected to moving load. This method combines the advantages of the mode decomposition and the precise time integration method. Numerical results are presented for a simply supported beam. The deflections of the beam on Winkler foundation, as a special case, are found to be in excellent agreement with the theoretical analysis. The proposed method is applied to obtain the response time histories in which a more economical computer effort and higher precision could be achieved with the same time step of integration compared with the Newmark method. At the same time, for the beam on viscoelatic foundation, the effects of some important parameters, such as the foundation stiffness, the damp ratio, traveling speed and the length of beam, have also been studied. The comparisons and calculations prove that this method can be used effectively foe forced vibration analysis of beams on foundation.

## References

1. Hetenyi M. In, *Beams on elastic foundations, Scientific series, vol. XVI,* (The University of Michigan Press, University of Michigan Studies, Ann Arbor, 1946).
2. Andersen L, Nielsen SRK and Kirkegard PH, Finite element modeling of infinite Euler beams on Kelvin foundations exposed to moving loads in convected co-ordinates, J. Sound Vib. 578 (2001).
3. Thambiratnam D P and Zhuge Y, Dynamic analysis of beams on an elastic foundation subject to moving loads, J. Sound Vib. 149 (1996).
4. Lu Sun, Analytical dynamic displacement response of rigid pavements to moving concentrated and line loads, Int. J. Solids and Struct. 1830 (2005).
5. Mengfu Wang and F.T.K. Au, Assessment and improvement of precision time step integration method, J. Comp. and Struct. 779 (2006).
6. W.X. Zhong and F.W. Williams. A precise time step integration method-Proc Inst Mech Engrs, Part C: J Mech. Eng. Sci. 427 (1994).
7. X. Q. ZHU and S. S. LAW. Precise time-step integration for the dynamic response of a continuous beam under moving loads, J. Sound Vib. 962 (2001).
8. Bin Tang, Combined dynamic stiffness matrix and precise time integration method for transient forced vibration response analysis of beams, J. Sound Vib. 1 (2007).
9. Kenny J T, Steady-state vibration of beam on elastic foundation for moving load, J. Appl. Mech. **21**, 359 (1954).
10. Violeta Medina Andres, BS, Thesis, Massachusetts Institute of Technology, 2006.
11. A.K. Chopra, Dynamic of Structures: *Theory and Applications to Earthquake engineering, second ed.* (Prentice-Hall, New Jersey, 2005).

# Part D

# Engineering Applications
# and Case Studies

# TRIMMING LINE DESIGN USING AN INCREMENTAL DEVELOPMENT METHOD AND A FINITE ELEMENT INVERSE METHOD

YOUN-JOON SONG[1]
*Hyundai-motor Company, Ulsan, KOREA,*
*yjsong01@hyundai-motor.com*

CHUN-DAL PARK[2],
*Daegu Machinery Institute of Components and Materials, Daegu, KOREA,*
*pcd0727@dmi.re.kr*

YOUNG-HO HAHN[3]
*Dept. of Mechanical Engineering, Konkuk University, Seoul, KOREA,*
*yhhahn@konkuk.ac.kr*

WAN-JIN CHUNG[4†]
*Department of Die and Mould Engineering, Seoul National University of Tech. Seoul, KOREA,*
*wjchung@snut.ac.kr*

Received 15 June 2008
Revised 23 June 2008

In automobile panel manufacturing, the design of a feasible trimming line is crucial in obtaining an accurate edge profile after flanging. In this study, an effective method which combines the robust incremental development method and finite element inverse method is presented. The finite element inverse method is used to analyze the flanging process. In using the finite element inverse method, the main obstacle is the generation of the initial guess. An improved incremental development method is presented to handle badly-shaped element, various element sizes, and undercut parts. This method incrementally develops a 3D triangular mesh onto the drawing tool surface by layer-wise development and smoothing technique. The effectiveness of this method is verified by two numerical examples.

*Keywords*: Trimming line; finite element inverse method; incremental development method.

## 1. Introduction

After drawing of an automobile panel, trimming is generally performed before flanging. Finding an optimal trimming line is essential in obtaining a precise product edge profile after flanging. An inaccurate edge profile can cause troubles in the subsequent fabrication process.

---

†Corresponding Author.

In the section-based method, widely used in industries for the design of trimming lines, the designer generates a tool section and product section by defining a section plane containing both the drawing tool surface and the product shape after flanging. Then, the product section is wrapped on the tool section while maintaining the length. The end point of the wrapped product section is defined as the trimming point of this section. The result is dependent strongly on the choice of sections because the section plane constrains the motion of the sheet. Especially, for parts with considerable deformation, this method shows a large deviation from the target edge profile.

Many researchers have tried to make a robust automatic development technique from a final product shape using area constancy[1-2]. However, these geometry-based methods have encountered serious problems in producing reliable results since they can generate reliable results only for regular mesh. Recently, an incremental development method by authors[3] and a mosaic development method[4] are reported. However, these methods show limitations in handling irregular mesh. Thus, more robust development method is required for the irregular mesh with bad-shaped elements. In the previous work[3], the finite element inverse method[5-6] is used to find a more accurate trimming line using an initial guess obtained from the incremental development method.

In this study, new development method using the consistent layer-wise development strategy and an enhanced energy-based smoothing is presented. Using an initial guess obtained by the incremental development, the finite element inverse method finds the initial configuration on the drawing tool surface. Trimming lines created by the present method and section-based method are compared using incremental FEM simulation for numerical examples.

## 2.  Trimming-Line Design Process

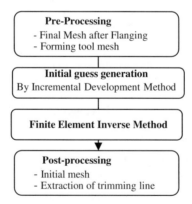

Fig. 1. Flow chart for design of trimming Line.

The overall flow chart for finding the trimming line is presented in Fig. 1. To express an arbitrarily-shaped surface, we used triangular elements throughout the whole process. The desired final shape after flanging can be easily determined based on the product shape considering the subsequent process, such as hemming. From input data, this program develops the final mesh onto the drawing tool mesh by the incremental development method. The finite element inverse method improves the accuracy of a developed mesh by considering mechanics. Finally, the post processor extracts a trimming line by distinguishing the outer edges from the developed mesh.

## 3. An improved Incremental Development Method

Before the finite element inverse simulation, a sound initial guess without inverted elements should be supplied.

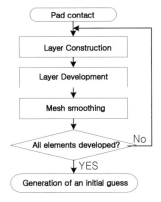

Fig. 2. Procedure of initial guess generation.

By taking advantage of the characteristics of the flanging process, we devised the procedure of the incremental development method as shown in Fig. 2. At first, the pad area is identified according to the conformity of shape between the final mesh and drawing tool mesh surface and is excluded from the development because this region usually has little deformation in flanging. The outer edge of the pad area is defined as the first development front. The first layer, the group of elements that contains the nodes in the first development front, is constructed. The next layer is constructed in the same way after the development of the first layer. This procedure is repeated until the whole mesh is developed. Because the final mesh usually has a large aspect ratio and various element sizes, mesh may be severely distorted and even inverted during development. Therefore, energy–based smoothing is utilized to develop mesh with low distortion. During energy minimization, the motions of every node are constrained to slide on the tangent plane orthogonally to the normal vector of the surface.

### 3.1. *Layer-wise development*

When developing elements in a layer, the algorithm to define the locations of new nodes development is required. Fig. 3 demonstrates a development algorithm. A development front is defined as the outer edges of the pre-developed elements. Fig 3(a) shows an example of development front. Elements in a layer are classified into two groups. One group (A-type) has two nodes in a development front. The other group (B-type) has one node in a development front. At first, A-type elements are developed as in Fig. 3(b). As shown in Fig. 3(c), a node may be defined in a variety of ways by several A-type elements. When m elements are involved, the node position is calculated by area-weighted averaging, as in the following equation:

$$X_i = \sum_{k=1,m} A_k x_i^k \qquad i = 1,2,3$$

(1)

where $X_i$ is the coordinate of the node after averaging. $A_k$ and $x_i^k$ are the area and the coordinate of the developing node of k-th element, respectively.

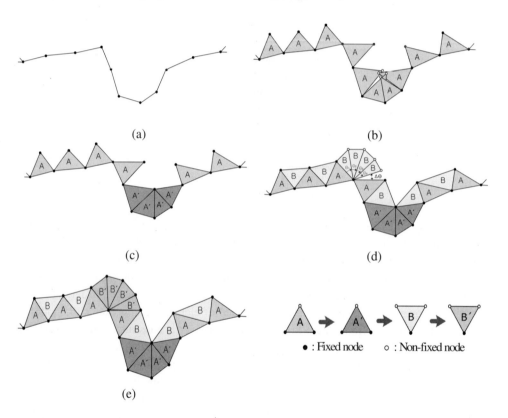

Fig. 3.  Procedure of a development method.

After all the A-type elements are developed, B-type elements, which have two adjacent A-type elements, are automatically defined as in Fig 3(d). Still, a group of B-type elements between A-type elements should be located. These elements are denoted as B'-type elements. In general, the development of B'-type elements is accompanied with large area change so that B'-type elements are difficult to develop perfectly at first. After all the B'-type elements are developed, a virtual triangular element is created to compensate the area change. The approximate total area, $A_{total}$, which B`-type elements can occupy is calculated by the sum of areas of the B`-type elements and the virtual element as in equation (2-a). At k-th iterative development, areas of B`-type elements are modified by the following proportional relation to consider an area change, as in equation (2-b).

$$A_{total} = \sum_{i=1,n} A_i + A_V \text{ (2-a)}, \qquad A_i^{k+1} = A_i^k \left( \frac{A_{total}}{\sum\limits_{i=1,n} A_i} \right) \text{ (2-b)}, \qquad \frac{A_v}{A_{total}} \langle \varepsilon \quad \text{(2-c)}$$

Iteration is terminated if the ratio of the area of a virtual element to the total area decreases below a pre-defined small value, $\varepsilon$, as in equation (2-c). In this way, the development of a layer is completed as shown in Fig 3(e).

In the present study, a triangular element was projected rotationally to handle undercut. At first, an edge of a developing element is attached to a pre developed common edge and the remaining point is rotated until it comes into contact with the drawing tool mesh surface.

### 3.2. *Smoothing using a length and area based energy*

In the previous study[3], minimization of a length and area-based energy is suggested to map the final mesh onto the tool surface with low distortion. In this study, the modified form of an energy as in equation (3) is suggested.

$$\Pi = \sum_{k=1,m} \Pi_e^k \qquad \text{(3-a)} \qquad\qquad \Pi_e = \alpha \Pi_l + (1-\alpha) \Pi_A \qquad\qquad \text{(3-b)}$$

$$\Pi_l = \sum_{i=1}^{3} \left( \frac{l_i^2 - L_i^2}{L_i^2} \right)^2 \qquad \text{(3-c)} \qquad\qquad \Pi_A = \left( \frac{A - A_0}{A_0} \right)^2 \qquad\qquad \text{(3-d)}$$

where $\Pi_e$ is the energy of a triangular element. In equation (3-c), $\Pi_l$ is a length-based energy that preserves length and is formulated to have the same dimension of the area based energy. In equation (3-d), $\Pi_A$ is defined using the difference of a signed element area. If a triangular element is inverted during calculation, the energy increase causes it to have the right orientation. Normalization is chosen so that this method works well for various mesh sizes. Since equation (3-a) is nonlinear, a Newton-Raphson iteration method is utilized to find the minimum energy.

## 4. Numerical Examples

### 4.1. *Flanging of L-shaped cup*

The effectiveness of proposed method is investigated by predicting an initial blank for the product with the shape shown in Fig. 4(a). The product has flanges with two protruded ears and has corners which have the deformation characteristics of cup drawing. The thickness of sheet is 0.8 mm and the stress-strain relationship is given as the following.

$$\bar{\sigma} = 645(0.023 + \bar{\varepsilon}_p)^{0.238} \tag{4}$$

Trimming lines are obtained by both the present method and conventional section-based method. For two trimming lines, incremental analyses using dynamic explicit FEM are carried out. Along the 77 sections shown in Fig. 4(a), the predicted heights are compared with those of target product. In Fig. 4(b), it is clearly shown that the present method give a more accurate result than the section-based method. For the straight flanges, the predicted heights are in close agreement with the target heights. Even for the ear with height of 25.0 mm, the error of height is around 0.26 mm. However, for the corner flanges with height of 10.0 mm, the error of height increases considerably due to the severe compression around the corner. Thus, for the areas with large deformation, the present method reveals some limitations since the complex external work during the flanging is not considered in the present method.

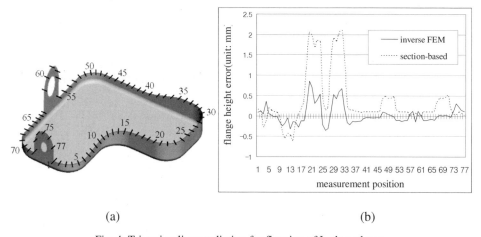

(a)                                                         (b)

Fig. 4. Trimming line prediction for flanging of L-shaped cup.

### 4.2.  *Flanging of a front fender*

The trimming line of a front fender is considered. The shape of the front fender is shown in Fig. 5(a). Sheet thickness is 0.7mm, and stress-strain relation is given by the following equation.

$$\bar{\sigma} = 663(0.01455 + \bar{\varepsilon}_p)^{0.191} \tag{5}$$

With section-based design and simulation-based design, we simulated the flanging process by using a dynamic explicit FEM and compared those results with the target edge profile. In Fig. 5(b) and Fig. 5(c), the present method is clearly shown to give a more accurate result than the section-based method. The outer edge profile simulated by the present method is in very close agreement with the target edge profile.

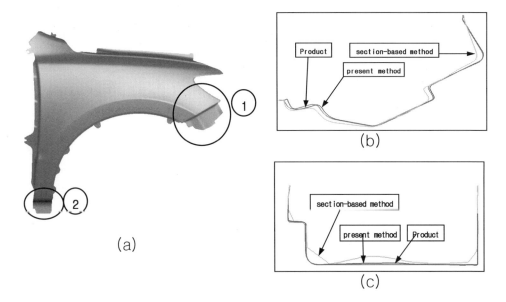

Fig. 5. (a) shape of developed fender, (b) comparison of edge profile for part-(1), (c) comparison of edge profile for part-(2).

## 5. Conclusion

For the design of the trimming line, an improved design method is suggested. Final mesh after flanging is developed on the drawing tool mesh surface by using a new incremental development technique. The improved layer-wise incremental development technique is devised to effectively develop mesh, which often have large aspect ratio, various element size, and undercut. Also, in order to overcome mesh distortion during development, an energy minimization technique is utilized. By using the finite element inverse method, the trimming line is found from the final mesh after flanging. From numerical comparisons using dynamic explicit FEM, it is proved that the present method can predict a trimming line more accurately than the section-based method.

## Acknowledgments

This research was financially supported by the Ministry of Knowledge Economy(MKE) and Korea Industrial Technology Foundation (KOTEF) through the Human Resource Training Project for Strategic Technology.

## References

1.  R. Sowerby, J.L. Duncan, E. Chu, Int. J. Mech. Sci. **28,** 415-430 (1986).
2.  G.N. Blount and B.V. Fischer, Int. J. of Prod. Res. **33**, 993-1005 (1995).
3.  W. J. Chung, C. D. Park and D. Y. Yang, Proceedings of NUMISHEET 2005, Detroit, USA, pp. 837-842.
4.  H. B. Shim, J. Engineering Manufacture, **220,** 1-10 (2006).
5.  K. Chung, O. Richimond,  Int. J. Mech. Sci., **34** 617-633 (1992).
6.  C.H. Lee and H. Huh, Journal of material processing technology,  **80-81** 76-82 (1998).

# EFFECTS OF HEAT TREATMENTS ON THE ON-LINE SERVICE LIFE OF A PRESS DIE MANUFACTURED BY W-EDM

KYE-KWANG CHOI[1]

*School of Mechanical & Automotive Engineering, Kongju National Univ.,*
*275 Budae-dong, Chonan-si, Chungnam, 330-717, KOREA,*
*ckkwang@kongju.ac.kr*

YONG-SHIN LEE[2†]

*Corresponding Author, School of Mechanical & Automotive Engineering, Kookmin Univ.,*
*861-1 Jungnung-dong, Sungbuk-gu, Seoul, 136-702, KOREA,*
*yslee@kookmin.ac.kr*

Received 15 June 2008
Revised 23 June 2008

Effects of heat treatments on the on-line service life of a press die manufactured by W-EDM are studied. In this work, four manufacturing processes for a press die are considered: (1) milling and then grinding, (2) wire-cut electric discharge machining (W-EDM), (3) low temperature heat treatment after W-EDM, and (4) high temperature heat treatment after W-EDM. On-line punching experiments for an automobile part of BL646-chain are performed. The amount of wear of the die and punch, roll-over and burnish depth in the punched chain are measured every 1,000 strokes. Overall productivities are carefully compared. Finally, it is concluded that heat treatment after W-EDM for a press die can enhance its on-line service life. Especially, high temperature heat treatment after W-EDM is very attractive as a fast and cheap manufacturing method for a press die.

*Keywords*: W-EDM; heat treatment; press die; on-line service life.

## 1. Introduction

Recent public demand requires a manufacturing system adequate to produce various kinds of goods with a small quantity. Correspondingly, new manufacturing technology for a die has been developed to replace a traditional machining method such as milling. For instance, Wire-cut electro-discharge-machining (W-EDM) enables machining even the high strength, low ductility material proper for a die. The big advantage of W-EDM is to require a short machining time satisfying a fast delivery, compared to a traditional machining process which needs a special tool or technique. However, W-EDM results in

---

[†]Corresponding Author.

several surface defects, hear affected zone, and relatively rough surface which may require additional processes. Thereby, a lot of works to solve such problems have been reported. For instance, many efforts made to reduce the amount of heat affected zone[1-5], which limit the service life of a product because the mechanical properties in that zone are deteriorated. However, heat generation during W-EDM is inherent and the existence of heat affected zone can not be avoided. The others [6-10] are related to examine the effects of process parameters such as the moving speed of the electrode, gap between the work piece and the electrode, EDM purse energy.

Most recently, the effects of heat treatments on the surface machined by W-EDM are carried out by the authors[11-12]. Those previous researches, which were on the effects of tempering heat treatments on the die steel STD11 machined by W-EDM, are extended to systematically examine the effects of heat treatments on the on-line service life of a press die. In this work, systematic on-line punching experiments for an automobile part of BL646 chain are planned with several sets of die and punch, prepared by W-EDM with/without heat treatments, and a traditional milling and grinding. The quality of products and overall productivities are carefully examined together with the on-line service life of a die.

## 2. Experiments

The previous works by the authors are not sufficient in terms of the effects of heat treatment on the on-line service life of a press die for the BL646 chain. Here, more experiments are systematically planned to examine both the low and high temperature heat treatments. The W-EDM machine used in this work is ACE535 made by Daewoo Heavy Industry. The diameter of a brass wire electrode is 0.25mm. Details of experimental and measuring apparatus are given in the previous paper[11-12].

### 2.1. *Die and punch preparation*

Since it is not easy to manufacture a complicated geometry with a milling, a simple shape of work piece such as an automobile part of BL646 chain is selected. The punch and die materials are SDT11 while the chain material is a carbon steel of S45C. Based on the previous works, four different manufacturing methods for the die and punch are examined, and are given in Table 1. Fig.1 shows detailed dimensions of a chain product considered in this work.

### 2.2. *On-line service life experiments*

On-line service life and productivity of the die sets by four manufacturing methods given in Table 1 are examined. Whenever every 1000 outputs of a leaf chain BL646 are produced, two products are examined. First, four dimensions of a leaf chain as shown in Fig. 1 are measured with a digital caliper. The roll-over and burnish depth in the punched surface as well as the concave part of a chain product are examined with image pictures made by a photomicroscope. If any defect in the products is observed, wear in the die is carefully examined to see if repair is needed. Otherwise, production was continued.

Table 1. Four manufacturing methods for punch and die block.

| Type | Punch | Die Block |
|------|-------|-----------|
| A | Milling + Profile Grinding | Milling + Jig Grinding |
| B | W-EDM | |
| C | W-EDM + Low Tempering(200℃) Tempering  2cycles | |
| D | W-EDM + High Tempering(510℃) Tempering 3cycles | |

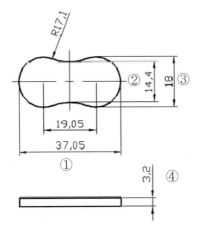

Fig. 1. Drawing of the leaf chain BL646 (unit: mm).

## 3.  Results and Discussion

First, the results in the previous works[11-12] are briefly summarized here. The surface roughness of the specimen by Type A or milling and grinding was lower than any other ones by Type B, C, and D. However, the heat treatment improves the surface roughness. Moreover, the heat affected zone produced while W-EDM could be removed completely by high temperature heat treatment while that zone was reduced quite a bit by the low temperature tempering as shown in Fig. 2.

Table 2 shows the measured values of the chain products at several strokes until the replacement of die set is needed. In general, the changes in product dimensions and burr increase abruptly once the production quantity becomes beyond its capacity determined by the quality of die. The common behavior observed in the dies by four types is that the dimensional increase of ② is faster than those of the other points or ①, ③, and ④. In other words, the wear at the concave part or ② is bigger than those at the other parts. It should be noted that the wear of a die increases suddenly when the stroke is close to the time for the replacement of die.

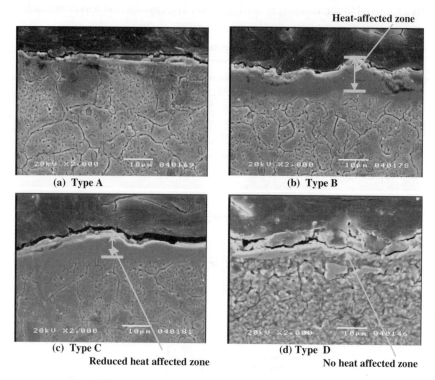

Fig. 2.  Microstructure of the surface machined by four types.

Table 2. Measured dimensions of chain products with the dies of type A, B, C, D  (unit : mm).

|  | Type A | | | | Type B | | | |
|---|---|---|---|---|---|---|---|---|
|  | No. of Stroke *1000 | | | | No. of Stroke *1000 | | | |
|  | 1 | 50 | 90 | 97 | 1 | 20 | 30 | 31 |
| ① 37.05 | 37.04 | 37.07 | 37.10 | 37.11 | 37.05 | 37.08 | 37.11 | 37.12 |
| ②14.40 | 14.36 | 14.38 | 14.44 | 14.64 | 14.32 | 14.36 | 14.47 | 14.66 |
| ③ 18.00 | 17.96 | 17.98 | 18.01 | 18.02 | 17.94 | 17.97 | 18.02 | 18.04 |
| ④ 3.20 | 3.26 | 3.28 | 3.30 | 3.34 | 3.25 | 3.31 | 3.34 | 3.36 |
|  | Type C | | | | Type D | | | |
|  | No. of Stroke *1000 | | | | No. of Stroke *1000 | | | |
|  | 1 | 20 | 40 | 45 | 1 | 50 | 90 | 92 |
| ① 37.05 | 37.04 | 37.06 | 37.08 | 37.10 | 37.05 | 37.08 | 37.10 | 37.11 |
| ② 14.40 | 14.37 | 14.42 | 14.48 | 14.61 | 14.35 | 14.37 | 14.45 | 14.63 |
| ③ 18.00 | 17.93 | 17.96 | 17.99 | 18.01 | 17.95 | 17.98 | 18.01 | 18.03 |
| ④ 3.20 | 3.27 | 3.32 | 3.34 | 3.35 | 3.26 | 3.31 | 3.35 | 3.38 |

From Table 2, it is observed that the wear of die by Type C with the low temperature tempering is smaller than that by Type B, which was machined only by W-EDM without any heat treatment. It is because the tempered martensite in the heat affected zone was stabilized although the heat affected zone was not removed completely, as was reported previously[11-12]. As shown in Table 2, the replacement strokes for the dies by Type A and Type D are similar while those for the dies by Type B and C are considerably small. It implies that wear resistance of a die by Type D is increased up to that by Type A. The reason is that the heat affected zone in the die by W-EDM could be completely removed by high temperature tempering.

The number of strokes until the die replacement is needed could be interpreted as productivity in some sense if the other factors for productivity such as number of outputs per coil of work piece, time for repairing die surface etc. are neglected. Fig. 2 shows comparisons of productivities with the dies prepared by four manufacturing types. In general, the service life of a press die manufactured by W-EDM is less than half of that by milling and grinding. The service life of a die by Type C with the low temperature tempering after W-EDM increased up to 145% of that by Type B, while the service life of a die by Type D is comparable to that by Type A. Thereby it can be concluded that heat treatment after W-EDM increased the on-line service life of a press die by W-EDM. Especially, the high temperature heat treatment could increase the service life of a press die manufactured by W-EDM as long as that of a die by the traditional manufacturing processes such as milling and grinding.

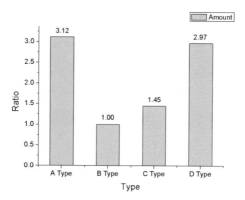

Fig. 3. Comparison of productivities of dies by four types A, B, C and D. of

## 4. Conclusions

In order to systematically examine the effects of heat treatments on the on-line life of a press die manufactured by W-EDM, on-line punching experiments for a leaf chain BL646 were performed. The punch and die are prepared by four different manufacturing methods - Type A: milling and then grinding, Type B: wire-cut electric discharge machining only, Type C; low temperature heat treatment after W-EDM, and Type D;

high temperature heat treatment after W-EDM.    Based on experiments, following conclusions can be drawn;

(1)  In general, heat treatment after W-EDM increases the service life of a press die.

(2)  The on-line service life of a press die by low temperature heat treatment after W-EDM (Type C) was increased up to 145% of that by W-EDM only (Type B).

(3)  From on-line experiments, the quality of press die by high temperature heat treatment after W-EDM(Type D) could be as good as that by the traditional manufacturing processes such as milling and grinding.

## References

1.   H. Yokoi, H. Hiraoka, *Annals of the CIRP*. Vol. 36 (1987) pp. 1~4.

2.   S. banerjee, B. V. prasad and P. K. Mishra, *J. Mater. Process. Tech.* Vol. 65 (1977) pp. 134~142.

3.   G. Spur, J. Chonbeck, *Annals of the CIRP*. Vol. 42 (1993) pp. 253~256.

4.   Y.S. Liao, Y.Y. Chu and M. T. Yan, *Int. J. Mach. Manufact.* Vol. 37 (1997) pp. 555~567.

5.   Y.S. Liao, J.T.Huang, Y.H.Chen, *J. Mater. Process. Technol.* 149 (2004) 165-171.

6.   C.A. Huang, C.C. Hsu, and H.H. Kuo, *J. Mater. Process. Technol.* Vol. 140 (2003) pp. 298-302.

7.   Y.F. Luo, *J. Mater. Process. Technol.* Vol. 94 (1999) pp. 208~215.

8.   H.T. Lee, T.Y. Tai, *J. Mater. Process. Technol.* Vol. 142 (2003) pp. 676~683.

9.   L.C. Lee, and L.C. Lim, and V. Narayanan, and V.C. Venkatesh, *Int. J. Machine Tools Manuf.* Vol. 28 (1988) pp. 359~372.

10.  J.C. Rebelo, A. Morao Dias, D. Kremer and J.L. Lebrun, *J. Mater. Process. Technol.* Vol. 84 (1998) pp. 90~96.

11.  K-K Choi, Y-S Lee, *Transactions of Materials Processing*, Vol. 15 (2006) pp. 539~543.

12.  K-K Choi, W-J Nam, Y-S Lee, *J. Mater. Process. Technol.* Vol. 201, (2008) pp. 580~584.

# CALCULATION OF THE ULTIMATE BEARING CAPACITY OF SOIL SLOPE BASED ON THE UNIFIED STRENGTH THEORY[*]

HONGJIAN LIAO[1]

*Department of Civil Engineering Xi'an Jiaotong University, Xianning West Road 28,*
*710049, Xi'an Shannxi, P. R. CHINA,*
*Key Lab. for the Exploitation of Southwestern Resources & the Environmental Hazard Control Engineering of*
*Chinese Ministry of Education, Chongqing University,Chongqing, 400044, P. R. CHINA,*
*hjliao@mail.xjtu.edu.cn*


ZONGYUAN MA[2]

*Department of Civil Engineering Xi'an Jiaotong University, Xianning West Road 28,*
*Xi'an Shannxi, 710049, P. R. CHINA,*
*Mazhongyuan9049@sohu.com*


LIJUN SU[3†]

*School of Civil Engineering, Xi'an University of architecture and technology,*
*13 Yanta Road, Xi'an Shannxi, 710055, P. R. CHINA,*
*sulijun1976@163.com*

Received 15 June 2008
Revised 23 June 2008

At present, the failure criteria used in calculating the ultimate bearing capacity of soil slope are the Tresca and Mohr-Coulomb criteria. But the results are conservative and the potential strength of soil mass cannot be utilized sufficiently because these two criteria do not take into account the effect of the intermediate principal stress. In this paper the unified strength theory was used to analyze the ultimate bearing capacity of soil slope. The formula for calculating the ultimate bearing capacity of soil slope using the unified strength theory was established. At the end, a case history was analyzed and it indicated that the result of the unified strength theory is larger than that of the Mohr-Coulomb criterion. This indicates that calculation of ultimate bearing capacity of soil slope with the unified strength theory can sufficiently exploit the strength of material. Therefore, the calculation of ultimate bearing capacity of the soil slope based on the unified strength theory will be of great significance in future applications.

*Keywords*: Unified strength theory; Intermediate principal stress; Soil slope; Ultimate bearing capacity.

[*]Financial support from the national natural science foundation of China (50379043) and key laboratory for the exploitation of southwestern resources & the environmental hazard control engineering (Chongqing university) of Chinese ministry of education are acknowledged.
[†]Corresponding Author.

## 1.  Introduction

At present, calculation of the ultimate bearing capacity and the limit analysis of soil slope are mainly based on Tresca (used for sand slope, $\varphi = 0$) and Mohr-Coulomb (used for cohesive soil slope, $\varphi \neq 0$) criteria (see Refs. 1-3). However, they do not take into account the effect of the intermediate principal stress, but experiment results indicate that the intermediate principal stress may affect the yield and failure of soil mass[1]. Therefore the results based on Tresca or Mohr-Coulomb criterion are conservative and cannot exploit the potential strength of soil mass sufficiently. The unified strength theory[1] takes into account the effect of the intermediate principal stress and uses a unified mechanical model and a simple unified mathematical expression to describe strength characteristics of material. The Terzaghi's ultimate bearing capacity was derived by using the unified strength theory (see Ref. 6). The plane-strain form of the unified strength theory was established and formula for the ultimate bearing capacity was derived by using assumptions of the Terzaghi's ultimate bearing capacity theory (see Ref. 7). In this paper, the unified strength theory was used to analyze the ultimate bearing capacity of soil slope. Mathematical expression of the unified strength theory is as follows:

$$
\begin{cases}
F = \sigma_1 - \dfrac{\alpha}{1+b}(b\sigma_2 + \sigma_3) = \sigma_t \, (\sigma_2 \leq \dfrac{\sigma_1 + \alpha\sigma_3}{1+\alpha}) \\[3mm]
F = \dfrac{1}{1+b}(\sigma_1 + b\sigma_2) - \alpha\sigma_3 = \sigma_t \, (\sigma_2 \geq \dfrac{\sigma_1 + \alpha\sigma_3}{1+\alpha})
\end{cases}
\tag{1}
$$

where $\sigma_t$ and $\sigma_c$ are the tensile and compressive strength respectively, b is a parameter of the unified strength theory which reflects the influence of the intermediate principal stress on failure of materials, namely, the intermediate principal stress impact factor, and $\alpha = \sigma_t/\sigma_c$ is the tensile-compressive strength ratio of a material. Fig. 1 shows the limit state line of unified strength theory in $\pi$ plane.

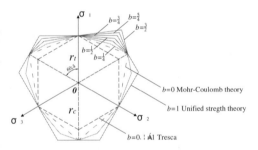

Fig. 1.  Limit state line of unified strength theory in $\pi$ plane.

## 2.   Derivation of Ultimate Bearing Capacity Formula of Soil Slope

### 2.1.  *Basic assumption*

(1) The whole slip zone of the slope shown in Fig. 2 is the same as that of the Prandtl's failure mechanism[2]. Wedge (*A'AB*) is an active Rankine zone. The failure mechanism of the slope[1] is shown in Fig. 2, while the Prandtl's failure mechanism is shown in Fig. 3.

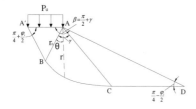 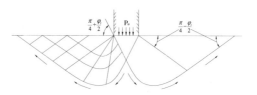

Fig. 2. Failure mechanism of soil slope.      Fig. 3. Prandtl's failure mechanism.

(2) The whole soil mass in the slip zone is in plastic state except for the wedge of $A'AB$. $ABC$ is a radial shear zone and $ACD$ is a passive Rankine zone. The surface $BC$ is a logarithmic spiral, which can be formulated as [3]:

$$r = r_0 e^{\theta \tan \varphi_t} \tag{2}$$

where $r_0$ is the initial radius of the radial shear zone, $\theta$ is the angle between any radius in the radial shear zone and the initial radius; $\varphi_t$ is the internal friction angle of the unified strength theory, the relationship between $\varphi_t$ and natural internal friction angle $\varphi_0$ is given by [1]:

$$\varphi_t = \arcsin \frac{(b-bm)+(2+bm+b)\sin \varphi_0}{2+b+b\sin \varphi_0} \tag{3}$$

where $m$ is the intermediate principal stress parameter and $\sigma_2$ can be defined as:

$$\sigma_2 = \frac{m}{2}(\sigma_1 + \sigma_3) \tag{4}$$

in which the value of $m$ can be determined by analytical or experimental method. When the soil mass is in elastic state, $m<1$; when the soil mass yields, $m$ is close to 1.

(3) The angle between the $A'B$ and $A'A$ is $\frac{\pi}{4}+\frac{\varphi}{2}$. The stress state of the wedge $A'AB$ is shown in Fig. 4.

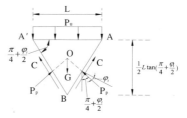

Fig. 4. Stress state of wedge in the upper slope.

## 2.2. Derivation of the formula

According to the basic assumption and the equilibrium condition of the triangular wedge shown in Fig. 4, the pressure on the top of slope is

$$Q_u = 2P_p \cos\left(\frac{\pi}{4}-\frac{\varphi_t}{2}\right) + c_t L \tan\left(\frac{\pi}{4}+\frac{\varphi_t}{2}\right) - \frac{1}{4}\gamma L^2 \tan\left(\frac{\pi}{4}+\frac{\varphi_t}{2}\right) \tag{5}$$

where

$$P_p = P_{pc}+P_{p\gamma} \;,\; c_t = \frac{2(1+b)c_0 \cos \varphi_0}{(2+b+b\sin \varphi_0)\cos \varphi_t} \tag{6}$$

where $c_t$ is the cohesion of the unified strength theory, $c_0$ is the natural cohesion, $P_P$ is the resultant force of passive earth pressures acting on boundaries $A'B$ and $AB$ of the wedge $A'AB$, $P_{pc}$ is the passive earth pressure yield from cohesion, $P_{p\gamma}$ is the passive earth pressure yield from gravity. The $P_P$ is given by[5]

$$P_p = \frac{L}{2\cos^2 \varphi_t}\left(c_t k_{pc} + \frac{1}{4}\gamma L \tan \varphi_t k_{p\gamma}\right) \tag{7}$$

where $k_{pc}$ is the passive earth pressure coefficient of cohesion. $k_{p\gamma}$ is the passive earth pressure coefficient of gravity. $k_{pc}$ can be written as

$$k_{pc} = \cos^2 \varphi_t \left[e^{\pi \tan \varphi_t}(1+\sin \varphi_t)-1\right] / \cos\left(\frac{\pi}{4}+\frac{\varphi_t}{2}\right)\sin \varphi_t \tag{8}$$

$k_{p\gamma}$ must be determined by try-and-error method.

When the slope reaches the state of limit equilibrium, the pressure on the top of the slope is the ultimate load of the slope. Substituting Eqs. (7) and (8) into Eq. (5) and simplifying, the formula of the ultimate load can be obtained as follows:

$$P_u = \frac{Q_u}{L} = c_t N_c + \frac{1}{2}\gamma L N_\gamma \tag{9}$$

where

$$N_c = \tan\left(\frac{\pi}{4}+\frac{\varphi_t}{2}\right) + \frac{\cos\left(\frac{\pi}{4}-\frac{\varphi_t}{2}\right)}{\cos\left(\frac{\pi}{4}+\frac{\varphi_t}{2}\right)\sin \varphi_t}\left[e^{\pi \tan \varphi_t}(1+\sin \varphi_t)-1\right] \tag{10}$$

$$N_\gamma = \tan\left(\frac{\pi}{4}+\frac{\varphi_t}{2}\right)\left[k_{p\gamma}\cos\left(\frac{\pi}{4}-\frac{\varphi_t}{2}\right)-\cos \varphi_t \cos\left(\frac{\pi}{4}+\frac{\varphi_t}{2}\right)\right] / 2\cos \varphi_t \cos\left(\frac{\pi}{4}+\frac{\varphi_t}{2}\right) \tag{11}$$

the $N_\gamma$ proposed by Terzaghi is[5]:

$$N_\gamma = 1.8\left[e^{\pi \tan \varphi_t}\tan^2\left(\frac{\pi}{4}+\frac{\varphi_t}{2}\right)-1\right]\tan \varphi_t \tag{12}$$

## 3.  Application to an Example

The example is a soil slope, with 120-150m in length and 58m in height, on top of which a utility area of an oil field was built. The slope angle is 30-50°. The loads of the buildings may affect the stability of the slope. Considering the security of the utility area, the ultimate bearing capacity of the slope was calculated. An overview of the slope is shown in Fig. 5. In order to show the whole shape and dimension of the slope clearly, a figure of 3D diagrammatic view for the slope is drawn by 3DMAX, which is shown in Fig. 6. The geological section of the slope is shown in Fig 7. The measured physical and mechanical properties for each layer of the soil are shown in table 1.

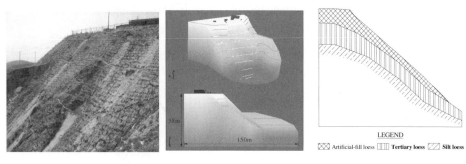

Fig. 5. Overview of the slope.  Fig. 6. 3D diagrammatic view of the slope.  Fig. 7. Geological section of the slope[7].

Table 1. Physical and mechanical properties for each layer of the soil[7].

|  | Statistic data | | Test Amount | Range of Results | Averaged Results |
|---|---|---|---|---|---|
| First layer | Density $\rho$ (g/cm³) | | 85 | 1.36-1.73 | 1.53 |
|  | Quick shear | cohesion (kPa) | 15 | 13-29 | 18.83 |
|  |  | internal friction angle (°) | 15 | 16.9-26.2 | 20.09 |
| Second layer | Density $\rho$ (g/cm³) | | 85 | 1.4 -1.67 | 1.52 |
|  | Quick shear | cohesion (kPa) | 7 | 25.1-32 | 30.29 |
|  |  | internal friction angle (°) | 7 | 19.3-25.6 | 22.56 |
| Third layer | Density $\rho$ (g/cm³) | | 18 | 1.46-2.18 | 1.69 |
|  | Quick shear | cohesion (kPa) | 18 | 36.2-58.6 | 52.78 |
|  |  | internal friction angle (°) | 18 | 19.1-23.5 | 21.09 |

These experimental results show that, for the average value of the strength indices from the first to the third layer of the soil, increase in cohesion is larger than increase in internal friction angle. In order to have a representative value, the strength indices of the second layer was selected in the calculation (density $\gamma$ =15.2kN/m³, cohesion $c_0$=30.29kPa, internal friction angle $\varphi_0$=22.56°). The foundation load of the buildings is simulated by a uniform load with 10m in length. The model for the calculation is shown in Fig. 8.

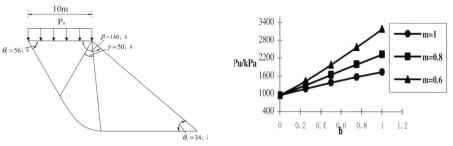

Fig. 8. Model for the calculation.          Fig. 9. Relationship between $P_u$ and $b$.

The relationship of the ultimate bearing capacity $P_u$, the internal friction angle $\varphi$ and the cohesion $c$ with the intermediate principal stress impact factor $b$ are shown in Figs. 9-11 respectively. On the basis of the calculated results, it can be concluded that the ultimate bearing capacity of the slope, the cohesion and the internal friction angle significantly increase with the increase in the intermediate principal stress impact factor (shown in Figs. 9-11). These findings confirm that the intermediate principal stress strongly affects the strength of soil mass and the ultimate bearing capacity of the slope.

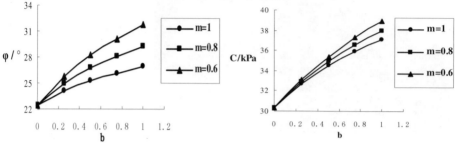

Fig. 10. Relationship between $\varphi$ and $b$.    Fig. 11. Relationship between $c$ and $b$.

## 4. Conclusion

In this paper, the formula of the ultimate bearing capacity of soil slope was established based on the unified strength theory. An example was analyzed and the results showed that the ultimate bearing capacity of the slope, cohesion and internal friction angle increase with the increase in the intermediate principal stress impact factor. These findings confirm that the intermediate principal stress strongly influences the strength of the soil mass and the ultimate bearing capacity of the slope. Moreover, the results show that the formula of the slope ultimate bearing capacity based on Mohr-Coulomb theory (the case of $b$ =0 in the unified strength theory) is conservative because it does not take into account the effect of the intermediate principal stress. Therefore, using the unified strength theory to calculate the ultimate bearing capacity of soil slope can exploit the potential strength of soil mass sufficiently and it will be of great significance in future applications.

## References

1. Yu Mao-hong. Unified strength theory and its applications (Springer, Beilin, 2004).
2. R.F. Craig. Soil mechanics (Van nostrand reinhold, New York, 1983).
3. W.F. Chen. Limit analysis in soil mechanics (Elsevier Science, Amsterdam AN-1991).
4. J.Salecon. Applications of the theory of plasticity in soil mechanics (A Wiley-Interscience Publication.1976).
5. T. William Lambe, Robert V. Whitman. Soil Mechanics (John Wiley & Sons, New York, 1969).
6. Xiaoping Zhou, Yongxing Zhang. *Study on the Terzaghi ultimate Bearing Capacity of Foundation Based on the unified strength theory.* (Journal of Chongqing university, Chongqing, 2004, Vol.27, No.9), p.133-136.
7. Wen Fan, Xiaoyu Bai, Maohong Yu. *Formula of ultimate bearing capacity of shallow foundation based on unified strength theory.* (Rock and soil mechanics, Wuhan, 2005, Vol.26, No.10), p.1617-1622.
8. Yanxun Song. *Analyse of high loess filling slope stability.* MA.Eng. Thesis, Univ. of Chang'An, Xi'an, 2005.

# TOP JOINT STUDY ON TEMPERATURE STRESS FOR SUPER-LONG SLAB-COLUMN STRUCTURE

MINGHAI DONG[1], LI SONG[2][†]

*Department of Civil Engineering, Xi'an Jiaotong University,*
*Xi'an 710049, P. R. CHINA,*
*minghaidong@hotmail.com, songli0125@tom.com*

YING SHAO[3]

*Urban Construction Institute, The First Survey and Design Institute of China Railway*
*Xi'an 710043, P. R. CHINA,*
*shawy8210@163.com*

Received 15 June 2008
Revised 23 June 2008

In this paper, top joint method is proposed to solve a practical engineering problem of temperature stress and temperature crack of super-long slab-column structure bearing temperature difference. From the study, it is shown that as for super-long slab-column structure undergoing temperature difference of inside and outside, joint located in bottom stories nearly has no influence on temperature stress and deformation while joint in top stories can significantly reduce temperature stress and deformation of super-long slab-column structures. In addition, comparison of joints located in top one story, top several stories and from bottom to top stories indicates that influences of them on temperature stress and deformation are similar. As for top joint method, among which cantilever plate method, double column method and corbel method are discussed and results indicate that influence effects of these methods on structures are similar.

*Keywords*: Super-long structure; slab-column; temperature stress; joint.

## 1. Introduction

In recent years due to unobstructed and concise space, flexible layout, even slab bottom and convenient construction, super-long slab-column structures have been widely used in non-seismic regions or regions with low requirement for seism. In some countries, i.e. England, France, Japan, America, there is no strict demand for expansion joint and only necessary measurement for temperature stress is required when structure exceeds certain length. As for China, maximum expansion joint is stipulated re. "Concrete structure design code"[1]. Stress induced by factors such as temperature difference and concrete contraction is the main reason of super-long structure cracking. Presently temperature effect is not considered in Chinese code[1,2] and only construction requirements are demanded for super-long structures. Distribution law, analysis model and calculation theory are not well studied[3]. In this paper, temperature stress of super-long concrete

---

[†]Corresponding Author.

slab-column structure bearing internal and external temperature difference of solar radiation is calculated, and top joint method is proposed which can significantly reduce temperature stress, well control temperature crack and barely influence using function of structure.

## 2. Calculation Method

### 2.1. *Temperature action and cases*

Seasonal temperature difference (uniform temperature difference) and sunshine temperature difference (internal and external temperature difference) are chosen as temperature cases in this paper according to classification and characteristic of temperature.

Uniform temperature difference is calculated with discount coefficient method of elastic modulus[4]. Shrinkage strain of concrete is expressed as

$$\varepsilon_y(t) = 3.24 \times 10^{-4} \cdot (1 - e^{-0.01t}) \cdot M_1 \cdot M_2 \cdot M_3 \cdots M_n \tag{1}$$

where $M_1, M_2, M_3, M_n$ are correction coefficients considering nonstandard conditions[1].

Creep coefficient is calculated by

$$\varphi(t, t_0) = \varphi(\infty, t_0) \beta_c(t - t_0) \tag{2}$$

where $\beta_c(t - t_0)$ is cerrp variation coefficient varying with stress duration calculated as

$$\beta_c(t - t_0) = \left[ \frac{(t - t_0)}{\beta_H + (t - t_0)} \right]^{0.3} \tag{3}$$

$$\beta_H = 1.5 \left[ 1 + \left( 1.2 \frac{RH}{100} \right)^{1.8} \right] \frac{2A_c}{u} + 250 \leq 1500 \tag{4}$$

where $f_c$ is axial compressive strength of concrete ($N \cdot mm^{-2}$), $A_c$ is section area of member($mm^2$), $u$ perimeter length of section contacting with atmosphere($mm^2$), $\beta(f_c)$ parameter related to $f_c$ ($N \cdot mm^{-2}$), $\beta(t_0)$ parameter determined by load age $t_0$, $RH$ environmental humility, $\beta_H$ determined by relative humility and member dimension. T-B method is widely used in creep calculation of concrete. Discount coefficient of elastic modulus for concrete is

$$\gamma_\varphi = \frac{E_\varphi(t, t_0)}{E_c(t_0)} = \frac{1}{1 + \chi(t, t_0) \cdot \varphi(t, t_0)} \tag{5}$$

Aging coefficient is derived as follows using creep constitutive theory and elastic continuation and plastic flow theory considering age influence:

$$\chi(t,t_0) = \frac{1}{1-e^{\left[-0.665\varphi-0.107\left(1-e^{-3.1314}\right)\right]}} - \frac{1}{\varphi} \tag{6}$$

Substituting Eq.(6) into Eq.(5), discount coefficient of elastic modulus is obtained as

$$\gamma_\varphi = \frac{1}{1+\left\{\dfrac{1}{1-e^{\left[-0.665\varphi-0.107\left(1-e^{-3.1314}\right)\right]}} - \dfrac{1}{\varphi}\right\}\cdot\varphi} \tag{7}$$

In this paper, age-adjusted effective modulus method is replaced with elastic modulus method of concrete. Then creep and relaxation of concrete can be successfully solved using elastic method for temperature effect.

As for internal external temperature difference, thermal convection is used to simulate sunlight action. Indoor temperature is assumed constant in this paper. Commercial software ANSYS as well as equation and assumption above is used in this paper to analyze temperature action. Simulation and calculation results are shown in Sec. 3.

## 2.2. Calculation model

In this paper temperature difference between outdoors and air conditioning indoors of diurnal changing in Xi'an China in summer are studied. A five-story super-long concrete slab-column structure with column network dimension $7.5m \times 7.5m$, longitudinal span number 9, transverse span number 4. Longitudinal total length of this structure is $67.5m$ which is greater than the limited joint distance $55m$. Section size of column is $600mm \times 600mm$ and plate thickness is $240mm$. Flat plate floor design of the structure is adopted with 4.8m story height and C40 concrete. Linear expansion coefficient of concrete is $1.0 \times 10^{-6}$. Solar radiation intensity changes with daily temperature. Indoors and outdoors daily temperature of Xi'an in summer is calculated according to Civil Architecture Heat Engineering Design Regulation (GB50045-95), listed in Table1.

Table 1. Roof and column outdoor side temperature of Xi'an in summer (℃).

| Direction | Outdoors | Roof | Southward column | Northward column | Eastward and westward column | indoors |
|-----------|----------|------|------------------|------------------|------------------------------|---------|
| 6 o'clock | 29.3 | 32.9 | 30.2 | 33.7 | 30.2 | 25 |
| 9 o'clock | 33.1 | 57.9 | 39.7 | 37.6 | 37.6 | 25 |
| 12 o'clock | 39.0 | 73.8 | 51.7 | 44.8 | 44.8 | 25 |
| 15 o'clock | 38.6 | 63.4 | 45.2 | 43.1 | 60.4 | 25 |
| 18 o'clock | 33.5 | 37.1 | 34.4 | 37.9 | 45.7 | 25 |

Temperature distributions of roof plate and side span column along section height in different time are analyzed, shown in Fig.1 and Fig.2.It can be obtained from the two figures above that solar radiation greatly influences roof plate and southward column with very uneven temperature distribution and large temperature difference. Temperature difference of structure on inside and outside surface of members reaches maximum value at 12 o'clock. So in this paper temperature field at 12 o'clock is adopted as the one to calculate temperature stress of structure.

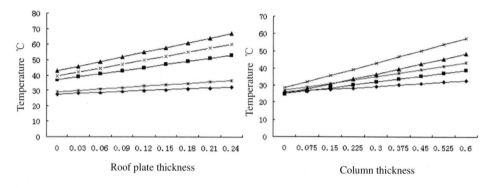

Fig. 1. Temperature distribution of roof plate.   Fig. 2. Temperature distribution of southward column.

## 3. Calculation Results and Analysis

### 3.1. *Top joint methods*

As for super-long slab-column structure, the main temperature effect is induced by axial deformation of horizontal members. According to temperature stress distribution of super-long structure, joint is set in part of the stories which only partially influence using function and structural integrity but can significantly decrease temperature stress of the structure. There are three feasible methods for top joint: (1) cantilever plate method; (2) double column method; (3) corbel method. From the viewpoint of architecture, top joint method is very flexible; moreover temperature deformation and stress can be greatly reduced. In addition internal forces become uniform which are within the limited range of normal use.   Requirements for top joint are: (1) one or more joints are set at the middle of structure considering the length and lateral stiffness of structure; (2) floor at joint should be separated apart and expansion joint requirement should be satisfied.

### 3.2. *Results and Analysis*

Ten joint forms shown in Table 2 are analyzed.

Table 2. Top joint structure forms.

| Numbering | Joint form | Numbering | Joint form |
|---|---|---|---|
| 1 | No joint | 6 | 2 joints at $2^{nd}$ and $3^{rd}$ stories |
| 2 | 1 joint at $3^{rd}$ story | 7 | 3 joints at $2^{nd}$ and $3^{rd}$ stories |
| 3 | 2 joints at $3^{rd}$ story | 8 | 1 joint from bottom to top |
| 4 | 3 joints at $3^{rd}$ story | 9 | 2 joints from bottom to top |
| 5 | 1 joint at $2^{nd}$ and $3^{rd}$ stories | 10 | 3 joints from bottom to top |

Three super-long concrete slab-column structures are adopted here for analysis with 3 stories; 4 spans in lateral direction; 15 spans (112.5$m$), 21 spans (157.5$m$) and 27 spans (202.5$m$) in longitudinal direction respectively. Calculation results are shown in Fig.3~Fig.8[*].

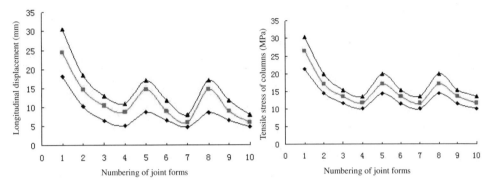

Fig. 3. Comparison of Lateral displacements in longitudinal direction.   Fig. 4. Comparison of tensile stresses of columns.

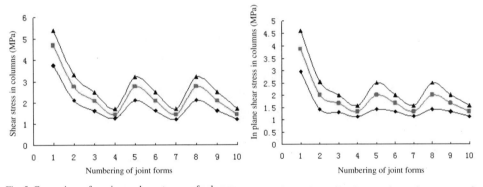

Fig. 5. Comparison of maximum shear stresses of columns.   Fig. 6. Comparison of in-plane maximum shear stresses of columns.

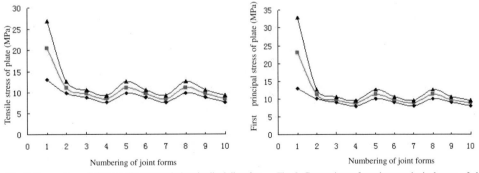

Fig. 7. Comparison of plate tensile stresses in longitudinal direction.   Fig. 8. Comparison of maximum principal stress of plates.

[*]◆ represent cases of 15spans, ■ 21spans and ▲27spans.

It can be found from Fig.3~Fig.8 that as for slab-column structures of different lengths, internal forces and deformations vary like wave with the change of joint forms. In addition, joint forms of 1~3 joints of $3^{rd}$ floor, 1~3 joints of $2^{nd}$ and $3^{rd}$ floors, 1~3 joints throughout bottom to top show similar effect on structures. So for super-long slab-column structures bearing inside and outside temperature differences, joint effects for joint set in top floor are nearly the same compared with that set throughout the whole height of structure.

Internal forces and deformations decrease rapidly with the increase of joint number. From the figures above it shows that for 21-span structure with 1~3 joints set in top floor, longitudinal displacement of the three cases reduces to 59.6%、42.8%、35.3% respectively of the value without joint; tensile stress of column reduces to 64.4%、51.1%、44.3% and in-plane first principal stress of plate reduces to 48.5%、42.4%、38%. As a result it is proposed that for super-long slab-column structure, more joints set at the top floor will efficiently reduce internal force and deformation of the structure.

## 4. Conclusion

From the study in this paper, the following conclusions can be drawn:

As for super-long slab-column concrete structure bearing internal external temperature difference, joint set in bottom stories nearly have no influence on temperature stress and deformation of structure while joint set in top stories can greatly decrease temperature stress and deformation. As for top joint method, influence effects of cantilever plate method, double column method and corbel method on structures are similar. And from the calculation and analysis it is shown that more joints set at the top floor will efficiently reduce internal force and deformation of the structure.

## References

1.  T. M. Wang. *Engineering structure crack control* (China Building Industry Press, Beijing, 1997).
2.  B. F. Zhu, *Temperature stress and its control of mass concrete structure* (China Electric Power Press, Beijing, 1999).
3.  R. F. Wang and G. R. Chen. *Temperature field and temperature stress* (Science Press, Beijing, 2005).
4.  C. Q. Han and S. P. Meng. *Research on Temperature Stress in Large Area Concrete Beam-Slab Structure*, Architectural Technology (Chinese Journal), **31**, 820(2000).
5.  X.W. Wang, J.Y. Pan. *Evaluating Aging Coefficient x in Age-adjusted Effective Modulus Method*, China Railway Science, **17**, 12(1996).
6.  Sami A.Al-Sanea. *Thermal performance of building roof element*. Building and Environment. **37**, 665(2002).
7.  B.A.Price, T.F.Smith. *Thermal response of composite building envelopes account for thermal radiation*. Energy Convers Mgmt., **36**, 23(1995).

# A PROCESS MAP FOR SEQUENTIAL COMPRESSION-BACKWARD EXTRUSION OF AZ31 MG ALLOYS AT WARM TEMPERATURS

DUK-JAE YOON[1], EUNG-ZOO KIM[2]

*New Forming Lab, KITECH, Songdo-dong, Yunseo-gu,*
*Incheon, 406-840, KOREA,*
*ydj@kitech.re.kr, ezkim@kitech.re.kr*

CHONG-DU CHO[3]

*Dept. of Mechanical Engineering, Inha Univ., Yonghyun-dong, Nam-gu*
*Incheon, 402-751, KOREA,*
*cdcho@inha.ac.kr*

YONG-SHIN LEE[4†]

*Corresponding Author, School of Mechanical & Automotive Engineering, Kookmin Univ.,*
*861-1 Jungnung-dong, Sungbuk-gu, Seoul, 136-702, KOREA,*
*yslee@kookmin.ac.kr*

Received 15 June 2008
Revised 23 June 2008

This paper is concerned with development of a process map for sequential compression-backward extrusion of bulk AZ31 Mg alloy at the warm temperatures. In experiments, metal flows and crack initiations are carefully investigated and formability is examined systematically for various forming conditions such as forming temperature, punch speed, and a gap between a specimen and die. Then, a process map for the sequential compression-backward extrusion of bulk AZ31 Mg alloy at the warm temperature is proposed. In order to further understand deformation behaviors and damage evolution during warm forming process, thermo mechanical finite element analyses coupled with damage evolutions are carried out. In general, finite element predictions support experimental observation. Finally, it is concluded that the process map, proposed for the sequential compression-backward extrusion of AZ31 Mg alloy, is valid.

*Keywords*: Warm backward extrusion; AZ 31 Mg alloy; Damage evolution, Finite element analysis.

## 1. Introduction

Recent advances in industrial technology require new engineering materials with high strength and low weight. Mg alloy is one of the lightest one among the practical structural materials with high strength and hardness, which are important characteristics for

---

†Corresponding Author.

structural materials. For instance, automobile industry expects Mg alloy as future material which can further reduce the weights of automobiles.[1, 2] There are, however, a couple of reasons why Mg ally has not been used widely. For instances, easy oxidation of Mg alloy requires a specific atmosphere during manufacturing, and corresponding surface treatment is necessary. Moreover, its formability is not so good that the technology for forming of Mg alloy at low and warm temperatures has not been well developed. Although there have been some efforts[3-9] to disclose forming characteristics of Mg alloys, further researches are still necessary in order to design a forming process of Mg alloy.    Most recently, Yoon et al.[10] examined several models to predict the damage evolution during warm backward extrusion of AZ31 Mg alloy and reported that Lee & Dawson's damage evolution model was the most effective one among those models.

In this study, both experimental studies and finite element analyses are performed to identify the proper process conditions for the sequential compression-backward extrusion of AZ31 Mg alloy at warm temperatures. Based on the previous works[8-10], forming temperature, punch speed, and gap between a specimen and die are chosen as most important process parameters.  In finite element simulations, thermo-mechanical analyses for metal flow are coupled with damage evolution. Eventually, a process map, which can identify the successful process conditions for the sequential compression-backward extrusion at warm temperatures, is proposed.

## 2.  Warm Sequential Compression-Backward Extrusion

A schematic drawing of sequential compression-backward extrusion of AZ31 Mg alloy is given in Fig. 1. Since the diameter of a specimen is smaller than that of a die, a specimen is first compressed until it contacts the inside wall of die. Then, backward extrusion is following. The specific chemical compositions of commercial wrought magnesium alloy AZ31 can be found elsewhere[8]. In order to get the homogeneous microstructure and mechanical properties of the specimen AZ31, the cast ingot was extruded, annealed at 400 ℃ for 10 hours, and cooled down to room temperature in the furnace. $MoS_2$ was applied as a lubricant on the surface of work piece and dies. The temperature of AZ31 billet was raised by keeping in the heated dies for approximately 10 minutes due to its low heat capacity.

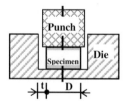

| Diameter of a specimen (D) | Gap(t) |
|---|---|
| φ29.7 | 0.15 |
| φ28.0 | 1.0 |
| φ26.0 | 2.0 |
| φ22.0 | 4.0 |

Fig. 1.  Schematic of sequential compression-backward extrusion with specific dimensions.

The process parameters examined in this research are temperature, forming speed, and the gap between a die and work piece. The forming temperatures are 180℃ and 200℃. The forming speed varies from 2mm/s to 20mm/s. The gap varies according to a specimen size because the diameter of a die is fixed as 30mm. The relations between the gap and specimen size are given in Fig. 1. The maximum forming load and punch speed of the forming machine developed in this work are 900kN and 25mm/sec, respectively.

## 3.  Finite Element Analysis Coupled with Damage Evolution

Finite element simulations for the above sequential compression-backward extrusion of AZ31 Mg alloy were performed using a commercial package DEFORM with a user subroutine, which was developed specifically in this work to predict the accumulated material damage during warm extrusion process. The material properties for finite element simulation are obtained by performing both tensile and compression tests, and are reported in the previous paper, Ref. 9. It must be noted that the tensile tests at the temperature of 175℃ showed the mode of ductile fracture. Here, the die and punch are assumed as rigid. The Coulomb friction is assumed with the coefficient of 0.3. The finite element formulation for the thermo-mechanical analysis of hardening viscoplastic materials like AZ31 Mg alloy has been a standard one, and the detail numerical derivation is available elsewhere[11-12].

The damage evolution model proposed by Lee & Dawson was adopted in this work. Their model was developed by analyzing growth of a spherical void in a larger cylinder of strain hardening, viscoplastic matrix. The specific form of their evolution model is given in the form of;

$$\dot{\phi} = C_0 \frac{\phi}{1-\phi} Exp[C_1 \frac{\sigma_m}{\sigma^*}] \cdot d_e \tag{1}$$

Here, $C_0$ and $C_1$ are material constants. In the above equation, the evolution of damage depends on the mean stress, effective strain rate as well as damage itself. The special feature of the above model is the scaling of a mean stress by a hardening state variable, $\sigma^*$, which never goes to zero even when shearing plastic deformation is negligibly small. In this model, the evolution rate depends on damage itself unlike other damage models. It is quite reasonable because the damaged material can be fractured faster than the undamaged one. Since Lee & Dawson's model did not include the nucleation of a micro crack, the initial value of damage must be given. The specific value of initial damage is 0.0011, as was used in Ref. 10. $C_0$ and $C_1$ are taken from Ref. 10, and those are 5 and 0.5, respectively.

## 4.  Results and Discussions

The maximum plastic resistance of AZ31 Mg alloy occurs around the strain of $0.2^{10}$. In sequential compression-backward extrusion, a work piece is extruded after compression. Thus, the low flow stress during extrusion is relatively low such that the overall forming load could be smaller than that for purely backward extrusion process. Thereby, it is needed to find the optimal amount of compression before backward extrusion. That is why the gap between the work piece and die diameter is selected as an important process parameter beside the punch speed and forming temperature. In general, the smaller specimen size causes the lower maximum forming load. During the first compression stage, the larger portion of Mg alloy work piece undergoes the maximum flow stress as the work piece size becomes smaller. In the following backward extrusion stage, the most deformation occurs near the punch corner. The work piece near the punch corner already experienced strain hardening when a work piece size was small, and its flow stress during extrusion could be smaller than the maximum flow stress shown in a stress–strain curve. Thereby, the maximum forming load during backward extrusion stage becomes lower with the smaller size of work piece.

Deformation characteristics such as the distributions of flow stress and effective strain can be predicted by finite element simulations. The most severe changes in work piece shape occurred during backward extrusion. The overall deformation behaviors during backward extrusion are the complex ones of shear and rotation near the punch corner, and the compression just below the punch. Once the work piece is extruded passing the punch corner completely, then it is pushed rigidly between the punch and die. The detail deformation characteristics for backward extrusion are available elsewhere[10]. In this work, the metal flow around a punch corner and the initiation of micro cracks are examined in detail, since the failure of a work piece begins there.  In Fig. 2, the metal flow near the punch corner is shown in detail for the work piece with the specimen size of 26 mm, under the punch speed of 20mm/s and the forming temperature of 200℃. The distribution of effective strain predicted by finite element analysis matches qualitatively well with metal flow by experiments. It should be also noted that the micro cracks initiated when the work piece was passing around the punch corner as shown in micro photos.  When the punch speed was reduced to 2mm/sec at the same forming temperature of 200℃, all extruded products were fine without any apparent defects regardless of a specimen size. However, many extruded products formed at the temperature of 180℃ had some defects regardless of a punch speed, except when the gap was as small as 0.15mm or when the specimen diameter was 29.7mm.

Fig. 2. Contour of effective strain, micro photos showing metal flow and crack initiation:
Specimen diameter 26.0mm, Punch speed 20 mm/s, Forming temperature 200℃.

Fig. 3 shows the extruded products with and without defects under the punch speed of 2mm/s and the forming temperature of 180℃. Distributions of accumulated damage normalized by an initial damage, damage ratio ($\phi/\phi_i$), predicted by finite element simulations for the same processes, are also shown. Although the damage evolution model did not define the criterion for the crack initiation, it gives an idea about the possible region and time of a defect creation. As mentioned earlier, only the extruded work piece with the specimen diameter of 29.7mm was fine and the maximum damage ratio predicted by finite element simulation for this case was about 13. The second one with a diameter of 28.0mm had many defects on the outside wall, for which the maximum damage ratio predicted by finite element simulation was about 35 and occurred at the outside wall. In general, finite element predictions for damage evolution matched experimental observations.

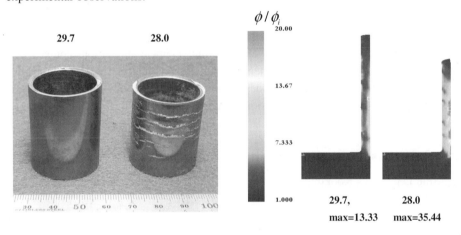

Fig. 3. Extruded specimens and distribution of accumulated damage predicted by FEM.
Punch speed of 2mm/s, Forming temperature of 180℃.

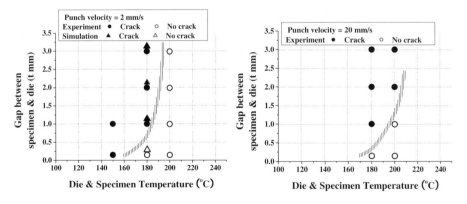

Fig. 4. Process maps for the sequential compression-backward extrusion:
(a) Punch speed of 2mm/sec, (b) Punch speed of 20mm/sec.

Finally, two process maps for the punch speed of 2mm/s and 20mm/s were proposed based on experimental observations as shown Fig. 4. Finite element predictions were also marked with the assumption that a crack would initiate when the damage ratio reaches about 25. But, the prediction for the damage evolution by finite element simulations still needs further verifications although it supports experimental observation qualitatively. It must be noted that such process maps can identify the forming conditions for the successful processes.

## 5. Conclusions

Both experimental studies and finite element simulations for the sequential compression-backward extrusion of bulk AZ31 Mg alloy at the warm temperatures were performed. In experiments, metal flows and crack initiations are carefully investigated for various forming conditions such as forming temperature, punch speed, and a gap between a specimen and die. Based on experimental observations, two process maps for the punch speed of 2mm/s and 20mm/sec were developed. Finite element simulations coupled with damage evolution supported the experimental observations. Finally, it was concluded that the proposed process maps, which can identify the forming conditions for the successful sequential compression-backward extrusion of AZ31 Mg alloy at the warm temperatures, are valid.

## Acknowledgments

Yong-Shin Lee thanks Kookmin University for the supports of the Research Fund 2008.

## References

1. S. Kamado, H. Ohara, and Y. Kojima, Advanced manufacturing technology of Magnesium alloy(CMC Publishing Co. 2005).
2. H.E. Friedrich and B.L. Mordike, Magnesium technology (Springer 2006).
3. N. Ogawa, M. Shiomi, K. Osakada, *Int. J. Mach. Tool. Manufac.*, Vol. 42 (2002) pp. 607~614.
4. P. Maier, K. U. Kainer, *62$^{nd}$ Annual World Conference Proceedings* (2005) pp.99~104.
5. H.E. Friedrich and S. Schumann, *J. Materials Process. Tech.*, Vol. 117 (2001) pp. 276~281.
6. E. Doege and K.Droder, *J. Materials Process. Tech.* Vol. 115, (2005).
7. D.J. Yoon, Y. W. Seo, C. Cho, H. J. Choi, and K. H. Na, *Advanced Technology of Plasticity (2005)-Proceedings of the 8$^{th}$ ICTP*, (2005) pp. 691~692.
8. D.J. Yoon, S.J. Lim, E-Z Kim, and C.D. Cho, *Trans. of Mater. Process.*, Vol. 15 (2006) pp. 597-602.
9. D.J. Yoon, K.H. Na, C.D. Cho, *Material Science Forum*, Vol. 539-543 (2007) pp. 1818-1823.
10. D.J. Yoon, E-Z Kim, Y-S Lee, *Trans. of Mater. Process.*, Vol. 16 (2007) pp. 614-620.
11. Y-S Lee and P.R. Dawson, *Mechanics of Materials*, Vol. 15 (1993) pp. 21-34.
12. Y-S Lee, *Metals and Materials International*, Vol. 12 (2006) pp. 161-165.

# A NUMERICAL SIMULATION OF STRIP PROFILE IN A 6-HIGH COLD ROLLING MILL[*]

XIAOZHONG DU[1†], QUAN YANG[2]

*National Engineering Research Center for Advanced Rolling Technology, University of Science and Technology Beijing, Beijing, 100083, P. R. CHINA,* [‡]
*xiaozhong_du@hotmail.com; yangquan@nercar.ustb.edu.cn*

CHENG LU[3], ANH KIET TIEU[4]

*School of Mechanical, Materials & Mechatronic Engineering, University of Wollongong, Wollongong, 2500, AUSTRALIA,*
*cheng_lu@uow.edu.au; ktieu@uow.edu.au*

SHINIL KIM[5]

*Technical Research Laboratories, POSCO, Gumho-dong, Gwangyang-si, Jeonnam, 545-090, KOREA,*
*shinil@uow.edu.au*

Received 15 June 2008
Revised 23 June 2008

Shape control is always a key issue in the six-high rolling mill, in which the shifting of the intermediate roll and the work roll have been used to enhance the shape control capability. In this paper, a finite element method (FEM) model has been developed to simultaneously simulate the strip deformation and the roll stack deformation for the six-high rolling mill. The effects of the work-roll bending, the shifting of the intermediate roll and the work roll on the strip crown and edge drop are discussed in details. Results have shown that both higher bending force and more roll shifting will significantly reduce the strip crown. The edge drop is also reduced with the bending force and the roll shifting.

*Keywords:* Rolling; shape control; crown; edge drop; HC; FEM.

## 1. Introduction

The control of the strip profile and flatness still remains unsatisfactory in tandem cold mill (TCM) with more stringent demand on the quality of cold rolled strip. A number of technologies, including continuous variable crown (CVC) roll, pair cross (PC) roll, six-high rolling mill, variable crown (VC) roll etc., have been developed to improve the strip profile and flatness. The six-high rolling mill, also known as high crown (HC) rolling

---

[†]Corresponding Author.

mill, has been widely used in TCM. According to statistics, there were 19 TCMs with the yield of over 1 million  tons built before 2007 in China, 16 of which have been entirely or partly equipped with HC mill.[1,2] In a HC rolling mill, a pair of intermediate rolls between work rolls and backup rolls can be shifted axially in opposite directions. A shifting of the intermediate rolls can increase the bending ability of the work rolls and the flatness control ability of the mill.[3] Other advantages of HC rolling mill include high reduction with small work roll diameter, zero work roll crown and small edge drop.[4,5]

The HC rolling mill exhibits a more complex roll stack deformation behaviour than a four-high rolling mill. Research effort has been focused on the simulation of roll stack deformation of a six-high rolling mill to obtain a better model to control the strip profile and flatness. Pawelski et al.[6] and Wang[3] have developed analytical models to analyze the roll stack deformation for a six-high rolling mill. Allwood[7] has proposed a roll stack deformation model which considers the effect of horizontal roll offset. Hacquin et al. [8] have modeled the elastic deflection and flattening of the work roll using the finite element method (FEM). However, a detailed model of roll stack deformation for a six-high rolling mill, especially coupled with the strip deformation, is still lacking. In this paper, we have developed a finite element model to simulate the roll stack deformation coupled with strip deformation for the HC mill with work roll shifting, also called UCM mill. The effect of the intermediate roll shifting, work roll shifting and work roll bending on the strip profile will be discussed.

## 2.  FEM Simulation

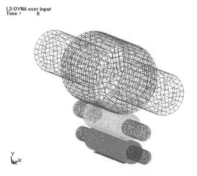

Fig. 1. FEM model for a six-high rolling mill.

The FEM model has been developed with the FEM package ANSYS/LS-DYNA. The mesh of the simulation model is shown in Fig.1. The total elements for strip, work roll, intermediate roll and backup roll are 80000, 85180, 8092 and 3700, respectively. The element type is SOLID164. Symmetric boundary conditions are applied to the central plane of the strip. The bending force is acted at the distance of 500mm from the edge of the work roll neck. The simulation parameters are shown in Table 1. To investigate the control capability of the strip profile in the six-high rolling mill, shifting of intermediate roll and work roll, and bending force have been investigated in the simulation. The adjustment ranges of these parameters are detailed in Table 1.

Table 1. Simulation Parameters.

| Parameters | Value |
|---|---|
| Diameter of work roll (mm) | 425 |
| Length of work roll (mm) | 1500 |
| Diameter of intermediate roll (mm) | 490 |
| Length of intermediate roll (mm) | 1510 |
| Diameter of back up roll (mm) | 1330 |
| Length of back up roll (mm) | 1500 |
| Bending force of work roll (kN/Chock) | +360/-180 |
| Shifting of intermediate roll (mm) | 200 |
| Shifting of work roll (mm) | 100 |
| Work roll speed (m/min) | 300 |
| Entry thickness (mm) | 2.5 |
| Exit thickness (mm) | 1.85 |

## 3. Results and Discussion

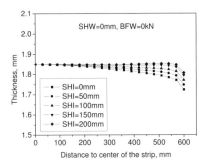
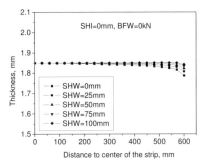

Fig. 2. Effect of the intermediate roll shifting (SHI) on the strip profile.

Fig. 3. Effect of the work roll shifting on the strip profile.

Fig.2 shows the effect of the intermediate roll shifting (SHI) on the strip profile without the bending force (BFW) and the work roll shifting (SHW). As expected, the typical convex profile is obtained for the case without shifting. The strip profile (thickness) decreases from the center to the edge of the strip due to deflection of the roll stack. A significant reduction of the profile can be observed near the edge of the strip. This phenomenon is called edge drop. The edge drop is mainly caused by a lateral mass flow and a smaller roll flattening near the strip edge. It is obvious that the intermediate roll shifting reduces significantly the thickness variation and can produce a flatter strip profile. The dependence of the strip profile on the work roll shifting is shown in Fig.3. With increased shifting, the profile in the middle of the strip tends to be flatter across the width, while the edge drop still exists although its value has been reduced. The effect of the work roll shifting is similar to the intermediate roll shifting. As the work roll shifting increases, the thickness variation become smaller. The simulation results clearly indicate that both the intermediate roll shifting and the work roll shifting are effective tools to control the strip profile.

Fig.4 depicts the effect of work roll bending on the strip profile. A positive bending force increases the roll crown, while the negative value reduces it. When the bending force is zero, the strip thickness decreases from the center to the edge of the strip due to deflection of the roll stack. With negative bending forces, the strip thickness variation increases with bending force. When the positive bending force is applied, the profile becomes concave. The profile deviation to the center increases as the bending force increases. Figs.2-4 demonstrate that the bending force has a larger adjustment capability than the shifting of the intermediate roll and work roll. The thickness at the distance of 40mm to the strip edge can be increased from 1.589mm to 1.971mm when the bending force is increased from −180kN to 360kN, while it is increased from 1.771mm to 1.846mm for intermediate roll shifting in Fig.2 and from 1.816mm to 1.839mm for work roll shifting in Fig.3.

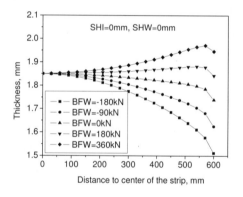

Fig. 4. Effect of the work roll bending force on the strip profile.

The strip crown and edge drop are often used to characterize the strip profile in the middle part of the strip and near the edge, respectively. The strip crown used in this paper is defined as the strip thickness difference between the thickness at the strip center and that at the distance of 40mm from the edge. The edge drop is defined as the strip thickness difference between the thickness at the distance of 100mm from the edge and that at the distance of 15mm. Figs.5(a) and 5(b) show the effect of intermediate roll shifting and the work roll shifting on strip crown and edge drop for three different bending forces (BFW=-180kN, 0kN and 360kN), respectively. The effect on the strip crown and the edge drop exhibits a similar pattern. It is clear that the strip crown and the edge drop decrease as the bending force increases. They also decrease with intermediate roll shifting and work roll shifting. For zero bending force, the effect of intermediate roll shifting and work roll shifting is close. When bending force is applied, the magnitudes of the strip crown and the edge drop caused by intermediate roll shifting is larger than those caused by work roll shifting. This indicates that the combination of the intermediate roll shifting and the work roll bending force provides a higher control capability of the strip profile than the work roll shifting with the work roll bending force.

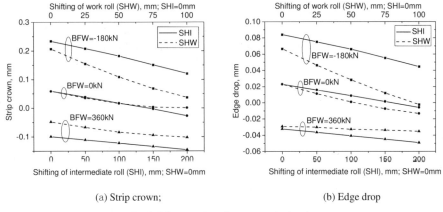

(a) Strip crown;　　　　　　　　　　　　　(b) Edge drop

Fig. 5. Effects on the strip crown and edge drop.

## 4. Field Application

Based on the simulation results, a technical procedure was proposed and an online shape control model has been developed. It has been applied successfully to the five-stand UCM tandem cold mills in Wuhan Iron&Steel Corporation with very good results.

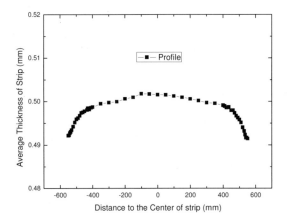

Fig. 6. Transverse Distribution of Average Strip Thickness.

According to the statisticed results for 28 steel coils, the percentage of coils with an average edge drop less than 7 $\mu m$ increases from 65.70% to 89.87%; and that with average edge drop less than 10 $\mu m$ increases from 83.6% to 98.13%. The profile of finished product after 20mm side trim is shown in Fig.6.

## 5. Conclusion

A FEM model has been developed to investigate the control capability of the strip profile in a six-high rolling mill. The strip deformation and the roll stack deformation were simultaneously simulated. The effects of work-roll bending, intermediate roll shifting and work roll shifting on the strip crown and edge drop have been discussed. The results show that both higher bending force and more intermediate roll shifting significantly reduce the strip crown and the edge drop. The roll bending force plays a dominant role in the control of the strip profile. The roll shiftings help increase the profile control capability. Comparing two roll shifting approaches, the intermediate roll shifting exhibits a higher capability than the work roll shifting.

## Acknowledgments

The authors gratefully acknowledge the Anshan Iron&Steel Corporation and Wuhan Iron&Steel Corporation for the support to this work. In addition, special thanks are extended to Chongtao WANG (Senior Engineer) for his cooperation and assistance.

## References

1. S. Q. Chen, *Encyclopedia of Chinese Cold Rolling Strip and Plate* (Metallurgy industry press, Beijing, 2002).
2. Q. Yang, *Cold rolling mills in China, Report,* University of Science and Technology Beijing, 2007.
3. G. D. Wang, *Shape control and shape theory* (Metallurgy Industry Press, Beijing, 1986).
4. Akio Suzuk, *Recent Progress in the Rolling Mills Part I,* Transactions ISIJ, 24(1984), 228-249.
5. V. B. Ginzburg., *Steel Rolling Technology, Theory and Practice* (Marvel Dekker, New York, 1989).
6. O. Pawelski, W. Rasp, and J. Rieckmann, *A mathematical model for predicting the influence of elastic and plastic deformations on strip profile in six-high cold rolling.* Proceedings of 4th International Steel Rolling Conference, Deauville, France 1987, pp. E.3.l-6.
7. J. M. Allwood, *Model-based evaluation of the effect of horizontal roll offset on cross-directional control performance in cold-strip rolling,* IEE Proc.-Control Theory Appl.. 149(2002), No. 5, 463-370.
8. P. M. Hacquin, J. P. Guillerault, *A three-dimensional semi-analytical model of rolling stand deformation with finite element validation,* Eur. J. Mech, A/Solids. 17(1998), 79-106.

# DYNAMIC SIMULATION OF THE TAILING PROCESS IN HOT FINISHING MILL

SHINIL KIM[1†]

*Technical Research Laboratories, POSCO, KOREA,*
*shinil@posco.com*


CHENG LU[2]

*School of Mechanical, Materials & Mechatronic Engineering, University of Wollongong, AUSTRALIA,*
*chenglu@uow.edu.au*


XIAOZHONG DU[3]

*National Engineering Research Center for Advanced Rolling Technology, University of Science and Technology Beijing, Beijing, P. R. CHINA,*
*xiaozhong_du@hotmail.com*


ANH KIET TIEU[4]

*School of Mechanical, Materials & Mechatronic Engineering, University of Wollongong, AUSTRALIA,*
*ktieu@uow.edu.au*

Received 15 June 2008
Revised 23 June 2008

In this paper an explicit dynamic finite element method model has been developed to investigate the strip deformation behavior between two adjacent stands in hot finishing mill. The effect of the roll speed ratio of second stand to first stand on tension and the tailing behavior of the strip has been discussed in details. It has been found that the strip accumulation occurs if the roll speed ratio is small. The tensile stress increases with the roll speed ratio. During the tailing process the accumulated strip caused by the small roll speed ratios knocks onto the roll, while the swing of the strip tail occurs for the large roll speed ratios and it strikes the roll as well. Both tailing phenomena will result in the strip tail pincher or roll damage in the real operation.

*Keywords*: Tailing; hot finishing mill; finite element method.

## 1. Introduction

The tail pincher, also called tail crash, refers to the strip tail rolled in a folded form. It is a serious operation occurrence in hot finishing mill. There are two major causes. One is due to the asymmetric rolling condition, which will lead to the snake motion of the strip,

---

[†]Corresponding Author.

namely the strip moving to one side. When the strip tail leaves the previous stand, the tension suddenly disappears. The tail will knock at the side guide, and then it is folded and rolled with increased thickness. Another cause is attributed to improper control strategies of the speed and looper controls. With improper controls, the strip tail will seriously swing up and down once the strip loses the tension from the previous stand, especially for thin, wide and hard strips. Serious swinging will then cause the tail folding. The swinging may also cause other problems, such as roll damage and strip jamming in the stand. The former cause has been experimentally and numerically investigated [1-4]. The research on the latter cause is still lacking.

In this paper an explicit dynamic finite element model has been developed to study the deformation behavior of the strip between two stands. The effect of the roll speed ratio of second stand to first stand on tension and the tailing behavior of the strip has been discussed in details. It has been found that tension increases with the roll speed ratio. The roll speed ratio should be controlled in a certain range. Larger or smaller speed ratio will result in the serious contact pressure between the strip and the roll outside the roll bite.

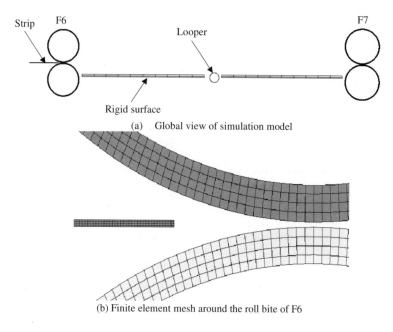

(a)    Global view of simulation model

(b) Finite element mesh around the roll bite of F6

Fig. 1. FEM simulation model.

## 2.  Dynamic Finite Element Method Simulation

To analyze the tailing process between last two stands (F6 and F7) of hot finishing mill, an explicit dynamic finite element method (FEM) model has been developed using the commercial FEM package LS_DYNA [5]. The simulation model is shown in Fig. 1. In the simulations, the strips traveled at an initial speed of 0.5 into the roll bite of F6. After the deformation in F6, the strip headed for F7. When the strip was deformed in F7 the continuous rolling between two stands was setup. There was a looper located at inter-stand position, which is used to control the mass flow in the operation. In the simulation once the head of the strip left the roll bite of F7, the looper started to move up. The relation between the simulation time (t) and the position of the looper is shown in Figure 2. With the looper moving up, the strip between stands was lifted. In the FEM model two rigid surfaces were used to simulate the delivery table, preventing the strip from falling down.

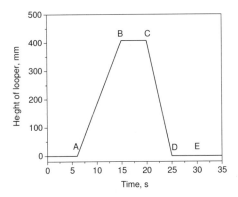

Fig. 2. Height of the looper against time in the simulation.

Since only the relative roll speed of stands F6 and F7 affects the strip deformation behavior between stands, the F6 roll speed was set to 1 in the simulation. The ratio (α) of F7 roll speed ($v_7$) to F6 roll speed ($v_6$) varied from 1.375-1.475. The detailed process parameters used in the simulations are shown in Table 1.

Table 1. Parameters used in the simulation.

| Parameters | Values | Parameters | Values |
|---|---|---|---|
| Roll diameter of F6 | 634mm | Entry thickness of F6 ($h_5$) | 8.16mm |
| Roll diameter of F7 | 676mm | Exit thickness of F6 ($h_6$) | 5.71mm |
| Looper diameter | 184mm | Exit thickness of F7 ($h_7$) | 4mm |
| Distance between F6 and F7 | 5800mm | Ratio of F6 speed to F7 speed(α) | 1.375-1.475 |
| Distance between F6 and looper | 2345mm | Yield stress of strip | 250MPa |
| Length of the simulated strip | 20m | Total elements | 44680 |

## 3. Results and Discussion

Fig.3 shows the deformation behavior of the strip at the time (t) of 14.7s, when the looper just contacts the strip. The strip accumulation, which is called the loop in the following context, can be observed for low speed ratios (α=1.375 and 1.4). This is attributed to the unbalanced mass flow between the exit of F6 roll bite and the entry of F7 roll bite. When the speed ratio equals to or exceeds 1.425, the inter-stand strip accumulation is hardly seen. This indicates the tension has been soundly established. The mass flow at F6 exit is $v_6*(1+f_6)$, where $f_6$ is the forward slip of the stand F6. The mass flow at F7 entry is $v_7*(1+f_7)*h_7/h_6$, where $f_7$ is the forward slip of the stand F7, $h_6$ and $h_7$ are the exit strip thicknesses of F6 and F7 respectively. $h_7/h_6$ is 0.7 in the simulation. It can be roughly assumed that f6 equals to $f_7$. Therefore, the roll speed ratio to maintain the constant inter-stand mass flow is $\alpha=1.429$. When the speed ratio is less than this critical value, the strip accumulation occurs. Otherwise the strip accumulation disappears and the tension can be established. The critical speed ratio $\alpha=1.429$ is determined based on the assumption of the equality of $f_6$ and $f_7$. It may change because $f_6$ and $f_7$ are slightly different.

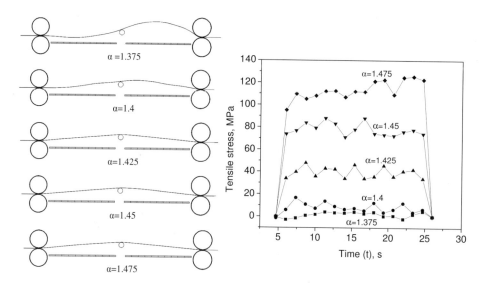

Fig. 3. Strip deformation behavior at t=14.7s.          Fig. 4. Tensile stress again simulation time.

The calculated tensile stress against the simulation time (t) is plotted in Fig.4. For $\alpha \leq 1.4$, the tension is very small. Though the lopper position has been increased to 408mm at t=15s as shown in Fig.2, the tensile stress has not been significantly enhanced. This is due to the effect of tension on the forward slip. Tension can be calculated by

$$\sigma = E\left(\frac{L'}{L} - 1\right) \tag{1}$$

$$L = L_0 + \int_0^t (v_6(1+f_6) - v_7(1+f_7)h_7/h_6)dt \tag{2}$$

where $\sigma$ is the inter-stand strip stress, E is Young's modulus, $L'$ is the strip length between F6 and F7, $L_0$ is the distance between F6. $L'$ increases with the looper position. The lift of the looper lengthens $L'$ and in turn increases the tension. However, the increased tension consequently increases the forward slip of F6 and decreases the forward slip of F7. This leads to a reduced tension, which compensates for the increase of tension caused by the elongated $L'$. Therefore, tension does not significantly change as the looper position increases. It can be seen in Fig.4 that the tension can be effectively established as the speed ratio is over 1.425 and it significantly increases with the speed ratio. This demonstrates that the speed ratio of two adjacent stands plays more important effect on tension than the looper position.

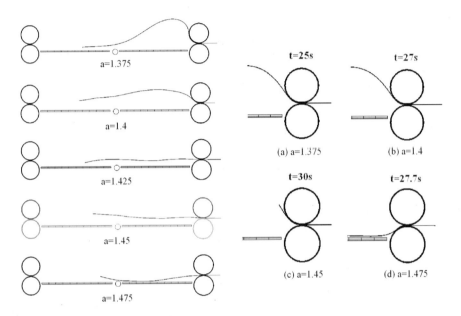

Fig. 5. Strip deformation behavior at t=27s.　　　Fig. 6. Strip deformation for maximum contact pressure.

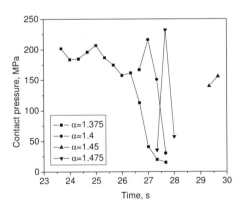

Fig. 7. Contact pressure again simulation time.

Fig.5 shows the strip deformation behavior at the time of 27s after the strip tail has left the first stand. The strip deformation behavior at this stage is strongly dependent on the strip deformation behavior before the strip tail leaves F6. For low speed ratios (1.375 and 1.4), the loop does not disappear during tailing. As the rolling in F7 proceeds, the loop moves toward the roll. It knocks onto the roll during a certain period as shown in Figs.6(a) and 6(b). This will cause the tail folding or roll damage in the real operation. For $\alpha=1.425$, there is no loop before the strip tail leaves F6. The strip tail travels steadily between stands. It does not contact the roll outside the roll bite. However, the strip tail swings up and down for higher speed ratios (1.45 and 1.475) due to the sudden loss of large tension established by large roll speed ratio. The swing tail will knock onto the roll when the tail reaches the stand F7 as shown in Figs.6(c) and 6(d).

The contact pressure between the roll and the strip outside the roll bite for various F7 roll speeds is shown in Fig.7. It can be seen that if the loop is large, such as the case with $\alpha=1.375$, the strip will knock onto the roll for a long period from t=23.7s to 27.7s. The maximum contact pressure is about 215MPa, which is 86% of the yield stress of the strip. As the loop is reduced ($\alpha=1.4$), the contact period decreases, while the maximum contact pressure still remains as high as the large loop. For the tail swing cases ($\alpha=1.45$ and 1.475), the contact period is short. However, the contact pressure is still quite high. The above simulation results have shown that the small speed ratios may cause a large strip accumulation, which results in a severe impact on the roll during the tailing process. However, if the speed ratio is too high the tail swing will occurs, leading to the large contact pressure as well. Therefore, the speed ratio should be carefully controlled in the real operation in order to reduce the operation occurrences.

## 4.  Conclusion

To analyze the tailing process between last two stands (F6 and F7) of hot finishing mill, an explicit dynamic finite element method (FEM) model has been developed using the commercial FEM package LS_DYNA. The effect of the roll speed ratio of second stand to first stand on tension and the tailing behavior of the strip has been discussed in details. The simulation results show that the strip accumulation occurs if the roll speed ratio is small. The tensile stress increases with roll speed ratio. During the tailing process the accumulated strip caused by small roll speed ratios knocks onto the roll, while the swing of the strip tail occurs for the large roll speed ratios and it strikes the roll as well. Therefore, the speed ratio should be carefully controlled in the real operation in order to reduce the operation occurrences.

## References

1.  H. Nakashima, et al., Proc. of Spring Conf. of Japan Soc. for Tech. of Plasticity, pp. 61-64, 1980.
2.  T. Kiyota and H. Matsumoto, *Proceedings of the American Control Conference* (Denver, Colorado, 2003), pp. 3049-3054.
3.  A. Nilsson, Journal of Materials Processing Technology, 80–81, 325–329, (1998).
4.  Y. Okamura and I. Hoshino, *Control. Eng. Practice,* 5, 1035-1042, (1997).
5.  J. O. Happquist, *LS-DYNA theory manual*, Livermore Software Technology Corporation, 2005.

# FACTOR STUDY FOR THE SEPARATOR PLATE OF MCFC HAVING UNIFORM STIFFNESS AT ELEVATED TEMPERATURE

SANG-WOOK LEE[1†]

*Department of Mechanical Engineering, Soonchunhyang University, Asan, Chungnam 336-745, KOREA,*
*swlee@sch.ac.kr*


JUNG-HYUN KIM[2]

*R&D Center, Yura Corporation, Hwasung, Gyeonggi 445-913, KOREA,*
*jhkim014@yura.co.kr*


JOONG-HWAN JUN[3]

*Fuel Cell Project, RIST, Pohang, Gyeongbuk 790-330, KOREA,*
*junjh@rist.re.kr*

Received 15 June 2008
Revised 23 June 2008

A molten carbonate fuel cell (MCFC) is composed of several stacks of unit cells. A unit cell is composed of two electrodes and a matrix that is inserted between separator plates. Separator plates should properly contact the electrodes to reduce the electricity loss arising from contact resistance. To this end, a pressure of about 2 kgf/cm$^2$ is usually applied on the top of the stack, which results in the separator plates being somewhat compacted. Furthermore, the stiffness of the separator plates becomes degraded at elevated temperatures due to softening of the plate materials. Therefore, a non-uniform temperature distribution across the separator plates induced by exothermic reactions of the oxidant and reactant gases leads to a non-uniform plate stiffness. This study has firstly evaluated the change in separator plate stiffness as temperature changes by applying pressure to the plates. Secondly, using the Taguchi method, several design factors that affect stiffness have been investigated to determine which has the most influence. Based on these results, a new design for the separators, which allows for uniform stiffness at elevated temperatures, has been proposed.

*Keywords*: MCFC; Separator plate; Elevated temperature; Taguchi method.

## 1. Introduction

Molten carbonate fuel cells (MCFCs) are a promising future energy source. They operate through electrochemical reactions between the oxidant and reactant gases and the molten carbonate being used as an electrolyte. An MCFC stack is composed of several unit cells stacked in layers. A unit cell is composed of three main components: electrodes, matrices, and separators. The main roles of the separators are to maintain good contact between the

---

[†]Corresponding Author.

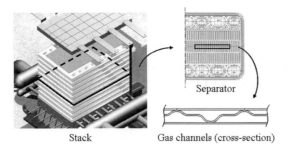

Fig. 1. An MCFC stack composed of several separators. Gas channels are formed in the active area of separator.

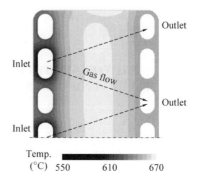

Fig. 2. Temperature gradient in separator due to the exothermal reactions.

electrodes and matrices, to provide passages for the gases, and to collect the electricity produced by the electrochemical reactions.

Separator plates should be in good contact with the electrodes and matrices to minimize electricity loss due to contact resistance. To this end, a pressure of about $2 \text{ kgf/cm}^2$ is applied to the stack head after stacking is completed.

MCFC stacks normally operate at 650 °C. The heat generated by exothermal reactions spreads into the separators resulting in a temperature gradient. Temperatures in the range of normally 550 °C to 700 °C are built up in the plane of the separator. The separator stiffness decreases at high temperatures, so a stiffness gradient induced by the temperature gradient occurs across the separator. This may cause a non-uniformity in the contact across the separator.

Most work regarding MCFC separators has been dedicated to determining corrosion behaviors over the long-term operation of the stack,[1,2] slurry coating effects,[3] a new design for the separator,[4] and the temperature distribution during stack operation.[5]

In this paper, we will evaluate the temperature dependence of the separator stiffness, and suggest a separator design that provides a uniform stiffness distribution over its active area at the working temperatures of an MCFC stack. The most effective design factor will be selected based on an assessment of the results using the Taguchi method.

## 2. Evaluation of Separator Stiffness at Elevated Temperatures

Figure 1 shows that a separator plate is one of the main components of an MCFC stack. The electrochemical reactions take place on the active area of the separator which is composed of two flat skin plates and one inner core. They are made of stainless steel SUS 310S and manufactured with a sheet metal press.

A separator plate is normally designed to have a uniform compressive stiffness over its active region at room temperature. However, as shown in Fig. 2, a temperature gradient is built up across the separator during use because the electrochemical reactions occur along the gas paths. Therefore a uniform stiffness across the active area at elevated temperatures cannot be expected, and could cause the contact between unit cells to loosen.

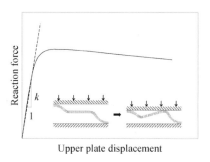

Fig. 3. Evaluation method for stiffness $k$ of separator using the displacement-force curve.

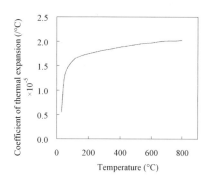

Fig. 4. Variation of coefficient of thermal expansion of stainless steel SUS 310S according to temperature.

Table 1. Stress-plastic strain curves of stainless steel SUS 310S.

| Temp (°C) | $E$ (GPa) | $\sigma_y$ (MPa) | $\sigma = K(\varepsilon_0 + \varepsilon_p)^n$ | | |
|---|---|---|---|---|---|
| | | | $K$ (MPa) | $\varepsilon_0$ | $n$ |
| 25 | 174.66 | 207.838 | 1153.975 | 0.024 | 0.459 |
| 550 | 118.137 | 149.342 | 1129.725 | 0.022 | 0.529 |
| 600 | 115.348 | 145.156 | 1052.935 | 0.021 | 0.512 |
| 650 | 105.119 | 142.504 | 734.095 | 0.021 | 0.467 |
| 700 | 84.195 | 133.638 | 501.220 | 0.007 | 0.266 |
| 750 | 55.433 | 123.898 | 321.985 | 0.003 | 0.164 |

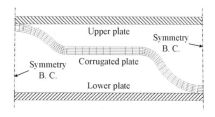

Fig. 5. Finite element modeling for simulations of stiffness evaluation of separator.

It is therefore necessary to investigate the variation in compressive stiffness at elevated temperatures.

We suggest that the initial slope measured from the displacement-force curve in Fig. 3 can be used as a measure of the separator stiffness. Finite element analysis has been performed to obtain displacement-force curves at elevated temperatures. The material properties of SUS 310S at room temperature and at elevated temperatures are shown in Table 1 for the stress-plastic strain curves and in Fig. 4 for the coefficient of the thermal expansion curve.

Figure 5 shows a finite element modeling for one repetitive segment of the separator. The corrugated plate was modeled using plane strain solid elements, and the two flat plates were modeled using rigid elements because no computation of internal stresses was needed. Both ends of the corrugated plate were assigned symmetric boundary conditions, which did not allow them to move inward during forming.

A commercial finite element code, ABAQUS,[6] was used for the simulations. The analysis procedures were as follows:
 (i) Thermal deformation analysis during a change from room temperature to a certain elevated temperature.
 (ii) Sufficient movement of the upper plate down to obtain displacement-force curves.
 (iii) Determination of the stiffness, $k$.

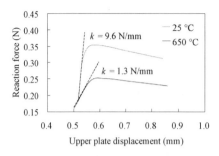

Fig. 6. Degradation of the stiffness of separator.

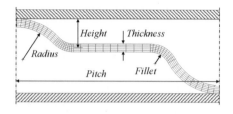

Fig. 7. Five design factors selected used in the Taguchi method.

Table 2. Factors and their levels set in the Taguchi method.

| Factors | Values at each level (mm) | | |
|---|---|---|---|
| | 1 | 2 | 3 |
| Thickness | 0.3 | 0.4 | 0.5 |
| Radius | 2.4 | 2.7 | 3.0 |
| Height | 1.0 | 1.2 | 1.4 |
| Pitch | 8 | 10 | 12 |
| Fillet | 1.0 | 1.2 | 1.4 |

Table 3. 18 runs arranged in the $L_{18}$ table and computed stiffness results.

| No. | Factors | | | | | k |
|---|---|---|---|---|---|---|
| | Thick. | Radius | Height | Pitch | Fillet | (N/mm) |
| 1 | 1 | 1 | 1 | 1 | 1 | 0.795 |
| 2 | 1 | 2 | 2 | 2 | 2 | 0.579 |
| 3 | 1 | 3 | 3 | 3 | 3 | 0.454 |
| 4 | 2 | 1 | 1 | 2 | 2 | 0.970 |
| 5 | 2 | 2 | 2 | 3 | 3 | 0.805 |
| 6 | 2 | 3 | 3 | 1 | 1 | 7.743 |
| 7 | 3 | 1 | 2 | 1 | 3 | 10.47 |
| 8 | 3 | 2 | 3 | 2 | 1 | 2.996 |
| 9 | 3 | 3 | 1 | 3 | 2 | 1.116 |
| 10 | 1 | 1 | 3 | 3 | 2 | 0.401 |
| 11 | 1 | 2 | 1 | 1 | 3 | 0.850 |
| 12 | 1 | 3 | 2 | 2 | 1 | 0.593 |
| 13 | 2 | 1 | 2 | 3 | 1 | 0.765 |
| 14 | 2 | 2 | 3 | 1 | 2 | 7.411 |
| 15 | 2 | 3 | 1 | 2 | 3 | 1.037 |
| 16 | 3 | 1 | 3 | 2 | 3 | 2.933 |
| 17 | 3 | 2 | 1 | 3 | 1 | 1.095 |
| 18 | 3 | 3 | 2 | 1 | 2 | 12.25 |

Table 4. Results of the ANOVA analysis for stiffness.

| Source of variation | Sum of squares | DOF | Mean square | $F_0$ | Sig. prob. |
|---|---|---|---|---|---|
| Model | 233.019 | 10 | 23.302 | 16.933 | 0.55e-3 |
| Thick. | 61.842 | 2 | 30.921 | 22.470 | 0.90e-3 |
| Radius | 7.959 | 2 | 3.980 | 2.892 | 0.12 |
| Height | 36.388 | 2 | 18.194 | 13.221 | 0.42e-2 |
| Pitch | 120.099 | 2 | 60.049 | 43.637 | 0.11e-3 |
| Fillet | 6.730 | 2 | 3.365 | 2.445 | 0.16 |
| Error | 9.633 | 7 | 1.376 | | |
| Total | 242.652 | 17 | | | |

Figure 6 represents the analysis results of the displacement-force curves obtained at room and elevated temperatures. The stiffness value at stack-operation temperature, 650 °C, was 1.3 N/mm, which is only 14 % of the stiffness value at room temperature, 9.6 N/mm. From this result, we know that the separator can have a stiffness that is significantly lower at operational temperatures than its designed, room temperature value.

## 3. Determination of Design Factors Controlling the Stiffness

To minimize the non-uniformity of stiffness across the separator it is necessary to find the design factor that most influences stiffness and use it to redesign a separator with uniform stiffness across the active region. The Taguchi method, which is widely used in engineering applications for design of experiments, is introduced.

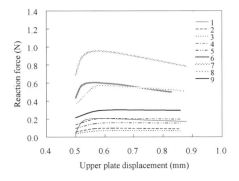
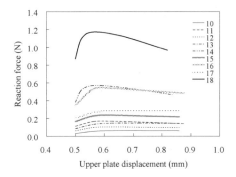

Fig. 8*a*. Results of displacement-force curves for 18 runs (Run 1 to 9).

Fig. 8*b*. Results of displacement-force curves for 18 runs (Run 10 to 18).

Figure 7 shows the design factors that are likely to most affect the separator stiffness: *Thickness*, *Radius*, *Height*, *Pitch*, and *Fillet*. Three kinds of levels for each factor are shown in Table 2. These levels were determined from design experience.

The orthogonal array, $L_{18}$, is one of most useful arrays in the Taguchi method because it can show the major effects of the design factors after only 18 runs. The design factors and their levels were entered into $L_{18}$ as shown in Table 3, and then 18 numerical analysis runs were performed. The resulting displacement-force curves are shown in Figs. 8*a* and 8*b*. The stiffness values obtained from the curves are shown in the last column in Table 3.

Based on these results, ANOVA (analysis of variance) was carried out over the stiffness values. Table 4 shows the ANOVA results. The model we set up using the five chosen factors explains 96 % of the total stiffness variation. *Pitch*, *thickness*, and *height*, in this order, were shown to strongly influence the stiffness, while *radius* and *fillet* had a weaker effect. The five factors can be ordered by the strength of influence: *Pitch* > *Thickness* > *Height* >> *Radius* > *Fillet*.

## 4. A Separator Having Uniform Stiffness at Elevated Temperature

In an MCFC stack, non-uniform temperature distributions are inevitable due to the electrochemical exothermic reactions in the unit cells. Temperature gradients across the separator plates can induce stiffness gradients as described in previous sections. If we control the *pitch* factor across the separator, we can achieve uniform stiffness across the separator at elevated temperatures.

We have investigated the stiffness variation after reducing the *pitch* from the existing design value of 10 mm to 7 mm at intervals of 0.5 mm. Figure 9 shows the results. As expected, a smaller *pitch* led to greater stiffness. Figure 9 also shows that when we reduced the *pitch* to around 7 mm, the stiffness at 650 °C was similar to that at room temperature.

The stiffness gradient and the temperature distribution in the separator are compared in Fig. 10. To achieve a uniform stiffness, the *pitch* size in the high temperature zone (the central region) must be tuned down and in the low temperature zone (the inlet and outlet

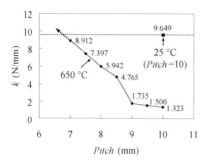

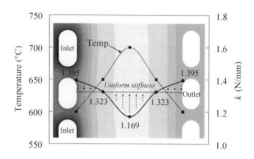

Fig. 9. The increase of stiffness according to the decrease of the factor *pitch* at elevated temperature.

Fig. 10. The line for *uniform stiffness* is drawn over the stiffness of 650 °C.

regions) it must be tuned up. Based on the results in Table 5, at 650 °C, the *pitch* size must be decreased by 5% near the central region and increased by 2% near the inlets and outlets to design a separator with uniform stiffness.

## 5. Conclusions

(1) The compressive stiffness of a separator in an MCFC stack at 650 °C is degraded to only 14 % of that at room temperature. This indicates the importance of evaluating the stiffness of the separator at the working temperature of the MCFC.

(2) Among the five chosen design factors, the one most strongly affecting the stiffness of the separator was shown to be the *pitch*, according to the results of the Taguchi method. The order of factor influence was found to be *pitch*, *thickness*, *height*, *radius*, and *fillet*.

(3) A separator plate with uniform stiffness over its active area can be created by controlling the *pitch* appropriately. Based on the existing design, the *pitch* size should be decreased by 5 % near the high temperature regions (the central region) and increased by 2 % near the low temperature regions (the inlet and outlet regions).

## Acknowledgments

Financial support by the Korean Ministry of Knowledge and Economy through the Electric Power Industry Technology Evaluation & Planning Center is gratefully acknowledged (Project no. R-2004-1-142).

## References

1.  Youngjoon Moon, Dokyol Lee, *J. Power Sources*, **115**, 1 (2003).
2.  M. Keijzer, K. Hemmes, P. J. J. M. Van Der Put, J. H. W. De Wit, J. Schoonman, *Corrosion Sci.*, **39**(3), 483 (1997).
3.  F. J. Perez, D. Duday, M. P. Hierro, C. Gomez, A. Aguero, M. C. Garcia, R. Muela, A. Sanchez Pascual, L. Martinez, *Surface and Coatings Tech.*, **161**, 293 (2002).
4.  Li Zhou, Huaxin Lin, Baolian Yi, Huamin Zhang, *Chem. Eng. Journal*, **125**, 187 (2007).
5.  F. Yoshiba, T. Abe, T. Watanabe, *J. Power Sources*, **87**, 21 (2000).
6.  ABAQUS User's Manual (ver. 6.5), ABAQUS Inc., Richmond, USA (2004).

# A STUDY ON THE IMPROVEMENT OF THE ACCURACY OF A SHAVING PROCESS USING A PROGRESSIVE DIE

WAN-JIN CHUNG[1], KU-BOK HWANG[2], JONG-HO KIM[4†]

*Department of Die and Mould Engineering, Seoul National University of Technology*
*172 Gongneung2-Dong, Nowon-Gu, Seoul, 139-743, KOREA,*
*wjchung@snut.ac.kr, kbh4717@chol.com, jhkim365@snut.ac.kr*

HO-YEUN RYU[3]

*Center for Transportation & Machinery Parts Service Center, Korea Institute of Industrial Technology*
*#209 BS-DITIC, 1274, Jisa-Dong, Gangseo-Gu, Busan, 618-230, KOREA,*
*hyryu@kitech.re.kr*

Received 15 June 2008
Revised 23 June 2008

Shaving is widely used as a finish process to generate a smooth sheared surface from a sheared surface originally obtained by blanking or piercing. In the sheet metal forming industry, progressive operations are commonly used to increase productivity. However, it is difficult to utilize shaving in a progressive die because a blank is dropped down after blanking. In this study, a method that addresses this issue is introduced, and the two-stage progressive operation of half-blanking and blank-shaving is suggested. For this study, a square part made by blanking and piercing is considered, and a progressive die that includes piercing, pierce-shaving, half-blanking and blank-shaving is prepared for a specimen of cold rolled steel sheet. Experiments are carried out involving the process parameters of the shaving allowance, shaving clearance and the half-blanking penetration depth. It was found that the proposed method using a progressive die offers a completely smooth sheared surface along with feasible working conditions for a cold rolled steel sheet with a thickness of 1.56 mm.

*Keywords*: Progressive die; Pre-piercing; Pierce-shaving; Half-blanking; Blank-shaving.

## 1. Introduction

Due to the strong competition in today's industry, parts with smooth sheared surfaces are greatly desired. Fine blanking and shaving, as precision shearing processes, are widely used to meet these needs. Fine blanking[1,2] is typically carried out using an expensive exclusive-use hydraulic apparatus in a single operation. On the other hand, the tooling costs for fine blanking are usually high owing to the demands for a high level of dimensional accuracy. For moderate and large parts, fine blanking is a highly effective

---

[†]Corresponding Author.

process that obtains a high-quality sheared surface. However, currently, as complex functions are in strong demand in small-size products, the gaps between the holes have become very close to one another. Thus, the use of fine blanking techniques is a challenge due to the low strength of the tools.

In industry, an inexpensive machine press is commonly used to obtain a smooth sheared surface via a shaving process[3,4]. Most shearing processes such as piercing and blanking yield an uneven surface with fractures that develops from such processes. In order to obtain a smooth surface without fractures, shaving is used in a subsequent process using the principles of a cutting process. However, a two-stage operation of shearing and shaving introduces problems in mass production because it requires a high level of positioning accuracy between the two separate stages.

Especially to produce a completely smooth sheared surface without fractures in the small and precise parts that comprise cameras, computers and communication devices, an additional expensive process such as grinding is required.

In this study, a method using a progressive die for high productivity is suggested, and optimal process conditions that lead to a smooth sheared surface without fractures are investigated. In this method, pierce-shaving and blank-shaving are applied to obtain a smooth sheared surface on both the inner sheared surface and outer sheared surface. In the pierce-shaving process, optimal process conditions for shaving are studied. Blanking is replaced with half-blanking as the part is separated from the strip. In the half-blanking process, the half-blanking depth that avoids cracking is studied. In the blank-shaving process, the effect of clearance is studied using the optimal condition found in the study of the pierce-shaving process.

## 2.  Experiment

### 2.1.  *Materials and process design*

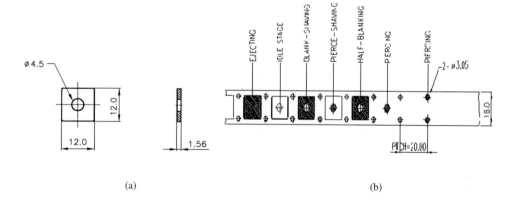

Fig. 1. (a) Shape of the product and (b) strip layout.

A square-shaped part with a circular center hole for digital camera is considered. The shape of this product is shown in Fig. 1(a). In order to obtain a smooth sheared surface without fractures, the part was finished by grinding. The materials used for the experiment is a cold-rolled steel sheet with a thickness of 1.56mm. The progressive strip layout for the experiment is shown in Fig. 1(b). A corresponding tool was designed and manufactured. The progressive die has seven stages: piercing of two pilot pin holes, piercing of a center hole, half-blanking of a square profile, pierce-shaving, blank-shaving, an idle stage, and ejecting. The positioning accuracy is maintained within 0.01 mm with the aid of a precise feeding device and pilot pins.

## 2.2. *Experimental method*

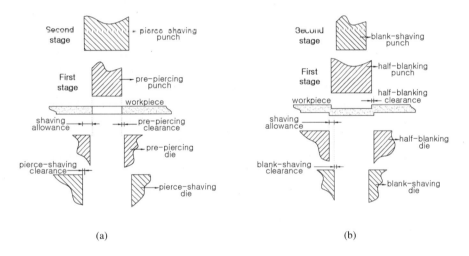

(a)                                                        (b)

Fig. 2. (a) Piercing and subsequent pierce-shaving, (b) Half-blanking and subsequent blank-shaving.

The two-stage operation of piercing and pierce-shaving is illustrated in Fig. 2(a). Piercing of a center hole is carried out using insert punches with diameters of 3.0 mm, 3.5 mm, 4.1 mm and 4.6 mm. Mating die inserts are manufactured with 5% clearance in terms of thickness. The punch diameter of the pierce-shaving process corresponds to the diameter of the center hole of the product. Thus, the corresponding shaving allowance ($\Delta a$) is half of the punch diameter difference between the piercing and the pierce-shaving stages. For the pierce-shaving process, mating die inserts are manufactured with four different clearances (0.3, 1.3, 5 and 8%) in terms of thickness. In this paper, a smooth sheared surface without fracture is considered as an effective sheared surface. All experimental process conditions are shown in Table 1. In the piercing-shaving process, the proportion of an effective sheared surface is examined according the above process parameters and the optimal process conditions are explored.

Table 1. Experimental conditions for pierce-shaving.

| Shaving allowance ($\Delta a$, mm) | Clearance (%) | Remark |
|:---:|:---:|:---:|
| 0.1 | 0.3 | Pre-piercing Clearance: 5% (0.08 mm) |
| 0.2 | 1.3 | |
| 0.5 | 5 | |
| 0.75 | 8 | |

The two-stage operation of half-blanking and blank-shaving is presented in Fig. 2(b). Except for the piercing process, as part is separated from the strip after blanking, blanking should be replaced by half-blanking in which blanking is performed via partial punch penetration that is less than the thickness of the sheet. This is shown in Fig. 2(b). After the blank-shaving process, the part is separated from the strip. The die dimension of blank-shaving corresponds to the dimension of the outer profile of the product. Thus, the corresponding shaving allowance is half of the difference of the die dimension between blanking and blank-shaving. The shaving allowance is the optimal value obtained in the study of the pierce-shaving process. It should be noted that only a blanked surface in the half-blanking process is processed by blanking and shaving. Therefore, an un-blanked surface in the half-blanking process is processed only by blanking and may have an unsmooth surface. Thus, in order to improve the smoothness of the surface, the proportion of blanked surface in the half-blanking process should be maximized. Therefore, the occurrence of cracking is investigated according to the half-blanking depth ( $\Delta t$ ) listed in Table 2. After the blank-shaving process, the percentage of an effective sheared surface is investigated according to the clearance shown in Table 3.

Table 2. Experimental condition for the investigation of crack occurrences.

| Half-blanking depth (% in terms of thickness) | Remark |
|:---:|:---:|
| 32 | |
| 45 | |
| 57 | $\Delta a$ : 0.2 mm |
| 64 | Clearance : 5 % |
| 70 | |
| 77 | |

Experiments were carried out for 20 specimens. A specimen was cut by wire-cutting process along the center line, and the percentage of an effective sheared surface is measured using a tool microscope. The average value for 20 specimens was chosen as the representative value. The occurrence of cracking was also observed the by tool microscope.

Table 3. Experimental conditions for blank-shaving ( $\Delta t \, / \, t$ : relative punch penetration depth).

| Clearance (%) | 0.3 | 1.3 | 5 | 8 | Remark |
|---|---|---|---|---|---|
| Punch Size | 11.99 | 11.96 | 11.84 | 11.74 | $\Delta a$ : 0.2 mm<br>$\Delta t \, / \, t \times 100\%$ : 70% |

## 3. Results and Discussion

In the pierce-shaving experiment, the proportion of an effective sheared surface was compared according to the clearance and the shaving allowance. As shown in Fig. 3, with a clearance of 0.3%, the proportion of an effective sheared surface was 100% with all shaving allowances ($\Delta a$). At a clearance of 1.3%, with the exception of a shaving allowance of 0.75mm, the proportion of an effective sheared surface was 100%. With clearances of 5% and 8%, a fractured surface was observed for all shaving allowances. With shaving allowances of 0.1mm and 0.2mm, nearly the same proportion of an effective sheared surface was obtained. On the other hand, with a shaving allowance larger than 0.5mm, a smaller proportion of an effective sheared surface was observed. In addition, a fractured surface occurred in piercing remains in some cases due to the positioning and fabrication inaccuracies at a shaving allowance of 0.1mm.

Thus, considering the difficulty of manufacturing and the stability of production, the most feasible pierce-shaving process conditions are considered to be a shaving allowance of 0.2 mm and a clearance of 1.3%.

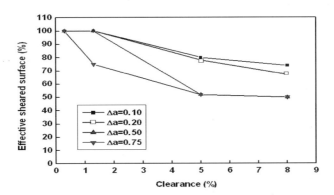

Fig. 3. Proportion of the effective sheared surface with different clearances and shaving allowances.

In the blank-shaving experiment, a square-shaped part with a side length of 12mm was half-blanked before a subsequent blank-shaving process. In the half-blanking process, the maximum half-blanking depth without cracking should be determined in order to maximize the proportion of blank-shaving. The occurrence of cracking is observed at a half-blanking depth of 77% (1.2mm) in terms of thickness. On the other hand, at half-blanking depth of 70% and less, cracking was not occurred. If a sheared surface has crack, the part may separate from the strip and blank-shaving cannot be carried out in a progressive die. As shown in Fig. 4, at half-blanking depth of 64% and less, fracturing of the surface was observed after the subsequent blank-shaving process. Thus, a half-blanking depth of 70% ($\Delta t/t = 0.7$) was applied in the following half-blanking and blank-shaving processes.

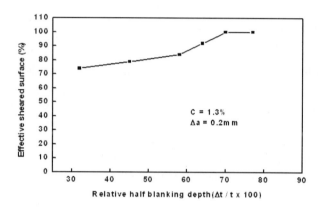

Fig. 4. Proportion of the effective sheared surface according to the half-blanking depth.

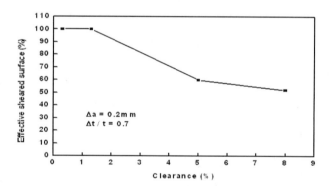

Fig. 5. Proportion of the effective sheared surface according to the clearance in the blank-shaving process.

In the subsequent blank-shaving process, the optimal shaving allowance of 0.2 mm determined from the pierce-shaving investigation was applied. Fig. 5 shows the proportion of an effective sheared surface after blank-shaving according to the clearance given in Table 3. For all clearances, it was found that the proportion of an effective sheared surface by blank-shaving is less than that by pierce-shaving, as only part of the sheared surface is generated by blanking and shaving due to the use of half-blanking. At a clearance of 5%, a fractured surface of 40% is observed and the proportion of fractured surface increased as the clearance increased. At clearances of 0.3% and 1.3%, the proportion of an effective sheared surface was 100%. However, at a clearance of 0.3%, the blank-shaving force tends to increase considerably; hence, the possibility of die failure increased and a fractured surface was observed in several specimens. Thus, a clearance of 1.3% and a half-blanking depth of 70% are the most feasible process conditions within the context of this experiment.

## 4. Conclusion

In this paper, a method that obtains a smooth sheared surface using a progressive die is proposed in the shearing of a cold-rolled steel sheet. In addition, the optimal process conditions of this method are studied experimentally.

In progressive operations, piercing and blanking are followed by pierce-shaving and blank-shaving to obtain a smooth sheared surface. The blanking process is replaced by a half-blanking process in order to realize a progressive operation. It was found that a completely smooth sheared surface can be obtained by the proposed method. The optimal process conditions for a 1.56 mm cold-rolled steel sheet are given below.

In the study of piercing and pierce-shaving, a shaving allowance of 0.2mm and a clearance in terms of thickness of 1.3% were found to be most feasible conditions to achieve a completely effective sheared surface.

In the investigation of blanking and blank-shaving, with a shaving allowance of 0.2mm, it was found that a half-blanking depth of 70% in a half-blanking process and a clearance of 1.3% in the blank-shaving process are the most feasible process conditions to achieve a completely smooth sheared surface.

Finally, it was found that the proposed method involving the use of a progressive die provides an effective means to ensuring a completely smooth sheared surface. This method can be used in place of methods that are more expensive.

### Acknowledgments

This research was financially supported by the Ministry of Knowledge Economy (MKE) and Korea Industrial Technology Foundation (KOTEF) through the Human Resource Training Project for Strategic Technology

### References

1. J. Kim, J. Ryu, K. Shim, D. Kim, Y. Jang and J. Lee, *J. of Korean Society for Precision Engineering* **14**(6), 15 (1997).
2. K. Hirota, K. Kondo and K. Maeda, *J. of the Japan Society for Technology of Plasticity* **35**(386), 67 (1996).
3. M. Murakawa, S. Thipprakmas and M. Jin, *J. of the Japan Society for Technology of Plasticity* **44**(513), 1049 (2003).
4. S. Thipprakmas, M. Jin, M. Murakawa, *Proceeding of the 7th APCMP*, 41 (2006).

# APPLICATION OF FORMING LIMIT CRITERIA BASED ON PLASTIC INSTABILITY CONDITION TO METAL FORMING PROCESS

SEONG-CHAN HEO[1], TAE-WAN KU[2], JEONG KIM[3], BEOM-SOO KANG[4†]

*Dept. of Aerospace Engineering, Pusan National University, Busan, 609-735, KOREA,*
*pnuhsc@pusan.ac.kr, longtw@pusan.ac.kr, greatkj@pusan.ac.kr, bskang@pusan.ac.kr*

WOO-JIN SONG[5]

*ILIC, Pusan National University, Busan, 609-735, KOREA,*
*woodysong@pusan.ac.kr*

Received 15 June 2008
Revised 23 June 2008

Metal forming processes such as hydroforming and sheet metal forming using tubular material and thin sheet metal have been widely used in lots of industrial fields for manufacturing of various parts that could be equipped with mechanical products. However, it is not easy to design sequential processes properly because there are various design variables that affect formability of the parts. Therefore preliminary evaluation of formability for the given process should be carried out to minimize time consumption and development cost. With the advances in finite element analysis technique over the decades, the formability evaluation using numerical simulation has been conducted in view of strain distribution and final shape. In this paper, the application of forming limit criteria is carried out for the tube hydroforming and sheet metal forming processes using theoretical background based on plastic instability conditions. Consequently, it is confirmed that the local necking and diffuse necking criteria of sheet are suitable for formability evaluation of both hydroforming and sheet metal forming processes.

*Keywords*: Hydroforming; Plastic instability; Forming limit criteria.

## 1. Introduction

Nowadays, hydroforming and sheet metal forming processes are commonly used as manufacturing method for various mechanical components in many industrial fields.[1] It is difficult to design the processes properly because there are various design variables that affect on the formability of the parts. In tube hydroforming process, most parts are produced by the combination of two or more loading forces such as internal pressure for material expansion and axial feeding for prevention of excessive thinning. Therefore, it could cause unexpected failures or defects on the parts such as bursting, buckling, folding and wrinkling unless proper loading paths are given.[2, 3] In sheet metal forming processes, similarly, tearing is frequently occurred during the processes because the sheet material

---

†Corresponding Author.

experiences local deformation due to discontinuity in distribution of forming loads transferred by contact between die tools and material. Both bursting and tearing are the most frequently occurred failure modes similar with plastic deformation behavior in view of necking. With the advances in finite element analysis technique over the decades, the formability evaluation using numerical simulation has been conducted in view of strain distribution and final shape. In this study, derivation and application of several forming limit criteria based on plastic instability conditions were carried out for an automobile engine cradle and a dimple type heat transfer tube as examples of hydroforming and sheet metal forming parts, respectively. Consequently, it is confirmed that the local necking and diffuse necking criteria of sheet are suitable for formability evaluation of both hydroforming and sheet metal forming processes.

## 2. Analytical Forming Limit Criteria

Strain based forming limit diagram (FLD) is a generally used criterion for formability evaluation on various sheet metal forming processes. However, it is well known that FLD is normally available for specific forming cases which have linear loading path and linear strain path thus it is dependent on strain path of the deformable body.[4] On the other hand, stress based forming limit diagram (FLSD) found by Kleemola[5] is independent on strain path. Therefore, it is more useful to evaluate formability of more general sheet metal forming processes. From this point of view, theoretical forming limit curves based on plastic instability conditions are derived.

### 2.1. *Plastic instability conditions*

Assuming that principal stresses and strain increments maintain uniform ratios as constants $\alpha = \sigma_2 / \sigma_1$ and $\beta = d\varepsilon_2 / d\varepsilon_1$, respectively. In order to derive the critical strains and stresses for forming limit curve analytically, three different necking criteria are

Table 1. Representative values for analytical forming limit curves based on three different plastic instability conditions.

| Classification | Local necking condition for sheet (abbreviation : L) | Diffuse necking condition for sheet (abbreviation : S) | Diffuse necking condition for tube (abbreviation : T) |
|---|---|---|---|
| Onset of plastic instability | $d\sigma_1 = \sigma_1\left(d\varepsilon_1 + d\varepsilon_2\right)$ & $d\sigma_2 = \sigma_2\left(d\varepsilon_1 + d\varepsilon_2\right)$ | $dF_1 = 0$ & $dF_2 = 0$ | $dp = 0$ & $dF_{axial} = 0$ |
| $d\sigma_1, d\sigma_2$ | $\dfrac{d\bar\varepsilon}{\bar\sigma}\left[\dfrac{\alpha+1}{2}\right]\sigma_i^2,\ \dfrac{d\bar\varepsilon}{\bar\sigma}\left[\dfrac{\alpha(\alpha+1)}{2}\right]\sigma_i^2$ | $\dfrac{d\bar\varepsilon}{\bar\sigma}\left[1-\dfrac{\alpha}{2}\right]\sigma_i^2,\ \dfrac{d\bar\varepsilon}{\bar\sigma}\left[\alpha\left(\alpha-\dfrac{1}{2}\right)\right]\sigma_i^2$ | $\dfrac{d\bar\varepsilon}{\bar\sigma}\left[\dfrac{3+\alpha}{2}\right]\sigma_i^2,\ \dfrac{d\bar\varepsilon}{\bar\sigma}\left[1-\alpha+\alpha^2\right]\sigma_i^2$ |
| $d\bar\sigma$ | $\bar\sigma d\bar\varepsilon\left[\dfrac{1+\alpha}{2(1-\alpha+\alpha^2)^{1/2}}\right]$ | $\bar\sigma d\bar\varepsilon\left[\dfrac{\alpha(2\alpha-1)^2+(2-\alpha)^2}{4(1-\alpha+\alpha^2)^{3/2}}\right]$ | $\bar\sigma d\bar\varepsilon\left[\dfrac{\alpha(2\alpha-1)^2+2(2-\alpha)(1+\alpha)}{4(1-\alpha+\alpha^2)^{3/2}}\right]$ |
| Sub-tangent | $\dfrac{1}{Z_L} = \dfrac{1}{\bar\sigma}\dfrac{d\bar\sigma}{d\bar\varepsilon} \leq \dfrac{(1+\alpha)}{2(1-\alpha+\alpha^2)^{1/2}}$ | $\dfrac{1}{Z_S} = \dfrac{1}{\bar\sigma}\dfrac{d\bar\sigma}{d\bar\varepsilon} \leq \dfrac{\alpha(2\alpha-1)^2+(2-\alpha)^2}{4(1-\alpha+\alpha^2)^{3/2}}$ | $\dfrac{1}{Z_T} = \dfrac{1}{\bar\sigma}\dfrac{d\bar\sigma}{d\bar\varepsilon} \leq \dfrac{\alpha(2\alpha-1)^2+2(2-\alpha)(1+\alpha)}{4(1-\alpha+\alpha^2)^{3/2}}$ |

adopted in this study. These criteria consist of local necking criterion proposed by Hill[6], and diffuse necking criteria by Swift.[7] The onset conditions of plastic instability and intermediately derived values are summarized in Table 1.

### 2.2. *Forming limit curves based on plastic instability conditions*

Critical principal strain components $\varepsilon_1^c$ and $\varepsilon_2^c$ are defined to obtain forming limit curves with regard to the three plastic instability conditions with an work-hardening model, $\bar{\sigma} = K(\varepsilon_0 + \bar{\varepsilon})^n$, which has initial strain, $\varepsilon_0$. Sub-tangent $Z_i$ in Table 1 can be written in a general form as follows:

$$1/Z_i = (1/\bar{\sigma})(d\bar{\sigma}/d\bar{\varepsilon}) = n/(\varepsilon_0 + \bar{\varepsilon}) = \Psi_i/\Omega_i \quad (i = L, S, T) \tag{1}$$

Assuming the proportional loading condition, critical principal strain components for the plastic instability conditions from Eq. (1) can be expressed as follows:

$$\varepsilon_1^c = \left[(\Omega_i/\Psi_i)n - \varepsilon_0\right]/\Theta, \quad \varepsilon_2^c = \beta\varepsilon_1^c \quad (i = L, S, T) \tag{2}$$

where $\Theta$ means $2\sqrt{1+\beta+\beta^2}/\sqrt{3}$.

From Eq. (2), critical principal stress components can be obtained as follows:

$$\sigma_1^c = \frac{4}{3}\left[K\left\{\left(\varepsilon_0 + \Theta\varepsilon_1^c\right)^n\right\}/\bar{\varepsilon}\right]\left(\varepsilon_1^c + \frac{1}{2}\varepsilon_2^c\right), \quad \sigma_2^c = \frac{4}{3}\left[K\left\{\left(\varepsilon_0 + \Theta\varepsilon_1^c\right)^n\right\}/\bar{\varepsilon}\right]\left(\varepsilon_2^c + \frac{1}{2}\varepsilon_1^c\right) \tag{3}$$

From Eq. (2) and (3), forming limit diagrams based on strain and stress for evaluation of failure on a thin material can be drawn.

### 3.  Numerical Simulations of Hydroforming and Sheet Metal Forming Processes

### 3.1. *Hydroforming: Engine cradle*

In this paper, an automobile engine cradle part was selected as a hydroforming part which is made by complicate loading conditions such as internal pressure and axial feedings at both ends of the tube. Fig. 1 shows numerical simulation model for the engine cradle

(a) Hydroforming dies and axial feeding punches          (b) Preformed tube model
Fig. 1. Numerical simulation model for tube hydroforming process analysis.

(a) Effective strain distribution          (b) Effective stress distribution
Fig. 2. Distributions of effective strain and stress at the end of hydroforming process simulation.

constructed by ANSYS pre-processor. Forming tools consist of upper and lower die, and two punches which feed the tubular material into the die cavity to prevent excessive thinning (See Fig. 1(a)). Mechanical behavior of the material was assumed that it obeys work-hardening law, $\bar{\sigma} = K(\varepsilon_0 + \bar{\varepsilon})^n$, with the strength coefficient, $K$=559.43MPa, work-hardening exponent, $n$=0.243 and initial strain $\varepsilon_0$ =0.0292, which are obtained from free bulge tests using straight tube. Friction coefficient between tools and material was assumed as 0.05. Loading path for the engine cradle part has maximum feeding amount, 30mm, and calibration pressure, 200MPa. Finite element simulation was carried out by ANSYS LS-DYNA solver which is based on explicit scheme. Fig. 2 shows simulation results in view of strain and stress

### 3.2. *Sheet metal forming: Dimple type heat transfer tube*

As an example of sheet metal forming process, dimpled heat transfer tube was selected in this paper. The tube has a lot of dimples on its surface along the longitudinal direction with repeated pattern as shown in Fig. 3. In order to minimize the simulation time, the partial model was used as shown in Fig. 4. In this figure, finite element analysis model was composed of 6 and 1/2 dimples that can be considered as a representative of the dimples because center located dimples are definitely enclosed with other dimples each other. Material properties of the sheet material were obtained from uni-axial tensile tests for work-hardening model, $\bar{\sigma} = K\bar{\varepsilon}^n$, with the strength coefficient, $K$=1796.5MPa and work-hardening exponent, $n$=0.591. Sheet blank which has 0.5mm of thickness and dies which were assumed as rigid material were modeled using solid and shell element, respectively. Friction coefficient between dies and material was set as 0.05. Similarly, finite element simulation was also carried out by ANSYS LS-DYNA solver. Fig. 5(a) and (b) depict the results such as equivalent strain and stress distributions.

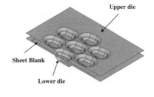

Fig. 3. Example of dimple type heat transfer tube model with repeated pattern.

Fig. 4. Simplified model for dimple forming analysis based on stamping process using dies.

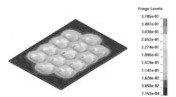

(a) Effective strain distribution        (b) Effective stress distribution

Fig. 5. Distributions of effective strain and stress for dimple type heat transfer tube.

## 4.  Application of Formability Evaluation based on Forming Limit Diagrams

### 4.1.  *Evaluation of formability: Hydroformed engine cradle*

As a premise, it was confirmed that the actual hydroformed part could be manufactured soundly in practical processes. Fig. 6(a) and (b) show forming limit curves for the engine cradle and the distributions of principal strain and stress results obtained from the simulation, respectively. As shown in this figure, only diffuse necking criterion for tube considerably underestimates the forming limit. In contrast, necking criterion for sheet estimates the formability of the tubular material properly. It means that the necking phenomenon can be considered as deformation in an infinitesimal planar material. It can be seen more clearly in the Fig. 6(b) that upper margin of the distributed stress data has good accordance with local and diffuse necking criteria with regard to sheet. From this figure, it is noted that the local necking criterion predicts the formability more suitably in the negative minor strain region, $-1 \le \beta \le 0$, ( $-1 \le \alpha \le 0.5$ in FLSD, where $\alpha = (2\beta + 1)$ $/(2 + \beta)$ ) and the diffuse necking criterion predicts well with regard to the positive minor strain region, $0 \le \beta \le 1$, ( $0.5 \le \alpha \le 1$ in FLSD). However, some data points are distributed over the forming limit diagram, thus bursting may be occurred at those points on the part after forming process. Therefore, several experiment such as free bulge test should be carried out to adjust the forming limit curves considering the experiment results.

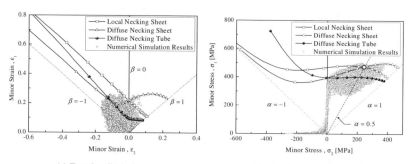

(a) Forming limit diagram          (b) Forming limit stress diagram

Fig. 6. Formability evaluation of engine cradle through analytical forming limit curves.

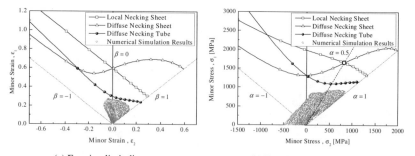

(a) Forming limit diagram          (b) Forming limit stress diagram

Fig. 7. Formability evaluation of dimple tube through analytical forming limit curves.

### 4.2. *Evaluation of formability: Dimple type heat transfer tube*

Fig. 7 (a) and (b) show the application of forming limit curves to dimple tube forming process. Despite of underestimation of diffuse necking criterion for tube, all data of strain and stress are distributed below the all of forming limit curves as shown in the figures. In this case, the sheet blank material experiences no much strain comparing with its elongation and fracture properties. In view of minor and major strain region, these results also shows similar tendency with tube hydroforming process for engine cradle. Consequently, it is expected that the dimple type heat transfer tube can be manufactured without any defect or failure.

### 5. Concluding Remarks

In this paper, strain and stress based forming limit diagrams using three different plastic instability conditions are introduced and applied to hydroformed engine cradle and the dimple tube parts. As results, forming limit curves for each material is obtained and the formability of the parts can be evaluated using the numerical simulation results. From the applications of the FLD and FLSD, it is confirmed that forming limit criterion considering deformation of tubular material that could be thought as deforming shape with arbitrary curvatures during the forming process underestimates the forming limit in comparison with the other criteria based on sheet. In other words, the formability evaluation criteria for the hydroformed parts should be derived with the planar material assumption. In addition, it is noted that the local necking and diffuse necking criteria for sheet show good accordance in the negative and positive minor strain region, respectively. Consequently, analytical forming limit diagrams can be used as criteria to predict the bursting or tearing failure with experiment based modification.

### Acknowledgments

This work was supported by the Korea Science and Engineering Foundation (KOSEF) NRL Program grant funded by the Korea government (MEST) (No. R0A-2008-000-2 0017-0). Also the last author would like to acknowledge the partial support of the fostering project of the Laboratory of Excellence by MOE, MOCIE and MOLAB and thank the partial support by grants-in-aid for the National Core Research Center program from MOST/KOSEF (No. R15-2006-022-02002-0).

### References

1. B. S. Kang, B. M. Son, J. Kim, *Int. J. Mach. Tools Manuf.*, **44**, 87 (2004).
2. F. Dohmann and C. Hartl, *J. Mater. Process. Technol.*, **71**, 174 (1997).
3. N. Asnafi and A. Skogsgardh, *Mater. Sci. Eng. A*, **279(1-2)**, 95 (2000).
4. T.B. Stoughton and X. Zhu, *Int. J. Plasticity*, **20**, 1463 (2004).
5. H.J. Kleemola and M.T. Pelkkikangas, *Sheet Metal Industries*, **63**, 591 (1977).
6. R. Hill, *The mathematical theory of plasticity* (Oxford University Press, New York, 1983).
7. H.W. Swift, *J. Mech. Phys. Sol.*, **1**, 1 (1952).

# TOOL PATH DESIGN OF INCREMENTAL OPEN-DIE DISK FORGING USING PHYSICAL MODELING

SUNG-UK LEE[1], DONG-YOL YANG[2†]

*Department of Mechanical Engineering, KAIST,*
*Daejeon, 305-701, KOREA,*
*silillo@kaist.ac.kr, dyyang@kaist.ac.kr*

Received 15 June 2008
Revised 23 June 2008

A small-batch product of large-sized parts is usually manufactured using incremental open-die forging. In order to control the overall change in the shape of a part, it is essential to be able to predict the shape changes that occur during each step. This paper addresses shape changes of a material according to the forging path. Rapid prediction of metal flows for continuing incremental deformation using theoretical methods is difficult. Accordingly, instead of a theoretical approach, an experiment that tests the tendency of the metal flow for development of forming processes is required. For the sake of convenience, simulative experiments are carried out using plasticine at room temperature. In present study, the tool movement is dominant parameters to with respect to changing the shape of the workpiece.

*Keywords*: Incremental open-die disk forging; physical modeling; tool path.

## 1. Introduction

Open-die forging, wherein a workpiece is manipulated manually and shaped under a hammering action, is carried out incrementally with only a part of the work-piece being deformed at each step. The system comprises a set or sets of flat or curved dies that can be produced economically. The tools are small compared to the overall sizes of the forgings. The resultant reduced forging load implies that a lower load capacity and thus lower investment on forging equipment can be anticipated in producing a given part. The principle of such an incremental forging process is simply compression or upsetting of the material step-by-step until the final target shape is attained[1].

The physical modeling method is an important technique in that it yields useful and practical information without requiring elaborate facilities and tooling. It is a method that plant operators can easily learn as it is straightforward and expedient. It must be recognized, however, that it is difficult to achieve complete similarity between a model and the actual deformation processes, as the similarity criteria for the mechanical properties, temperature and boundary conditions cannot be satisfied completely[2-3]. Nonetheless, it is now well established that the physical modeling approach, despite some physical disadvantages, is a powerful tool for improving the existing processes and

---

[†]Corresponding Author.

products and developing new processes. In the present work, through intensive use of physical modeling (especially plasticine modeling)[4-7], as an approach to investigate hot steel deformation problems, the metal deformation behavior in incremental open-die disk forging is investigated using an plasticine.

This work shows how a change in the final shape can be produced by specific control of the flat-bar forging parameters in a multi-pass design. The parameters can include, the upper die width, overlapping, staggering, configuration, and the billet shape. It is clear that the existing practice and empirical equations currently in use are not useful for disk forming (an increased ingot cross-section). Hence, the present work was executed in an effort to provide a forging guideline that can accurately predict metal flow. The associated flow pattern is produced in translational movement, incremental reduction and 90° rotation operations. The experimental results are used to derive the relationship between process variables and the change in the forming shape.

## 2. Design of Experiments

The quality of the forgings under given metallurgical conditions depends on the shape of the forging tool paths and the execution of the individual forging operation. The most important considerations for selection of the tools and forging sequence are related to the material flow. Generally, Incremental forging experiments were conducted to investigate the effects of the bite ratio, tool overlapping, tool staggering, ingot shape, tool speed, and quantity of stroke per step in the multiple-stroke pass of forged ingots. In the present work, the effects of tool width and tool path are considered. Experiments were carried out to obtain information regarding the final shapes and dimensions of the forged billet. To examine the flow patterns, a plasticine billet was drawn with 16 lines on the side of the cylinder ( diameter : 100mm, height : 60mm ) along the circumference. The reduction of height and the tool speed in each test were controlled by the use of UTM.

### 2.1. Tool overlapping and staggering

In an incremental open-die forging process the prima aims are to achieve consolidation and a uniform structure throughout the ingot cross-section and length with a lower forming load. However, as this is a discontinuous process, each stroke causes inhomogeneous displacement of the material under the surface of the tool. For a more homogeneous distribution of the deformation, the strain history should be controlled so that deformation is as uniform as possible. This can be achieved by means of proper assignment of tool overlapping and staggering[8].

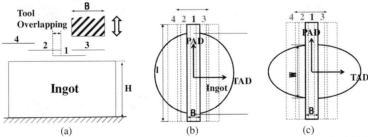

Fig. 1. Schematic representation of tool overlapping and staggering. (a)front, (b) top view for initial shape, and (c) top view of deformed shape.

In the experiment, tool strokes were overlapped by constant amounts for each pass. The amount of tool overlap considered was 5mm. Figure 1 illustrates one of the path designs with application of both upper die staggering and overlapping. The flat tool moves translationally from the billet center toward the edge. TAD is defined as the tool advance direction, and PAD is defined as the perpendicular direction to the tool advance direction in Figure 1.

### 2.2.   *Tool geometry and tool paths*

In incremental open-die disk forging, the workpiece is placed on a lower flat die to be forged by a narrower upper die. The width of the upper die, B is one of the most important parameters in forging operations, having a substantial influence on the distribution of the internal deformation, the magnitude of elongation, and finally, the material shape. Plasticine billets, employed as simulative ingots, were forged using a flat conventional tool corresponding to a length (l) of 180mm and a width (B) of 20mm or 30mm. The tool speed was limited to 30mm/min in all cases.

Table 1.   Tool path schedule.

|  | Stroke/ Stage [mm] | Tool width [mm] | Tool Direction ( ↔: x–dir., ↕: y–dir) | | |
|---|---|---|---|---|---|
|  |  |  | Initial→Stage 1 | Stage 1→2 | Stage 2→3 |
| Case 1 | 10mm | 20mm | ↔ | ↕ | ↔ |
| Case 3 | 10mm | 30mm | ↔ | ↕ | ↔ |
| Case 3 | 10mm | 20mm | ↔ | ↔ | ↔ |

Following the first path (initial → stage 1), the tool was advanced in the x-direction, and a subsequent second path was conducted according to the selected schedule in Table 1. The value of stroke was applied to the same for each stage during the process.

### 3.   Experimental Results and Discussion

The initial circle shape on the upper surface was changed into various elliptical forms according to the tool direction and the side shape also was changed in incremental open die disk forging.

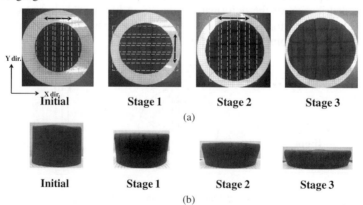

(a)

(b)

Fig. 2. Step-by-step shapes in Case 2 (a) top, (b) side view.

As shown in Figure 2, the size of the upper surface is larger than that of the lower surface and this difference increases with further steps of the process. This situation results from incremental forming. Unlike the upsetting process, incremental forming, whereby local deformation is induced, mainly deforms the upper flat part. Therefore, asymmetrical bulge shapes are seen in the flank.

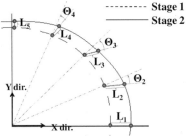

Fig. 3. Definition of difference angle ($\theta$) and elongated length (L).

The results measured through the experimental procedure were classified by their geometrical characteristics ($\theta$, L) in figure 3. The difference angle, $\theta$ is defined as the difference between the actual stretching direction and the radial direction of initial marked point. The elongated length (L), meanwhile, is an extended value of a point from the prior stage to the current stage. As points 1 and 5 are located on the axis line, their deformation angles have little effect on the tool path. Thus, the values of the difference angle for points 1 and 5 were not measured.

Observing the deformation aspect from the initial position to the final location, there exist two axes of symmetry. Therefore, the deformation path is drawn for one-quarter of the section in Figure 4.

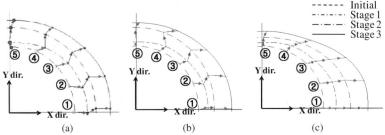

Fig. 4. Shapes of the upper surface (quarter section) (a) Case 1, (a) Case 2, (a) Case 3.

Figure 4 shows that the tool direction mainly affects the spread direction of the points for each stage. Hence, the spread direction inclines to TAD (tool advance direction). Furthermore, the elongated length for each point is also affected by the tool direction, because the elongated length accumulates the values of spread toward TAD.

### 3.1. *Effect of tool width*

In the case of conventional flat bar forging, the spread in PAD (perpendicular direction) increases with a reduction of height and also increases with an increase in $B/H_0$, where B is the upper die width and $H_0$ is the initial height. Meanwhile, the spread in TAD (tool advance direction) decreases with an increase in $B/H_0$[9-10]. In other words, the spread in

PAD increases relatively, whereas the spread in TAD decreases with an increase in the die width. Moreover, the effect of spread in TAD is more dominant than that in PAD if W, when is the breadth of the specimen in Figure 1, is larger than B. these characteristics of conventional flat bar forging (single-pass forming) show that the accumulated spread in TAD increases as the width of tool decreases in case of the incremental forging process (multi-pass forming).

Table 2 shows the area according to stage for each case. The results show that the area in the case of the small die width is larger than that in the case of the large die width. In addition, it is observed in Figure 5 that the elongated length of Case 1 is larger than that of Case 2 in general. Accordingly, this shows that smaller width of the upper die results in larger elongated length value. But, the difference in the experimental results according to the tool width is small.

Table 2.  Area of upper surface for each case and stage.

| | Area of upper surface[mm²] | | | |
|---|---|---|---|---|
| | Initial | Stage 1 | Stage 2 | Stage 3 |
| Case 1 | 7745.97 (100%) | 9929.50 (128%) | 13101.98 (169%) | 16850.11 (218%) |
| Case 2 | 7747.89 (100%) | 9788.20 (126%) | 12648.32 (163%) | 16471.87 (212%) |
| Case 3 | 8058.91 (100%) | 10322.90 (128%) | 13272.02 (165%) | 17226.72 (214%) |

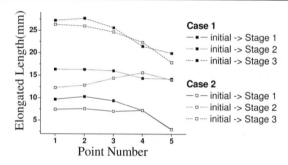

Fig. 5.  Elongated length (L) for each point number between Case 1 and Case 2.

## 3.2. *Results of tool paths*

Figure 6 reveals that tool paths affect the shape of the upper surface. When the TAD is crossing the prior TAD (stage 2 of case 1), the shape of the upper surface shows a circle. So the elongated length is around 15 and the difference angle is around zero. The present situation means that the total deformation is isotropic on the upper surface, when the TAD is crossed. If the TAD per stage is the same (Case 3), the shape of the upper surface will be an ellipse, which is wider in TAD than PAD. That is because spread in the TAD is accumulated during translational movement of the flat tool. Therefore, the difference between the length of the major axis (TAD) and minor axis (PAD) increases, as the forging process proceeds.

Generally, the quantity of spread decreases as $W/H_0$ decreases for $1/2 \leq W/H_0 \leq 2$.[10] Since the tool proceeds from the center to the edge of the ellipse, W is decreased by the geometric character of the ellipse. Accordingly, because the quantity of accumulated spread decreases as the tool proceeds from the center to the edge, the gradient of the elongated length decreases when the tool advances from the center to the edge.

Thus, the final shape on the upper surface is either elliptical or circular depending on the translational tool path, as the value of the spread is accumulated according to TAD (tool advance direction) and the initial shape is a circle.

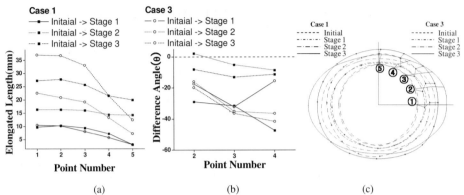

(a)                              (b)                              (c)

Fig. 6. Comparison of Case 1(solid square) with Case 3(hollow circle). (a) elongated length, (b) difference angel, and (c) surface shape.

## 4. Conclusions

Physical modeling using plasticine has been employed to predict the metal flow in incremental open-die disk forging. In this work it was found that smaller tool width results in slightly increased elongated length and area. However, the path of upper die is the dominant parameter to change the shape of the material.

## References

1. B. Aksakal F.H. Osman t, A.N. Bramley, "Upper-bound analysis for the automation of open-die forging", *Journal of Materials Processing Technology*, 71 (1997), pp. 215-223.
2. H. Suzuki, "The law of similarity in deformation processes", *Trans. Jpn. Soc. Mech. Eng.*, XVII(59) (1951), pp. 132.
3. S. Kobayashi, Dynamic similitude in materials processing, unpublished review on model tests in forming, (1966).
4. Kotaro Y., Hidehiko T.,Tsuneo E., Shunji O., and Junichi I., "A Study of Simulative Model Test for Metal Forming Using Plasticine", *MTB 91*, (1974), pp. 1-11.
5. R.L. Bodnar, D.C. Ronemus, B.L. Bramfitt and D.C. Shah, "Physical Modeling of Hot-Deformation Processes-Using Plasticine", *Trans. ISS.* (1986), pp. 35-46.
6. S.F. Wong, P.D. Hodgson, C.-J. Chong, P.F. Thomson, "Physical modelling with application to metal working, especially to hot rolling", *Journal of Materials Processing Technology*, 62 (1996), pp. 260-274.
7. Kaximierz Swiatkowski, "Physical modeling of metal working processes using wax-based model materials", *Journal of Materials Processing Technology*, 72(1997), pp. 272-276.
8. E. Erman, N.M. Medei, A.R. Roesch and D.C. Shan, "Physical modeling of the blocking process in open-die press forging", *Journal of Mechanical Working Technology*, 19 (1989), pp. 165-194.
9. A. Tomlison, A. Met., A.I.M and J.D. Stringer, B.Sc. (Tech.), "Spread and elongation in flat tool forging", *Journal of The Iron and Steel Institute*, (1959), pp. 157-162.
10. Baraya, G. L. and Johnson, W., "Flat Bar Forging", *Proc. of 5th Int. M.T.D.T. Conference*, Birmingham, (1964), pp. 449-469.

# NUMERICAL OPTIMIZATION OF SHEET METAL FORMING PROCESS USING NEW FRACTURE CRITERION

A. HIRAHARA[1†]

*Graduate school of Hiroshima University,*
*1-4-1, Kagamiyama, Higashi-Hiroshima, Hiroshima, 739-8527, JAPAN,*
*d081152@hiroshima-u.ac.jp*

R. HINO[2], F. YOSHIDA[3]

*Department of Mechanical System Engineering, Hiroshima University,*
*1-4-1, Kagamiyama, Higashi-Hiroshima, Hiroshima, 739-8527, JAPAN,*
*rhino@hiroshima-u.ac.jp, fyoshida@hiroshima-u.ac.jp*

V. V. TOROPOV[4]

*School of Civil and Mechanical Engineering, University of Leeds,*
*Leeds, LS2 9KG, UK,*
*v.v.toropov@leeds.ac.uk*

Received 15 June 2008
Revised 23 June 2008

A numerical optimization system for sheet metal forming process has been developed based on a combination of response-surface-based optimization strategy with finite element simulation. The most important feature of the optimization system is introduction of a new fracture criterion to predict fracture limit under non-proportional deformation. In addition, a sheet-edge fracture criterion is also introduced to predict fracture limit under stretch-flanging deformation. The numerical optimization system is developed using the fracture criteria as accurate fracture constraints to avoid sheet breakage. The developed optimization system is applied to the optimum blank design for a square-cup deep drawing process of perforated blank. The optimum blank design, which minimizes the amount of material and avoids the sheet fracture, is obtained successfully. The effect of definition of fracture constraints on optimization calculation is also discussed.

*Keywords*: Numerical optimization; FE analysis; forming limit diagram; non-proportional deformation; square-cup deep drawing.

## 1. Introduction

Nowadays, sheet-stamping industries actively replace stamping experiments with numerical simulations. Their next target is numerical optimization of sheet stamping

---

[†]Corresponding Author.

processes based on numerical simulation[1-3]. For successful optimization of sheet metal forming process, accurate fracture prediction is essential to avoid sheet breakage.

Conventionally, sheet thickness reduction or FLD (forming limit diagram) under proportional deformation (hereafter called proportional FLD) has been used for fracture prediction. However, such simple criteria do not provide accurate fracture prediction under complex deformation which is often observed in real products. To overcome this problem, a new fracture criterion, which can predict FLD under non-proportional deformation path (hereafter called non-proportional FLD), is necessary.

The aim of this study is to develop a numerical optimization system for sheet metal forming process in consideration of sheet breakage under non-proportional deformation. For that purpose, a new fracture criterion to predict the non-proportional FLD is introduced[4]. In addition, a sheet-edge fracture criterion is also introduced since the FLD can not predict the edge fracture in stretch flanging deformation. The optimization system is constructed using the above-mentioned fracture criteria as accurate fracture constraints.

The developed optimization system is applied to the optimum blank design for a square-cup deep drawing process of a perforated blank. The optimization problem is solved using several types of definitions of fracture constraints and the effect of the fracture-constraint definition on the optimization is also discussed.

## 2. Fracture Criteria Used as Constraints for Optimization

### 2.1. *Criterion based on non-proportional FLD*

Figure 1(a) shows the basic idea for determination of the non-proportional FLD[4]. Here, point O' is the current strain state of pre-strained sheet and $\theta$ shows a direction of the subsequent strain increment. Then point A shows the subsequent forming limit for the pre-strained sheet in $\theta$ direction, while point B shows the forming limit for a virgin sheet metal under proportional deformation in $\theta$ direction. Since the equivalent plastic strain integrated along the non-proportional path O-O'-A is equal to that under the proportional path O-B, if the proportional FLD is given in advance, the non-proportional FLD can be determined using the pre-strain and the direction of the strain increment[4].

Based on the above idea, a fracture parameter $f_{\text{FLD}}$ to judge forming severity and fracture occurrence under non-proportional deformation is defined as follows:

$$f_{\text{FLD}} = \overline{\varepsilon}_{(i)} / \overline{\varepsilon}_{lim}(\theta) \tag{1}$$

Here, $\overline{\varepsilon}_{(i)}$ denotes the current equivalent plastic strain, and $\overline{\varepsilon}_{\text{lim}}(\theta)$ is the equivalent plastic strain at the forming limit under the proportional strain path in $\theta$ direction as shown in Fig. 1(b). Sheet fracture takes place when the parameter $f_{\text{FLD}}$ exceeds 1.

## 2.2.  *Criterion for edge fracture*

In stretch flanging deformation, sheet fracture is often observed on its edge under an almost proportional strain path. However, in such a case, fracture limit strain is much larger than that given by FLD[5]. Neither the proportional FLD nor the non-proportional one is capable for predicting the edge fracture. Therefore, another fracture criterion based on hole-expanding tests is introduced in addition to the non-proportional FLD.

Severity of stretch flanging deformation and occurrence of the edge fracture is evaluated by the following fracture parameter $f_{\text{edge}}$ :

$$f_{\text{edge}} = \varepsilon_1 / \varepsilon_{\text{edge}} \tag{2}$$

Here, $\varepsilon_{\text{edge}}$ is the edge fracture limit strain obtained by the hole-expanding test and $\varepsilon_1$ is the major strain along sheet edge, respectively. Sheet edge fracture takes place when edge fracture parameter $f_{\text{edge}}$ exceeds 1.

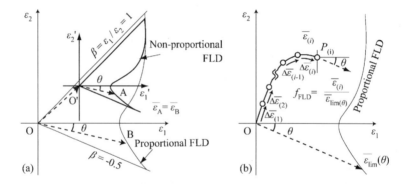

Fig. 1. Schematic illustration of (a) determination method of non-proportional FLD and (b) evaluation of fracture limit under non-proportional deformation.

## 3.  Optimization System for Sheet Metal Forming Problem

An optimization system for sheet metal forming problem is constructed based on a combination of an optimizer and an FE solver. In the present work, a response-surface-based optimizer, HyperOpt of Altair Engineering Inc. and a dynamic-explicit solver, PAM-STAMP 2G of ESI Group are utilized as the optimizer and the solver, respectively.

Fig. 2.  Schematic illustration of deep drawing process of perforated square blank.

## 4. Application to Optimum Blank Design for Deep Drawing Process

### 4.1. *Deep drawing process of perforated blank*

A deep drawing process of a perforated square blank (see Fig. 2) is investigated as an example problem for application of the developed optimization system. A square mild-steel blank with a center hole is drawn by a square punch. Then outer and inner flanges of the drawn box are trimmed off to obtain the target shape and the trimmed material is wasted. The optimization problem is formulated to determine the optimum blank design that minimizes the amount of material including the waste without any sheet breakage.

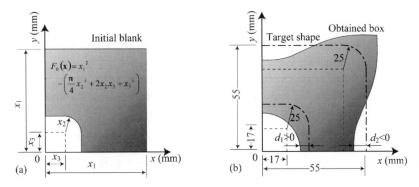

Fig. 3. Optimization problem setting; (a) Blank, design variables and objective function, (b) Drawn box, target shape of product and evaluation of profile constraints.

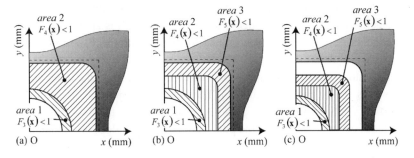

Fig. 4. Three types of fracture constraint definitions (a), (b) and (c) using fracture parameters based on edge-fracture criterion ($f_{edge}$ in area 1) and non-proportional FLD ($f_{FLD}$ in areas 2 and 3).

### 4.2. *Optimization problem formulation*

Figure 3 shows the optimization problem setting. The blank design is determined by three design variables, $x_1$, $x_2$ and $x_3$ as shown in Fig. 3(a). The lower and the upper bounds of the design variables are defined as follows:

$$85 \le x_1 \le 115, \quad 3 \le x_2 \le 10, \quad 2 \le x_3 \le 11 \text{ (in mm)} \tag{3}$$

The object of the present problem is to minimize the amount of material. Therefore the area of the initial blank is treated as the objective function as shown in Fig. 3(a). The definition of the objective function is as follows:

$$F_0(\mathbf{x}) = x_1^2 - \left(\frac{\pi}{4}x_2^2 + 2x_2x_3 + x_3^2\right) \tag{4}$$

There are four or five constraints including profile constraints and fracture constraints. The first and second constraints are profile constraints aiming to prevent the outline of the obtained box from penetrating the target line.

$$F_1(\mathbf{x}) = \text{The minimum } d_1 > 0 \tag{5}$$

$$F_2(\mathbf{x}) = \text{The minimum } d_2 > 0 \tag{6}$$

Here, $d_1$ and $d_2$ are distances from the outline to the target line, which are defined to become negative when the penetration takes place, as shown in Fig. 3(b).

The rest constraints are fracture constraints using fracture parameters $f_{edge}$ and $f_{FLD}$. The maximum value of the parameter $f_{edge}$ or $f_{FLD}$ must be smaller than a certain limit value $f_{lim}$. These constraints are described as follows:

$$F_3(\mathbf{x}) = \text{The maximum } f_{edge}/f_{lim} < 1 \text{ in area 1} \tag{7}$$

$$F_4(\mathbf{x}) = \text{The maximum } f_{FLD}/f_{lim} < 1 \text{ in area 2} \tag{8}$$

$$F_5(\mathbf{x}) = \text{The maximum } f_{FLD}/f_{lim} < 1 \text{ in area 3} \tag{9}$$

In this study, the limitation $f_{lim}$ is taken as 0.85 or 1.

As shown in Fig. 4, there are three types of fracture constraint definitions (a), (b) and (c). In the definition type (a), the edge-fracture constraint (7) is applied to the area 1 (i.e. the hole-edge part) and the non-proportional fracture constraint (8) to the area 2 (i.e. the bottom part of the drawn box except the hole-edge), respectively. In the definitions (b) and (c), the bottom part of the drawn box is divided into two, i.e. the inner area 2 and the outer area 3, as shown Figs. 4(b) and (c). Then the non-proportional fracture constraints (8) and (9) are applied to the areas 2 and 3, respectively. In the definition (b), the outer area 3 includes the bent portion on the punch corner, while it does not in the definition (c).

## 4.3. Optimization result and discussions

The results of the optimization are summarized in Table 1 and Fig. 5 where the fracture constraint boundary $f_{lim}$ is taken as 1.0. Table 1 shows the design variables, the objective/constraint functions and iteration number of simulation. Figure 5 shows distribution of the fracture parameter in the bottom part of the drawn product and position of a material element which has the maximum fracture parameter for the optimum solution. According to the results, two solutions under the fracture constraint definitions (a) and (b) seem to be considerably similar to each other, but the latter solution requires less computing time for optimization compared to the former one. This result shows that the fracture constraint definition affects the convergence of the optimum solution.

In those two solutions, fracture is likely to occur at the punch corner portion as shown in Figs. 5(a) and (b). However, according to the experimental observation, fracture had never occurred at that part but around the hole edge. In that point of view, the third solution under the definition (c) is found to be the most accurate one (see Fig. 5(c)). In addition, the amount of material is minimized in the third solution as shown in Table 1.

These results indicate that appropriate definition of fracture constraints helps to obtain an accurate solution in short computing time.

Table 1. Results of optimization where $f_{lim}$ is 1.0.

| Constraint definition | (a) | (b) | (c) |
|---|---|---|---|
| Design variables $x_1 / x_2 / x_3$ [mm] | 100.8 / 9.28 / 8.84 | 100.7 / 10.0 / 8.03 | 100.5 / 9.30 / 9.13 |
| Objective [mm²] | 9850 | 9834 | 9774 |
| Constraints $F_1 / F_2$ [mm] | 11.8 / 0.100 | 11.6 / 0.290 | 10.9 / 0.251 |
| $F_3 / F_4 / F_5$ | 0.665 / 0.945 / - | 0.721 / 0.861 / 0.986 | 0.705 / 0.860 / 0.769 |
| Iteration | 40 | 22 | 33 |

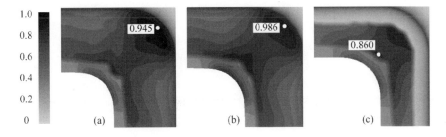

Fig. 5. Distribution of fracture parameter and position of element which has the maximum fracture parameter in bottom part of drawn product under the optimum condition.

## 5. Conclusions

A design optimization system for sheet metal forming process is constructed using a response-surface-based optimizer and an FE solver. In the optimization system, a new fracture criterion based on the non-proportional FLD and an edge fracture criterion are introduced as accurate fracture constraints. The developed optimization system is applied to the optimum blank design of a deep drawing process of a perforated square blank. The optimization problem is formulated to determine the optimum blank design that minimizes the amount of material without sheet fracture. It is confirmed that the optimum blank design is determined successfully. It is also found that appropriate definition of fracture constraints helps to obtain an accurate solution in short computing time.

## References

1. T. Ohata, E. Nakamachi and N. Omori, *J. Mat. Process. Technol.*, **60** (1996), pp. 543-548.
2. Y.Q. Guo, J.L. Batoz and H. Nauceur, *Computers and Structures*, **78** (2000), pp. 133-148.
3. R. Hino, F. Yoshida and V.V. Toropov, *Archive of Applied Mechanics*, **75** (2006), pp. 679-691.
4. R. Matsui, T. Naka, R. Hino and F. Yoshida, *Proc. 37th Japanese Spring Conf. Technol. Plasticity*, (2006), pp. 55-56.
5. M.J. Worswick and M.J. Finn, *Int. J. Plasticity*, **16** (2000), pp. 701-720.

# STUDY FOR BLADE CERAMIC COATING DELAMINATION DETECTION FOR GAS TURBINE

CHOUL-JUN CHOI[1]

*Researcher/DCC Business Center/Gwangju Regional Headquarter/Korea Electronics Technology Institute, No. 210 1110-11, Oryoung-dong, Buk-gu, Gwangju, 500-480, KOREA, cjchoi@keti.re.kr*

SEUNG HYUN CHOI[2], JAE-YEOL KIM[3†]

*Department of Mechatronics Engineering Graduate School of Chosun University, 375, Seosuk-Dong, Dong-Gu, Gwangju, 501-759, KOREA, cshddubi@hanmail.net, jyklm@chosun.ac.kr*

Received 15 June 2008
Revised 23 June 2008

The component of the hot gas path in gas turbines can survive to very high temperatures because they are protected by ceramic Thermal Barrier Coating (TBC); the failure of such coating can dramatically reduce the component life. A reliable assessment of the Coating integrity and/or an Incipient TBC Damage Detection can help both in optimizing the inspection intervals and in finding the appropriate remedial actions.

This study gives the TBC integrity; so other methods are required, like thermography to obtain indications of TBC delamination. Pulsed Thermography detects coating detachments and interface defects, with a large area of view but a spatial resolution of few mm. The mentioned techniques as a whole constitute a powerful tool for the life assessment of thermal barrier coating.

*Keywords*: Thermal Barrier Coating (TBC); delamination; Pulsed Thermography.

## 1. Introduction

Today, power demand increases rapidly as industry grows rapidly. Optimum generation facilities for smooth supply of demanded power and for minimizing environmental pollution can be gas turbine facilities that require short time for construction and use clean fuel. Since such gas turbines for generation are operated under high temperature, high pressure and corrosive environment for a long time, oxidization there of causes serious damages. In particular, since damages to the turbine blade may cause unexpected stop, it is very important to find measures for preventing accidents. Therefore, it is required to regularly diagnose, inspect, maintain or replace hot parts in a gas turbine in order to keep wholesomeness during operation. For this purpose, a life cycle test technique for predicting a maintenance and replacement cycle is needed.

---

†Corresponding Author.

In special, thermal barrier coating is applied to protect the material of blade. However, physical changes in coating and defects on coating layer happen during the cycle of use, causing difficulties in operating a gas turbine. This study is intended to examine the possibility of non destructive testing for delamination and defects on coating layer.

Conventional inspection methods used for such regular maintenance times require much expenses and time and reliable inspection is not applied. To solve such problems, new inspection methods have been presented. One of them is to use an infrared thermography camera. This method couldn't be applied to testing wholesomeness of the coating layer of a gas turbine blade since, while this method achieves quick diagnostics time and reduced expenses, it is difficult to quantitatively analyze the test results, and the tested object is heated. Therefore, in this essay, it is described to construct reference data for applying this method to testing wholesomeness of the coating layer of a gas turbine blade.

## 2. Configuring Test Equipment

### 2.1. *Infrared thermography camera.*

An infrared thermography camera is a measuring machine for detecting an infrared wavelength reflected from a measured object from an external light source. With an equation of temperature and a wavelength by the Stefan-Boltzmann Law(1), the infrared wavelength detected from the infrared detector is represented as a function of temperature to show high or low temperatures by graphic images.

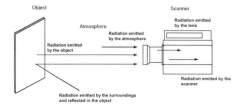

Fig. 1. Measurement situation of thermography.

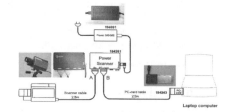

Fig. 2. Infrared thermography camera.

$$I_m = I(T_{abj}) \cdot \tau \cdot \varepsilon + \tau(1-\varepsilon) \cdot I(T_{amb}) + (1-\tau) \cdot I(T_{atm}) \tag{1}$$

As shown in Fig.1, a total standard radiant energy can be represented according to Equation (1). The first term is a radiation emitted by the surroundings and reflected in the object. The second term is a radiation emitted by the object. The second term is a radiation emitted by the atmosphere. The infrared thermography camera used in the test is Thermo-vision 900 SW/TE available from AGEMA. The cooling method of the infrared thermography camera is thermal electric (TE). The detection showed two splits, serial scanning, 2 to 5.4 micron spectrum response. The temperature ranged from -10℃ to 500℃ (up to 2000℃ can be detected when using a high temperature filter is used).

Sensitivity was 0.1 ℃ at 30 ℃, and spatial resolution was 104elements/line (50% modulation). The IR line frequency was 3.5 KHz. Fig.2 shows Research Package 900 that allows the Thermo-vision 900 system operated in X-Windows OS(operation system) to be connected to a Microsoft windows OS based laptop computer.

### 2.2.  Infrared thermography(Pulsed Thermography)

Pulsed thermography is one of the most popular thermal stimulation methods in thermography. One reason for this popularity is the quickness of the test relying on a short thermal stimulation pulse, with duration going from about 3 ms for high conductivity material testing(such as metal parts) to about 4s for high conductivity specimens(such as plastics and graphite epoxy laminates). Such quick thermal stimulation allows direct deployment on the plant floor with convenient heating sources, Moreover, the brief heating (generally a few degrees above initial component temperature) prevents damage to the component and pulse duration varies from about 3 ms to 2s.

Basically, pulse thermography consists of briefly heating the specimens and then recording the temperature decay curve. Qualitatively, the phenomenon is as follows. The temperature of the material changes rapidly after the initial thermal pulse because the thermal front propagates, by diffusion, under the surface. The presence of a discontinuity reduces the diffusion rate so that when observing the surface temperature, discontinuities appear as area of different temperatures with respect to surrounding sound area once the thermal front has reached them. Consequently, deeper discontinuities will be observed later and with a reduced contrast. In fact, the observation time t is a function(in a first approximation) of the square of the depth z and the loss contrast C is proportional to the cube of the depth.

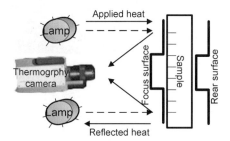

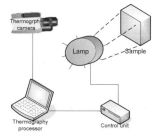

Fig. 3.  Principle of Pulse Thermography.          Fig. 4.  Principle of lockin thermography.

## 3.  Method and Result of the Test

### 3.1.  *Making a sample*

Usually, thermal barrier coating consists of both ceramic top coating of $ZrO_2$-(6-8wt.%)$Y_2O_3$ that is excellent for thermal barrier effect and bond coating of MCrAlY (M=Ni and/or Co) that is applied between the material of the blade and top coating, and has reportedly been most excellent in durability. Ceramic top coating is usually done by porous structure in order to reduce the damages from heat stress. Therefore, oxygen is input through porous ceramic layer at higher temperature and the gas combines with Al on MCrAlY bond coating layer, making up TGO of $AL_2O_3$. The TGO layer acts as barrier which prevents defective layer from interface oxidization. However, when the layer propagates too rapidly and densely, so reducing cohesion and increasing interface stress between layers, it suffers cracks or delamination. Many studies have been conducted to decrease these defects and to preserve coating layer as longest as possible.

For a sample for testing, a one layered used blade of GE's 7FA gas turbine was used. To analyze defects caused by dual coating layers, this was done by making artificial defects between blade and bond coating, between bond coating and top coating.  To obtain exact test data, a heat source was applied to cooling holes in the sample, with the testing conducted in a dark room.   The test device was configured as shown in figure 6.

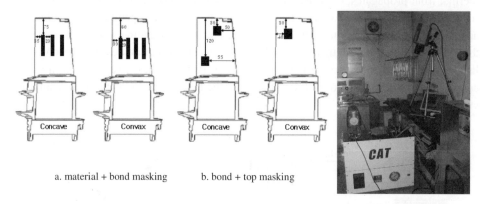

a. material + bond masking          b. bond + top masking

Fig. 5.  Artificial defect Sample.                    Fig. 6. Experiment device.

### 3.2.  *Aritificial Defect*

Tests were done with a new blade, a used one, and a defected sample under the same conditions and spot temperature data were obtained and analyzed. For a new blade, it had the best coating layer, emitting heat very well and recording the highest temperature. For a defected sample, heat conductivity was not well done in artificial air layer between the blade and coating layer, recording lower temperature. This was found in a testing with other kinds of defected samples.

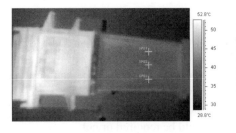

Fig. 7.  IR image of Heating system (Convax).         Fig. 8.  IR image of Heating system (Concave).

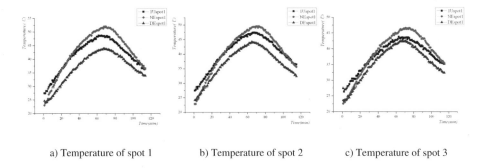

a) Temperature of spot 1          b) Temperature of spot 2          c) Temperature of spot 3

Fig. 9.  Temperature of each spot of IR image.

### 3.3.  *Coating Delamination*

While the blade is in operation, delamination often happens on coating layer. The testing methods were done with naked eyes. Tests were conducted to devise how to find defects. Delamination on coating layer could be enough seen with naked eyes. Thermal barrier effect against delamination was also tested through IR device. Defects were found at the ends of the blade as shown in figure 11. The coating layer was delaminated and the exposed blade was found to have higher temperature than those on coating layers as shown in figure 12. For electric appliances parts in relation with IR image, when the difference in temperature shows 4~5□, they turn out to have defects and are recommended to be repaired or replaced with new ones.

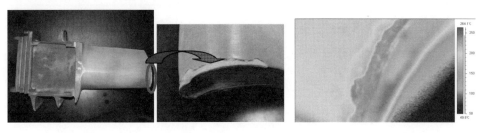

Fig. 10. 7FA+eBlade.        Fig. 11. Actuality image of defect.       Fig. 12. IR image of defect.

## 4. Conclusion

The test was conducted by examining coating delamination and interface defects on GE 7FA+e  with IR thermography camera and concluded as follows:

1. An examination was made on whether IR device could be used for non destructive testing of interface defects on coating layer and showed a possibility.

2. Thermal load by coating delamination on the blade could be seen, and objective data was presented to identify whether maintenance should be required or not.

3. Through IR device, any novice could determine whether gas turbine should undergo maintenance test or not.

## Acknowledgment

This study was supported by research funds from Chosun University, 2005

## References

1.  D.J. Yang, C.J. Choi , J.Y. Kim, Evaluation Method of Gas Turbine Blades Covering Integrity by IR Camera, Int. J. Modern Physics. Vol.20 (2006), 4329-4334.

2.  C.J. Choi, J.Y Kim, D.J. Yang, K.S. Song, Y.S Ahn, A Study on the Application Infrared Thermography Camera for the 7FA Gas Turbine Blade Covering Crack Detecting, Key Engineering Materials, Vol. 340-341 (2007) 483-488.

3.  Chan K.S, Cheruvu N.S, Leverant G.R, Coating Life Prediction under Cycle Oxidation Condition, J.of Eng. For Gas Turbine and Power (1998).

4.  Antonelli G, Tirone G, Condition Assessment of Service Degraded High Tmeperature Blades during Gas Turbine Overhauls, Power. Gen. Europe, Helsinki, Finland (2004).

5.  Marinetti S, Vavilov V, Bison P.G., Grinzato E., Cernuschi F., Quantitative Infrared Thermographic Nonestructive Testing of Thermal Barrier Coating, Mater. Eval., (2003).

6.  Gell M, Sridharan S, Wen M, Photoliminescence Piezospecroscopy : A Multi-purpose Quality Control and NDI Technique for Thermal Barrier Coating, Int. J. Appl. Ceram. Technol., (2004).

7.  Chen X.,Newaz G., Han X., Damage Assessment in Thermal Barrier Coating using Thermal Wave Image Technique, Int. Mech. Eng. Congress and Exposition Proc. IMECE, New York, (2001).

# STUDY ON CORRELATION BETWEEN SHEAR WAVE VELOCITY AND GROUND PROPERTIES FOR GROUND LIQUEFACTION INVESTIGATION OF SILTS

AILAN CHE[1], XIANQI LUO[2], JINGHUA QI[3†]

*School of Naval Architecture, Ocean and Civil Engineering, Shanghai Jiaotong University, Room 2302 Haoran Building, 1954 Hushan-Road, Shanghai 200030, P. R. CHINA,*
*alche@sjtu.edu.cn, luoxianqi@sjtu.edu.cn, qijinghua@sjtu.edu.cn*

DEYONG WANG[4]

*Key Laboratory of Geological Hazards on Three Gorges Reservoir Area, Ministry of Education, China Three Gorges University, 8 Daxue-Road, Xiling-qu, Yichang 443002, P. R. CHINA,*
*wangdeyong200218@yahoo.com.cn*

Received 15 June 2008
Revised 23 June 2008

Shear wave velocity ($V_s$) of soil is one of the key parameters used in assessment of liquefaction potential of saturated soils in the base with leveled ground surface; determination of shear module of soils used in seismic response analyses. Such parameter can be experimentally obtained from laboratory soil tests and field measurements. Statistical relation of shear wave velocity with soil properties based on the surface wave survey investigation, and resonant column triaxial tests, which are taken from more than 14 sites within the depth of 10 m under ground surface, is obtained in Tianjin (China) area. The relationship between shear wave velocity and the standard penetration test $N$ value (SPT-$N$ value) of silt and clay in the quaternary formation are summarized. It is an important problem to research the effect of shear wave velocity on liquefaction resistance of saturated silts (sandy loams) for evaluating liquefaction resistance. According the results of cyclic triaxial tests, a correlation between liquefaction resistance and shear wave velocity is presented. The results are useful for ground liquefaction investigation and the evaluation of liquefaction resistance.

*Keywords*: Silt; shear wave velocity $V_s$; SPT-$N$ value; liquefaction resistance; correlation analyses.

## 1. Introduction

Soil liquefaction is one of the important causations of site damage in earthquake engineering and geotechnical engineering. Many studies for problem of factors that affect the probability of sand liquefaction and the experiential expressions of sand liquefaction have been advanced, [1] but most of these are applicable for sandy soils. Silt is a special soil, which is made up of three particles such as sand particle, silt particle and clay particle. It is easy to be liquefied during earthquake, [2] and it is still deserved to study the liquefaction strength and dynamic character of silt.

[†]Corresponding Author.

Shear wave velocity ($V_s$) is an important parameter in evaluation of dynamic properties for a site, which uses as a parameter in the field of geotechnical earthquake engineering. To determine liquefaction potential in-site, several viable approaches are used, which are the Standard Penetration Test (SPT), measurement of in-situ shear wave velocity ($V_s$). [3] As known it is an important problem to research the effect of shear wave velocity on liquefaction resistance of saturated sands (silts) for evaluating its liquefaction resistance. Statistical relation of shear wave velocity with soil properties based on the investigation is obtained in Tianjin (China) Wangqingtuo area.

As a part of the China South to North Water Diversion Mid Line Project, an emergency water supply reservoir (Wangqingtuo reservoir) is planned to be constructed in the west of Tianjin City. An effective ground investigation method is required to provide information for liquefaction countermeasures, Since the Tangshan Earthquake in 1976 caused serious ground liquefaction in Tianjin area. The characteristic of soil is studied through laboratory soil tests. The soil stratums are investigated using the Standard Penetration Test (SPT) and the surface wave survey.

## 2.  Wangqingtuo Reservoir in Tianjin Area, China

Wangqingtuo reservoir is constituted of two parts located at the side of Jin Bao High Way, which is connected by the inverted siphon. The area and perimeter of the north part are 3.03 km², 6838.87 m; and those of the south part are 1.50 km², 5163.77 m (Figure 1). The dead storage level and design water level are 3.50 m, 12.80 m respectively. The average altitude of bottom stratum is 2.50 m, and that of gantry crane is 14.10 m.

The site is composed of Quaternary Period Holocene Seri upside terrestrial alluvial stratum ($alQ_4^3$), which consisted of sandy loam, silt, silty sand, loam; Quaternary Period Holocene Seri midsection limnetic facies sediment stratum ($l+hQ_4^2$), which consisted of thick layer clay and loam; Quaternary Period Holocene Seri midsection marine deposit stratum ($mQ_4^2$), which consisted of thick layer loam; Quaternary Period Holocene Seri subside terrestrial alluvial stratum ($alQ_4^1$), which consisted of thick layer loam and clay; Quaternary upper Pleistocene Series terrestrial alluvial stratum ($alQ_3$), which consisted of thick layer loam. The depth of groundwater level is 1.9~6.4 m.

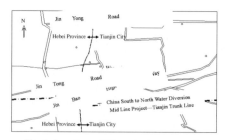

Fig. 1. Location of Wangqingtuo reservoir.

## 3.  Surface Wave Survey Investigation

Surface wave survey technique is one of seismic exploration methods.[4] At Wangqingtuo reservoir area, the surface wave survey is conducted surrounding the axis of the dam. Sledgehammer is used as a vertical source for data acquisition, and four shot gathers with source offsets for 4 m, 8 m, 12 m, and 16 m are recorded. Data are acquired with 1ms sample rate, and total of 2048 measurements are recorded. First, the cross-spectrum of each trace with a reference trace is calculated to form a cross-spectrum gather for each shot location. Then, in order to construct a 2D profile over the line, spatial windows of 24 traces are analyzed and moved along at 1 m intervals. Finally, the $V_s$ structure by inverting the phase velocity data is estimated.[5] Figure 2 shows the borehole log results and the estimated 2D shear-wave velocity model from the phase velocity inversion. The model given by the multi-mode inversion is in better agreement with soil types. At the Wangqingtuo reservoir area, the $V_s$ section shows 3 clear velocity boundaries inferring that the soil stratums consists of 4 layers, which shear wave velocities ($V_s$) are under 120 m/s, 130-160 m/s, 170-190 m/s, over 200 m/s, respectively.

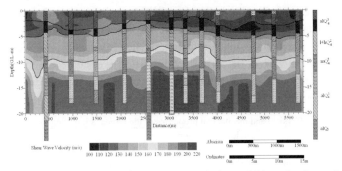

Fig. 2. Borehole log results and the estimated 2D shear-wave velocity model from the phase velocity inversion.

## 4.  Dynamic Triaxial Tests

Silt is a special type of soil which property differs from sand and clay. Table1 shows the physical and mechanical properties of the typical silt in the research area.

Table 1.  Physical and mechanical properties of the typical silt in the research area.

| Particle composition | | | Physical properties in natural state | | | | Permeability coefficient | |
|---|---|---|---|---|---|---|---|---|
| Sand | Silt | Clay | Density | Dry density | Pore ratio | Plasticity index | Temperature 20 ℃ | |
| 0.25- | 0.074- | <0.005 | | | | | | |
| 0.074mm | 0.005mm | mm | $\rho_0$ | $\rho_d$ | $e_0$ | $I_P$ | $K_v$ | $K_h$ |
| % | % | % | g /cm$^3$ | g/cm$^3$ | | | cm/s | cm/s |
| 13.5 | 75.5 | 11 | 1.83 | 1.46 | 0.85 | 6.9 | 4.75E-06 | 8.38E-06 |

### 4.1.  *Resonant column triaxial tests*

The resonant column triaxial tests are carried out for the dynamic shear modulus $G_0$ and the damping ratio $D$ of unsaturated silts (Figure 3). Experiment research results indicate that dynamic shear modulus of silts decrease with the increase of strain; damping ratio of silts increase with strain. Confining stress has much effect on dynamic modulus and damping ratio. Dynamic modulus increases while confining stress increasing and damping ratio decreases while confining stress increasing.

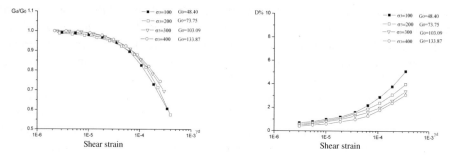

Fig. 3. Relationship of the dynamic shear modulus and the damping ratio of the silts.

## 4.2. *Cyclic triaxial tests*

The cyclic triaxial tests on saturated and undisturbed silts are carried out for the cyclic liquefaction resistance (Figure 4). The cyclic liquefaction resistance $N_{12}$ (the dynamic shear stress ratio when the equivalent vibration number is 12) and $N_{30}$ of silts in the research area and the effects of particle composition are studied. The characteristic of silts is demonstrated that cyclic liquefaction resistance increasing while silt and clay content increasing (Table 2).

Table 2.  Physical and mechanical properties of the typical silt in the research area.

| Number of sample | Particle composition | | | Cyclic liquefaction resistance | Cyclic liquefaction resistance |
| | Sand 0.25~0.075 mm | Silt 0.075~0.005 mm | Clay < 0.005mm | | |
| | % | % | % | $N_{12}$ | $N_{30}$ |
|---|---|---|---|---|---|
| QZK5-1 | 4.5 | 91.9 | 3.6 | 0.360 | 0.310 |
| QZK2-2 | 4.0 | 90.4 | 5.6 | 0.280 | 0.268 |
| QZK4-1 | 23.5 | 69.8 | 6.7 | 0.420 | 0.400 |
| QZK2-1 | 9.0 | 83.5 | 7.5 | 0.260 | 0.245 |
| QZK7-1 | 2.0 | 86.9 | 11.1 | 0.300 | 0.290 |

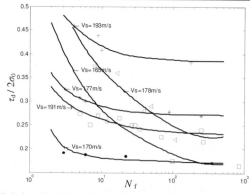

Fig. 4. Relationship of the cyclic liquefaction resistance versus vibration number.

## 5. Correlation Analysis Between SPT-$N$ Value and Shear Wave Velocity $V_s$

By using large amount of field tests, the correlation between the shear wave velocity $V_s$ and SPT-$N$ value is studied (Figure 5). An empirical correlation between shear wave velocity of soil and SPT-$N$ value in research region is provided as Eq. (1) and Eq. (2). From the results we find that shear-wave velocity $V_s$ correlates well with SPT-$N$ value of silt and clay in the research area. Shear-wave velocity $V_s$ increase monotonously with increasing of SPT-$N$ value. The relationship between shear-wave velocity $V_s$ and SPT-$N$ value shows power function relation. This result will provide a valuable way to estimate in-situ liquefaction resistance of granular soil stratum.

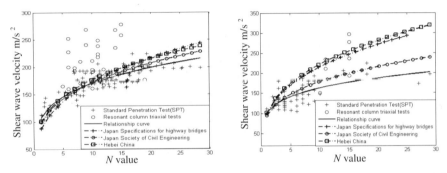

Fig. 5. Correlation between $V_s$ and SPT-$N$ value of silt. Fig. 6. Correlation between $V_s$ and SPT-$N$ value of clay.

$$V_s = 127.7N^{0.140} \qquad \text{(Silt)} \qquad (1)$$
$$V_s = 99.13N^{0.221} \qquad \text{(Clay)} \qquad (2)$$

## 6. Effect of Shear Wave Velocity on Liquefaction Resistance

It is shown that the cause of liquefaction is closely related to shear wave velocity $V_s$, so it is feasible to determine liquefaction potential using shear wave velocity $V_s$.[6] Figure 7 shows the correlation between cyclic liquefaction resistances of undisturbed saturated silts and its SPT-$N$ value, compared with the results of Japan Specifications for highway bridges.[7] It was seen that the trend is basically the same, and there exists a liquefied determination line, where different sand under different density and confining stress should fall on.

Figure 8 shows the correlation between cyclic liquefaction resistances of undisturbed saturated silts and its shear wave velocity $V_s$. The experimental approximate curve is shown that $V_s$ correlates well with liquefaction resistance of saturated silts within a liquefied determination line, which trend is basically the same as that with SPT-$N$ value. It can be considered that liquefaction strength is a form of its shear strength, and during dynamic load with the sand particles becoming density the pore water pressure increases and effective stress is reduced, as the results the shear strength gradually decreases till its liquefaction closing to zero. The results supported that the initial shear wave velocity of saturated silts is a key parameter controlling its cyclic liquefaction resistance.

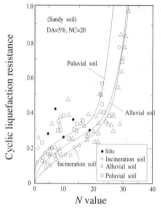

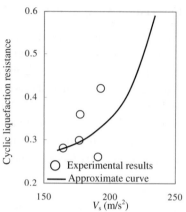

Fig. 7. Correlation between cyclic liquefaction resistances and SPT-$N$ value.

Fig. 8. Correlation between cyclic liquefaction resistances and shear velocity $V_s$.

## 7. Conclusions

(1) The characteristic of silts is demonstrated that cyclic liquefaction resistance increasing while silt and clay content increasing.

(2) The power function relation between shear wave velocity $V_s$ and the standard penetration test $N$ value (SPT-$N$ value) of silt and clay are obtained.

(3) The correlation between cyclic liquefaction resistances of undisturbed saturated silts and its shear wave velocity $V_s$ is shown that shear-wave velocity $V_s$ correlates well with liquefaction resistance of saturated silts within a liquefied determination line, its trend is basically same as that with SPT-$N$ value.

(4) The results supported that the initial shear wave velocity of saturated silts is a key parameter controlling its cyclic liquefaction resistance.

## References

1. Tokimatsu K., Yamazaki T. and Yoshimi Y., in *Soil liquefaction evaluation by elastic shear moduli*, (Soils and Foundations, 1986), p. 25-35.
2. Wang Shuyun, Lu Xiaobin, in *Effects of grain size distribution and structure on mechanical behavior of silty sand*, (Rock and Soil Mechanics, 2005), p. 1029-1032.
3. Shi Zhaoji, Yu Shousong, Feng Wanling, in *Evaluation of soil liquefaction potential with shear wave velocity method*, (Journal of Geotechnical Engineering, 1993), p. 74-80.
4. Yamanaka H., and Ishida H., in *Application of genetic algorithm to an inversion of surface-wave dispersion data*, (Bulletin of the Seismological Society of America, 1996), p. 436-444.
5. Shaokong Feng, Takeshi Sugiyama, Hiroaki Yamanaka, in *Effectiveness of multi-mode surface wave inversion in shallow engineering site investigations*, (Butsuri-Tansa Vol. 58, No.1, Mulli-Tamsa Vol. 8, No.1, Exploration Geophysics 36, 2005), p. 26–33.
6. SEED H. B, IDRISS I. M., *Soil module and damping factors for dynamic response analyses*, (Calif. U. C. Berkeley, Calif. 1970).
7. *Japan Specifications for highway bridges, Part 5 Seismic Design*, (Japan Road Association, 2002).

# A STUDY ON THE STABILITY OF EARTH DAM SUBJECTED TO THE SEISMIC LOAD

JINGHUA QI[1†], AILAN CHE[2], XIURUN GE[3]

*School of Naval Architecture, Ocean and Civil Engineering Shanghai Jiaotong University, Room 2302 Haoran High Technique Building, 1954 Hushan-Road, Shanghai, 200030, P. R. CHINA,*
*qijinghua@sjtu.edu.cn; alche@sjtu.edu.cn, xiurunge@public.wh.hb.cne*

Received 15 June 2008
Revised 23 June 2008

For ensuring the earth dam's stability of Wangqingtuo reservoir when silt liquefaction happens during Tangshan earthquake, a large amount of laboratory soil tests and field measurements have been performed to obtain the mechanic properties of the soil and silt dynamic parameters. On the basis of the soil tests, the equivalent linear constitutive model is employed in the dynamic numerical simulation of the typical dam and the results indicate that the shear deformation is induced by the foundation liquefaction with the help of the geo-slope software. Moreover, the stability analysis is performed using the finite element elasto-plastic model that is considered the Mohr-Coulomb failure criteria to calculate the stability factor. The factors indicate the local instability would take place because of the shear action. At last, the measures are introduced to the designers for preventing the dam from the instability.

*Keywords*: China South to North Water Diversion Project; earth dam; liquefaction; stability.

## 1. Introduction

Wangqingtuo reservoir is a regulative and on-line emergency water supply reservoir of the China South to North Water Diversion Mid Line Project. It is about 20 kilometers away from Tianjin city. Once the reservoir was destroyed during the earthquake, there would be the abnormal loss for the city. Since 1964, there have been a lot of earth dam failure examples because of the sand liquefaction, such as San Fernando dam[1] during the Great Alaskan earthquake in USA and many little earth dams in Japan Nigata Tyuubu earthquake. In 1976, the M7.8 earthquake stroke Tangshan-Tianjin area and caused extensive sandy soil liquefaction. Therefore, the great attention had been paid to the sandy soil liquefaction.

Over last decades, the study on the soil liquefaction engineering has focused on the assessment of liquefaction potential[2], evaluation of the deformation and settlement induced by the liquefaction[3], clay content influence on the liquefaction and the liquefaction-induced construction failure, etc. A few research works approached the post-liquefaction behavior of soil[4] and post-liquefaction failure[5]. Since the finite element

---

[†]Corresponding Author.

method has been introduced to geo-technical engineering, the liquefaction problem has also been focused on within the numerical analysis[6]. Although the achievements have been made in the aspects above, there are few research works on the silt liquefaction and the dam's local instability induced by the liquefaction.

The dynamic behavior of silt is acquired by laboratory tests, and on the basis of the tests, the dynamic analysis is performed. Then, the local instability of the dam is discussed for the liquefaction induced the non-uniform settlement and level sloping displacement in the paper.

## 2.  Wangqingtuo Reservoir (Tianjin city, China)

Wangqingtuo reservoir is the earth dam built of the surrounding soil to be constructed in the west of the Tianjin city. It is approximately 12000 square meters and has a storage capacity of 40000,000 cubic meters. It is constituted of two parts located at the sides of Jinbao High Way, which is connected by the inverted siphon. According to the construction plan, the box dam is to be made of the fill from the earth excavations inside the reservoir. The maximum height of the dam is about 14.1 meters. The upstream slope of the dam is $3\ h : 1\ v$ and horizontal above the water level, but $3.5\ h : 1\ v$ under the water level, and the downstream slope is $3\ h : 1\ v$. The dead storage water level and design water level are 3.50 meters and 12.40 meters respectively. For seepage prevention and dam landslide after first filling, the weight platform and the catchwater are constructed following the downstream slope.

The fill materials generally consist of medium dense loam with clay and silt ingredients. During geo-technical investigation, field explorations' programs consisting of 140 borings and a series of laboratory tests were completed to develop the subsurface conditions and evaluate the soil properties and engineering parameters for the site-specific seismic evaluations and seismic stability analyses. The site explorations, such as surface wave survey and micro-tremor survey, measured the shear wave velocity and the seismic predominant period of the ground. Based on the borings, the topsoil, whose components are composed of loam and silt, is the liquefying soil layer. The typical section and soil profile are shown in figure 1. The conventional mechanic properties are evaluated, as shown in table 1. Although the thin muddy clay layer exists at the bottom of the liquefying layer, it is more possible that liquefaction occur in the dam foundation. So it is necessary that the special study on the liquefaction in the dam foundation should be pursued.

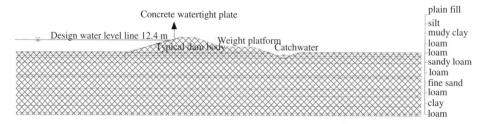

Fig. 1. Typical dam and soil layers distribution under the dam body.

Table 1. Mechanic properties of soil.

| Layer | Soil layer | Initial water content (%) | Initial void ratio | Unit weight Natural (kN/m³) | Unit weight Dry | Permeability coefficient (cm/s) | Cohesion (kPa) | Internal friction angle (°) |
|---|---|---|---|---|---|---|---|---|
| 1 | Plain fill | 21.9 | 0.621 | 17.2 | 13.8 | 4.95E-07 | 37 | 29.5 |
| 2 | Silt | 17.1 | 0.613 | 17.2 | 14.6 | 4.39E-04 | 14.4 | 28 |
| 3 | Muddy clay | 40.8 | 0.722 | 17.6 | 12.5 | 5.13E-05 | 35.9 | 5.7 |
| 4 | Loam | 30.5 | 0.666 | 18.8 | 14.5 | 1.84E-05 | 30.8 | 11.3 |
| 5 | Loam | 31.5 | 0.623 | 18.7 | 14.2 | 4.02E-05 | 21.8 | 21.4 |
| 6 | Sandy loam | 24.8 | 0.617 | 19.5 | 15.7 | 1.79E-04 | 15.7 | 28.8 |
| 7 | Loam | 24.3 | 0.498 | 19.7 | 15.9 | 3.22E-05 | 23.9 | 18.9 |
| 8 | Fine sand | 20.6 | 0.522 | 20.2 | 16.7 | 2.34E-04 | 15 | 30.8 |
| 9 | Loam | 23.9 | 0.507 | 19.7 | 15.9 | 2.72E-05 | 26.3 | 20.4 |
| 10 | Clay | 24.8 | 0.547 | 19.9 | 16 | 6.02E-07 | 47.5 | 13.7 |
| 11 | Loam | 22.1 | 0.555 | 20.2 | 16.6 | 5.66E-06 | 33.6 | 17.3 |

## 3. Dynamic Analysis of Dam

Considering the practical engineering, the equivalent linear equation is used to simulate the silt dynamic behavior in the finite element model. In the equation, the typical parameters of the silt dynamic properties, such as dynamic shear modulus, damping ratio and the cyclic liquefaction resistance were acquired by the dynamic triaxial tests. The micro-tremor surveys in the reservoir area showed that the seismic predominant period ranged from 0.9s to 1.5s. The surface wave survey gave the soil profile of the shear velocity. Based on the data from the soil tests, the typical dam section's two-dimensional model is simulated in the finite element analysis.

### 3.1. *Dynamic parameters of soil*

On the basis of investigation of earthquakes having taken place in reservoir area, earthquake intensity of 7 is the design seismic fortification intensity with the help of Chinese map of seismic intensity zoning[7]. According to the Chinese national code, a peak ground acceleration of 0.15g is used to evaluate the liquefaction potential analysis. For assessing the liquefying soil dynamic properties, the resonant column test and the vibration liquefaction test were performed in laboratory. In virtue of these dynamic tests, the soil dynamic properties, such as dynamic shear modulus ($G$), the damping ratio ($D$) obtained via the resonant column test, and the dynamic shear stress ratio ($\tau / 2\sigma_3$) by the vibration liquefaction test. The liquefying silt's dynamic behaviors are shown in figure 2. From the figure, the silt relationship between $G$ and $\gamma$, acquired by the spine fit method, present the typical tendency of $G$ modulus decreasing with the $\gamma$ increasing, however, silt damping ratio ($D$) increase with $\gamma$ increasing (figure 2). The cycle number ($N$) has influence on the shear stress ratio ($\tau / 2\sigma_3$), as shown in figure 3, and with the increase of $N$, the ratio lessens gradually. In addition, the confining stress has great effect on the $G$, $D$

and $\tau$. The table 2 shows the soils' initial dynamic mechanic properties in finite element analysis according the tests.

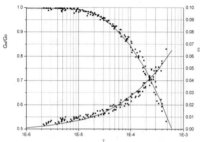

Fig. 2. $G \sim \gamma$ and $D \sim \gamma$ curves of silts.

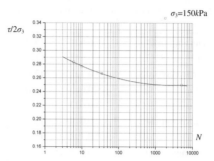

Fig. 3. $\tau / 2\sigma_3 \sim N$ curve of silts.

Table 2. Initial dynamic mechanic properties of the dam.

|          | Plain fill | Silt  | Muddy clay | Loam  | Loam  | Sandy loam | Loam | Fine sand |
|----------|-----------|-------|------------|-------|-------|-----------|------|-----------|
| $G$ (kPa) | 17.89     | 29.22 | 24.15      | 33.51 | 41.19 | 61.04     | 61.8 | 64.93     |
| $D$      | 0.001     | 0.001 | 0.008      | 0.008 | 0.007 | 0.007     | 0.007 | 0.007    |

### 3.2.  Model of the typical section

With the help of the Quake/w module of the Geo-slope software (Version 5.12) [8], the finite element model was established to simulate the typical dam section's behavior. The 8-node quadrangle elements and the 6-node triangle elements are chosen in the model, and reduced integration formulation is used in the dynamic analysis. During the dynamic analysis, the $G \sim \gamma$ curve, $D \sim \gamma$ curve and $\tau / 2\sigma_3 \sim N$ curve (Figure 2 and Figure 3) are inputted as the reduction functions.

### 3.3.  Tangshan earthquake motion

Consider that the reservoir be working when earthquake strikes the place where the reservoir lies, so the reservoir is subjected to homothetic dynamic load for the dynamic calculation. By the rule of similar geological condition, Tangshan earthquake acceleration wave record (Figure 4) is chosen as the input ground motion, which max peak value and duration are 0.66 m/s² at 1.65th second and 13.2 seconds respectively.

### 3.4.  Results of the dynamic analysis

In order to assess fully the reservoir's safety, two cases are discussed in the paper. The first case is that the concrete watertight plate works normally to prevent

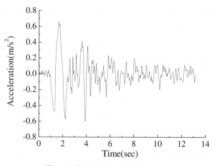

Fig. 4. Tangshan earthquake motion.

the seepage from the dam. Against the first case, the second case is that the plate is in failure because of the earthquake shaking. By the Cholesky Factorization technique, dynamic response of the system in finite element formulation is solved. The deformation is evaluated for the above critical cross section to predict the potential non-uniform settlement and level displacement for the significant deformation of the dam due to the earthquake shaking and the liquefaction induced by it (Figure 5). The picture shows that the level displacement is greater than the vertical settlement, and the maximum vertical displacement is presented at the foot of the reservoir foundation where the liquefaction would take place during the earthquake shaking. The results also indicate that the concrete watertight plate influences the displacement distribution. So based on the dynamic study, it is necessary to evaluate the reservoir dam's stability.

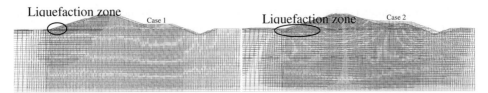

Fig. 5. Segment of displacement deformation (Magnification: 50, Red: deformable, Black: original).

## 4. Post-Liquefaction Stability Analysis

Whether the reservoir is in work or not during the earthquake shaking is the critical problem for hydraulic engineering. So in condition of the dam foundation liquation induced by the earthquake, further assessing the reservoir's safety is gone on in Slope/w finite element software. In the slope/w, the finite element stress method is applied to assess the dam stability factors. The factor of the slope by the method is defined as the ratio of the summation of the available resisting shear force $S_r$ along a slip surface to the summation of the mobilized shear $S_m$ along a slip surface. So the stability factor $k_s$ is expressed as[8]:

$$k_s = \frac{\sum S_r}{\sum S_m} \qquad (1)$$

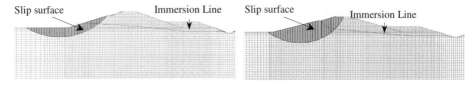

Fig. 6. Slip surface of case one ($k_s$=1.207).          Fig. 7. Slip surface of case two ($k_s$=0.931).

Based on the results of dynamic numerical analysis, the dam stability study is performed to assess the possibility of local instability. In the course of the slope analysis, the Mohr-Coulomb elasto-plasitic model is used and the parameters of the model is adopted from the table 1. Slope analysis of two cases is presented in the paper the same as the analysis above. Figure 6 showed the instability happened when the concrete watertight plate was in function as the imperious layer. The minimum of the stability is 1.207 when the earthquake acceleration at the peak value, therefore, the stability factor is lesser than the design factor 1.5 Figure 7 showed the instability of the case two. It illuminates that it would be possible that instability happen in the upstream slope of the dam when the watertight plate is in failure during the earthquake shaking. The result indicates that the slope instability extension of case two is much larger than the extension of case one.

## 5.  Conclusions and Discussion

Based on large numbers of soil tests, the dynamic analysis and the stability analysis are carried out. During Tangshan earthquake shaking, the ground of silts of Wangqingtuo earth dam is liquefied, and the earth dam shows shear behaviors. The results indicate that the shear deformation causes local instability on the side of the upstream dam. Compared to two cases' stability factors, the instability in case 2 happens more sensitively than case 1 because of dam's shear deformation.

For preventing the reservoir's dam from the shear local instability, preliminary remediation for seepage failure and simple foundation treatment for non-uniform settlement are insufficient to resist the shear local instability induced by the liquefaction during earthquake. Therefore, some suggestions are made taking the other measures, such as building up the little weight platform at the foot of the upstream slope dam for instability control and thickening the concrete plate or introducing the high-strength concrete to ensure the plate's safety when it works.

## References

1.  Seed, H.B., Lee, K.L., Idriss, I.M. and Makadisi, F.I. *The Slides in the San Fernando Dams during the Earthquake of February 9, 1971*, (ASCE, Journal of the Geotechnical Engineering Division, GT7), pp. 651-688.
2.  Ishihara k. and Yoshimine M. *Evaluation of settlements in sand deposits following liquefaction during earthquakes*, (Soils and Foundations, 32 (1)), pp. 173-188.
3.  Risheng (Park) Piao, Arlan H. Rippe, BarryMyers and Kim W. Lane. *Earth dam liquefaction and deformation analysis using numerical modeling*, (ASCE 2006, Geocongras).
4.  D. Porcino and G. Caridi, *Pre- and post-liquefaction response of sand in cyclic simple shear*, (ASCE 2007, Dynamic Response and Soil Properties), pp. 160-170.
5.  Scott Michel Olson, Ph.D. Thesis, Univ. of Illinois, 2001.
6.  Jinchi Lu, Ahmed Elgamal and Zhaohui Yang, *Pilot 3d numerical simulation of liquefaction and countermeasures*, (Earthquake engineering and soil dynamics, 2005), pp.133-144.
7.  *Chinese Code for seismic design of buildings(GB50011-2001)*, (Ministry of construction, PRC, 2001).
8.  *Geo slope version 5 help documentation*, (Geo-slope international Ltd., 2001).

# A STUDY OF THE RECONSTRUCTION OF ACCIDENTS AND CRIME SCENES THROUGH COMPUTATIONAL EXPERIMENTS

S. J. PARK[1], S. W. CHAE[2†]

*Korea University, Seoul, KOREA,*
*park134679@korea.ac.kr, swchae@korea.ac.kr*

S. H. KIM[3], K. M. YANG[4], H. S. CHUNG[5]

*National Institute of Scientific Investigation, Seoul, KOREA,*
*kshgp@nisi.go.kr, ykmoo@nisi.go.kr, hschung7@nisi.go.kr*

Received 15 June 2008
Revised 23 June 2008

Recently, with an increase in the number of studies of the safety of both pedestrians and passengers, computer software, such as MADYMO, Pam-crash, and LS-dyna, has been providing human models for computer simulation. Although such programs have been applied to make machines beneficial for humans, studies that analyze the reconstruction of accidents or crime scenes are rare. Therefore, through computational experiments, the present study presents reconstructions of two questionable accidents. In the first case, a car fell off the road and the driver was separated from it. The accident investigator was very confused because some circumstantial evidence suggested the possibility that the driver was murdered. In the second case, a woman died in her house and the police suspected foul play with her boyfriend as a suspect.

These two cases were reconstructed using the human model in MADYMO software. The first case was eventually confirmed as a traffic accident in which the driver bounced out of the car when the car fell off, and the second case was proved to be suicide rather than homicide.

*Keywords*: Accident reconstruction; crime scene; computational experiments; MADYMO.

## 1. Introduction

Computer simulations are commonly used to analyze safety belts and design interiors to reduce injuries of drivers and passengers. Recently, many studies have investigated methods for reducing injury, such as reduced bumper strength, reduced elasticity of the engine hood, and a higher engine hood structure with a large gap in the inner structure, all of which enable effective deformation. Computer software, such as MADYMO, Pam-Crash, and Ansys, has generally been used in these analyses. Several computer programs, like PC-Crash, HVE, are used by forensic scientists or engineers to analyze accident reconstructions but these programs do not have sufficient resolution compared with FE (Finite Element) based programs, such as MADYMO, PAM-Crash, and Ansys. Although PC-Crash and HVE (Human Vehicle Environment) programs have generally been used for

---

†Corresponding Author.

traffic accidents, are user-friendly, and generate simple and powerful results for traffic accidents, they cannot solve every case.

Regarding human injury, HIC[1] (Head Injury Criterion, Versace, 1971) is widely used as a criterion of head injury and is expressed in Eq. (1) below; this definition has been evaluated in experiments for corpses and animals.

$$HIC = \max_{T_0 \le t_1 \le t_2 \le T_E} \left[ \frac{1}{t_2 - t_1} \int_{t_1}^{t_2} R(t) dt \right]^{2.5} (t_2 - t_1).$$  (1)

In Eq. (1), $T_0$ is the starting time of the simulation, $T_E$ is the ending time, $R(t)$ is the resultant head acceleration (measured in $g$) over the time interval $T_0 \le t \le T_E$, and $t_1$ and $t_2$ are the initial and final times (in seconds) of the interval during which HIC attains a maximum value.

Bostrom[2] introduced a widely used criterion, NIC (Neck Injury Criterion), for neck injury. NIC focuses strictly upon the relative acceleration between the top and the bottom of the cervical spine. The equation for NIC is as follows.

$$NIC(t) = 0.2 \times a_{rel}(t) + (v_{rel}(t))^2.$$  (2)

In Eq. (2), $a_{rel}(t) = a_x^{T1}(t) - a_x^{Head}(t)$ and $v_{rel}(t) = v_x^{T1}(t) - v_x^{Head}(t)$. The value of NIC is calculated using the relative acceleration and velocity between T1 (the first thoracic spine) and the head of the occupant. Also, the value of NIC is defined mostly as a criterion of a sprain injury on the cervical vertebrae, and it has been reported that the sprain injury in the cervical vertebrae happens in over 15. In the area of the reconstruction of traffic accidents, criteria, such as HIC and NIC, have been applied mainly for analyzing injuries of pedestrians and passengers. However, the present study has reconstructed a mysterious case using computational experiments and analyzed the possibility of injuries on a body in the situation. In most cases, computational experiments are used when a suspect is evidently at fault or the importance of a crime is to be judged either during or after an investigation. However, in the present case, computational experiments are used at the beginning of the investigation and help the investigation procedure.

## 2. Case Report

The first case involved a driver of a passenger car in winter 2005. The body and car were found on the ground about 45m from the road. Figure 1 is a photograph taken the following day. The photograph reveals scratches on the ground and shows the car lying at an angle of 180° with its direction of travel, with the driver located outside and to the left of the car.

The questions in this case were how the driver died while the car lay upright, and how the driver was thrown from the car when the driver's door was closed and its window was not broken. These two issues led to many assumptions. Some police suggested that the driver had opened the door and gotten out of the car but had then been killed by another person. The uncertainties in this case perplexed the investigators in terms of whether the case was one of murder or merely a traffic accident.

The second case was that of a woman who died in her house. Figure 2 shows the scene at the beginning of the investigation. The body lay on the floor with the face up, the neck showed signs of pressure or strangulation, and the room was in disorder, all of which suggested the possibility of homicide. However, opinion about the cause of death differed among forensic examiner, in particular about whether the direction of the mark behind her neck or behind and above her neck. This caused confusion over whether the investigation was one of homicide or suicide: if someone else tied up her neck behind her, the mark on her neck would be on the nape of the neck. In the present study, these two cases are reconstructed using MADYMO, which is very effective in the analysis of human dynamics.

Fig. 1. Accident scene for the first case (the driver's body was located on the left side of the car).

Fig. 2. Crime scene for the second case.

## 3. Autopsy Findings

In the first case, the male body was 170cm tall, weighed 68kg, and had a laceration between the eyebrows and abrasion on the right side of the face. There were fractures of the cranial, the cervical, and the thoracic vertebrae. The main cause of death was concluded to be fractures of the cranial and cervical vertebrae.

In the second case, the 160cm tall, 55kg woman had died by suffocation. The ligature mark, including several circular patterns, ran obliquely from the front of the neck to the external protuberance of the occipital bone.

## 4. Reconstruction of Crime Scenes

### 4.1. *Reconstruction of the first case*

#### 4.1.1. *The physical marks and analysis*

The accident car showed that the front bumper and substructure of the radiator had been deformed by about 30cm. Deformation marks on the left-rear fender and other marks arising due to pressure from the roof structure were found. A pressure mark was also found at the right-rear of the roof structure. These facts suggested that the car had overturned.

Several pieces of evidence, such as the deformation of the front bumper and of inner structures, pieces of tissue including hair attached to the inner right side of front window, and the deformed shift lever that pointed to the right, presented the possibility that the head of the driver had crashed against the right side of the front window. In conclusion, it was inferred that the car had slipped on the road, fallen on the ground and overturned; the final deduction was that the driver, who had not been wearing a seatbelt, had been separated from the car at the last stage of the overturn. However, these methods of reconstruction, which are based on the marks on the car and the evidence at the accident scene, are limited in forming a complete reconstruction of the accident.

### 4.1.2. *Analysis using computational experiments*

MADYMO 6.4[3] was applied to analyze this accident. The car model was remodeled to fit the properties (mass, size, etc.) of the actual vehicle. The Hybrid 50th-percentile Dummy was used to describe the driver. The width of the road was modeled as 8m and the slope of the road was modeled as 5.5m (horizontal) to 4.7 m (vertical). The deformation of the front bumper was adjusted so that the model would experience a similar extent of deformation as the actual vehicle in the accident. The friction coefficient between the ground and the front bumper was estimated to be 0.45. When the car fell off, it was modeled as being directed at an angle of 20°with the direction of travel, and was analyzed at a variety of speeds and angular velocities.

Firstly, to make the car crash against the ground, the speed of the car was required to be 60km/h. At lower speeds, the front of the car crashed against the slope, whereas at higher speeds, the front of the car did not overturn.

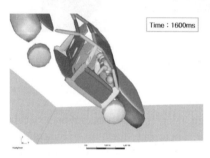

Fig. 3. Crashing motion against the ground.

Secondly, the simulation with a variety of angular velocities indicated that the car had slipped at a counter-clockwise angular velocity of 0.419rad/s. This indicated that the car had slipped in a counter-clockwise direction on the road and had fallen off the ground.

The driver trace showed that the driver had hit the upper part of the front window and the instrument panel due to inertia, as shown in Figure 3. As the car turned to the right, as shown in Figure 4(a), the driver moved against the right door and when the right portion of the car fell off, the driver began to exit the car. Figure 4(b) shows the graph of NIC, which occurred when the driver collided with the front window at about 1500ms~1600ms (as shown in Figure 3). At this moment, the value of HIC was maximized at about 3000 and the value of NIC was maximized at 700. If we compare these values with the criteria of EuroNCAP[4] for head injury, which corresponds to a risk of 20 percent that the injury is greater than AIS3, and for neck injury, namely, HIC of 1000 and NIC of 15, we find that the estimated values are much greater than the stipulated values.

The actual nature of the injury can slightly differ owing to the condition of contact (e.g., the stiffness of the material of contact) and coefficients of friction between the human model and the car interior and/or the ground. Nevertheless, we can easily reconstruct the driver's movements and explain, with certainty, the reason why the driver's neck and head were fractured.

Figure 5 shows the final results of the simulation. The distance from the road is 47m, which is similar to the distance of the accident at 45m. It was therefore concluded that the car slipped on an icy road and overturned, at which time the driver was ejected from the front right window of the car.

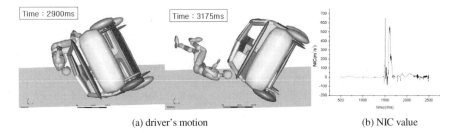

| (a) driver's motion | (b) NIC value |

Fig. 4. The driver was ejected from the front right just before the final position and the NIC and HIC values were maximized at about 1500~1600ms.

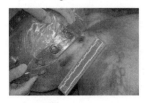

Fig 5 Comparison between the result of the simulation and the accident scene.

## 4.2. Reconstruction of the second case

### 4.2.1. Analysis of marks on the dead body at the scene.

The marks on the neck of the dead body in Figure 6 were concluded to have been caused by a belt. The marks were on both cervical sides toward the occipital region, suggesting that the dead body had received uniform loads. In the absence of any fracture on the hyoid bone and thyroid cartilage, the manner of death was considered to be suicide from hanging, rather than homicide through ligature strangulation. Therefore, the cause of death was considered to be hanging rather than homicide.

A very small amount of gold foil was found at the corner of the upper end of the door where the waistband may have been hooked; the gold foil was the same shape and color as the coating on the waistband.

Therefore, if the woman had stepped on a chair and kicked the chair behind her, she would have been found hanging from the upper corner of the door. In spite of this inference, there were many opposing opinions, e.g., her neck could have been strangled from behind in a manner that resulted in a similar outcome. Moreover, if the waistband had been hung at the top part of the door, it remained to be explained why the hinged door was shut.

Fig. 6. The waistband marks.

### 4.2.2. Analysis using computational experiments

The mid-sized male human model was used for the analysis. Under the assumption of hanging, the required parameters are basically a door, its framework, and a floor which contacted the dead body; the joints of the door are modeled as hinges. The effects of the door's crashing against its framework, when the door was closed, were examined. The waistband was omitted in the simulation because it was thought that it would not affect the simulation results for the analysis after the belt had been torn.

As soon as the human model began falling due to gravity, the knees and hip began to bend. The knees and hip moved both forward and backward. As the head moved forward, the right side of the face contacted the door. Parts of the knee hit the door and closed it, after which the human model turned left.

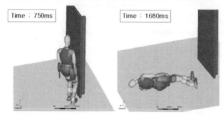 

Fig. 7. After the knee hit the door, the human model turned to the left, and the head crashed against the floor.    Fig. 8. Comparison of the final postures of the body.

After the door was closed, the human model fell down and then it turned slowly to the prostrate position shown in Figure 7. Finally, its posture became similar to that of the dead body in Figure 8. Therefore, this simulation confirmed that all of the evidence found at the death scene could be explained by hanging. Thus, this case was concluded to be a suicide rather than a homicide.

## 5. Conclusion

To summarize the first case, the car fell off the road, at which time the driver was thrown to the side of the passenger seat (since he did not wear a seatbelt) and crashed against the front window, which caused severe injury to the cervical spine. After that, when the car overturned, the driver was ejected from the front right window, which caused injuries to the chest. Both the cause of death and the progress of the accident were satisfactorily explained by the computational experiments. In the second case, a cord was broken when the body was against the door, after which the body fell and the knees folded. The door closed slowly as the body collapsed. The simulation results presented a similar posture to that of the dead body. This result indicated suicide rather than murder.

Any analysis that is based solely upon the evidence available at the crime or accident scene will be insufficient to explain the complete process of the accident and will be inadequate for convincing other people. Therefore, we have demonstrated that computational experiments can be a highly persuasive tool that shows great promise for application in the areas of accident reconstruction and crime scene analysis.

## References

1. John Versace, *A Review of the Severity Index, SAE 710881.*
2. Bostrom, O. et al. *IRCOBI Conference*, 1996.
3. TNO Automotive. MADYMO Model Manual Version 6.4.
4. EuroNCAP Assessment Protocol and Biomechanical Limits, Ver. 4.1.

# PREDICTION OF ROLLING FORCE USING AN ADAPTIVE NEURAL NETWORK MODEL DURING COLD ROLLING OF THIN STRIP

H.B.XIE[1†], Z.Y. JIANG[2], A.K. TIEU[3]

*School of Mechanical, Materials and Methatronic Engineering, University of Wollongong,*
*Wollongong NSW 2522, AUSTRALIA,*
*hx899@uow.edu.au, jiang@uow.edu.au, ktieu@uow.edu.au*

X.H. LIU[4], G.D WANG[5]

*The State Key Laboratory of Rolling and Automation, Northeastern University,*
*Shenyang, Liaoning 110004, P.R. CHINA,*
*liuxh@mail.neu.edu.cn, wanggd@mail.neu.edu.cn*

Received 15 June 2008
Revised 23 June 2008

Customers for cold rolled strip products expect the good flatness and surface finish, consistent metallurgical properties and accurate strip thickness. These requirements demand accurate prediction model for rolling parameters. This paper presents a set-up optimization system developed to predict the rolling force during cold strip rolling. As the rolling force has the very nonlinear and time-varying characteristics, conventional methods with simple mathematical models and a coarse learning scheme are not sufficient to achieve a good prediction for rolling force. In this work, all the factors that influence the rolling force are analyzed. A hybrid mathematical roll force model and an adaptive neural network have been improved by adjusting the adaptive learning algorithm. A good agreement between the calculated results and measured values verifies that the approach is applicable in the prediction of rolling force during cold rolling of thin strips, and the developed model is efficient and stable.

*Keywords:* Intelligent prediction; adaptive neural network; rolling force; cold rolling; thin strip.

## 1.    Introduction

Customers in the sheet metal industry require good accuracy of the strip dimensions and good uniformity of the mechanical properties. Steel manufacturers are therefore under pressure to improve their productivity by automating as many as possible and to optimize their process parameters to maximum efficiency and quality. Presently, the process control is mainly based on mathematical and statistical models. The development of mathematical preset models is difficult because the process of cold rolling is a non-linear system, in which lots of processing parameters such as rolling force, roll speed, frictional force, temperature, roll damage, materials, etc. have been interacted. Recently, neural-network-based adaptive control technique has attracted increasing attentions, because it has provided an efficient and effective way in the control of complex nonlinear or ill-defined systems. Neural networks have an excellent ability to learn and to predict

---

[†]Corresponding Author.

the complicated relationships between input and output variables. The relationship can be "learned" by a neural network through adequate training from the experimental data. It cannot only make decisions based on incomplete and disorderly information, but can also generalize rules from those cases on which there were trained and apply these rules to new cases. The applications of neural networks in rolling cover a wide field of applications ranging from flatness control to the prediction of mechanical properties of the rolled materials [1-3].

In this paper, a dynamic parameters adaptive BP neural networks (ABPNNs) model is proposed to predict the rolling force during cold rolling of thin strip. The model can optimize the input nodes, hidden nodes, transfer function, weights and bias of BP networks dynamically and adaptively. The simple architecture (less input and hidden nodes) of networks model is constructed in order to improve networks' adaptation and generalization ability, and to greatly reduce the subjective choice of structural parameters. Simulation results demonstrate that the proposed ABPNNs system with structure adaptation algorithm can achieve favorable tracking performance even unknown the control system dynamics function. The developed neural networks model is capable of providing good prediction accuracy of rolling force during cold rolling of thin strip.

## 2.    Structure of Neural Networks Control System

A conventional rolling force model has the following form:

$$P = \sigma_y \times B \times L_c \times Q_s \tag{1}$$

where $P$ is the rolling force, kN, $\sigma_y$ is the average yield stress, $B$ is the strip width, $L_c$ is the roll contact length, and $Q_s$ is the geometric term. The yield stress model is dependent on material temperature, strain, strain rate, material chemistry, and microstructure states.

However, there have not been any formulas that can describe the real rolling force exactly due to the complexity of the rolling process. There are too many factors affecting the rolling force, such as friction factor, the work-hardness of material, the front and back tension and so on. The rolling force model contains many inherent errors caused by the simplified model structure, the identification errors, disturbances, etc. Thanks to their powerful modelling and flexible self-learning capability, neural networks (NNs) have found successful applications in rolling process modelling and control in the steel industry [4]. The learning algorithm can improve the predictability of rolling force regardless of the change of working conditions, and the algorithm approach can optimize the network structure, transfer function and weights value [5, 6]. The rolling force has been predicated by a system in combination of conventional mathematical models and neural network. The model optimizes the input nodes, hidden nodes, transfer function, weights and bias of BP networks. Fig. 1 shows the architecture of optimization mechanism.

Selecting the important input variables from all possible input variables is important for data driven modeling. Incorporating only the important variables into a model learning process provides a simpler and more reliable model than the model that comprises all possible input variables [7]. The input nodes include material original thickness H; output thickness and input thickness at each pass h and $h_0$; the front and back tension $T_f$ and $T_b$ respectively; the basic material strength K; strip width B; and working roll radius R. The friction factor is not considered as an input node due to the emulsion used in the rolling mills does not change frequently. The structure of the BP network of rolling force is shown in Fig. 2 and the neural network has one hidden layer.

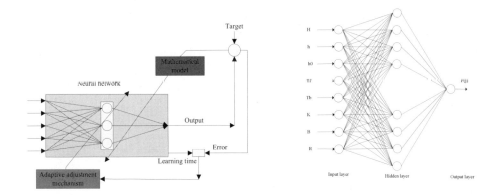

Fig. 1. Architecture of optimization mechanism.   Fig. 2. The structure of the BP network of rolling force.

In order to determine the numbers of the hidden nodes, both the model accuracy and calculation complexity are considered. Test shows that when the hidden nodes decrease from 18 to 15, the increase of model error is remarkable as shown in Table 1, and the hidden nodes are chosen to be 18. The output layer just has one node, i.e., the rolling force.

Table 1. Result of rolling force error testing (average square difference).

| Item | Hidden nodes | Train | Check | Test |
|------|--------------|-------|-------|------|
| Error ( kN) | 20 | 5.63 | 6.38 | 5.11 |
| | 18 | 6.13 | 6.76 | 6.58 |
| | 15 | 9.63 | 8.94 | 10.25 |

The weight is adjusted as below,

$$w_{j_i}{}^m(t+1) = w_{j_i}{}^m(t) + \alpha[w_{j_i}{}^m(t) - w_{j_i}{}^m(t-1)] + \eta \delta_j y_i^{(m-1)} \qquad (2)$$

where $\alpha$ and $\eta$ are positive constants, and 0.1 and 0.01 respectively are used,. The weights of each layer in the neural network will be changed after trained by the measured value off-line. Consequently, the precision of network's prediction will be higher and the predicted value will approach to the measured value.

## 3.   Results and Discussion

During the training of the neural network, the training samples need some data pre-treatments such as the dimension regulation and the input and output parameters normalization of the neural network. The original data is divided into three subsets: training subset, check subset and testing subset. The training subset is used for computing the gradient and updating the network weights and biases. The error on the validation subset is monitored during the training process. The comparison of rolling force controlled by conventional model based on physical process and neural networks are shown in Figs. 3-5 for three subset. The rolling force predicted using neural network approach is much closer to the measured values.

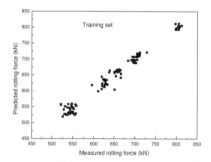

Fig. 3. Comparison of rolling force between measured and predicted values (training set).

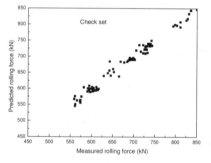

Fig. 4. Comparison of rolling force between measured and predicted values (check set).

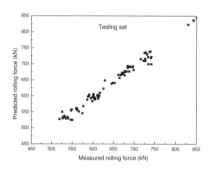

Fig. 5. Comparison of rolling force between measured and predicted values (testing set).

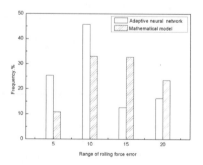

Fig. 6. Comparison of rolling force error between neural network and conventional model.

The comparison of predicted rolling force error using the proposed neural network and conventional model based on physical process is shown in Fig. 6, and the error is the range deviation. It can be seen that the proposed simulated method can improve the control accuracy significantly.

## 4. Conclusions

In this paper, the predicted model of rolling force using adaptive neural networks is proposed to improve the accuracy of rolling force control. A combination of self adaptive algorithm for adjusting the hidden node parameters as well as the weights and a neural was conducted successfully. Simulation results provided show the validity and applicability of the proposed neural networks. This confirms that the neural network is very effective in significantly improving the model prediction accuracy of rolling force. A hybrid mathematical roll force model and an adaptive neural network have been improved by adjusting the adaptive learning algorithm. A good agreement between the calculated results and measured values verifies that the approach is applicable in the prediction of rolling force during cold rolling of thin strips, and the developed model is efficient and stable.

## References

1. J. Larkiola, P. Myllykoski, A.S. Korhonen, L. Cser, The role of neural networks in the optimization of rolling process, Journal of Material Processing Technology 80-81(1998) 16-23.
2. P. Korczak, H. Dyja, E. Labuda, Using neural network models for predicting mechanical properties after hot plate rolling processed, Journal of Material Processing Technology 80-81 (1998) 481-486.
3. J.S. Son, D.M. Lee, I.S. Kim, S.K. Choi, A study on genetic algorithm to select architecture of a optimal neural network in the hot rolling process, Journal of Material Processing Technology 153-154 (2004) 643-648.
4. F. Janabi-Sharifi, A neuro-fuzzy system for looper tension control in rolling mills, Control Engineering Practice 13 (2005) 1-13.
5. Martin Schlang, B. Lang, T. Poppe, T. Runler, K. Weinzier, Current and future development in neural computation in steel processing, Control Engineering Practice 9 (2001) 975-986.
6. D.M. Lee, S.G. Choi, Application of on-line adaptable neural network for rolling force set-up of a plate mill, Engineering applications of artificial intelligence 17(5)(2004) 557-565.
7. H.B. Xie, Z.Y. Jiang, X.H. Liu, G.D. Wang, A.K. Tiue, Prediction of coiling temperature on run-out table of hot strip mill using data mining, Journal of Material Processing Technology 177 (2006) 121-125.

# AN INFLUENCE FUNCTION METHOD ANALYSIS OF COLD STRIP ROLLING

Z.Y. JIANG[1†], D.W. WEI[2], A.K. TIEU[3]

*School of Mechanical, Materials and Mechatronic Engineering, University of Wollongong, Northfields Avenue,*
*Wollongong NSW 2522, AUSTRALIA,*
*jiang@uow.edu.au, dwei@uow.edu.au, ktieu@uow.edu.au*

Received 15 June 2008
Revised 23 June 2008

An influence function method has been developed to simulate the asymmetric cold rolling of thin strip with work roll kiss at edges. The numerical simulation model was obtained based on the deformation compatibility of the roll system in rolling and lateral directions. The strip plastic deformation has been considered in the formulation, which is significantly different from the traditional theory of metal rolling. The rolling mechanics and crown of the strip with work roll edge kiss, which are new findings for cold rolling of thin strip, are obtained. A comparison of the rolling force, roll edge kiss force and the strip crown after rolling has been conducted for various cross shear regions in the roll bite. Results show that the calculated strip crown is in good agreement with measured value, and the rolling force and strip crown decrease with an increase of cross shear regions, as well as the work roll edge kiss force and edge wear decrease. The friction also has an influence on the profile of the rolled thin strip.

*Keywords*: Work roll edge kiss; influence function method; rolling force; roll edge kiss force; strip crown.

## 1. Introduction

Thin strip is produced on a tandem cold mill where the work rolls are flattened [1] to a non-circular deformed shape [2]. The produced thin strip has a wide application, especially in the electronic and instrument industries. Given in Fig. 1, an asymmetric strip rolling with work roll speed difference, $R_1$ and $R_2$ are the radii of upper and lower work rolls respectively, $x_{n1}$ and $x_{n2}$ the neutral points of upper and lower work rolls respectively, $\mu_1$ and $\mu_2$ the friction coefficients of upper and lower work rolls respectively, $h_0$ and $h_1$ the strip thickness before and after rolling respectively, $V_1$ and $V_2$ the speeds of upper and lower work rolls respectively. There is a cross shear region existing in the roll bite, so it will result in a significant decrease (up to 40 %) of rolling force [3], and can also reduce the strip thickness significantly in cold strip rolling [4-7].

As a thinner strip is rolled, some problems will be increased during the rolling, such as the strip shape, profile and flatness. A good strip shape and profile is based on the

---

[†]Corresponding Author.

mathematical models applied to the cold strip rolling. Researchers have investigated the mechanics of cold rolling of thin foil [8,9], and the deformation of thin strip, its shape, profile and flatness have also been investigated [10-12]. A solution on how to improve the strip shape, profile and flatness, and dimensional accuracy has always been of major interest to the steel manufacturers and researchers. These are rolling processes where the work rolls do not contact each other and their speeds are the same when relatively thick strip is rolled.

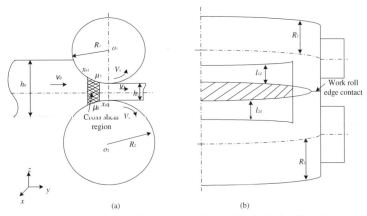

Fig. 1. Asymmetric strip rolling (a) and work roll edges kiss for half roll barrel length (b).

In cold rolling of thin strip, including the temper rolling, it has been found that the edges of the work rolls kiss and deform (see Fig. 1b). Work roll edge kiss forms a new deformation feature, especially for the asymmetric rolling of thin strip. In this case, the model of deformation mechanics is significantly different from the traditional analysis of strip rolling. When the work rolls kiss beyond edges of the strip, it will significantly change the crown of the rolled strip and work roll wear. In this study, an influence function method has been developed to simulate this special rolling process. Based on the numerical simulation, the effects of the work roll cross shear region on the rolling force and the crown of rolled strip are obtained. Effect of friction on the rolled strip crown is also discussed.

## 2. Influence Function Analysis of Asymmetric Rolling of Thin strip

Due to symmetry, the calculation involves only one half of a work roll system along the roll barrel as shown in Fig. 2. The profiles of the deformed work roll and backup roll (above the upper work roll and down the lower work roll) are obtained by calculating roll deflections due to the bending and shear force. Local deformations due to the flattening in the contact region between the work roll and backup roll, the work roll and strip, and the upper and lower work rolls are added to the roll deflections [13].

Refer to Ref. [13], the load is divided into a number of concentrated loads in the middle of each element, the deformation of the beam at an element $i$ caused by an arbitrary load distribution along the beam can be calculated by

$$y(i) = \sum_{j}^{m} g(i,j) p_j \qquad (1)$$

where the influence function $g(i,j)$ is defined as the deflection in the middle of element $i$ due to a unit load in the middle of element $j$. The deformation $y(i)$ in Eq. (1) not only stands for the deflection of the rolls, but also includes the flattening in the contact zone.

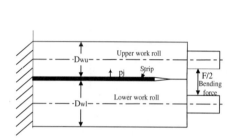

Fig. 2. Roll deflection model of the asymmetric rolling.

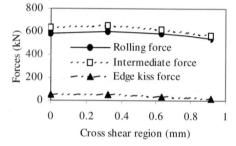

Fig. 3. Effect of cross shear region on forces.

Under the rolling load, the deformation of the work roll, backup roll and the strip are described in Ref. [13]. Compatibility for contact of the work roll and backup roll varies with the sum of the contour of the deformed work roll and backup roll and the local roll flattening. It can be calculated by Eqs. (2) and (3).

$$y_{wbu}(i) = y_{wbu}(0) + y_{bu}(i) - y_{wu}(i) - m_{bu}(i) - m_{wu}(i) \qquad (2)$$

$$y_{wbl}(i) = y_{wbl}(0) + y_{bl}(i) - y_{wl}(i) - m_{bl}(i) - m_{wl}(i) \qquad (3)$$

where the parameters can refer to Ref. [13], $u$ and $l$ stand for upper and lower rolls respectively.

The contour of the work roll surface in contact with the strip is determined by the combined influence of the rolling load, machined and thermal crown and the local flattening between the work roll and the strip. The exit thickness of the strip at any point is the same as the loaded gap height at that point. Therefore, the compatibility for the contact of the work roll and the strip can be expressed by Eq. (4).

$$h(i) = h(0) + y_{wsu}(i) - y_{wsu}(0) + m_{wu}(i) - y_{wu}(i) + y_{wsl}(i) - y_{wsl}(0) + m_{wl}(i) - y_{wl}(i) \qquad (4)$$

where $h(0)$ is the central thickness of the strip.

When rolling a thin strip, the edges of the work rolls beyond the strip may kiss and deform. The compatibility Eq. (5) for edge contact of upper and lower work rolls is calculated from the deformed work roll profile and the centreline value of the flattening between the work roll and strip.

$$y_{ww}(i) = y_{wsu}(0) + y_{wu}(i) - m_{wu}(i) + y_{wsl}(0) + y_{wl}(i) - m_{wl}(i) - h(0)$$  (5)

where $u$ and $l$ stand for upper and lower work rolls respectively.

## 3.   Results and Discussion

Based on an analysis of the influence function method, a simulation program for the asymmetric rolling of thin strip was developed. For different cross shear regions and the friction, the rolling pressure, work roll edge kiss force and the strip crown can be obtained. The deformation resistance of strip can be described by the following equation,

$$k_s = 740 \cdot (\varepsilon_m + 0.01)^{0.23} \quad (\text{MPa})$$  (6)

where $\varepsilon_m$ is an average reduction that can be calculated by entry and exit strip thickness.

### 3.1.  *Effect of cross shear region*

Due to asymmetric rolling of thin strip, a cross shear region exists in the roll bite. When the entry and exit thickness of strip are 0.2 and 0.12 mm (reduction 40%) respectively, the roll bending force, front and back tensions are 0 kN. Based on the parameters of Hille 100 rolling mill, the radius of the upper work roll is 31.5 mm, diameters of the upper and lower backup rolls 228 mm, barrel length of the upper and lower work rolls 249 mm, barrel length of the upper and lower backup rolls 249 mm, crown of the upper and lower work rolls 0 mm, crown of the upper and lower backup rolls 0 mm, strip width 160 mm, friction coefficient at the upper and lower rolls contact region 0.1. The radii ratios of the lower and upper work rolls $R_2/R_1$ are 1.0, 1.1, 1.2 and 1.3 ($R_1$ = 31.5 mm) respectively. The relevant cross shear regions are 0, 0.32, 0.63 and 0.92 mm respectively. Fig. 3 shows the effect of the cross shear region on the rolling force, intermediate force and roll edge kiss force. It can be seen that the rolling force, intermediate force and roll edge kiss force decrease with an increase of the cross shear region. The decrease of the rolling force is up to 7.7 %, intermediate force 12.0 % and roll edge kiss force 60.1 % if the cross shear region increases from 0 to 0.92 mm. It can also be seen that the difference between the rolling force and intermediate force decreases when the cross shear region increases. This result demonstrates that the rolling load reduces for asymmetric rolling when the work rolls edges kiss, which can result in the decrease of the roll deformation, and reduces the length of roll edge kiss. As a result, the work roll edge kiss force decreases significantly. Therefore, the work roll edge wear will be reduced, which can extend the work roll service life.

The effect of cross shear region on the crown of the rolled strip is shown in Fig. 4. It can be seen that the crown of the rolled strip reduces when the cross shear region increases, and the decrease rate becomes significant when the cross shear region is larger than 0.3 mm. Therefore, if the work roll edge kiss, the profile of the rolled strip becomes better when the cross shear region increases.

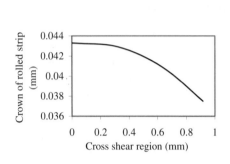

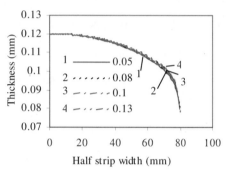

Fig. 4. Effect of cross shear region on crown of rolled strip.

Fig. 5. Effect of friction at upper and lower contact region on strip crown.

### 3.2.  *Effect of friction at upper and lower contact regions*

One of the key issues in cold rolling of thin strip is the strip crown. When the reduction is 40%, friction coefficients of upper and lower contact regions are 0.05, 0.08, 0.1 and 0.13 respectively, the peripheral speed ratio of the lower and upper work rolls is 1.2, other parameters are the same as the above. The effect of the friction coefficient on the rolled strip crown is shown in Fig. 5. It can be seen that the strip crown reduces slightly when the friction coefficient increases. This indicates that the friction coefficient can improve the strip crown. Simulation also shows that the length of the cross shear region varies with a change of friction coefficient in the roll bite.

In order to verify the simulation results, an experimental verification was carried out in the laboratory. When the rolling speed is 0.272 m/s, entry thickness of strip is 0.5 mm, strip width 100 mm, and work roll peripheral speed ratio 1.2, a low carbon steel was rolled on Hille 100 rolling mill, the oil lubricant was applied on sample surface, the friction coefficient is 0.1. Fig. 6 shows a product of the rolled thin strip. Based on the results, a comparison of the calculated strip crown with the measured value is shown in Fig. 7. It can be seen that the calculated strip crown is close to the measured result. This indicates that the developed model of calculating the strip crown is effective and applicable.

### 4.   Conclusions

The influence function method analysis of asymmetric cold rolling of thin strip with work roll edge kiss was effectively carried out. The results show that rolling force increases with the length of cross shear region in the roll bite, as well as the work roll edge kiss force decreases significantly. As a result, the work roll edge wear reduces, and the work roll service life can be extended. The crown of the rolled strip also reduces significantly when the cross shear region increases, and a better strip profile can be obtained. Higher friction coefficient can improve the strip shape and profile. The calculated strip crown is consistent with the measured value. The developed model of calculating the strip crown is effective and applicable in the asymmetric cold rolling of thin strip.

Fig. 6.  Product of the rolled ultra thin strip.

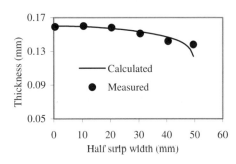

Fig. 7. Comparison of calculated strip crown with measured value.

## Acknowledgments

This work is supported by an Australian Research Council (ARC) Discovery Project grant including Australian Research Fellowship. The author would like to thank Dr H.T. Zhu's assistance in modelling, and to Mr. Joe Abbot, Mr. Kelvin Chong, Mr. Chong Pink Chuah and Ms Limei Yang in assisting the experimental work.

## References

1.  P. Montmitonnet, E. Massoni, M. Vacance, G. Sola, P. Gratacos, *Ironmaking & Steelmaking* **20**, 254-260(1993).
2.  J. Shi, D.L.S. McElwainand, T.A.M. Langlands, *Int. J. Mech. Sci.* **43**, 611-630(2001).
3.  D. Pan, D.H. Sansome, *J. Mech. Work. Technol.* **16**, 61-371(1982).
4.  Y.M. Huang, G.Y. Tzou, *Int. J. Mech. Sci.* **39**(3), 289-303(2001).
5.  G.Y. Tzou, *J. Mater. Proc. Technol.* **86**, 271-277(1999).
6.  G.Y. Tzou, M.N. Huang, *J. Mater. Proc. Technol.* **119**, 229-233(2001).
7.  M. Salimi, F. Sassani, *Int. J. Mech. Sci.* **44**(9), 1999-2023(2002).
8.  L.C. Zhang, *Int. J. Machine Tools Manufact.* **35**(3), 363-372(1995).
9.  N.A. Fleck, K.L Johnson, M. Mear, L.C. Zhang, *Proc. Inst. Mech. Engrs Part B: Eng. Manufact.* **206**, 119-131(1992).
10. Z.Y. Jiang, A.K. Tieu, *J. Mater. Proc. Technol.* **130-131**, 511-515(2002).
11. Z.Y. Jiang, A.K. Tieu, *Tribology Int.* **37**, 185-191(2004).
12. Z.Y. Jiang, A.K. Tieu, X.M. Zhang, C. Lu, W.H. Sun, *J Mater. Proc. Technol.* **140**(1-3), 544-549(2003).
13. Z.Y. Jiang, H.T. Zhu, A.K. Tieu, W.H. Sun, *J. Mater. Proc. Technol.* **155-156**, 1280-1285(2004).

# EFFECT OF ASYMMETRICAL STAND STIFFNESS ON HOT ROLLED STRIP SHAPE

DIANYAO GONG[1], JIANZHONG XU[2†]

*State Key Laboratory of Rolling technology and Automation, Northeastern University,*
*Shenyang, Liaoning Province, 110004, P.R.CHINA,*

*dygong1976@gmail.com, xjzyyh@263.net*

ZHENGYI JIANG[3]

*Faculty of Engineering, University of Wollongong,*
*Wollongong NSW 2522, AUSTRALIA,*
*jiang@uow.edu.au*

XIAOMING ZHANG[4], XIANGHUA LIU[5], GUODONG WANG[6]

*State Key Laboratory of Rolling technology and Automation, Northeastern University,*
*Shenyang, Liaoning Province, 110004, P.R.CHINA,*

Received 15 June 2008
Revised 23 June 2008

The difference of elastic springs between the operating side (OS) and driving side (DS) of rolling mill has a significant influence on the strip shape not just the strip thickness. Based on the slit beam and roll deformation theories, the roll force distribution was analysed considering the asymmetric stiffness of the OS and DS of rolling mill, and the work roll and backup roll deformation equations were deduced respectively, and the thickness distribution in lateral direction of the hot rolled strip at exit was discussed. Using the roll elastic deformation analysis software which was developed previously based on the influence coefficient method, the roll flattening distribution, roll pressure distribution and the rolling force distribution caused by the asymmetric stand stiffness were calculated and analysed, and the exit strip profile of the rolling mill was also presented. The relationship between the mill stiffness difference and the strip wedge shape or single wave was obtained. Effect of the upstream asymmetric mill on strip crown and flatness of the downstream stands was discussed.

*Keywords:* Influence coefficient method; Asymmetric stiffness; Hot rolled strip; Strip shape; 4-high mill.

In theory, effect of stiffness on strip flatness was neglected, symmetric model was built to resolve deflection of the roll based on slit beam theory or freely supported beam theory.

---

[†]Corresponding author.

Based on these theory, no matter how to describe the deflection of roll, there is at least one single point of the roll is motionless. In fact, the difference of mill stiffness between DS and OS will effect not only strip thickness but also strip lateral thickness distribution. Currently many important theoretical and practical flatness problems are resolved using Influence Coefficient Method, therefore, based on the roll elastic deformation theory, using Influence Coefficient Method, effect of the asymmetrical stiffness of rolling mill on strip crown and flatness are researched.

## 1.   Performance of Asymmetrical Stiffness

So-called stiffness of mill is that the mill unit bouncing generated by rolling force, it means the elastic deformation resistance capacity of mill, the greater stiffness of mill, the better stability. Analysis based on flatness in theory considering mill operating side and the driving side of the stiffness is the same, strip shape is presented to be symmetrical joints or symmetrical distribution points. In fact, there are some strip rolling plants on the two sides has different stiffness, when the difference exceeds a certain limit, the impact of stiffness difference of two sides of plant on the strip shape must be considered. Fig. 1   shows the stiffness difference between driving side and operation side measured in certain factory.

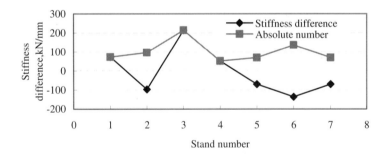

Fig. 1.  Stiffness difference between DS and OS of finishing mill in certain domestic factory.

For the hot rolled strip flatness control, mill such as PC, CVC and the associated flatness control strategies and mathematical model are used, these strip shape devices limit the scope of control in the flatness defect for quadratic curve and aquatic curve.[1-3] For such reasons of the equipment itself, it can do nothing to control waves caused by asymmetric condition. Because of flatness online measurement technical limitations, strip shape control relay on operatives visual experience and judgment together with forward and feedback auto control of bending force.

Considering above situation, analysis of the strip thickness distribution and the effect on exit profile of adjacent downstream stand under the conditions of asymmetric stiffness has very important theoretical and practical value[4].

## 2.   Force Analysis of 4-High Mill

When the force distribution of four-high mill is analyzed in theory, it is often assumed that the stress of roll, steel and the plant are symmetrical. The ordinary four-high mill can be supposed as axial symmetric, to CVC mill, it can be supposed as *o* bit symmetric. Rolling force distributes along with the width of steel, plus the bending force acts on the both ends of work roll, roll force generated between work roll and backup roll.

When there is difference of stiffness between two sides of stand, the symmetric state is broken therefore the stress of rolling stand must be updated. To analyzing the elastic deformation of rolls during strip rolling, the backup roll and work roll are regarded as two slit beams whose fixed ends are near the centre of roller. The deflection status of roll considered as slit beam with the roll center regarded as fixed support point when the BR and WR are assumed to be spring base. The free ends at both side of slit beam is kept for a kind of dynamic balance of torque created by rolling force and bending force. The deflection of the free end of slit beam content the elastic deformation caused by unit width rolling force, the deflection caused by stand spring, the deflection caused by WR bending force and BR bending force, as shown in Fig. 3.

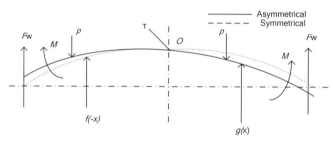

Fig. 3.   Roll deflection with asymmetrical stand stiffness.

## 3.   Calculation of Roll Elastic Deformation

The roll gap is asymmetric in lateral direction when strip is rolling since the asymmetric stiffness of stand of operating side and driving side, then the rolling force along with strip width and roll force between BR and WR along with roll barrel length are asymmetric, with the result that the roll elastic deformation is asymmetric thus the strip thickness along with strip width is asymmetric. The Influence Coefficient Method is a practical method to calculate roll elastic deformation which is shown in Ref.5 and Ref.6.

For modeling the deflection and elastic deformation of rolls, firstly, the rigid motion of rollers during rolling are assumed as linear. The rigid motion of the fixed point of slit beam is shown in Eq. 1, Eq. 2 and Eq. 3.

$$JWW : \Omega = S_0 + (K_D + K_O)(P + F_w) \tag{1}$$

$$JWB : \Omega = S_0 + (K_D + K_O)P \tag{2}$$

$$JWW : \Omega = S_0 + (K_D + K_O)(P + F_b) \tag{3}$$

Where, $\Omega$ is the rigid motion of the slit beam(rolls' center); $P$ is half of rolling force; $K_D$ and $K_O$ is the stiffness of DS side and OS side of stand respectively; $F_W$ is the bending force of WR; $F_b$ the is bending force of BR. *JWW*, *JWB* and *JBB* indicate the status when considering work roll bending force, no bending force and backup bending force is used respectively.

The rigid motion of the free end of the slit beam is shown in Eq. 4, Eq. 5 and Eq. 6.

$$JWW : \begin{cases} \Omega_D = S_0 + K_D (P + F_W / 2) \\ \Omega_o = S_0 + K_o (P + F_W / 2) \end{cases} \tag{4}$$

$$JWB : \begin{cases} \Omega_D = S_0 + K_D P \\ \Omega_D = S_0 + K_O P \end{cases} \tag{5}$$

$$JWB : \begin{cases} \Omega_D = S_0 + K_D (P + F_b / 2) \\ \Omega_D = S_0 + K_O (P + F_b / 2) \end{cases} \tag{6}$$

The rigid motion of any point at roll barrel can be obtained by linear interpolation. The equation of work roll elastic deformation is:

$$\begin{cases} Y_{WD} = \sum_{j=1}^{n} q(j) g(i, j) - \sum_{j=1}^{n} \phi(j) p(j) q_w (i, j) + Z_w (i) - \Omega_{WD} (i) \\ Y_{WO} = \sum_{j=1}^{m} q(j) g(i, j) - \sum_{j=1}^{m} \phi(j) p(j) q_w (i, j) + Z_w (i) - \Omega_{WO} (i) \end{cases} \tag{7}$$

Where, $\phi(j)$ is variable, $\phi(j) = 0 \, (j = 1 \sim k)$, $\phi(j) = 1 \, (j > k)$; $Z_w (i)$ is the press amount of unit $i$ of work roll caused by the contact press between work roll and backup roll; $\Omega_{WD} (i)$ and $\Omega_{WO} (i)$ is the rigid motion of drive side and operation side of work roll.

The equations of backup roll deformation are shown as follows:

$$\begin{cases} Y_{BD} = -\sum_{i=1}^{n} q(j) g(i, j) - Z_b (i) - \Omega_{BD} \\ Y_{BO} = -\sum_{i=1}^{m} q(j) g(i, j) - Z_b (i) - \Omega_{BO} \end{cases} \tag{8}$$

$$\begin{cases} Y_{BD} = -\sum_{j=1}^{n} q(j) g_b (i, j) - Z_b (i) - \Omega_{BD} + e(i) \\ Y_{BO} = -\sum_{j=1}^{m} q(j) g_b (i, j) - Z_b (i) - \Omega_{BO} + e(i) \end{cases} \tag{9}$$

Where, $Z_b (i)$ is the elastic bruise caused by rolls contact; $\Omega_{BD}$ and $\Omega_{BO}$ is the rigid motion of drive side and operation side of backup roll; $e_b (i)$ is the deformation of backup roll caused by force couple.

## 4.    Calculation Result and Discussion

Considering stand stiffness were 2450kN/mm(operating side) and 2550kN/mm(driving side), the difference of stiffness is 100kN/mm. The roll elastic deformation, rolling force distribution, roll pressure distribution between the work roll and backup roll during strip rolling were calculated, the calculating result are shown in Fig. 4, Fig. 5. Fig. 4 shows the distribution of rolling force and roll force between roll pressure along with BR barrel. When the rolling force value turns negative, there the start of strip edge. At the edge end of strip, the roll force between rolls shows a sudden salient, the value of the peak can be 20.796 kN/mm, then decreases smoothly to the center of strip, at the other side, the roll force between rolls put up a smoothly salient (10.297kN/mm) then decrease smoothly to the center of strip. The rolling force is almost linear increasing from the small stiffness side to the bigger stiffness side.

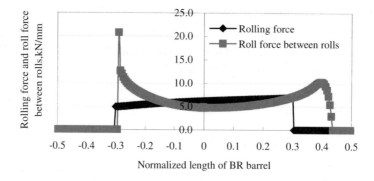

Fig. 4.    The rolling force and roll force between WR and BR along with BR barrel.

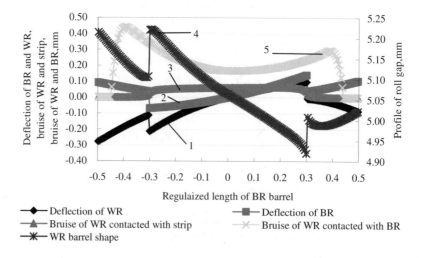

Fig. 5.    The elastic deflection curves and profile of roll gap along with BR barrel.

Fig. 5 shows the elastic deformation curves of work roll and backup roll, includes the deflection of backup roll, the deflection of work roll, the bruise of work roll contacted with strip, the bruise of work roll contacted with backup roll. The curves in Fig. 5 are indicated with both legend and serial number. There are sharp transition in curve 1, curve 2, and curve 4 at the strip edge end. When the value of roll deformation is negative, that means the direction of elastic deformation is toward the strip, when the value is positive, the direction of elastic deformation is reverse. The absolute value of deflection of WR and BR are decreasing from the OS (low stiffness side) to DS (high stiffness side) of stand. There are sharp change of force at the sharp transition points, which caused by asymmetrical stiffness of stand. The sharp decrease of WR deflection is 0.1mm occurred at the 0.3 point (lateral axial) of regulative length of strip width, at the same place, the sharp decrease of BR deflection is 0.09mm, the WR shape curve has a sharp increase, the value is 0.12mm. At the other side of WR, the curves have the same changing trend. These sharp transition point shows that there are discontinued elastic deformation along with the roll, the discontinued deformation is dangerous to the rolls. Although the stiffness is smaller at OS, the WR bruise is greater than the DS side as shown in Fig. 5, curve 5.

## 5.  Conclusion

The force distribution under asymmetrical stiffness condition is analysed and the elastic deformation of work roll and backup roll are calculated using Influence Coefficient Method. The analysed and calculation result show that:

1)  There are sudden salient of roll force between work roll and backup roll at the both place edge end of strip during rolling. The sudden salient is very great and harmful to the backup roll.

2)  The rolling force at the lower stiffness is smaller, the rolling force at the higher stiffness is higher, the rolling force curve along with backup roll barrel keep linear.

3)  There are sudden salient at elastic deformation curve of work roll and backup roll, the sudden salient at both side is almost equal, can get about 0.1mm. The elastic deformation curves are asymmetrical.

## References

1)  Zhou S.X, Funke P, Zhong. *J. Steel Reasearch*, 1996, 67(5): 200-204 (in Chinese).

2)  Kitahama. Masanori. *Iron and Steel Engineer*, 1987, 64(11): 34-43.

3)  Wood G. E, Kenneth Humphries, Arnold P. W, et al. MPT, 1989, 12(5): 92-96.

4)  Shimizhu Y, Iwatani J, Yamasaki Y, et al. *Technology Review-Mitsubishi Heavy Industries*, 2000, 37(2): 45-47.

5)  Wang G D. Beijing: *Metallurgical Industry Press*, 1986. 289-306 (in Chinese).

6)  Xu Jianzhong, Zhang Fengqin, *Gong Dianyao etal. Steel Rolling*, 2003, 20(2): 8-11.

# RIGID-PLASTIC SEISMIC DESIGN OF REINFORCED CONCRETE SHEAR WALL

CHANG LIN FAN[1†], SHAN YUAN ZHANG[2]

*College of Architecture and civil Engineering, TaiYuan University of Technology,*
*No.79 West Yingze Street, Taiyuana, 030024, Shanxi, PR CHINA,*
*Fancl7538@126.com, luguoyun@tyut.edu.cn*

Received 15 June 2008
Revised 23 June 2008

Basing the displacement-capacity design method and capacity spectrum method, a new rigid-plastic seismic design procedure is proposed to describe the behavior of shear wall structure under strong earthquakes. Firstly the concept of rigid-plastic hinge is used to choose a collapse mechanism of shear wall, then according to the dynamic performance criterion the yield load of structure is determined through rigid-plastic response spectrum. This procedure is used in 11-story reinforced structure shear wall design, the results of comparison with refined Non-Linear Time-History Analysis showing good agreement. .

*Keywords*: Rigid-plastic model; the theory of plasticity; Virtual work principle; Shear wall structure.

## 1. Introduction

Reinforced concrete (RC) shear wall to withstand the effects of strong earthquake motion should preferably be designed to have localized plastic behavior, where an appropriate level of damage is accepted and the safety of a structure depends on its capacity to withstand large plastic displacements in these plastic regions.[1] In fact, for strong earthquakes, elastic deformations are very small when compared with the magnitude of plastic deformations, making their contribution to dynamic response, including resonance effects, negligible.[2] determination of the seismic demand on inelastic system is necessary and rather complicated. This paper presents a new procedure for seismic design of RC shear wall buildings for strong ground motion based on simplified Non-Linear Time History (NLTHA), the Rigid-Plastic Seismic Design (PRSD) method.

## 2. Rigid-Plastic Dynamics

The RSPD method deals with rigid-plastic based on the following assumptions:
 (i) The behavior of hinges is rigid-plastic and the rest of the structures remain in the rigid domain. Pinching is taken into account in a simple manner see Fig.1.
 (ii) Masses of structures are geometrically lumped at the height of floors.

---

[†]Corresponding Author.

(iii) All yield moments in the plastic hinge at the end of the coupling beams have the same value and relate to the yield moment at the base of shear wall by a factor $\beta$.
there are axial force $N$ shear force $Q$ and bending moment $M$ in base section of shear wall. Only stress constituting internal work is called generalized stress. Thus, $M$ and $Q$ in section of shear wall are generalized stresses. The perfectly plastic stress condition is expressed by the Eq. (1), [3]

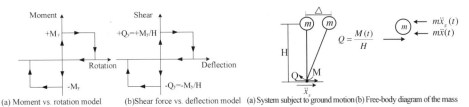

| (a) Moment vs. rotation model | (b)Shear force vs. deflection model | (a) System subject to ground motion (b) Free-body diagram of the mass |
|---|---|---|

Fig. 1. Rigid-plastic hysteretic relationship.　　Fig. 2. Model of SODF system.

$$\Psi p - \tilde{m}^2 + \tilde{q}^2 - 1$$

Where $\qquad\qquad \tilde{m} = M/M_p = Q/Q_{Mp} \quad \tilde{q} = Q/Q_p \qquad\qquad$ (1)

$M_p$ , $Q_p$ are the full plastic bending moment and the full plastic shear force of the shear wall base section, respectively. For the rectangular section,

$$Q_P = \tau_p \cdot bh \, , Q_{Mp} = \sigma_p bh^2/(4H) \qquad (2)$$

Where $\tau_p = \sigma_p/2$ , [3] $\tau_p$ , $\sigma_p$ are shear yield strengthen, and uniaxial tensile yield strength, respectively. $b$ , $h$ are the width, height of the shear wall base section, respectively. $Q_{Mp}$ is shear force of the shear wall base section, when the base moment reaches $M_p$ .

For perfect rigid-plastic model, the plastic flow lows are[3]

$$\dot{\theta}^P = \mu \frac{\partial \Psi_p}{\partial M} \quad \dot{V}^P = \mu \frac{\partial \Psi_p}{\partial Q} \qquad (3)$$

Where $\mu$ is a positive scalar constant. One can obtain

$$\theta^P/V^P = (2H \cdot \beta)/h \cdot (1+\beta) \qquad (4)$$

Where $\theta^P$ and $V^P$ are shearing and rotational plastic displacement respectively.

## 2.1. Rigid-plastic dynamics

Consider the case of the rigid-plastic oscillator in Fig.2 (a). Using D'Alembert's principle, dynamic equilibrium may be expressed by Eq. (5)

Rigid behavior: $\qquad\qquad \dot{x}(t) = 0 \qquad\qquad$ if $\psi_p < 0$ $\qquad$ (5.1)

Plastic behavior: $\qquad\qquad \ddot{x}(t) \pm Q_y/M = -\ddot{x}_g(t) \qquad$ if $\psi_p = 0$ $\qquad$ (5.2)

Slip behavior: $\qquad\qquad \ddot{x}(t) = \; -\ddot{x}_g(t) \qquad\qquad$ if $\dot{x}(t) \bullet x(t) < 0$ $\quad$ (5.3)

In the Eq. (5.2) the sign +or-should be taken depending on whether $\ddot{x}(t) + \ddot{u}_g(t)$ is greater or less than zero, respectively.

For a rigid-plastic MDOF structure chosen to develop a collapse mechanism under severe ground motion may be treated as an assemblage of rigid bodies where the internal deformations take place at the plastic zones only.[4] The relative displacement of each mass, $m_i$ is given by the product of a displacement shape vector, $\phi_i = h_i / H$ and a given amplitude, e.g. the top of displacement, $\Delta$. In Fig.3 a sketch of the structure is shown when the collapse mechanism in the positive direction is activated. During a virtual displacement $\delta$ as shown in Fig.3, using Virtual Work Principle, one can obtain Eq. (4)

$$\begin{cases} -\sum_{k=1}^{2n+2} M_k \cdot \left|\frac{\delta^m}{H}\right| - \sum_{i=1}^{n+1} m_i \varphi_i^2 \ddot{x}(t)\delta^m - \sum_{i=1}^{n} m_i \varphi_i \ddot{x}_g(t)\delta^m - \sum_{k=1}^{2} V_k \cdot \eta\delta^m - \sum_{i=1}^{n} m_i \ddot{x}(t)\cdot\eta\delta^m - \sum_{i=1}^{n} m_i \ddot{x}_g(t)\cdot\eta\delta^m = 0 (if \delta < 0) \\ -\sum_{i=1}^{2n+2} |M_k| \cdot \frac{\delta^m}{H} + \sum_{i=1}^{n} m_i \varphi_i^2 \ddot{x}(t)\delta^m + \sum_{i=1}^{n+1} m_i \varphi_i \ddot{x}_g(t)\delta^m + \sum_{k=1}^{2} V_k \cdot \eta\delta^m + \sum_{i=1}^{n} m_i \ddot{x}(t)\cdot\eta\delta^m + \sum_{i=1}^{n} m_i \ddot{x}_g(t)\cdot\eta\delta^m = 0 (if \delta > 0) \end{cases}$$

Where $\eta = h \cdot \beta / [2H^2 \cdot (1+\beta)]$, $\delta^v = \eta\delta^m$, $\delta^m = \delta^\theta \cdot H$ \hfill (6)

Here, $M_k$ is the bending moment at the $k$th plastic hinge, $\delta^v$ and $\delta^m$ are the virtual shearing and rotational displacement respectively, for instance (cf. Fig.3), if $\delta > 0$ the bending moment at left of the coupling beam is positive yield moment, whereas at the right of the coupling beam is the negative one.

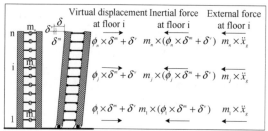
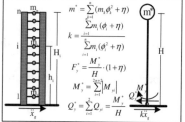

Fig. 3. MDOF system with mass $m_i$ external forces displacement in each floor.

Fig. 4. Transition from a MDO system to equivalent SDOF (ESDOF) system.

Introducing the terms defined in Fig. 3, one may rearrange Eq. (6)

Plastic behavior
$$\begin{cases} \ddot{x}(t) = -\frac{F_y^*}{m^*} - k \cdot \ddot{x}_g(t) & \text{if } \dot{x}(t) > 0 \quad and \quad x(t) > 0 \\ \ddot{x}(t) = +\frac{F_y^*}{m^*} - k \cdot \ddot{x}_g(t) & \text{if } \dot{x} < 0 \quad and \quad x(t) < 0 \end{cases}$$

Slip behavior    $\ddot{x}(t) = k \cdot \ddot{x}_g(t)$ if $\dot{x}(t) \cdot x(t) < 0$ or $M_K = 0$ with $k \in [1, n]$

Rigid behavior    $\ddot{x}(t) = 0$ if $\dot{x}(t) = 0$ or $|k \cdot x_g(t)| < 0$ \hfill (7)

In Eq. (6), Eq. (7), $\ddot{x}(t)$, $\dot{x}(t)$ and $x(t)$ are the relative acceleration velocity and displacement of the top floor, respectively. $\ddot{x}_g(t)$ is ground acceleration.

## 2.2. Rigid-plastic spectra

The dynamic response of rigid-plastic oscillators with lateral force $F_y$ and mass $m$ under a given earthquake depend on the quantity $F_y/m$ designated $a_y$. Any rigid–plastic system with a certain value of $a_y$ will develop the same dynamic response.[3] thus the plastic spectrum can be constructed. Fig.5 (a) and (b) show the rigid-plastic spectrum in terms of the relative peak of displacement denoted by the parameter $R_{max}$. The rigid-plastic spectrum of a ground motion scaled by a factor $\alpha$ is the same as the spectrum of the original ground motion with the axes $R_{max}$ and $a_y$ scaled by a factor $\alpha$. This may be demonstrated by multiplying the Eq. (5) by the factor $\alpha$.

According to Eq. (7), the dynamic response of a specific MODF system subject to a ground motion is $k$ times the dynamic response of SODF system with $a_y = F^*/km^*$ for the same ground motion. This means that the rigid-plastic spectrum can be used to estimate the seismic demand of a MDOF structure. In the following, the general rigid-plastic spectrum associated with the rigid-plastic oscillator is termed GRPS, and the one corresponding to a specific MDOF structure with chosen collapse mechanism is termed SRPS, Peak Ground Acceleration is termed PGA

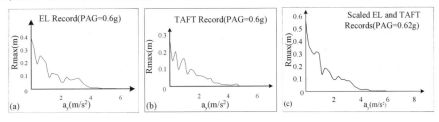

Fig. 5. Example of RPS in terms of relative peak displacement, $d_{max}$.

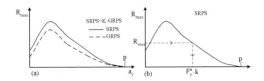

Fig. 6. (a) Determination of SRPS for MDOF structure from the GRPS, (b) Determination of $F_y$ using SRPS.

As shown in Fig.6 (a) one may determined the general strength $F_y^*$, using Eq. (7)

$$F_y^* = k \cdot a_y \cdot m^* \qquad (8)$$

The lateral force at each degree of freedom at any time $t$, $F_i(t)$ is given by Eq. (9)

Plastic behavior $\qquad F_i(t) = -m_i \cdot [\pm\varphi_i ka_y + \ddot{x}_g(t) \cdot (1 - \varphi_i k)]$

Slip behavior $\qquad F_i(t) = -m_i \ddot{x}_g(t) \cdot (1 - \varphi_i k) \qquad (9)$

## 3.   The Rigid-Plastic Seismic Design Procedure

The design procedure is as follows: Step1. Choose a suitable collapse mechanism. The guidelines leading to a suitable collapse was discussed detailedly in Ref.5. Step2. Choose the dynamic performance criterion, $R_{max}$ . The extent of damage is related to plastic deformations in the plastic zones, which in turn are related to displacements. Thus, the displacement based on the parameters should be chosen.Step3. Using the rigid-plastic spectrum defines the seismic demand. One should transform GRPS to SRPS, Subsequently, determine the value of $a_y$ from the $R_{max}$ in the SRPS. Finally the generalized yield force $F_y$ is determined by Eq.(16). Step4. Determine safe internal stress field using the extreme loading scenarios approach.[3] Step5. Design the structure, the required strength derived from the previous step should multiply an over-strength factor to ensure that a chosen collapse mechanism takes place.

## 4.   Design Example

The structure shear walls of an 11-story were designed. $b = 0.3$ m, $h = 2.65$ m. PGA is up to 0.62g. The length of coupling beam is 6m. Story heights are 3m. Mass of each floor $m_i$ is 42 ton.

Step1. The most suitable collapse mechanism for this structure is the one in Fig.3. Using the expressions in Fig.4, One can get $m^* = 175.65$ Ton, $k = 1.33$ Step2. Due to the shape of the collapse mechanism, the dynamic performance criterion may be written in terms of the maximum displacement at the top, $d_{max} = 0.33$ m. Step3. Here scaling the GRPS of Fig.5(c) by $k$ , which is 1.33, the SRPS is defined and given by the solid line in Fig.7. When $d_{max} = 0.33$m, $a_y = 1.2$m/s$^2$. If $\beta = 4$ one can get $\eta = 9.7 \times 10^{-4}$ , $M_y^* = 9241$ kNm. Step4. Using the extreme loading scenarios approach[11]and replacing $a_y$ by PAG, which is 0.62g, in Eq.9, one reaches the internal forces $F_i$ shown in table1.

Table 1. External force for design purpose.

| floor | Plastic behavior | | | | Slip behavior | |
|---|---|---|---|---|---|---|
| | Mechanism in the positive direction | | Mechanism in the negative direction | | | |
| | +PGA | -PGA | +PGA | -PGA | +PGA | -PGA |
| 11 | 153 | -18.9 | 18.9 | -153 | 86 | -86 |
| 10 | 115 | 6.5 | -6.5 | -115 | 55 | -55 |
| 9 | 78 | 32 | -32 | -78 | 23 | -23 |
| 8 | 40 | 57 | -57 | -40 | -9 | 9 |
| 7 | 2.5 | 83 | -83 | -2.5 | -40 | 40 |
| 6 | -35 | 108 | -108 | 35 | -72 | 72 |
| 5 | -72 | 133 | -133 | 72 | -103 | 103 |
| 4 | -110 | 159 | -159 | 110 | -134 | 134 |
| 3 | -148 | 184 | -184 | 148 | -166 | 166 |
| 2 | -185 | 210 | -210 | 185 | -197 | 197 |
| 1 | -223 | 235 | -235 | 223 | -228 | 228 |

According to assumption (iii) and the factor $\beta$, which is 4, one can obtain the all yield moments in the plastic hinges at the ends of coupling beams, $\sum M_y^b = 105$ kNm, and the yield moments at the base shear wall, $\sum M_y^w = 3696.4$ kNm. The required flexural strengths in the shear wall are shown in table2.

Table 2. Required flexural strength in the shear wall for $\beta = 4$ (kNm).

| Floor | | 1 | 2 | 3 | 4 | 5 | 6 | 7 | 8 | 9 | 10 | 11 |
|---|---|---|---|---|---|---|---|---|---|---|---|---|
| Cross | Top | 1420 | 1038 | 711 | 441 | 336 | 393 | 412 | 392 | 335 | 239 | 105 |
| section | Bottom | 3696 | 1315 | 933 | 606 | 236 | 231 | 288 | 307 | 287 | 230 | 134 |

Step5. Here, the tensile reinforcement ratio of shear wall was set at 0.3%, and that of coupling beam was set at 0.8%. The over-strength factor is 1.4. The axial compressive strength of concrete is 14.3MPa and the yield strength of reinforcement of steel is 300MPa. The detailing of the different cross-sections in the structure according to the value of $\beta$ is shown in table3.

Table 3. Cross-section detailing.

| Element name | Cross section(m) | | Reinforcement steel area(mm$^2$) | | | |
|---|---|---|---|---|---|---|
| | Height | Width | Plastic zones | | Outside plastic zones | |
| | | | Tensile | Shear | Tensile | Shear |
| Shear wall | 2.65 | 0.3 | 3525 | 2370 | 2518 | 3318 |
| Coupling beam | 0.8 | 0.3 | 941 | 1500 | 941 | 2180 |

## 5. Discussions

Refined NLTHA was performed for the structure using the scaled acceleration of EL record to illustrate the accuracy of the design method. The structure was modeled using integral model of RC Solid (solid65) element of compute program ANSYS.

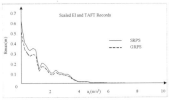
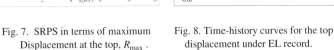

Fig. 7. SRPS in terms of maximum Displacement at the top, $R_{max}$ .

Fig. 8. Time-history curves for the top displacement under EL record.

From Fig.8 it appears that the refined NLTHA yields a maximum top displacement only 10% higher than the one estimated by the RSPD method, which is a small deviation for civil engineering. Paglietti[1] and Bento[3] explained the discrepancy. The RPSD follows performance-based seismic design and bases on the Theory of Plasticity. Rigid-plastic structures may be considered as rigid-plastic oscillators. Rigid-plastic spectra can be constructed. GRPS can be scaled up or down according to the PGA expected at the implementation site. SRPS is simply defined by multiplying the ordinates by $k$ . Because the method is base on the principles of rigid-plastic theory, any elastic behavior of structures is unable to be considered. Final design should be adjusted considering higher mode effects and serviceability limit stats against more frequent and moderate earthquake by means of other methods for dynamic analyses based on elastic theory. Examples illustrated the method provides a simple and rational tool that can facilitate the task of designing structures to survive strong earth.

## Acknowledgments

This work was supported by the National Natural Science Foundation of China under Award Numbers 10772129 to the authors, and was carried out during the graduate studies of the first author.

## References

1. Tjen N. Tjhin, Mark A. Aschheim. *Yield displacement-based seismic design of RC wall buildings*. Engineering Structures, 2911, 2946(2007).
2. Paglietti A, *Rigid-plastic approximation to predict plastic motion under strong earthquakes*. 30,115(2001).
3. Tong xi Yu. *Dynamic Models for Structural Plasticity*. (Peking University, Beijing China, 2002).
4. Domingues Costa, R. Bento. *Rigid-plastic seismic design of reinforced concrete structures*. Earthquake Engng Struct Dyn.; 36, 55(2007).
5. Taewan Kima, and Douglas A. Foutch. *Application of FEMA methodology to RC shear wall buildings governed by flexure*. Engineering Structures 29, 2514(2007).
6. Paulay T, Priestley MJN. *Seismic Design of Reinforced Concrete and Masonry Buildings*. (Wiley, New York, 1992).

# Part E
# Experimental and Numerical Techniques

# NUMERICAL STUDY ON THE SHAPED CHARGES

TIANBAO MA[1†], CHENG WANG[2], JIANGUO NING[3]

*State Key Laboratory of Explosion Science and Technology, Beijing Institute of Technology, 5 South Zhongguancun Street, Haidian District, Beijing, 100081, P.R. CHINA,*
*madabal@bit.edu.cn, wangcheng@bit.edu.cn, jgning@bit.edu.cn,*

Received 15 June2008
Revised 23 June2008

This paper developed a hydro-elasto-plastic method to describe the mechanical behavior of the metal material under explosion load. Based on this method, the EXPLOSION-2D hydrocode was developed, and the shaped charges with steel liner and copper liner were numerically simulated, respectively. The simulation results agree well with experimental ones, which show that the proposed method is efficient for engineering design.

*Keywords*: Shaped charge jet; numerical simulation; hydro-elasto-plastic method.

## 1. Introduction

A shaped charge consists in its simplest form, of an explosive with a conical cavity, lined with metal. In a more elaborate form, a detonator and an aluminum or steel casing are added[1]. With the ignition of the explosive a detonation front sweeps the liner causing it to collapse owing to the very high pressure in the gaseous detonation products. The collapse of the liner under the high pressures produces a hypervelocity jet, the fastest parts of which can reach typical speeds of 8-12km/s. This hypervelocity jet is used to penetrate targets in both military and industrial applications.

As the physics of shaped charges are very complex, numerical simulation becomes one of the important means for the study of this kind of problems. This paper developed a hydro-elasto-plastic method to describe the mechanical behavior of the metal material under explosion load. Based on this method, the EXPLOSION-2D hydrocode was developed, and the shaped charges with steel liner and copper liner were numerically simulated.

## 2. Numerical Methods

The mechanical behavior of the metal material under explosion load is very complicated. In order to provide an efficient method to describe this mechanical behavior in a hydrocode, the following numerical model was adopted.

---

[†]Corresponding Author.

The stress tensor $\sigma_{ij}$ is defined by

$$\sigma_{ij} = -P\delta_{ij} + S_{ij} \tag{1}$$

where $P$ is the hydrostatic pressure; $S_{ij}$ is the stress deviator tensor; $\delta_{ij}$ is the kronecker delta function.

Within the range of small deformations, stress deviator tensor follows the general Hooke's law,

$$
\begin{aligned}
\overset{\triangledown}{S}_{ij} &= \dot{S}_{ij} + \Omega_{ij} \cdot S_{ij} - S_{ij} \cdot \Omega_{ij} \\
&= 2G\left(\dot{\varepsilon}_{ij} - \dot{\varepsilon}_{kk}\right) + \Omega_{ij} \cdot S_{ij} - S_{ij} \cdot \Omega_{ij}
\end{aligned} \tag{2}
$$

where, the strain rate tensor and spin rate tensor are

$$\dot{\varepsilon}_{ij} = \frac{1}{2}\left(\frac{\partial u_i}{\partial x_j} + \frac{\partial u_j}{\partial x_i}\right) \tag{3}$$

$$\Omega_{ij} = \frac{1}{2}\left(\frac{\partial u_i}{\partial x_j} - \frac{\partial u_j}{\partial x_i}\right) \tag{4}$$

where $u_i$ and $u_j$ are velocity components.

The plastic flow regime is determined by the von-Mieses criterion when the equivalent stress, $J_2$, exceeds the known flow stress, $Y_0$. The individual deviators are then brought back to the flow surface.

$$S_{ij} = S_{ij}\sqrt{\frac{Y_0^2}{3J_2}} \tag{5}$$

in which,

$$J_2 = \frac{1}{2}S_{kl} \cdot S_{kl} \tag{6}$$

The flow stress $Y_0$ changes with the specific internal energy as shown in Eq.(7).

$$Y_0 = \begin{cases} Y_0\left(1 - e/e_m\right), & e < e_m \\ 0, & e \geq e_m \end{cases} \tag{7}$$

When the specific internal energy $e$ exceed $e_m$, the flow stress $Y_0$ is set to zero. Accordingly, the individual deviators are changing to zero. In this case, the metal is treated as fluid.

The above mentioned method is hydro-elasto-plastic model, which is very efficient to describe the mechanical behavior of the metal material under explosion load.

For the hydrostatic pressure $P$, the Mie-Gröneisen equation of state for metal liner can be expressed as

$$P = \begin{cases} \dfrac{\rho_0 C^2 \mu \left[1 + \left(1 - \dfrac{\gamma_0}{2}\right)\mu - \dfrac{a}{2}\mu^2\right]}{\left[1 - (S_1 - 1)\mu - S_2 \dfrac{\mu^2}{\mu+1} - S_3 \dfrac{\mu^3}{(\mu+1)^2}\right]^2} + (\gamma_0 + a\mu)E & (\mu \geq 0) \\[4mm] \rho_0 C^2 \mu + (\gamma_0 + a\mu)E & (\mu < 0) \end{cases} \tag{8}$$

where $E$ is the internal energy per unit volume; $\mu = \rho / \rho_0 - 1$, $\rho_0$ is the initial density; $C$ is the intercept of the $u_s$-$u_p$ curve; $S_1$, $S_2$, $S_3$ are the coefficients of the slope of the $u_s$-$u_p$ curve; $\gamma_0$ is the Grüneisen gamma, and $a$ is the first order volume correction to $\gamma_0$.

Based on this model, we developed a multi-material Eulerian hydrocode named EXPLOSION-2D[2]. In this code, the operator splitting method is employed, that is to say that the calculation for a given time step involves two phases. The first phase is a Lagrangian phase in which the mesh is allowed to distort with the material. In the second advection phase, transport of mass, internal energy and momentum across mesh boundaries is computed. An adaptive grid subdivision algorithm based on the Youngs' interface reconstruction algorithm[3] is developed to treat the multi-material interfaces.

For the simulation of shaped charge jet, it involves metal liner, explosive and air in the computational region. For the air, the ideal gas equation of state is adopted. For explosive products, the following JWL (Johns-Wilkins-Lee) equation of state is

$$P = A(1 - \frac{\omega}{R_1 V})e^{-R_1 V} + B(1 - \frac{\omega}{R_2 V})e^{-R_2 V} + \frac{\omega E}{V} \tag{9}$$

where $E$ is the internal energy of per unit volume of explosive products, $V$ is the relative volume of explosive products, $A$、$B$、$R_1$、$R_2$、$\omega$ are the known constants.

For the detonation of high explosive, the programmed burn technique, which is a simple method and frequently used in hydrocode calculations, is adopted in this paper[4].

## 3. Experimental Configuration

In the experiment, the high explosive Comp-B was used for the charge. The length of the charge was 110mm, with a diameter of 60mm. The liner materials were steel and copper, respectively. The conical angle of the liner was 60°, with a thickness of 2.4mm and diameter of 60mm, as shown in Fig.1.

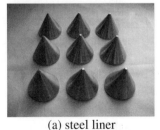

(a) steel liner               (b) copper liner

Fig. 1. Photos of the shaped charge liner.

The facility was placed in an experimental shelter. Two channels of flash X-ray systems were equipped to take images of the jet in the locus of its flight at different positions. The experimental setup of shaped charge jet was shown in Fig.2.

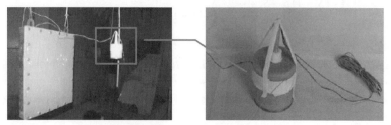

Fig. 2. An experimental setup of shaped charge jet.

## 4.  Numerical Results

Fig.3 and Fig.4 illustrate the processes of expansion of detonation products and deformation of the liner by the VISC-2D code[5]. It can be seen from the figure that, when detonation front reaches the liner, the liner is subjected to the intense pressure of the front and begins to collapse. The upper region of the cone has collapsed and collided on the axis of symmetry. This collision results in liner material being extruded. The extruded material is called the jet. Thus, the cone collapses progressively from apex to base, and a part of liner flows into the jet, while other part of the liner forms the slug that is the large massive portion at the rear of the jet.

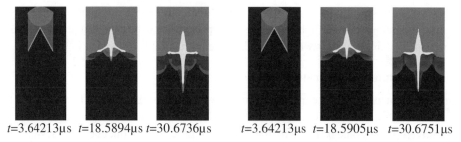

$t=3.64213\mu s$  $t=18.5894\mu s$ $t=30.6736\mu s$                $t=3.64213\mu s$   $t=18.5905\mu s$   $t=30.6751\mu s$

Fig. 3. Image of the shaped charge jet with steel liner.              Fig. 4. Image of the shaped charge jet with copper liner.

The comparison of the experimental images by radiography and the numerical results are shown in Fig.5 and Fig.6. It can be seen from these images that the shapes of the numerically simulated jets are consistent with the experimental ones.

The comparison of the movement distance of the tip of the jet between the above two recorded moments is listed in table.1. It can be shown that the numerical results are in good agreement with the experimental data, which indicates that the numerical method proposed in this paper is efficient for the simulation of shaped charge jet.

$t=41.4\mu s$   $t=25.5\mu s$

Image by radiography          numerical results

Fig. 5. Numerical results vs. experimental results
          (steel liner).

 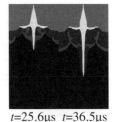

$t=25.6\mu s$   $t=36.5\mu s$

Image by radiography          numerical results

Fig. 6. Numerical results vs. experimental results
          (copper liner).

Table 1. Comparison of tip velocity with different liner material and different liner conical angle.

| | | $t_1(\mu s)$ | $t_2(\mu s)$ | the tip's movement distance (mm) | average tip velocity (m/s) | error |
|---|---|---|---|---|---|---|
| steel liner | experimental data | 25.5 | 41.4 | 83.9 | 5276.7 | 1.53% |
| | numerical results | | | 85.2 | 5358.5 | |
| copper liner | experimental data | 25.6 | 36.5 | 60.3 | 5532.1 | 1.49% |
| | numerical results | | | 61.2 | 5614.7 | |

Furthermore, the shaped charge jets with different liner conical angle were numerically simulated. The numerical data at the moment of the jet's tip reaching to the position two times of the charge diameter were selected to be compared. The images of the shaped charge jet at this moment are shown in Fig.7 and Fig.8, respectively. The tip velocity and movement time are listed in table.2 and table.3, respectively. It can be seen from the tables that with the increase of the conical angle of the liner, the tip velocity of the jet is decreased, so the movement time is longer, which is consistent with the jet formation theory. This can be explained that the initial pressure on the surface of the liner is related to the included angle between the detonation wave front and the surface of liner. With the decrease the liner conical angle, the included angle will decrease, and the detonation load will increase, which leads to the increase of the collapse velocity of the liner element. Therefore, smaller liner conical angle can get higher jet velocity.

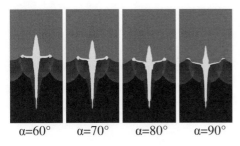

$\alpha=60°$      $\alpha=70°$      $\alpha=80°$      $\alpha=90°$

Fig. 7. Image of steel liner jets with different conical
          angle of liner.

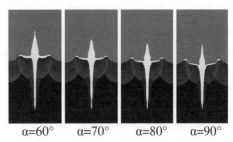

$\alpha=60°$      $\alpha=70°$      $\alpha=80°$      $\alpha=90°$

Fig. 8. Image of copper liner jets with different
          conical angle of liner.

Table 2. Comparison of tip velocity with different liner material and different liner conical angle.

| liner material | tip velocity (m/s) | | | |
|---|---|---|---|---|
| | 60° | 70° | 80° | 90° |
| steel | 4902.44 | 4290.57 | 3829.09 | 3403.9 |
| copper | 4925.45 | 4413.01 | 3882.35 | 3446.14 |

Table 3. Comparison of movement time with different liner material and different liner conical angle.

| liner material | movement time (μs) | | | |
|---|---|---|---|---|
| | 60° | 70° | 80° | 90° |
| steel | 35.9383 | 37.509 | 39.039 | 42.3604 |
| copper | 36.6428 | 38 | 39.3231 | 42.5191 |

## 5.  Summary and Conclusions

This paper developed a hydro-elasto-plastic numerical method to describe the mechanical behavior of the metal material under explosion load. Based on this method, the EXPLOSION-2D hydrocode was developed, and the shaped charges with different liner material and conical angle were numerically simulated, respectively. The simulation results agree well with experimental ones, which show that the proposed method is efficient and can be used for engineering design.

## Acknowledgments

The support of National Natural Science Foundation of China under grant number 10625208 is gratefully acknowledged.

## References

1. J. F. Molinari, Finite Elements in Analysis and Design, 38, 921–936(2002).
2. T. B .Ma, Ph.D. Thesis, Beijing Institute of Technology, Beijing, 2007.
3. D. L. Youngs, in Numerical Methods for Fluid Dynamics, ed. K. W. Morton et al. (Academic Press, 1982), pp. 273–285.
4. J. G. Ning, C. Wang, T. B. Ma, International Journal of Nonlinear Sciences and Numerical Simulation, 7 (1), 71–78(2006).
5. W. Y. Zhang, J. G. Ning, Z. G. Zheng, Journal of Beijing Institute of Technology (English Edition), 8(4), 424–429(1999).

# EXPERIMENTAL STUDY ON MECHANICAL PROPERTY OF STEEL REINFORCED CONCRETE L-SHAPED SHORT COLUMNS

ZHE LI[1][†], HAO QIN[5]

*Department of Civil Engineering, Xi'an University of Technology, Shaanxi, Xi'an, 710048, CHINA,*
*lizhe009@163.com, qinhao9095@163.com*


HUI DANG[2]

*Department of Civil Engineering, Xi'an University of Technology, Shaanxi, Xi'an, 710048, CHINA,*
*Shaanxi Taier Industry Co., Ltd, 710600, CHINA,*
*Danghui0129@126.com*


HUI LI[3]

*Department of Civil Engineering, Xi'an University of Technology, Shaanxi, Xi'an, 710048, CHINA,*
*Shenzhen-Construction Company Limited, 518005, CHINA,*
*lihui@163.com*


JIAN-SHAN ZHANG[4]

*China Ji Kan Geotechnical Institute, 710048, CHINA,*
*zhangjs@jk.com.cn*

Received 15 June 2008
Revised 23 June 2008

The horizontal press performance of column is deteriorated because of its special-shaped section. Moreover, because the antiseismic performance of column is worse, special-shaped column is only used in regions where seismic intensity is lower. So the main problem is to enhance the ductility and shear capacity. This test study on mechanical performance has been carried out through 14 SRCLSSC and 2 RCLSSC. The study focuses on the impacts of test axial load ratio ($n_t$), hooped reinforcement ratio ($\rho_v$), shear span ratio ($\lambda$) and steel ratio ($\rho_{ss}$) on the shear strength and the antiseismic performance of SRCLSSC. It can be concluded that the shear strength of SRCLSSC is increasing with the increasing of $n_t$ and $\rho_{ss}$, but the degree of increasing is small when $n_t$ is a certainty value, and that the shear strength of SRCLSSC is decreasing with increasing of $\lambda$; The shear resistance formula of L-shaped column is derived through tests, the calculated results are in correspondence with those of the tests. It also can be concluded that the hysteretic loops of the SRCLSSC are full and the hysteretic behaviors are improved; the displacement ductility is increasing with increasing of $\rho_v$ and $\rho_{ss}$, but decreasing with the increasing of $n_t$; the degree of variety in high axial load ratio is larger than that in low axial load ratio. If steel bars are added, the shear strength and displacement ductility of SRCLSSC are increased in a large degree.

*Keywords*: Steel reinforced concrete L-shaped short column (SRCLSSC); Reinforced concrete L-shaped short column (RCLSSC); Displacement ductility; shear strength.

[†]Corresponding Author.

## 1. Introduction

Special-column structure emerged in the late 70s of the 20th century. Recent years witness its rapid development. Related researches have born considerable fruits. While, being weak in web plate, flange plate, and asymmetric in section, special-column structure is prevented from wide application for its poor supporting capacity and weak ductility. Particularly in higher seismic intensity regions, wide application of special-column is severely restricted. The regulation[1] stipulates the height of portal frame structure is no more than 12m and the height of frame-shearing for a wall structure is less than 28m in 8 ⁰seismic intensity areas. Refs. 2, 3 show that ductility of steel reinforced T-shaped short columns could be optimized in a large degree when steel bars are added in. Therefore, it is necessary to conduct experimental study on the mechanical properties of steel reinforced L-shaped short columns.

## 2. Test General Situation

### 2.1. *Design of members and test parameters*

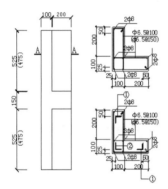

The section of the experimental column is shown in Fig1. According to the analytical results of some steel reinforced concrete short columns of special-shape, the parameters of

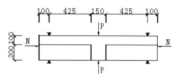

Fig. 1. Dimensions and reinforcement details of specimen.

Fig. 2. Test "loading plan".

section design can be given as: limb width is 300mm and limb thickness is 100mm; specimen adopts the compound stirrups, and the interval of them is 75mm, 100mm and 150mm; the steel in the experimental column adopts angle steel of rolling; the models of angle steel are L40×40×5,L50×50×5 and L75×50×5;The length of column is fixed as 1100mm and 1200mm. Considering the main factors influencing on short column and the demands of code, the main parameters and their change ranges are decided: $\lambda=H/2h_0=$ 1.62 ,1.78; $\rho_{ss}$=2.75% (as figure 1 shows: the steel L40×40×5 is situated in ①and the steel L50×50×5 is situated in ②), $\rho_{ss}$=3.20% （three pieces of steel L50×50×5 are situated in ① and ②） and 3.79%(the steel L75×50×5 is situated ① and the steel

L50×50×5 is situated in ②); $n_t = N / (f_c A_c + f_{ss} A_{ss})$ = 0.24, 0.36, 0.48;stirrups ratio per unit volume $\rho_v$=0.62%, 0.93%,1.24%. Considering the demands of code, concrete intensity is C40, and specimen adopts P.042.5R silicate cement, 5mm~15mm macadam and medium sand in Ba river; Mixed ratio is 1:1.49:2.76 (cement: sand: cobblest), and the ratio of water to plaster is 0.48; The parameter of specimens is shown in table.1.

## 2.2. *Loading device and system*

Simply-supported beam loading model is adopted in this test. First, the maximal axial load is located in the center of section with geometry method, and then the low cyclic horizontal load is exerted. Test "loading plan" is shown in Fig.2. Loading system adopts variable-amplitude and variable- displacement systems, and the horizontal load circulates for three times under every fixed displacement. Loading control is adopted before the crack occurs. The loading doesn't stop until the first crack appears. The method of displacement control is adopted after the crack occurs. Under the control of different times of ⊿, loadings are repeated and loading doesn't stop until the horizontal strength decreases to 85 percentage of the maximal strength in a certain cycle. The specimen is considered to be damaged at that time.

## 3. Analysis on Test Results

### 3.1. *Analysis on failure*

The test results show two kinds of failure models of steel reinforced concrete L-shaped short columns: shear-compression failure and shear-coherent failure.

(1)     Shear-compression failure: This kind of failure occurs under the lower test axial load ratio. As shown in fig.3.

(2)     Shear-coherent failure: This kind of failure occurs under the higher test axial load ratio and lower hooped reinforcement ratio. Cracks in shear-coherent failure are shown in Fig.4.

Fig. 3. Shear-compression failure.     Fig. 4. Shear-coherent failure.

### 3.2.  Ductility of short L-shaped steel reinforced concrete columns

#### 3.2.1. Definition of ductility3

#### 3.2.2. The main influential factors on the ductility

(1)  Axial load ratio: Axial load ratio not only influences the characteristics of failure and deformation capability, but also influences supporting capacity. Meanwhile, it is one of the main factors which influence displacement ductility of steel reinforced concrete L-shaped columns. The fig.5 illustrates the relationship between ductility ratio ($\mu_r$) and axial load ratio ($n$) under different steel ratios. According to these figures, the displacement ductility ratio is decreasing with the increasing of axial load ratio; the extent of decreasing is larger under low axial load ratio than that under high axial load ratio.

(2)  Steel ratio: According to comparison of RCLSSC with SRCLSSC, the displacement ductility ratio of reinforced concrete L-shaped short columns has been improved after adding steel into columns. As figure 5 shows: with the increase of steel ratio, ductility ratio increases.

(3)  Hooped reinforcement ratio: Stirrup can bind up concrete around steel and make concrete cooperate with steel to increase ductility of frame. Also, the increasing of hooped reinforcement ratio can improve the strength and deformation capability of member. According to influence curve of $\rho_v$ in Fig.6, under the same condition, the displacement ductility is increasing with $\rho_v$ increasing. But under the high axial load ratio condition, the influence of $\rho_v$ on displacement ductility is smaller than that under the low axial load ratio condition.

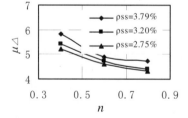

Fig. 5. Influence curves of axial load ratio $n$ on ductility coefficient $\mu_r$.

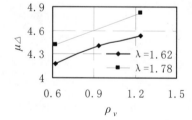

Fig. 6. Influence curves of hooped reinforcement ratio $\rho_v$ on ductility coefficient $\mu_r$.

## 4. The Analysis on Influential Factors on Shear Resistance of SRCLSSC

### 4.1. *The main influential factors on the shear capacity*

(1) Steel ratio .It can be concluded from the experiments that steel contributes a lot to the shear capacity. Shear capacity of SRCLSSC is improved with the increase of steel ratio. Fig.7 illustrates that shear capacity of L-shaped section short columns increases evenly about 12% when steel ratio rises from 2.75% to 3.79%. Based on this data, we can conclude that shear capacity is improved dramatically with the increase of steel ratio.

(2) Axial load ratio. Axial load ratio exerts great influence on shear capacity of column, meanwhile demonstrating considerable influence on its failure models. Evidently, shear capacity of special-shaped column is affected subtly by the change of axial load ratio. The experiments show that within two kinds of steel ratio, shear capacity of experimental column increases to some extent with the increase of axial load ratio. As figure 8 shows.

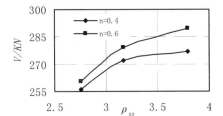
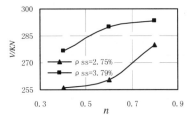

Fig. 7. Influence curves of steel ratio $\rho_{ss}$ on shear strength V.

Fig. 8. Influence curves of axial pressure ratio n on shear strength V.

(3) The influence of shear span ratio. Shear span ratio determines the failure model of the short steel reinforced concrete special-shaped column. The shear strength of SRCLSSC is decreasing with the increasing of $\lambda$ and the degree of decrease is larger.

### 4.2. *The calculated formula for shear capacity*

Based on intensity superimposition theory and related special-column Standard,[4, 5, 6] the author comes up with suggested calculated formula for the shear capacity. The shear capacity formula for L-shaped steel reinforced column is listed below:

$$V \leq V_u = \frac{1.75}{\lambda+1.0} f_t b_c h_{c0} + f_{yv} \frac{A_{sv}}{s} h_{c0} + 0.07 N_{rc} + V_{sw} \tag{1}$$

Where: $V$ stands for shear design value of L-shaped steel reinforced concrete; $V_u$ is the shear capacity of steel reinforcement concrete L-shaped short columns. $N_{rc}$ signifies the axial design value of reinforced concrete, $N_{rc}=Nf_cA_c/ (f_cA_c+f_{ss}A_{ss})$, when N>0.3$f_cA_c$, N=0.3$f_cA_c$; when $\lambda$<1, $\lambda$=1.0, and when $\lambda$>3.0, $\lambda$=3.0; $b_c$ is section thickness of calculated column limb, $h_{c0}$ is section height of calculated column limb. $V_{sw}$ stands for the shear capacity of some steel. Formula (2) is applied in operation:

$$V_{sw} = \frac{1.0}{\lambda + 2.0} f_{ss} t_w h_w \qquad (2)$$

Where: $V_{sw}$ is shear capacity undertaken by web plate of angle steel; $f_{ss}$ is design value of anti-pull intensity undertaken by web plate of angle steel; $t_w$ is thickness of web plate of shaped steel; $h_w$ is height of web plate of angle steel which lies in the same direction with loading direction.

Angle steel of rolling is adopted in the tests. Based on code, formula (2) is applied in the calculation of shear capacity. According to the suggested formula derived from the tests, the calculated results of section shear capacity of steel reinforced concrete L-shaped short columns are in accordance with those of the tests. The test value of steel reinforced concrete L-shaped short column and suggested calculated formula are shown in table1.

Table1. Parameters of specimen and contrast of test results and calculated results.

| Specimen number | $f_{cu}$ /Mpa | $\lambda$ | $\rho_v\%$ | $n_t$ | $\rho_{ss}\%$ | $V_{cr}$ | $V_u$ | $V_j$ | $V_u/V_j$ | $N(KN)$ | $\mu_\Delta$ | Failure pattern |
|---|---|---|---|---|---|---|---|---|---|---|---|---|
| L93-08-37 | 49.8 | 1.62 | 0.93 | 0.48 | 3.79 | 250 | 293.3 | 271.8 | 1.079 | 931 | 4.71 | Shear-coherent |
| L93-06-37 | 43.8 | 1.62 | 0.93 | 0.36 | 3.79 | 253.2 | 289.8 | 265.5 | 1.092 | 715 | 4.88 | Shear-compression |
| L93-04-37 | 44.5 | 1.62 | 0.93 | 0.24 | 3.79 | 233.4 | 276.7 | 266.2 | 1.039 | 640 | 5.83 | Shear-compression |
| L93-08-27 | 43.8 | 1.62 | 0.93 | 0.48 | 2.75 | 226.5 | 280 | 240.6 | 1.164 | 931 | 4.32 | Shear-coherent |
| L93-06-27 | 47.6 | 1.62 | 0.93 | 0.36 | 2.75 | 216.7 | 260.7 | 244.6 | 1.066 | 715 | 4.61 | Shear-coherent |
| L93-04-27 | 44.5 | 1.62 | 0.93 | 0.24 | 2.75 | 193.3 | 256.2 | 241.3 | 1.062 | 640 | 5.21 | Shear-compression |
| L93-08-32 | 49.8 | 1.62 | 0.93 | 0.48 | 3.2 | 193.3 | 280.1 | 257.7 | 1.087 | 931 | 4.41 | Shear-coherent |
| L²93-08-32 | 47.6 | 1.62 | 0.93 | 0.48 | 3.2 | — | 285.6 | 255.4 | 1.118 | 931 | 4.72 | Shear-coherent |
| L93-04-32 | 44.5 | 1.62 | 0.93 | 0.24 | 3.2 | 203.3 | 272.1 | 252.1 | 1.079 | 640 | 5.42 | Shear-compression |
| L62-06-32 | 43.8 | 1.62 | 0.62 | 0.36 | 3.2 | 216.7 | 279.4 | 226.4 | 1.234 | 715 | 4.18 | Shear-compression |
| L12-06-32 | 49.8 | 1.62 | 1.24 | 0.36 | 3.2 | 250 | 282.3 | 282.6 | 0.999 | 715 | 4.55 | Shear-coherent |
| $L_t$12-06-32 | 47.6 | 1.78 | 1.24 | 0.36 | 3.2 | 210 | 263.1 | 274.4 | 0.959 | 715 | 4.83 | Shear-coherent |
| $L_t$62-06-32 | 47.6 | 1.78 | 0.62 | 0.48 | 3.2 | 246.7 | 260 | 224.5 | 1.158 | 715 | 4.42 | Shear-coherent |
| L93-06-25 | 43.8 | 1.62 | 0.93 | 0.36 | 2.52 | 300 | 322.2 | 239.9 | 1.343 | 715 | 5.33 | Shear-compression |
| L²93-06-0 | 44.5 | 1.62 | 0.93 | 0.36 | 0 | — | 230 | 175.2 | 1.313 | 715 | 3.56 | Shear-compression |
| L93-06-0 | 49.8 | 1.62 | 0.93 | 0.36 | 0 | 200 | 227.3 | 180.7 | 1.273 | — | — | Shear-compression |

Note：(1) $V_{cr}$, $V_u$ and $V_j$ represent the shear strength as skew crack appearing, the test result and calculated result of the shear strength (kN); The average of $V_u/V_j$ is $x = 1.06$, $\sigma=0.089$; (2) The first number 93 of T93-08-37 represents hooped reinforcement ratio $\rho_v$ =0.93%, the second number 08 represents axial load ratio ($n=1.66n_t$, $n_t$ represents experimental axial load ratio) $n=0.8$, the third number 37 represents steel ratio $\rho_{ss}$=3.79%. N represents axial pressure of members; $\mu_\Delta$ represents displacement ductility ratio.

## 5.  Conclusions

(1)  Steel reinforced concrete L-shaped short column shows two kinds of failure models: the shear-compression failure and the shear-coherent failure.

(2)  When steel is added, the ductility performance and shear capacity of SRCLSSC improve greatly.

(3)  Shear span ratio, as well as hooped reinforcement ratio, exerts some influence on shear capacity. Under the same parameters, the specimen with great shear span ratio shows great ductility and poor shear capacity. Furthermore, as the increase of hooped reinforcement ratio, ductility, in addition to shear capacity, improves to some extent. So, in design, shear span ratio and hooped reinforcement should draw our attention.

(4)  The shear strength formula of SRCLSSC is derived through test, and the calculated results are in good agreement with those of the test. It can provide the scientific foundation to apply the special-shaped column to the higher intensity regions

## References

1. JGJ149-2006 (2006).
2. Z.Li, X.F.Zhang.Int.J.Civil Eng., **40**, 1(2007).
3. Z.Li. Ph.D. Thesis, Xi'an Univ. of Tech, 2007.
4. L.G.Wang Lianguang, L.X. Li Int.J. Build Struct., **31**, 23(2001).
5. YB9082-97 (1998).
6. Z.Li, X.F.Zhang, Z.Y.Guo, Int.J. K.Eng.Mat (2007).

# MICROSTRUCTURE AND CRYSTALLOGRAPHIC TEXTURE OF STRIP-CAST FE-3.2%SI STEEL SHEET

Y.B.XU[1†], Y.M.YU[2], G.M.CAO[4], C.S.LI[5], G.D.WANG[6]

*State Key Laboratory of Rolling and Automation, Northeastern University,*
*Shenyang, Liaoning Province, 110004, P. R. CHINA,*
*xuyunbo@mail.neu.edu.cn*

Z.Y. JIANG[3]
*Faculty of Engineering, University of Wollongong,*
*Wollongong NSW 2522, AUSTRALIA,*
*jiang@uow.edu.au*

Received 15 June 2008
Revised 23 June 2008

Fe-3.2%Si steel strips were produced using vertical type twin casting process, and the changes of microstructure and texture trough thickness direction were analyzed. The equiaxed grains of approximately 44.6μm were observed in the center layer, the great mass of columnar dendrite was formed near the surface, and the dendrite truck mainly developed in the transverse direction with respect to the casting direction of about 45° or less. From the subsurface to the center, the volume fraction of the Goss texture (110)[001] gradually decreases. The Goss {110} <001> components at the surface are two times those at the center layer, the {001} <100> components are three times those at the center layer, and the overall texture components are similar to that of the hot-rolled oriented silicon steel strip. The minor α texture could be found from the $\varphi_2=45°$ sections of ODF, and there is no remarkable composition segregation of Si element in the thickness direction of thin strip.

*Keywords*: Fe-3.2%Si; twin-roll strip casting; microstructure; texture; equiaxed grain.

## 1. Introduction

Silicon steel is a soft magnetic material that is used in electric power transformers, motors and generators. It can often be divided into oriented silicon steel and non-oriented silicon steel according to production technology. Although conventional manufacturing process has been well established, the process is still complicated, which would lead to a high cost of production. To shorten the production steps, strip-casting technology has been proposed.[1] This process would eliminate the continuous casting and hot-rolling processes in the conventional manufacturing and could offer steel sheets with the same

---

†Corresponding Author.

thickness and width as hot rolled products.[2] In the strip-casting process, however, it is of most importance to control the microstructure and texture of materials.

Grain-oriented silicon steel that is used for non-rotating applications, i.e. transformers, is characterized by a pronounced Goss texture, i.e. a {110} <001> preferred crystal orientation.[3] Since grain oriented silicon steel sheet was invented in 1935 by Goss[4], a lot of effort has been made to improve the sharpness of the Goss texture. In the conventional manufacturing process, the (110) [001] preferred orientation often forms in the layer as deep as 1/5 to 1/4 thickness of hot-rolled sheets, and hence the texture must be controlled strictly in each processing step.[5] The solidified and deformed conditions during strip casting are quite different from the conventional case, and thus, it is essential to study the evolution of microstructure and texture of twin-roll casting strip. Raabe *et al.*[6] have reported the variation of texture in stainless steel strip produced by the twin-roll casting process.

In the present study, Fe-3.2% Si strips were processed using the vertical type twin-roll strip caster, the microstructure and texture trough thickness direction were studied and the texture of subsurface was compared with those of hot-rolled strip.

## 2. Material and Experimental Procedure

Table 1. Chemical composition of as-cast strip (wt %).

| C | Si | Mn | Cu | S | Al | N | Fe |
|------|-----|------|------|--------|-------|---------|------|
| 0.042 | 3.2 | 0.14 | 0.02 | <0.025 | 0.078 | <0.0075 | Bal. |

Table 1 shows the chemical composition of the silicon steel used in this investigation. The experiments were carried out in Ar gas atmosphere by a laboratory twin-roll caster with internally water-cooled steel rolls of 500mm diameter and 300mm width. The thickness and width of the strip were 2.56mm and 250mm, respectively.

The specimens for metallographic examination were etched with 4%Nitric acid and 96% ethanol. The solidification structure through the thickness of cast strip was observed from the transverse and longitudinal directions by LEICA DMIRM optical microscope and the Si distribution in the thickness direction of as-cast strip was investigated by scanning electron microscopy (SEM).

The texture evolution through the thickness of the as-cast strip was observed, particularly at the subsurface (approximately 1/4~1/5 thickness of the sheet) and center layer. The crystallographic texture was quantitatively examined by measuring the three incomplete pole figures {110}, {200}, {211} at the PW3040160 X'Pert Pro MRD. The orientation distribution function (ODF) was calculated from the measured pole figures by two-steps method. The micro-texture of the as-cast Fe-3.2%Si strip from subsurface to center layer was examined by the electron back-scattered diffraction (EBSD).

## 3. Results and Discussion

### 3.1. *Microstructure*

The procedure of twin-roll strip casting mainly includes primary cooling (crystal solidification) and secondary cooling. Primary cooling parameters, such as casting temperature, casting rate, flow of cooling water and casting-strip thickness, have a significant influence on microstructure. As an example, the solidification morphology, size and distribution of structure can be varied by adjusting the cooling rate. Fig. 1 shows the schematic diagram of melt pool in twin-roll casting process. The melt pool is generally divided into three parts, such as solid layer, mushy zone and liquid pool. The solidifying shells, which are formed during the contact with rolls, meet at the solidification end point. This point is dependent on the casting temperature and rolling speed. The station time and phase transformation occurring during secondary cooling affects the final microstructure. In addition, the parameters determining the microstructure in the cast strip also comprise the superheat, the casting speed, the heat transfer characteristic between the solidifying strip and the rolls, and the flow characteristics of the incoming liquid metal. [7]

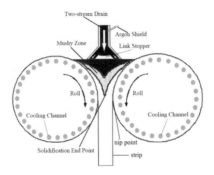

Fig. 1.  Schematic diagram of melt pool in twin-roll strip casting process.

Fig.2 shows the optical micrograph of the strip-cast 3.2%Si steel on the longitudinal and transverse sections throughout the thickness from surface down to center of the sheet. The strip-cast silicon steel sheet displays an inhomogeneous microstructure through the thickness direction. The equiaxed grains of approximately 44.6μm in the center layers and the columnar grains near the surface region were observed. This is closely related to the twin-roll casting process. When the liquid metal contacts with rolls, the liquid metal firstly nucleates at surfaces of rolls due to chilling, and then the columnar dendrites grows towards the maximal heat-flux direction. But the solidification takes place in a narrower region as columnar dendrites increase, which would lead to the variation in cooling intensity and temperature gradient of dendrite front. It is clearly that the heat-transfer coefficient of strip surfaces abruptly decreases after the strip leaves the roll nip, which would provide an opportunity for the formation of equiaxed-grains. That is to say, the unsolidified layer solidifies under slow cooling rate to form equiaxed crystal zone in the strip center by the preferential growth of free crystal ahead of the dentritic solidification front.[8, 9] The presence of relatively coarser equiaxed grains in solidification structure is considered to be associated with the relatively larger initial roll gap (1.9mm). This can be explained by the fact that the equiaxed zone generally increases greatly with increasing the initial roll gap.[10]

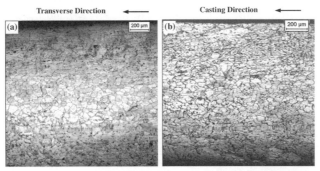

Fig. 2. Fe-3.2% Si steel solidification structure of cast strip on the transverse (a) and longitudinal (b) sections.

It can also be seen from Fig.1 that, the columnar dendrites region is symmetric, which is due to identical twin-roll conditions, such as cooling rate and rough face etc. The dendrites are known to grow with an inclination towards the upstream of molten metal flow parallel to the solidification front. In the present case, the dendrite truck developed in the transverse direction with respect to the casting direction of about 45° or less. Twin-roll casting is a dynamic process, and dendrite directions are determined by various factors, such as liquid flowing, heat transfer characteristic, mass transfer and deformation conditions etc, in which the effect of liquid flowing should be of most significance.

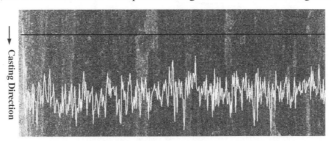

Fig. 3. SEM line profile of Si as-cast Fe-3.2%Si strip.

Fig.3 shows the SEM line profile along the thickness direction, indicating that no remarkable segregation of Si in the cross section of the as-cast strip. Twin-roll strip casting is usually regarded as a sub-quick solidification procedure at the cooling rate of approximately $10^3$ k/s. Therefore, with the progress of solidification the equilibrium distribution coefficients of solute exhibit a deviation from equilibrium state, but, whether these values are greater than 1 or not, the actual partition coefficients of solute would approach unity, that is, the segregation would decrease with increasing the solidification rate.

### 3.2. Texture

Fig.4 shows crystal orientation map and inverse pole figure of the as-cast Fe-3.2%Si strip. Columnar crystals of dendritic structure mainly develop on the surface, and equiaxed crystals are mostly observed near the center thickness. The transverse direction is parallel to the <113>, the rolling direction is parallel to the <112>, and the minor $\alpha$ texture is observed, which is consistent with the ODF of the subsurface and center layers.

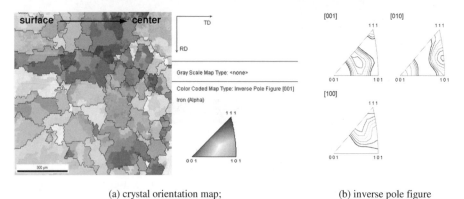

(a) crystal orientation map;                                    (b) inverse pole figure

Fig. 4. EBSD results of the as-cast Fe-3.2%Si strip from the surface to the center layer.

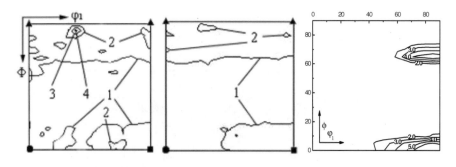

(a) subsurface of as-cast strip;    (b) center layer of as-cast strip;    (c) subsurface of hot-rolled strip

Fig. 5. $\varphi_2=45°$ ODF sections of the as-cast and hot-rolled 3.2% Si steel strips.

▲ = {001}<110>;■={110}<001>;● = {110}<110>

Fig.5(a-b) shows the measured ODFs at two positions in the thickness of the as-cast 3.2%Si steel strip. The volume fractions of the Goss {110} <001>components and brass texture components are two times those of the center layer and the {001}<100> components are about three times that of the center layer. S texture doesn't occur in the center layer and the minor $\alpha$ texture is observed in the two layers. It is well known that {100} <0vw> fiber component is the typical solidification texture, but the Goss {110}

<001> component is the typical shear texture of BCC metals. The α and γ-fibers are the typical rolling textures according to plane strain deformation. The texture components in the subsurface are similar to that of the hot-rolled sheet, but the textures of the former are stronger. Fig.5(c) shows $\varphi_2$=45° ODF section in the subsurface of the hot-rolled Fe-3.2%Si steel. As we known, the inhomogeneous texture gives the origin of Goss texture in final products of silicon steel and is closely connected with the heterogeneous deformation along the thickness due to shear deformation. This formation reason of the heterogeneous deformation is markedly different for as-cast strip and hot-rolled strip. During strip casting, the temperature gradient through the thickness direction gives rise to the strength gradient and finally the large shear deformation.[8] But, the shear deformation in hot-rolled strips is mainly related to the heavy friction between the rolls and strips.

## 4. Conclusion

The inhomogeneous microstructure distribution through the thickness of strip cast 3.2%Si steel has been examined. The equiaxed grains of about 44.6μm in the center layers and the columnar grains near the surface region were observed, the dendrite truck developed in the transverse direction with respect to the casting direction of about 45°or less. The minor α texture was observed from crystal orientation maps and inverse pole figures, which is consistent with the ODFs of the subsurface and center layer. The $\varphi_2$=45° sections of ODF in the subsurface is similar to that of hot-rolled strip. The volume fractions of the Goss{110}<001>components and brass texture components are two times those of the center layer, the {001}<100> components are three times that of the center layer, S texture doesn't occur in the center layer and the minor α texture is observed in both layers. The SEM line profile along the thickness direction shows that, no remarkable segregation of Si in the cross section of the as-cast strip.

## Acknowledgments

This work was financially supported by the National Natural Science Foundation of China under contract No. 50504007 and No.50534020.

## References

1. H. Litterscheidt, R. Hammer, C. Schneider, R. W. Simon, D. Senk, R.Kopp and B. Hehl, *Stahl Eisen*, **111**, 61 (1991).
2. K.Günther, G.Abbruzzese, S.Fortunati, *the Iron and Steel of World*, **5**, 1(2005).
3. D.Dorner, S.Zaeffer, L.lahn, D.Raabe, *J.Magn.Magn.Mater.*, **304**, 183(2006).
4. N.P.Goss,*Trans. Am. Soc. Met.*, **23**, 511(1935).
5. M.Matsuo, *ISIJ Int.*, **29**, 809 (1989).
6. D. Raabe, F. Reher, M. Hölscher, and K. Lücke, *Scripta Metall.*, **29**, 113 (1993).
7. J.Y.Park, K.H.Oh and H.Y.Ra, *ISIJ Int.*, **40**, 1210(2000).
8. J.Y.Park, K.H.Oh and H.Y.Ra, *Scripta Mater.*, **40**, 881(1999).
9. J.Y.Park, K.H.Oh and H.Y.Ra, *ISIJ Int.*, **41**, 70(2001).
10. T.Mizoguchi and K.Miyazawa, *ISIJ Int.*, **35**,771(1995).

# FINITE ELEMENT SIMULATION OF PORE CLOSING DURING CYLINDER UPSETTING

MIN CHEOL LEE[1], SUNG MIN JANG[2], JU HYUN CHO[3]

*School of Mechanical and Aerospace Engineering, Gyeongsang National Universit, 900 Gajwa-dong, Jinju-City, GyeongNam, 660-701, KOREA,*

MAN SOO JOUN[4†]

*School of Mechanical and Aerospace Engineering, RRC/Aricraft Parts Technology, Gyeongsang National University, 900 Gajwa-dong, Jinju-City, GyeongNam, 660-701, KOREA,*
*msjoun@gnu.ac.kr*

Received 15 June 2008
Revised 23 June 2008

Three-dimensional precision simulation of pore closing during cylinder upsetting is carried out in this paper. A matrix of pores on the longitudinal section of a cylindrical material is traced at the same time. Various ratios of pore-to-cylinder diameter are tested to reveal their effect on the pore closing phenomena. Hydrostatic pressure and effective strain as well as the order of pore closing are investigated in detail to find out major factors affecting pore closing. It is shown that the effective strain has a strong relationship with the pore closing phenomena and that the finishing time of pore closing increases as the ratio of pore-to-cylinder diameter increases but that the pore size effect becomes negligible if the ratio exceeds a critical value. An intelligent metal forming simulator AFDEX 3D is used.

*Keywords*: Precision simulation; pore closing; cylinder upsetting; intelligent remeshing.

## 1. Introduction

In open die forging of large mechanical parts including ship engine parts, wind power generator parts and the like, the cast material, that is, the cast ingot, is progressively formed to the desired shape by a series of upsetting or cogging processes. The major purposes of the open die forging include improvement of product quality as well as material saving. In general, a large cast ingot has many cavities or defects especially near its central region. These may cause deterioration of product quality or decisive damage of the system. For this reason, the pore or cavity closing has attracted interests of the researchers on the open die forging of the large mechanical parts for a long time. In spite of its significance, the related research works are not sufficient enough to be a guide for process design engineers. The reason lies in that both experimental approaches[1-4] and analytical or numerical approaches[5-14] have their own limitations. That is, experiments of

---

†Corresponding Author.

large-scaled materials are very costly and difficult and most conventional simulation techniques are also poor at describing the pore closing phenomena due to their poor remeshing capability.

Tomlinson and Stringer[1] and Kopp and Ambaum[2] carried out an experimental study on the pore closing phenomena. Erman et al.[3,4] studied a physical modeling for the experiments of metal flow during open die forging. Tanaka et al.[5] conducted analysis of pore closing during cogging process and proposed a pore closing criterion. Stahlberg et al.[6] studied rectangular cavity closing by the upper-bound method and they compared the predictions with experiments. From the late 1980s, finite element methods[7-14] have been used to study the pore closing phenomena in open die forging. Shah et al.[7] applied three-dimensional rigid-plastic finite element method to simulating open die forging processes without considering the pore closing phenomena. In the early years of finite element method based researches on pore closing, most researchers had used two-dimensional approaches[8,10,11,12] in which the pore shapes were considered as pipes in plane strain approach or rings in axi-symmetric approach. On the contrary, in recent years, most researchers on these topics have used three-dimensional finite element methods[9,13,14]. However, it is not easy to find research works on precise simulation of pore closing of small spherical pores even though several researchers have tried to apply them to solving the pore closing phenomena. The reason has caused from some difficulties in remeshing because geometries of the pores during being closed are too complex to be automatically remeshed.

In recent years, Lee et al.[15] developed an intelligent mesh generation technique and applied it to an intelligent metal forming simulator AFDEX 3D[16], based on a rigid-thermoviscoplastic finite element method and tetrahedral MINI-elements. In this paper, the pore closing phenomena is investigated using AFDEX 3D.

## 2.  Simulation of Pore Closing in Cylinder Upsetting

Detailed pore closing phenomena during cylinder upsetting in open die hot forging of large mechanical parts have not fully revealed until now, even though its major purpose is to improve product quality by closing the internal pores in a mechanical way. Large size of a material has discouraged the researchers to approach empirically to the problem and poor remeshing capabilities have also limited application of the simulation technologies to solving the problem.

Recently Lee et al.[12] developed an intelligent remeshing technique and applied it to developing an intelligent metal forming simulator, called AFDEX 3D. The intelligent remeshing technique is based on various surface mesh quality optimization schemes[17,18], which is essential to solve the pore closing phenomena with higher accuracy. We thus applied AFDEX 3D to solving the pore closing phenomena in this study.

Figure 1 shows the initial material of which diameter and height are all 300 mm. In the figure, all pores are located equally on the twelve planes of symmetry, evenly spaced by the circumferential angle of 30° and a matrix of ten pores having the same diameters is located on each plane. The diameters of the pores investigated involve 2.5 mm,

5.0 mm, 7.5 mm, 10.0 mm, 12.5 mm and 15.0 mm. The flow stress of the material was assumed as $\bar{\sigma} = 64.4\dot{\bar{\varepsilon}}^{0.2}$ MPa and the friction factor as 0.6. The upper die and lower dies moved in the opposite direction with the constant speed of 300 mm/sec.

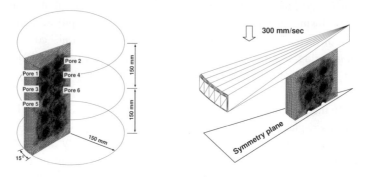

Fig. 1.  Schematic description of the workpiece.      Fig. 2.  Analysis model of the process.

Using the symmetry of the process, only one forty-eighth of the cylinder was considered as the solution domain as shown in Figure 2. During simulation, the number of tetrahedral elements was controlled to be less than 40000. When mesh densities were calculated for remeshing, not only state variables including the effective strain and effective strain-rate but also geometrical features including the surface curvature were considered[15]. However, user-intervention during the whole simulation and mesh density control using special user-defined mesh density functions[18] were excluded.

During automatic simulation, when a node on the material surface penetrated into the material, then the related region was considered as closed. Overall procedure of pore closing and disappearing for the 15 mm pore diameter case is shown in Figure 3 and detailed pore closing history of Pore 1 is shown in Figure 4. As shown in these figures, mesh densities near the pores were well distributed and thus even small shrunk pores did not disappear artificially during remeshing. Throughout the whole simulation, 65 remeshings were conducted to obtain the solutions shown in Figure 3.

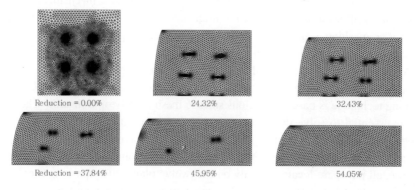

Fig. 3.  Pore closing history of the 15.0 mm pore diameter case.

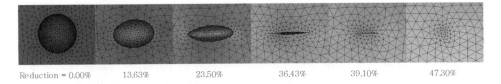

Reduction = 0.00%      13.63%      23.50%      36.43%      39.10%      47.30%

Fig. 4. Detailed pore closing history of Pore 1 for the 15.0 mm pore diameter case.

Figures 5(a) and 5(b) show variation of the hydrostatic pressure and effective strain at all the pore locations with the stroke or reduction for the 15.0 mm pore diameter case, respectively. The lowest point of the pore was traced to measure those state variables. Figure 6 shows the reductions at which each pore disappeared for all the cases of six different pore diameters.

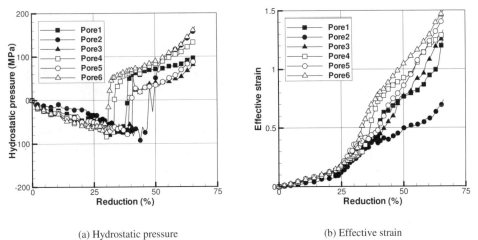

(a) Hydrostatic pressure                    (b) Effective strain

Fig. 5. Variations of the hydrostatic pressure and effective strain with the reduction for the 15.0 mm pore diameter case.

Figure 5(a) shows that hydrostatic pressure changed drastically from negative to positive during pore closing. It can be seen from the figure that the hydrostatic pressures at all pores were narrowly distributed before pore closing started while they were relatively widely scattered to the contrary after pore closing finished. It is noted that the hydrostatic pressure at Pore 2 was relatively high in comparison with the other pores except Pore 6 even though Pore 2 closed last as can be seen in Figure 5(a) and Figure 6(a), indicating that the hydrostatic pressure has no direct influence on pore closing.

On the contrary, the order of pore closing that can be seen in Figure 6(b) for the 15 mm pore diameter case is nearly the same with the order of pores having higher effective strain in Figure 5(b), implying that the effective strain has a strong and direct influence on pore closing. It can be seen from Figures 5(b) and 6(a) for the 15.0 mm pore diameter case that all pores but Pores 3 and 5 closed when their effective strains reached around 0.5 while Pores 3 and 5 closed when they reached around 0.6. This fact might be associated with the effect of hydrostatic pressures because they are lowly ranked as seen in Figure 5(a). The hydrostatic pressure may also play an important role in determining the order of pore closing if the effective strains of two pores are nearly the same, as can be seen from Figures 5 and 6(a) for Pores 1 and 5. It is thus concluded that the hydrostatic pressure plays an indirect but non-negligible role in pore closing.

It can be also seen from Figure 6(a) that the finishing time of pore closing increased on average as the pore diameter increased but that the pore size effect became negligible when the pore diameter exceeded 7.5 mm.

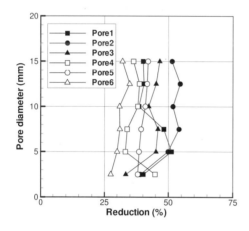
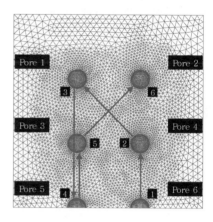

(a) Reductions at which pore closing took place     (b) Order of pore closing, 15.0 mm pore diameter

Fig. 6.  Reductions when pore closing took place and the order of pore closing for the 15.0 mm pore diameter case.

## 3.  Conclusions

In this paper, three-dimensional finite element simulations of pore closing in cylinder upsetting was carried out using AFDEX 3D, a general-purpose intelligent metal forming simulator based on the rigid-thermoviscoplastic finite element method and tetrahedral MINI-elements. Various ratios of pore-to-cylinder diameter, that is, 2.5/300.0, 5.0/300.0, 7.5/300.0, 10.0/300.0, 12.5/300.0 and 15.0/300.0, were considered and a matrix of pores located on the longitudinal section of the cylindrical material was investigated.

It was observed that the finishing time of pore closing increases as the ratio of pore-to-cylinder diameter increases but that the pore size effect becomes negligible if the ratio exceeds 7.5/300.0. It was found out that the effective strain has strong and direct relationship with the pore closing phenomena but that the hydrostatic pressure has indirect and secondary influence on pore closing.

## Acknowledgments

This work was supported by grant No. RTI04-01-03 from the Regional Technology Innovation Program of the Ministry of Commerce, Industry and Energy (MOCIE).

## References

1. A. Tomlinson and J. D. Stringer, *J. Iron Steel Inst.*, March, (1958), pp. 209–217.
2. R. Kopp and E. Ambaum, *Stahl und Eigen*, **96**, (1976), pp. 1004–1009.
3. E. Erman, N. M. Medei, A. R. Roesch and D. C. Shah, *J. Mech. Working Tech.* **19**, (1989), pp. 165–194.
4. E. Erman, N. M. Medei, A. R. Roesch and D. C. Shah, *J. Mech. Working Tech.* **19**, (1989), pp. 195–210.
5. M. Tanaka, S. Ono, M. Tsuneno and T. Iwadate, *Proc. 2nd Int. Conf. Adv. Tech. Plast.* **2**, (1987), pp. 1035–1042.
6. U. Stahlberg, H. Keife, M Lundberg and A. Melander, *J. Mech. Working Tech.* **4**, (1980), pp. 51–63.
7. K. N. Shah, B. V. Kiefer and J. J. Gavigan, *Adv. Manuf. Proc.* **1**, (1986), pp. 501–516.
8. S. P. Dudra and Y. T. Im, *J. Mat. Proc. Tech.* **21**, (1990), pp. 143–154.
9. B. V. Kiefer and K. N. Shah, *ASME Trans., J. Eng. Mat. and Tech.*, **112**, (1991), pp. 477–485.
10. C. Y. Park, J. R. Cho, D. Y. Yang, D. J. Kim and I. S. Park, *Trans. Korean Soc. Mech. Engnrs.* **16** (10), (1992), pp. 1877–1889.
11. J. R. Cho, D. K. Kim, Y. D. Kim and B. Y. Lee, *Korean Soc. Tech. of Plast.* **5** (1), (1996), pp. 18–26.
12. C. Y. Park and D. Y. Yang, Proc. Korea Soc. Prec. Eng. 1996 Spring Annual Meeting, (1996), pp. 819–823.
13. M. S. Chun, J. S. Ryu and Y. H. Moon, *Korean Soc. Tech. Plast.* **13** (2), (2004), pp. 148–153.
14. Y. C. Kwon, J. H. Lee, S. W. Lee, Y. S. Jung, N. S. Kim and Y. S. Lee, *Trans. Mat. Proc.* **16** (4), (2007), pp. 293–298.
15. M. C. Lee, M. S. Joun and June K. Lee, *Finite Elem. Anal. Des.* **43** (10), (2007), pp. 788–802.
16. http://www.afdex.com.
17. M. C. Lee and M. S. Joun, *Adv. in Soft. Eng.* **39** (1), (2008), pp. 25-34.
18. M. C. Lee and M. S. Joun, *Adv. in Soft. Eng.* **39** (1), (2008), pp. 35-46.
19. M. S. Joun, H. K. Moon, J. S. Lee, S. J. Yoo and J. K. Lee, *ASME Trans., J. Eng. Mat. and Tech.* **129**, (2007), pp. 349-355.

# TENSILE TEST BASED MATERIAL IDENTIFICATION PROGRAM AFDEX/MAT AND ITS APPLICATION TO TWO NEW PRE-HEAT TREATED STEELS AND A CONVENTIONAL Cr-Mo STEEL

MAN-SOO JOUN[1†]

*School of Mechanical and Aerospace Engineering, RRC/Aircraft Parts Technology,*
*Gyeongsang National University, 900 Gajwadong, Jinju, Gyeongnam, 660-70 KOREA,*
*msjoun@gnu.ac.kr*

JAE-GUN EOM[2], MIN-CHEOL LEE[3], JEONG-HWI PARK[4]

*Graduate School, Gyeongsang National University, Jinju, 660-701 KOREA,*

DUK-JAE YOON[5]

*Digital production processing and forming team, Korea Institute of Industrial Technology (KITECH),*
*Incheon, 406-840 KOREA,*

Received 15 June 2008
Revised 23 June 2008

A material identification program AFDEX/MAT is presented in this paper. The program is based on the method for acquiring true stress-strain curves over large range of strains using engineering stress-strain curves obtained from a tensile test coupled with a finite element analysis. In the method, a tensile test is analyzed using a rigid-plastic finite element method combined with a perfect analysis model for its associated simple bar to provide the information of deformation. An initial reference true stress-strain curve, which predicts the necking point exactly, is modified iteratively to minimize the difference in tensile force between the experiments and predictions of the tensile test. It was applied to identifying the mechanical behaviors of two new pre-heat treated steels of ESW95 and ESW105 and a conventional Cr-Mo steel of SCM435. The predictions are compared with the experiments for the tensile test of the three materials, showing an excellent similarity.

*Keywords*: Flow stress; large strain; stress-strain curve; tensile test; pre-heat treated steel.

## 1. Introduction

A true stress-strain curve is affected by the manufacturing history, metallurgical treatments, and chemical composition of a material. Therefore, metal-forming simulation engineers require true stress-strain curves that reflect special conditions of their materials. However, it is difficult to obtain material properties from experiments and very limited information about true stress-strain curves can be found in the literature. Most metal forming simulation engineers use the material properties supplied by software companies, which are very limited and sometimes unproven.

---

†Corresponding Author.

It is general that commercial metals show the strain-hardening characteristics during cold working. These characteristics have been being quantified in terms of strain-hardening exponent. At the room temperature, the flow stresses of the common commercial metals are usually represented by the Hollomon's law, which has been widely used for metal forming simulation. Joun et al.[1, 2] were the first to obtain accurate finite element solutions that satisfied the Considère criterion[3] exactly in an engineering sense using a perfect tensile test analysis model, that is, a cylindrical specimen consisting of a simple bar model without any imperfections. They used the flow stress curve described by the Hollomon's law with the strain hardening exponent calculated from the true stain at the necking point, which was called a reference stress-strain curve. However, a great deal of discrepancy in tensile load was observed between the predictions and experiments after the necking point when mild steel was examined. To reduce the discrepancy by improving iteratively the true stress-strain curve, Joun et al.[4] presented a finite element method based algorithm of predicting the exact engineering stress-strain curve.

In this paper, a tensile test based material identification program AFDEX/MAT was introduced and applied to identifying the mechanical behaviors of two new pre-heat treated steels of ESW95 and ESW105[5] as well as a conventional Cr-Mo steel of SCM435.

## 2. Material Identification Algorithm and AFDEX/MAT

Let us first introduce the reference stress-strain curve[2] described by the Hollomon's law, which is formulated by the strength coefficient $K$ and strain hardening exponent $n$, that is, $\sigma = K\varepsilon^n$. The true strain $\varepsilon_N$ at the necking point in a tensile test of a cylindrical bar for a material is considered as its reference strain hardening exponent $n_N$ used for describing the reference stress-strain curve, that is, $n = n_N = \varepsilon_N$. Then the corresponding reference strength coefficient $K_N$ is calculated by making the reference stress-strain curve predict the true stress at the necking point exactly in simulating the tensile test. Of course, the true stress at the necking point can be obtained from the experiments with the incompressibility condition.

After necking occurs in the tensile test, non-uniformity of the true strain increases rapidly in the longitudinal direction. The maximum strain occurs at the minimum cross-section where the shear stress is free due to symmetry and the non-uniformity of the strain is comparatively low. Therefore, it is relatively easy to define a representative strain at the minimum cross-section.

Through finite element simulation of the tensile test, the minimum cross-section of the tensile test specimen can be traced at a set of $N$ sampled elongations. The representative strain of the minimum cross-section at the elongation $\delta^i (i = 1, 2, \cdots, N)$, denoted as $\varepsilon_R^i$, can be calculated from finite element solutions of the tensile test. The difference between the measured load $F_t^i$ and the predicted load $F_e^i$ at the elongation $\delta^i$ can be reduced by modifying the true stress $\sigma_R^i$ corresponding to the representative strain $\varepsilon_R^i$, which is defined by the following average area scheme in this study:

$$\varepsilon_R^i = \int_{A^i} \overline{\varepsilon} \, dA \, / \, A^i \,. \tag{1}$$

where $A^i$ indicates the area of the minimum cross-section of the tensile test specimen at the sampled elongation $\delta^i$. The current true stress $\sigma_{R,old}^i \equiv \sigma_R^i$ at $\varepsilon_R^i$ is modified to give the new true stress $\sigma_{R,new}^i$ by multiplying the current true stress by $F_t^i / F_e^i$ as follows:

$$\sigma_{R,new}^i = \sigma_{R,old}^i \, F_t^i \, / \, F_e^i \,. \tag{2}$$

An iterative algorithm based on the above idea is proposed to obtain an optimized true stress-strain curve. The reference stress-strain curve is used before the necking occurs. After the necking, a true stress-strain relationship is interpolated linearly using the sampled points $(\varepsilon_R^i, \sigma_R^i)$ defined at the elongation $\delta^i$. The detailed procedure used to calculate the improved sampled points $(\varepsilon_R^i, \sigma_R^i)$ at the sampled elongation $\delta^i$ is seen in Figure 1. In the algorithm, $\varepsilon_R^{i,j}$ and $\sigma_R^{i,j}$ are the j-times modified strain and stress, respectively, at the sampled elongation $\delta^i$.

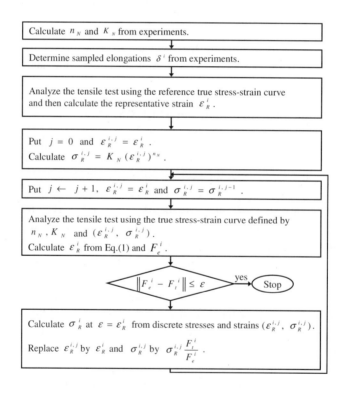

Fig. 1. Flow chart of the algorithm.

More details on the approach can be found from the Reference[4]. Based on the proposed algorithm together with the rigid-plastic finite element simulator AFDEX[6], we developed a material identification program AFDEX/MAT, which is one of the modules of AFDEX, an intelligent metal forming simulator[7]. It needs the diameter and gage length of a cylindrical tensile test specimen and a series of experimental tensile forces at the sampled elongations as the input data. The outputs are either a series of true stresses and strains or the strength coefficients at the sampled elongations. These discrete outputs are interpolated to give the optimized true stress-strain curve.

## 3. Application to New Pre-Heat Treated Steels and Conventional Cr-Mo Steel

The material identification program AFDEX/MAT was applied to two new pre-heat treated steels of ESW95 and ESW105 and a conventional Cr-Mo steel of SCM435 to reveal their mechanical behaviors. The tensile test specimens were prepared according to the Korean standard for tensile test. The screw grip was adopted to minimize the end effect and to increase the reliability of the tensile test.

Five specimens for each material were tested and a representative tensile test was selected considering the measured yield and tensile strengths. Thus obtained engineering stress-strain curves are shown in Figure 2.

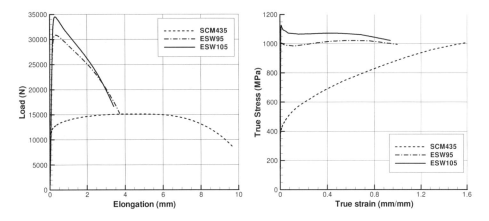

Fig. 2. Experimental load-elongation curves.          Fig. 3. Predicted true stress-strain curves.

The optimized flow stress curves for ESW95, ESW105 and SCM435 were obtained after five iterations by AFDEX/MAT and are shown in Figure 3. The Cr-Mo steel of SCM435 shows a typical strain-hardening behavior up to the fracture point while the pre-heat treated steels of ESW95 and ESW105 show a typical softening behavior after the strains reach 0.58 and 0.45, respectively.

It is interesting to note that ESW95 and ESW105 have very high initial yield stresses compared with SCM435 and that they do not have the distinct strain-hardening behavior as a whole. As a consequence the flow stress of SCR435 at the strain of 1.5 is nearly the same with that of ESW95 at the initial strain-free state.

The tensile test was simulated for the three materials using AFDEX 2D and the flow stress information depicted in Figure 3. Figure 4 compares the final predicted shapes at the fracture point, showing that plastic deformation for the pre-heat treated steels takes place only near the necked region. To the contrary, the tensile test specimen of SCM435 was relatively much elongated as a whole before it fractured. Figure 5 compares the predicted and measured engineering strain-stress curves. The comparison shows a close similarity between the experiments and predictions especially for the plastic region, emphsizing the applicability of the material identification program AFDEX/MAT. The maximum error of the predicted load relative to the measured load after the necking point was less than 0.5%. It is interesting to note that the flow stress of SCM435 was calculated up to a relatively high strain of 1.5.

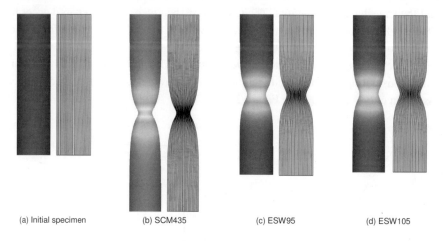

(a) Initial specimen          (b) SCM435          (c) ESW95          (d) ESW105

Fig. 4. Predicted deformed shapes with effective strain and metal flow lines at the fracture point.

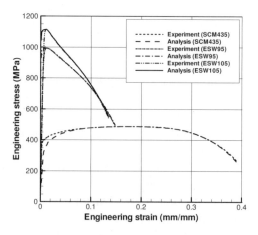

Fig. 5. Comparison of experiments with predictions of the tensile tests.

## 4. Conclusions

A tensile test based material identification program AFDEX/MAT was presented in this paper. The algorithm of the program was based on the method for acquiring true stress-strain curves over large range of strains using engineering stress-strain curves obtained from a tensile test coupled with a finite element analysis. The program was applied to two new pre-heat treated steels of ESW95 and ESW105 and a conventional Cr-Mo steel of SCM435 to reveal their mechanical behaviors involving plastic flow stresses.

It is worthy of emphasizing that the flow stress of SCM435 was calculated up to the considerably high strain of 1.5. In addition, the comparison between experiments and predictions of the tensile test showed very close similarity, revealing the appropriateness of the approach and the applicability of the program AFDEX/MAT.

It has been known from the flow stress information that the initial yield stresses of the pre-heat treated steels tested are quite high but that they have so weak strain-hardening behaviors as to be treated as perfectly plastic materials in the engineering sense. It is believed that their unique feature can contribute not only to reducing the post-work such as straightening and heat treatment but also to minimizing the weight of products in structural design. Of course, it incurs the change of manufacturing process design because their initial yield stresses are considerably high compared to the conventional steels.

## Acknowledgments

This research was financially supported by the second stage BK21 project of Korea. Experiments were supported by TIC(Technology Innovation Center) of Gyeongsang National University, KITECH(Korea Institute of Industrial Technology) and KIMM(Korea Institute of Machinery and Materials). Samhwa Steel Co. supplied the tensile test specimens of the materials tested.

## References

1. M. S. Joun, I. S. Choi, J. G. Eom and M. C. Lee, *Proc. Korean Soc. Mech. Eng.* (Spring Annual Meeting, Jeju Island, Korea, 2006).
2. M. S. Joun, I. S. Choi, J. G. Eom and M. C. Lee, *Comput. Mater. Sci.* **41** (2007) 63.
3. M. Considère, *Annales des Ponts et Chaussées* **9** (1885) 574.
4. M S. Joun, J. G. Eom and M. C. Lee, *Mech. Mater.* **40** (2008) 586.
5. S. T. Ahn and Y. Yamaoka, *Patent No. 10-2001-0056917* (2001).
6. http://www.afdex.com
7. M. S. Joun, M. C. Lee, J. G. Eom, I. S. Choi, J. H. Cho, H. T. Kim and J. H. Park, submitted to *ICTP 2008*.

# AN ELASTO-PLASTIC ANALYSIS OF SOLIDS BY THE LOCAL MESHLESS METHOD BASED ON MLS

Y.T. GU[1†]

*Queensland University of Technology, Brisbane, AUSTRALIA,*

*yuantong.gu@qut.edu.au*

Received 15 June 2008

Revised 23 June 2008

A pseudo-elastic local meshless formulation is developed in this paper for elasto-plastic analysis of solids. The moving least square (MLS) is used to construct the meshless shape functions, and the weighted local weak-form is employed to derive the system of equations. Hencky's total deformation theory is applied to define the effective Young's modulus and Poisson's ratio in the nonlinear analysis, which are obtained in an iterative manner using the strain controlled projection method. Numerical studies are presented for the elasto-plastic analysis of solids by the newly developed meshless formulation. It has demonstrated that the present pseudo-elastic local meshless approach is very effective for the elasto-plastic analysis of solids.

*Keywords*: Material nonlinearity; Elasto-plastic analysis; Meshless method; Moving Least Squares.

## 1. General Appearance

The finite element method (FEM) is currently the dominated numerical simulation tool for the analysis of material behaviors in elastic and elasto-plastic ranges. A group of numerical techniques based on FEM have been developed so far to solve the elasto-plastic problems. Owen and Hinton[1] provided finite element computer implementation of elasto-plastic problems based on incremental theory. Seshadri[2] developed a GLOSS method based on two-linear elastic finite element analysis which is used to evaluate the approximate plastic strains at certain local regions. Desikn and Sethuraman[3] developed a so-called pseudo-elastic method for the determination of inelastic material parameters.

However, the nonlinear stress-strain relationship and the loading path dependency in the plastic range make the analysis tedious. On the other hand, some shortcomings of FEM are revealed in the elasto-plastic analysis including difficulty for adaptive analysis and poor accuracy of the stress field. These shortcomings are inherent of numerical methods formulated based on predefined meshes or elements. In recent years, some meshless (or meshfree) methods have been proposed to overcome the shortcomings of FEM. According to the classifications of Liu and Gu[4], meshless methods can be largely grouped into three different categories: the meshless method based on strong-forms, the

---

[†]Corresponding Author.

meshless method based on weak-forms, and the meshless method based on the combination of weak- and strong-forms. The famous smooth particle hydrodynamics (SPH)[5] belongs to the first category, and the element-free Galerkin (EFG) method[6] and the point interpolation method (PIM)[7] belong to the second category. The advantages of the meshless method include: 1) no mesh used; 2) high accuracy, and 3) good performance for adaptive analysis. Therefore, the meshless technique provides a big freedom for numerical modeling and simulation. However, the freedom does not present without cost (e.g., some undetermined parameters and worse computational efficiency).

In the family of meshless methods, the meshless method based on the local weak-form is a well-developed technique including the meshless local Petrov-Galerkin (MLPG) method[8,9] and the local radial point interpolation method (LRPIM)[10]. Due to the stability and accuracy of MLS, the local meshless method based on the moving least squares (MLS) is becoming a robust numerical tool for the practical analysis. However, almost all current researches and applications of this type of meshless methods (based on MLS) for solids are limited to linear elasticity, and few research for the material nonlinear analysis is reported. In this paper, a pseudo-elastic local meshless formulation is developed to solve elasto-plastic problems in solids. The locally weighted residual method is used to derive the meshless system of equations and MLS is applied to construct the meshless shape functions. The Hencky's total deformation theory with the iterative manner is used to define effective material parameters. Numerical examples are studied to demonstrate the effectivity of the newly developed pseudo-elastic local meshless formulation for the elasto-plastic analysis.

## 2. The local Memeshless Formulation

Consider the following two-dimensional solid problem:

$$\sigma_{ij,j} + b_i = 0 \quad \text{in} \quad \Omega \tag{1}$$

The corresponding boundary conditions are

$$u_i = \overline{u}_i \, ; \quad t_i = \sigma_{ij} n_j = \overline{t}_i \tag{2}$$

In the local meshless method, a local weak-form is constructed over a sub-domain $\Omega_s$ bounded by $\Gamma_s$. Using the locally weighted residual method, the generalized local weak-form of Eqs. (1) and (2) for a field node, $I$, can be written as

$$\int_{\Gamma_{si}} w_I t_i d\Gamma + \int_{\Gamma_{su}} w_I t_i d\Gamma + \int_{\Gamma_{st}} w_I \overline{t}_i d\Gamma - \int_{\Omega_s} (w_{I,j} \sigma_{ij} - w_I b_i) d\Omega = 0 \tag{3}$$

where $w_I$ is the weight function, which is constructed based on the node $I$.

The problem domain and boundaries are discretized by arbitrarily distributed field nodes. To approximate the displacement function $u(\mathbf{x})$ in $\Omega_s$, a finite set of $\mathbf{p}(\mathbf{x})$ called basis functions is considered in the space coordinates $\mathbf{x}^T = [x, y]$. The moving least squares (MLS) interpolant $u^h(\mathbf{x})$ is defined in the domain $\Omega_s$ by

$$u^h(\mathbf{x}) = \sum_{j=1}^{m} p_j(\mathbf{x}) a_j(\mathbf{x}) = \mathbf{p}^T(\mathbf{x}) \mathbf{a}(\mathbf{x}) \tag{4}$$

where $m$ is the number of basis functions, and the coefficient $a_j(\mathbf{x})$ is also a functions of $\mathbf{x}$. $\mathbf{a}(\mathbf{x})$ is obtained at a point $\mathbf{x}$ by minimizing a weighted discrete $L_2$ norm of:

$$J = \sum_{i=1}^{n} w(\mathbf{x} - \mathbf{x}_i)[\mathbf{p}^T(\mathbf{x}_i)\mathbf{a}(\mathbf{x}) - u_i]^2 \tag{5}$$

where $n$ is the number of nodes in the neighborhood of $\mathbf{x}$ for which the weight function $w(\mathbf{x}-\mathbf{x}_i)\neq 0$, and $u_i$ is the nodal value of $u$ at $\mathbf{x}=\mathbf{x}_i$. The stationarity of $J$ with respect to $\mathbf{a}(\mathbf{x})$ leads to the following linear relation between $\mathbf{a}(\mathbf{x})$ and $\mathbf{u}$:

$$\mathbf{A}(\mathbf{x})\mathbf{a}(\mathbf{x}) = \mathbf{B}(\mathbf{x})\mathbf{u} \tag{6}$$

In which $\mathbf{A}(\mathbf{x})$ and $\mathbf{B}(\mathbf{x})$ are the interpolation matrices defined by coordinates and the weight functions. Solving $\mathbf{a}(\mathbf{x})$ from Eq. (6) and substituting it into Eq. (4), we have

$$u^h(\mathbf{x}) = \sum_{i=1}^{n} \phi_i(\mathbf{x})u \text{ , in which } \phi_i(\mathbf{x}) = \sum_{j=1}^{m} p_j(\mathbf{x})(\mathbf{A}^{-1}(\mathbf{x})\mathbf{B}(\mathbf{x})) \tag{7}$$

where $\phi_i(\mathbf{x})$ is the MLS shape function. It should be mentioned here that the MLS shape function obtained above does not have the Kronecker delta function properties[4].

Substituting the displacement expression given in Eq. (7) into the local weak- form Eq. (3) and applying this local weak-form for all filed nodes, we have the following discretized system of equations,

$$\mathbf{KU} = \mathbf{F} \tag{8}$$

where $\mathbf{K}$ is the stiffness matrix and $\mathbf{F}$ is the force vector, i.e.

$$\mathbf{K}_I = \int_{\Omega_s} \mathbf{v}_I^T \mathbf{D}_e \mathbf{B} d\Omega - \int_{\Gamma_{si}} \mathbf{w}_I \mathbf{N} \mathbf{D}_e \mathbf{B} d\Gamma - \int_{\Gamma_{su}} \mathbf{w}_I \mathbf{N} \mathbf{D}_e \mathbf{B} d\Gamma \tag{9}$$

$$\mathbf{F}_I = \int_{\Omega_s} \mathbf{w}_I b d\Omega + \int_{\Gamma_{st}} \mathbf{w}_I \bar{\mathbf{t}} \ d\Gamma \tag{10}$$

In Eq. (9), $\mathbf{D}_e$ is the effective material matrix that is obtained from the effective constitutive equation, i.e.:

$$\mathbf{D}_e(\mathbf{x}_Q) = \frac{E_e(\mathbf{x}_Q)}{1 - v_e^2(\mathbf{x}_Q)} \begin{bmatrix} 1 & v_e(\mathbf{x}_Q) & 0 \\ v_e(\mathbf{x}_Q) & 1 & 0 \\ 0 & 0 & \dfrac{1 - v_e^2(\mathbf{x}_Q)}{2} \end{bmatrix} \tag{11}$$

where $E_e$ and $v_e$ are effective Young's modulus and Poisson's ratio, which will be discussed in the following section.

It should be mentioned here that to get the matrix $\mathbf{K}$ in Eq. (9), Gauss quadrature is used, and it means that $\mathbf{K}$ is obtained based on all quadrature points. Hence, $\mathbf{D}_e$ is the material parameter matrix at the Gaussian quadrature point $\mathbf{x}_Q$.

## 3. Effective Material Parameters

The strain-stress relationship can be taken in the form[3] of $\varepsilon_{ij} = f(\sigma_{ij})$. $\varepsilon_{ij}$ is the total strain tensor which is the summation of conservative elastic $\varepsilon_{ij}^e$ and nonconservative plastic part $\varepsilon_{ij}^p$, i.e.,

$$\varepsilon_{ij} = \varepsilon_{ij}^e + \varepsilon_{ij}^p \tag{12}$$

The elastic strain tensor is related to the stress tensor and is given by Hooke's law[11] for isotropic material. The plastic strain tensor is related to the deviatoric part of stress tensor and is given by Hencky's total deformation relation of $\varepsilon_{ij}^p = \Psi S_{ij}$, where $\Psi$ is a scalar valued function, given by

$$\Psi = \frac{3\varepsilon_{\text{equivalent}}^p}{2\sigma_{\text{equivalent}}} = \frac{3}{2}\frac{\sqrt{2\varepsilon_{ij}^p \varepsilon_{ij}^p / 3}}{\sqrt{3S_{ij}S_{ij}/2}}; \quad S_{ij} = \sigma_{ij} - \frac{1}{3}\sigma_{kk}\delta_{ij} \tag{13}$$

Hence, from Equation (12), we can get

$$\varepsilon_{ij} = \varepsilon_{ij}^e + \varepsilon_{ij}^p = \frac{1+v}{E}\sigma_{ij} - \frac{v}{E}\sigma_{kk}\delta_{ij} + \Psi S_{ij} \tag{14}$$

$$= \left(\frac{1+v}{E} + \Psi\right)\sigma_{ij} - \left(\frac{v}{E} + \frac{1}{3}\Psi\right)\sigma_{kk}\delta_{ij}$$

The above equation can be re-written as

$$\varepsilon_{ij} = \left(\frac{1+v_e}{E_e}\right)\sigma_{ij} - \left(\frac{v_e}{E_e}\right)\sigma_{kk}\delta_{ij} \tag{15}$$

where $E_e$ and $v_e$ are the equivalent Young's modulus and Poisson's ratio, which are given by

$$E_e = 1\bigg/\left(\frac{1}{E} + \frac{2\Psi}{3}\right); \quad v_e = \left(\frac{v}{E} + \frac{\Psi}{3}\right)\bigg/\left(\frac{1}{E} + \frac{2\Psi}{3}\right) \tag{16}$$

Eq. (15) is the effective constitutive equation for the analysis of material nonlinearity. It should be mentioned here that the effective material parameters are functions of the final state of stress fields, which are usually unknown. Because the system of equations is constructed based on the Gauss quadrature points, the effective material parameters should be also calculated for Gauss quadrature points. In addition, the direct method is unable to lead to the final solution, and the following iteration method based on the projection technique[3] is used.

A linear elastic analysis is firstly carried out to get the initial stress field. To determine whether a material enters the plastic range, the Von Mises yield criterion, which compares the equivalent stress with the yield stress, is used. If the equivalent stress calculated from linear analysis is smaller than the yield stress, $\sigma_0$, the computing is finished because the material still satisfies the linear elasticity; if the equivalent stress is larger than the yield stress, it means the deformation already enters the plastic region, and the following iteration computing will be performed.

From the initial linear elastic analysis, the strain value is kept unchanged (i.e. strain controlled), and is projected on the experimental uniaixial $\sigma - \varepsilon$ curve. Based on the projection point, the effective value of Young's modulus, $E_e^{(1)}$, for the next iteration is obtained from the slope, and then the effective Poisson's ratio, $v_e^{(1)}$, can also be obtained from Equation (16). Using the new effective materials parameters $E_e^{(1)}$ and $v_e^{(1)}$, the next linear elastic meshless analysis is carried out to get the next point, its projection point, and further to obtain $E_e^{(2)}$ and $v_e^{(2)}$, similarly. This iterative procedure is repeated until all the effective material parameters converge and equivalent stress falls on the experimental uniaxial stress-strain curve. However, if the applied loading is too big, the computing may not converge, and it means that the material is already failure, and this certain loading is called the critical failure loading which is also an important parameter for solids and structures.

## 4.  Numerical Example

To illustrate the effectiveness of the presented pseudo-elastic local meshless formulation for the material nonlinear problems, several cases of elasto-plastic analyses have been studied, and good results have been obtained. Following, the results for the uniaxial tension of a bar is presented and discussed in details.

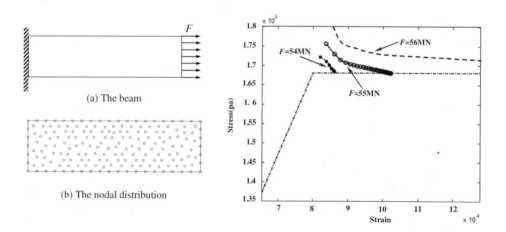

(a) The beam

(b) The nodal distribution

Fig. 1. A cantilever bar under uniaxial tension.

Fig. 2. The convergence path for the bar with the elastic-perfectly plastic material (*F*=54MN, 55MN and 56MN).

As shown in Fig. 1, a cantilever bar with length 3m and height 0.3m is subjected to a uniform tensile pressure (the resultant force is *F*). The Young's modulus is assumed as $E = 2.1 \times 10^{11}$ Pa, the Poisson's ratio is $v = 0.3$, and the yield stress is $\sigma_0 = 1.68 \times 10^8$ Pa. The bar is assumed as being in a plane stress state. As shown in Fig. 1, 182 irregularly distributed field nodes are used to discretize the problem domain and boundaries.

The material is initially considered as elastic-perfectly plastic. The new developed pseudo-elastic local meshless method and the iterative projection technique are applied to get the results. Fig. 2 shows the convergence paths for different $F$. It can be seen that the present method using the projection technique can quickly produce convergent results. However, when $F$=56MN, the result is not convergent. It has been found that when $F$ is larger than a certain value, the results will become non-convergent, and the structure fails. This value is called the critical failure load, and it is $F$=55.5 MN for this problem. Comparing with the FEM and other method results[10,12], it can be seen that the results obtained by the present method are in good agreement with those obtained by other methods. It should be mentioned that the present method needs much less iteration steps than FEM. Therefore, it is more efficient than FEM.

A work-hardening material is also considered, as shown in Fig. 3. It clearly shows that the pseudo-elastic local meshless method gives convergent results for $F$=80 MN. It is reasonable that a work-hardening material has a much higher critical failure load than an elastic-perfectly plastic material.

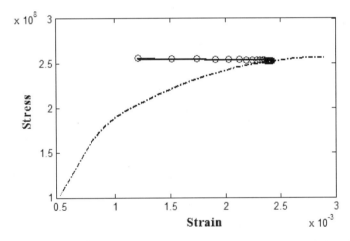

Fig. 3. The convergence path for the bar with the work-hardening material (the unit of stress is Pa and $F$=80MN).

## 5. Conclusions

A pseudo-elastic local meshless formulation is developed for solving elasto-plastic problems in solids. The moving least squares (MLS) is used to construct meshless shape function and the weighted local weak-form is used to derive the meshless system of equations. The Hencky's total deformation theory is utilized to define the effective material parameters, which are obtained in an iterative manner using strain controlled projection method. Numerical studies are presented for the elasto-plastic analysis of solids by the newly developed meshless formulation. It has demonstrated that the present pseudo-elastic local meshless approach is very effective for the elasto-plastic analysis of solids.

## References

1. D.R.J. Owen and E. Hinton, *Finite Elements in Plasticity: Theory and Practice* (Pineridge Press Limited, 1980).
2. R. Seshadri, ASME J. Pressure Vessel Technology **113**(1991), 219–227.
3. V. Desikn and R. Sethuraman, *ASME J. of Pressure Vessel Technology* **122**(2000), 457-461.
4. G.R. Liu and Y.T. Gu, An introduction to meshfree methods and their programming (Springer Press, Berlin, 2005).
5. R.A. Gingold and J.J. Moraghan, *Monthly Notices of the Royal Astronomical Society* **181**(1977), 375-389.
6. T. Belytschko, Y.Y. Lu and L. Gu, *Int. J. for Numer. Meth. in Eng.* **37**(1994), 229-256.
7. G.R. Liu and Y.T. Gu, *Int. J. for Numerical Methods in Eng.* **50**(2001), 937-951.
8. S.N. Atluri and T. Zhu, *Computational Mechanics* 22(1998), 117-127.
9. Y.T. Gu and G.R. Liu, *Computational Mechanics* **27(3)**(2001), 188-198.
10. Y.T. Gu, Q.X. Wang et al., *Engineering Analysis with Boundary Elements* **31** (2007), 771–782.
11. S.P. Timoshenko, J.N. Goodier, *Theory of Elasticity* (McGraw-hill, New York, 1970).
12. K.M. Liew,Y.C. Wu, G.P. Zou and T.Y. Ng, *Int. J. Numer. Meth. Engng* **55**(2002), 669–683.

# PREDICTION OF PLUG TIP POSITION IN ROTARY TUBE PIERCING MILL USING SIMULATION AND EXPERIMENT

HYOUNG WOOK LEE[1†]

*Dept. of Energy System Engineering, Chungju National University,*
*Daehak-ro 72, Chungju, Chungbuk, 380-702 KOREA,*
*hwlee@cjnu.ac.kr*


GEUN AN LEE[2], EUNG ZU KIM[3], SEOGOU CHOI[4]

*Manufacturing Process Research Division, Korea Institute of Industrial Technology,*
*7-47 Songdo-dong, Yeonsu-gu, Incheon City, 406-840, KOREA,*
*galee@kitech.re.kr, ezkim@kitech.re.kr, schoi@kitech.re.kr*

Received 15 June 2008
Revised 23 June 2008

Typical seamless tube production processes are an extrusion and a rotary tube piercing. The rotary piercing process is more competitive than the extrusion process from the viewpoint of flexibility. Main components of this equipment are twin rolling mills and a plug. Twin rolling mills are installed on a skew with proper angles in two directions. These angles are called the cross angle and the feed angle. The internal crack initiation and the growth at central area of the billet are gradually progressed due to the Mannesmann effects. The feed angle affects on the position of the crack initiation in the rotary tube piercing process. Adjustable design parameters in the equipment are the feed angle and the plug insertion depth.

In this research, the rotary tube piercing equipment was developed. Finite element analyses with the plug and without the plug were carried out in order to predict the internal crack initiation position under good calculation efficiency. Ductile fracture models with the one variable equation were utilized to crack initiation criteria. The forward distance of the plug was determined to be 5 mm through the analysis. It was testified by experiments of a carbon steel billet.

*Keywords*: Rotary piercing mill; seamless tube; ductile fracture criteria; feed angle.

## 1. Preface

There are many processes for manufacturing tubes and pipes, and each process has been developed and adapted for industrial production according to its own strengths. Among these processes, the rotary tube piercing process can target special steel tubes with the higher tensile strength. It is a superior process in practical uses because it can offer particular advantages in cost efficiency and productivity with flexibility for producing small quantities of a wide range of product types.

---

†Corresponding Author.

Rotary tube piercing, first devised by the Mannesmann brothers of Germany in 1886, is a process where a billet rotates and progresses between two or three rolls. The repetitive tensile and compression stresses caused during the forming process induce cracks at the central area of the billet. Material flows of the billet were controlled by a plug located near the front of crack tip for making seamless tubes or pipes. This process has a higher correlation between the process parameters and quality compare to the other rolling based processes. The important process parameters of the rotary tube piercing are as follows[1,2]: the temperature of billets, the cross angle, the feed angle, the area reduction ratio, the plug insertion depth, the shape of roll and plug, and the lubrication condition. The feed angle of twin roll mills affects to the position of a generation of crack in a billet. A proper position of a plug can give an easy material flow. The position of a plug tip is an important process parameter because when a distance between the crack and the plug tip is too large, the material may become oxidized, and if the distance is too small, there will be severe wears on the plug tip.

Hayashi and his group[3] researched material deformation patterns caused by stresses and surface twisting in the circumferential direction of the cones. Mori and his team[4] adopted a simplified 2D approach instead a three-dimensional analysis and validated their predictions using Oyane's ductile fracture criteria. Recently, in order to simulate the rotary tube piercing process in 3D, Pater's team[5] used MSC.SuperForm to perform a thermal-mechanical analysis of a process with Diescher's Mill, which is a disc-shape guide.

In the present work, finite element simulations and experiments are performed in order to determine the position of the plug tip. An analysis was carried out using DEFORM/3D to take the plug into consideration, and LS-NYNA3D was employed to consider the effects of the feed angle under the absence of the plug with the computing efficiency. Also, the ductile fracture criteria model was applied to design the appropriate position of the plug tip.

## 2.  Rotary Tube Piercing Equipment

In traditional Mannesmann piercing equipment, the shape of rolls is designed in barrel-form, and although it appears that the piercing process is carried out smoothly with a constant speed of progression from inlet to outlet, in fact its design makes it difficult for the pierced tube to rotate and progress. This makes a rotary forging effect that causes defects within the material. In order to improve this, Hayashi and others developed the "super piercer", which uses cone-type rolls. In this case, due to the circumferential length on the roll, tangential velocity increases towards the outlet.

In this work, rotary tube piercing equipment using cone-shaped rolls was developed as shown in Fig. 1(a). The inlet angle of the cone roll is set at 4 degrees, the outlet angle at 3 degrees, and the cross angle at 20 degrees, while the feed angle can be varied from 0 to 15 degrees. The piercing machine is capable of forming material with a diameter in a range of 20 mm to 60mm, and the guide can be either a straight type or rotary disc type.

Using specimens of pure aluminum, the tube manufactured from the equipment is shown in Fig. 1(b). Fig. 2 shows the schematic diagram of the cross angle and feed angle. An increase of the feed angle accelerates the material's movement, while a decrease in angle raises the material's rotation.

(a)                                                    (b)

Fig. 1. Developed rotary piercing machine with cone type roll mills and pierced pure aluminum billet by the developed machine. (a) developed equipment; (b) achieved material.

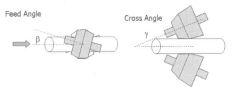

Fig. 2. Schematic diagram of the feed angle and cross angle in rotary tube piercing mill.

## 3. Determination of the Position of Plug Tip using Numerical Simulation

### 3.1. *Finite Element Analysis of Rotary Tube Piercing Process*

To simulate the piercing process, DEFORM/3D was used. The design and process parameters are a 20 degree the cross angle, a 4 degree feed angle, a 52 mm gap between the rollers, a 56 mm gap between the left and right guides, 50 rpm rotating speed of the rollers, and a 60 mm inlet material diameter. The material used was S45C at 1100 degrees centigrade, while the plug's forward insertion distance was 10 mm.

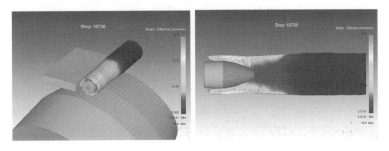

Fig. 3. Effective strain distribution of FE result from the DEFORM/3D.

The computing time of the analysis up to obtaining the shape shown in Fig. 3 was approximately 26 hours. The material was forced out around the plug and formed into a hollow shape. In this analysis, the size of finite elements was limited in consideration of the computing time of analysis using only re-meshing method, while excluding ductile fracture models. If ductile fracture models had been used for the analysis, it would have been necessary to reduce the size of elements to at least ¼ values. This would take the duration of the analysis to roughly one month, thus it makes difficult to obtain various tries in simulation. The commercial code of LS-DYNA3D was employed to analyze the effects of the feed angle while excluding the effects of the plug and the index value of ductile fracture criteria was calculated at the center line.

Fig. 4 (a) shows the distribution of the maximum principal stress from the analysis results in the sectional view. It shows the maximum principal stress, which affects the ductile fracture criteria, in the central region of the material, which leads to the prediction that cracks will first appear in the center of the material. Fig. 4 (b) shows that various index values of ductile fracture criteria, calculated at the center line, are in accordance with each other. The Brozzo ductile fracture index value, calculated according to the radial direction, is illustrated in Fig. 5. Since the maximum stress occurs on the interior portion of the material and cracks occur in the center of the material before the surface, it can be confirmed that a solid material can be formed into a hollow shape using a mandrel.

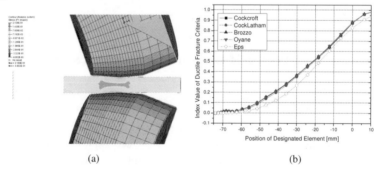

(a)                                              (b)

Fig. 4. Finite element modeling of rotary piercing process and results (a) principal stress distribution in sectional view from the LS-DYNA3D without a plug; (b) various index values of ductile fracture criteria at the center line.

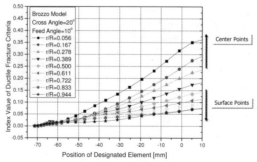

Fig. 5. Variation of Brozzo index values according to the radial position with respect to the material position.

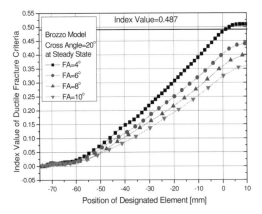

Fig. 6. Variation of Brozzo index values at central point of billet according to the feed angle with respect to the material position.

Fig. 7. Experimental result of the central crack initiation in the feed angle of 4 degrees.

## 3.2. *Analysis of Effects of Feed Angle*

Parametric studies were carried out with the calculation of the ductile fracture index value at the material center in order to analyze effects of the feed angle while excluding the plug. In the simulation, the cross angle is set to 20 degrees and the feed angle is varied from 4 to 10 degrees with the interval of 2 degrees. The calculated index values according to the center line of material are displayed in Fig. 6. It can be seen that the smaller the feed angle gives the larger the ductile fracture index value. Fig. 7 shows the experimental results at 4 degrees feed angle using S45C carbon steel without the plug.

The test results shown in Fig. 7 indicate that the size of the crack at the center is approximately 9.5% of the radius; when this is applied to Fig. 8, the index value is determined to be 0.487. From this result, we can predict the timing of crack occurrence per radial position in relation to the material position when the feed angle is set at 4 degrees. When a crack forms too rapidly at the material center, the material may become oxidized, and if it forms too slowly, there will be severe wear on the plug.

Therefore, the forward distance of the plug in Fig. 8 may be determined to be 5 mm from where cracks start to form at the material center. Fig. 9 shows a S45C tube specimen produced through this condition.

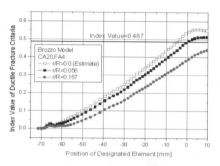

Fig. 8. Variation of index value of Brozzo ductile fracture criteria and the threshold line of the crack initiation with respect to the radial position.

Fig. 9. Experimental result of the S45C tube by rotary piercing machine with the feed angle of 4 degrees.

## 4. Conclusion

In this research, the timing of crack initiation at the material center was predicted based on changes in the feed angle through a finite element analysis. Simulation and experiment without the plug were performed in order to predict an internal crack initiation position with a good computing efficiency. The smaller feed angle gives the larger index values and the faster crack initiation time. The position of the plug was designed based on these results and experimental results. The forward distance of the plug may be determined to be 5 mm from where cracks start to initiate at the material center.

## Acknowledgments

This work was supported by SAC and MOCIE (Ministry of Commerce, Industry and Energy) of Korea and a grant from the Academic Research Program of Chungju National University in 2008.

## References

1.  C. Hayashi and T. Yamakawa, *ISIJ International*, **38**, 1255 (1998).
2.  M. Morioka, H. Oka, and T. Simizu, *Kawasaki Steel Technical Report*, **29**, 57 (1997).
3.  C. Hayashi and T. Yamakawa, *ISIJ International*, **37**, 146 (1997).
4.  K. Mori, H. Yoshimura, and K. Osakada, *J. Mater. Process. Technol.*, **80-81**, 700 (1998).
5.  Z. Pater, J. Kazanecki, and J. Bartnicki, *J. Mater. Process. Technol.*, **177**, 167 (2006).

# STUDY OF ACCURATE HIGH-SPEED TENSION TESTING METHOD

TSUTOMU UMEDA[1†], KOJI MIMURA[2]

*Division of Mechanical Engineering, Graduate School of Engineering, Osaka Prefecture University,*
*1-1, Gakuen-cho, Naka-ku, Sakai, Osaka 599-8531, JAPAN,*
*umeda@me.osakafu-u.ac.jp, mimura@me.osakafu-u.ac.jp*

Received 15 June 2008
Revised 23 June 2008

The high measurement accuracy in dynamic tension testing is required for designs and numerical simulations based on the accurate modeling of stress–strain relations at various strain-rates. The non-coaxial Hopkinson bar method (NCHBM) is one of the recently proposed methods for dynamic tension tests. In this study, the direct measurement of both stress and strain was conducted by using the strain gauges affixed on the special specimen with load cell part. The stress–strain relations obtained by this revised NCHBM were compared with those obtained by the normal NCHBM. In order to verify the experimental results, some numerical models were examined by using the FEM code LS-DYNA, in which the whole finite-element model of the apparatus and the plate-type specimen were made in detail. Furthermore, new appliances for fixing specimen were examined by experiment. The target material employed in this study is SUS316.

*Keywords*: Non-coaxial Hopkinson bar method; measurement accuracy; stress–strain relation.

## 1. Introduction

The importance of the measurement accuracy in dynamic material testing has increased in the fields of design and numerical simulation based on modeling the stress-strain relation at various strain-rates. The Hopkinson bar method (HBM)[1] is the most widely used experimental technique for the dynamic compression test, but is not used very much for the dynamic tension test, even though some special techniques been developed.[2, 3]

NCHBM is one of the recently proposed methods for dynamic tension testing and is based on HBM.[4, 5] NCHBM has the advantages of the simple configuration and the high performance at high strain rates. On the measurement accuracy of testing methods of this kind, there still remains of the problem concerning the transient vibration. In this paper, the direct measurement of both stress and strain was conducted by using the strain gauges affixed on the specimen. In order to understand the experimental results, some numerical models were examined by using the FEM code LS-DYNA.[6] Furthermore, new appliances were examined by experiment.

---

†Corresponding Author.

## 2.  Non-coaxial Hopkinson Bar Method

In NCHBM, the nominal stress $\sigma_N$ and the apparent nominal strain $\varepsilon_N^*$ (positive in tension) can be evaluated by the traditional process of HBM as follows:

$$\sigma_N = -\frac{\sigma_T + \sigma_I + \sigma_R}{2} \cdot \frac{A_B}{A_S} \quad \text{or} \quad -\sigma_T \cdot \frac{A_B}{A_S}, \tag{1}$$

$$\varepsilon_N^* = -\int_0^t \frac{\Delta V}{l} dt' = -\int_0^t \frac{\sigma_I - \sigma_R - \sigma_T}{\rho_B c_0 l} dt', \tag{2}$$

where $A_B$ and $A_S$ are the cross-sectional areas of the input or output bar and the specimen, respectively. $\sigma_T$, $\sigma_I$ and $\sigma_R$ are the transmitted, the incident and the reflected stresses, respectively. Here, $\sigma_N$ evaluated by the former formulation of Eq. (1) is greatly affected by the disturbance included in $\sigma_R$ .[5] Therefore, $\sigma_N$ was evaluated by the latter formulation. $\Delta V$, $\rho_B$ and $c_0$ are the relative velocity between the pins, the density of the bars and the velocity of stress propagation in the bars, respectively. $\varepsilon_N^*$ usually includes the errors due to extra elastic deformation, which should be eliminated by the following:

$$\varepsilon_N = \varepsilon_N^* - \frac{\sigma_N}{E'} \quad \text{with} \quad \frac{1}{E'} = \frac{1}{E^*} - \frac{1}{E}, \tag{3}$$

where $E^*$ is the apparent Young's modulus and $E$ is the true. Then, the true stress $\sigma$ and the true strain $\varepsilon$ are obtained by the following equations used under uniform deformation:

$$\sigma = \sigma_N (1 + \varepsilon_N) \quad \text{and} \quad \varepsilon = \ln(1 + \varepsilon_N). \tag{4}$$

## 3.  Overview of Apparatus

The basic configuration of the NCHBM system is shown in Fig. 1 (a).[5] NCHBM is realized by the simple idea of slightly shifting the axes of input and output bars from one another. Those of HBM system are set up in coaxial. The details around the installed specimen are shown in Fig. 1 (b). The input and output bars are the same one of the length 2000mm, the diameter 10mm, and made of tool steel. In Fig. 1 (a), two strain gauges are glued onto both sides (symmetrically with respect to this page) of each bar. Because those pairs of strain gauges are set over the points, which are placed on the neutral axis, the deviation of stress distribution in the cross section due to the offset of axes does not affect the measurement. The effects of the disturbance and deflection caused by the bending wave were also discussed, so that the validity was confirmed.[5]

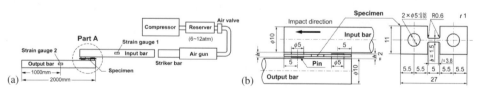

Fig. 1.   Configuration of NCHBM system. (a) Schematic. (b) Details around the Part A (unit: mm).

## 4. Examination of Obtained Stress–strain Relation

Fig. 2 shows example outputs for the specimen of the parallel-part length $l$ = 10mm, which is the largest one used in this study, with the impact velocity $v_i \cong$ 7.6m/s. The specimen material is SUS316. The output from the output bar is magnified by ten. The duration of the transmitted stress is longer than that of the incident stress because the bars keep moving after the duration of the incident stress, and the relative motion between them keep generating a wave-form signal until the fracture of specimen. In this case, the specimen was too long to rupture in the duration of the incident stress, so that the stress–strain relation could not be calculated after the duration. Figs. 3 (a) and (b) show the obtained stress–strain relations under the normal condition and fixed one at the boundary between specimen and pin with an epoxy adhesive (Araldite), respectively. In each figure, all results show good agreement. The specimen of $l$ = 10mm and the parallel-part breadth $b$ = 1mm causes a strong transient vibration under the normal condition, while the reinforcing the fixation of specimen with the adhesive considerably reduces the vibration.

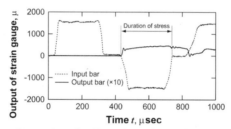

Fig. 2. Outputs from input and output bars ($l$ = 10mm, $b$ = 1mm, $h$ = 2mm, $v_i \cong$ 7.6m/s, $\dot{\varepsilon}_N \cong$ 1400s$^{-1}$, No. 3).

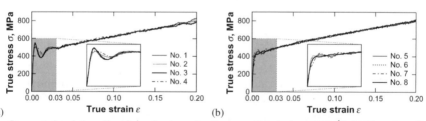

(a)                    (b)

Fig. 3. Stress–strain relations ($l$ = 10mm, $b$ = 1mm, $h$ = 2mm, $v_i \cong$ 7.6m/s, $\dot{\varepsilon}_N \cong$ 1400s$^{-1}$ ). (a) Normal condition. (b) With epoxy adhesive (Araldite).

### 4.1. *Experiment of direct measurement from specimen*

We have tried to obtain the stress and strain outputs through the strain gauges (E-02W-12T11W3, Minebea Co., Ltd.) glued on the specimen.[7] Fig. 4 (a) shows the specimen used in this experiment. Figs. 5 (a) and (b) show the strain time histories and the stress–strain relations. In Fig. 5 (a), the strain was recorded up to only about 1% from the gauge A, so that the subsequent history was obtained by taking away the offset $\overline{ab}$ from the corrected nominal strain $\varepsilon_N$ , which includes the contribution of the uneven contact between bars, the gap between specimen and pin, and so on. In Fig. 5 (b), the stress and strain

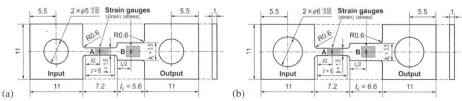

Fig. 4.    Specimen with load cell (unit: mm). (a) A ($l_L = 5.6$mm). (b) B ($l_L = 6.6$mm, using with new appliances).

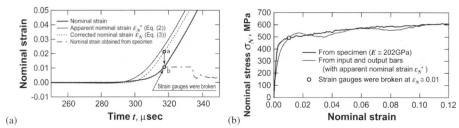

Fig. 5.    Experimental results. (a) Strain time histories. (b) Stress–strain relations.

within $\varepsilon < 0.01$ for the bold line were obtained from the gauges A and B, and in the subsequent part only the strain was exchanged for the corrected one, as mentioned above. As compared with the fine line, the bold line hardly shows the transient vibration.

### 4.2.  Material model

In the numerical simulation, SUS316 was modeled as an isotropic elastic/viscoplastic material based on the von Mises yield criterion. The dynamic yield stress or the flow stress $\sigma_y$ was represented in the form of the Johnson-Cook model[8] without thermal coupling:

$$\sigma_y = \left\{\sigma_0 + B\left(\varepsilon_{eq}^p\right)^n\right\}\left(1 + C\ln\frac{\dot{\varepsilon}_{eq}^p}{\dot{\varepsilon}_0}\right), \tag{5}$$

where $\sigma_0$ , $\varepsilon_{eq}^p$ and $\dot{\varepsilon}_{eq}^p$ are the initial yield stress at the reference strain-rate $\dot{\varepsilon}_0$ , the equivalent plastic strain and the equivalent plastic strain rate, respectively. The elastic and viscoplastic material constants are shown in Tables 1 and 2, respectively.

Table 1.  Basic material properties of tool steel and SUS316.

| Material | Density $\rho$ (kg/m³) | Young's modulus $E$ (GPa) | Poisson's ratio $\nu$ |
|---|---|---|---|
| Tool steel | 7860 | 189 | 0.3 |
| SUS316 | 7980 | 193 | 0.3 |

Table 2.  Viscoplastic material properties of SUS316.

| Material | $\sigma_0$ (MPa) | $B$ (MPa) | $n$ | $C$ | $\dot{\varepsilon}_0$ (s⁻¹) |
|---|---|---|---|---|---|
| SUS316 | 325 | 1050 | 0.78 | 0.0379 | 1 |

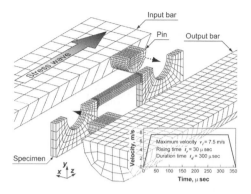

Fig. 6.   1/2 symmetric model ($l$ = 10mm, $b$ = 2mm, $h$ = 2.5mm, 31556 elements).

### 4.3.  Finite element model

Numerical modeling was carried out only for half of the system.[5, 7] The neighborhood of example specimen assembling with the bars is shown in Fig. 6. The pin and bar were 'tied' to each other.[6] The outer end of the output bar was fixed, while the velocity was applied to that of the input bar in **z**-direction. The coefficient of friction $\mu$ at the contact surface between specimen and pin was changed from the static value 0.20 to the kinetic value 0.14 according to the relative velocity (the exponential coefficient 1s/m).[6] The nodes on the symmetry plane were fixed in **y**-direction.

### 4.4.  Improvement of fixation of specimen

The numerically evaluated effect of reinforcing the specimen fixation with adhesive was compared with that evaluated by the experiment. The numerical analyses were conducted to reproduce the experimental boundary conditions (normal: contact with the friction, adhesive: tied contact). Figs. 7 (a) and (b) show the comparisons between the results of experiment and calculation, and they agree well under respective conditions. Furthermore, the effect of reinforcing the specimen fixation with thin nuts (two nuts and a pin were made in one body, as shown in Fig. 7 (d)) was numerically examined. In Fig. 7 (c), the stress difference ($\delta_\sigma$: the ratio of difference of the obtained true stress to that given by the material model[5]) evaluated with the epoxy adhesive or with the nuts becomes as small as that evaluated under the tied contact condition.

### 4.5.  New appliances for improvement of fixation and reproducibility

On the basis of the investigation mentioned above, we newly prepared additional appliances.

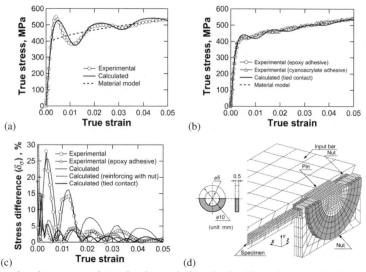

Fig. 7.   Comparison between experimental and numerical results ($l = 10$mm, $b = 1$mm, $h = 2$mm, $t_r = 30\mu$sec, $\dot{\varepsilon}_N \cong 1400$s$^{-1}$). (a) Stress–strain relations (normal condition). (b) Stress–strain relations (with adhesive). (c) Stress difference–strain relations. (d) Improved model with thin nuts ($h = 2.5$mm).

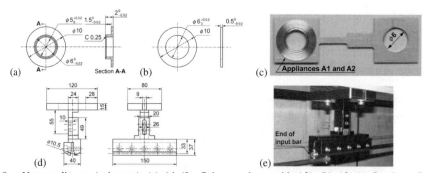

Fig. 8.   New appliances (unit: mm). (a) A1 (for fixing specimen with A2). (b) A2. (c) Specimen B with appliances A1 and A2. (d) B (for fixing input bar). (e) Installed appliance B.

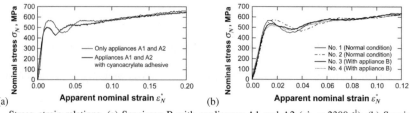

Fig. 9.   Stress–strain relations. (a) Specimen B with appliances A1 and A2 ($\dot{\varepsilon}_N \cong 2300$s$^{-1}$). (b) Specimen in Fig. 1 (b) with appliance B ($\dot{\varepsilon}_N \cong 3400$s$^{-1}$).

Figs. 8 (a) and (b) show a pair of appliances; A1 and A2, which are set in the specimen (see Fig. 8 (c)) and used with an adhesive, for the improvement of reproducibility and the easy installation. Fig. 8 (d) shows the appliance B for restraining the fluctuation of input bar, which is installed as shown in Fig. 8 (e). Fig. 9 (a) shows example stress–strain relations obtained for the specimen in Fig. 4 (b). The transient vibration is greatly decreased by using A1 and A2 with the adhesive. Fig. 9 (b) also shows the decrease of vibration by using B.

## 5. Summary

The direct measurement of both stress and strain was conducted by revising NCHBM with the strain gauges affixed on the specimen, and the measurement accuracy of stress–strain relation in the small-strain region was remarkably improved as compared with the normal NCHBM. The improvement caused by reinforcing the specimen fixation was also confirmed both by the experiment and by the FEM simulation. Furthermore, the effects of new appliances proposed through the simulations were confirmed by the experiments.

## References

1. H. Kolsky, *Proceedings of the Physical Society of London*, **62 B**, 676 (1949).
2. L.S. Costin and J. Duffy, *Trans. ASME, J. Eng. Mater. Technol.*, **101**, 258 (1979).
3. T. Nicholas, *Experimental Mechanics*, **21**, 177 (1981).
4. K. Miura, S. Takagi, O. Hira, O. Furukimi and S. Tanimura, *SAE Technical Paper Series*, **980952**, 23 (1998).
5. T. Umeda, H. Umeki and K. Mimura, *JSME International Journal*, **48-4**, Ser. A, 215 (2005).
6. Livermore Software Technology Corporation, *LS-DYNA Ver.960 USER'S MANUAL*, (2001).
7. T. Umeda and K. Mimura, *Proceedings of ATEM '07*, CD-ROM, (2007).
8. G.R. Johnson and W.H. Cook, *Proceedings of the 7th International Symposium on Ballistics*, 541 (1983).

# DYNAMIC MATERIAL PROPERTIES OF
# THE HEAT-AFFECTED ZONE (HAZ) IN RESISTANCE SPOT WELDING

JI-WOONG HA[1], JUNG-HAN SONG[2], HOON HUH[3†]

*School of Mechanical, Aerospace & System Engineering, KAIST, 335, Gwahangno, Deadoek Science Town,
Daejeon, 305-701 KOREA,
hajiwoong@kaist.ac.kr, jhsong_me@kaist.ac.kr,hhuh@kaist.ac.kr*


JI-HO LIM[4],SUNG-HO PARK[5]

*Automotive Steel Research Center, POSCO, 699, Gumho-dong, Gwangyang-si, Jeonnam, 545-090 KOREA,
jiholim@posco.com, sunghopark@posco.com*

Received 15 June 2008
Revised 23 June 2008

This paper is concerned with a methodology to identify the dynamic material properties of the heat-affected zone (HAZ) near the base metal in a resistance spot weld process at various strain rates. In order to obtain the dynamic material properties of the HAZ in the spot-welded steel sheet, specimens are prepared to have similar material properties, hardness and microstructure to the actual HAZ. Such thermally simulated specimens are fabricated with the material thermal cycle simulator (MTCS) and compared with the real one for the hardness and microstructure. Dynamic tensile tests are then conducted with a high speed material testing machine. Stress–strain curves of the thermally simulated HAZ are obtained at various strain rates ranged from 0.001/sec to 100/sec. Obtained material properties are applied to the finite element analysis of the spot-welded tensile-shear specimen in order to verify validity of the proposed testing methodology and obtained results. Analysis results demonstrate that the material properties obtained are appropriate for the FE analysis of spot-welded specimens.

*Keywords*: Resistance spot weld; material thermal cycle simulator (MTCS); thermally simulated HAZ; dynamic material properties.

## 1. Introduction

The electric resistance spot welding process is an indispensable assembling process of steel auto-panels in the automotive industries. As a modern auto-body contains several thousands of spot welds[1], the strength of spot welds becomes extremely important in the crashworthiness assessment of auto-body members. During the spot welding process, heat generated by electrical resistance induces inhomogeneous microstructures around

---

†Corresponding Author.

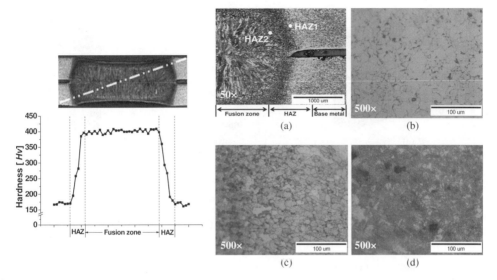

Fig. 1. Hardness distribution of the spot-welded SPRC340R.

Fig. 2. Optical micrograph for cross-section of the spot-welded SPRC340R: (a) overview; (b) base metal; (c) HAZ1; (d) HAZ2.

the welded nugget. These inhomogeneous properties should be considered in the FE analysis to characterize the actual mechanical behavior of a spot-welded specimen. In case of automotive simulation, material properties of inhomogeneous regions can be applied to a simplified FE modeling of the welded part described as a beam element instead of a rigid link to restrict two sheets. However, in most numerical simulations, inhomogeneous material properties are not considered due to the lack of both an effective testing methodology and available data for describing the material properties of the HAZ. Although Zuniga and Sheppard[2] proposed the relationship to approximate the yield and ultimate stress of HAZ as a function of a hardness value, few studies reported regarding appropriate testing methodology to acquire accurate stress–strain curves of the HAZ. Moreover, there have been little studies about the dynamic material properties of the HAZ, which are important to evaluate the dynamic failure load of a spot weld.

In this paper, testing methodology is newly proposed to identify the dynamic material properties of HAZ near the base metal in a resistance spot weld. At first, the thermally simulated HAZ specimen which has similar hardness and microstructure as the actual HAZ is fabricated using the material thermal cycle simulator (MTCS). In order to determine the thermal cycle of MTCS, the peak temperature was changed from 1000°C to 1300°C keeping other conditions. Dynamic tensile tests are then conducted with high speed material testing machine at various crosshead speed. Stress–strain curves of the thermally simulated HAZ are obtained at the strain rates ranged from 0.001/sec to 100/sec. Finally, the material properties obtained are applied to the finite element analysis of the spot-welded tensile-shear specimen in order to evaluate the validity of the proposed testing methodology and obtained results.

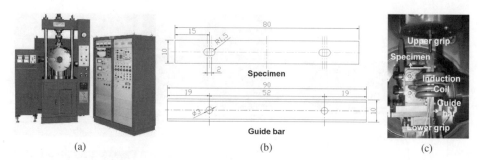

Fig. 3. Fabrication of the thermal simulated HAZ specimen: (a) material thermal cycle simulator (MTCS); (b) dimension of a specimen and a guide bar; (c) experimental set up for thermal simulation using MTCS.

## 2.   Thermal Simulation Tests for the Simulated HAZ

### 2.1.   *Classification of the spot-welded region*

The HAZ of a spot weld was investigated with a high strength steel sheet of SPRC340R whose thickness was 1.2 mm. Spot welding of steel sheets was performed using a static spot/projection welding machine. The welding current of 7.6 kA was imposed during the welding time of 12 cycles at 60 Hz with the holding force of 4.0 kN. Hardness distribution and optical micrographs for produced spot-welded regions are shown in Fig. 1 and Fig. 2, respectively. From the hardness profile and micro-structures shown in the figures, spot-welded region can be classified into three heterogeneous metallurgical zones as follows[3]: the base metal (BM) where no metallurgical transformation occurs; the heat-affected zone (HAZ) where regeneration and transformation zone conveying thermal and structural gradients between the melted zone and the base metal; the fusion zone or nugget (FZ) where melting and solidification occurs. The range of hardness values at BM, HAZ and FZ are 165~170 *Hv*, 170~400 *Hv* and 400~410 *Hv*, respectively. As the hardness profile explains, the HAZ is the transition zone between the FZ and the BM with steep gradients in the hardness. The varying mechanical properties of the HAZ are resulted from non-uniform thermal histories during the spot weld. The HAZ also can be divided into several subzones which have distinct microstructures and mechanical properties.

In this paper, the HAZ is divided into two subzones of the HAZ1 which is close to the BM and the HAZ2 which is close to the FZ. The material properties of the HAZ1 are examined in this paper since failure of the spot weld occurs at the interface between the HAZ and the BM when a large load is applied to the spot-welded components. The average grain size in the HAZ1 is 9.22 μm and the range of the hardness value is 170~190 *Hv*. Compared with the grain size of 12.54 μm in the BM, grains are refined due to the thermal histories of melting and fast solidification during the welding process. The grain size mentioned above was measured by Planimetric method[4] explained in ASTM E112 standard.

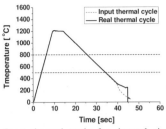

Fig. 4. Input thermal cycle for thermal simulation using MTCS.

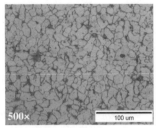

Fig. 5. Optical micrograph of the specimen after thermal simulation.

## 2.2. *Fabrication of the thermal simulated HAZ*

The goal of the simulated HAZ development effort is to generate tensile specimens of simulated HAZ materials that have similar hardness values and microstructures to those of the HAZ1 in actual spot weld and uniform microstructure throughout the tensile specimen. This preparation was carried out with the assumption that material properties would be the same as that of the HAZ1 when the simulated HAZ specimen has similar hardness values and grain sizes that match those of the HAZ1.

MTCS shown in Fig. 3(a) is used to fabricate the thermally simulated HAZ specimen. Although the MTCS has been widely adopted to fabricate the simulated HAZ in arc welding of thick plates, sufficient eddy currents to elevate the temperature up to melting points cannot be fully induced in the steel sheets. As a remedy to induce sufficient eddy current, two guide bars are attached onto the sheet specimen as shown in Fig. 3(b). The specimen with guide bars is mounted on MTCS to fabricate the thermally simulated HAZ specimen as shown in Fig. 3(c).

Fig. 4 represents the thermal cycles imposed on the specimen to fabricate simulated HAZ specimen. The peak temperature is assigned as 1200°C with the holding time of 5 sec considering the thermal histories of HAZ1 in actual spot welding process. Cooling rate from 800°C to 500°C is 35°C/sec. During the operation of the MTCS, a function generator transmits the thermal cycles into the feedback controller in the MTCS to control the temperature by comparing the measured temperature with the thermal cycles inputted. Microstructure and hardness distribution of the simulated HAZ specimen are shown in Fig.5 and Fig. 6, respectively. Average grain size obtained from the thermal simulation is 9.62 μm with the hardness of 186 *Hv*. Compared with the grain size and hardness value in HAZ1 of the actual spot weld, the simulated HAZ specimen can be fabricated with the thermal cycles shown in Fig. 4.

## 3. Dynamic Material Properties of the Simulated HAZ

### 3.1 *Dynamic tensile tests of the simulated HAZ specimen*

Using the simulated HAZ specimen, dynamic tensile tests were conducted at the strain rate ranged from 0.001/sec to 100/sec. A high speed material testing machine[5] is utilized in the dynamic tensile tests. The simulated HAZ specimen is mounted onto the high

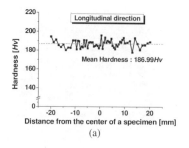
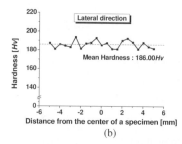

Fig. 6. Hardness distribution of the simulated HAZ specimen: (a) longitudinal direction; (b) lateral direction.

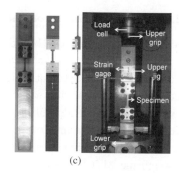

(a)                    (b)                    (c)

Fig. 7. Dynamic tensile tests of the simulated HAZ specimen: (a) high speed material test machine; (b) dimension of the specimen; (c) fixture set for mounting the specimen onto high speed material testing machine.

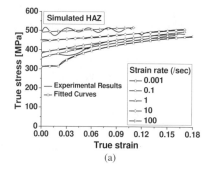
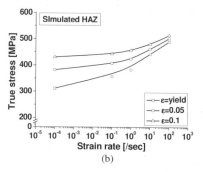

(a)                    (b)

Fig. 8. Experimental results of the simulated HAZ specimen: (a) true stress-strain curves at various strain rates; (b) strain rate sensitivity.

speed material testing machine using the fixture set as shown in Fig. 7.

The stress–strain curves of the simulated HAZ are obtained from the tests at various strain rates as shown in Fig. 8(a). Compared with the yield stress of 246.8 MPa for the BM at quasi-static states, the simulated HAZ shows higher strength due to the grain refinement caused by the thermal histories. The results also indicate that the flow stress gradually increases as the strain rate increases. The strain rate sensitivity of the flow stress is also plotted in Fig. 8(b) with different plastic strains. The figure represents that the strain rate sensitivity of the flow stress decreases with increasing plastic strain.

(a)         (b)

Fig. 9. FE modeling of the lap-shear specimen: (a) overview; (b) detailed FE model near the spot weld.

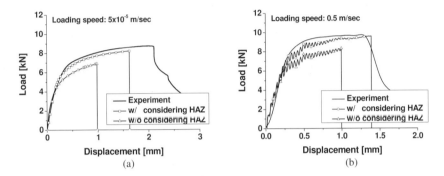

Fig. 10. Comparison of the FE analysis results to the experiment: (a) quasi-static loading; (b) dynamic loading.

## 3.2 *Application to FE analysis of the spot-welded specimen*

In order to verify the testing methodology and experimental results using the simulated HAZ, material properties obtained from the simulated HAZ are applied to the FE analyses of the spot-welded lap-shear specimen. A finite element model for lap-shear specimen is shown in Fig. 9(a). The specimen is modeled with three-dimensional solid elements. The HAZ is divided into five element groups. The HAZ close to BM is named as HAZ-A and that close to FZ is named as HAZ-E as shown in Fig. 9(b). The stress–strain curves shown in Fig. 8(a) are given to HAZ-A and the properties from HAZ-B to nugget are estimated with the assumption that stress in plastic deformation region is proportional to hardness profile.

Fig. 10 compares load–displacement curves obtained from the analyses to those from the experiments. The comparison represents that the results obtained from the analyses are close in coincidence with those obtained from the experiments. The discrepancies between analyses and experiments are less than 3% when the material properties of HAZ are considered in the FE analysis. Analysis results indicate that material properties obtained from the simulated HAZ specimens are valid for the FE analysis of the spot-welded specimen.

## 4. Conclusion

This paper proposes a methodology to identify the dynamic material properties of the HAZ in a spot weld. Using the MTCS, the simulated HAZ specimen is fabricated as the representative of the HAZ1 in actual spot weld. Stress–strain curves of the thermally simulated HAZ are obtained at various strain rates ranged from 0.001/sec to 100/sec. Compared with the BM, the simulated HAZ shows higher strength due to the grain refinement caused by the thermal histories. The material properties obtained from the experiment are applied to the FE analyses of the spot-welded lap-shear specimen. The analyses results fully demonstrate that testing method proposed are valid to evaluate the material properties of the HAZ.

### References

1.   Y. J. Chao, *J. Eng. Mater.- T. ASME.* **125** (2003) 125.
2.   S. M. Zuniga and S. D. Sheppard, *Modeling Simul. Mater. Sci. Eng.* **3** (1995) 391.
3.   S.-H. Lin, J. Pan, S.-R. Wu, T. Tyan and P. Wung, *Int. J. Solids Struct.* **39** (2002) 19.
4.   ASTM E112, Standard Test Methods for Determining Average Grain Size, (2006).
5.   H. Huh, S. B. Kim, J. H. Song and J. H. Lim, *Int. J. Mech, Sci.* **50** (2008) 918.

# AN INFLUENCE OF SELECTED MECHANICAL PARAMETERS OF MMC ON THE THERMAL SHOCK RESISTANCE

DARIUSZ RUDNIK[1†], ZBIGNIEW KOWALEWSKI[2], KRYSTYNA PIETRZAK[3], ANDRZEJ WOJCIECHOWSKI[4]

*Motor Transport Institute, ul. Jagiellonska 80, Warsaw, 03-301, POLAND,*

*dariusz.rudnik@its.waw.pl, zbigniew.kowalewski@its.waw.pl, krystyna.pietrzak@its.waw.pl*

*andrzej.wojciechowski@its.waw.pl*

Received 15 June 2008
Revised 23 June 2008

The results concerning the basic mechanical properties are presented for selected metal matrix composites (MMC). Three types of composites were produced by means of squeeze casting and next heat treatment. An influence of reinforcement type on the crack appearance in matrix and composite material was analyzed. Theoretical evaluation of the thermal shock resistance was proposed on the basis of conventional mechanical characteristics. Moreover, the qualitative metallographic investigations were carried out using optical microscopy method. A good consistence between the theoretical evaluation of thermal shock resistance and experimental data was achieved.

*Keywords*: Composite; mechanical properties; metallography; thermal shock resistance.

## 1. Introduction

For several years metal matrix composites (MMC) have been developed because of their attractive operating properties. The high level of mechanical properties and low weight of products fabricated from MMC are advantageous especially in aerospace and motor industry solutions. Introducing the MMC reinforced by hard particles or fibers, in total volume or locally, results in an improvement of mechanical properties of ready made products.[1-3] From the technological point of view it is important to obtain the products of the repeatable properties assessed by means of quantitative methods and considered with respect to material microstructure from which they were made. The phenomenon mechanisms in operating conditions as well as microstructure differences, revealed as the result of manufacturing processes, require development of good quality assessment methods. The research work can be treated as an attempt to create a new methodical solution in thermal shock resistance assessment. The aim of this reasoning was to obtain a good consistence between the theoretical evaluation of thermal shock resistance and experimental data.

---

[†]Corresponding Author.

## 2. Qualitative Analysis of an Influence of Reinforcement on the Crack Appearance

The resistance of a material to the thermal shock is closely connected with its mechanical strength, since when it is subjected to the alternating thermal field, the internal stresses responsible for the formation of cracks are generated. The application of the external heat source to the material causes a temperature gradient formation. For the determined value of thermal expansion linear coefficient the material begins to be deformed. A strain magnitude can be calculated from the formula:

$$\varepsilon = \alpha \cdot \Delta T \tag{1}$$

where: $\varepsilon$   - strain,
$\quad$ $\alpha$   - linear thermal expansion coefficient [m/m·K],
$\quad$ $\Delta T$ - temperature gradient [°C].

Knowing Young's modulus it is easy to determine the stresses inside the material due to the temperature gradient:

$$\sigma = E \cdot \alpha \cdot \Delta T \tag{2}$$

where: E   - Young modulus [GPa],
$\quad$ $\varepsilon$   - strain,
$\quad$ $\alpha$   - linear thermal expansion coefficient [m/ m·K],
$\quad$ $\Delta T$ - temperature gradient [°C].

In case of the material being subjected to the constant thermal field, the homogeneous temperature distribution is kept in it ($T_1 = T_2$) and, as a consequence, the deformation does not appear. An application of the external heat source changes the temperature distribution inside the material. A temperature gradient ($\Delta T \neq 0$) created leads to the internal stresses formation, and as a consequence, to the process of material deformation. A degree of deformation is dependent on the temperature gradient and linear thermal expansion coefficient. The frequency of temperature variations, the values of $T_1$ and the coefficient of the thermal conductivity $\lambda$ may have an influence on the temperature gradient ($\Delta T$). The material located inside the alternating thermal field is exposed to the tension-compression cycles. During the rapid cooling, the temperature on the surface of the material is lower than that inside it. Hence, relatively high decohesion stresses appearing, which are responsible for cracks generation. The materials of high value of the thermal conductivity coefficient are resistant to thermal shocks, since inside them, the temperature gradient does not appear, even at large differences between the temperatures $T_1$ and $T_2$. Hence the resistance of monolithic materials to the thermal shocks is dependent on the coefficient $\lambda$ mentioned earlier. The thermal conductivity is proportional to the electric conductivity coefficient, which allows to conclude that the introduction of alloy additions to the alloy solution lowers the electric (and thermal) conductivity and diminishes the resistance of the alloy to the thermal shock. Only an adequate heat treatment, leading to creation of the alloy additions as the separate phase, can increase thermal resistance of the alloy. In case of low thermal conductivity composites, their greater resistance can be explained by the occurrence of the reinforcing phase. This phase, representing about 10-times smaller coefficient of the thermal expansion, enables the composite enlargement during heating. Because of the rapid cooling stage, composite surrounding the shrinkage is subjected to the smaller tension stresses in comparison to those occurring on the surface. Such process influences beneficially the thermal resistance of composite. In case of variations observed in the material microstructure, the hypothetical model of the microcracks propagation can be divided in two groups. In case of two-phase monolithic (α, β) material the propagation of cracks can proceed in the following ways.

(1) Across the phase α and β,

(2) Along grain boundaries,

(3) Across shrinkage pores,

(4) Only across the phase α,

(5) Across the precipitations of the second phase.

Taking into account reinforced metal matrix composites, the additional ways of the crack propagation appear i.e.:

(6) Across particles (or fibers),

(7) Along by particles (fibers) - on the matrix border,

(8) Across the fibers.

By an adequate modification of the material microstructure it is possible to prevent efficiently the occurrence of the particular form of crack propagation. In case of monolithic materials by decreasing of the quantity of pores as well as by the improvement of strength properties of the component in the alloy it is possible to increase the strength of the alloy. Materials with the reinforcement demonstrate much better strength properties. An optimum-bond of phase borders, may be obtained by introducing an additional metal layers improving the wettability. In the optimum case, the bond on the phase should be of relatively high strength that enables microcrack crossing by the fiber. Due to the need of an additional data for modeling, the structural investigations were carried out using OLYMPUS PMG3 microscope. The representative microstructures are shown in Figs. 1-2. They enable visualization of the places where an initiation and then propagation of cracks created after the thermal shock test takes place in monolithic material (Fig. 1) and in composite material (Fig. 2).

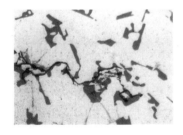

Fig. 1. Microstructure of the AK 12 alloy (the die casing) after thermal shock tests. Magn. 500x.

Fig. 2. Microstructure AK12/Al2O3 composite (after heat-treatment), after thermal shock tests of the material. Magn. 500x.

Microscopic observations of the tested materials confirmed predictions achieved using the propagation model. In case of monolithic material the microcracks are crossing along gas-pores and interphase borders. For the composite, with the weak connection between a fiber and matrix, the created cracks are passing around the fibers, which is consistent with the model proposed. A good example of the load transfer by reinforcing particles is shown in Fig. 2, because in this instance the fiber became torn. This case confirms the correctness of the rule of the reinforcement phase which can be treated as specific kind of mechanical barrier with relation to composite metal matrix.

## 2. Theoretical Estimation of the Composite Resistance to Thermal Shocks

The resistance to thermal shocks ($\sigma$) was evaluated for three composites, F3S.20S, THE ALPHA® (produced by squeeze casting and then heat treated) and AK12/Al$_2$O$_3$. In general the resistance to thermal shocks $\sigma$ can be expressed as a function of different properties of the material using the following formula:

$$\sigma = f\left[\frac{R_m^R, R_m^T, I, E}{\alpha, \lambda, K, A}\right] \tag{3}$$

where: $R_m$ - ultimate tensile strength at room temperature (MPa),
$R_m^T$ - ultimate tensile strength at elevated temperature T, (MPa),
$I$ - the coefficient characterizing the bond between reinforcement and matrix ($I>1$),
$E$ - Young's modulus (GPa),
$\alpha$ - coefficient of thermal expansion ($\mu$m/m$\cdot$K),
$\lambda$ - thermal conductivity (W/m$\cdot$K),
$K$ - shape factor,
$A$ - coefficient characterizing plastic proprieties.

In practice the resistance to thermal shocks is characterized by the temperature variation $\Delta T$ calculated on the basis of heating and cooling cycles. Such process leads to the stress creation, and as a consequence, to the crack initiation in the monolithic material. The $\Delta T$ can be determined using the following equation:

$$\Delta T = \frac{\sigma(1-v)}{E\alpha} \tag{4}$$

where $\sigma$ and $v$ are the cracking stress and Poisson's ratio, respectively. As a first attempt, the resistance to thermal shock can be assessed for the composite material from equation (4) putting in to it the measured or estimated values of E, $\alpha$, $v$ and $\sigma$.

Such approach was used in calculations of the resistance to thermal shocks for selected composites. The Poisson ratio and elasticity modulus of composite can be estimated as a mixing rule (RM):

$$v_c = v_m\phi + v_r(1-\phi), \ i \ E_c = E_m\phi + E_r(1-\phi) \tag{5}$$

where "φ" means the volume fraction of the reinforcing phase, while "c", "m" and "r" denote indices identifying composite, matrix, reinforcement, respectively. RM is extremely important for the reinforcement which is thermodynamically stable in the metal. In the case of Al-Fly-ash system such behavior is not observed. The fly-ashes are chemically active[1] and they are covered continuously by $Al_2O_3$ nano-particles according to the reaction:

$$6MeO + 4Al = 2Al_2O_3 + 6Me$$

where Me denotes metal-component of the oxide. The particles of $Al_2O_3$ reinforce additionally the metal matrix and increase the thermal shock resistance of composite containing fly-ashes.

The thermal expansibility $\alpha_c$ of composite was estimated from Turner's formula[5]:

$$\alpha_c = \frac{\alpha_m(1-\phi)\cdot K_m + \alpha_r\phi\cdot K_r}{(1-\phi)\cdot K_m + \phi K_r} \tag{6}$$

where $K_r$ and $K_m$ are the elasticity coefficients of reinforcement and matrix, respectively. The data used in all calculations, are put together in the Table 1.

Table 1.  Input data used in calculations of thermal shock resistance.[3]

| Material \ Properties | F3S.20S | AK12 | SiC | Al₂O₃ | Fly-ashes |
|---|---|---|---|---|---|
| | | | ΔT = 270°C | | |
| Young modulus E [GPa] | 75 [6] | 80 [4] | 400 | 285 | 304 [*] |
| Poisson's coefficient ν | 0.33 [6] | 0.33 [6] | 0.17 | 0.25 | 0.20 [*] |
| Coefficient of the thermal expansion, [μm/m.K] | 18.4 [4] | 18.4 [4] | 4.3 | ~5.5 | 5.79 [*] |
| Coefficient of volume elasticity K [**] [GPa] | 73.5 | 78.4 | 202.1 | 190 | 170 |

[*] Calculated values  [**] Calculated values according to the relationship: K = E/3(1-2ν).

A diagram showing that the crack resistance variations for three composites as a function of the volume fraction of reinforcing phase is presented in Fig. 3. The results confirm qualitatively a good consistency with the experimental data. The composite AK12/$Al_2O_3$ shows better thermal shock resistance than F3S.20S composite. The thermal shock resistance of composite with the SiC particles shows the local maximum, within the range of examined volume fraction. It was observed for the 35% SiC content.

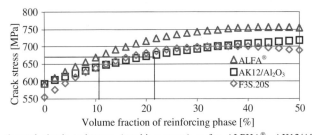

Fig. 3. Theoretical thermal shock resistance (cracking stress) σ for ALPHA®, AK12/Al₂O₃ and F3S.20S composites.

The ALPHA® composite is of higher crack resistance than F3S.20S composite within the range of total volume fraction of reinforcing phase (its crack stress for the 10.36% vol. content of the ash is equal to the crack stress of $AK12/Al_2O_3$ composite). It has to be noticed that the calculations presented here, characterize the static or thermal equilibrium state, only. They do not reflect the dynamic or temporary states during the cooling or heating. The large difference of the thermal conductivity between the matrix and reinforcement during heating (or cooling) leads to creation of the different thermal profiles in these two phases. Such differences in the temperature profile can change the thermal deformations in both phases and generate additional stresses.

## 3.  Conclusions

The following conclusions can be formulated:
*   The results of investigations confirm the beneficial effect of reinforcing phase and squeeze casting manufacturing methods, on the thermal shock resistance.
*   The proposed equation characterizing thermal shock resistance, takes into account the reinforcing phase. The final result depends on the strength properties, stiffness, volume fraction of reinforcing phase and metal matrix.
*   The results of thermal shock resistance estimation of selected composites are qualitatively in good consistency with the experimental observations.

## References

1.  J. Sobczak, *Kompozyty metalowe* (Motor Transport Institute, Foundry Institute, Warsaw, 2001), p. 314.
2.  A. Wojciechowski, J. Sobczak, *Kompozytowe tarcze hamulcowe pojazdow drogowych* (Motor Transport Institute, Warsaw, 2001), p. 169.
3.  D. Rudnik, J. Sobczak, *Tloki kompozytowe do silnikow spalinowych* (Motor Transport Institute, Warsaw, 2001), p. 127.
4.  N. Sobczak, J. Sobczak, P.K. Rohatgi, in *Proceedings of ECOMAP-98* (High-Temperature Society of Japan, Kyoto, Japan, 1998), p. 195.
5.  P.S. Turner, *J. Res. NBS* (1946), vol. 37, pp. 239.
6.  *Metals Handbook, vol. 2 Properties and Selection: Nonferrous Alloys & Special-Purpose Materials,* 10th edition (ASM International, 1990), pp. 152-177.

# NUMERICAL SIMULATIONS OF THE INFLUENCE OF STRIKER BAR LENGTH ON SHPB MEASUREMENTS

DONG WEI SHU[1†], CHUN QI LUO[2], GUO XING LU[3]

*School of Mechanical and Aerospace Engineering, Nanyang Technological University, 50 Nanyang Avenue,*
*639798, SINGAPORE,*
*mdshu@ntu.edu.sg, luoc0003@ntu.edu.sg, GXLu@ntu.edu.sg*

Received 15 June 2008
Revised 23 June 2008

Split Hopkinson Pressure Bar (SHPB) has become a frequently used technique for measuring uni-axial compressive stress-strain relationship of various engineering materials under high strain rates. The pulse shape generated in the incident bar is sensitive to the length of the striker bar. In this paper, a finite element simulation of a Split Hopkinson Pressure Bar is performed to estimate the effect of varying length of striker bar on the stress-strain relationship of a material. A series of striker bars with different lengths, from 200mm to 350mm, are employed to obtain the stress-strain response of AL6061-T6 in both simulation and experiment. A comparison is made between the experimental and the computed stress-strain curves. Finally the influence of variation of striker bar length on the sample's stress-strain response is presented.

*Keywords*: Hopkinson Bar; striker bar length; stress-strain; plastic.

## 1. Introduction

The high strain rate behavior of materials is of considerable importance in many applications related to the design of armor systems, aircraft, and high-speed transportation vehicles and machinery. With the knowledge of mechanical deformation of materials under high strain rate loading, new structures under extreme conditions can be designed and built to withstand high velocity and compressive impact. One of the most widely used experimental configurations for high strain-rate material measurement is the Split Hopkinson Pressure Bar[1]. Fig. 1 shows the experimental arrangement.

The stress,strain and strain rate of the specimen can be obtained as follows.

---

†Corresponding Author.

$$\sigma_{specimen} = E \frac{A}{A_s} \varepsilon_t . \tag{1}$$

$$\varepsilon_{specimen} = -2 \frac{C_0}{L} \int \varepsilon_r dt . \tag{2}$$

$$\dot{\varepsilon}_{specimen} = -2 \frac{C_0}{L} \varepsilon_r(t) . \tag{3}$$

Where L is the initial length of the specimen and $C_0$ is wave propagation velocity. A and $A_s$ are the cross sectional area of the pressure bar and specimen, respectively. E is the Young's modulus of the bar material.

This paper reports numerical and experimental studies of the influence of striker bar length on SHPB Measurements. P.S.Follansbee[2] studied the longitudinal collinear impact

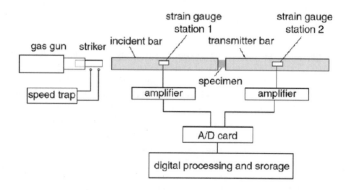

Fig. 1. Schematic diagram of the Hopkinson Bar set-up.

of two rods and pointed out that, the amplitude of strain in the incident bar is proportional to its impact speed and its duration is only related to length of the impact bar. Liu and Li[3] investigated how the shape of a striker bar could influence the rise time of the incident pulse, resulting in significant reduction of stress oscillation. In this paper, we vary the striker bar length from 200mm to 350mm to investigate the effect of changing the bar length on stress-strain relationship of aluminum specimen.

## 2.   Numerical Simulation and Analysis

The finite element code LS-DYNA is employed to analyze the SHPB testing system. A three-dimensional model is more appropriate due to its accuracy with the cost of longer computation time[4]. The reason for using a three-dimensional model rather than a two-dimensional model is that friction force exists at the contact interface of the specimen and the pressure bars and it could not be simplified into a plane-stress model. The finite element model includes the incident bar, specimen and the transmitted bar. Fig. 2 shows the finite element model for SHPB used in LS-DYNA.

The material parameters of the bars are taken as the same as in the experiments: Young's modulus E=200GPa, Poisson ratio $V$ =0.3,density $\rho = 7800 kg/m^3$ . The

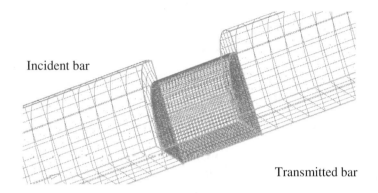

Incident bar

Transmitted bar

Fig. 2. Finite element model Specimen in LS-DYNA.

element type is selected as SOLID-164 and contact type is automatic-surface-to-surface which contains the effect of slip when the contact force is bigger than the friction force whilst the former is calculated automatically by LS-DYNA.The calculation parameters, such as damping coefficient, static friction and dynamic friction coefficients, are selected as 0.08,0.1 and 0.06,respectively,which are taken from the best results of numerical simulations compared with the experimental results. By taking advantage of symmetry[5], one quarter model with the corresponding symmetrical boundary condition is used for the calculation. In order to improve the accuracy, the symmetrical modeling technique and a mapped meshing technique are employed to define high-quality three-dimensional hexahedral elements. Considering both numerical precision and calculation efficiency, the sizes of the elements in the numerical simulations are 1.5mm for the bars and 0.5mm for the specimen.

The geometric parameters of the bars and specimen are presented as follows.

Table 1.  Geometric parameters of specimen and Hopkinson Bars.

| Item | Length(mm) | Diameter(mm) |
|---|---|---|
| Striker bar | 200,250,300,350,respectively | 12.7 |
| Specimen | 5 | 10 |
| Incident bar | 1000 | 12.7 |
| Transmitted bar | 1000 | 12.7 |

The empirical Johnson-Cook strength model is used to simulate material behavior of the aluminum 6061-T6 specimen at high strain rates[6].

$$\bar{\sigma} = [A + B(\bar{\varepsilon})^N] \left[ 1 + C \ln \left( \frac{\dot{\varepsilon}}{\dot{\varepsilon}_0} \right) \right] (1 - \theta^m). \tag{4}$$

The Johnson Cook coefficients for specimen AL6061-T6 used in this paper are summarized in Table 2.

Table 2. The Johnson Cook coefficients for specimen AL6061-T6.

| A | B | C | N | m | $\dot{\varepsilon}_0$ |
|---|---|---|---|---|---|
| 289.6MPa | 203.4MPa | 0.011 | 0.35 | 1.34 | $1.0\,S^{-1}$ |
| $\theta_{melt}$ | $\theta_{transition}$ | $\rho$ | $V$ | E | |
| 925.37K | 294.26K | 2700(kg/$m^3$) | 0.29 | $2.0 \times 10^{11}$ | |

In order to investigate the effect of variation of the striker bar length, A series of bars with varying lengths ranging from 200mm to 350 mm are used in simulation. The

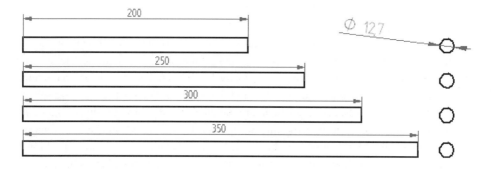

Fig. 3. Striker bars with varying length from 200mm to 350mm used in simulations and experiments.

dimensions of the striker bars are shown in Fig. 3.

## 3.  Experimental Setup

Experiments are performed to investigate the influence of changing the striker bar length. The experimental setup is shown in figure 1.The geometric parameters of the bars are summarized in Table 1. The striker bar is fired onto the incident bar by a pneumatic gun at impact velocities of 10-30m/sec. The specimen is sandwiched between the incident bar and transmitted bar[7] .A strain gauge installed on the incident bar is designed to measure both incident and reflected stress wave, while its counterpart on the transmitted bar measures the transmitted stress wave. A digital oscilloscope records these waveforms and then the data is transferred to a personal computer for further analysis. Fig. 4 shows a typical set of oscilloscope traces.

Four strikers bars with lengths of 200mm,250mm,300mm,350mm are employed to conduct the experiments under an identical striking speed 15m/s, respectively.

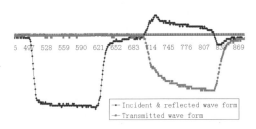

Fig. 4. Typical set of oscilloscope traces from SHPB test.

## 4. Results and Discussion

The aluminum specimen AL6061-T6 first undergoes elastic deformation before reaches its yield strength, after that the plastic region begins which is modeled in FEM using Johnson-Cook plastic strength model. The experimental results and simulation results are plotted in Fig 5.

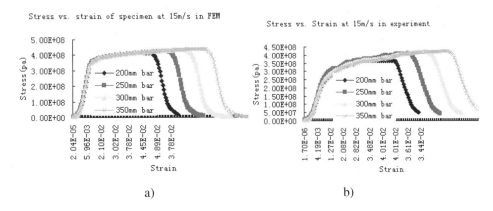

a)          b)

Fig. 5. Stress vs. Strain curves of specimen using 200mm, 250mm, 300mm, 350mm striker bar under impact speed 15m/s, a) in simulation b) in experiment.

A linear relationship of stress and strain is exhibited in Fig.5 at the initial part in which the stress of sample is within yield stress 300Mpa, the slope of the curve in this region denotes the Young's modulus of AL6061-T6 which is 70Gpa.

After the yield point, the plastic deformation region begins in which the stress grows slower than the former elastic deformation region. When the stress reaches the maximum strength of the specimen, the curve decreases drastically to zero.

Under an identical striking speed, the maximum stress that the curve can reach depends on the length of the striker bar. The longer the striker bar, the larger the maximum value. For 200mm, 250mm, 300mm and 350mm striker bars, the maximum stresses are 418Mpa, 427Mpa, 435Mpa, and 441Mpa respectively.

## 5.  Conclusion

It is concluded that the stress-strain curves obtained by using different length striker bars exhibit the same profile except that the maximum stress points are different. Under an identical striking speed, the maximum stress that the curve can reach depends on the length of the striker bar, the longer the striker bar, the larger the maximum value.

## References

1.  D. W. Shu, "Tensile Mechanical Properties of AM50A Alloy by Hopkinson Bar", Key Engineering Materials Vols. 340-341 (2007) pp. 247-254.
2.  P. S. Follansbee and C. Franz, Wave propagation in the split Hopkinson pressure bar. J. Engng Mater. Technol.105, 61-66 (1963).
3.  Liu, D. and Li, X. B. Dynamic Inverse design and experimental study of impact pistons, Chinese J. of Mechanical Eng., 34(4), 506-514. (1998).
4.  J. W. Tedesco, M. L. Hughes and B. P. O'Neil, Numerical analysis of dynamic direct tension and direct compression tests. Final Report, No. ESL-TR-91-41, Air Force Civil Engineering Support Agency, Tyndall AFB, FL (1992).
5.  Hallquist JO. LS-DYNA theoretical manual. Livermore Software Technology Corporation, May 1998.
6.  G. R. Johnson and W. H. Cook, A constitutive model and data for metals subjected to large strains, high strain rates, and high temperature. Proc.7th Int.Symp.on Ballistics, Hague, Netherland, April (1983).
7.  H. Kolsky, An investigation of the mechanical properties of materials at very high strain rates of loading. Proc. Phys Sot. Section B &2, 676-700 (1949).

# QUANTITATIVE RELATIONSHIPS BETWEEN MICROSTRUCTURAL AND MECHANICAL PARAMETERS OF STEELS WITH DIFFERENT CARBON CONTENT

KRYSTYNA PIETRZAK[1†], ADAM KLASIK[2], ZBIGNIEW KOWALEWSKI[3], DARIUSZ RUDNIK[4]

*Motor Transport Institute, Jagiellonska 80,*
*Warsaw, 03-301, POLAND,*
*krystyna.pietrzak@its.waw.pl, adam.klasik@its.waw.pl, zbigniew.kowalewski@its.waw.pl,*
*dariusz.rudnik@its.waw.pl*

Received 15 June 2008
Revised 23 June 2008

The steels of different carbon content were tested with respect to their microstructural and selected mechanical parameters variations. The original combinatorial method, based on the phase quanta theory, was applied in the quantitative metallographic assessment. Hence, the adequate geometrical parameters of steel microstructure were possible to be determined. Moreover, the selected mechanical parameters of the materials were evaluated using non-destructive and destructive methods. The results enabled to formulate some significant quantitative relationships between microstructure and mechanical parameters.

*Keywords*: Steel; phase composition; quantitative metallography; mechanical behaviour.

## 1. Introduction

A basic problem in fabrication of the final products, which are made from the various kinds of materials, is to obtain their high and also repeatable quality resultant from the set of required functional properties of materials used in this production. Continuous growth of requirements concerning the level of these properties forces an improvement of investigation methods of functional properties as well as the methods for microstructural characterization of the material. The elaboration of quantitative relationships between the selected functional properties and material microstructure, described by the set of stereological parameters, is the optimal solution of material characterization. Besides, the cognitive advantage of the relationships, the knowledge can be treated as the basis to the conscious formation of the microstructure by the specific technological processes, which have to be applied for obtaining the adequate compromises with regard to anticipated properties. On the basis of previous papers[1-11] it is easy to conclude that many researchers try to show the relationships between selected properties and a material on the basis of quantitative metallographic methods. Many efforts were used to develop the quantitative

†Corresponding Author.

447

metallographic methods of microstructure assessment, but still there are a lot of problems to be solved. The investigators usually choose a certain geometrical structural parameter and then correlate it with the chosen mechanical properties. In this way, the complex relationships cannot be obtained because the features of material microstructure are changing continuously. Therefore, there is a need to describe completely the microstructure by means of an adequate set of parameters to identify uniquely e.g. a phase/compound according to its content and morphological characteristic. This is possible by the use of combinatorial method, based on the phase quanta theory. The details of the combinatorial method are presented in ref 12 and also in Table 1 and Fig. 1.

Table 1. Computational formulas used for calculations of stereological parameters on the basis of combinatorial method.

| No | Stereological value | Theoretical parameter | Computational formula |
|---|---|---|---|
| 1 | Volume fraction $V_V$ [%] | $m$ | $V_V = \dfrac{m}{M} \times 100\%$ |
| 2 | Surface area of the interphase boundary in unit test volume $S_V$ [mm$^{-1}$] | $r$ | $S_V = 4\,r\ [\mathrm{mm^{-1}}]$ |
| 3 | Specific surface area $S_V/V_V$ [mm$^{-1}$] | $\dfrac{r}{m}$ | $\dfrac{S_V}{V_V} = \dfrac{4r\,M}{m}[\mathrm{mm^{-1}}]$ |
| 4 | Total surface curvature $K_V$ [mm$^{-2}$] | $\dfrac{r^2}{m}$ | $K_V = 2N_A = \dfrac{16M}{3} \times \dfrac{r^2}{m}[\mathrm{mm^{-2}}]$ |
| 5 | Mean free distance between the particles $\lambda$ [mm] | $\dfrac{M-m}{M\,r}$ | $\lambda = \dfrac{M-m}{M\,r}[\mathrm{mm}]$ |
| 6 | Mean chord $\bar{l}$ [mm] | $\dfrac{m}{r}$ | $\bar{l} = \dfrac{m}{r}a\,[\mathrm{mm}]$ |
| 7 | Number of interceptions of particles per unit area $N_A$ [mm] | $\dfrac{r^2}{m}$ | $N_A = \dfrac{8M}{3\pi} \times \dfrac{r^2}{m}[\mathrm{mm^{-2}}]$ |
| 8* | Number of particles in the unit test volume $N_V$ [mm$^{-3}$] | $\dfrac{r^3}{m^2}$ | $N_V = \left(\dfrac{4}{3}\right)^2 \dfrac{M^2}{\pi} \times \dfrac{r^3}{m^2}[\mathrm{mm^{-3}}]$ |
| 9* | Mean diameter of particles $D$ [μm] | $\dfrac{m}{r}$ | $D = \dfrac{3}{2} \times \dfrac{m}{r} \times \dfrac{10^3}{M}[\mu m]$ |

* concerns the polydispersion sphere set assumed

From the theoretical point of view any material microstructure can be treated as an arrangement of its elements (being subjected to quantitative metallographic assessment) in material matrix. In measuring practice this way minimizes the quantitative analysis to the determination of two basic estimators: the estimator of the phase/compound volume fraction ($L_L$ [%]) and the estimator of the surface area (phase/compound - matrix as $N_L$) [mm$^{-1}$]. Then, on the basis of adequate computational formulas (Table 1), any geometrical parameters of the microstructure can be determined.

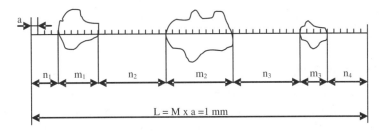

a   -   measurable microstructure element (phase quanta)
m   -   number of "a" elements of phase/component on the test line consisted of "M" elements
n   -   number of "a" elements of matrix on the test line consisted of "M" elements
r   -   number of segments (chords)

Fig. 1. Scheme showing the meaning of parameters determined by means of combinatorial method.

This simplifies significantly the problem of measurement using the quantitative metallographic method and assures the same accuracy of geometrical microstructure parameters determined because output data ($L_L$ and $N_L$), used for further calculations, are proceeded from the single measuring process only.

## 2. Materials

The chemical compositions of investigated steels are shown in Table 2.

Table 2. The chemical compositions of carbon steels in as-received state.

| Steel grade | C | Mn | Si | P | S | Cr | Ni | Mo | Al | Cu | As |
|---|---|---|---|---|---|---|---|---|---|---|---|
| | content [%] | | | | | | | | | | |
| C10 | 0.09 | 0.52 | 0.28 | 0.03 | 0.03 | 0.15 | 0.20 | 0.05 | 0.02 | 0.1 | 0.04 |
| C15 | 0.14 | 0.45 | 0.23 | 0.02 | 0.01 | 0.21 | 0.15 | 0.03 | 0.01 | 0.1 | 0.03 |
| C25 | 0.26 | 0.48 | 0.25 | 0.02 | 0.02 | 0.18 | 0.25 | 0.08 | 0.02 | 0.2 | 0.05 |
| C45 | 0.44 | 0.62 | 0.32 | 0.03 | 0.02 | 0.23 | 0.21 | 0.06 | 0.02 | 0.1 | 0.04 |
| C55 | 0.56 | 0.78 | 0.29 | 0.02 | 0.03 | 0.17 | 0.18 | 0.07 | 0.03 | 0.2 | 0.03 |

## 3. Results of Investigations

The results of analysis are presented in the form of relationships between selected mechanical and microstructural properties quantitatively assessed for steels of varying carbon content.

### 3.1. *Mechanical properties*

Mechanical properties were determined using the modified Low Cycle Fatigue method[13] adapted to practical purposes by Karamara and Maj[14], and damping capacity. Damping capacity was assessed by means of the Förster's elastomat using the method of free vibration damping.

## 3.2. *Microstructural properties*

The carbon steels C10, C15, C25, C45 and C55 in as-received state were subjected to tests in order to determine an influence of the phase composition of the steel matrix on the damping capacity of vibrations and other mechanical properties. The microstructures of these steels are shown in Fig. 2 a, b, c, d.

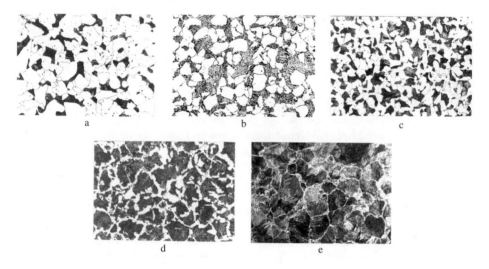

a                     b                     c

d                          e

Fig. 2.  Microstructures of unalloyed steels in as-received state: a) C10, b) C15, c) C25, d) C45, e) C55.

The correlation analysis revealed that there are the negative correlations between the damping coefficients ($\vartheta_n$), elongation ($A_5$) and the pearlite content ($L_L = A_A = V_v$) as well as between the damping coefficients ($\vartheta_n$), elongation ($A_5$) and the relative area pearlite-ferrite ($S_v = 4N_L$). The remaining mechanical parameters show the opposite tendency. Therefore, the synthetic variable consisted of both estimators was proposed as $\lambda^{-1/2}$, where $\lambda = (1-V_V)/N_L$ and denotes the mean free distance between pearlite-pearlite areas in unalloyed steels. Consequently, the stereological parameter $\lambda^{-1/2}$ describes the refinement of ferrite microstructure by pearlite areas in these steels. The results of investigations related to the proposed stereological parameter are summarized in Table 2. It has to be pointed out that, there are significant correlations between all mechanical parameters considered and the synthetic stereological parameter $\lambda^{-1/2}$ (even for the significance level $\alpha = 0,001$ - Table 3).

The variation tendency of the logarithmic decrements of vibrations ($\vartheta_n$) is opposite to the remaining mechanical properties, except the elongation $A_5$ which is connected with properties of the softer phase. In general, the damping capacity ($\vartheta_n$) depends also on the properties of the softer phase, and as a consequence, shows the same variation tendency as the elongation ($A_5$) does.

Table 3. Correlation coefficients ($\rho$) between the mechanical properties and stereological variables ($\lambda^{-1/2}$) and also their calculated ($t_o$) and theoretical ($t_t$) significances.

| Number of measurements - 12 | | Mechanical properties | | | | | | | | |
|---|---|---|---|---|---|---|---|---|---|---|
| Stereological parameter | $\lambda^{-1/2}$ | $\vartheta_n$ | $R_{0,02}$ | $R_{0,05}$ | $R_{0,10}$ | $R_{0,20}$ | $R_m$ | $Z_{go}$ | $A_5$ |
| Correlation coefficient | $\rho$ | -0,8367 | 0,8412 | 0,8881 | 0,9080 | 0,9082 | 0,9085 | 0,8468 | -0,7771 |
| Calculated significance | $t_o$ | 4,888 | 4,920 | 6,110 | 6,853 | 6,866 | 6,875 | 5,030 | 3,905 |
| Theoretical significance | $t_t$ | 2,228 for ($\alpha$=0,05)  3,169 (for $\alpha$=0,01)  4,578 for ($\alpha$=0,001) | | | | | | | |

The examples of determined quantitative relationships are presented in the form of diagrams, where for visualization of the best fit of curves the regression equations are shown in Figs. 3 and 4.

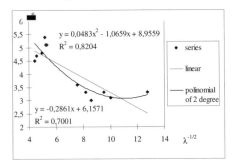 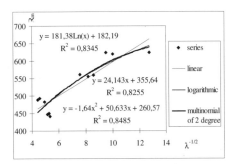

Fig. 3. Relationship between $\vartheta_n$ and $\lambda^{-1/2}$.    Fig. 4. Relationship between $R_m$ and $\lambda^{-1/2}$.

The results allow to formulate the following conclusions:
- the damping capacity and elongation of the unalloyed steels decrease significantly when the pearlite content ($L_L$) as well as the relative area pearlite-ferrite ($S_v$) increase,
- the refinement of ferrite by pearlite areas formulated by means of the synthetic parameter ($\lambda^{-1/2}$) affects significantly the mechanical properties of unalloyed steels,
- the quantitative relationships between the mechanical properties of unalloyed steels and the parameters of their microstructure can be formulated by means of the regression equations for the synthetic variable ($\lambda^{-1/2}$),
- the refinement of ferritic microstructure by pearlite areas affects significantly the mechanical properties such as: tensile strength ($R_m$), yield limits ($R_{0.05}$ $R_{0.10}$ $R_{0.2}$), estimated fatigue limit of rotary bending ($Z_{go}$) but the variation tendency is opposite to that connected with the damping capacity ($\vartheta_n$) and the elongation ($A_5$).

## 4. Concluding Remarks

Based on the results presented in this study it can be concluded that the combinatorial method used in the quantitative structure analysis makes it possible to investigate the quantitative relationships between selected mechanical properties and geometrical structure parameters. More importantly, the microstructural behaviour was described by the formulated synthetic parameter, which characterizes well the microstructure of the steels in question, giving volume fractions and morphological characteristics of elements existing in the microstructures investigated. The combinatorial method can be used to determine the microstructure of unalloyed steels in the complex way by means of the synthetic parameter. However, the results described in this study show that the variation tendency of some mechanical properties is opposite with respect to that for damping capacity and elongation observed. Such phenomenon depends on the synthetic stereological variable formulated as a result of the quantitative metallographic methods based on the phase quanta concept.

## References

1. J. Ryś J, *Metalurgia i Odlewnictwo* Tom 14, Zeszyt 3-4, (1988), pp. 417-433.
2. P. Clayton and D. Danks, *Wear,* 135, (1990), pp. 369-389.
3. A. Kulmburg, STERMAT'94 — Proc. 4th Int. Conference on Stereology in Materials Science, ed. L. Wojnar (Beskidy, 1994), pp. 269-276.
4. J. Majta, A. Bator, EUROMAT'2001 — Proc. 7th European Conference on Advanced Materials and Processes, (Rimini, 2001), pp. 265.
5. M. M Ura, A. F. Padilha, N. Alonso, *Associacao Brasileira de Metalurgia e Materais*, (1995), pp. 337-349.
6. M. Ohashi, *Experimental Mechanics*, vol. 38, no. 1, (1998), s. 13-17.
7. R. Barcik, STERMAT'94 — Proc. 4th Int. Conference on Stereology In Materials Science, ed. L. Wojnar (Beskidy, 1994), s. 319-323.
8. Z. Cvijovic, D. Mihajlovic, R. Manojlovic, *Acta Stereologica*, vol. 11, no. I-II, (1992), pp. 665-700.
9. Y. Kuroshima, H. Shimizu, K. Kawasaki, *Trans. Tech. Publications*, (1991), pp. 49-54.
10. D. Lazecki, K. J. Kurzydłowski, *Acta Stereologica*, vol. 15, no. 3, (1996), pp. 219-226.
11. J. Dankmeyer-Łączny, J. Buzek, STERMAT'94 — Proc. 4th Int. Conference on Stereology In Materials Science, ed. L. Wojnar (Beskidy, 1994), s. 339-344.
12. B. Kęsy, STERMAT'90 — Proc. 3rd Int. Conference on Stereology In Materials Science, (Szczyrk, 1990), pp. 226-231.
13. A. Karamara, Foundry Research Institute - Research, (1986).
14. A. Karamara, M. Maj, *Metalurgia,* 34, (1986), pp. 43-52.

# EXPERIMENT STUDY ON DYNAMIC STRENGTH OF LOESS UNDER REPEATED LOAD

ZHENGHUA XIAO[1][†], BO HAN[2], HONGJIAN LIAO[3]

*Department of Civil Engineering Xi'an Jiaotong University, Xi'an, P. R. CHINA,*
*xzh-70@163.com, han_bo961@163.com, hjliao@mail.xjtu.edu.cn*

AKENJIANG TUOHUTI[4]

*College of Architectural Engineering, Urumqi, P. R. CHINA,*
*jgxyakj@xju.edu.cn*

Received 15 June 2008
Revised 23 June 2008

A series of dynamic triaxial tests are performed on normal anisotropic consolidation and over anisotropic consolidation specimens of loess. Based on the test results, the variable regularity of dynamic shear stress, axial strain and pore water pressure of loess under dynamic loading are measured and analyzed. The influences of the dynamic shear strength and pore water pressure at different over consolidation ratio are analyzed. The relationship between dynamic shear strength and over consolidation ratio of loess is obtained. The evaluating standard of dynamic shear strength of loess is discussed. Meanwhile, how to determine the effective dynamic shear strength index of normal anisotropic consolidated loess is also discussed in this paper. Several obtained conclusions can be referenced for studying the dynamic shear strength of loess foundation.

*Keywords*: Over consolidation ration; Dynamic shear strength; Pore water pressure; Dynamic triaxial test.

## 1. Introduction

The hazards of foundation failure have been verified by many earthquakes. At the present time, China mainland is still in active high tide earthquake interval and loess area is one of meizoseismal area. A large numbers of liquefaction researches on sand foundations have been done, but the problems of dynamic shear strength of cohesive soils are not definitely proposed some specific judgments method and standards which are similar to sand foundation. The Hai Yuan large earthquake of 8.5 earthquake magnitude that happened at the edge of Gan Su province and Ning Xia province had caused a large damage in loess area.[1] Dinyi Xie (1986, 1988[2], 1994), Zhihui Wu (1986[3]) and Ruwen Duan (1979, 1990[4]) have done plentiful researches on loess stress-strain relationship, strength, elastic modular and damping ratio under constant amplitude sinusoid cycling load; Gongshe Liu (1994[4]) studied the varying regularity of pore pressure of saturated loess under dynamic load; Hongjian Liao (2001[6], 2003) studied the strength of volcanic cohesive soil under repeated loads, but the research findings on dynamic strength of

---

[†]Corresponding Author.

anisotropic consolidation loess are seldom. For this reason, a series of dynamic triaxial tests directed towards anisotropic consolidation loess have been done in this paper. The dynamic characteristics under dynamic load, variable regularity of pore water pressure and evaluating dynamic shear strength index are studied. The variable rule of dynamic shear stress, axial strain and pore water pressure of anisotropic consolidation loess are analyzed. The influences of dynamic shear strength and pore water pressure at different over consolidation ratio are analyzed. The relationship between dynamic shear strength and over consolidation ratio of loess is obtained. How to evaluate the dynamic strength parameters is also discussed. Some obtained conclusions from the tests can be referenced for studying the dynamic shear strength of loess foundations.

## 2. Experiment Condition

DDS-70 electromagnetic vibrating triaixial test system controlled by computer was adopted to study the dynamic characteristics of soils. The samples are made of loess obtained from excavating field in Xi'an in China. It is composed of sands (19.6%), silts (28.5%) and clay (51.9%). Its specific gravity $G_s$ is 2.71, natural water content $w_0$ is 16.8%, natural density is 1.75g/cm$^3$, liquid limit $w_L$ is 36.6%, plastic limit $w_P$ is 20.7%. Its plasticity index $I_p$ is 15.9, so it belongs to silty clay. First, the specimens are in anisotropic consolidation for 24 hours at static coefficient of earth pressure $K_0$=0.67 and axial effective pressure $\sigma_1'$=270, 180, 135, 90kPa respectively, corresponding point A, B, C, D are shown in Fig.1. Second, the axial effective pressure is decreased to 90kPa respectively, corresponding point D in Fig. 1, the over consolidation ratio $K_c$ of samples equals to 3, 2, 1.5, 1. Then, vibration tests are performed after 24 hours absorbing water to expand. The effective stress path of the specimens consolidated process is shown in Fig. 1, $q$ is the effective deviator stress, $p'$ is the average effective stress. The repeated load is 1Hz sine wave.

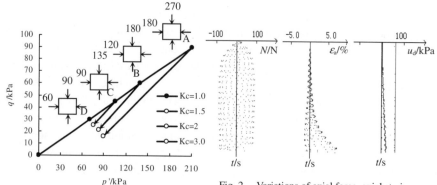

Fig. 1.    Effective stress path.

Fig. 2.    Variations of axial force, axial strain and pore water pressure with time.

The variable curves of initial axial force, axial strain and pore water pressure with time are shown in Fig. 2. The axial force can keep a constant; the variation of axial strain with time increases only at the side of compression till to failure at anisotropic consolidation; the value of pore water pressure is small and can't reach the confining pressure (Fig. 2).

## 3. Test Results and Analysis

### 3.1. *The relationship between dynamic shear strength and over consolidation ratio*

Fig. 3 is the dynamic triaxial test results of the anisotropic consolidation samples endure through different circulating load. The vibration times tends to less and the axial strain $\varepsilon_a$ tends to larger as the initial axial force is greater (Fig. 3a). While reaching the same axial strain, the times of vibration tends to more as the initial axial force is smaller and has the trend of rapid increasing. The dynamic triaxial test results of different over consolidation ratio specimens are shown in Fig. 3b. The times of vibration also have the tendency of rapid increasing as the over consolidation ratio increases.

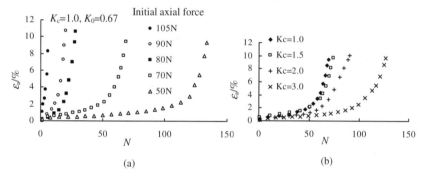

Fig. 3.   Relationship between axial stain and times of vibration.

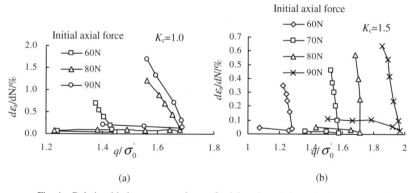

Fig. 4.   Relationship between speed rate of axial strain and dynamic shear stress ratio.

The ratio of axial strain increment $d\varepsilon_a$ and vibration times increment $dN$ is called the speed rate of axial strain $d\varepsilon_a/dN$. The relationship between $d\varepsilon_a/dN$ and ratio of dynamic shear stress $q/\sigma'_0$ (the effective confining pressure $\sigma'_0=\sigma_3-u_0$) is shown in Fig. 4. Whether to the normal anisotropic consolidation or the over anisotropic consolidation samples, they all have an apparent yield point $q_y$ with the raising speed rate of axial strain and usually occur at little speed rate of axial strain (Fig. 4). The yield point value tends to increase as the initial axial force increases. The yield point value is also larger as the over consolidation ratio is larger in Fig. 5.

When axial strain is equal to 5%, the relationship between ratio of dynamic shear stress and number of breakdown times is shown in Fig. 6. The ratio of dynamic shear stress is larger as the over consolidation ratio is larger; the ratio of dynamic shear stress is less as the number of failure times increases for the same specimen (Fig. 6).

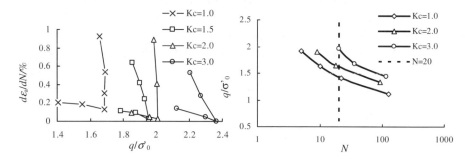

Fig. 5.    Dynamic shear stress ratio at different over consolidation ratio.

Fig. 6.    Relationship between dynamic shear stress ratio and failure vibration times.

The variable curves of yield point value $q_y$ and dynamic shear strength $q_d$ ($\varepsilon_d=5\%$, $N=20$) along with over consolidation ratio are shown in Fig. 7. In coordinate axis, $q_y$ is greater than $q_d$, they all tend to increase as the over consolidation ratio increases and relational expressions can be shown as follow:

$$q_y\big/\sigma'_0 = 1.6898K_c^{0.2879} \tag{1}$$

$$q_d\big/\sigma'_0 = 1.1696e^{0.1682K_c} \tag{2}$$

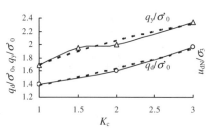

Fig. 7.    Variable curves for $q_y/\sigma'_0$-$K_c$ and $q_d/\sigma'_0$- $K_c$.

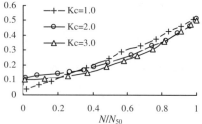

Fig. 8.    Relationship between ratio of pore water pressure and ratio of vibration times.

### 3.2. *The relationship between pore water pressure and over consolidation ratio*

The influences of pore water pressure caused by different over consolidation ratio are shown in Fig.8, vertical ordinate is the ratio of pore water pressure $u_{dN}/\sigma_3$ ($u_{dN}$ is the dynamic pore water pressure as the number of vibration times is N), horizontal ordinate is the ratio of vibration times $N/N_{50}$ ($N_{50}$ is the vibration times as $u_d=50\%\sigma_3$). Under the identical initial axial force, with the increasing of over consolidation ratio, the number of vibration times increase while reaching the same pore water pressure. In the vibrating process, the pore water pressure curves display monotone increasing and this is dissimilar to the dilatancy of sand (Fig. 8). The difference between loess and sand in dynamic pore water pressure curve may be caused by the adhesiveness and structural strength of loess.

### 3.3. *Determination of the index of effective dynamic shear strength of loess*

The confining pressure of normal anisotropic consolidation specimens is equal to 60kPa, 90kPa and 120kPa respectively. The consolidation undrained(CU) dynamic triaxial tests of these samples were done to determine the index of effective dynamic shear strength. The unified strength theory is adopted to determine the strength index in this paper. Base on the unified strength theory dynamic equilibrium condition line equation[7]:

$$q_u' = m + n\sigma_u' .  \tag{3}$$

Where $q'_u$ is the deviator stress, $\sigma'_u$ is the effective consolidation pressure. Interception of dynamic equilibrium condition line at vertical ordinate is $m$, its slope coefficient is $n$. Based on $m$ and $n$, the index of effective dynamic shear strength $\phi'_d$ and $c'_d$ can be determined, the equations are as follows:

$$\varphi_d' = \sin^{-1}(\frac{n}{n+2}) ,  \tag{4}$$

$$c_d' = m(\frac{1-\sin\varphi_d'}{2\cos\varphi_d'}) .  \tag{5}$$

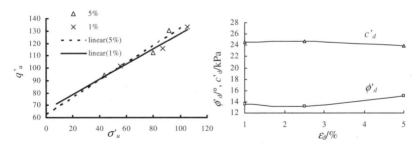

Fig. 9.   The relationship between deviator stress and effective consolidation pressure.

Fig. 10.   Variable curves for $\phi'_d$-$\varepsilon_d$ and $c'_d$-$\varepsilon_d$.

Two dynamic equilibrium condition lines of $\varepsilon_d$ =1% and 5% are shown in Fig. 9. The equation of $\varepsilon_d$ =5% is as follows:

$$q_u' = 0.6978\sigma_u' + 62.371 .  \tag{6}$$

Then $\phi'_d = 14.99°$, $c'_d = 23.93\text{kPa}$ are calculated. The equation of $\varepsilon_d = 1\%$ is as follows:

$$q'_u = 0.6106\sigma'_u + 66.447 . \tag{7}$$

Then $\phi'_d = 13.20°$, $c'_d = 24.72\text{kPa}$ are obtained. The index of effective dynamic strength, that are $\phi'_d$ and $c'_d$, is not a constant when the different dynamic strain is selected as failure standard (Fig. 9). This is not similar to static strength index. $\phi'_d\text{-}\varepsilon_d$ and $c'_d\text{-}\varepsilon_d$ variable curves are shown in Fig. 10. When dynamic strain $\varepsilon_d$ is less, the effective cohesive force has been full used and the internal friction angle is small. When dynamic strain $\varepsilon_d$ is large, the internal friction angle acts on and the effective cohesive force is decreased (Fig.10). This is verified by tests.

## 4. Conclusions

Based on the above study, several conclusions are obtained as follows:

(1) The yield strength value $q_y$ and dynamic shear strength $q_d(\varepsilon_a=5\%, N=20)$ all increase as the over consolidation ratio increases and $q_y$ is greater than $q_d$ and their expressions are obtained.

(2) Loess can't reach the complete liquefaction standard of sand under dynamic load. When the axial strain comes to a stable value, even if a larger deformation has been produced, the pore water pressure only reaches to a constant but can't reach the initial effective consolidated pressure. In the vibrating process, the pore water pressure curves display monotone increasing and this is dissimilar to the dilatancy of sand.

(3) The index of effective dynamic strength is not a constant according to different failure standard and this is not similar to static strength index. When $\varepsilon_d$ is smaller, the effective cohesive force has been full used.

## Acknowledgments

This paper is funded by scientific research program of the Higher Education Institution of Xinjiang (XJEDU2005S04) and Key Lab. for the Exploitation of Southwestern Resources & the Environmental Hazard Control Engineering (Chongqing University) of Chinese Ministry of Education, which are gratefully acknowledged.

## References

1. Lanmin Wang, *Loess Dynamics* (The Earthquake Publishing House, Beijing, 2003).
2. Dinyi Xie, *Soil Dynamics* (Xi'an Jiao Tong University Press, Xi'an, 1988).
3. Zhihui Wu, *in the Symposium of the National Earth Structure and Foundation of Earthquake-resistance Conference*, (1986) p.219.
4. Ruwen Duan, *Chinese Journal of North-West Earthquake*, **12**, (1990) p.72.
5. Gongshe Liu, *Chinese Journal of Industrial Construction*, **3**, (1994) p.40.
6. Hongjiang Liao, *Chinese Journal of Rock and Soil Mechanics*, **27**, (2001) p.16.
7. Hongjiang Liao, *Chinese Journal of Rock Mechanics and Engineering*, **22**, (2003) p.1994.

# SHEAR STRENGTH PREDICTION OF RC BEAMS WRAPPED WITH FRP

SUYAN WANG[1], YINGWU ZHOU[2], HONGNAN LI[3†]

*State Key Laboratory of Coastal and Offshore Engineering, Dalian University of Technology*
*Dalian, 2# Ligong Road, 116024, P. R. CHINA,*
*suyanwang1958@yahoo.com.cn, albert_williancs@163.com, hnli@dlut.edu.cn*

Received 15 June 2008
Revised 23 June 2008

During past decades, substantial studies on the external bonding of fiber reinforced polymer (FRP) strips to deficient reinforced concrete (RC) beams have been carried out for the well-known superior properties of the FRP. Several shear prediction models have been established by using the effective strain of the FRP or introducing an ultimate stress discount coefficient. And the latest design concept is the use of stress distribution factor. In this paper, an equivalent effective strain model of the FRP is presented, which contains the concepts of maximum strain, stress distribution factor and critical shear crack angle influences. To develop a simple and accurate approach for such equivalent effective strain, major influenced factors are investigated and analyzed by the statistical independent hypothetic tests on a database of 128 RC beams wrapped by the FRP. Finally, a simple and rational shear design proposal is given, which is more accurate than the existing models using the above database.

*Keywords*: FRP; RC beams; strengthening; effective strain; hypothetic test.

## 1. Introduction

The external bonding of fiber reinforced polymers (FRP) to the RC beam as additional web reinforcement can obviously enhance the shear resistance of the RC beam. Up to now, many studies on the shear strengthening of RC beams have been carried out, and some shear strength models have been established[1-6]. This paper focuses on the study of shear failure with the FRP rupture for the FRP reinforced beam. Existing methods are firstly reviewed in predicting the shear contribution of the FRP. A more rational shear strength model is presented, called as the equivalent effective strain model, which is on the base of analyses with the database of 128 RC beams wrapped by the FRP or failure with FRP rupture in literatures. The data are selected with considering the different concrete strength $f_{cu}$, the shear span-to-effective-depth ratio $\lambda$, the shear steel reinforcement ratio $\rho_v$ and the depth of the beam section $d$. The statistical independent hypothetic test on such database is applied to distinguish the major influenced factors. By neglecting the minors, the shear capacity of the RC beams reinforced with the FRP is

---

[†]Corresponding Author.

given. The results indicate that the model presented here is of better advantages in accuracy and simplicity compared to the existing shear strength models.

## 2.  Existing Shear Strength Models

In literatures[1-6], several shear strength models of RC beams strengthened with the FRP were proposed, which result in an identical expression given by

$$V_u = V_c + V_s + V_f \tag{1}$$

where $V_u$ is the total shear strength of the RC beam reinforced with FRP, and $V_c$, $V_s$ and $V_f$ are the shear contributions of the concrete, stirrups and FRP, respectively.

Derived from a truss model, Triantafillou[1] presented a model as follows

$$V_f = \frac{0.9}{\gamma_f} \rho_f E_f \varepsilon_{f,e} bd(1+\cot\beta)\sin\beta \tag{2}$$

where $E_f$ is the Young's modulus of the FRP; $\rho_f$ means the shear reinforcement ratio of the FRP; $b$ and $d$ are the width and the effective depth of the RC beam; $\beta$ denotes the bonding angle of the FRP; $\gamma_f$ implies the partial safety factor; and $\varepsilon_{f,e}$ is defined as an effective strain of the FRP at failure as:

$$\varepsilon_{f,e} = 0.0119 - 0.0205(\rho_f E_f) + 0.0104(\rho_f E_f)^2 \quad for \quad 0 \le \rho_f E_f \le 1 \tag{3a}$$

$$\varepsilon_{f,e} = -0.00065(\rho_f E_f) + 0.00245 \quad for \quad \rho_f E_f > 1 \tag{3b}$$

However, no distinction has been made between the FRP rupture and debonding until now. Triantafillou[2] further improved this model by distinguishing the two different types of failures. When the shear failure due to the CFRP rupture, the relation between $\varepsilon_{f,e}$ and the ultimate tensile strain $\varepsilon_{f,u}$ is:

$$\frac{\varepsilon_{f,e}}{\varepsilon_{f,u}} = 0.17(\frac{f_c^{2/3}}{E_f \rho_f})^{0.3} \tag{4}$$

All of the above models were established by the regression of limited experimental data and assumed with the critical shear crack angle equal to $45^0$. Yet, such an assumption sometimes results in a conservative prediction[4]. Thus, given a critical shear crack angle $\theta$, Chen & Teng[4] suggested:

$$V_f = \rho_f E_f \varepsilon_{f,e} bd(\cot\theta + \cot\beta)\sin\beta \tag{5}$$

where the effective strain $\varepsilon_{f,e}$ is evaluated by introducing the strain distribution factor $D_f$, a ratio reflecting the strain distribution in the FRP strips intersected by the shear crack:

$$\varepsilon_{f,e} = D_f \varepsilon_{f,u} \tag{6}$$

The use of the ultimate strength in Eq.(6) sometimes overestimates the contribution of the FRP strips as the observation in experiments indicates that the FRP strips usually rupture due to the local stress concentration so that the maximum strain or stress in the FRP strips intersected by the critical shear crack doesn't reach to its ultimate one. Moreover, a parabolic normalized strain distribution in the FRP along the critical shear crack is assumed to obtain the distribution factor $D_f$. However, no reasonable explanation of the assumption is provided in Ref. 4.

## 3.  Modified Shear Strength Model

### 3.1.  *Definition and derivation*

Considering the different failure patterns of the FRP (i.e. the FRP rupture and debonding), the strain distribution along the critical shear crack $D_f$, the maximum strain of the FRP strips intersected by the critical shear crack $\varepsilon_{f,max}$ and the diagonal shear crack angle $\theta$, the modified shear model is given by

$$V_f = \rho_f E_f \varepsilon_{f,e} bd(\cot\theta + \cot\beta)\sin\beta = \rho_f E_f \varepsilon_{f,e} bd\cot\theta(1+\eta\cot\beta)\sin\beta$$
$$= \rho_f E_f D_f \varepsilon_{f,max} bd\cot\theta(1+\eta\cot\beta)\sin\beta = \rho_f E_f D_f \alpha\varepsilon_{f,u} bd\cot\theta(1+\eta\cot\beta)\sin\beta$$
$$= \rho_f E_f \varepsilon_{f,ee} bd(1+\eta\cot\beta)\sin\beta$$
(7)

where $\varepsilon_{f,ee}=D_f \alpha \cot\theta \varepsilon_{f,u}$, is defined as an equivalent effective strain of the FRP; $\alpha=\varepsilon_{f,max}/\varepsilon_{f,u}$ and $\eta=1/\cot\theta$ will be further discussed as follows.

### 3.2.  *Discussion about η*

Since the combination of the FRP bonding angle $\beta$ and the shear crack angle $\theta$ is stochastic, the value of $\eta$ can be determined by the statistics. Known the distributions of $\beta$ and $\theta$, generating a series group of random numbers $(\beta, \theta)$, the corresponding values of $y=1+\cot\beta/\cot\theta=1+\eta\cot\beta$ are obtained; based on the generated random sample $(\beta, y)$, $\eta$ is then determined by the least square estimation approach. Suppose that $\beta$ obey the uniform distribution over the interval $[30^0, 90^0]$ and $\theta$ summarized in Table 1 obey the normal distribution, which has passed the Kolmogorov-Smirnov (K-S) test under the significant level of 0.05 with the mean of $39.2^0$ and standard deviation of $6.2^0$, and then $\eta$ has the best estimation value of 0.835 through the computing procedure as shown in Fig. 1.

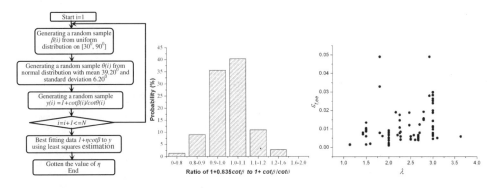

Fig. 1. Procedure for determining $\eta$.

Fig. 2. Probability vs. ratio of $1+0.835\cot\beta$ to $1+\cot\beta/\cot\theta$.

Fig. 3. $\varepsilon_{f,ee}$ versus $\lambda$.

In Fig. 2, it can be seen that such method is feasible since the probability of errors within ±10% has reached to 75.93%, and almost all of the errors are within ±20%. Thus, the modified shear model can be expressed as

$$V_f = \rho_f E_f \varepsilon_{f,ee} bd(1+0.835\cot\beta)\sin\beta \tag{8}$$

Eq. (8) is a precise model when $\beta$ is equal to $90^0$.

## 3.3. Parameter study and suggestion for $\varepsilon_{f,ee}$

To study the influenced parameters of the equivalent effective strain, $\varepsilon_{f,ee}$ is calculated based on 128 RC beams using Eq. (9) and the results are listed in Table 1.

Table 1. Experimental data on shear strengthening with FRP failure with FRP rupture.

| Ref. | Specimen | Crack angle $\theta$ | $\varepsilon_{f,ee}$ | Ref. | Specimen | Crack angle $\theta$ | $\varepsilon_{f,ee}$ | Ref. | Specimen | Crack angle $\theta$ | $\varepsilon_{f,ee}$ |
|---|---|---|---|---|---|---|---|---|---|---|---|
| Ref. 8 | LA1 | 35 | 0.007936 | Ref. 11 | LA-4 | 43 | 0.006057 | Ref. 4 | No.3 | 46 | 0.005555 |
| | LA2 | 39 | 0.007087 | | LA-9 | 38 | 0.019239 | | S-3 | 35 | 0.011348 |
| | LA3 | 46 | 0.006596 | | LA-10 | 41 | 0.015676 | | S-4 | 35 | 0.013707 |
| | LA5 | 34 | 0.009739 | | LB-4 | 41 | 0.014963 | | S200H | 35 | 0.010725 |
| | LA6 | 41 | 0.007737 | Ref. 12 | 3 | - | 0.004548 | | S200L | 35 | 0.010735 |
| | LA10 | 32 | 0.010898 | | AN-1/5Z-3 | - | 0.011293 | | BT4 | 45 | 0.006645 |
| | LA7 | 41 | 0.006529 | | AN-1/2Z-3 | - | 0.007067 | | CF-045 | 45 | 0.009088 |
| | LA8 | 44 | 0.003778 | | CN-1/LZ-2 | - | 0.008276 | | CF-064 | 45 | 0.009083 |
| | LA9 | 47 | 0.005187 | | CS2 | - | 0.004597 | | CF-097 | 45 | 0.009180 |
| | LA19 | 48 | 0.004827 | | AS2 | - | 0.029794 | | CF-131 | 45 | 0.009505 |
| | LA11 | 36 | 0.006958 | | CS4) | - | 0.028196 | | CF-243 | 45 | 0.006240 |
| | LA16 | 55 | 0.003685 | | CS1 | - | 0.006249 | | AF-060 | 45 | 0.010235 |
| | LB1 | 39 | 0.007714 | | CS3 | - | 0.011114 | | AF-090 | 45 | 0.009863 |
| | LC1 | 42 | 0.009113 | | AS3 | - | 0.014348 | | AF-120 | 45 | 0.013445 |
| Ref. 9 | BVI-3 | 45 | 0.008348 | | AB2 | - | 0.028335 | | S-2 | 45 | 0.009921 |
| | BVI-4 | 45 | 0.008348 | | AB3 | - | 0.025212 | | S-3 | 45 | 0.006227 |
| | BVII-2 | 45 | 0.004174 | | AB4 | - | 0.029824 | | S-4 | 45 | 0.005350 |
| | BVII-3 | 45 | 0.004174 | | AB5 | - | 0.023876 | | S-5 | 45 | 0.008206 |
| | BVII-4 | 45 | 0.005844 | | AB8 | - | 0.026320 | | No.2 | 40 | 0.012023 |
| | BVII-6 | 45 | 0.005844 | | AB9 | - | 0.019880 | | No.3 | 40 | 0.011300 |
| Ref. 4 | A1 | 35 | 0.005816 | | AB11 | - | 0.018734 | | No.7 | 40 | 0.013163 |
| | E1 | 35 | 0.006570 | | CF045 | - | 0.008607 | | No.8 | 40 | 0.020083 |
| | E2 | 35 | 0.005284 | | CF064 | - | 0.008667 | | No.2 | 35 | 0.008568 |
| | G1 | 35 | 0.004628 | | CF097 | - | 0.008802 | | SB5 | 45 | 0.005527 |
| | G2 | 35 | 0.004762 | | CF131 | - | 0.009126 | | SC2 | 45 | 0.004388 |
| | 45G1 | 35 | 0.003897 | | CF243 | - | 0.005987 | | SC3 | 45 | 0.004602 |
| Ref. 10 | PC1 | 35 | 0.017588 | | CF060 | - | 0.010314 | | BS2 | 30 | 0.013017 |
| | PC2 | 35 | 0.014657 | | CF090 | - | 0.009804 | | BS5 | 25 | 0.013702 |
| | PC3 | 35 | 0.010694 | | CF120 | - | 0.013402 | | BS6 | 25 | 0.019763 |
| | PC4 | 35 | 0.007732 | | S-2 | - | 0.005884 | | BS7 | 25 | 0.020411 |
| | SL12 | 45 | 0.006619 | | S-3 | - | 0.004283 | Ref. 2 | BS12 | - | 0.007560 |
| | SL18 | 45 | 0.001585 | | A | - | 0.001431 | | BS24 | - | 0.005580 |
| | SL19 | 45 | 0.005046 | | B | - | 0.001538 | | BM06 | - | 0.010530 |
| | S-GO-2-1 | 45 | 0.019014 | | D | - | 0.001973 | | BM12 | - | 0.008370 |
| Ref.5 | L2 | 26 | 0.027952 | | S4 | - | 0.004044 | | BM18 | - | 0.007020 |
| | L3 | 23 | 0.048917 | | G5.5-1L | - | 0.001803 | | BM24 | - | 0.005400 |
| | L6 | 43 | 0.032844 | | G5.5-2L | - | 0.001542 | | BL06 | - | 0.007560 |
| | L7 | 37 | 0.048917 | | G8-1L | - | 0.001803 | | BL12 | - | 0.007020 |
| | Bb | 29 | 0.011950 | | G8-2L | - | 0.001799 | | BMW06 | - | 0.007560 |
| | Bc | 34 | 0.018778 | | G16-1L | - | 0.002318 | | BMW12 | - | 0.006210 |
| | A2 | 40 | 0.007838 | | G24-2L | - | 0.001414 | | BMW24 | - | 0.004140 |
| | A3 | 38 | 0.012291 | | G24-3L | - | 0.001029 | | 2 | - | 0.010800 |
| | A5 | 41 | 0.007838 | | | | | | 3 | - | 0.009270 |

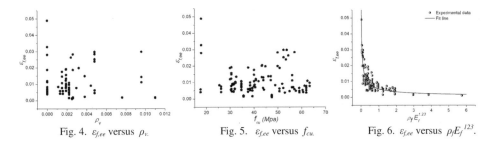

Fig. 4. $\varepsilon_{f,ee}$ versus $\rho_v$.  Fig. 5. $\varepsilon_{f,ee}$ versus $f_{cu}$.  Fig. 6. $\varepsilon_{f,ee}$ versus $\rho_f E_f^{123}$.

In Figs. 3-5, $\varepsilon_{f,ee}$ seems to be independent on the shear span-to-effective-depth ratio, $\lambda$, the shear steel reinforcement ratio, $\rho_v$, and the concrete strength, $f_{cu}$. Thereby, the statistical independent hypothetic test[7] is used to give a strict demonstration on their independences based on the 128 strengthened RC beams. The $\chi^2$-hypothetic test is adopted since the distributions of the samples are uncertain. Take the samples $\varepsilon_{f,ee}$ and $\rho_v$ for example, they are divided into several intervals as shown in Table 2.

Table 2. $\chi^2$-hypothetic test on $\varepsilon_{f,ee}$ and $\rho_v$.

| $\varepsilon_{f,ee}$ \ $\rho_v$ | [0, 0.0015] | [0.0015, 0.0030] | [0.0030, 0.012] | $n_{i.}$ |
|---|---|---|---|---|
| [0, 0.006] | $n_{11}=9$ | $n_{12}=16$ | $n_{13}=6$ | $n_{1.}=31$ |
| [0.006, 0.010] | $n_{21}=6$ | $n_{22}=17$ | $n_{23}=2$ | $n_{2.}=25$ |
| [0.010, 0.050] | $n_{31}=10$ | $n_{32}=10$ | $n_{33}=9$ | $n_{3.}=29$ |
| $n_{.j}$ | $n_{.1}=25$ | $n_{.2}=43$ | $n_{.3}=17$ | $n=85$ |

Grouping principle: $n_{.j}\,n_{i.}\,/n$ must be larger than 5

Based on Table 2, the test statistic value of $\chi^2$ can be calculated by

$$\chi^2 = \sum_{i=1}^{3}\sum_{j=1}^{3}\frac{[n_{ij}-(n_{i.}n_{.j})/n]^2}{(n_{i.}n_{.j})/n} = 7.0676 < \chi_{(4,0.95)}^2 = 9.488 \tag{9}$$

Under the significant level of 0.05, the result from Eq. (9) indicates that $\varepsilon_{f,ee}$ and $\rho_v$ are mutually independent. The same conclusions can be drawn on the relationship between $\varepsilon_{f,ee}$ and $\lambda$ and the relationship between $\varepsilon_{f,ee}$ and $f_{cu}$ when repeating the $\chi^2$- hypothetic test on the experimental data. Thus, a simple expression of $\varepsilon_{f,ee}$ is possibly obtained since the influenced parameters are significantly reduced to two major factors, that is, the FRP shear reinforcement ratio, $\rho_f$, and the Young's modulus of the FRP, $E_f$.

In Fig. 6, it can be found that $\varepsilon_{f,ee}$ is dependent on $\rho_f E_f$, thus their relationship can be obtained from the best fitting exponential equation where the unit of Gpa is used in the regression:

$$\varepsilon_{f,ee} = 0.005229(\rho_f E_f^{1.23})^{-0.6152} \tag{10}$$

Finally, the shear contribution of the FRP can be estimated by Eqs. (10) and (8).

## 4.  Comparisons with Experiment and modified model

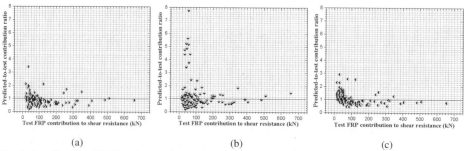

|                | (a)                    | (b)                    | (c)                    |

Fig. 7. Contribution of FRP to shear resistance: (a) Triantafillou (1998) (Eqs. (2) & (3));  (b) Cheng & Teng (Eqs. (5) & (6)); (c) Modified model (Eqs. (9) & (11)).

As can be seen from Fig. 7, the equivalent strain model presented here agrees quite well with the experiments and its accuracy in predicting the shear contribution of the FRP is better than those of Triantafillou's second order polynomial model (Eqs. (2) and (3)) and the Chen & Teng's model (Eqs. (5) and (6)). And this conclusion is further proved by the statistical properties of the three design data shown in Table 3.

Table 3. Statistical performances of the three proposals.

| Design methods | Triantafillou (Eqs. (2) & (3)) | Cheng & Teng (Eqs. (6) & (7)) | Suggested model (Eqs. (9) & (11)) |
|---|---|---|---|
| Average | 0.93 | 1.31 | 1.07 |
| Coefficient of variation % | 47.2 | 93.7 | 42.2 |

## 5.  Conclusions

In this paper, a new modified design model of shear capacity of RC beams strengthened with the FRP is presented when failed by the FRP rupture, which is called as the equivalent effective strain model taking into account the influences on the strain distribution along the critical shear crack, the actual maximum strain of the FRP intersected by the shear crack, and the diagonal shear crack angle. All of the factors are reflected in the equivalent effective strain $\varepsilon_{f,ee}$ that has been further determined based on the regression of 128 test data. It is validated that the presented model is a simple and rational shear strength model without necessity to consider the shear crack angle anymore, which is usually assumed to be $45^0$ in the most existing design methods. It is also an accurate model when compared to the experimental data and other existing design models.

## Acknowledgments

The authors gratefully acknowledge the financial support provided by the Natural Science Foundation of China (No. 50378711), the Key Laboratory Open Research

Foundation of Academy Affiliated to Beijing city (EESR200708) and the Province Key Laboratory Open Research Foundation of Shenyang Jian Zhu University (JG-200605).

## References

1. T. C. Triantafillou, *ACI Struct. J.* **95**, (1998) 107.
2. T. C. Triantafillou and P. A. Costas, *J. Comp. Constr. ASCE,* **4**, (2000) 198.
3. A. Khalifa, W. J. Nanni, et al, *J. Comp. Constr. ASCE,* **2**, (1998) 195.
4. J. F. Chen and J. G. Teng, *J. Struct. Eng. ASCE,* 12**9**, (2003) 615.
5. S. Y. Cao, J. F. Chen and J. G. Teng, *J. Comp. Constr. ASCE,* **9**, (2005) 417.
6. C. Deniaud and J. J. R. Cheng, *J. Comp. Constr. ASCE,* **8**, (2004) 425.
7. S. Z. Teng, *Mathematical statistics* (Publishing Company of Dalian University of Technology, China, 2000).
8. Y. W. Zhou, S. Y. Wang and H. N. Li, *Chinese Journal of Dalian University of Technology*, **48**, (2008) 235.
9. T. Zhao, J. Xie and Z. Q. Dai, *Chinese Building Structure*, 30, (2000) 21.
10. Y. W. Zhou, S. Y. Wang and H. N. Li, *Chinese Journal of Railway Science and Engineering*, **2**, (2005) 22.
11. Z. Hao, MPhil, Thesis. Southeast Univ., Nanjing, China, 2004.
12. A. Boussenlham and O. Chaallal, *ACI Struct. J.* **101**, (2004) 219.

# STUDY ON STRESS-STRAIN RELATIONSHIPSHIP OF LOESS BASED ON TWIN SHEAR UNIFIED DAMAGE CONSTITUTIVE MODEL

BO HAN[1†], HANGZHOU LI[3], HONG-JIAN LIAO[4]

*Department of Civil Engineering Xi'an Jiaotong University, Xianning West Road 28*
*Xi'an , Shannxi 710049,P. R. CHINA,*
*han_bo961@163.com, lihangzhou77@163.com, hjliao@mail.xjtu.edu.cn*

ZHENGHUA XIAO[2]

*Collegd of Architectural Engineering, Xinjiang University,*
*Urumqi, Xinjiang 830008, P. R. CHINA,*
*xzh-70@163.com*

Received 15 June 2008
Revised 23 June 2008

To investigate the change of loess stress state, a series of triaxial shear tests were performed on normal consolidation and over consolidation loess. From the test results, the stress-strain relationships of loess were obtained and discussed. Based on unified strength theory, the statistical damage constitutive equation was obtained under triaxial stress state assuming distribution statistical probability of micro-units strength. Then the proposed formulation was adopted to study on stress-strain constitutive relationships of loess and to simulate consolidation undrained triaxial test and consolidation drained triaxial test for normal consolidated and over-consolidated specimens. Compared between experimental and theoretical results, it was shown that the proposed constitutive model can well describe stress-strain relationship of loess, whatever the characteristic of strain softening or stain hardening.

*Keyword*: Stress-strain relationship; Triaxial shear test; Statistical Damage Constitutive model.

## 1. Introduction

Because natural sedimentary loess has structural properties, it shows different mechanical properties before and after structural failure. This leads to a complicated stress-strain relationship of loess. Damage mechanics is effective methods to determine the stress-strain relationship and it has been used in metals, geo-materials, etc. Shen Zhujiang proposed a binary-medium model for over consolidated clays. The results obtained from simulating triaxial test results show that the model can reflect the main features of the behavior of London clay. [1, 2] Xia Wangmin researched the stress-strain and deformation characteristics of $Q_1$ loess by means of conventional triaxial tests. The elastoplastic damage model of $Q_1$ loess is proposed based on the continuous damage

---

[†]Corresponding Author.

mechanics and the law of thermodynamics.[3] Lin Bin researched how to define reasonably the damage variable of loess and evaluated the damage tendency of loess[4]. However, none of these models considered the effect of intermediate principal stress. Great deals of true triaxial tests and plane strain tests had proved that strength and stress-strain relationship of geo-materials are not only related to the maximum and minimum principal stress, but also related to the intermediate principal stress.[5] Ever since professor Yu has proposed unified strength theory in 1991, it has been widely used.[6]

The main purpose of the present study was to investigate the change of loess stress state, a series of triaxial shear tests were carried out on both normal consolidated and over-consolidated loess specimens. The statistical damage constitutive model based on unified strength theory under triaxial stress state was adopted to study on stress-strain relationship of loess and to simulate results of consolidation undrained triaxial test (CU-test) and consolidation drained triaxial test (CD-test).

## 2. Preparation of Samples and Test Method

The samples were made of loess obtained from excavating field in Xi'an in China. The main physical index tests of loess were made in laboratory. Its specific gravity $G_s$ was 2.71, Natural water content $\omega_0$ is 16.8%, and natural density was 1.75g/cm. Then the CU-test and CD-test of normal consolidated and over consolidated samples were performed. To make over consolidated samples, the specimens were consolidated for 24 hours at confining pressure 400kPa firstly. Then the confining pressure was respectively decreased to 100,150,200,250kPa, so the over consolidation ratio $K_c$ equaled to 4, 2.67, 2, 1.6.

## 3. Test Results Analysis

### 3.1. *CU-test of Normal consolidation and Over-consolidation specimens*

Fig.1 and Fig.2 show the results of CU-test for normal consolidated and over consolidated specimens. It is shown all stress-strain curves had strain-softening tendency. The strain-softening tendency was much more obvious and the stress-strain relationship curves were relatively steep in original stage in Fig.1. With the increase stain, the deviator stress q was reduced gradually; the stress-strain relationships curves tended towards smooth and belonged to strain-softening curves.

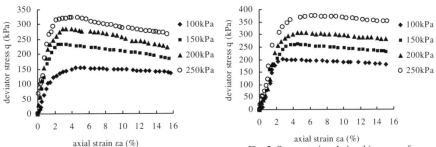

Fig. 1. Stress-strain relationship curves for normal consolidated soil of CU-test.

Fig. 2. Stress-strain relationship curves for over consolidated soil of CU-test.

### 3.2. CD-test of Normal consolidation and Over-consolidation specimens

Fig.3 and Fig.4 show the results of CD-test for normal consolidated and over-consolidated specimens. The stress-strain curves of normal consolidated and over consolidated specimens had a strain-hardening tendency. For over consolidated specimens, strain-hardening tendency was more obvious. Therefore, it belonged to strain-hardening curves, that is to say, the deviator stress $q$ was increased gradually with the increase strain.

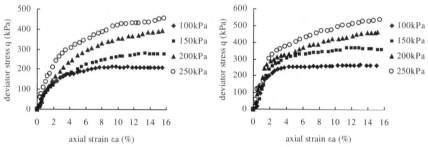

Fig. 3. Stress-strain relationship curves for normal consolidated soil of CD-test.    Fig. 4. Stress-strain relationship curves for over consolidated soil of CD-test.

## 4. Numerical Simulation and Determination of Model Parameters

### 4.1. Damage model based on unified strength theory

According to the hypothesis of equivalent strain, the effective stress is expressed as follows:

$$[\sigma] = \sigma^* (1 - D) \tag{1}$$

Where $\sigma$ is nominal stress, $\sigma^*$ is equivalent stress, and D is damage variable.

$$D = \frac{n}{N} \tag{2}$$

Where $n$ is the number of failure units, $N$ is the total number of the units.

When loading arrive at a certain level $F$, the amount of failure units is:

$$N[f(\sigma)] = N \int_0^{f(\sigma)} p[f(\sigma)] df(\sigma) = N\left\{1 - \exp\left[-\left(\frac{f(\sigma)}{F_0}\right)^m\right]\right\} \tag{3}$$

Where $f(\sigma)$ is distribution variable of micro-units strength random distribution; $F_0$, $m$ are Weibull's distribution parameters.

According to the definition of statistical damage variable, substituting equation (3) into equation (2), the damage variable is obtained, and its mathematic expression is:

$$D = 1 - \exp\left[-\left(\frac{f(\sigma)}{F_0}\right)^m\right] \tag{4}$$

Because observed stresses are nominal stresses in tests, according to generalized Hooke's law and equation (1), nominal stresses is expressed as follows:

$$\left. \begin{aligned} P &= \frac{(\sigma_1 + \sigma_3)E\varepsilon_1}{\sigma_1 - \mu(\sigma_2 + \sigma_3)} \\[2mm] R &= \frac{(\sigma_1 - \sigma_3)E\varepsilon_1}{\sigma_1 - \mu(\sigma_2 + \sigma_3)} \end{aligned} \right\} \tag{5}$$

$$\left. \begin{aligned} \sigma_1^* &= \frac{\sigma_1 E\varepsilon_1}{\sigma_1 - \mu(\sigma_2 + \sigma_3)} \\[2mm] \sigma_2^* &= \frac{\sigma_2 E\varepsilon_1}{\sigma_1 - \mu(\sigma_2 + \sigma_3)} \\[2mm] \sigma_3^* &= \frac{\sigma_3 E\varepsilon_1}{\sigma_1 - \mu(\sigma_2 + \sigma_3)} \end{aligned} \right\} \tag{6}$$

The unified strength theory is expressed as follows: [6]

$$R + P \sin \varphi_t = 2C_t \cos \varphi_t \tag{7}$$

Where $\varphi_t$, $C_t$ are defined as unified strength parameters.

Because micro-units obey unified strength theory, damage variable of loess is obtained from equations (5), (6) and (7).

$$D = 1 - \exp\left[ -\left( \frac{R + P \sin \varphi_t}{F_0} \right)^m \right] \tag{8}$$

Where : when $\mu_\sigma \leq \sin \varphi_0$, $\sin \varphi_t = \dfrac{2(1+b) \sin \varphi_0}{2 + b + b \sin \varphi_0 - (1 - \sin \varphi_0)b\mu_\sigma}$

when $\mu_\sigma \geq \sin \varphi_0$, $\sin \varphi_t' = \dfrac{2(1+b) \sin \varphi_0}{2 + b - b \sin \varphi_0 + (1 + \sin \varphi_0)b\mu_\sigma}$

Where b is a parameter that reflect influence of intermediate principal shear stress, $\varphi_0$ is shear strength parameters of loess, $\mu_\sigma$ is lode parameter

According to generalized Hooke's law, substituting equation (8) into equation (1), elastic damage constitutive equation is obtained as follows:

$$\sigma_1 = E\varepsilon_1 \exp\left[ -\left( \frac{R + P \sin \varphi_t}{F_0} \right)^m \right] + \mu(\sigma_2 + \sigma_3) \tag{9}$$

### 4.2. *Model parameters*

The proposed model contains seven parameters, i.e. elastic modulus $E$, Poisson's ratio $\mu$, internal friction angle $\varphi_0$, Weibull's parameters $F_0$, $m$, intermediate principal shear stress parameter $b$, and lode parameter $\mu_\sigma$. Weibull's parameters are obtained from triaxial tests. Constitutive equation is treated linearly by taking logarithm:

$$Y = mX + A \qquad (10)$$

where: $Y = \ln\{-\ln[\dfrac{\sigma_1 - \mu(\sigma_2 + \sigma_3)}{E\varepsilon_1}]\}$ ; $X = \ln(R + P\sin\phi_t)$ ; $A = -m\ln(F_0)$.

For the triaxial tests, elastic modulus $E$ is determined as follows:

$$E = Kp_a(\frac{\sigma_3}{p_a})^n \qquad (11)$$

Where $p_a$ is atmospheric pressure, parameters $K$, $n$, are test constant.

## 4.3. Comparison between calculated and experimental results

The stress-strain behavior of normal consolidated and over consolidated specimens is simulated using the proposed model for triaxial stress state. Model parameters are shown in table1. Theoretical and experimental results of normal consolidated and over consolidated specimens for CU-test are shown in Fig.5 and Fig.6. It is obtained that the theoretical results can well simulate experimental results, they are closed to experimental results, and especially the characteristic of strain softening of loess can well reflects using damage model based on unified strength theory.

Table 1. Model parameters.

| Test | Pressure (kPa) | Normal Consolidated specimens | | | | | Over Consolidated specimens | | | | |
|------|------|------|------|------|------|------|------|------|------|------|------|
| | | $E$ (MPa) | $m$ | $F_0$ | $\varphi_0(^0)$ | $\mu$ | $E$ (kPa) | $m$ | $F_0$ | $\varphi_0(^0)$ | $\mu$ |
| CU | 100 | 88.92 | 0.270 | 85.3 | 18.87 | 0.5 | 84.17 | 0.322 | 238.3 | 21.92 | 0.5 |
| | 150 | 147.25 | 0.267 | 119.5 | 18.87 | 0.5 | 143.67 | 0.274 | 159.5 | 21.92 | 0.5 |
| | 200 | 192.42 | 0.271 | 162.5 | 18.87 | 0.5 | 181.17 | 0.263 | 159.7 | 21.92 | 0.5 |
| | 250 | 265.00 | 0.259 | 151.7 | 18.87 | 0.5 | 225.42 | 0.251 | 153.0 | 21.92 | 0.5 |
| CD | 100 | 169.83 | 0.214 | 30.99 | 24.82 | 0.35 | 86.58 | 0.314 | 297.9 | 28.54 | 0.35 |
| | 150 | 123.46 | 0.223 | 56.39 | 24.82 | 0.35 | 134.67 | 0.273 | 227.7 | 28.54 | 0.35 |
| | 200 | 173.08 | 0.193 | 30.11 | 24.82 | 0.35 | 178.92 | 0.265 | 242.0 | 28.54 | 0.35 |
| | 250 | 201.92 | 0.236 | 124.4 | 24.82 | 0.35 | 221.08 | 0.220 | 96.16 | 28.54 | 0.35 |

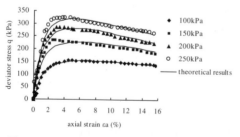
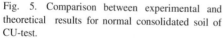
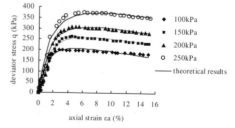

Fig. 5. Comparison between experimental and theoretical results for normal consolidated soil of CU-test.

Fig. 6. Comparison between experimental and theoretical results for over consolidated soil of CU-test.

The calculated results of stress-strain relationship of normal consolidated specimens and over consolidated specimens for CD-test are shown in Fig.7 and Fig.8. From the results, the calculated results are very close to the test results. The strain-hardening characteristic is well simulated also. It is shown that the proposed constitutive model can describe the stress-strain behavior curves and predict the strength trend of loess.

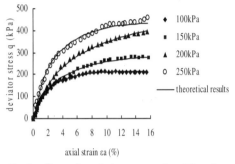

Fig. 7. Comparison between experimental and theoretical results for normal consolidated soil of CD-testl.

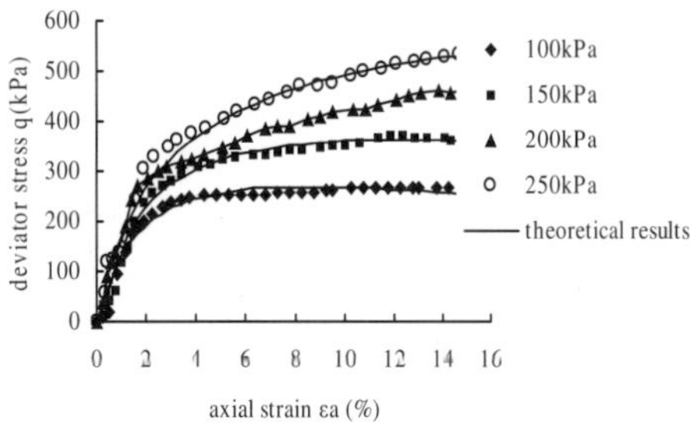

Fig. 8. Comparison between experimental and theoretical results for over consolidated soil of CD-test.

## 5. Conclusion

From a series of triaxial shear tests, the stress-strain relationships of loess were obtained, and were discussed on normal consolidated and over consolidated loess. It is shown that the stress-strain curves of loess have strain-softening and strain-hardening characteristic.

Based on unified strength theory, the statistical damage model is proposed, which can well describe characteristics of strain softening and strain hardening on loess. The proposed model can predict stress-strain relationships under various stress states.

## Acknowledgment

This paper is funded by scientific research program of the Higher Education Institution of XinJiang (XJEDU2005S04).

## References

1. Shen Zhujiang, Hu Zaiqiang, *Journal of hydraulic engineering*, **7**, 1(2003).
2. Shen Zhujiang, Den Gang, *Rock and Soil Mechanics*, **24**, 495(2003).
3. Xia Wangmin, Guo Zengyu, *Rock and Soil Mechanics*, **25**, 1423(2004).
4. Lin Bin, Zhao Fasuo, *Journal of Anhui university of Science and Technology*, **123**, 17(2006).
5. Xu Dongjun, Geng Naiguang, *Acta Mechanica Solida Sinica*, **6**, 72(1985).
6. Mao-hong Yu. *Unified Strength Theory and its Applications* (Springer-Verlag Berlin Heidelberg, Berlin, 2004).

# THE INFLUENCE OF STRAIN PATH ON BIAXIAL COMPRESSIVE BEHAVIOR OF AZ31 MAGNESIUM ALLOY

ICHIRO SHIMIZU[1†], NAOYA TADA[2]

*Graduate School of Natural Science and Technology, Okayama University,*
*3-1-1 Tsushima-naka, Okayama 700-8530, JAPAN,*
*shimizu@mech.okayama-u.ac.jp, tada@mech.okayama-u.ac.jp*

KOSUKE NAKAYAMA[3]

*OG Giken Co., Ltd, 1835-7 Miyoshi, Okayama 703-8261, JAPAN,*
*kousuke-nakayama@og-giken.co.jp*

Received 15 June 2008
Revised 23 June 2008

The strain path dependence of the compressive flow behavior of cast AZ31 magnesium alloy was investigated. Biaxial compression tests with linear strain paths were conducted using a unique biaxial compression device. It was found that the equivalent stress-strain relations varied according to the strain paths. The work contour for linear strain paths was well described by the Logan-Hosford yield criterion. Biaxial compressions with abrupt strain path change were also carried out to investigate the influences of the prestrain amplitude and angular relation of the sequential strain paths on the flow behavior. Rapid increase in the equivalent stress was observed just after the abrupt strain path change. These specific flow behaviors were discussed with regard to the plastic anisotropy, which showed rapid evolution in the early stage of the biaxial compressions.

*Keywords*: Plasticity; Biaxial compression; AZ31 magnesium alloy; Strain path; Plastic anisotropy.

## 1. Introduction

In recent years, magnesium alloys have become important industrial materials because of their low weight and high specific strength. This trend has attracted research on the plastic behavior of magnesium alloys, particularly for metal forming applications. Owing to the easy formation of magnesium alloys at high temperature, several researchers have investigated formability,[1,2] yield locus,[3] and effects of microstructure on the mechanical behavior[4,5] by focusing on warm and hot working. Meanwhile, since cold working is productive and economical, researches on plastic deformation at an ambient temperature have also been conducted.[6-9] The characteristics of the plastic behavior of magnesium alloys are mainly attributed to the limited slip systems of hexagonal close-packed structures. Thus, texture evolution and twinning are important factors affecting specific

---

[†]Corresponding Author.

472

flow behavior.[7–10] However, deformation modes utilized in most studies are uniaxial tension and compression. Although important for compressive metal forming processes such as forging and extrusion, information on the mechanical behavior in biaxial compressive stress conditions is highly limited due to experimental difficulties.

The primary objective of the present study is to provide fundamental information on the flow behavior of AZ31 magnesium alloy during biaxial compressions at an ambient temperature. Biaxial compressions were carried out using a unique device developed by the authors.[11] Tests were conducted for investigating the biaxial compressions of linear strain paths with different strain ratios; further, two types of biaxial compression tests with an abrupt strain path change without unloading were also conducted. Thus, the influence of strain paths on the biaxial compressive flow behavior was discussed.

## 2. Experiments

The material used was cast AZ31 magnesium alloy with a chemical composition of Mg-2.8wt%Al-0.9wt%Zn. Rectangular block specimens of $8 \times 8 \times 7$ mm$^3$ (in the $x$, $y$, and $z$ directions) were machined and annealed at 523 K for 2 h, the resulting grain size being 180 μm. After mechanical polishing, biaxial compression was applied to the specimens.

The compression tests with linear strain paths were performed under uniaxial, plane strain, and equibiaxial conditions to obtain fundamental flow behaviors. In addition, two types of biaxial compression tests with strain path change (shown in Fig. 1) were performed. In test A, for prestraining, plane strain compression was carried out in the $x$ direction until the preset strain $\varepsilon_{x0}$ was achieved, and then compression was carried out in the $y$ direction under the plane strain condition. In test B, the biaxial compression with an arbitrary strain ratio was carried out as the first path, until $\varepsilon_x + \varepsilon_y$ became −0.15. The compressive direction was then changed, and the subsequent linear strain path was selected to $(\varepsilon_x, \varepsilon_y) = (-0.15, -0.15)$. These strain paths in test B were characterized by the angle $\alpha$ between sequential strain paths in the $\varepsilon_x$-$\varepsilon_y$ strain plane. In all compression tests, a lubricant of silicone grease mixed with boron nitride powder was applied to the specimen/die interfaces to reduce friction. Hence, bulging of specimens was barely observed even at an equivalent strain of 0.3. All tests were performed at an ambient temperature at a slow strain rate of ~5 × 10$^{-4}$ s$^{-1}$. In order to verify the repeatability, several experiments were conducted under each set of conditions. For a direct

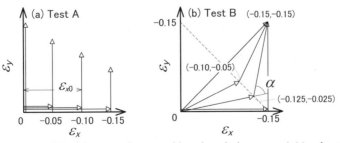

Fig. 1. Two types of biaxial compression tests with strain path change, test A (a) and test B (b).

comparison of the results obtained by various biaxial compressions, von Mises equivalent stress and equivalent strain were employed as a first approximation.

## 3. Results and Discussion

The true stress-strain curves and the variation in the strain hardening rate by uniaxial compressions in the $x$, $y$, and $z$ directions are shown in Fig. 2. The stress-strain curves for the three compressive directions were almost the same. Hence, the initial condition was regarded as isotropic. An interesting feature regarding the variation in the strain hardening rate is observed; this rate can be divided into three stages. Stage I is the elastic-plastic transition range in which the hardening rate decreases gradually. In stage II, the hardening rate becomes almost constant and finally, in stage III, it decreases again.

Fig. 3 shows the von Mises equivalent stress-strain relations by uniaxial, plane strain, and equibiaxial compressions. After yielding, the equivalent stress varied according to the strain paths, that is, equi-biaxial compression resulted in the maximum equivalent stress, while uniaxial compression resulted in the minimum equivalent stress. This result implied the rapid evolution of plastic anisotropy during biaxial compressions, as discussed later. The three stages of the hardening rate also appeared in the cases of plane strain and equibiaxial compressions. The constant or decreasing hardening rate in stage II for AZ31 alloys was reported only by uniaxial compression[7,9] and not by uniaxial tension.[1,5,6] These results imply that for the cast material employed, this phenomenon is also observed in the case of biaxial compressions. Jiang et al.[10] carried out a detailed observation of microstructural evolution during the uniaxial compressions of AZ31 and explained that the change in the hardening rate was due to twinning ($\{10\bar{1}2\}$ extension and $\{10\bar{1}1\}$ contraction twinning) and texture change induced by twin. Although these results are observed at moderate temperatures, twinning may be the dominant cause of changes in the hardening rate in the biaxial compressions at an ambient temperature as well.

The equivalent stress-strain relations due to biaxial compressions with strain path change are shown in Figs. 4(a) and (b) for tests A and B, respectively. In test A, the increasing rate of equivalent stress varied just after the strain path change, and subsequently, the equivalent stress became larger than that by monotonic plane strain

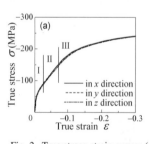 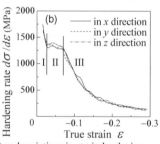 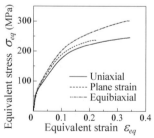

Fig. 2. True stress-strain curves (a) and variations in strain hardening rate (b) by uniaxial compressions. The hardening rate can be divided into stages I, II and III.

Fig. 3. Equivalent stress-strain curves by biaxial compressions with linear strain paths.

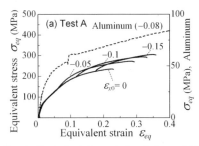 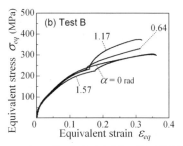

Fig. 4. Equivalent stress-strain relations of AZ31 magnesium alloy by test A (a) and test B (b). The dotted curve in (a) is a sample curve of pure aluminum.

compression. In test B, a sudden increase in the equivalent stress just after the strain path change was also observed. However, the increasing rate did not vary uniformly with the angle $\alpha$ between sequential strain paths, and the maximum increasing rate was observed for an $\alpha$ value of 1.17. A similar variation in the increasing rate of equivalent stress just after the strain path change was also observed in the biaxial compression of pure aluminum (as shown in Fig. 4(a)), mainly due to the latent hardening effect.[11] However, the variation in the equivalent stress for AZ31 was different from that for pure aluminum, and a larger equivalent stress accompanied by the strain path change was observed until fracture. Thus, it is supposed that the larger equivalent stress is caused mainly by the evolution of plastic anisotropy during prestraining.

In order to understand the specific equivalent stress-strain relations of AZ31 shown in Figs. 3 and 4, we focused on the change in work contours, namely, contours of the equal expenditure of plastic work, introduced by Hill et al.[12] Fig. 5(a) shows work contours obtained by biaxial compressions with linear strain paths. The equi-work points were determined by evaluating plastic work corresponding to 0.2% proof stress, $W_{0.2}$.[3,12] Since the equi-work points deviated from the von Mises yield criterion, the Logan-Hosford yield criterion was introduced.[13] Assuming symmetry with respect to the line $\sigma_x = \sigma_y$, the Logan-Hosford criterion for plane strain condition becomes

$$\bar{\sigma}^M = \frac{1}{1+R}\left(\sigma_x^M + \sigma_y^M\right) + \frac{R}{1+R}\left(\sigma_x - \sigma_y\right)^M , \qquad (1)$$

where $M$ is a material constant and $R$ is a Lankford value, which represents plastic anisotropy. The calculated Logan-Hosford criterion with $M = 10$ and $R = 1.0$ was found to agree well with the experimental equi-work points, as shown in Fig. 5(a). By this result, $M = 10$ was used throughout the present study. Fig. 5(b) shows the work contours for $W = 0.2$, 5, and 10 MPa. It was found that a large $R$ value (>4) was necessary for fitting the Logan-Hosford curve to the work contours. Although such a high value of $R$ as the Lankford value appears unrealistic, the increasing $R$ indicates the evolution of plastic anisotropy. Thus, $R$ can be used as a parameter that represents the degree of anisotropy.

Fig. 6 shows the variation in $R$ with plastic work. The value of $R$ was evaluated by fitting the Logan-Hosford curve to the equi-work points for each plastic work. The true

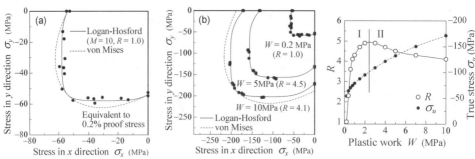

Fig. 5.  Plastic work contours $W$ corresponding to 0.2% proof stress (a) and for $W = 0.2$, 5.0, and 10.0 MPa.  The Logan-Hosford curves were derived from Eq. (1) with $M = 10$.

Fig. 6.  Variations of $R$ and corresponding true stress $\sigma_u$ by uniaxial compressions, with plastic work.

stress by uniaxial compressions for each plastic work is also shown in Fig. 6.  The value of $R$ was found to increase rapidly in the initial stage of plastic deformation, namely, in stage I, and then $R$ decreased slightly in stage II.  It should be noted that $R$ is not necessarily equivalent to the Lankford value in the present results because the Logan-Hosford curve was fitted to the equi-work contour and $M = 10$ is not necessarily adequate for the AZ31 alloy.  However, the result implies that the plastic anisotropy increases rapidly in biaxial compressions.  Considering the results by Jiang et al.,[10] it is supposed that the rapid growth of anisotropy is due to the interactions between primary twinning and slip deformation in stage I, and the anisotropy becomes stable or decreases slightly with the onset of secondary twinning in stage II.  Such a variation in the anisotropy in the initial stage of deformation is also in qualitative agreement with the texture changes shown by Yi et al.[7] and Brown et al.,[14] in which the texture evolved by a small tensile or compressive strain less than 0.1.

   Fig. 7 shows the variations in stress points on the $\sigma_x$-$\sigma_y$ plane for biaxial compressions with strain path change in tests A and B.  In test A, the direction of the stress path in the first compression changed rapidly when the compression direction was switched.  After the strain path change, the deviation of the work contour from the von Mises curve became significant, and the deviation increased with a larger prestrain $\varepsilon_{x0}$.  A distortion of work contours after the strain path change was also observed for test B.  Furthermore, though the final dimensions of specimens were almost the same, the final stress points

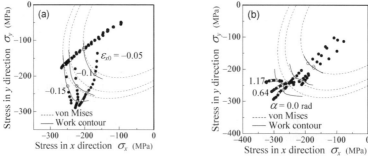

Fig. 7.  Variation in stress paths and work contours during biaxial compressions in test A (a) and test B (b).

varied according to the strain paths. These specific stress paths and the resultant distortion of work contours in tests A and B indicate a strong influence of strain path change on the flow behavior, due to the evolution of plastic anisotropy during prestraining.

## 4. Conclusions

The flow behavior of AZ31 magnesium alloy was investigated by biaxial compressions with various strain paths at ambient temperature. The conclusions of the present study are summarized as follows.

- The von Mises equivalent stress-strain relations varied according to the strain paths, and an increase in equivalent stress was observed just after the abrupt strain path change. These phenomena were explained by the rapid evolution of plastic anisotropy.

- Variation in the hardening rate was observed in not only uniaxial compression but also biaxial compressions. It was divided into three stages: (1) an elastic-plastic transition stage, where the hardening rate decreased gradually, (2) a stage with an almost constant hardening rate, (3) and a stage in which the hardening rate decreased again. In the case of biaxial compressions with linear strain paths, the equi-work points determined by plastic work corresponding to 0.2% proof stress agreed well with the Logan-Hosford yield criterion. The parameter $R$ of the Logan-Hosford criterion was used to assess plastic anisotropy; thus, the anisotropy was found to grow rapidly in the initial stage of biaxial compressions.

- The work contours deviated from the von Mises curve after the strain path change. The deviation increased with an increase in the prestrain in the sequential plane strain compressions. Meanwhile, even under the same strain conditions, the final stress ratios varied according to strain paths. These results imply that the evolution of plastic anisotropy influences the flow behavior after the strain path change.

## References

1. M. Kohzu, F. Yoshida, H. Somekawa, M. Yoshikawa, S. Tanabe and K. Higashi, *Mater. Trans.* **42**, 1273 (2001).
2. Q.F. Chang, D.Y. Li, Y.H. Peng and X.Q. Zeng, *Int. J. Machine Tool. Manuf.* **47**, 436 (2007).
3. T. Naka, T. Uemori, R. Hino, M. Kohzu, K. Higashi and F. Yoshida, *J. Mater. Process. Technol.* **201**, 395 (2008).
4. A.G. Beer and M.R. Barnett, *Mater. Sci. Eng.* **A423**, 292 (2006).
5. S.B. Yi, S. Zaefferer and H.G. Brokmeier, *Mater. Sci. Eng.* **A424**, 275 (2006).
6. A. Jäger, P. Lukáč, V. Gärtnerová, J. Haloda and M. Dopita, *Mater. Sci. Eng.* **A432**, 20 (2006).
7. S.B. Yi, C.H.J. Davies, H.G. Brokmeier, R.E. Bolmaro, K.U. Kainer and J. Homeyer, *Acta Mater.* **54**, 549 (2006).
8. R.H. Wagoner, X.Y. Lou, M. Li and S.R. Agnew, *J. Mater. Process. Technol.* **177**, 483 (2006).
9. Y.N. Wang and J.C. Huang, *Acta Mater.* **55**, 897 (2007).
10. L. Jiang, J.J. Jonas, A.A. Luo, A.K. Sachdev and S. Godet, *Mater. Sci. Eng.* **A445**, 302 (2007).
11. I. Shimizu and N. Tada, *Key Eng. Mater.* **340-341**, 883 (2007).
12. R. Hill, S.S. Hecker and M.G. Stout, *Int. J. Solids Struct.* **31**, 2999 (1994).
13. R.W. Logan and W.F. Hosford, *Int. J. Mech. Sci.* **22**, 419 (1980).
14. D.W. Brown, S.R. Agnew, M.A.M. Bourke, T.M. Holden, S.C. Vogel and C.N. Tomé, *Mater. Sci. Eng.* **A399**, 1 (2005).

# 3D FINITE ELEMENT MODELLING OF COMPLEX STRIP ROLLING

Z.Y. JIANG[†]

*School of Mechanical, Materials and Mechatronic Engineering, University of Wollongong, Northfields Avenue,*
*Wollongong NSW 2522, AUSTRALIA,*
*jiang@uow.edu.au*

Received 15 June 2008
Revised 23 June 2008

A main feature of complex strip - ribbed strip is the significant local residual deformation on a flat strip, resulting in the pulling down of rib height. The interesting issue in this paper shows a developed three-dimensional finite element model of the complex strip rolling, coupling the use of an extremely thin array of elements which is equivalent to the calculation of the additional shear deformation work rate occurred by the velocity discontinuity in the deformation zone. The 3D finite element modelling includes the consideration of the special rib inclined contact surface boundary condition has been carried out on a computer. An examination of the equivalent stress field, forward slip and rib height demonstrates the effectiveness of the developed model. The computed forward slip and rib height are in good agreement with the measured values. The effect of the rib inclined angle on the pulling down of rib height is also discussed.

*Keywords*: Local residual deformation; complex strip; finite element modeling; extremely thin elements; pulling down of rib height.

## 1. Introduction

In order to meet the demand for various steel markets in the globe, the manufacturers have been focusing on the development of new products and its applications. A complex strip shown in Fig. 1 ($\theta$ is the rib inclined angle, $h$ is the rib height, $E_1$ is the distance between two ribs, I, II and III are the deformation zones) has continuous ribs on the surface of strip that involves a significant local residual deformation [1], resulting in the pulling down of rib height. It has been widely used in various engineering fields including electrical equipment, automobile industry, offshore platform construction, ship structures and thermal transfer industry. An important application of the strip, for example, is as pipe piles in civil construction, when it is made into internal spiral pipe with the internal ribs providing a much stronger bond between the pipe and concrete than that with smooth-surfaced ones [2]. The higher the rib height is, the stronger the bond [3] is. Thus this pile leads to a significant savings of both the steel pipes and concrete[1, 2]. The

---

[†]Corresponding Author.

application of this type of strips is therefore expected to increase significantly in civil construction, which can modify the current civil construction structure greatly.

For complex strip rolling, due to significant local residual deformation across the width of strip in the roll bite, the modelling of mechanics and deformation nearby the ribs is an interesting issue, which involves the treatment of the formulations of a special rib inclined contact surface boundary conditions, singular point [4] and friction in the roll bite. In this paper, the author concentrates on the analysis of rib deformation of the ribbed strip in the roll bite using a three-dimensional thermo-mechanical model during the ribbed strip rolling taking into account friction variation model [5]. The theoretical analysis is based on the utilization of the finite element flow formulation with boundary conditions to characterize the material flow, coupling the use of an extremely thin array of elements which is equivalent to the calculation of the additional shear deformation work rate due to the velocity discontinuity in the deformation zone. The effective treatments of the inclined contact surface, the first singular points and friction in the roll bite for this local residual deformation have been conducted. The rib height, forward slip and pulling down of rib height [1] during the ribbed strip rolling are discussed.

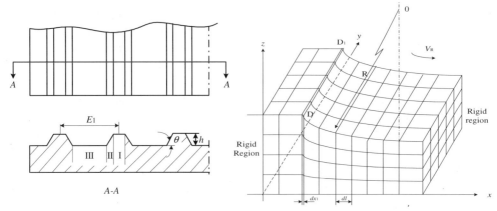

Fig. 1. Cross-section and plan view of the ribbed strip.

Fig. 2. A method to treat the first singular point occurring in steady rolling by 3D rigid plastic FEM.

## 2. Theoretical Analysis

### 2.1. *Extremely thin elements*

In general, there is a point at which the velocity of metal flow changes its direction in the metal forming processes. As shown in Fig. 2, R is the roll radius and $V_R$ is the roll speed, and metal at points $DD_1$ flows along rolling direction $x$ before it contacts the work roll, but it flows along the roll tangential direction when D contacts the work roll. The point D is known as the first singular point [4]. First singular point can be dealt with a thin array of elements [6].

Based on the practical feature of the discontinuous surface of velocity, thin elements were selected such that the thickness $dx_1 < 1$ % of general element size [6], see Fig. 2. For a 3D problem, the sizes of the thin elements are *max* $(dx_1/dy_1, dx_1/dz_1, dx_1/dl) < 0.01$, where

$dx_1$, $dy_1$ and $dz_1$ are the average element length along the rolling direction $x$, thickness $y$ and width $z$ directions respectively, $dl$ is the length in the $x$ direction for a general element. Shear deformation mainly occurs in the extremely thin elements. In particular, for very thin elements near the entry of the deformation zone, the calculated result shows that the rate of shear velocity is high, which makes the equivalent strain rate $\dot{\bar{\varepsilon}}$ to be higher than that of the general elements [6]. The introduction of the extremely thin elements can simplify the programming and improve the convergence of simulation [7].

### 2.2. FE formulations

Many rolling processes have been simulated by finite element method. Hartley et al. [8] and Liu et al. [9] have used an elastic-plastic finite element method to analyse the flat rolling and slab rolling, and Wen et al. [10-12] also investigated the section rolling. Osakada et al. [13] developed a slightly compressible material formulation for the analysis of metal forming by rigid plastic FEM. Mori et al. [14] has used the rigid plastic FEM with a slightly compressible material to analyse the plane-strain rolling. Other researchers have used the rigid plastic/visco-plastic FEM to solve the strip rolling [15, 16], shape rolling [17] and the edge rolling [18-20]. Jiang et al. [1, 5] have investigated the special strip rolling by rigid plastic/visco-plastic FEM. However, the deformation mechanics of the ribs with friction variation model in the roll bite and the effect of the rib inclined contact surface on the pulling down of rib height have not been studied in the modelling of ribbed strip rolling by a 3D rigid plastic/visco-plastic FEM.

For a 3D rolling process, according to variational principle [15], the real velocity field must satisfy the functional at the minimum. In theory, the set up of thin elements is equivalent to consider the additional shear work rate [6]. In order to improve the simulation accuracy, a friction variation model [21, 22] has been used in the hot strip rolling. In the study of the ribbed strip rolling, a friction model used is as follows:

$$\tau_f = \pm \frac{m_1 \sigma_s}{\sqrt{3}} \frac{V_g}{\sqrt{V_g^2 + k_1^2}} \tag{1}$$

where $m_1$ is friction factor, $\sigma_s$ yield stress, $k_1$ a small positive constant, $V_g$ the relative slip velocity between the rolled material and roll. The signs "+" and "–" are for forward and backward slip zones respectively. Detailed distribution of the frictional shear stress can refer to Ref. [23]. The rib height can be determined from Eq. (2):

$$h = Z_i - 0.5 h_2 \tag{2}$$

where $Z_i$ is the $z$ average coordinate of nodes on rib top at exit and $h_2$ the thickness of the ribbed strip. The forward slip can be obtained from Eq. (3):

$$S = \frac{\sum_{i=1}^{m_2} v_{xi}}{m_2 V_R} - 1 \tag{3}$$

where $v_{xi}$ is the node velocity at exit of the strip, and $m_2$ is the number of nodes at the exit of the strip.

## 3. Simulation and Results

A three-dimensional thermo-mechanical model was employed in this paper [24], coupling the use of an extremely thin array of elements. The FE mesh used in the simulation is shown in Fig. 3. Isoparametric hexahedral elements with eight gauss points were used throughout the ribbed strip rolling, $m_1$ is 0.5 and $k_1$ is 0.01. Process data and material flow stress model utilised in the theoretical analysis of the ribbed strip rolling are shown in Table 1. Eight ribs were produced on a rolling mill. It should be noted that if the ribs do not fill up the grooves of the roll, the PP'O'O, QQ'R'R, TT'S'S and UU'V'V are free surfaces.

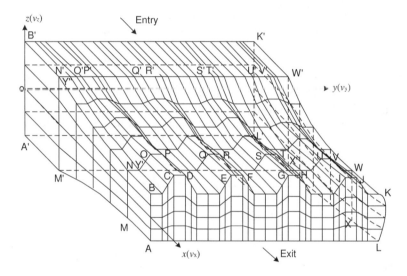

Fig. 3. FE mesh used in the simulation.

Table 1. Process data and material flow stress model.

| Material | Aluminum |
|---|---|
| Slab thickness | |
| Before rolling | 5.0 – 5.5 mm |
| After rolling | 1.8 – 3.0 mm |
| Slab width | 80 mm |
| Rib height | 1.0 – 1.5 mm |
| Two-high rolling mill, diameter of roll | 120 mm |
| Flow stress model [16] | $\sigma = 31.70 + 113.77\,\bar{\varepsilon}^{0.5539}$ |

In the simulation, the inclined contact surface, OO'Y"Y' etc is a non-steady surface. The determination of the inclined contact surface is based on the calculation stability of the rib heights during the iterations. A criterion was used to determine the stability of rib heights across the strip width [23]. Then the inclined contact area of the rolled strip can be obtained, which is used to determine the ribbed strip deformation and mechanics.

The numerical simulation has been carried out. Fig. 4 shows the distribution of equivalent stress of the side rib and strip side shape. The pattern of the equivalent stress field at the side (the right hand part in Fig. 4) is different from that between the ribs (the left hand part in Fig. 4), because of the lower temperature at the sides of the ribbed strip. The local residual deformation offers a large strain gradient near the sides of the ribs. Therefore, the stress gradient is the largest at the sides of the ribs. It can also be seen that a lateral flow of metal exists (see right hand side of Fig. 4). A comparison of the calculated forward slip with measured value is shown in Fig. 5, it can be seen that the calculated forward slip is in good agreement with the measured value. This demonstrates the suitability of this method. Fig. 6 shows the relationship between the reduction and rib height. It can be seen that the rib height increases significantly when the reduction $\varepsilon <$ 40%, but the increasing rate of rib height reduces as $\varepsilon > 40\%$. It can also be seen that the calculated results are in good agreement with the measured values.

A comparison of the calculated and measured pulling down of rib height [5] is shown in Table 2. It can be seen that the pulling down of rib height decreases when the inclined angle of the rib $\theta$ increases. The width of the zone II (see Fig. 1) reduces as the rib inclined angle increases. As a result, more metal in zone III flows to zone I to form ribs, which results in an increase of rib height. However, the pulling down of rib height increases significantly when the reduction increases due to the increase of non-uniform deformation across the strip width. The calculated pulling down of rib height is close to the measured value.

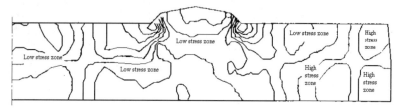

Fig. 4. Distribution of equivalent stress of the side rib and strip side shape.

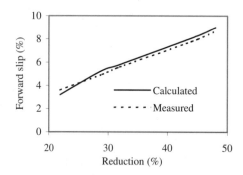

Fig. 5. Comparison of calculated forward slip with measured value.

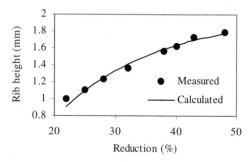

Fig. 6. Comparison of calculated rib height with measured value.

Table 2. Comparison of the calculated and measured pulling down of rib height.

| $\theta$, ° | | 60 | | 75 | |
|---|---|---|---|---|---|
| | | Calculated (mm) | Measured (mm) | Calculated (mm) | Measured (mm) |
| $\varepsilon\%$ | 20.5 | 0.40 | 0.39 | 0.30 | 0.28 |
| | 46.0 | 0.73 | 0.70 | 0.64 | 0.61 |

## 4.  Conclusions

A 3D finite element method model of the complex strip rolling has been developed in this study, coupling the use of an extremely thin array of elements, which is equivalent to the calculation of the additional shear deformation work rate. The rib inclined contact surface, first singular point and friction in the roll bite have been treated effectively, which make a significant scientific contribution for rolling theory with local residual deformation and control model development. The rib height increases significantly when the reduction $\varepsilon < 40\%$, but the increasing rate of rib height reduces as $\varepsilon > 40\%$. The calculated forward slip, rib height and pulling down of rib height are in good agreement with the experimental values. The pulling down of rib height decreases when the rib inclined angle increases, but it increases significantly with reduction.

## Acknowledgments

The financial support from Australian Research Council (ARC) for the present study is greatly appreciated.

## References

1. Z.Y. Jiang, S.W. Xiong, X.H. Liu, G.D. Wang, Q. Zhang, *J. Mater. Proc. Technol.* **79**(1-3), 109-112 (1998).
2. S. Masude, T. Hirazawa, I. Taniguchi, *Nippon Kokan Technical Report* **107**, 31-37 (1985).
3. Z.Y. Jiang, X.H. Liu, G.D. Wang, Q. Zhang, *J. Welded Pipe and Tube* **18**(2), 17-19 (1995).
4. X.H. Liu, *Rigid-Plastic FEM and Its Application in Steel Rolling* (Metallurgical Industry Press, Beijing, 1994).
5. Z.Y. Jiang, X.L. Liu, X.H. Liu, G.D. Wang, *Int. J. Mech. Sci.* **42**, 693-703 (2000).
6. Z.Y. Jiang, A.K. Tieu, *J. Mater. Proc. Technol.* **117**(1-2), 146-152 (2001).
7. S.W. Xiong, J.M.C. Rodrigues, P.A.F. Martins, *Finite Elements in Analysis and Design* **39**(11), 1023-1037 (2003).
8. P. Hartley, C.E.N. Sturgess, C. Liu, G.W. Rowe, *Int. Materials Reviews* **34**(1), 19-34(1989).
9. C. Liu, P. Hartley, C.E.N. Sturgess, G.W. Rowe, *Int. J. Mech. Sci.* **29**(4), 271-283(1987).
10. P. Hartley, S.W. Wen, I. Pillinger, C.E.N. Sturgess, D. Petty, *Ironmaking & Steelmaking* **20**(4), 261-263(1993).
11. S.W. Wen, P. Hartley, I. Pillinger, C.E.N. Sturgess, Proceedings of the Institution of Mechanical Engineers Part B-J. Engrg Manufacture **206**(B2), 133-141(1992).
12. S.W. Wen, P. Hartley, I. Pillinger, C.E.N. Sturgess, Proceedings of the Institution of Mechanical Engineers Part B-J. Engrg. Manufacture **211**(B2), 143-158(1997).
13. K. Osakda, J. Nakano, K. Mori, *Int. J. Mech. Sci.* **24**(8), 459-468(1982).
14. K. Mori, K. Osakada, T. Oda, *Int. J. Mech. Sci.* **24**(9), 519-527(1982).

15. S. Kobayashi, S. Oh, I.T. Altan, *Metal Forming and the Finite Element Method* (Oxford University Press, New York, 1989).
16. X.H. Liu, *Rigid-Plastic FEM and Its Application in Steel Rolling* (Metallurgy Industry Press, Beijing, 1994).
17. J.L. Chenot, P. Montmitonnet, P. Buessler, F. Fau, *Engineering Computations* (Swansea, Wales) **8**(3), 245-255(1991).
18. H. Nikaido, T. Naoi, K. Shibata, T. Kondo, K. Osaka, K. Mori, *J. Japanese Society Technology Plasticity* **25**(277), 129-135(1984).
19. S.W. Xiong, X.H. Liu, G.D. Wang, Q. Zhang, *J. Materials Engineering and Performance* **6**(6), 757-765(1997).
20. H.J. Huisman, J. Huetink, *J. Mech. Working Tech.* **21**, 333-353(1985).
21. S.M. Hwang, M.S. Joun, *Int. J. Mechan. Sci.* **34**(12), 971-984(1992).
22. S.M. Hwang, M.S. Joun, Y.H. Kang, *Trans. ASME J. Eng. Ind.* **115**(3), 290-298(1993).
23. Z.Y. Jiang, Forming Theory and Practice of the Strip with Ribs and the Internal Spiral-ribbed Tube (Metallurgical Industry Press, Beijing, 1998).
24. Z.Y. Jiang, A.K. Tieu, C. Lu, W.H. Sun, X.M. Zhang, X.H. Liu, G.D. Wang, X.L. Zhao, *Finite Elements in Analysis and Design* **40**, 1139-1155(2004).

# Part F
# Molecular Dynamics

# THE STUDY ON THE GAS PERMEABILITIES OF THE ETHYLENE/1-HEXENE COPOLYMER BY MOLECULAR DYNAMICS SIMULATION

SIZHU WU[1], JUN YI[2], LISHU ZHANG[3], LIQUN ZHANG[4†]

*Key Laboratory of Beijing City on Preparation and Processing of Novel Polymer Materials,*
*Beijing University of Chemical Technology, Beijing 100029, P. R. CHINA,*
*wusz@mail.buct.edu.cn, yijun0516@163.com, lishu917@yahoo.com.cn, zhanglq@mail.buct.edu.cn*

JAMES E. MARK[5]

*Department of Chemistry and the Polymer Research Center,*
*The University of Cincinnati, Cincinnati, OH 45221-0172, USA*
*markje@email.uc.edu*

Received 15 June 2008
Revised 23 June 2008

In this research, molecular dynamics(MD) simulations were used to study the transport properties of small gas molecules in poly(ethylene-*co*-1-hexene) copolymer. The condensed-phase optimized molecular potentials for atomistic simulation studies (COMPASS) forcefield was applied. The diffusion coefficients were obtained from MD (NVT ensemble). The results indicated that the diffusion coefficient of oxygen increased with increasing 1-hexene content in copolymer membrane.

*Keywords*: Molecular dynamics; ethylene/1-hexene; diffusion coefficient; diffusion selectivity.

## 1. Introduction

Gas separation through a membrane is one of the most attractive industrial processes because of the low energy consumption and easy operation.[1] The transportation of small molecules in polyolefins is a hot research objective in both technical and commercial areas. Therefore, the study on the diffusive characteristics of gases, especially oxygen, is of great importance.[2] Since experimental techniques are not easy to give details at the molecular level, the transport of gas molecules through membranes is still not completely understood.[3] In recent years, atomistic simulation techniques have been proven to be a useful tool for understanding of structure-property relationships of materials, and the molecular dynamics (MD) can be used particularly for the detailed descriptions of the complex morphologies and the transport mechanisms.[4] In this work, the transport properties of ethylene/1-hexene copolymer were studied by MD simulation.

†Corresponding Author.

## 2.  Description of the Simulations

The simulations were performed with Material Studio (MS) software of Accelrys Inc. Materials. Visualizer module was used to construct the copolymer chains, and then the chains were subjected to energy minimization. Subsequently the packing models were constructed with Amorphous Cell Module, which implement a modification of the rotational isomeric state(RIS) of Theodorous and Suter method.[5,6] The simulation parameters were shown in Table 1, which were used to study the relationship between diffusion coefficient of oxygen and the 1-hexene content in the copolymer.

Table 1. The parameters of the simulation systems of gas transportation.

| System | $f^a_{hex}$ (mol%) | DP | $N_{atom}$ | Number of polymer in cell | Number of oxygen | $\rho_{sim}$(g/cm$^3$) | $\rho^b_{exp}$(g/cm$^3$) | Cell lengths(Å) |
|---|---|---|---|---|---|---|---|---|
| PE | 0 | 200 | 1216 | 1 | 4 | 0.9452 | 0.9499 | 21.61 |
| CEH2.0 | 2.0 | 50 | 1264 | 4 | 4 | 0.9257 | 0.9206 | 22.10 |
| CEH4.2 | 4.2 | 50 | 1312 | 4 | 4 | 0.9154 | 0.9088 | 22.40 |
| CEH7.0 | 7.0 | 50 | 1408 | 4 | 4 | 0.8719 | 0.8962 | 22.30 |
| CEH11.0 | 11.0 | 50 | 1456 | 4 | 4 | 0.8651 | 0.8842 | 23.62 |

*a* $f_{hex}$ was the content of 1-hexene in ethylene/1-hexene copolymer. *b* Ref.[2]

Amorphous cells were initially constructed at reduced density of 0.4g/cm$^3$. The increased free volume at the lower density was sufficient for the packing algorithm. Following the cell construction at the initial density, NPT (NPT means the condition of constant particle number $N$, constant pressure $P$, and constant temperature $T$) dynamics was performed to bring the cells to the experimental density which was shown in Table 1. When the system density stabilized at a constant value, an NVT (NVT means the condition of constant particle number $N$, constant volume $V$, and constant temperature $T$) dynamics of 500ps was performed at the temperature of 300K. The trajectories were recorded at 1ps intervals for further analysis. In NPT process, the pressure was controlled by Berendsen's method.[7] The cutoff distance for the non-bonded interactions was selected as 9.5Å and the time step was 1fs and the velocity Verlet algorithm was used.

The gas diffusion coefficient can be calculated from the Einstein equation as following:[8]

$$D = \tfrac{1}{6N}\lim_{t\to\infty}\frac{d}{dt}\sum_{i=1}^{N_\alpha}\left\langle\left[r_i(t)-r_i(0)\right]^2\right\rangle. \tag{1}$$

where $r_i$ is the position vector of particle. $N_\alpha$ is the number of particle of type $\alpha$. $\left[r_i(t)-r_i(0)\right]^2$ is the mean squared displacement (MSD) and the angular brackets denote averaging over all choices of time origin.

To obtain the gas diffusion coefficient $D$, it is needed to plot MSD vs. time and calculate the slope of the line of the best fit. Since the value of the MSD has already been averaged over the number of atoms $N$, then Eq. (1) can simplify to:

$$D=a/6. \tag{2}$$

where *a* is the slope of the line. Then, according to Eq. (2), the gas diffusion coefficient *D* can be calculated from the plot of MSD vs. time.

Studies of the microscopic free-volume properties at molecular and atomic scales can provide a basic understanding of the mechanical and physical properties of polymers.[9] Free-volume theories of the glass transition assume that, if the conformational changes of the backbone are to take place, there must be a space available for molecular segment to move into. As the temperature is lowered from a temperature well above $T_g$ (glass transition temperature), the free-volume eliminate because of conformational rearrangement. When the temperature approaches $T_g$, the molecular motions becomes so slow that the molecules cannot rearrange within the time-scale of the experiment. At temperatures well below $T_g$, the elastic deformation that comes from bonds stretch and bond angles change, can be understood by assuming that all the conformational changes of backbone of the molecule are freezing out. Therefore, the free-volume size and distribution have a vital influence on the transport properties of penetrants. Recent studies showed that the positron annihilation lifetime spectroscopy was used to measure the free-volume.[10] But it usually consumes long time (almost 1 week) to measure a sample. However, MD could easily calculate the free-volume.[11] In order to study the free-volume size and distribution in bulk simulated molecular models, the hard probe method was applied,[12] where the van der Waals radius of oxygen and hydrogen atoms are 1.52Å and 1.2Å respectively.[13] When the probe with its radius $R_P$ moved along the van der Waals surface, the Connolly surface could be calculated,[14] which was shown in Fig. 1. Then the fraction of free-volume (FFV) could be obtained by dividing system volume with free-volume.

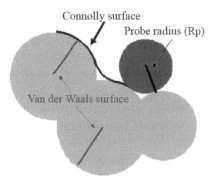

Fig. 1. Definition of Connolly surface.

## 3. Results and Discussion

Oxygen and hydrogen molecules were chosen as the penetrants in ethylene/1-hexene random copolymer systems in Table 1. One of the MSD curve of oxygen molecules in the ethylene/1-hexene copolymer cell was shown in Fig. 2, where the slope *a* of MSD vs.

$t$ could be obtained. Then the diffusion coefficient $D$ of oxygen and hydrogen could be calculated by Eq. (2). The results were listed in Table 2, which indicated that the diffusion coefficient usually increases with increasing 1-hexene content and the one of hydrogen molecule was larger than that of oxygen molecule.

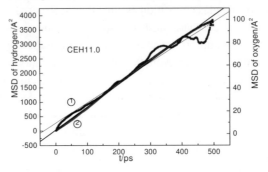

Fig. 2. MSD vs. time for the diffusion of oxygen.

Table 2. Calculated and Experimental Diffusion coefficients of $H_2$ and $O_2$ in the copolymer systems.

| System | Oxygen | | hydrogen |
|---|---|---|---|
| | $D^a_{exp} \times 10^7$ (cm$^2$s$^{-1}$) | $D_{sim} \times 10^7$ (cm$^2$s$^{-1}$) | $D_{sim} \times 10^7$ (cm$^2$s$^{-1}$) |
| PE | | 8.60 | 882.80 |
| CEH2.0 | 4.80 | 6.72 | 999.37 |
| CEH4.2 | 5.51 | 11.15 | 1690.08 |
| CEH7.0 | 12.36 | 26.46 | 1038.01 |
| CEH11 | 11.56 | 29.00 | 1318.95 |

The calculation results of free volume were very sensitive to the probe radius $R_P$. Table 3 listed the calculated value of FFV in the systems with different probe radius. It showed that FFV decreased rapidly with the increasing $R_P$, which indicated that the gas molecules have less volume to motion with the increasing radius of gas molecules. This is the main reason that the diffusion coefficient of hydrogen molecule is larger than that of oxygen molecule. Figure 3 showed the spatial morphology of the free volume in different probe radius. The free-volume decreases as the probe radius increased from Fig. 3(a) to 3(c).

Table 3. The calculated value of FFV in systems with different probe radius.

| System | $R_p$=0.00Å | $R_p$=1.20Å | $R_p$=1.52Å |
|---|---|---|---|
| PE | 0.316 | 0.030 | 0.008 |
| CEH2.0 | 0.335 | 0.055 | 0.019 |
| CEH4.2 | 0.346 | 0.065 | 0.022 |
| CEH7.0 | 0.355 | 0.092 | 0.048 |
| CEH11.0 | 0.360 | 0.084 | 0.035 |

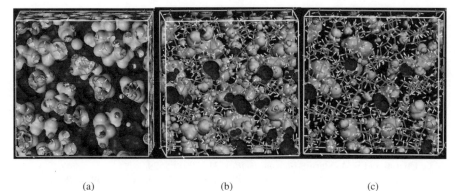

(a)                              (b)                              (c)

Fig. 3. Spatial morphology of the free volume (a) $R_p$=0.00Å; (b) $R_p$=1.20Å; (c) $R_p$=1.52Å.

Figure 4 showed the FFV of each system calculated at $R_p$=0.00Å. The FFV increased as the increasing 1-hexene content in the copolymer backbone and reached to a largest value at $f_{hex}$ = 11.0. It indicated that the changes in copolymer backbone structure have an important influence on the transport properties. Figure 5 was the diffusion coefficient of oxygen and FFV in poly(ethylene-co-1-hexene) with different content of 1-hexene. It was concluded that diffusion coefficient increased as the FFV increases.

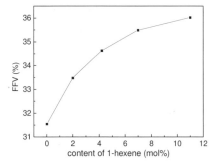

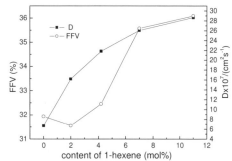

Fig. 4. Effect of $f_{hex}$ on FFV.

Fig. 5. The diffusion coefficient of oxygen and FFV in poly(ethylene-co-1-hexene) with different content of 1-hexene.

## 4. Conclusion

In this work, MD simulation was used to study the transport properties of ethylene/1-hexene copolymer and diffusion selectivity. It indicated that the diffusion coefficient was strongly influenced by FFV. The diffusion coefficient of oxygen increased with increasing 1-hexene content in the ethylene/1-hexene copolymer because of larger FFV. Temperature also presented an important effect on the diffusion coefficient.

# References

1. E. F. Vansant and R. Dewolfs, *Gas separation Technology* (Elsevier Publishing Company: Amsterdam, Netherlands, 1990).
2. M. F. Laguna, M. L. Cerrada, R. Benavente et al., *J. Membr. Sci.* **212** (2003) 167.
3. E. Tocci, E. Bellacchio, N. Russo and E. Diroli, *J. Membr. Sci.* **206** (2002) 389.
4. Q. L. Liu and Y. Huang, *J. Phys. Chem. B.* **110** (2006) 17375.
5. D. N. Theodorou, U. W. Suter, *Macromolecules.* **18** (1985) 1467.
6. D. N. Theodorou, U. W. Suter, *Macromolecules.* **19** (1986) 379.
7. E. Kucukpinar, P. Dorukera, *Polymer.* **44** (2003) 3607.
8. S. G. Charati and S. A. Stern, *Macromolecules.* **31** (1998) 5529.
9. J. Asaad, E. Gomaa, I. K. Bishay, *Materials Science and Engineering A.* **490** (2008) 151.
10. C. M. McCullagh, Z. Yu, A. M. Jamieson, J. Blackwell and J. D. McGervey, *Macromolecules.* **28** (1995) 6100.
11. S. Lee and W. L. Mattice, *Comput. Theor. Polym. Sci.* **9** (1999) 57.
12. D. Rigby and R. J. Roe, *Macromolecules.* **23** (1990) 5312.
13. Accelrys Inc., San Diego, CA, USA (Materials studio 4.0.0, POLYMERIZER, DISCOVER, AMORPHOUS_CELL, BUILDER and RIS Modules).
14. F. S. Pan, F. B. Peng and Z. Y. Jiang, *Chem. Eng. Sci.* **62** (2007) 703.
15. R. A. Hussein, S. Z. Wu, L. S. Zhang and J. E. Mark, *J. Inorg. Organomet, Polym.* **18** (2008) 100.
16. J. Han and R. H. Boyd, *Polymer.* **37** (1996) 1797.

# DYNAMICS BEHAVIOR ANALYSIS OF GLOBULAR PARTICLE IN RHEOLOGY MATERIAL WITH CONTROLLED SOLID FRACTION

KI-YOUNG KWON[1]

*National Core Research Center for Hybrid Materials Solution, Pusan National University,*
*Busan 609-735, KOREA,*
*kkybug@pusan.ac.kr*

CHUNG-GIL KANG[2†]

*School of Mechanical Engineering, Pusan National University,*
*Busan 609-735, KOREA,*
*cgkang@pusan.ac.kr*

SANG-MAE LEE[3]

*Engineering Research Center for Net Shape and Die Manufacturing, Pusan National University,*
*Busan 609-735, KOREA,*
*smlee@pusan.ac.kr*

Received 15 June 2008
Revised 23 June 2008

Semi-solid forging process has many advantages such as long die life, good mechanical properties and energy savings. But rheology material has complex characteristics, i.e., thixotropic behavior. Also, difference of the particle velocity between solid and liquid phase in the semi-solid state material causes a liquid segregation and specific stress variation. A number of simulation tools have been attempted for analyzing these behaviors of rheology material. However, general plastic or fluid dynamic analysis is not suitable. Therefore, we set up the stress equation to include viscosity, in order for investigating on how the moving behavior the solid particle in the rheology material during forging process is affected by viscosity, temperature, and solid fraction. In this study, a dynamics simulation was performed for the control of liquid segregation and the prediction of stresses on particles by changing forming velocity and viscosity in a compression experiment as part of a study on the analysis of the rheology aluminum forming process.

*Keywords*: Molecular dynamics; Rheology material; Solid fraction; Compression velocity.

## 1. Introduction

Recent attention has been paid to the nature of the semi-solid process as a new material processing technology because it can produce the complicate shape component through less forming force than does the solid forging process due to the small deformation resistance of the material [1]. In addition, semi-solid processing is carried out at low temperature and thus thermal fatigue of dies can be reduced. Rheology material exhibits

---

†Corresponding Author.

characteristics of fluidity, while the slurry has a shear thinning ability. Recently, the study to predict separation of solid-liquid phases of rheological material during semi-solid processing, by employing rheology theory, has been attempted. Because each solid and liquid phase can be analyzed and compared through rheological analysis considering ideal flow of rheological material, there is advantage, i.e., characteristics of semi-solid forming process can be found out.

At present, studies in the area of molecular dynamics have focused on pure materials [2, 3] and analyzed areas which are nano scale cutting, crack growth, and transfer of dislocation[4,5]. However, no studies that are concerned with particle behavior in a microstructure scale have been undertaken when forming the crystal grain controlled aluminum alloy in a semi-solid state. Therefore, this study demonstrates the development of a method for predicting particle behavior that occurs when compressing a rheological material, using molecular dynamics.

## 2.  Principle of Molecular Dynamics

### 2.1.  *The potential function*

The interparticle potential model is the typical Derjaguin-Landau-Verwey-Overbeek model which is widely used in colloidal science [6, 7]. The pair potential, P is the sum of an attraction component, $P_A$ and a repulsive component, $P_R$ as follows,

$$P = P_A + P_R = -\frac{a_1}{12\pi r^2} + 2\pi a_2 r_{ij} exp(-a_3 r) \tag{1}$$

$$\left( P_A = -\frac{a_1}{12\pi r^2}, \ P_R = 2\pi a_2 r_{ij} \exp(-a_3 r) \right)$$

where $a_{1-3}$ are constants and $r_{ij}$ represents the distance between the particles i and j. The parameter r is the distance between particles and R is the particle radius. A typical plot of a potential P is given in Fig. 1.

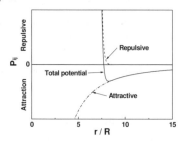

Fig. 1. Typical Derjaguin-Landau-Verwey-Overbeek inter-particle potential.

## 2.2.  Force subjected to single sphere particle

When performing a compression test, a sphere particle can be deformed into an elliptical particle. Thus, the floating force associated with an elliptical particle is given by,

$$F_a = (-cv)K' = -(6\pi\eta av)K' \tag{2}$$

where a is the length of the short axis and $K'$ is the shape factor. The shape factor is given as follows:

$$K' = \frac{\frac{4}{3}(B^2 - 1)}{\left[\frac{B(B-2)}{(B^2-1)^{1/2}}\tan^{-1}(B^2-1)^{1/2} + B\right]} \qquad \left(B = \frac{b}{a}\right) \tag{3}$$

$$K' = \frac{\frac{8}{3}(B^2 - 1)}{\left[\frac{B(3B^2-2)}{(B^2-1)^{1/2}}\tan^{-1}(B^2-1)^{1/2} + B\right]} \qquad \left(B = \frac{b}{a}\right) \tag{4}$$

Eq. (3) represents movement of the oblate ellipsoid along the short axis in an elliptical circle, while Eq. (4) represents movement of the oblate ellipsoid along the long axis.

As presented by Langevin, the one-dimensional motion equation of a sphere particle can be expressed.

$$m\frac{dv}{dt} = m\frac{d^2x}{dt^2} = F_a + F_T \tag{5}$$

The resultant force $F_T$ is given by,

$$F_T = mg - V\rho g + F_e \tag{6}$$

where V is the volume of the sphere particle and m is the mass of the solid sphere particle. The force subjected to an individual particle and a sphere particle was obtained through Eq. (5), by using an interaction according to the interparticle distance, which is calculated by the DLVO potential function, as described in Eq. (1).

## 2.3.  Numerical analysis technique

In order to numerically solve Eq. (5) in this study, the Verlet algorithm was used, as is frequently used in molecular dynamics. The modified algorithm was used, as follows, in order to calculate the Stokes floating force. And than Solving by using the geometrical coefficient, c and a(t)=F(t)/m=b-c´v(t), position and velocity vectors are represented by the equation given below

$$x(t + \Delta t) = \frac{2x(t) - x(t - \Delta t)\left(1 + \frac{c'\Delta t}{2}\right) + b\Delta t^2}{\left(1 + \frac{c'\Delta t}{2}\right)} \tag{7}$$

$$v(t) = \frac{x(t + \Delta t) - x(t - \Delta t)}{2\Delta t} \tag{8}$$

Thus, Eqs. (7) and (8) determine the velocity and location of the particle after the time increment, $\Delta$t. However, the initial location and velocity are required for numerical integration. The initial location of the particle is determined on the basis of the atom structure of the object material that is being analyzed, while the initial velocity is determined by Maxwell dispersion, where the velocity of the system reaches zero.

### 2.4. Viscosity equation

The viscosity equation for rheology material proposed by Okano[8]

$$\eta_a = \eta_{La}\left(1 + \frac{\alpha \rho_m C^{1/3} \dot{\gamma}^{-4/3}}{2\left(\dfrac{1}{f_s} - \dfrac{1}{0.72 - \beta C^{1/3} \dot{\gamma}^{-1/3}}\right)}\right) \tag{9}$$

Where the parameters are defined as follows: $\alpha = 2.03^2\left(\frac{x}{100}\right)^{1/3}$, $\beta = 19.0\left(\frac{x}{100}\right)^{1/3}$ , X = 0.6%, $\eta_{La} = 0.0045 Pa \cdot sec$,

Where C is cooling rate, $\dot{\gamma}$ is shear rate, $f_s$ is solid fraction and X is composition percentage of Si.

## 3.  Analysis Results and Discussion

### 3.1. Analysis Model

The conditions for the rheology forging simulation are summarized in Table 1. Each computer simulation was performed by using Okano viscosity equation.

Table 1. Simulation conditions (Al 7075).

| Test | Initial shape | Particle array | Solid Fraction | Tool velocity (m/s) |
|---|---|---|---|---|
| 1 | Rectangular | Square | 0.5 | 0.5 |
| 2 | Rectangular | Square | 0.6 | 0.5 |
| 3 | Rectangular | Square | 0.5 | 1.0 |

### 3.2. Analysis Results

Figs. 3 and 4 show the solid particle behavior with 20, 35, and 50% compression under test conditions for tests 1 and 2. Particles crowding begin as compression proceeds. Because there is no restriction along the x-axis, rheological solid particle exits through the right and left sides and consequently the workpiece approaches an elliptical circle.

When the particle size is identical and solid fraction is large, size in the entire shape become small. However, exception of shape size, the effect of solid fraction on the particle behavior is not observed.

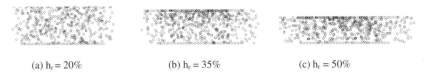

(a) $h_r$ = 20%　　　　(b) $h_r$ = 35%　　　　(c) $h_r$ = 50%

Fig. 3. Simulation result of square array particles behavior during compression by test 1 (Okano viscosity, rectangular tool, square array, $f_s$= 0.5, $v_t$= 0.5 m/s).

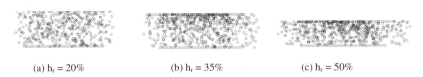

(a) $h_r$ = 20%　　　　(b) $h_r$ = 35%　　　　(c) $h_r$ = 50%

Fig. 4. Simulation result of square array particles behavior during compression by test 2 (Okano viscosity, rectangular tool, square array, $f_s$ = 0.6, $v_t$ = 0.5 m/s).

It was known that through tests 1 and 2, when solid fraction of rheology material was high, the right and left movement of the particle were free but the up and down movement were restricted. When the interparticle gap narrows and the upper tool compresses, the upper located particle exits. This phenomenon is significant with solid fraction of 60% having the initial narrow interparticle distance. Fig. 5 shows location of selected particles. To predict particle behavior according to compression velocity, the velocity and force vector of particles that were selected between the 300th and the 330th particles, are shown in Figs. 6 and 7. When compressing with the compression velocity of 1.0 m/s, location of the particle was crowded at the upper of the material. When compressing with the velocity of 0.5 m/s, the particle distributed over the entire material and exited largely to the edge of the material as compared to compression by 1.0 m/s of compression velocity. It was observed that the velocity and force vectors' direction were different. This may be due to the fact that viscosity hindering movement of the particle applies to the opposite direction of the particle velocity.

Fig. 5. Selected area and particle numbers 301-330 for measurement of particle velocities and reversed forces.

(a) Velocity vectors of selected particles　　　　(b) = Force vectors of selected particles

Fig. 6. Velocity and force vector of selected particles (301-330) at $h_r$ = 50% (Okano viscosity, rectangular tool, square array model, $f_s$ = 0.5, $v_t$ = 0.5 m/s).

| (a) Velocity vectors of selected particles | (b) = Force vectors of selected particles |

Fig. 7. Velocity and force vector of selected particles (301-330) at $h_r$= 50% (Okano viscosity, rectangular tool, square array model, $f_s$ = 0.5, $v_t$ = 1.0 m/s).

## 4.  Conclusion

Through a molecular dynamic simulation of rheology material, a number of predictions were performed in this study.

(1) Simulating molecular dynamics for rheology material by employing Okano's viscosity, it was confirmed that movement of the particle at solid fraction of over 0.5 was restricted by the reaction force subjected to the particle.

(2) It was confirmed that the solid particle flow differs by varying a solid fraction. In the case of high solid fraction of the material, the particle crowding at the upper of the material causes movement of the particle to the right and left hand sides. Movement in up and down direction of the particle is restricted and thus the particle distributes at the upper of the material.

(3) This study confirmed that when the compression ratio are identical, the compression velocity significantly influences the behavior of particle movement. When compression velocity was high, the crowding rate of the particles at the upper part of the workpiece became significant because the compression time shortens and tool velocity was rapid. Furthermore, the reverse distance between particles at the lower part of the workpiece increased and a consequent liquid segregation occurred.

## Acknowledgments

This work was supported by a grant-in-aid for the National Core Research Center Program from MOST and KOSEF (No. R15-2006-022-01001-0).

## References

1. A. M. Mullis, Toward a microstructural model of semi-solid rheology, Proc. 5th Int. Conf. on Semi-Solid Processing of Alloys and Composites, pp.265-272, 1998.
2. G. Yosiyuki, Molecular dynamics and Monte Carlo computing simulation, Chap. 2, p.12, 1996.
3. D. C. Rapaport, The art of molecular dynamics simulation, Cambridge university press, Chap. 2, pp.12-41, 1995.
4. C. I. Kim, S. H. Yang, and Y. S. Kim, Computer Simulation of Nano Material Behavior using Molecular Dynamics, Transaction of Material Processing Vol. 12, No. 3 pp.171-183, 2003
5. D. U. Kim, Y. K. Son, S. H. Rhim, and S. I. Oh, Molecular Dynamic Simulation of Nano Indentation and Phase Transformation, Proceedings of the Korean Society for Technology of Plasticity Conference, 49 issues, pp.339-346, 2003.

6. B. V. Derjagin and L. Landau, Theory of the stability of strongly charged lyphobic sols the adhesion of strongly charged particles in solutions of electrolytes, Acta Physiochem. URSS (14), pp.633-662, 1941.

7. E. J. W. Verywey, J.Th.G. Overbeek, The theory of the stability of lyophobic colloids, Elsevier, 1948.

8. S. Okano, Research activities in rheo-technology, The 3rd int. conf. on semi-solid processing of alloy and composites, pp.7-13, Tokyo, June, 1994.

# EFFECT OF HELICITY ON THE BUCKLING BEHAVIOR OF SINGLE-WALL CARBON NANOTUBES

JEEHYANG HUH[1], HOON HUH[2†]

*KAIST, Daejeon, KOREA,*
*utopias62@kaist.ac.kr, hhuh@kaist.ac.kr*

Received 15 June 2008
Revised 23 June 2008

Simulations of single-wall carbon nanotube(SWCNT)s having a different chiral vector under axial compression were carried out based on molecular dynamics to investigate the effect of the helicity on the buckling behavior. Calculation was performed at room temperature for (8,8) armchair, (14,0) zigzag and (6,10) chiral single-wall carbon nanotubes. The Tersoff potential was used as the interatomic potential since it describes the C-C bonds in carbon nanotubes reliably. A conjugate gradient (CG) method was used to obtain the equilibrium configuration. Compressive force was applied at the top of a nanotube by moving the top-most atoms downward with the constant velocity of 10m/s. The buckling load, the critical strain, and the Young's modulus were calculated from the result of MD simulation. A zigzag carbon nanotube has the largest Young's modulus and buckling load, while a chiral carbon nonotube has the lowest values.

*Keywords*: Molecular dynamics; single-wall carbon nanotube; buckling behavior; helicity.

## 1. Introduction

After discovered first by S. Iijima[1] in 1991, carbon nanotubes (CNTs) are widely used in Micro-Electro-Mechanical Systems (MEMS) and Nano-Electro-Mechanical Systems (NEMS)[2] since they have good mechanical,[3,4] electrical,[5] and magnetic[6] properties.

It is essential to obtain the mechanical properties of CNTs, such as the Young's modulus and the buckling load, for its application to mechanical systems. Many researchers have obtained the mechanical properties of CNTs through experimental measurements or computer simulations. Quantum or molecular level simulation has been of great interest to obtain the properties with computer simulations, since continuum level simulation does not reproduce the atomistic behavior correctly. Yakobson *et al.*[4] observed the mechanical behavior of single-wall carbon nanotubes (SWCNTs) under large deformation by molecular dynamics (MD). Ozaki *et al.*[7] used a tight binding (TB) method to investigate the behavior of SWCNTs under axial load. Buehler *et al.*[8] studied buckling behavior of SWCNTs with various aspect ratios and Dereli *et al.*[9] investigated a (10,10) SWCNT under compressive loading with a TB method. Most of the researchers

---

[†]Corresponding Author.

focused mainly on the behavior of armchair CNTs and few studies investigated the effect of the helicity. A study on the effect of the helicity can be meaningful since CNTs have different electric properties according to their helicity.[10] Ozaki *et al.*[7] reported that elastic moduli are insensitive to the size and the helicity. However, Lu[11] found that the 0K stress of SWCNTs is remarkably sensitive to the helicity. Since those two results are contrary to each other, further investigation on the effect of the helicity on the mechanical behavior of CNTs is required.

This study observes the mechanical behavior of SWCNTs having different helicity under axial compression through molecular dynamics simulations with the Tersoff potential function.[12] The Young's modulus, the critical strain, and the buckling load are calculated for each tube comparing with other literatures.

## 2. Theoretical Backgrounds

### 2.1. *Carbon nanotubes*

A carbon nanotube (CNT) is a wrapped graphene sheet that has a diameter of a few nanometers. There exist CNTs with various dimensions. A CNT can be classified into two groups: single-wall nanotubes (SWCNT) and multi-wall nanotubes (MWCNT). SWCNTs consist of one graphite sheet, whereas MWCNTs consist of more than one SWCNT piled up into a single tube.

A SWCNT can be classified into armchair, zigzag and chiral nanotubes according to their helicity, or the chiral vector.[10] The chiral vector $\mathbf{C_h}$ is a pair of indices $(n,m)$ which represents the way the graphene sheet is wrapped as in Fig. 1

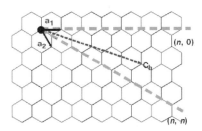

Fig. 1.[10] Schematic of a two-dimensional graphene sheet illustrating lattice vectors $\mathbf{a_1}$ and $\mathbf{a_2}$, and the roll-up vector $\mathbf{c_h} = n\mathbf{a_1} + m\mathbf{a_2}$.

The integers $n$ and $m$ denote the number of unit vectors along two directions in the honeycomb crystal lattice of graphene. Nanotubes are called "zigzag" if $m=0$, "armchair" if $n=m$, and "chiral" otherwise.

### 2.2. *Molecular dynamics*

Molecular dynamics (MD) is a type of computer simulation to predict the motion of atoms and molecules by time integration of the known equations of motion. Conventional MD uses the Hamiltonian equation (Eq. (1))[13] as the equations of motion.

$$\dot{\mathbf{p}}_i = -\frac{\partial H}{\partial \mathbf{q}_i}, \quad \dot{\mathbf{q}}_i = -\frac{\partial H}{\partial \mathbf{p}_i} \tag{1}$$

where $H$ is Hamiltonian of the system, $\mathbf{q}_i$ and $\mathbf{p}_i$ are the generalized position vector and the generalized momentum vector respectively. Hamiltonian contains the kinetic energy and the potential energy.

A different material has different potential, so it is very important to select an appropriate potential function[14] that is appropriate for each material. Among various potentials, density dependent potential functions[15] or multi-body potential functions[12,16] are generally used for metals and ceramics. In this paper, the Tersoff potential function[12], a type of multi-body potential functions, was used.

## 3. Simulation Methods

Buckling simulation of SWCNTs under axial compression was carried out based on molecular dynamics simulation to investigate how the buckling behavior of nanotubes changes according to the helicity. MD simulation was performed for (8,8) armchair, (14,0) zigzag, (6,10) chiral SWCNTs. Chiral vectors are carefully selected that those three nanotubes have similar dimensions, that is, a diameter of around 11 Å and a length of around 80 Å. Positions of atoms were generated by setting the initial C-C bond length to be 1.42 Å. Initial dimensions of each nanotube are shown in Table 1.

Table 1. Initial dimensions of the nanotubes.

| Chiral vector | D(Å) | L(Å) | Aspect ratio(L/D) |
|---|---|---|---|
| (8,8) | 10.8 | 79.934 | 7.4 |
| (14,0) | 10.94 | 76.325 | 6.977 |
| (6,10) | 10.96 | 78.8087 | 7.19 |

The interaction among atoms was described by the Tersoff potential function,[12] which is known to describe the C-C bond in CNTs reliably.[8] The Tersoff potential is described as Eq. (2):

$$V_{ij} = f_c(r_{ij})[a_{ij} f_R(r_{ij}) + b_{ij} f_A(r_{ij})] \tag{2}$$

where $f_R$ is the repulsive potential, $f_A$ is the attractive potential, $f_C$ is a smooth cut-off function, and $a_{ij}$, $b_{ij}$ are the coefficient to consider the effect of the bond order.

LAMMPS[†], a classical molecular dynamics code, was used to carry out the MD simulation and VMD[‡], a visualization program, was used for visualization. The time step was set to be 1fs, and the simulation was carried out at room temperature. Firstly, energy minimization was carried out with the conjugate gradient (CG) method. Secondly, the atoms at the top were moved downwards with the constant velocity of 10m/s to apply axial compression while the atoms at the bottom were fixed. Force-strain curves were obtained after finishing a total of 70,000 steps of simulations. Force was calculated by

[†]Large-scale Atomic/Molecular Massively Parallel Simulator, http://lammps.sandia.gov/
[‡]Visual Molecular Dynamics, http://www.ks.uiuc.edu/Research/vmd/

summing up all the forces applied to the top-layer-atoms. The buckling load and the critical strain were depicted as the load and strain of the point at which the curve starts to drop. The Young's modulus was computed from the slope of the linear part.

## 4. Result and Discussion

The dimensions of each CNT after energy minimization are shown in Table 2 and the shapes of nanotubes are shown in Fig. 2.

Table 2. Dimensions of CNTs after energy minimization.

| Chiral vector | D(Å) | L(Å) | Aspect ratio(L/D) |
|---|---|---|---|
| (8,8) | 11.28 | 82.12 | 7.28 |
| (14,0) | 11.414 | 78.2552 | 6.856 |
| (6,10) | 11.5804 | 80.95 | 6.99 |

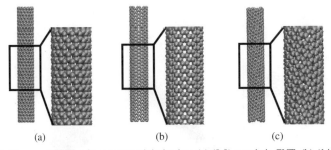

(a)　　　　　　　　　(b)　　　　　　　　　(c)

Fig. 2. Shapes of CNTs after energy minimization; (a) (8,8) armchair CNT, (b) (14,0) zigzag CNT, (c) (6,10) chiral CNT.

The length of each CNT increases after minimization as C-C bond length increases from 1.42 Å to 1.45 Å. The length of the C-C bond in a CNT measured by Wildoer *et al.*[17] is 1.42 Å, and the discrepancy between the measurement and simulation results is 2%, which signifies that the Tersoff potential simulates the C-C bond in CNTs reliably.

Force-strain curves obtained from the buckling simulation are shown in Fig. 3. Forces increase linearly to the strain of around 0.05, then drops abruptly, which implies that current simulation describes buckling behaviors properly. The maximum force is the buckling load, and the corresponding strain is the critical strain. The buckling load and the critical strain are represented in Table 3. The Young's modulus was calculated from the slope of the linear part of the curves. The stress was calculated by dividing the force by the section area of the CNTs as in Eq. (3). $D$ is the diameter of CNTs and the layer thickness $t$ was assumed to be 3.4 nm.[18]

$$A = Dt\pi \tag{3}$$

Table 3. Critical strain, Buckling Load, and Young's modulus of CNTs.

| Chiral vector | Critical strain($10^{-2}$) | Buckling Load(nN) | E(TPa) |
|---|---|---|---|
| (8,8) | 5.34 | 87.1 | 1.38 |
| (14,0) | 5.37 | 89.5 | 1.39 |
| (6,10) | 4.70 | 75.8 | 1.32 |

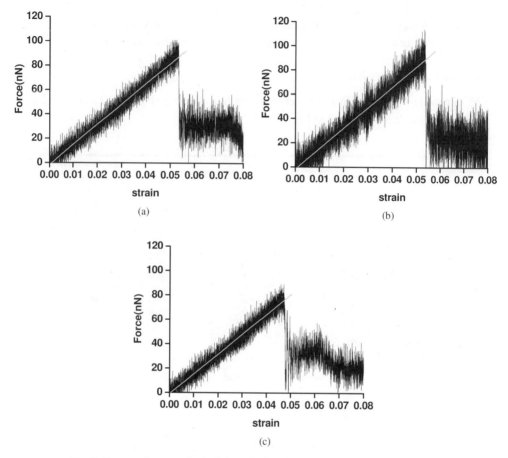

Fig. 3. Force-strain curve obtained from the buckling simulation, gray lines indicate linear fitting of the force-strain curves; (a) (8,8) armchair CNT, (b) (14,0) zigzag CNT, (c) (6,10) chiral CNT.

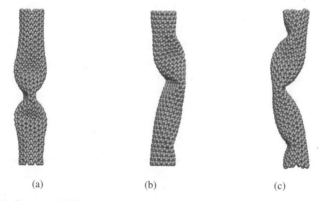

Fig. 4. Shapes of CNTs under axial loading at the strain of 0.06; (a) (8,8) armchair CNT, (b) (14,0) zigzag CNT, (c) (6,10) chiral CNT.

A zigzag CNT has the highest value of the critical strain, the buckling load and the Young's modulus, and a chiral CNT has the lowest of them, which was shown by Lu[11] as well. The differences in the Young's modulus among the nanotubes are approximately 5%. However, the difference in the buckling load among them is around 15%. The effect of helicity on the buckling load is three times larger than that on the Young's modulus. The Young's modulus for nanotubes is around 1.3TPa, which is consistent with the value reported by other researchers.[18] The shapes of the nanotubes at the strain of 0.06 are shown in Fig. 4. An armchair CNT shows a diamond buckling mode with two lobes, where as a zigzag CNT shows global bending with concentrated local deformation, and a chiral CNT squashed entirely. Different buckling modes can be the reason for the discrepancy in the buckling load of CNTs.

## 5. Conclusion

Molecular dynamics simulations of CNTs under axial compressive load were carried out for SWCNTs with different chiral vector to investigate the effect of the helicity on the buckling behavior of SWCNTs. The buckling load, the critical strain and the Young's modulus were calculated from the force-strain curves. The effect of the helicity of a SWCNT on the buckling load and the critical strain were newly investigated. SWCNTs with around 1.1nm thickness and 80nm length have the buckling load around 85nN, the critical strain around 0.05, and the Young's modulus about 1.3TPa. A zigzag nanotube has the largest Young's modulus, critical strain, and buckling load, followed by an armchair, and a chiral nanotube. The differences in the Young's modulus among the SWCNTs with different chiral vectors are slight, but the differences in the buckling load are up to 15%. The shapes of the CNTs at the strain of 0.06 are different from each other. The discrepancy in the buckling load may result from the different buckling mode.

## References

1. S. Iijima, *Nature* **354**, 56 (1991).
2. M. Meyyappan, *Carbon Nanotubes: Science and Applications* (CRC Press, 2004).
3. M.M.J. Treacy, T.W. Ebbesen, and J.M. Gibson, *Nature* **381**, 678 (1996).
4. B.I. Yakobson, C.J. Brabec, and J. Bernholc, *Phys. Rev. Lett.* **76**, 2511 (1996).
5. N. Hamada, S. Sawada, and A. Oshiyama, *Phys. Rev. Lett.* **68**, 1579 (1992).
6. J.P. Lu, *Phys. Rev. Lett.* **74**, 1123 (1995).
7. T. Ozaki, Y. Iwasa, and T. Mitani, *Phys. Rev. Lett.* **84**, 1712 (2000).
8. M.J. Buehler, Y. Kong, and H. Gao, *J. Eng. Mater. Technol.* **126**, 245 (2004).
9. G. Dereli and C. Özdoğan, *Phys. Rev. B* **67**, 035416 (2003).
10. T.W. Odom et al., *Nature* **391**, 63 (1998).
11. J.P. Lu, *Phys. Rev. Lett.* **79**, 1297 (1997).
12. J. Tersoff, *Phys. Rev. B* **37**, 6991 (1988).
13. J.M. Haile, *Molecular dynamics simulation: elementary methods* (Wiley, 1992).
14. Z. Yao et al., *Computational Materials Science* **22**, 180-184 (2001).
15. M.S. Daw and M.I. Baskes, *Phys. Rev. B* **29**, 6443 (1984).
16. M.W. Finnis and J.E. Sinclair, *Philosophical Magazine A* **50**, 45 (1984).
17. J.W.G. Wildoer et al., *Nature* **391**, 59 (1998).
18. S. Akita and M.N.A.Y. Nakayama, *Jpn. J. Appl. Phys.* **45**, 5586-5589 (2006).

# Part G
# Nano, Meso, Micro and Crystal Plasticity

# ENERGY OF ARMCHAIR NANOTUBE USING THE MODIFIED CAUCHY-BORN RULE

S. LU[1], C.D. CHO[2†], L. SONG[3]

*Department of Mechanical Engineering, Inha University, Incheon, KOREA,*
*lxlusheng@hotmail.com, cdcho@inha.ac.kr, songli0125@hotmail.com*

Received 15 June 2008
Revised 23 June 2008

Due to the difference of nanotube diameters, the single-walled carbon nanotubes (SWCNTs) show the different energy and mechanical properties. In order to take the effect of the curvature of nanotubes into account in the modeling of those structures, the present paper proposes an atomistic based continuum model with using a type of modified Cauchy-Born to link the continuum strain energy to the interatomic potential. This modified Cauchy-Born is developed by incorporating the concept of differential mean value theorem into the standard Cauchy-Born rule. The present model not only can bridge the microscopic and macroscopic length scales, but also can investigate the curvature effect of a single layer film on the continuum level. Application of the current model to armchair carbon nanotubes and graphite shows an excellent prediction of the size dependent strain energy which are compared in a good agreement with the existing experimental and theoretical results.

*Keywords*: Carbon nanotubes; Cauchy-Born rule; Interatomic potential; Curvature.

## 1. Introduction

Since multi-walled carbon nanotubes (MWCNTs) were discovered[1], carbon nanotubes (CNTs) have attracted much attention in science and engineering, owing to their many unique and novel properties: low density, specific stiffness, specific strength, excellent electrical and thermal conductivities.

In order to develop a direct link between the continuum analysis and atomistic simulation, the concept of atomistic-based continuum is gradually developed. Tadmor et al.[2] first proposed a quasi-continuum model to link atomistic simulation with continuum analysis. Gao and Klein[3] developed a virtual internal bond (VIB) model to construct the constitutive model of solids from atomic level. With CNTs, Zhang et al.[4] incorporated the interatomic potential directly into a constitutive model based on the modified Cauchy-Born rule. Jiang et al.[5] took the effect of tube radius into account. Arroyo and Belytschko[6, 7] proposed the exponential Cauchy-Born rule by introducing the exponential map to extend the standard Cauchy-Born rule to describe the deformation of CNTs much

---

†Corresponding Author.

more accurately. Guo et al.[8] studied the mechanical properties of CNTs by higher order Cauchy-Born rule to have a better description of the deformation of CNTs.

Depending on this kind of concept, the present study concentrates on the modification of the Cauchy-Born rule, then the differential geometry concept of inscribed surface is introduced into CNTs. Finally a modified and extended Cauchy-Born rule, which makes the relationship between deformation and curvature much more clearly, is developed. This method can not only bridge the microscopic and macroscopic length scales, but also make this connection more accurate.

## 2. The Interatomic Potential for Carbon

Tersoff-Brenner interatomic potential for carbon[9, 10] is introduced as follows:

$$V(r_{ij}) = V_R(r_{ij}) - B_{ij} V_A(r_{ij}) . \tag{1}$$

In Eq. (1), $i$ and $j$ represent the respective carbon atoms at the ends of one covalent bond; $r_{ij}$ is the distance between $i$ and $j$ ; the repulsive interaction $V_R$ and attractive interaction $V_A$ between two carbon atoms are given by:

$$V_R(r) = f_c(r) \frac{D_e}{S-1} e^{-\sqrt{2S}\beta(r-r_e)} \qquad V_A(r) = f_c(r) \frac{D_e S}{S-1} e^{-\sqrt{2/S}\beta(r-r_e)} , \tag{2}$$

where the parameters $D_e = 6.000eV$, $S = 1.22$, $\beta = 21nm^{-1}$ and $r_e = 0.1390nm$ . And the multi-body coupling term $B_{ij}$ represents the coupling between bond $ij$ and the local environment of atom $i$ , which is given by:

$$B_{ij} = \left[ 1 + \sum_{K(\neq i,j)} G(\theta_{ijk}) f_c(r_{ik}) \right]^{-\delta} , \tag{3}$$

where $\delta = 0.50000$, $r_{ik}$ is the distance between atom $i$ and its local environmental atom $k$ and $\theta_{ijk}$ is the angle between bond $ij$ and $ik$ .The smooth cutoff function $f_c$ restricts the pair potential to nearest neighbors with $r_1 = 0.17nm$ and $r_2 = 0.2nm$ , is given by

$$f_c(r) = \begin{cases} 1 & r < r_1 \\ \frac{1}{2} \left\{ 1 + \cos[\frac{\pi(r-r_1)}{(r_2-r_1)}] \right\} & r_1 \leq r \leq r_2 \\ 0 & r > r_2 \end{cases} . \tag{4}$$

The function $G$ in Eq. (3) is given by

$$G(\theta) = a_0 \left[ 1 + \frac{c_0^2}{d_0^2} - \frac{c_0^2}{d_0^2 + (1 + \cos\theta)^2} \right] , \tag{5}$$

with $a_0 = 0.00020813$, $c_0 = 330$ and $d_0 = 3.5$ .

With this set of parameters, this potential function is applied into the hexagonal atomic structure. The equilibrium bond length can be determined by:

$$\frac{\partial V}{\partial r_{ij}} = 0 . \tag{6}$$

Eq. (6) gives the equilibrium bond length $a_{cc} = 0.145068nm$, which is in good agreement with that of graphite (0.144nm).

## 3.  The Modified Cauchy-Born Rule

The Cauchy-Born rule[11,12] plays an important role in the atomistic-based continuum mechanics, which is the fundamental kinematics assumption that establishes a connection between the deformation of an atomic system and that of a continuum. It is widely accepted as the form of:

$$a = FA , \qquad (7)$$

where $F$ denotes a two-point deformation gradient tensor, $A$ and $a$ denote one lattice vector on the respective undeformed and deformed crystals. In the absence of diffusion, phase transition, slip, lattice defect and other non-homogeneities, Eq. (7) is good at describing the deformation of bulk material.

Noticeably, there are two limitations to solid materials in the application of the standard Cauchy-Born rule: one is that the solid material must have a centrosymmetric atomic structure, the other is that the material must be a type of bulk material, or equivalently, space-filling material. Unfortunately, the structure of CNTs does not satisfy these two limitations exactly. From the continuum mechanics point of view, the linear transformation between the two line segments is given by

$$d x = F d X , \qquad (8)$$

where $d x$ and $d X$ denote one infinitesimal line segment joining two material particles in the deformed and undeformed configurations, respectively; $F$ is the deformation gradient tensor. Integrating Eq. (8) by taking the finite length of the lattice vectors $a$ and $A$ into account, Eq. (8) can be written as:

$$\int_{x}^{x+a} d x = \int_{X}^{X+A} F d X . \qquad (9)$$

The corresponding deformed lattice vector $a$ can be expressed as:

$$a = \int_{X}^{X+A} F d X . \qquad (10)$$

If the deformation gradient tensor $F$ is smooth enough on the closed interval $[X, X + A]$, it is natural to introduce the mean value theorem in calculus. To be brief, there exists a certain point $X_a \in [X, X + A]$ such that

$$a = F( X_a ) A . \qquad (11)$$

## 4.  Apply the Modified Cauchy-Born Rule into CNTs

Apply the standard Cauchy-Born rule into the modeling of CNTs, there is a challenge as pointed out by Arroyo and Belytschko[8], "The deformation gradient on a surface gives the tangent behavior, whereas the deformation of the bonds depends on the behavior of the chords to the surface". Thus how to use the deformation gradient in a tangent space to describe the deformation of the chords of the surface, the differential geometry concept of the inscribed surface can be adopted.

As shown in Fig. 1, generally, the CNTs is viewed as the cylindrical surface as the outer red surface, owing to the position of atoms. So it can be called as atomic surface, which is denoted by $C_1$ . If the standard Cauchy-Born rule is applied to the atomic surface, it means that the deformation of the bond *BD* is described by tangent behavior along the direction of *AK* . So errors will be inevitably introduced. However, the chord *BD* is in coincidence with the tangent of the inscribed surface $C_2$ at point *E* ; in other words, the bond *BD* is the tangent of the inscribed surface $C_2$ at point *E* . That is to say, the variations of the bonds (the chords of the atom surface $C_1$ ) can be investigated by the deformation gradient of the inscribed surface $C_2$ at each tangent point. Moreover, the CNTs structure possesses axial symmetry at the midpoint of every bond, so it can be determined that the tangent point is the midpoint of the bond.

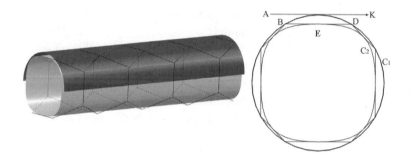

Fig. 1.   The relationships between the atom surface and the inscribed surface in CNTs structures. The inner configurations are the inscribed surfaces, and the outer configurations are the atom surface.

So the inscribed surface of one crystalline film can describe the variations of the lengths and directions of bonds by the deformation gradient in the process of deformation. Obviously, for curved crystalline films, applying the Cauchy-Born rule on the inscribed surface is much justifiable.

The inscribed surface introduced in the above can be combined with the modified Cauchy-Born rule. Eq. (11) can be used in the single layer crystalline films by its inscribed surface. In Eq. (11), **A** and **a** denote one lattice vector in the undeformed and deformed crystal, respectively; $\mathbf{F}(\mathbf{X}_a)$ denotes deformation gradient of a tangent point on the inscribed surface. So the method of using the deformation gradient of the tangent point on the inscribed surface to describe the deformation of the lattice vector of a single layer crystalline film is developed.

## 5.  Calculation and Result

For the non-centrosymmetric atomic structure as CNTs, an inner shift vector $\boldsymbol{\lambda}$ should be taken into consideration since the inner displacement can occur without disobeying Cauchy-Born rule. See Refs. 4-7 for more details. After introducing the inner shift vector $\boldsymbol{\lambda}$ , the bond vector can be expressed as

$$\mathbf{r}_{ij} = \mathbf{F}(\mathbf{X}_a)\,\mathbf{r}_{ij}^0 + \boldsymbol{\lambda} . \tag{12}$$

where vector $\mathbf{r}_{ij}^0$ is used to express a bond between atom *i* and atom *j* on the undeformed configuration and the vector $\mathbf{r}_{ij}$ denotes the bond *ij* on the deformed configuration.

The mappings of the rolling energy surface and the atom surface are shown in Fig. 2. The mapping of inscribed surface can be expressed as:

$$x_1 = X_1, \quad x_2 = eX_2, \quad x_3 = aX_2^{\ 3} + bX_2^{\ 2} + cX_2 + d, \tag{15}$$

where $(x_1, x_2, x_3)$ are the point coordinates in current configuration, $(X_1, X_2)$ are the point coordinates at the undeformed configuration. The coefficients $(a, b, c, d, e)$ can be determined from geometric relations, which are functions of nanotube chiral $(m, n)$ and the inner shift $\lambda$. Then, the deformation gradients can be obtained.

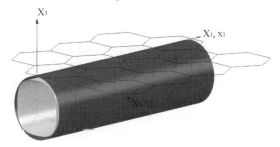

Fig. 2.  Illustration of rolling a graphite sheet into one armchair naotube (2, 2).

The energy $\overline{W}$ per atom can be calculated with the use of

$$\overline{W}[\mathbf{F}(\mathbf{X}_\alpha), \lambda] = [V(r_{i1}) + V(r_{i2}) + V(r_{i3})]/2. \tag{13}$$

By minimizing the strain energy of each carbon atom with respect to $\lambda$, the inner shift vector can be obtained by

$$\left. \frac{\partial \overline{W}[\mathbf{F}(\mathbf{X}_\alpha), \lambda]}{\partial \lambda} \right|_{\lambda = \bar{\lambda}} = 0. \tag{14}$$

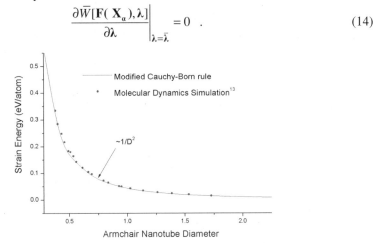

Fig. 3.  The strain energy (relative to graphite sheet) per atom in an armchair nanotube versus the tube diameter.

Based on the present model, the energy per atom in both graphite sheet and armchair nanotube can be calculated. For the graphite sheet, the energy per atom is -7.3756 eV, which is corresponding to the molecular dynamics simulation of Robertson et al.[13] For the CNTs, if the graphite sheet is taken as the ground state, the energy variations from this ground state to the nanotube state can be defined as the strain energy. The strain energy per atom in armchair nanotube versus the tube's diameter D can be expressed as $C/D^2$ as shown in Fig. 3, where C is a constant. This result is in a good agreement with a series of well known simulation results by molecular dynamics[13], tight binding method[14] and exponential Cauchy-Born rule[7].

## 6.  Conclusion

A methodology to construct continuum models for nanoscale crystalline films has been developed in the present study. The proposed model is a hyper-elastic inscribed surface constitutive model by introducing the inscribed surface into the modified Cauchy-Born rule. With the use of the present model and the Tersoff-Brenner atomic potential for carbon, the size dependent strain energy of armchair nanotube is predicted. The obtained results are in a good agreement with those obtained by other experimental, atomic modeling and continuum concept based studies.

## Acknowledgments

This work was supported by Inha University.

## References

 1. S. Iijima, *Nature.*  **354**, 56 (1991).
 2. E. B. Tadmor, R. Phillips and M. Ortiz, *Langmuir.* **12**, 4529 (1996).
 3. H. Gao and P. Klein, *J. Mech. Phys. Solids.* **46**, 187 (1998).
 4. P. Zhang, Y. Huang, P.H. Klein and K.C. Hwang, *Int. J. Solids. Struct.* **39**, 3893 (2002).
 5. H. Jiang, P. Zhang, B. Liu, Y. Huang, P. H. Geubelle, H. Gao and K. C. Hwang, *Comp. Mater. Sci.* **28**, 429 (2003).
 6. M. Arroyo and T. Belytschko, *J. Mech. Phys. Solids.* **50**, 1941 (2002).
 7. M. Arroyo and T. Belytschko, *Phys. Rev.* **B69**, 115415-1 (2004).
 8. X. Guo, J.B. Wang and H. W. Zhang, *Int. J. Solids. Struct.* **43**, 1276 (2006).
 9. J. Tersoff, *Phys. Rev.* **B37**, 6991 (1988).
10. D. W. Brenner, *Phys. Rev.* **B42**, 9458 (1990).
11. M. Born and K. Huang, *Denamical Theory of the Crystal Lattice* (Oxford University, Oxford, 1954).
12. F. Milstein, *J. Mater. Sci.*  **15**, 1071 (1980).
13. D. H. Robertson, D.W. Brenner and J.W. Mintmire, *Phys. Rev.* **B45**, 12592 (1992).
14. C. Goze, L. Vaccarini, L. Henrard, P. Bernier, E. Hernándz and A. Rubio, *Synthetic. Met.* **103**, 2500 (1999).

# FABRICATION OF POLYMER MASTER FOR ANTIREFLECTIVE SURFACE USING HOT EMBOSSING AND AAO PROCESS

HONG GUE SHIN[1†], JONG TAE KWON[2], YOUNG HO SEO[3], BYEONG HEE KIM[4]

*Division of Mechanical Engineering and Mechatronics, Kangwon National University, Kangwondaehakgil 1, Chunchon, 200-701, KOREA,*
*rushhong @kangwon.ac.kr*

Received 15 June 2008
Revised 23 June 2008

A simple method for the fabrication of polymer master for antireflective surface is presented. In conventional fabrication methods for antireflective surface, coating method with low refractive index have usually been used. However, it is required to have a high cost and a long processing time for mass production. In this paper, antireflective surface was fabricated by using hot embossing process with porous anodized aluminum oxide. Through multi-AAO and etching processes, nano patterned master with high aspect ratio was fabricated at the large area. Size and aspect ratio of nano patterned master are about 175±25nm and 2~3, respectively. In order to replicate nano patterned master, hot embossing process was performed by varying the processing parameters such as temperature, pressure and embossing time etc. Finally, antireflective surface can be successfully obtained after etching process to remove selectively silicon layer of AAO master. Optical and rheological characteristics of antireflective surface were analyzed by using SEM, EDX and spectrometer inspection. Antireflective structure by replicating hot embossing process can be applied to various displays and automobile components.

*Keywords*: AAO process; Hot embossing; Antireflective surface.

## 1. Introduction

Recently, antireflective (AR) films have been widely used for industrial technique to minimize the reflectance of the incident light. In addition to, AR films improve the transmittance and the contrast of display panel [1-2]. It is usually used conventional methods by coating a single layer or multi-layer with a material of specific refractive index [1]. The main advantage is capable to remove the reflectance of the incident light without affecting the substrate. AR films are used in a wide range of applications such as the front panels of liquid crystal displays (LCD) [3], light-emitting diode (LED) [4], photovoltaic solar cells [5]. However, it is required to have a high cost and a long processing time for mass production [3].

†Corresponding Author.

In recent years, many researchers have been attempted by the UV imprinting and injection molding using nano-patterned masters fabricated by UV and hologram lithography, electron-beam etching and nano-probe lithography [6-7]. However, several conventional approaches for the fabrication of the nano-patterned master are not only expensive but also locally fabricated and limited to small areas.

This paper presents a simple method for the fabrication of antireflective surface using hot embossing process. A porous nano-patterned master with high aspect ratio was fabricated by multi-anodic aluminum oxidation, etching and pore widening process at a 4-inch silicon wafer. In order to replicate the antireflective surface, the nano-patterned master was used as the mold stamp on the polymethyl methacrylate (PMMA) substrate in the hot embossing process.

## 2.  Fabrication of AR Master

In order to fabricate a porous nano-patterned master, anodic aluminum oxidation (AAO) was used. AAO process is one of the most alternative and promising technique for nanotechnology [8]. Due to a flexibility, a high structural controllability and a simple procedure at low cost, AAO has anticipated for used in a broad range of potential nano-applications such as carbon nano tube (CNT), nano template, electronic and optoelectronic devices, field emission displays and data storage [9].

Generally, most of AAO structures have been fabricated by the muti-step anodizing and etching process at optimal acidic electrolyte environment for long time from 4 to 24hours. Furthermore, the pore size (diameter and height) and regularity in AAO structure can be determined by various electrolytes (sulphuric ($H_2SO_4$), oxalic ($C_2H_2O_4$) , phosphoric ($mP_2O_5nH_2O$)), anodizing voltage (40-60V), temperature (0-20°C) and anodizing time (4~24h). However, the conventional AAO process for the fabrication of ordered nano-structures with high aspect ratio, such as CNT or nano-pillars, has to be required for a long processing time and no variation of voltage and temperature (mild anodization). In this paper, AAO experiments were performed by using the multi-step anodizing process with Al thin film, which has a thickness of 1μm on silicon wafer, for 2 min (hard anodization).

Figure 1 shows the schematic diagram of AAO procedure. It consists of 1'st anodizing, alumina removal, 2'nd anodizing and pore widening. Especially, the initial pore size is depend on the anodizing voltage in the 1'st anodizing process.

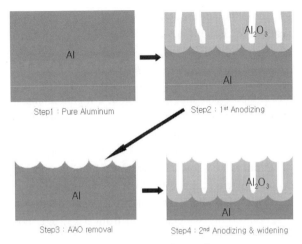

Fig. 1. Schematics of AAO procedure.

## 2.1. *Preparation for AAO process*

For the fabrication of porous AAO layer, it consists of a power supply, a magnetic stirrer and a heater, a chiller, an electrolytic bath and a thermocouple. Al thin film with a thickness of 1μm, which was used as an anodic material, deposited on silicon wafer by using sputtering process. In case that thickness of Al thin film is over 1μm, not only the surface roughness decreases but also the sputtering cost increases remarkably. Al thin film was fabricated by using DC magnetron sputtering process. For a surface roughness under 50nm, Al thin film was used instead of bulk aluminum (purity 99.999%). Furthermore, Al thin film has no need of electropolishing process. A platinum (Pt) was used as a cathode material. Power supply and chiller can controlled the anodic voltage from 0V to 200V and temperature from 0°C to 100°C, respectively. Oxalic acid was used as an electrolyte.

## 2.2. *1'st anodizing*

In order to fabricate the porous nano-patterned master, 1'st anodizing process was performed in 0.04M oxalic acid solution at 4°C for 2 min. Anodizing voltage and distance between Al and Pt were 90V and 8cm, respectively. AAO growth length per minute was about 240nm/min. As shown in Fig. 2(a), initial pore nucleation was generated on the Al thin film.

## 2.3. *Alumina removal*

After 1'st anodizing, the alumina layer (1'st AAO layer) was removed in the etching solution of 1.8wt% chromic acid and 6wt% phosphoric acid at 65°C for 60 min. As indicated in Fig. 2(b), hemispheric nano patterns were placed on the surface of Al thin film. In our earlier study, the hemispheric nano patterns, which have aspect ratio of 0.5, were used as a master of the hot embossing process.

### 2.4.  2'nd anodizing and pore widening

For improvement of the pore size and aspect ratio, 2'nd anodizing was performed in the same conditions of 1'st anodizing. However, anodizing voltage was decreased 60V due to the limitation of film thickness as mentioned in section 2.1. In 2'nd anodizing, AAO growth ratio (length per minute) was about 150nm/min. After 2'n anodizing, nano pores with a diameter of 80nm were shown in Fig. 2(c). Finally, pore widening process was performed in 0.1M phosphoric acid at 30°C for 120 min to increase the pore diameter. As shown in Fig. 2(d), the pore diameter and aspect ratio increased from 80nm (2'nd anodizing) to 170~200nm (pore widening) and from 0.5 (2'nd anodizing) to 2~3 (pore widening), respectively.

Figure 2(e) shows a SEM image of cross-section for the porous nano-patterned master. It has small pores (diameter; 90±10nm, height; 225±25nm) in the large pore (diameter; 135±15nm, height; 90±10nm). Because it is different from the variation of anodizing voltage in the 1'st anodizing and 2'nd anodizing process. As shown in Fig. 2(e), it used as a mold stamp of hot embossing process.

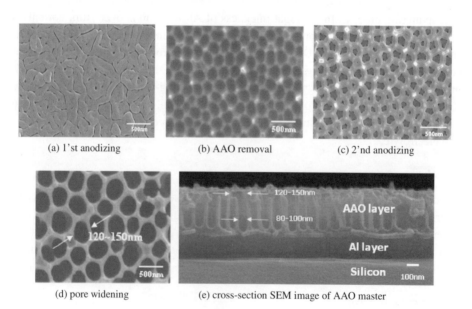

(a) 1'st anodizing          (b) AAO removal          (c) 2'nd anodizing

(d) pore widening          (e) cross-section SEM image of AAO master

Fig. 2. SEM images with respect to AAO procedure.

## 3. Replication of AR Nano-Patterns

### 3.1. *Hot embossing*

Nano hot or thermal embossing is one of the most promising fabrication techniques for micro-nano molding of thermoplastic polymer [10]. Three major steps of the hot embossing process are shown in Fig. 3. Typically, a thermoplastic material is fed into the embossing machine (Fig. 3 (a)); heated above the glass transition temperature ($T_g$); and pressed by the stamper with high pressure (Fig. 3(b)). After the temperature of the stamper decreases to below $T_g$, generally it takes several minutes, the product is released from the stamper as shown in Fig. 3(c). The nano hot embossing system consists of an air compressor, two heating and cooling blocks, several thermocouples and controllers. A porous nano-patterned stamper (20mm x 20mm), which was fabricated by AAO process, was fixed on the bottom of the upper block and a PMMA (4" wafer type) was placed on the lower block. The upper and lower blocks could be heated separately.

Figure 4 shows the variation of temperature and pressure during the hot embossing process. The temperature of the stamper increases to a little higher than $T_g$ during heating time; then the temperature of embossing is maintaining for tens of seconds; and after cooling process for few minutes, the stamper is opened and the product is released. Pressure increases to designated value at the start of the embossing process and is maintained constantly during embossing and cooling. In addition to, pressure increases to tens of bars during cooling time in order to improve the filling ratio of the nano patterns. In order to compensate of the temperature of the stamper and substrate (PMMA), not only several thermocouples but also an infrared camera (NEC, Thermo Tracer, TH5104) were used.

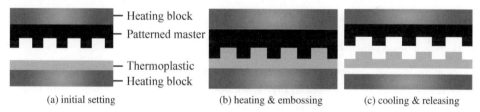

(a) initial setting      (b) heating & embossing      (c) cooling & releasing

Fig. 3. Three steps of hot embossing process.

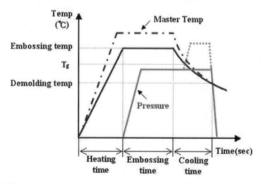

Fig. 4. Temperature and pressure variation during hot embossing process.

### 3.2. *Replication for fabrication of antireflective structure*

A porous AAO layer which was fabricated by multi-anodizing and etching process (including pore widening) was used as a mold for the hot embossing process to replicate the antireflective structures on a polymethyl methacrylate (PMMA) substrate. Table 1 is several experimental conditions for hot embossing process. A porous nano-patterned master and a PMMA substrate were embossed each other at the force of from 3kN to 7kN and from 115°C to 140°C for 2 min, and then were released after cooling under 80°C for 5min. Embossing pressure per square centimeters is about from 7.5MPa to 17.5MPa (a force between AAO stamper and PMMA / square areas of AAO stamper).

Table 1. Experimental conditions for hot embossing.

| Stamper | Material | | Porous alumina layer |
|---|---|---|---|
| | Size | Depth | 280~350nm |
| | | Width | 120~150nm |
| Substrate | Material | | PMMA |
| | Diameter | | 4 inch |
| Processing conditions | Temperature [°C] | | 105, 110, 120, 130 |
| | Force [kN] | | 3, 5, 7 |
| | Em. Time [sec] | | 120 |
| | Releasing temp [°C] | | 80 |
| | Cooling time [min] | | 5 |
| Temp. compensation; T. of stamp = T. of PMMA + 2°C | | | |

## 4.  Results and Discussion

### 4.1. *Replication characteristic*

For the analysis of antireflective structures, field emission scanning electron microscope (FE-SEM) was used. As shown in Fig. 5, we obtained several SEM images with respect to the temperature and the pressure of embossing process. The more the temperature and pressure of hot embossing were increased, the better the replication of antireflective structures was improved. However, it was difficult to release of a stamp and a substrate over the temperature of 110°C. In case of the embossing temperture of 110°C, the seperation of master and substrate was impossible due to high aspect ratio of AAO master and mechanical locking structure, regardless the variation of the embossing pressure. In order to release effectively AAO master and PMMA substrate, tetramethylammonium hydroxide (TMHA, (CH3)4NOH) was used as an anisotropic etchant of silicon. It is also used as a basic solvent in the development of acidic photo-resist in the photolithography process. In addition to, the PMMA can resist the TMHA etchant. For etching of silicon using TMHA etchant, we performed etching experiment at 80°C for 20h. As shown in Fig. 5(d), Fig. 5(e) and Fig. 5(f), we obtained clearly several good results without damages of antireflective structures by using the TMHA etchants.

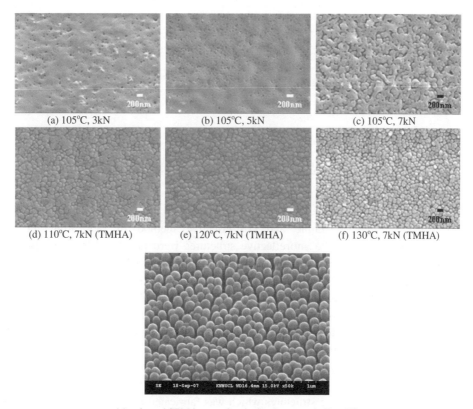

(a) 105°C, 3kN          (b) 105°C, 5kN          (c) 105°C, 7kN

(d) 110°C, 7kN (TMHA)   (e) 120°C, 7kN (TMHA)   (f) 130°C, 7kN (TMHA)

(g) enlarged SEM image of antireflective areas in Fig. 6(f)

Fig. 5. SEM images of replication results with respect to temperature, pressure and releasing type.

### 4.2. *Optical properties*

For the valuation of the optical characteristics of antireflective structures, a spectrometer (UV-2450 spectrometer, SHIMAZU Scientific Instruments, wavelength ($\lambda$) range; 200~800nm) and an ellipsometer ($\lambda = 405~750$nm) were used. By using a spectrometer and an ellipsometer, the transmittance of AR surface was measured as shown in Fig. 6. The transmittance of pure PMMA was about 85~90%. The transmittance of antireflective structures was also kept to 93% in the broad range of the wavelength. Furthermore, its contrast was improved together with the transmittance. In the optical analysis, we confirmed the optical validity of our antireflective structures which was fabricated and replicated by using the AAO process and the hot embossing process. Consequently, the simple fabrication process of metal mold with porous nano-patterns was demonstrated and the antireflective surface was successfully replicated on the PMMA substrate by the hot embossing process.

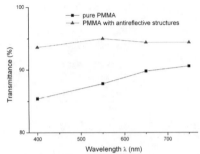

Fig. 6. Transmittance curves of periodic nanopillar arrays on PMMA.

## 5. Conclusions

This paper presents a simple and cost-effective method of fabricating for the antireflective structures. The antireflective structures were fabricated by using the hot embossing on the anodized aluminum substrate. Porous nano-patterned master used as a mold stamper of hot embossing has a diameter of 135±15nm and a height of 300±20nm. Porous nano-patterned master has a high aspect ratio of 2~3. Various experimental conditions of hot embossing process were also successfully replicated antireflective structures with respect to temperature, pressure and embossing time. The transmittance of PMMA with AR structures increased from 85~90% to 93% at the broad range of the wavelength (350~750nm). The contrast of PMMA with antireflective structures was improved together with the transmittance. Consequently, we confirmed the optical validity of our antireflective structures which was fabricated and replicated by the AAO process and the hot embossing process, respectively.

## Acknowledgment

This research was supported by a grant (06K1401-00511) from the Center for Nanoscale Mechatronics & Manufacturing, one of the 21st Century Frontier Research Programs, which are supported by Ministry of Science and Technology, Korea. Also, this work was partly supported by the 2nd stage of BK21 project funded by the Ministry of Education & Human Resources Development, Korea.

## References

1. T. Hoshino and etc., *Appl. Opt.* **46(5)**, 648 (2007).
2. R. Leitel and etc., *Plasma Prcess. Polym.* **4**, 878 (2007).
3. J. Ho and etc., *Appl. Opt.* **44**, 6176 ( 2005).
4. M. Fujita and etc., *Appl. Phys. Lett.* **85**, 5769 (2004).
5. J. Zhao and etc., *IEEE Transactions on Electron Devices* **41(9)**, 1592 (1994).
6. V. Piotter and ect., *Microsyst. Technol.* **73(3)**, 99 (2001).
7. J. Steigert and ect., *J. Micromech. Microeng.* **17**, 333 (2007).
8. H. Masuda, *Appl. Phys. Lett.* **71 (19)**, 2770 (1997).
9. J. Suh and J. Lee, *Appl. Phys. Lett.* **75**, 2047 (1999).
10. M. Heckele and ect., *Sensor. Actuat. A-Phys.* **83(1)**, 130 (2000).

# TEXTURE AND FORMABILITY DEVELOPMENT OF ASYMMETRY ROLLED AA 3003 AL ALLOY SHEET

INSOO KIM[1†], SAIDMUROD AKRAMOV[2], HAE BONG JEONG[3]

*Department of Information and Nano Materials Science and Engineering, Kumoh National Institute of Technology, Gumi, Gyung Buk, 730-701, KOREA,*
*iskim@kumoh.ac.kr, saidmurod23@yahoo.com, jjn7nmr@dreamwiz.com*

Received 15 June 2008
Revised 23 June 2008

The physical, mechanical properties and formability of sheet metal depend on preferred crystallographic orientations (texture). In this research work, we investigated texture development and formability of AA 3003 aluminum alloy sheets after asymmetry rolling and subsequent heat treatment. After asymmetry rolling, the specimens showed fine grain size. We also investigated the change of the plastic strain ratios after asymmetry rolling and subsequent heat-treating condition. The plastic strain ratios of asymmetrically rolled and subsequent heat treated samples are 1.5 times higher than the initial AA 3003 Al alloy sheets. These could be attributed to the formation of ND//<111> texture component through asymmetry rolling in Al sheet.

*Keywords*: Texture; AA 3003 Al alloy; Asymmetric rolling; Plastic Strain Ratio; R-value; Formability; γ-fiber component.

## 1. Introduction

Aluminum alloys have good potentials to replace the low carbon steels in automotive industry. However most of aluminum alloys have lower r-value (Lankford parameter or plastic strain ratio) than low carbon steels in fully annealed conditions[1,2]. The texture of fully annealed aluminum alloy sheet is mainly consist of cube component {001}<100> which has low r-values, whereas γ-fiber is a major texture with high r-value after full annealing in low carbon steels[3]. The r-value is known to be related to the formability of sheet metals. Many researchers devoted their works to improve the r-values and formabilities of aluminum alloys and only few of them were successful[4-6].

It has been found that rotated cube {001}<110> texture and γ-fiber component appear after severe shear deformations in fcc metal sheets. This leads to increase the r-value[5]. However, due to the high stored energy in the deformed state and the existences of structural heterogeneities, annealing conditions are critical and may form unwanted structures and textures[5,6].

---

[†]Corresponding Author.

In this paper, development of textures and plastic strain ratios after the asymmetric rolling without lubricant condition and subsequent heat treatment in AA 3003 Al sheet was observed.

## 2.  Experimental Procedure

AA 3003 aluminum alloy sheets with initial thickness of 3mm were used in present study. The sheet samples, with dimensions of 60mm x 40mm x 3mm, were prepared from a rolled sheet along the rolling direction. Then these plates were annealed at 500 °C for 2 hours to homogenize the initial grain size through thickness (named initial Al sheet). The initial Al sheets were asymmetrically rolled to 90% reduction in thickness on a laboratory rolling mill with rolls diameter ratio of 1.5. To obtain high friction coefficient between rolls and samples, without lubricant condition was used during rolling process. Samples were kept in air furnace at 100°C for 3-4 minutes between each pass. Change of texture was investigated after the asymmetrical rolling. Change of texture was also investigated after subsequent heat treatment at the temperature of 300°C for 20 minutes and 275°C for 60 minutes in a salt bath solution. The microstructure of the transverse direction (TD) side surfaces of the asymmetrically rolled specimens was investigated by using an optical microscopy. Texture measurements were performed at one-tenth and center layer of thickness of asymmetrical rolled sheets. (111), (200), and (220) pole figures were obtained by using the Cu-Kα radiation with the Schultz reflection method and the grain orientations were represented in the orientation distribution function (ODF) calculated by the harmonic method[7].

Table 1.  Chemical compositions of AA 3003 aluminum alloy.

| Si | Fe | Cu | Mn | Zn | Others | | Al |
|----|----|----|----|----|--------|--------|----|
| | | | | | Each | Total | |
| 0.6 | 0.7 | 0.2 | 1.0-1.5 | 0.1 | 0.05 | 0.15 | Bal |

Table 1 is shown chemical compositions of AA 3003 aluminum alloy which having the 0.6% Si, 0.7% Fe, 0.2% Cu, Zn 0.1% and aluminum.

R-value is one of the most important parameters in textured sheet metals forming. The r-value, defined as the ratio of the width-to-thickness strain changes in uni-axial tension, influences the deep drawability or the limit drawing ratio (LDR).

In the present study, tensile test specimens were taken along 0°, 45° and 90° to the rolling direction from the sheet and the r-value were measured as a function of strain in each direction. Due to the small thickness of specimen, width-to-length measurements were conducted instead of width-to-thickness measurements.

The average r-value (r -value), and Δr-value was obtained from the measured and calculated r-value data. The average measured r-value (r -value) was calculated by the use of $r = (r_0 + 2r_{45} + r_{90}) / 4$, and $\Delta$ r-value was calculated by the use of $\Delta r = (r_0 - 2r_{45} + r_{90}) / 2$. Here, the $r_{0, 45, \text{ and } 90}$ means the r-value of along the angles of 0°, 45°, and 90° to rolling direction, respectively.

Furthermore, the measured r-values by tensile tests were compared with those predicted by r-value prediction theories[7].

## 3. Results and Discussion

Fig. 1 shows the microstructures of the AA 3003 aluminum alloy sheet at the different processing conditions. The microstructure of initial Al sheet is presented on Fig. 1(a). Fig. 1(b) shows the microstructure of 90% asymmetric rolled sample. After the asymmetric rolling, the microstructure is consist of a highly elongated grains which are characteristic of a rolled material. The elongated grains sizes become smaller with increase in reduction. Recrystallization does not occurs after heat treatment at 275°C for 60 minutes in salt bath condition (Fig. 1c), recrystalization microstructure shows in heat treated Al sheet at 300°C for 20 minutes in Fig. 1(d).

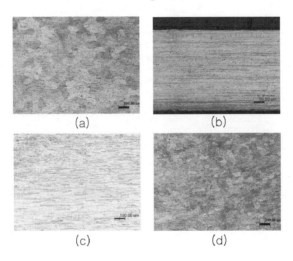

Fig. 1. Optical micrographs obtained from the TD section surfaces of AA 3003 Al alloy sheet; (a) initial Al sheet, (b) 90% reduction in thickness, and (c) 90% reduction in thickness and heat treated at 275°C/60min. (d) 90% reduction in thickness and heat treated at 300°C/20min.

After homogenizing heat treatment, weak cube textures observed on initial samples of one tenth surface and center layers in Fig. 2(a). After 90% asymmetrical rolling, pole figures of one tenth surface layer showed week γ-fiber ND//<111> texture components in Fig. 2(b). After subsequent heat treatment of samples at 275°C for 20 minutes, strong γ-fiber ND//<111> texture components showed on one-tenth surface layer in Fig. 2(c). After subsequent heat treatment of samples at 300°C for 20 minutes, strong γ-fiber ND//<111> and rotate Cube {001}<110> texture components observed on one-tenth surface layer in Fig. 2(c).

In order to analyze the detailed texture components, ODFs were calculated and presented in Fig. 3. The ODF cutouts at $\Psi_2 = 0°$, 25°, 45°, 65° are presented in Fig. 3. The initial AA 3003 aluminum alloy sheets showed cube textures in Fig. 3(a).

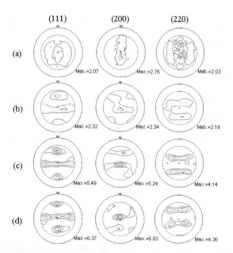

Fig. 2. (111), (200) and (220) pole figures obtained from one-tenth surface layer of the AA 3003 Al alloy sheet; (a) initial Al sheet, (b) 90% reduction in thickness, and (c) 90% reduction in thickness and heat treated at 275°C/60min. (d) 90% reduction in thickness and heat treated at 300°C/20min.

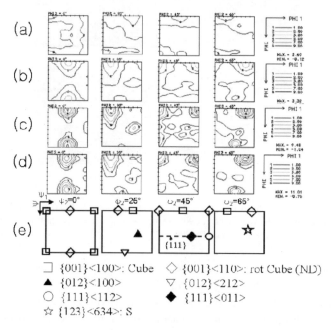

Fig. 3. ODFs obtained from one-tenth layer surface of the AA 3003 Al alloy sheet; (a) initial Al sheet, (b) 90% reduction in thickness, (c) 90% reduction in thickness and subsequently heat treated at 275°C/60min., (d) 90% reduction in thickness and subsequently heat treated at 300°C/20min., (e) the prominent orientations in Euler space angel with constant $\psi_2$ section.

After asymmetric rolling with 90% reduction in thickness, rotated Cube texture {001}<110> texture component was observed in Fig. 3(b). Asymmetrically rolled and subsequently heat treated at 275°C for 60 minutes also observed the rotated Cube texture in Fig. 3(c). After asymmetry rolling and subsequent heat treatment at 300°C for 20 minutes, sample also showed the rotated Cube {001}<110> texture component in Fig. 3(d).

Table 2. Variation of the f(g) values of major texture components.

| | {001}(100) | {001}(110) | Betta fiber | | | Gamma fiber | |
|---|---|---|---|---|---|---|---|
| | | | {112}(111) | {123}(634) | {001}(211) | {111}(110) | {111}(112) |
| Initial | 2.1 | 1.4 | 0.9 | 1.5 | 1.5 | 0.1 | 0.1 |
| 90% reduced | 0.85 | 3.34 | 1.1 | 1.6 | 3.1 | 1.2 | 0.8 |
| 90% reduced and 275°C heat treated for 60min | 0.9 | 9.2 | 1.2 | 2.3 | 9.2 | 1.9 | 1.1 |
| 90% reduced and 300°C heat treated for 20min | 0.86 | 7.6 | 1.86 | 1.75 | 8.15 | 0.1 | 1.2 |

Table 2 shows variation of the f(g) value of major texture components. {001}<100> Cube texture component shows the highest intensity in initial Al sheet. After asymmetric rolling, Cube texture decreased and rotated Cube {001}<110> and beta fiber textures were observed. After heat treatment at 275°C for 60 minutes, rotated Cube {001}<110> texture component were increased as well as {001}<211> beta and gamma fiber texture component. But in case of heat treatment at 300°C for 20 minutes, intensity of {001}<211> component remained high and {111}<110> texture component was decreased as in Table 2.

Table 3. The comparison of measured r-value, $\bar{r}$ and |Δr| of the 90% frictionally rolled Al sheet data.

| Conditions of samples | r-value | | | $\bar{r}$ | |Δr| |
|---|---|---|---|---|---|
| | 0° | 45° | 90° | | |
| Initial specimen (500°C/2 hr) (Measured) | 0.463 | 0.633 | 0.387 | 0.529 | 0.161 |
| 90% frictionally rolled (275°C/60min) (Calculated from surface layer data) | 0.118 | 1.69 | 0.152 | 0.912 | 1.55 |
| 90% frictionally rolled (300°C/20min) (Measured) | 0.405 | 0.988 | 0.661 | 0.760 | 0.035 |

The average r-values were calculated using by ODF data as shown in Table 3 and Fig. 4. The measured r-value of initial Al sheet is 0.529, after asymmetric rolling and subsequent heat treatment at 275°C for 60 minutes, calculated r-value is 0.912 (Fig. 4b and Table 3), after asymmetric rolling and subsequent heat treatment at 300°C for 20 minutes, measured r-values was 0.760 (Fig. 4a and Table 3). After asymmetric rolling and subsequent heat treatment of AA 3003 Al sheet, the measured average r-value was 1.5 times higher than that of the initial Al sheet. The increasing of r-value after asymmetrical rolling and subsequent heat treatment may be related to the formation of ND//<111> texture.

The |Δr| of initial sample was about 0.161. After asymmetrical rolling and subsequent heat treatment at 275°C for 60 minutes, it increased to 1.55. after asymmetrical rolling and subsequent heat treatment at 300°C for 30 minutes it was 0.035. The change of |Δr|

after asymmetrical rolling and subsequent heat treatment may be related to rotated the formation of Cube {001}<110> texture component[1].

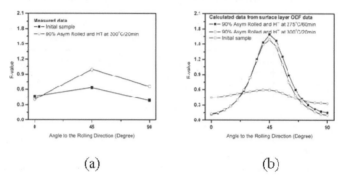

(a)                                    (b)

Fig. 4. Measured (a) and calculated (b) r-values of 90% asymmetrically rolled and subsequently heat treated at 300°C for 20 minutes and 275°C for 60 minutes AA 3003 aluminum alloy sheet.

## 4. Summary

Formability and texture development after asymmetric rolling of AA 3003 aluminum alloy sheet was investigated in this work. From the results it was found that:

(1) The $\gamma$-fiber ND//<111> and rotated Cube {001}<110> texture components were observed after asymmetry rolling and subsequent heat treatment of AA 3003 Al sheet.

(2) After asymmetric rolling and subsequent heat treatment of AA 3003 Al sheet, the average r-value was 1.5 times higher than that of the initial Al sheet.

## Acknowledgment

This paper was supported by the research fund of Kumoh National Institute of Technology (2006) and Korea Research Foundation Grant Funded by the Korean Government (MOEHRD) (KRF-2005-D00372).

## References

1.  PH. Lequeu and J. J. Jonas: *Metal. Trans.* **A 19A**, 105 (1988).
2.  M. Hatherly, W.B. Hutchinson, *An Introduction to Textures in Metals* (Institution of Metallurgists, London, 1979).
3.  J. Hirsch, K. Lucke, *Mechanism of deformation and development of rolling textures in polycrystalline F.C.C. metals—simulation and interpretation of experiments on the basis of Taylor-tape theories, Acta Metall.* **36**, 2883 (1988).
4.  A.K. Vasudevan, R.D. Doherty, *Aluminum Alloys—Contemporary Research and Applications, Treatise on Materials Science and Technology* 31 (Academic Press, 1989).
5.  K.J. Kim, H.-T. Jeong, K.S. Shin and C.W. Kim, *Texture evolution of rolled AA 5052 alloy sheets after annealing, J. Mater. Process. Tech.* **187-188**, 578 (2007).
6.  S. Akramov, I. Kim and N.J. Park, *Texture and formability of frictionally rolled AA 1050 aluminum alloy sheets, Advanced Mater. Research* **26-28**, 393 (2007).
7.  H.J. Bunge, *Texture Analysis in Material Science* (Butterworth, Guildford, UK, 1982).

# NUMERICAL SIMULATION OF DEEP DRAWING PROCESS OF ALUMINUM ALLOY SHEET USING CRYSTAL PLASTICITY

JUNG GIL SHIM[1†]

*Graduate School, Hanyang University, Seoul, 133-791, KOREA,*
*nasim79@nate.com*

YOUNG TAG KEUM[2]

*Department of Mechanical Engineering, Hanyang University, Seoul, 133-791, KOREA,*
*ytkeum@hanyang.ac.kr*

Received 15 June 2008
Revised 23 June 2008

In this study, the FEM material model based on the crystal plasticity is introduced for the numerical simulation of deep drawing process of A5052 aluminum alloy sheet. For calculating the deformation and stress in a crystal of aluminum alloy sheet, Taylor's model is employed. To find the texture evolution, the crystallographic orientation is updated by computing the crystal lattice rotation. In order to verify the crystal plasticity-based FEM material model, the strain distribution and the draw-in amount are compared with experimental measurements. The crystal FEM strains agree well with measured strains. The comparison of draw-in amount shows less 1.96% discrepancy. Texture evolution depends on the initial texture.

*Keywords:* Crystal Plasticity; Texture, Anisotropy; Deep Drawing; Aluminum Alloy Sheet.

## 1. Introduction

Due to the enhancement of the computer technology and the development of numerical methods, the simulation of sheet forming process has been greatly advanced. However, the continuum based FEM has many difficulties in predicting the fractures, spring-backs, and wrinkles because sheet forming process includes microscopically complicate deformation behaviors. Furthermore, the sheet metal is poly-crystally structured with the textures where the grains are preferentially orientated by the rolling process so that the material properties of the sheet are greatly influenced by the textures. In order to perform the accurate simulation of sheet metal forming process, therefore, the textural behavior has to be considered.[1]

For the microscopic simulation of sheet forming process, the crystal plasticity based on the crystallographic characteristics has been recently investigated by many researchers. Gambin et al.[2] modeled a deformation texture evolution based on strain rate-independent

---

[†]Corresponding Author.

crystal plasticity and performed the simulation considering the rolling textures of silver, copper and aluminum. Hosford3, 4 calculated the shape of yield locus from the information of texture and slip system using Taylor model and showed more accurate locus than Hill's quadratic theory. Nakamachi et al.[5] introduced the homogenization method to performed the multi-scale simulation of sheet forming process. Takahashi et al.[6] predicted the anisotropy evolution of aluminum sheet employing the successive accumulation method. Dawson et al.[7] solved constitutive models that incorporate the effects of plasticity and slip gradients. the rate of texture evolution in simple shear of a magnesium polycrystal is examined by Neale et al.[8]

In this study, the deep drawing process of A5052 aluminum alloy sheet is numerically simulated using the FEM material model based on the crystal plasticity. And thickness strain and draw-in amount are compared with experimental measurements. In addition, the texture evolution during the drawing process is traced with pole figures to present anisotropy developments.

## 2. Crystal Plasticity

### 2.1. *Crystal material model*

The plastic deformation theory led to texture evolution in the crystalline kinematics is employed in this study.[9] When the sheet plastically deforms, the velocity gradient tensor $\dot{L}_{ij}^{P}$ is written as follows:

$$\dot{L}_{ij}^{P} = \dot{\varepsilon}_{ij}^{P} + \bar{\omega}_{ij}^{P} \tag{1}$$

where $\dot{\varepsilon}_{ij}^{P}$ and $\bar{\omega}_{ij}^{P}$ are plastic strain rate tensor and spin tensor, respectively, which are defined as follows:

$$\dot{\varepsilon}_{ij}^{P} = \sum_{S} P_{ij}^{S} \dot{\gamma}^{S} \tag{2}$$

$$\bar{\omega}_{ij}^{P} = \sum_{S} M_{ij}^{S} \dot{\gamma}^{S} \tag{3}$$

where $P_{ij}^{S}$ and $M_{ij}^{S}$ are symmetric and asymmetric parts of Schmid tensor, respectively. $\dot{\gamma}^{S}$ is shear strain rate of slip system.

The critical resolved shear stress of slip system $\tau^{S}$ is computed as follows:

$$\tau^{S} = P_{ij}^{S} \sigma_{ij} \tag{4}$$

where $\sigma_{ij}$ is the microscopic stress of the grain calculated by projecting the macroscopic stress to the crystal lattice orientation.

Following the Taylor's isotropic hardening assumption that $k^{S}$ s in all slip systems are the same[10], the hardening function $k^{S}$ can be written as follows: [4, 12]

$$k^{S} = \frac{K}{M^{n+1}} (\dot{\Gamma}^{S})^{n} \tag{5}$$

where $K, n, M$ and $\dot{\Gamma}^{S}$ are strength coefficient, strain hardening exponent, Taylor factor, and the shear strain rate of the grain obtained by summing $\dot{\gamma}^{S}$, respectively.

Yield condition of the slip system can be described as follows:

$$sign(\tau^s)\,\dot{\gamma}^s > 0 \quad \text{when} \quad \tau^s = k^s$$
$$sign(\tau^s)\,\dot{\gamma}^s = 0 \quad \text{when} \quad \tau^s < k^s \tag{6}$$

In equation (6), shear strain rate of the slip system $\dot{\gamma}^s$ is computed as follows: [6]

$$\dot{\gamma}^s = (\tau^s - sign(\tau^s)k^s)/4G \tag{7}$$

where $G$ is shear modulus.

## 2.2. Texture evolution

Velocity gradient tensor $\dot{L}_{ij}^P$ in equation (1) is expressed as follows: [11]

$$\dot{L}_{ij}^P = \dot{\beta}_{ij} + \dot{\Omega}_{ij} \tag{8}$$

where $\dot{\beta}_{ij}$ and $\dot{\Omega}_{ij}$ are plastic spin tensor and lattice rotation tensor, respectively. Plastic spin tensor $\dot{\beta}_{ij}$ is defined as follows:

$$\dot{\beta}_{ij} = \sum_s m_{ij}^s \dot{\gamma}^s \tag{9}$$

where $m_{ij}^s$ is Schmid tensor.

In order to calculate new crystal orientation after plastic deformation, namely lattice rotation tensor $\dot{\Omega}_{ij}$, the Velocity gradient tensor $\dot{L}_{ij}^P$ obtained from equation (1) and the plastic spin tensor $\dot{\beta}_{ij}$ calculated from equation (9) are substituted into equation (8). Deformation texture is then calculated from the Euler angle of the grain by substituting lattice rotation tensor $\dot{\Omega}_{ij}$ into the following equations. [13, 14]

$$\dot{\varphi}_1 = (\dot{\Omega}_{23}\sin\varphi_2 + \dot{\Omega}_{31}\cos\varphi_2)/\sin\Phi$$
$$\dot{\Phi} = \dot{\Omega}_{23}\cos\varphi_2 - \dot{\Omega}_{31}\sin\varphi_2 \tag{10}$$
$$\dot{\varphi}_2 = \dot{\Omega}_{12} - \dot{\varphi}_1\cos\Phi$$

In this study, the deformation texture during the deep drawing is calculated by assigning initial textures to the integration points of finite elements and is represented by the pole figures showing the texture evolution. [14]

## 3. Deep Drawing Simulation

### 3.1. Modeling

For a numerical simulation of deep drawing process of A5052 aluminum alloy sheet, due to the 4-fold symmetry, the first quadrant of the circular sheet with a radius 45mm is modeled. Finite element model employs 331 nodes and 300 4-node shell elements. Integration points of the finite element are regarded as crystal aggregates and 200 texture information are employed in a crystal aggregate. Table 1 shows the material property of A5052 aluminum alloy sheet. Blank holding force and initial thickness of the sheet are 0.3ton and 1.0mm, respectively. Fig. 1 shows the section view of the tooling.

## 3.2.  *Simulation result and comparison with experimental measurement*

The distances of edge nodes from the center to see the ear shape of sheet edge are compared with measurements when punch travels 20mm (see Fig. 2). The overall trend of draw-ins agrees well between numerical simulation and experimental measurement and there is averagely 1.96% discrepancy. There is a maximum discrepancy 6.52% at an edge node located in the direction of 22.5° from the x-axis. Therefore, the current FEM material model based on the crystal plasticity depicts well the anisotropy behavior of A5052 aluminum alloy sheet in the deep drawing process.

Fig. 3 shows, at the punch travel of 20mm, the formed shape of the sheet and the comparison of thickness strain distribution in the x-axis and y-axis directions with experimental measurements. The biggest strain found 30~35mm far away from the center, where the fracture occurs, agrees well with the experimental observation. Experimentally measured thickness strain distribution shows that thickness strain in x-axis direction is smaller than that in y-axis direction except 2 positions: one position 26mm far away from the center and boundary of contact and non-contact regions with the punch, the other position 35mm and diffusion necking region.

Table 1. Material properties of
        A5052 aluminum alloy sheet.

| Property | Value |
|---|---|
| Young's modulus | 69GPa |
| Poisson ratio | 0.33 |
| Friction coefficient | 0.08 |
| Taylor factor | 3.21 |

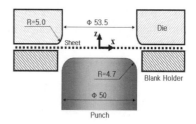

Fig. 1. Section view of the tooling.

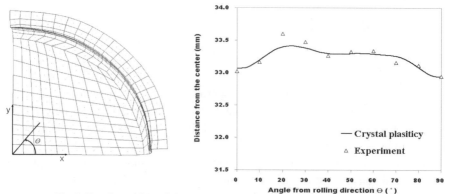

Fig. 2. Top view of formed sheet and distance of edge nodes from the center.

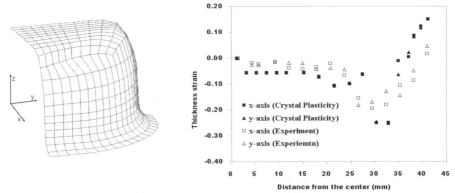

Fig. 3. Side view of formed sheet and thickness strain distribution.

### 3.3.  *Texture analysis*

As shown in Fig. 4, the initial textures randomly distributed and pre-oriented to certain directions are used to simulate the deep drawing process and to find the pole figures showing deformation texture in {100}, {110}, and {111} directions. In the {100} direction of random initial textures, the poles concentrated in [110] and [101] directions are evolved to the [100] fiber texture centered, and the pole in [011] direction is randomly distributed in {110} direction. In {111} direction, the pole focused in the center concentrated to the [101] direction. In the case of the pre-oriented initial texture, the pole figures in {100} and {111} directions show that the poles significantly scattered to the random directions. In The {110} pole figure, the poles of Goss component often seen in FCC rolled sheet are focused to the center. Therefore, the initial textures of the sheet influence deformation behavior of the grain so that, for the accurate simulation of sheet forming process like a deep drawing, the initial textures evolved in the rolling process have to be properly considered.

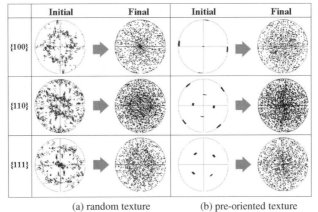

(a) random texture          (b) pre-oriented texture

Fig. 4. Texture evolution during the deep drawing process.

## 4.  Conclusion

In order to accurately simulate the deep drawing process of A5052 aluminum alloy sheet, the FEM material model based on the crystal plasticity is introduced and the numerical simulation is performed. The conclusions derived in this research are listed as follows:

(1) The accuracy of the FEM material model based on the crystal plasticity is verified by comparing thickness strains and draw-in amounts with those experimentally measured.

(2) The ear and anisotropy behavior of the sheet can be accurately predicted by the crystal plasticity based FEM.

(3) The texture evolution expressed with pole figures depends on the initial texture.

## Acknowledgments

The authors would like to thank Prof. E. Nakamachi for discussing texture evolution algorithms and constitutive equation. This work was supported by the Korea Research Foundation Grant funded by the Korean Government (MOEHRD, Basic Research Promotion Fund) (KRF-2007-313- D00046).

## References

1. H. Takahashi, *Polycrystal Plasticity* (Corona Publishing Co, Tokyo, 1999).
2. W. Gambin and F. Barlat, *Int. J. Plasticity*, **13** (1997) 75.
3. W. F. Hosford, *Mater. Science and Eng.*, **A257** (1998) 1.
4. W. F. Hosford, The Mechanics of Crystals and Textured Polycrystals (Oxford University Press, 1993).
5. E. Nakamachi, C. L. Xie, and H. Morimoto, *Int. J. Mechanical Sciences*, **43** (2001) 631.
6. H. Takahashi, H. Motohashi, and S. Tsuchida, *Int. J. Plasticity*, **12** (1996) 935.
7. J. M. Gerken and P. R. Dawson, *Computer method in applied Mech. Eng.*, **197** (2008) 1343.
8. B. Beausir, L. S. Toth, and K. W. Neale, *Int. J. Plasticity*, **23** (2007) 227.
9. R. J. Asaro and A. Needleman, *Acta Metal*, **33** (1984) 923.
10. G. I. Taylor, *J. Inst. Metals*, **62** (1938) 307.
11. S. H. Choi and K. H. Oh, *Metals and Mater.*, **3** (1997) 252.
12. G. Lin, S. Kerry, and Havner, *Int. J. Plasticity*, **12** (1996) 695.
13. U. F. Kocks, *Acta Metal*, **6** (1958) 85.
14. H. J. Bunge, *Texture Analysis in Materials Science* (Butterworths, London, 1982).

# AN INTEGRATED MICROMECHANICS MODELLING APPROACH FOR MICRO-FORMING SIMULATION

W. ZHUANG[1]

*College of Automobile Engineering, Jilin University, 5988 Renmin Street,*
*Changchun, 130022, P. R.. CHINA,*
*zhuangwm@jlu.edu.cn*

J. LIN[2†]

*Department of Mechanical Engineering, Imperial College London, South Kensington,*
*London, SW7 2AZ, UK*
*Jianguo.lin@imperial.ac.uk*

Received 15 June 2008
Revised 23 June 2008

An effort has been made to create an integrated Crystal Plasticity FE (CPFE) system. This enables micro-forming process simulation to be carried out easily and the important features in forming micro-parts can be captured. Firstly, based on Voronoi tessellation and the probability theory, a VGRAIN system is created for the generation of grains and grain boundaries for micro-materials. Numerical procedures have been established to link the physical parameters of a material to the control variable in a gamma distribution equation. An interface has been created, so that the generated virtual microstructure of the material can be inputted in the commercial FE code, ABAQUS, for mesh generation. Secondly, FE analyses have been carried out to demonstrate the effectiveness of the integrated system for the investigation of uncontrollable curvature and localized necking in extrusion of micro-pins and hydro-forming of micro-tubes.

*Keywords*: Crystal plasticity; Finite element; micro-mechanics modeling; micro-forming.

## 1. Introduction

In conventional metal forming processes, the size of the workpiece is usually much larger compared with the grain size of the metal. Macro-mechanics-based FE techniques can be used for the process simulation and the material appears homogeneous and different samples of the same material in the same thermo-mechanically treated condition, exhibit the same properties. At a micro-level, grain size can be similar to that of the part being formed. Thus, Crystal Plasticity (CP) theories, which assume that the crystalline slip is a predominant deformation mechanism, need to be employed to describe the inelastic behaviour of crystallined materials. Finite element (FE) methods are commonly considered as standard analysis tools for obtaining sub-grain stress and strain fields[1, 2]. However, to construct an FE model of a workpiece, which contains a number of grains in random shapes and random material orientations, using just the commercially available

---

[†]Corresponding Author.

FE software, e.g. ABAQUS, itself is very time-consuming and sometimes impossible[3, 4]. Many micro-mechanics analyses were carried out using the microstructures with a few grains with very simple grain shapes and grain boundaries [5, 6]. Although it can be used to study particular deformation and damage mechanisms of crystal materials, it does not represent typical microstructure of real materials and is not suitable for structural and micro-forming simulations.

The main objective of this research is to introduce an integrated numerical approach for crystal plasticity finite element (CPFE) analysis for micro-forming applications. This CPFE system includes virtual microstructure generation and orientation assignment for individual grains based on probability theories, integration between the virtual grain generation system (VGRAIN) and ABAQUS/CAE, where further FE pre-processing can be carried out. Two case studies will be carried out: one is the extrusion of micro pins and the other is hydro-forming of micro tubes.

## 2. Integrated Numerical Procedure for CPFE Analysis

Fig. 1 shows the structure of the integrated numerical procedure for micro-mechanics modelling using ABAQUS. The system is split into three parts and the last one, post-processing, is standard, which will not be discussed here. The only specific feature in the second part is the implementation of crystal plasticity equations, which has been given everywhere by Taylor[7], Hill & Rice[8]. The equations have been calibrated for a stainless steel and the material constants are listed for the material by Harewood *at al*[9]. The CP equation set describes slipping strain rates for the 12 slipping systems of the FCC material, including self and latent hardening moduli. The crystal plasticity equations have

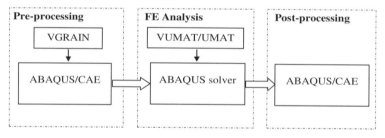

Fig. 1.   Flowchat of integrated numerical procedures for micro-mechanics modelling.

been implemented into ABAQUS via user defined subroutine VUMAT following the same procedure of Huang[10].

The specific feature of the system is within the first part, pre-processing. A virtual grain generation system, known as VGRAIN, has been developed. Polygon grains within a defined area are generated using Voronoi tessellation methods[11]. The shape of grains is controlled by a regularity parameter introduced by Zhu *et al*[12] and the Gamma distribution function is used here. Cao *et al*[13] developed a technique to link the regularity and gamma distribution parameters with the physical parameters of a material: average, minimum and maximum grain sizes. Having generated virtual grains, orientations can be assigned to individual grains according to probability theories, which have been embedded into the VGRAIN system.

The output of VGRAIN is a "*jnl*" file, which contains the grain boundary and grain orientation information. Each grain is considered as a block, which contains specific material properties, such as orientation, Young's modulus, etc. This "*jnl*" file can be read into ABAQUS/CAE for further pre-processing, such as generating elements, adding boundary and loading conditions, creating tools and defining contact pairs, etc. The integration between VGRAIN and ABAQUS/CAE enables CPFE analysis to be carried out efficiently and human interaction for the pre-processing has been minimized.

## 3. Forming Micropins

The quality of extruded micropins is affected by the grain size, grain orientation and grain distributions of the material and the geometric defects cannot be captured using conventional continuum-based FE forming simulation techniques. The established CPFE technique can be used for the process modelling. Fig. 2(a) shows the micropins of 570μm in diameter extruded using the material with average grain sizes of 211μm, i.e. about 2 ~ 3 grains across the diameter of the extruded micropins, uncontrollable bending and curvature of the extruded pins are experimentally observed[14]. The uncontrollable curvature feature can not be modelled, if macro-mechanics-based FE analysis is used[14]. This clearly shows the grain size effect for the micro-forming process and the CPFE technique needs to be employed to predict such a feature.

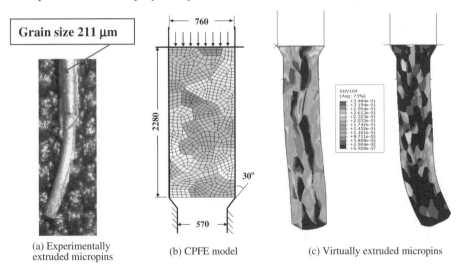

(a) Experimentally extruded micropins    (b) CPFE model    (c) Virtually extruded micropins

Fig. 2. Comparison of the shape of (a) experimentally (in Ref.14) and (c) virtually extruded micropins. (b) shows the FE model (dimension in μm) with the grains.

Numerical investigations have been carried out using the developed integrated CPFE simulation system. Random distributed grains and their orientations with an average grain size of 211μm were generated using the VGRAIN system (Fig. 2(b)).

The dimensions of the workpiece and the die, shown in Fig. 2(b), are the same as those given in Ref. 14, apart from that a plane strain analysis model is created here for simplicity. The FE mesh was created using ABAQUS/CAE with quad-dominated elements (CPE4R). A displacement of 2280µm is applied on the extrusion punch. A friction coefficient of 0.1 between the die and the work piece is assumed. This value is commonly used for cold extrusion processes. The CPFE model is shown in Fig. 2(b), where the grain orientation is assigned randomly.

Fig. 2(c) shows the virtually "formed" micropins and the contour shows the cumulative shear strain distribution for two CPFE analyses results. The same FE model and the grain structure are used for both FE analyses and the only difference is that the grain orientations are assigned to the grain using the same probability theories twice. Thus the grain orientations are different from the two CPFE models. It can be seen clearly that geometric errors for the extruded micropins are different. This could confirm the results obtained experimentally that if the ratio of the diameter of micropins and the grain size of the material is small, uncontrollable bending and curvature of extruded micropins are the major geometric defects. It can also be observed from the figure that the maximum cumulative shear strains occur locally along grain boundaries, which is induced by strong mis-matches of orientations among grains. The uncontrollable curvature feature (Fig. 2a & c) can not be modelled, if continuum FE analysis is used. This clearly shows the grain size effect during the micro-forming process and CPFE needs to be employed to predict such feature.

## 4. Hydro-Forming of Micro-Tubes

Hydro-formed micro tubes with the wall thickness of 30-50 µm are used in electronic and medical devices due to an increasing integration of micro-system technology into products. This case study will demonstrate the CPFE analysis should be used for the process modelling and to predict the non-uniform thinning of the formed micro-tubes.

Since the tube is much larger compared with grain size of the metal from which it is constituted, the conventional macro-mechanics FE modelling is effective for the modelling of hydro-forming processes [15-17]. Fig. 4 shows an example of a hydro-formed stainless steel tube. The outer diameters of the initial and the formed tube are 800µm and 1030µm, respectively. Cracking occurs, as shown in Fig. 4, due to the localized thinning of the material in the hydro-forming process. Many tubes were formed using the same hydro-forming parameters and it was found that the cracking takes place at different locations. This random cracking feature in hydro-forming of micro-tubes can not be captured using the conventional macro-mechanics based process modelling technique.

A plane strain CPFE model is created using the technique described in the previous sections, which is shown in Fig. 5. The minimum, average and maximum grain sizes of the material are 25, 30 and 40 µm, respectively. Hence it is about 1~2 grains across the thickness of the tube section in average. The grains and their orientations of the material are generated using the VGRAIN system, which are read into ABAQUS/CAE for further mesh generation, boundary and loading definitions. The die is defined in ABAQUS/CAE as well. The maximum applied loading pressure is 400MPa, this high pressure ensure workpiece is deformed to the die. A friction coefficient of 0.1 is used when the workpiece and the die is in contact in the forming process. For simplicity, a 2D plane strain CPFE analysis was carried out here.

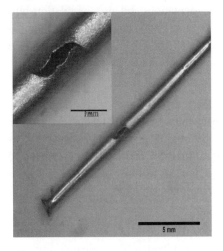

Fig. 4. Localized failure of formed micro-tube.

Fig. 5. CPFE model for micro-tube hydro-forming. Grains and grain boundaries are also shown in the figure.

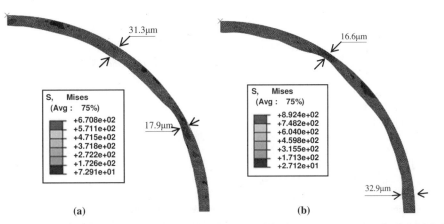

Fig. 6. Comparison of predicted thinning features of the part with microstructures generated twice ((a) & (b)) using the same control parameters of the material with VGRAIN based on embedded probability theory.

As mentioned before, the grain structures and grain orientations are generated using the VGRAIN system automatically based on the embedded probability theories. To simulate the deformation and thinning behaviour of hydro-forming of two micro-tubes, which are cut from the same piece of materials, the grain structures and orientations are generated twice using the same microstructure control parameters defined above. This indicates that the microstructures and grain orientations may be different from the two

CPFE models, although they are within the range of the material specification. The results of the virtually hydro-formed micro-tubes are shown in Fig. 6. It can be seen that the minimum and maximum values of the wall thickness of the formed tubes, shown in Fig 6 (a) (17.9 µm and 31.3 µm) and (b) (16.6 µm and 32.9 µm), are different for the two cases studied. This is due to the grain size and their orientations and also the relationships with their neighbouring grains. This complicated relationship and localized thinning features cannot be captured using the conventional macro-mechanics based FE techniques. It can also be seen that the localized thinning occurs at different locations.

## 5. Conclusions

In traditional metal forming processes, macro-mechanics based FE techniques can be used for the process simulation. However, in the forming of micro components, such as extrusion of micro-pins and hydro-forming of micro-tubes, CPFE analysis may need to be used for real cases. Otherwise, the important geometric features of the formed parts due to the microstructure variation of the material can not be captured.

## Acknowledgments

The support from the European Commission for conducting the research into the "Integration of Manufacturing Systems for Mass-manufacture of Miniature/Micro-Products (MASMICRO)" is acknowledged.

## References

1. A. Needleman, *Acta Mater.* **48**, 105 (2000).
2. G. Cailletaud, S. Forest, D. Jeulin, F. Feyel, I. Galliet, V. Mounoury, and S. Quilici, *Comput. Mater. Sci.* **27**, 351(2003).
3. E. Nakamachi, K. Hiraiwa, H. Morimoto, and M. Harimoto, *Int. J. Plast.* **16**, 1419 (2000).
4. M. Kovac, and C. Leon, *Nucl. Eng. Des.* **235**, 1939 (2005).
5. M. Dao, and M. Li, *Philos. Mag. A*, **8**, 1997 (2001).
6. Z. Y. Ren, and Q. S. Zheng, Mech. Mater. **36**, 1217 (2004).
7. G. I. Taylor, *J. Inst. Met.* **62**, 307 (1938).
8. R. Hill, J. R. Rice, *J. Mech. Phys. Solids*, **20**, 401 (1972).
9. F. J. Harewood, and P. E. McHugh, *Comput. Mater. Sci.* **39**, 481 (2007).
10. Y. Huang, *Harvard University Report*, MECH No. 178 (1991).
11. K. Kobayashi, and K. Sugihara, *Future Gener. Comp. Sy.* **18**, 681 (2002).
12. H. X. Zhu, S. M. Thorpe, and A. H. Windle, *Philos. Mag. A* **81**, 12, 2765 (2001).
13. J. Cao, W. Zhuang, S. Wang, K. Ho, N. Zhang, J. Lin, and T.A. Dean, To appear in *Int. J. of Multiscale Modelling* (2008).
14. K. Krishnan, j. Cao and K. Dohda, *J. Manuf. Sci. E.-ASME* **129**, 669 (2007).
15. Ch. Hartl, J. Mater. Process. Technol. **167**, 283 (2005).
16. Y. Aue-u-lan, G. Ngaile, and T. Altan, *J. Mater. Process. Technol.* **146**, 137 (2004).
17. J. Kim, W. J. Kim, and B. S. Kang, *Int. J. Mech. Sci.* **47**, 1023 (2005).

# STRETCHING-INDUCED ORIENTATION TO IMPROVE MECHANICAL PROPERTIES OF ELECTROSPUN PAN NANOCOMPOSITES

X. X. HOU[1], X. P. YANG[2†], F. ZHANG[3], S. Z. WU[4]

*Beijing University of Chemical Technology, Beijing, P. R. CHINA,*
*hxx198525@tom.com, yangxp@mail.buct.edu.cn, zfipis@163.com, wusz@mail.buct.edu.cn*


E. WACLAWIK[5]

*Queensland University of Technology, Queensland, AUSTRALIA,*
*e.waclawik@qut.edu.au*

Received 15 June 2008
Revised 23 June 2008

Partially aligned and oriented polyacrylonitrile(PAN)-based nanofibers were electrospun from PAN and SWNTs/PAN in the solution of dimethylformamide(DMF) to make the carbon nanofibers. The as-spun nanofibers were hot-stretched in an oven to enhance its orientation and crystallinity. Then it were stabilized at 250 □ under a stretched stress, and carbonized at 1000 □ in $N_2$ atmosphere by fixing the length of the stabilized nanofiber to convert them into carbon nanofibers. With this hot-stretched process and with the introduction of SWNTs, the mechanical properties will be enhanced correspondingly. The crystallinity of the stretched fibers confirmed by X-ray diffraction has also increased. For PAN nanofibers, the improved fiber alignment and crystallinity resulted in the increased mechanical properties, such as the modulus and tensile strength of the nanofibers. It was concluded that the hot-stretched nanofiber and the SWNTs/PAN nanofibers can be used as a potential precursor to produce high-performance carbon composites.

*Keywords*: Nanofiber; electrospinning; polyacrylonitrile; orientation.

## 1. Introduction

Electrospinning from polymer solutions can produce fibers having the diameters in the nanometer to micron range, with the sizes depending on the electrostatic force applied.[1] Polyacrylonitrile (PAN) nanofiber is a common precursor of general carbon fibers. However, these applications of PAN nanofiber were hindered by the poor strength, attributed to their small diameters and the unoptimized molecular orientation and crystallinity in the fibers .[2-3]

Recently, carbon nanofibers have received increased attention due to their potential uses in composite reinforcing fillers, heat-management materials, high-temperature catalysts

---

[†]Corresponding Author.

and components in nanoelectronics etc.[1,3] To enhance the strength of the nanofibers, the single-walled carbon nanotubes (SWNTs) has considered as promising reinforced materials for modification of polymer by hybrid or nanocomposites due to their excellent mechanical and electronic properties.[4] Some researchers also tried hot-stretched to improve the molecular orientation and the crystallinity of the nanofibers.[5-8] Therefore, in this study, PAN and PAN/SWNTs nanofiber composites were electrospun, and then hot-stretched which demonstrated good dispersion of SWNTs and the high orientation of PAN molecules.

## 2.   Experiment

### 2.1.  *Electrospinning*

The required PAN solution was prepared from a commercial PAN fiber (UK Courtaulds Ltd.) composed of PAN/methyl acrylate/itaconic acid (93:5.3:1.7 w/w, average molecular weight of 100,000 g/mol). The solvent used was N,N-dimethylformamide (DMF, Beijing Chemical Plant Co.). To uniformly disperse the SWNTs in the organic polymer matrix, the SWNTs were modified to form an individually polymer-wrapped structure.[9] SWNTs were dispersed in DMF at 40□ with a fixed concentration for 12h in bath sonication (KQ-250DB, Kunshan Ultrasonic Instrument Co., Ltd). Then PAN at a concentration of 12wt% was added to the solution while stirring to obtain the well-dispersed solution. SWNTs concentration was 0.25%, 0.5%, 0.75%, 1% of the mass of the dissolved PAN respectively.

### 2.2.  *Stretching*

Relatively aligned PAN nanofiber and PAN/SWNTs nanofiber composites are obtained by electrospinning with the rotating drum at high-spun speed. However, the whirlpool jet from the pinhead to the collector made it difficult to get unidirectional alignment in a large-area sheet. Therefore, the electrospun nanofiber sheet needed a subsequent hot-stretching to improve the fiber alignment in the sheet.

In this study, the PAN nanofibers and PAN/SWNTs nanofiber composites were hot-stretched according to the method proposed by Phillip and Johnson.[10,11] Both ends of the sheet were clamped with pieces of graphite plates. Then one end was fixed to the ceiling of the oven and the other end was weighted by 75g of metal poise to give a desired tension and elongation in the temperature-controlled oven at about 135°C for 5 minutes. The stretching ratio, $\lambda$, was calculated from $\lambda = L / L_0$, where $L_0$ and $L$ are the lengths of nanofiber sheet before and after the hot-stretching, respectively.

### 2.3. *Characterization*

The unidirectional alignment and the diameter change of the PAN nanofibers and PAN/SWNTs nanofiber composites were observed by scanning electron microscope (SEM, HITACHI S-4700 FEG-SEM). The cross sectional view of a nanofiber was obtained by embedding a sheet in resin, immersing it in liquid nitrogen, and fracturing it perpendicular to the fiber alignment direction to expose the cross-section area.

The crystallinities of the as-electrospun and the hot-stretched PAN nanofiber were investigated with X-ray diffractometer (XRD, Rigaku D/max 2500VB2+ /PC), operated at 40 kV and 200 mA to produce CuK$\alpha$ radiation ($\lambda = 1.54$ Å). The percent crystallinity was obtained by extrapolation of the crystalline and amorphous parts of the diffraction pattern.[12] The *Hermans* orientation coefficient, $f$, was determined using the primary equatorial arcs from the *(100)* reflection at $d \approx 5.3$Å according to the following Eq.[12,13]

$$f = \frac{\int_0^{90} \frac{(3\cos^2 \beta - 1)}{2} I \sin \beta d\beta}{\int_0^{90} I \sin \beta d\beta} \tag{1}$$

Where, $\beta$ is the azimuthal angle between the axis of the molecular segment and the fiber alignment, and $I$ is the scattering intensity of the *(100)* reflection at that angle. The crystallite size was calculated by using the formula $L_c = k\lambda / (\beta\cos\theta)$, with $k = 0.89$ and $\lambda = 1.54$Å respectively.[12]

Mechanical properties of PAN nanofibers and PAN/SWNTs nanofiber composites after hot-stretched were measured using an LR30K Testing Machine. The samples were prepared in 5mm width and 20mm length respectively. The cross section areas of the nanofibers were calculated through the weights of the samples and the densities of PAN and SWNTs, where the density of PAN nanofibers and PAN/SWNTs nanofiber composites are 1.2 g/cm.$^2$

### 3.  Results and Discussion

Fig. 1. SEM micrographs: (a) as-spun nanofibers, (b) hot-stretched nanofibers, (c) cross sectional views of as-spun nanofibers and (d) hot-stretched nanofibers.

Fig. 1(a) showed SEM observation of the as-spun PAN nanofibers in a sheet. The enlarged photo showed the rough surface of nanofiber. Fig. 1(b) showed the hot-stretched nanofibers, proving the excellent alignment along the sheet axis. However, the fiber surface was still rough, possibly as a result of imposed tension. If the tension increased, the stretching ratio and the alignment may increase, but too much increase may resulted in the rupture of the sheet. Fig. 1(c) and (d) showed the cross-sectional views of as-spun and hot-stretched nanofibers in the sheet. The remarkable changes were the decrease in fiber diameter.

The morphology and distribution of fiber diameter of PAN/ SWNTs nanofiber composites were examined using scanning electron micrographs (SEM). Fig. 2 showed the SEM of samples with different SWNTs concentration. There was no conglutination in the composite fibers, which proved that the SWNTs were dispersed in the solution well. When the concentration of the SWNTs increased, the surface of the composite fibers became rough, which indicated that at high concentration some SWNTs might not completely embedded into the nanofiber matrix.

Fig. 2. SEM micrographs: (a)pure PAN, (b)0.5% SWNTs and (c)1% SWNTs.

Fig. 3(a) and (b) showed *X*-ray diffraction (XRD) patterns from the as-spun and the oriented nanofibers sheet respectively. There were two diffraction peaks at 17 and 29° which were corresponding to the $d \approx 5.3$Å from the (*100*) and the $d \approx 3.03$Å from the (*110*) reflections respectively.[14] The diffraction pattern of the as-spun nanofiber showed one weak peak with a value of $2\theta$ at 17°. This indicated that electrospinning of the nanofibers onto a rotating drum generates limited crystallinity. In contrast, the oriented

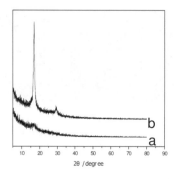

Fig. 3. *X*-ray diffraction patterns for the nanofibers: (a) as-spun, (b) highly oriented.

nanofibers also showed two diffraction peaks indexed with values of 2 $\theta$ of 17 and 29°. The former, very strong peak corresponded to the (*100*) diffraction planes of the crystallite, and the latter to the (*110*) crystal planes.[14] The degree of orientation was usually expressed by orientation coefficient, which can be calculated from the half-width of the corresponding peak in the XRD curve.

Table 1 presented values of the percent crystallinity and the orientation coefficient for pure PAN nanofiber. The percent crystallinity of the hot-stretched nanofiber increased about 3 times and the orientation coefficient increased about 21.5% in comparison with those of as-spun nanofiber. The crystallite size also decreased about 10%.

Table 1. Percent crystallinity and orientation coefficient of nanofibers obtained from *X*-ray diffraction curves.

| Nanofiber | Percent Crystallinity % | Orientation Coefficient % | Crystallite size Å |
|---|---|---|---|
| As-spun | 7.92 | 45.6 | 2.51 |
| Hot-stretched | 31.8 | 55.4 | 2.26 |

Fig. 4(a) and (b) showed the typical stress–strain curves of the PAN nanofibers before and after hot-stretching respectively. The stress of the as-spun fiber sheet increased gradually to a maximum value of approximately 100MPa at around 10% strain, and then decreased with further increase in the strain. In contrast, the hot-stretched nanofiber sheets showed stresses that increased rapidly to the maximum value of about 220MPa.

Fig. 5 showed the typical stress–strain curves of the PAN and PAN/SWNTs nanofiber composites after hot-stretched. SWNTs improved the modulus and tensile strength of the nanofiber. The tensile strength 128.76MPa at about 0.75% SWNTs is increased with 58.9%. And also the tensile modulus showed a peak value of 4.62GPa at about 0.75% SWNTs by weight with 66.8% improvement. The (e) curve is departure from the trend, which might be the uncompleted uniform dispersion of SWNTs in high concentration.

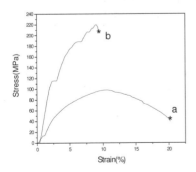

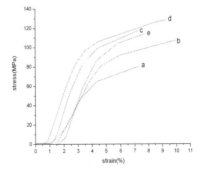

Fig. 4. Stress–strain curves for the nanofiber sheets: (a) as-spun, (b) highly oriented. The star symbols locate the rupture points of the samples.

Fig. 5. Stress–strain curves for PAN and PAN/SWNTs Nanofiber: (a) pure PAN, (b) 0.25%SWNTs, (c) 0.5%SWNTs, (d) 0.75%SWNTs, (e) 1%SWNTs.

## 4.   Conclusions

PAN nanofiber and PAN/SWNTs nanofiber composites were prepared by electrospinning from PAN/N,N-dimethylformamide solution. Hot-stretched method was used to increase the degree of crystallinity and molecular orientation of PAN nanofiber and PAN/SWNTs nanofiber composites. Improved unidirectional alignment and decrease of the fiber diameter were observed by scanning electron microscopy in the hot-stretched nanofiber sheet. The crystallinity of the stretched sheet confirmed by *X*-ray diffraction has enhanced 3 times in comparison with those of as-spun sheets. The improved fiber alignment and crystallinity resulted in the increased modulus and tensile strength. Thus, the hot-stretched nanofiber sheet and the nanofiber composites with the component of SWNTs can be used as the potential precursor to produce high-performance carbon nanofibers. The mechanical properties of the PAN nanofibers and PAN/SWNTs nanofiber composites can be improved more by extensive studies of electrospinning and hot-stretched conditions.

## References

1.  E. Hammel, X. Tang, M. Trampert, T. Schmitt, K. Mauthner, A. Eder and P. Pötschke, Carbon **42** (2004) 1153.
2.  H. Q. Hou, J. J. Ge, J. Zeng, Q. Li, D. H. A. Greiner, Z. D. Stephen, S. Z. D. Cheng, Chem. Mater. **17** (2005) 967.
3.  J. J. Ge, H. Hou, Q. Li, M. J. Graham, A. Greiner, D. H. Reneker, F. W. Harris and S. Z. D. Cheng, J. Am. Chem. Soc., **126** (2004) 15754.
4.  S. Chand, J. Mater. Sci. **35** (2000) 1303.
5.  J. T. McCann, M. Marquez and Y. Xia, J. Am. Chem. Soc. **128** (2006) 1436.
6.  Y. Dror, W. Salalha, R. L. Khalfin, Y. Cohen, AL. Yarin and E. Zussman, Langmuir **19** (2003) 7012.
7.  S. Maensiri, W. Nuansing, Mater. Chem. Phys. **99** (2006)104.
8.  W. Watt, B. V. Perov, Handbook of Composites (Amsterdam, HOLAND, 1985).
9.  W Eric, B. J. John, G. S. G. Roland, M. Anthony and M. Nunzio, Proc. of SPIE, **6036** (2006) 603607-4.
10. J. Johnson, L. N. Phillips and W. Watt, Brit Patent 1,110,790 (1965).
11. J. Johnson, W. Watt, L. N. Phillips, R. Moreton, Brit Patent 1,166,251 (1966).
12. W. H. Norman, W. H. Cheetham and L. Tao, Carbohydrate Polymers **36** (1998) 277.
13. C. P. Lafrance, M. Pezolet and R. E. Prud'homme, Macromolecules **24** (1991) 4948.
14. S. F. Fennessey, R. J. Farris, Polymer, **45** (2004) 4217.

# THE BENDABILITY AND FORMABILITY OF COLD ROLLED AND HEAT TREATED AZ31 MG ALLOY SHEETS

INSOO KIM[1†], SAIDMUROD AKRAMOV[2]

*Department of Information and Nano Materials Science and Engineering, Kumoh National Institute of Technology, Gumi, Gyung Buk, 730-701, KOREA,*
*iskim@kumoh.ac.kr, saidmurod23@yahoo.com*

Received 15 June 2008
Revised 23 June 2008

In this study, strong {0002} textured (or strong ND//<0002> textured) AZ31 Mg alloy sheets were prepared. These AZ31 Mg alloy sheets were cut along the angles of 0, 12.5, 25, 30 and 37.5 degrees to the rolling direction or {0002} texture. Prepared samples with different angles to the rolling direction were rolled at room temperature and after subsequent heat treatment, the bendability, ultimate tensile stress, elongation and texture are investigated using the 3-points bending tester, tensile tester and x-ray diffractometer, respectively. The specimen having along the angles of 0 degree to the rolling direction shows the highest load and 45 degrees specimen shows the highest displacement among any other specimens in the bending test. The specimen having along the angles of 12.5 and 25 degrees to the rolling direction shows highest elongation and of 0 degree specimen shows the highest ultimate tensile stress among any other specimens after cold rolling and subsequent heat treatment at 150°C for 30 minutes in bending test.

*Keywords*: AZ31 Mg alloy; Bendability; Formability; Cold Rolling; Microstructure; Texture; HCP; Basal Plane; 3-Points Bending Test; Tensile Test.

## 1. Introduction

During the last one decade, many researchers have made very significant efforts to study the formability of Mg alloy sheet for automobile applications for weight savings. However the number of possible applications of magnesium alloys is limited due to low ductility and high anisotropy of mechanical properties[1]. Magnesium alloy sheets have also poor formability, because only few available slip systems in the hexagonal close packed (HCP) crystal structure at room temperature and it is attributed to a strong planar anisotropy where the {0002} basal plane tends to be set parallel to the rolling direction (RD)[2,3]. Severe deformation processes are one of useful tools for improving formability through the grain refinement[4,5].

Many studies have been carried out on the deformation of Mg alloys. The results of these studies showed that slip system occurs only on the basal plane {0002} texture of the HCP crystal structure at room temperature and other slip systems occur over about 200 -

---

[†]Corresponding Author.

$250°C^{6-8}$. In this work, AZ31 Mg alloy sheets having strong basal plane {0002} texture. These sheets were cut along the angles of 0, 12.5 and 25, 37.5 degrees to the RD or basal plane {0002} and were used for 3-points bending test and tensile test at room temperature to compare the bendability, ultimate tensile stress and elongation with the change of texture.

## 2. Experimental Procedure

The chemical compositions of AZ31 alloy sheet are shown in Table 1. Strong basal plane {0002} textured (or ND//<0002> textured) AZ31 Mg alloy sheets were cut along the angles of 0, 12.5, 25°, 37.5 and 45 degrees to the rolling direction or basal plane {0002} texture with dimensions of 20mm x 10mm x 0.7mm in Fig. 1. Mg alloy sheets having the angles of 0, 12.5, 25°, 37.5 and 45 degrees to the RD were symmetrically cold rolled with different reduction ratios ranging from 0 to 45% on a laboratory rolling mill at room temperature and without lubricant condition. The sheet samples were reduced about 5% in thickness at each pass through rolling mill. Before and after cold rolled AZ31 Mg alloy sheets having the angles of 0, 12.5, 25, 30, 37.5 and 45 degrees to the RD or basal plane {0002} texture were investigated bendability, ultimate tensile stress, elongation and texture using 3-points bending tester, tensile tester and x-ray diffractometer, respectively. Before tensile test, 45% cold rolled samples have a heat treatment at 150°C for 30 minutes.

Table 1. Chemical composition of AZ31 magnesium alloy.

| Al | Zn | Mn | Si | Fe | Ni | Cu | Ca | Mg |
|------|------|------|------|------|-------|-------|----|-----|
| 2.89 | 0.96 | 0.31 | 0.31 | 0.15 | 0.002 | 0.011 | – | BAL |

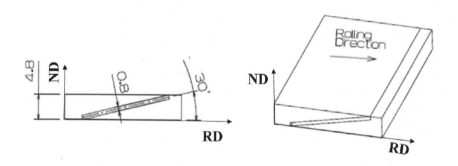

Fig. 1. The method of initial sample preparation.

## 3. Results and Discussion

Fig. 2 shows the bending characteristics of initial samples with various angles to the RD or basal plane {0002} texture; where is 1- initial sample with 0° to the RD, 2 - initial sample with 12.5° to the RD, 3 - initial sample with 25° to the RD, 4 - initial sample with 30° to RD, 5 - initial sample with 37.5° to the RD and 6 - initial sample with 45° to the RD. Sample with 0° degree to the RD shows the lowest flexural bending strain and the highest flexural bending stress, with increasing degrees of prepared samples to the RD, stresses were decreased and strains were increased. Sample having 45 degrees to the RD or basal plane {0002} texture observed the highest strain and the lowest stress value in Fig. 2.

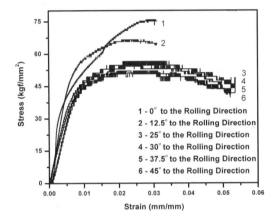

Fig. 2. The result of 3-points bending test corresponding to: 1- initial sample with 0° to the RD, 2 – initial sample with 12.5° to the RD, 3 - initial sample with 25° to the RD, 4 - initial sample with 30° to the RD, 5 initial sample with 37.5 to the RD and 6 – initial sample with 45° to the RD.

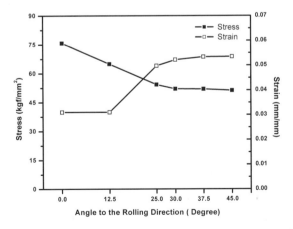

Fig. 3. Flexural bending stress and flexural bending strain change versus angle to the RD.

Fig. 3 shows stress-strain change based on Fig. 2 after the 3-points bending test. Initial sample with 0° to the RD shows the lowest displacement, but the highest load. Sample with 45° to the RD shows the highest displacement, but the lowest load. It is attributed to the change of texture with the angles of 0, 12.5, 25, 30, 37.5 and 45 degrees to the RD.

The ultimate tensile stress and elongation of the initial and 45% rolled and subsequent heat treated at 150°C for 30min specimens were measured at room temperature and shown in Fig. 4. All the initial specimens with the angles of 0, 12.5, 25, 37.5 and 45 degrees to the RD show the maximum stress at about 50 MPa. The Ultimate tensile stress of the 45% rolled and subsequently heat treated samples at 150°C for 30 minutes show about 35-40 MPa in Fig. 4. The ultimate tensile stress of initial specimens with 0 degree to the RD is higher than that of the angles of 12.5, 25, 37.5 and 45 degrees. Based on the data of Fig. 4, Fig. 5 is created, which shows the change of ultimate tensile stress and elongation with the angles of 0, 12.5, 25, 30, 37.5 and 45 degrees to the RD in initial and 45% rolled and subsequently heat treated samples at 150°C for 30 minutes.

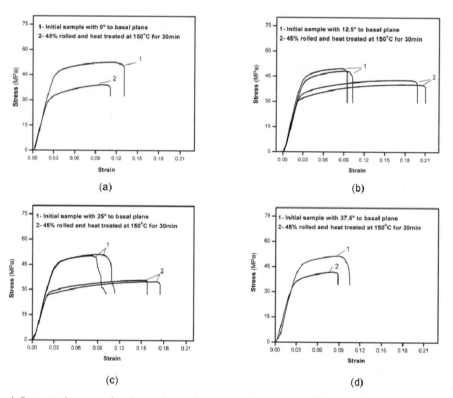

Fig. 4. Stress–strain curves after the tensile test from samples having the angles of (a) 0° to RD, (b) 12.5° to RD (c) 25°to RD, (d) 37.5° to RD.

The ultimate tensile stresses of initial and 45% rolled and subsequent heat treated at 150°C for 30min specimens are similar to each others in Fig. 5.

The higher elongation observed on the samples of cold rolled and subsequent heat treated at 150°C for 30 minutes with the angles of 12.5 and 25 degrees to the RD in Figs. 4 and 5. Elongation of samples with the angles of 0 and 37.5 degrees to the RD slightly decreased after 45% reduction in thickness and subsequent heat treatment at 150°C for 30 minutes.

In order to understand the change of the ultimate tensile stress and elongation with the angles to the RD, pole figures of all samples were conducted. Fig. 6 shows measured {0002} plane pole figures on initial samples with different angles to the RD and after 45% cold rolling and subsequent heat treatment at 150°C for 30 minutes. After 45% cold

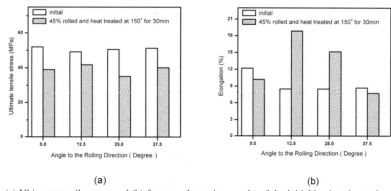

(a)                                                        (b)

Fig. 5. (a) Ultimate tensile stress and (b) fracture elongation graphs of the initial having the angles of 0°, 12.5°, 25° and 37.5°to the RD, and 45% cold rolled and subsequent heat treated 150° C for 30 min. having angles of 0°, 12.5°, 25° and 37.5°to the RD.

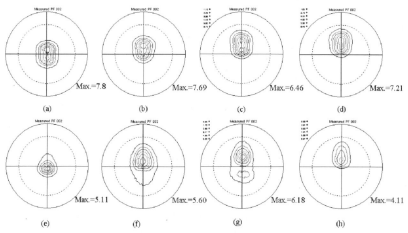

Fig. 6. {0002} Pole figures corresponding to: (a) initial sample with 0° to the RD, (b) initial sample with 12.5° to the RD, (c) initial sample with 25° to the RD, (d) initial sample with 37.5° to the RD, (e) 45% cold rolled sample with 0° degree to the RD and heat treated at 150°C for 30 min, (f) 45% cold rolled sample with 12° to the RD and heat treated at 150°C for 30 min., (g) 45% cold rolled sample with 25° to the RD and heat treated at 150°C for 30 min., (h) 45% cold rolled sample with 37.5° to the RD and heat treated at 150°C for 30 min.

rolling and subsequent heat treatment at 150°C for 30 minutes, intensity of sample with 0 degree to the RD remained same before and after cold rolled and heat treatment. {0002} poles of samples with the angles of 12.5, 25 and 37.5 degrees to the RD rotated toward to the normal direction (ND) after cold rolling and subsequent heat treatment. This may be explained with recrystalization during cold rolling and subsequent heat treatment. The higher elongation observed on the samples of cold rolled and subsequent heat treated at 150°C for 30 minutes with the angles of 12.5 and 25 degrees to the RD, but it can not be perfectly explained by texture analysis in Fig. 6, which needs more studies in the future.

## 4. Conclusion

(1) Initial sample with 0° to the RD shows the lowest flexural bending strain, but the highest flexural bending stress and sample with the angle of 45° to the RD shows the highest flexural bending strain, but the lowest flexural bending stress.

(2) After 45% cold rolling and subsequent heat treatment at 150°C for 30 minutes, elongation of the specimens having the angles of 12.5 and 25 degrees to the RD is higher than that of any other specimens.

(3) The ultimate tensile stresses are similar with each other under different texture and same process conditions.

## Acknowledgment

This paper was supported by the research fund of Korean Research Foundation Grant Funded by the Korean Government (MOEHRD) (KRF-2005-D00372).

## References

1. M. Avedesian and H. Baker, *Magnesium and Magnesium Alloys, ASM Specialty Handbook* (ASM International, 1999).
2. S.R. Agnew, O¨. Duygulu, *Int. J. Plast.* **21**, 1161 (2005).
3. I. Kim and S. Akramov, *The development of the microstructure and texture in the cold rolled AZ31 Mg alloy sheet, Morden Phys. Letter B* (2008) will be published.
4. W.J. Kim, S.I. Hong, Y.S. Kim, S.H. Min, H.T. Jeong and J.D. Lee, *Texture development and its effect on mechanical properties of an AZ61 Mg alloy fabricated by equal channel angular pressing, Acta Mater.* **51**, 3293 (2003).
5. H.T. Jeong and W.J. Kim, *Mechanical properties and texture evolution of AZ31 Mg alloy during equal channel angular pressing, Mater. Sci. Forum* **475–479**, 545 (2005).
6. H.S. Kim, H.T. Jeong, H.G. Jeong and W.J. Kim, Grain refinement and texture evolution in AZ31 during ECAP process and their effects on mechanical properties, *Mater. Sci. Forum* **475–479**, 549 (2005).
7. L. Chabbi, W. Lehnert and R. Kawalla In: K.U. Kainer, Editor, *Magnesium Alloys and Their Applications* (Wiley-VCH-Verlag, Weinheim, p 590, 2000).
8. E. F. Emley: *Principle of Magnesium Technology* (Pergamon, Elmsford, NY, p 483, 1966).

# THE DEVELOPMENT OF THE MICROCTRUCTURE AND TEXTURE IN COLD ROLLED AZ31 MG ALLOY SHEETS

INSOO KIM[1†], SAIDMUROD AKRAMOV[2]

*Department of Information and Nano Materials Science and Engineering, Kumoh National Institute of Technology, Gumi, Gyung Buk, 730-701, KOREA,*
*iskim@kumoh.ac.kr, saidmurod23@yahoo.com*

Received 15 June 2008
Revised 23 June 2008

Formability is very important parameter of magnesium alloy sheets and it would be related to the texture of sheet metals. In this study, magnesium alloy sheets with strong {0002} texture were cut along the angles of 0, 12.5, 25 and 37.5 degrees to rolling direction (RD). Prepared samples were rolled at room temperature condition. Cold rolled AZ31 magnesium alloy sheets along the angles of 0, 12.5, 25, 37.5 and 45 degrees to rolling direction were investigated microstructure and texture with optical microscopy and x-ray diffractometer, respectively.

*Keywords*: Cold Rolling; Microstructure; Pole Figure; Texture; AZ31 Mg alloy; HCP; Basal Plane; {0002} Texture; Formability.

## 1. Introduction

In recent years, many researchers have made very strong efforts to study the formability of Mg alloy sheet in automobile applications and in casings of mobile electronics for weight savings and prevention of electromagnetic waves, respectively. However the number of possible applications of Mg alloys is limited due to low ductility and high anisotropy of mechanical properties[1,2]. Mg alloy sheets have also poor formability, because only few available slip systems in the hexagonal close packed (HCP) crystal structure at room temperature and it is attributed to a strong planar anisotropy where the {0002} basal plane tends to be set parallel to the rolling direction (RD)[3]. Therefore, the {0002} texture needs to be changed to other textures to enhance the formability of Mg alloy sheets. Some researchers have been tried to change the strong {0002} texture of Mg alloy sheets by the various types of processing methods[4-8].

Many studies have been carried out on the deformation of Mg alloys. The results of these studies showed that slip system occurs only on the basal planes {0002} texture of the HCP crystal structure at room temperature and other slip systems operate over about 200 - 250°C temperature[9-10]. AZ31 Mg alloy sheets having strong basal plane {0002} texture were prepared in this work. These sheets were cut along the angles of 0, 12.5, 25,

†Corresponding Author.

37.5 and 45 degrees to the RD or basal plane {0002} and were used to investigate the change of microstructures and texture after cold rolling.

## 2.  Experimental Procedure

Strong {0002} // ND textured AZ31 Mg alloy sheets were used in this research work. The chemical compositions of AZ31 alloy sheet are shown in Table 1. These specimens were cut along the angles of 0, 12.5, 25, 37.5 and 45 degrees to rolling direction(RD) or basal plane {0002} texture with dimensions of 20mm x 10mm x 0.7mm (Fig.1). The prepared Mg alloy sheets were rolled symmetrically to different reductions ranging from 0 to 45% on a laboratory rolling mill in room temperature and without lubricant condition. The sheets samples were reduced about 5% in thickness at each pass through rolling mill. Cold rolled AZ31 Mg alloy sheets with different angles to the rolling direction were investigated the microstructure and texture with the optical microscope and x-ray diffractometer, respectively. It was investigated that the microstructures of the sectional plane which is perpendicular to transverse direction of specimens and measured that the pole figures of the sectional plane which is perpendicular to normal direction of specimens which were cut along the angles of 0, 12.5, 25, 37.5 and 45 degrees to rolling direction(RD) or basal plane {0002} texture.

Table 1. Chemical composition of AZ31 Mg alloy.

| Al | Zn | Mn | Si | Fe | Ni | Cu | Ca | Mg |
|------|------|------|------|------|-------|-------|-----|-----|
| 2.89 | 0.96 | 0.31 | 0.31 | 0.15 | 0.002 | 0.011 | – | BAL |

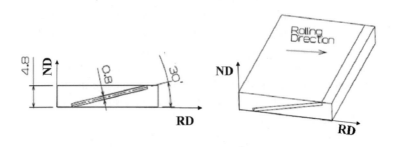

Fig. 1.  Initial sample preparation method.

## 3.  Results and Discussion

Fig. 2 (a, b, c and d) show the microstructures of the initial specimens which were prepared along the angles of 0, 12.5, 25, 37.5 and 45 degrees to RD. The microstructures are consisted of small grains and twin boundaries. In case of samples with along the angles of 37.5 degrees to RD, their average grain size was decreased. It may be related to the elongated grains after cold rolling of Mg alloy sheets. Fig. 2 (e, f, g, and h) show the microstructures of 45% reduction rolled sheets in thickness. The microstructures are also consisted of small grains and twin boundaries.

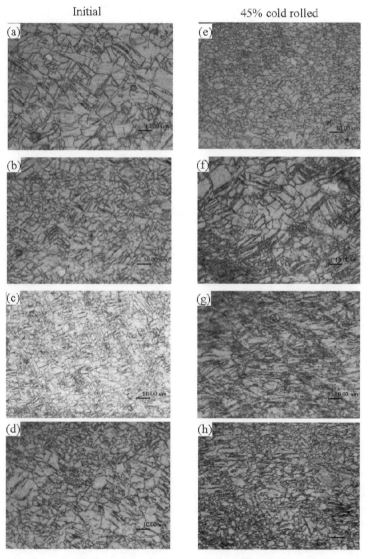

Fig. 2. Optical micrographs corresponding to: (a) initial sample with 0 degree to RD, (b) initial sample with 12.5 degrees to RD, (c) initial sample with 25 degrees to RD, (d) initial sample with 37.5 degrees to RD, (e) 45% cold rolled sample with 0 degree to RD, (f) 45% cold rolled sample with 12.5 degrees to RD, (g) 45% cold rolled sample with 25 degrees to RD, (h) 45% cold rolled sample with 37.5 degrees to RD.

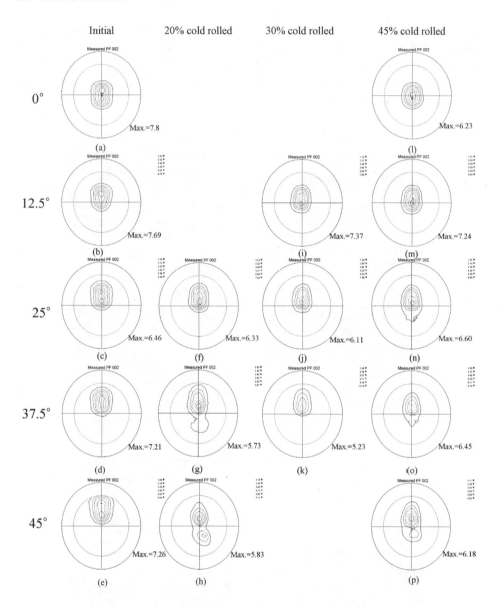

Fig. 3. {0002} Pole figures from one tenth thickness layer: (a) initial sample with 0 degree to RD, (b) initial sample with 12.5 degrees to RD, (c) initial sample with 25 degrees to RD, (d) initial sample with 37.5 degrees to RD, (e) initial sample with 45 degrees to RD, (f) 20% cold rolled sample with 12.5 degrees to RD, (g) 20% cold rolled sample with 37.5 degrees to RD, (h) 20% cold rolled sample with 45 degrees to RD, (i) 30% cold rolled sample with 12.5 degrees to RD, (j) 30% cold rolled sample with 25 degrees to RD, (k) 30% cold rolled sample with 37.5 degrees to RD, (l) 45% cold rolled sample with 0 degree to RD, (m) 45% cold rolled sample with 12.5 degrees to RD, (n) 45% cold rolled sample with 25 degrees to RD, (o) 45% cold rolled sample with 37.5 degrees to RD, (p) 45% cold rolled sample with 45 degrees to RD.

After 45% reduction rolling in thickness, the average grain size of samples with angles of 0 and 37.5 degrees to RD was decreased. It may also be related to dynamic recrystalization during the cold rolling process. During the cold rolling, the dynamic recrystalization will be occurred in the deformed and elongated grains, the recrystalized grains will be rotated to equilibrium state. But the recrystalized grains were not remained with equi-axed shape.

Texture measurements were performed at one tenth (S=0.9) thickness of the symmetrically rolled Mg alloy sheets. The {0002} pole figures were obtained by using the Co-Kα radiation with the Schultz reflection method. Figure 3 shows {0002} pole figures of the initial AZ31 Mg alloy sheets and after symmetrical cold rolled sheets. Strong basal plane {0002} texture was observed in initial Mg alloy sheet in Fig 3(a). After cold rolling, strong basal plane {0002} texture was not changed in initial Mg alloy sheet with angles of 0 degree to RD in Fig 3(l), but the intensity of x-ray diffraction of sample decreased about 10-20% than that of the initial sample. It may be related to dynamic recrystalisation during cold rolling process. Fig. 3 (b, c, d and e) show the pole figures of initial samples having the angles of 12.5, 25, 37.5 and 45 degrees to the RD, respectively. Fig. 3 (b, c, d and e) show tilted basal plane {0002} texture. After cold rolling, {0002} poles of samples having the angles of 12.5 and 25 degrees to the RD also moved to the normal direction in Fig. 3(f, i, j, m and n). After 45% reduction cold rolling, the {0002} poles of samples having along the angles of 37.5 and 45 degrees to the RD move toward to the normal direction, but tilted basal plane {0002} texture were maintained in Fig. 3(g, h, k, o and p). Also, intensities of x-ray diffraction of almost all the cold rolled samples were decreased in Fig. 3. These may be explained with the dynamic recrystalization during the cold rolling process of AZ31 Mg alloy sheet.

## 4. Conclusion

(1) The grain sizes of AZ31 Mg alloy sheets were changed in all kinds of directional cut specimens due to the dynamic recrystalization during the cold rolling process.

(2) The basal plane {0002} poles of specimens having the angles of 0, 12.5 and 25 degrees to the rolling direction were easily moved toward the normal direction after the cold rolling.

(3) The basal plane {0002} poles of specimens having the angles of 37.5 and 45 degrees to the rolling direction were not moved to the normal direction after the cold rolling.

## Acknowledgment

This paper was supported by the research fund of Korean Research Foundation Grant Funded by the Korean Government (MOEHRD) (KRF-2005-D00372).

## References

1. R.E. Brown, *Proceedings of 59th Annual World Magnesium Conference* (International Magnesium Association, VA, USA, p 25, 2002).
2. M. Avedesian and H. Baker, *"Magnesium and Magnesium Alloys" ASM Specialty Handbook* (ASM International, 1999).
3. S.R. Agnew, O¨. Duygulu, *Int. J. Plast.* **21**, 1161 (2005).
4. W.J. Kim, S.I. Hong, Y.S. Kim, S.H. Min, H.T. Jeong and J.D. Lee, Texture development and its effect on mechanical properties of an AZ61 Mg alloy fabricated by equal channel angular pressing, Acta Mater. **51**, 3293 (2003).
5. L.W.F. Mackenzie and M. Pekguleryuz, The influences of alloying additions and processing parameters on the rolling microstructures and textures of magnesium alloys, Mater. Sci. and Eng. A **480**, 189 (2008).
6. M.-Y. Zhan, Y.-Y. Li and W.-P. Chen, Improving mechanical properties of Mg-Al-Zn alloy sheets through accumulative roll-bonding, Trans. of Nonferrous Metals Society of China **18**, 309 (2008).
7. X. Huang, K. Suzuki, A. Watazu, I. Shigematsu and N. Saito, Mechanical properties of Mg-Al-Zn alloy with a tilted basal texture obtained differential speed rolling, Mater. Sci. and Eng. A **488**, 214 (2008).
8. I. Kim and S. Akramov, The bendability and formability of cold rolled and heat treated AZ31 Mg alloy sheets, Morden Phys. Letter B (2008) will be published.
9. L. Chabbi, W. Lehnert and R. Kawalla In: K.U. Kainer, Editor, *Magnesium Alloys and Their Applications* (Wiley-VCH-Verlag, Weinheim, p 590, 2000).
10. E. F. Emley: *Principle of Magnesium Technology* (Pergamon, Elmsford, NY, p 483, 1966).

# TEXTURE DEVELOPMENT AND DRAWABILITY OF FRICTIONALLY ROLLED AA 5052 AL ALLOY SHEET

INSOO KIM[1†], SAIDMUROD AKRAMOV[2], HAE BONG JEONG[3], TAE KYOUNG NO[4]

*Department of Information and Nano Materials Science and Engineering, Kumoh National Institute of Technology, Gumi, Gyung Buk, 730-701, KOREA,*
*iskim@kumoh.ac.kr, saidmurod23@yahoo.com, jjn7nmr@dreamwiz.com, hyun5a@naver.com*

Received 15 June 2008
Revised 23 June 2008

The microstructure, pole figure and r-value of the frictionally rolled and subsequently heat treated AA 5052 Al sheets were investigated by optical microscopy, x-ray diffractometer and tensile tester, respectively. Frictionally rolled AA 5052 Al specimens showed a fine grain size. After subsequently heat treated specimens, the ND//<111> texture component was increased. The r-values of the frictionally rolled and subsequently heat treated Al alloy sheets were about two times higher than those of the original Al sheets. These could be related to the formation of ND//<111> texture components through frictional rolling in and subsequent heat treatment of AA 5052 Al sheet.

*Keywords*: Frictional rolling; Shear deformation; Pole figure; Texture; Microstructure; Tensile test; Plastic strain ratio; R-value; Drawability; ND//<111> texture.

## 1. Introduction

Severe deformation processes are one of the useful tools in the development of the r-value and drawability characteristics of aluminum alloys. It has been found that the formation of the <111> plane parallel to the sheet surface (ND//<111> texture) by the severe deformation in fcc metal sheets helps to increase the r-value (Lankford parameter or plastic strain ratio) and to decrease the $\Delta$r-values[1-3]. It has been found that rotated cube {001}<110> texture and ND//<111> texture or γ-fiber component appear after severe shear deformations in fcc metal sheets. This leads to increase the r-value[4]. The r-value is known to be related to the drawability of sheet metals. However, due to the high energy in the deformed state and the existence of structural heterogeneities, annealing conditions are critical and unwanted structures may form[5-7]. In the present paper, it is investigated that the effects of the frictional rolling[8] on texture, r-value and drawability of AA 5052 Al alloy sheet.

---

[†]Corresponding Author.

## 2. Experimental Procedure

Table 1 shows chemical compositions of AA 5052 Aluminum alloy which having 0.45% Si+Fe, 0.1% Cu, 2.2-2.8% Mg, 0.15-0.35% Cr, 0.1% Zn and aluminum. AA 5052 aluminum alloy sheets with initial thickness of 3.4mm were used in the present study.

Table 1. Chemical composition of AA 5052 aluminum alloy.

| Si | Fe | Cu | Mn | Mg | Cr | Zn | Others | | Al |
|----|----|----|----|----|----|----|--------|------|----|
| 0.45Si+Fe | | 0.1 | 0.1 | 2.2–2.8 | 0.15–0.35 | 0.1 | Each | Total | Bal |
| | | | | | | | 0.05 | 0.15 | |

The sheet samples, with dimensions of 60mm x 40mm x 3.4mm, were prepared from a rolled sheet along the rolling direction. Then these plates were annealed at 500 °C for 2 hours to homogenize the initial grain size through thickness (named initial Al sheet). The initial Al sheets were then frictionally rolled to range from 0 to 90% on a laboratory rolling mill with rolls of 150 mm in diameter. Hence, we will observe 90% frictionally rolled samples. To obtain a high friction coefficient between rolls and sample, without lubrication condition was used during rolling process. After the frictional rolling, to measure the plastic strain ratio of the frictionally rolled Al sheet, samples were subsequently heat treated at the temperature of 300°C for 5, 20 and 60 minutes in a salt bath. The microstructure of the transverse direction (TD) side surfaces of the frictionally rolled specimens were investigated by using an optical microscopy. Texture measurements were performed at one tenth and half thickness of the frictionally rolled sheets. (111), (200), and (220) pole Figures were obtained by using the Cu-K$\alpha$ radiation with the Schultz reflection method and the grain orientations were represented in the orientation distribution function (ODF) calculated by the harmonic method[9].

R-value is one of the most important parameters in textured sheet metals forming. The r-value, defined as the ratio of the width-to-thickness strain increments in uni-axial tension, influences the deep drawability or the limit drawing ratio (LDR). In this study, tensile test specimens were taken along 0°, 45° and 90° to the rolling direction from the sheet and the r-value were measured as a function of strain in each direction. Due to the small thickness of the specimen, width-to-length measurements were conducted instead of width-to-thickness measurements.

The average r-value ($\bar{r}$ -value) and $\Delta$r-value were obtained from the measured and calculated r-value data. The average measured r-value ($\bar{r}$ -value) was calculated by the use of $r = (r_0 + 2r_{45} + r_{90}) / 4$, and $\Delta$ r-value was calculated by the use of $\Delta r = (r_0 - 2r_{45} + r_{90}) / 2$. Here, the $r_{0, 45, and 90}$ means the r-value along the angles of 0°, 45°, and 90° to rolling direction, respectively.

Furthermore, the measured r-values by tension tests were compared with those predicted by r-value prediction theories[9].

## 3. Experimental Results and Discussion

Fig.1 shows the microstructures of the AA 5052 aluminum alloy sheet at different conditions. The initial Al sheet microstructure is presented in Fig.1 (a). Fig. 1 (b) showsthe microstructure of 90% frictionally rolled sample which is indicative of homogeneous deformation. Equi-axis grains become elongated with the rolling reduction. The elongated grains sizes become smaller with the increase in reduction ratio. After heat treatment at 300°C for 5, 20 and 60 minutes in the salt bath condition, deformed grains were recrystallized as in Fig. 1(c).

After frictional rolling, surface layer contains a severe deformed texture. Surface layer can not become the representative layer for r-value calculation which is based on pole figure data. Therefore pole figures of one tenth and center layers are measured in this paper. After homogenizing heat treatment, strong cube textures were observed on initial samples of one tenth surface and center layers in Fig. 2(a, d). After frictional rolling, pole figures of one tenth surface and center layer were measured in Fig. 2(b, e).

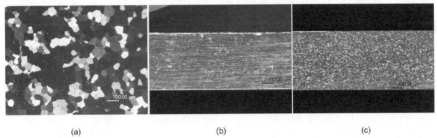

(a)  (b)  (c)

Fig. 1. Optical micrographs obtained from the transverse section of surfaces of AA 5052 Al alloy sheet; (a) initial Al sheet, (b) 90% reduction in thickness and (c) 90% reduction in thickness and heat treated at 300°C for 20min.

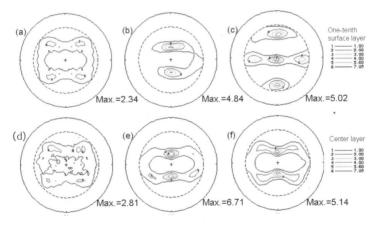

Fig. 2. (111) Pole figures from one-tenth surface layer and center layer of AA 5052 aluminum alloy sheet: (a) initial Al sheet from one-tenth surface layer, (b) 90% frictionally rolled from one-tenth surface layer, (c) 90% frictionally rolled and subsequent heat treated at 300°C for 20 min from one-tenth surface layer, (d) initial Al sheet from center layer, (e) 90% frictionally rolled from center layer, (f) 90% frictionally rolled and subsequent heat treated at 300°C for 20 min from one-tenth surface layer.

After subsequent heat treatment of samples at 300°C for 20 minutes, textures on one-tenth surface and center layers were changed as in Fig. 2(c, f). After subsequent heat treatment on the one-tenth surface thickness layer, rolling texture changed to rotated Cube texture and ND//<111> texture components in Fig.2 (c). On the center layer, strong rolling texture was remained and a Cube texture was observed as a result of recrystalization texture as in Fig. 2 (f).

Based on the obtained (111), (200) and (220) pole figures on the one-tenth surface thickness layers, ODFs were calculated. The ODF cutouts at $\Psi_2 = 0°, 25°, 45°, 65°$ are presented in Fig. 3. After homogenizing heat treatment, strong cube textures observed on initial samples of one tenth surface and center layers in Fig. 3(a). S component {123}<634> was observed in Fig. 3 (b) after 90% frictionally rolling. After subsequent heat treatment at 300°C for 20 minutes, strong rotated cube texture and the γ-fiber ND//<111> texture components were observed on one-tenth surface layer and S component {123}<634> was observed on center layer in Fig. 3 (c).

To make the quantity analysis of texture components development, f(g) values were calculated from ODF results after frictional rolling and subsequent heat treatment conditions. Table 2 shows the f(g) values variation of major texture components.

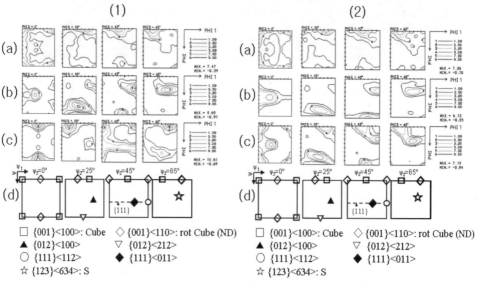

Fig. 3. ODF cutouts of AA 5052 aluminum alloy sheet (1) one-tenth surface layer, (2) center layer measurement results: (a) initial Al sheet, (b) 90% frictionally rolled, (c) 90% frictionally rolled and 300°C subsequent heat treated for 20 min, (d) the prominent orientations in Euler space angel with constant ψ2 section.

Initial samples show high Cube texture, {001}<100> component, but 90% frictionally rolled and subsequent heat treated samples increased γ-fiber ND//<111> texture components. Especially, 90% frictionally rolled and subsequently heat treated for 20 min samples show strong {011}<211>, {111}<110> and {111}<112> texture components on the one tenth surface layer and {001}<100> and β-fiber texture on the center layer.

Table 2. Variation of f (g) value in major texture components.

| | Beta-fiber | | | | Gamma-fiber <111>//ND | |
|---|---|---|---|---|---|---|
| | {001}<100> | {112}<111> | {123}<634> | {011}<211> | {111}<110> | {111}<112> |
| Initial s=0.9 | 7.6 | 1.6 | 7.5 | 7.5 | 0.75 | 0.11 |
| Initial s=0 | 7.3 | 2.2 | 7.3 | 7.3 | 0.1 | 0.1 |
| 90% frictionally rolled S=0.9 | 0.5 | 4.2 | 4.1 | 6.4 | 0.1 | 0.15 |
| 90% frictionally rolled S=0 | 0.4 | 9.8 | 8.5 | 5.6 | 0.1 | 0.7 |
| 90% frictionally rolled(300°C/20min) S=0.9 | 1.2 | 1.2 | 3.3 | 10.2 | 4.2 | 3.9 |
| 90% frictionally rolled(300°C/20min) S=0 | 5.0 | 6.3 | 7.2 | 5.8 | 0.1 | 0.7 |

Table 2 shows comparison of measured and calculated r-values at various heat treatment conditions. In the initial AA 5052 aluminum alloy sheet, measured average r-value was 0.663. After frictionally rolling and subsequent heat treatment at 300°C for 5, 20 and 60 minutes, measured average r-values were 1.104, 0.862 and 0.854, respectively.

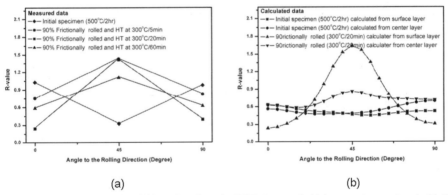

(a)                                              (b)

Fig. 4. Measured (a) and calculated (b) r-values from the ODFs (one tenth thickness and center layer) of initial and 90% frictionally rolled and subsequently heat treated at 300°C during 5, 20 and 60 minutes in AA 5052 aluminum alloy sheet.

Table 3. The Comparison of measured r-value, $\bar{r}$, and $|\Delta r|$ of the 90% frictionally rolled and subsequently heat treated in AA 5052 aluminum alloy sheet.

| Conditions of samples | r-value | | | $\bar{r}$ | $|\Delta r|$ |
|---|---|---|---|---|---|
| | 0° | 45° | 90° | | |
| Initial specimen (500°C/1 hr) | 1.029 | 0.324 | 0.975 | 0.663 | 0.679 |
| 90% frictionally rolled (300°C/5min) | 0.754 | 1.422 | 0.821 | 1.104 | 0.634 |
| 90% frictionally rolled (300°C/20min) | 0.242 | 1.408 | 0.392 | 0.862 | 0.261 |
| 90% frictionally rolled (300°C/60min) | 0.584 | 1.106 | 0.621 | 0.854 | 0.503 |

The measured and calculated r-values are coincided with each others as shown in Fig. 4. The r-value of the frictionally rolled and subsequently heat treated at 300°C for 5 minutes in AA 5052 aluminum alloy sheet is 2 times higher than that of the initial Al specimen. The increasing of r-value after frictional rolling and subsequent heat treatment may be related to ND//<111> texture formation.

## 4. Summary

The drawability and texture development after frictional rolling of AA 5052 aluminum alloy sheet was investigated in this work. From the results it was found that:

(1) The ND//<111> texture, γ-fiber was observed after frictional rolling and subsequent heat treatment in AA 5052 aluminum alloy sheet.

(2) The r-value of the frictionally rolled and subsequently heat treated at 300°C for 5 minutes in AA 5052 aluminum alloy sheet was 2 times higher than that of the initial Al specimen.

## Acknowledgment

This paper was supported by the research fund of Kumoh National Institute of Technology (2008) and Korean Research Foundation Grant Funded by the Korean Government (MOEHRD) (KRF-2005-D00372).

## References

1.  P.H. Lequeu and J. J. Jonas: *Metal. Trans.* **A 19A**, 105 (1988).
2.  M. Hatherly, W.B. Hutchinson, *An Introduction to Textures in Metals* (Institution of Metallurgists, London, 1979).
3.  J. Hirsch, K. Lucke, *Mechanism of deformation and development of rolling textures in polycrystalline F.C.C. metals-simulation and interpretation of experiments on the basis of Taylor-tape theories, Acta Metall.* **36**, 2883 (1988).
4.  K.J. Kim, H.-T. Jeong, K.S. Shin and C.W. Kim, *Texture evolution of rolled AA 5052 alloy sheets after annealing, J. Mater. Process. Tech.* **187-188**, 578 (2007).
5.  A.K. Vasudevan, R.D. Doherty, *Aluminum Alloys-Contemporary Research and Applications, Treatise on Materials Science and Technology*, **31** (Academic Press, 1989).
6.  S. Akramov, M.G. Lee, I. Kim, D. Y. Sung and B.H. Park, *The texture of 1050 Al sheet produced by equal channel angular pressing, Mater. Sci. Forum* **475-479**, 417 (2005).
7.  S. Akramov, I. Kim, M.G. Lee and B.H. Park, *Sheet formability of AA 1050 Al alloy sheet by equal channel angular pressing of route C type, Solid State Phenomena* **116-117**, 324 (2006).
8.  S. Akramov, I. Kim and N.J. Park, *Texture and formability of frictionally rolled AA 1050 aluminum alloy sheets, Advanced Mater. Research* **26-28**, 393 (2007).
9.  H.J. Bunge, *Texture Analysis in Material Science* (Butterworth, Guildford, UK, 1982).

# GRAIN-SIZE DEPENDENT YIELD BEHAVIOR UNDER LOADING, UNLOADING AND REVERSE LOADING

NOBUTADA OHNO[1†], DAI OKUMURA[2]

*Department of Mechanical Science and Engineering, Nagoya University,*
*Chikusa-ku, Nagoya 464-8603, JAPAN,*
*ohno@mech.nagoya-u.ac.jp, okumura@mech.nagoya-u.ac.jp*

TOMOYUKI SHIBATA[3]

*Department of Computational Science and Engineering, Nagoya University,*
*Chikusa-ku, Nagoya 464-8603, JAPAN,*
*shibata@mml.mech.nagoya u.ac.jp*

Received 15 June 2008
Revised 23 June 2008

Two types of higher-order stresses work-conjugate to slip gradient in single crystals are investigated to analyze the grain-size dependent yield behavior of polycrystals. The first is the higher-order stress due to the self-energy of geometrically necessary dislocations (GNDs). The second higher-order stress examined is an extension based on the non-recoverable energy that is postulated to be proportional to the accumulated density of incrementally defined GNDs. It is shown that the second higher-order stress is co-directional with the in-plane gradient of *slip-rate* and consequently causes isotropic hardening, whereas the first higher-order stress is co-directional with the in-plane gradient of *slip*. These higher-order stresses are incorporated into a strain gradient plasticity theory of single crystals. Subsequently, using a finite element method, 2D model polycrystals are analyzed to demonstrate the influence of the two types of higher-order stresses on the grain-size dependent yield behavior under loading, unloading and reverse loading.

*Keywords*: Polycrystals; yield behavior; grain-size dependence; geometrically necessary dislocations.

## 1. Introduction

To analyze the grain-size dependence of initial yield stress of polycrystals, Ohno and Okumura[1] considered the self-energy of geometrically necessary dislocations (GNDs). Firstly, they showed that the self-energy of GNDs provides one type of higher-order stress work-conjugate to slip gradient in single crystals; this higher-order stress changes stepwise as a function of slip gradient on each slip system. Secondly, by incorporating the higher-order stress into the single-crystal strain gradient plasticity theory of Gurtin[2], they analyzed two-dimensional (2D) and three-dimensional (3D) model crystal grains to obtain a closed-form evaluation of the initial yield stress of polycrystals. Subsequently, using published experimental data, it was demonstrated that the self-energy of GNDs

---

[†]Corresponding Author.

successfully explains the grain-size dependence of initial yield stress in the submicron to several-micron range.

The self-energy of GNDs is induced if slip gradient occurs under loading. The self-energy of GNDs is completely recovered, when the slip gradient returns to zero under reverse loading. The self-energy, however, cannot necessarily be recovered if reverse loading induces opposite-sign dislocations to form statically stored dislocations. Consequently, it is worthwhile to extend the theory so that the self-energy of GNDs cannot be recovered under reverse loading.

In this study, two types of higher-order stresses are considered to analyze the grain-size dependent yield behavior of polycrystals. The first is the higher-order stress derived previously using the self-energy of GNDs.[1] The second higher-order stress investigated is an extension based on the non-recoverable energy that is postulated to be proportional to the accumulated density of incrementally defined GNDs. Using these two types of higher-order stresses, 2D model polycrystals are analyzed to demonstrate their influence on the grain-size dependent yield behavior under loading, unloading and reverse loading.

## 2. Theory

In this section, single crystals at small strains will be considered. Cartesian coordinates $x_i$ ($i = 1, 2, 3$) will be used for the index notation of vectors and tensors; $(\ )_{,i}$ and $(\dot{\ })$ will indicate differentiation with respect to $x_i$ and time $t$, respectively, and the summation convention will be employed.

### 2.1.  *Higher-order stress due to the self-energy of GNDs*

The density of GNDs on slip system $\beta$ is expressed as[3,4]

$$\rho_{\mathrm{G}}^{(\beta)} = b^{-1} \left( \gamma_{,i}^{(\beta)} \gamma_{,i}^{(\beta)} \right)^{1/2},$$   (1)

where $b$ denotes the magnitude of the Burgers vector, $\gamma^{(\beta)}$ the slip on slip system $\beta$, and $\gamma_{,i}^{(\beta)}$ the gradient of $\gamma^{(\beta)}$ on this slip plane. From here on, $\gamma_{,i}^{(\beta)}$ will be referred to as the in-plane gradient of $\gamma^{(\beta)}$.

A single dislocation in an infinitely large single crystal has the following self-energy per unit length, if the crystal is assumed to be isotropic:[5]

$$E_0 = a\mu b^2,$$   (2)

where $a$ is a coefficient, and $\mu$ indicates the elastic shear modulus. Therefore, except for the interaction among GNDs, the strain energy density $\psi^{(\beta)}$ of GNDs on slip system $\beta$ is proportional to $\rho_{\mathrm{G}}^{(\beta)}$, i.e.,

$$\psi^{(\beta)} = E_0 \rho_{\mathrm{G}}^{(\beta)}.$$   (3)

Using Eqs. (1) and (3), the increment of $\psi^{(\beta)}$ is expressed as follows:[1]

$$\mathrm{d}\psi^{(\beta)} = \xi_i^{(\beta)} \mathrm{d}\gamma_{,i}^{(\beta)} = \xi_i^{(\beta)} \mathrm{d}\gamma_{,i}^{(\beta)},$$   (4)

where

$$\xi_i^{(\beta)} = b^{-1}E_0 v_i^{(\beta)}, \tag{5}$$

$$v_i^{(\beta)} = \gamma_{,\_i}^{(\beta)} / \left( \gamma_{,\_k}^{(\beta)} \gamma_{,\_k}^{(\beta)} \right)^{1/2}. \tag{6}$$

Here, $v_i^{(\beta)}$ is the unit vector indicating the direction of in-plane slip gradient $\gamma_{,\_i}^{(\beta)}$. Eq. (4) allows $\xi_i^{(\beta)}$ to be interpreted as a higher-order stress work-conjugate to slip gradient $\gamma_{,i}^{(\beta)}$. As shown in Eqs. (5) and (6), $\xi_i^{(\beta)}$ is co-directional with $\gamma_{,\_i}^{(\beta)}$ and has a magnitude equal to $b^{-1}E_0$ if $\gamma_{,\_i}^{(\beta)} \neq 0$. Consequently, $\xi_i^{(\beta)}$ changes stepwise as a function of $\gamma_{,\_i}^{(\beta)}$ (Fig. 1(a)).

## 2.2. *Higher-order stress based on incrementally defined GNDs*

We can also consider the following energy density $\bar{\psi}^{(\beta)}$ as an extension of the self-energy density $\psi^{(\beta)}$ described in the preceding subsection:

$$\bar{\psi}^{(\beta)} = E_0 \bar{\rho}_G^{(\beta)}, \tag{7}$$

where $\bar{\rho}_G^{(\beta)}$ indicates the accumulated density of incrementally defined GNDs, i.e.,

$$\bar{\rho}_G^{(\beta)} = b^{-1} \int_0^t \left( \dot{\gamma}_{,\_i}^{(\beta)} \dot{\gamma}_{,\_i}^{(\beta)} \right)^{1/2} dt. \tag{8}$$

Because $\bar{\psi}^{(\beta)}$ is proportional to $\bar{\rho}_G^{(\beta)}$ and $\bar{\rho}_G^{(\beta)}$ monotonically increases, $\bar{\psi}^{(\beta)}$ does not decrease. Therefore, $\bar{\psi}^{(\beta)}$ is *non-recoverable in contrast to* $\psi^{(\beta)}$.

Differentiating Eq. (7) with respect to time $t$ gives

$$\dot{\bar{\psi}}^{(\beta)} = b^{-1}E_0 \left( \dot{\gamma}_{,\_i}^{(\beta)} \dot{\gamma}_{,\_i}^{(\beta)} \right)^{1/2}. \tag{9}$$

This equation is rewritten as

$$d\bar{\psi}^{(\beta)} = \bar{\xi}_i^{(\beta)} d\gamma_{,\_i}^{(\beta)} = \bar{\xi}_i^{(\beta)} d\gamma_{,i}^{(\beta)}, \tag{10}$$

where

$$\bar{\xi}_i^{(\beta)} = b^{-1}E_0 \bar{v}_i^{(\beta)}, \tag{11}$$

$$\bar{v}_i^{(\beta)} = \dot{\gamma}_{,\_i}^{(\beta)} / \left( \dot{\gamma}_{,\_k}^{(\beta)} \dot{\gamma}_{,\_k}^{(\beta)} \right)^{1/2}. \tag{12}$$

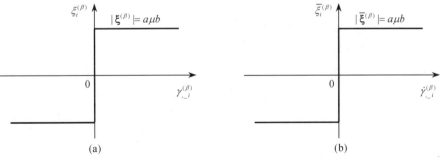

Fig. 1. Variation in higher-order stress; (a) $\xi_i^{(\beta)}$ versus $\gamma_{,\_i}^{(\beta)}$ relation, and (b) $\bar{\xi}_i^{(\beta)}$ versus $\gamma_{,\_i}^{(\beta)}$ relation.

As seen from Eqs. (10) – (12), $\overline{\xi}_i^{(\beta)}$ is also a higher-order stress work-conjugate to $\gamma_{,i}^{(\beta)}$, and $\overline{\xi}_i^{(\beta)}$ is co-directional with the in-plane gradient of $\dot{\gamma}^{(\beta)}$ and changes stepwise as a function of $\dot{\gamma}_{,i}^{(\beta)}$ (Fig. 1(b)). This co-directionality means that $\overline{\xi}_i^{(\beta)}$ causes isotropic hardening. The co-directionality of higher-order stress and in-plane slip-rate gradient was empirically assumed to introduce dissipative higher-order stress by Gurtin and others[6].

Let us consider monotonic proportional loading followed by unloading and reverse loading. Under monotonic proportional loading, $\overline{\psi}^{(\beta)}$ increases identically to $\psi^{(\beta)}$, and consequently $\overline{\xi}_i^{(\beta)}$ is identified with $\xi_i^{(\beta)}$. Under unloading and reverse loading, on the other hand, $\overline{\psi}^{(\beta)}$ continues to increase, whereas $\psi^{(\beta)}$ decreases as $\gamma_{,i}^{(\beta)}$ returns to zero under reverse loading. Therefore, $\overline{\xi}_i^{(\beta)}$ can be considerably different from $\xi_i^{(\beta)}$ under unloading and reverse loading.

## 3. Analysis of 2D Model Polycrystals

A 2D model periodic volume element of polycrystals was analyzed by incorporating the two types of higher-order stresses described in Section 2 into the single-crystal plasticity theory of Gurtin[2]. The periodic element consisted of 16 grains, to which double slip systems were randomly assigned (Fig. 2(a)). The periodic element was subjected to uniaxial tension followed by unloading and reverse loading in the $x_2$-direction. The periodicity of perturbed displacement was imposed at the boundary of the periodic element.[7,8] The grains were hexagonal and had an identical side length equal to $L/2$. The microclamped condition of slip, $\gamma^{(\beta)} = 0$, was imposed at all grain boundaries. Each grain was divided into finite elements using eight-node quadratic elements (Fig. 2(b)). Elastic isotropy and the following slip-rate equation were assumed:

$$\dot{\gamma}^{(\beta)} = \dot{\gamma}_0 \frac{k^{(\beta)}}{k_0} \left| \frac{k^{(\beta)}}{k_0} \right|^{(1/n)-1}, \tag{13}$$

where $\dot{\gamma}_0$, $k_0$ and $n$ are material parameters, and $k^{(\beta)}$ is the slip-rate driving stress resulting from the microforce balance equation of Gurtin[2]. The material parameters listed

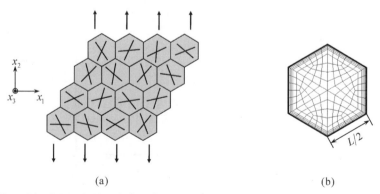

|       | (a)  |       | (b)  |
|-------|------|-------|------|

Fig. 2.  2D model polycrystal; (a) periodic volume element consisting of 16 grains with double slip-systems, and (b) finite element mesh of each grain (234 elements, 739 nodes).

Table 1. Material parameters.

| Elastic shear modulus, $\mu$ | 45.4 [GPa] | Reference slip-resistance, $k_0$ | 60 [MPa] |
|---|---|---|---|
| Poisson's ratio, $v$ | 0.33 | Reference slip-rate, $\dot{\gamma}_0$ | $10^{-3}$ [s$^{-1}$] |
| Magnitude of Burgers vector, $b$ | $2.5 \times 10^{-10}$ [m] | Slip-rate sensitivity, $n$ | 0.05 |
| Coefficient in Eq.(2), $a$ | 1.0 | | |

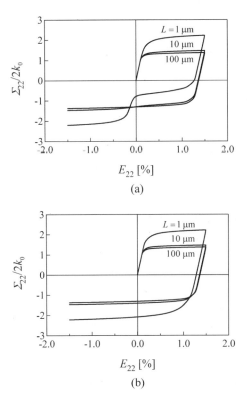

Fig. 3. Macroscopic stress versus strain relations at $\dot{E}_{22} = 0.1$ %/s; (a) based on $\xi_i^{(\beta)}$ expressed as Eq. (5), and (b) based on $\bar{\xi}_i^{(\beta)}$ expressed as Eq. (11).

in Table 1 were employed to assume a copper for the grains. The analysis was performed using the finite element method developed by Okumura and others[9].

Figs. 3(a) and 3(b) show the macroscopic stress versus strain relations obtained using the two types of higher-order stresses in the analysis. In the figures, the macroscopic stress in the loading direction, $\Sigma_{22}$, is non-dimensionalized with $2k_0$ and plotted against the macroscopic strain in the loading direction, $E_{22}$. The following features are evident from the figures. (1) The two types of higher-order stresses provided the same results under monotonic tensile loading (see also Figs. 4(a) and 4(b)). (2) The higher-order stress based on the self-energy of GNDs gave a sigmoid $\Sigma_{22}$ versus $E_{22}$ relation under reverse loading (i.e., an unusual Bauschinger effect), when $L = 1$ μm (Fig. 3(a)). (3) In contrast,

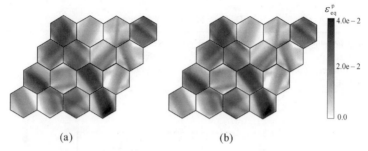

(a)                                    (b)

Fig. 4. Distribution of Mises equivalent plastic strain $\varepsilon_{eq}^{p}$ just before unloading ($L = 1$ μm, $\dot{E}_{22} = 0.1$ %/s); (a) based on $\xi_i^{(\beta)}$ expressed as Eq. (5), and (b) based on $\bar{\xi}_i^{(\beta)}$ expressed as Eq. (11).

the higher-order stress based on the accumulated density of incrementally defined GNDs did not cause such an unusual Bauschinger effect (Fig. 3(b)).

## 4. Conclusions

Two types of higher-order stresses were considered to analyze the grain-size dependent yield behavior of polycrystals. The first was the higher-order stress derived previously using the self-energy of GNDs. The second higher-order stress examined was an extension based on the non-recoverable energy that was postulated in this study to be proportional to the accumulated density of incrementally defined GNDs. By analyzing 2D model polycrystals, it was demonstrated that the extended higher-order stress does not provide any unusual Bauschinger effect in contrast to the higher-order stress based on the self-energy of GNDs even when the grain size is very small. The extended higher-order stress can therefore be appropriate under unloading and reverse loading, though the two types of higher-order stresses are identical under monotonic loading.

## Acknowledgments

This study was supported in part by the Japan Society for the Promotion of Sciences under a Grant-in-Aid for Scientific Research B (No. 19360048).

## References

1. N. Ohno and D. Okumura, *J. Mech. Phys. Solids* **55**, 1879 (2007).
2. M. E. Gurtin, *J. Mech. Phys. Solids* **50**, 5 (2002).
3. M. F. Ashby, *Philos. Mag.* **21**, 399 (1970).
4. N. A. Fleck, G. M. Muller, M. F. Ashby and J. W. Hutchinson, *Acta Metall. Mater.* **42**, 475 (1994).
5. J. P. Hirth and J. Lothe, *Theory of Dislocations, 2nd ed.* (Wiley, New York, 1982).
6. M. E. Gurtin, L. Anand and S. P. Lele, *J. Mech. Phys. Solids* **55**, 1853 (2007).
7. N. Ohno, D. Okumura and H. Noguchi, *J. Mech. Phys. Solids* **50**, 1125 (2002).
8. D. Okumura, N. Ohno and H. Noguchi, *J. Mech. Phys. Solids* **52**, 641 (2004).
9. D. Okumura, Y. Higashi, K. Sumida and N. Ohno, *Int. J. Plasticity* **23**, 1148 (2007).

# FINITE ELEMENT MODELLING OF MICRO-CUTTING PROCESSES FROM CRYSTAL PLASTICITY

Y. P. CHEN[1], W. B. LEE[2†], S. TO[3], H. WANG[4]

*Department of Industrial and Systems Engineering, The Hong Kong Polytechnic University,*
*Hung Hom, Kowloon, Hong Kong, P. R. CHINA,*
*mfypchen@inet.polyU.edu.hk, mfwblee@inet.polyU.edu.hk, Sandy.To@inet.polyU.edu.hk,*
*solidvc@gmail.com*

Received 15 June 2008
Revised 23 June 2008

In ultra-precision machining (UPM), the depth of cut is within an extremely small fraction of the average grain size of the substrate materials to be cut. Polycrystalline materials commonly treated as homogeneous in conventional machining have to be considered as heterogeneous. The cutting force, one of the dominant factors influencing the integrity of the machined surface in UPM, is observed to strongly depend on the grain orientations. To accurately capture the intrinsic features and gain insight into the mechanisms of UPM of single crystals, the crystal plasticity constitutive model has been incorporated into the commercial FE software Marc by coding the user material subroutine Hypela2 available within it. The enhanced capability of the FE software will be adopted to simulate factors influencing the micro-cutting processes, such as grain orientation variation, the tool edge radius and the rake angle. The simulation results will provide useful information for the optimization of critical processing parameters and enhancement of quality of machined products.

*Keywords*: Ultra-precision machining (UPM); user subroutine Hypela2; crystal plasticity.

## 1. Introduction

To meet the growing demand for the miniaturization of products and three-dimensional features with dimensions ranging from less than a micron to a few tens of microns in a wide range of materials, mechanical micro-machining methods including ultra-precision machining are commonly used.[1] The characteristic of these processes is that a very small amount of material is removed, in the order of a few microns. At these scales of length in the removal of material, the features of material microstructure, such as the size and crystallographic orientation of grain of the substrate material being cut will exert an appreciable influence on the micro-machining process and the mechanical behaviors of the work-piece. As the depth of cut in ultra-precision cutting is usually less than the average grain size of a polycrystalline aggregate, cutting is generally performed within a grain. The polycrystalline work-piece material, which may be considered to be an isotropic and homogeneous continuum in the conventional analysis, must be treated as a series of single crystals which possess random or preferred orientations. Therefore the

---

†Corresponding Author.

machined surface quality and the mechanism of chip formation depend very much on the crystallographic orientation of the material being cut. The effect of crystallographic orientation on the surface quality cannot be avoided in the cutting of polycrystalline materials. More investigation on this topic is needed to minimize the crystallographic origin of variations in cutting force. Numerical simulations of machining processes by FEM were initiated in 1970s. The updated Lagrangian formulation was adopted to simulate the chip formation from the incipient stage to the steady stage of orthogonal cutting.[2] Huang et al.[3] made an analysis of different chip separation criteria for the FE simulation of cutting. The constitutive model adopted in FE simulation of machining is decisive in influencing the modeling results. However, the review[4] on the related research over the last decade shows that most of them employ phenomenological plasticity model and no account has been taken of the micro-structural features of the work-piece.

Recent years witness a continued interest in the research and application of continuum crystal plasticity constitutive model on account of its potential to delve into the mechanical response from the micro-structural features, such as the orientation of grain, which constitutes the main advantage over the phenomenological counterpart in applications. Thus, the present investigation will apply the crystal plasticity to the simulation of micro-machining process with the focus on the effect of grain orientation on the variation of cutting force.

## 2. Crystal Plasticity

The present constitutive time integration algorithm is based on that proposed by Zhang et al.[5] with some modifications in implementation details.

To take into account the effect due to rigid body rotation in finite deformation, the resultant stress is updated as follows:

$$^{t+\Delta t}\sigma = \Delta R \cdot {}^t\sigma \cdot \Delta R^T + \Delta\sigma \tag{1}$$

where $\sigma$ is Cauchy stress. The first term on the RHS of Eq.(1) represents the stress increment due to the rigid body rotation $\Delta R$ from $t$ to $t+\Delta t$ and the second term is solely determined by the constitutive laws. We establish two coordinate systems: One is the lattice coordinate system and the other the global fixed one. The orientations of a single crystal at time $t=0$, $t$ and $t+\Delta t$ are described by the rotation matrices ${}^0T^{crys}$, ${}^tT^{crys}$ and ${}^{t+\Delta t}T^{crys}$, respectively. They admit the following relationship,

$$^{t+\Delta t}T^{crys} = \Delta R^e \cdot {}^tT^{crys} = {}^{t+\Delta t}R^e \cdot {}^0T^{crys} \tag{2}$$

For simplicity, ${}^{t+\Delta t}T^{crys}$ will be denoted as $T^{crys}$ hereafter.    The assumption of small elastic deformation yields the following relation expressed with respect to the fixed global coordinate system,

$$\Delta\sigma_{ij} = {}^4C_{Gijkl}\,\Delta\varepsilon^e_{kl} \tag{3}$$

where ${}^4C_{Gijkl}$ is the 4th-order elastic modulus and $\Delta\varepsilon^e_{kl}$ the elastic logarithmic strain increment. The additive decomposition of strain increment is adopted as follows;

$$\Delta\varepsilon = \Delta\varepsilon^e + \Delta\varepsilon^p \tag{4}$$

The flow law governing the plastic strain increment of a single crystal is

$$\Delta\varepsilon^p = \sum_{\alpha=1}^{N} P^{(\alpha)*} \Delta\gamma^{(\alpha)} \tag{5}$$

where Schmid tensor is

$$P^{(\alpha)*} = (1/2)(m^{(\alpha)*} \otimes s^{(\alpha)*} + s^{(\alpha)*} \otimes m^{(\alpha)*}) \tag{6}$$

The two vectors $m^{(\alpha)*}$ and $s^{(\alpha)*}$ are expressed in the updated crystal orientation by means of the transformation matrix $T^{crys}$ and $\Delta\gamma^{(\alpha)}$ is the shear strain increment of the slip system $\alpha$. The power law for describing slip system hardening is employed,

$$\dot{\gamma}^{(\alpha)} = \dot{\gamma}_0 \mathrm{sgn}(\tau^{(\alpha)}) \left| \tau^{(\alpha)} / g^{(\alpha)} \right|^{1/m} \tag{7}$$

where $\dot{\gamma}_0$ is the reference shear rate, $g^{(\alpha)}$ the function characterizing the strain hardening state and $m$ the rate sensitivity. The resolved shear stress on the slip system $\alpha$, $\tau^{(\alpha)}$, is obtained by the Schmid law:

$$\tau^{(\alpha)} = P^{(\alpha)*} : \sigma \tag{8}$$

The evolution of $g^{(\alpha)}$ is described as
$$\dot{g}^{(\alpha)} = \sum_{\beta=1}^{N} h_{\alpha\beta} \left| \dot{\gamma}^{(\beta)} \right| \tag{9}$$

where $h_{\alpha\beta}$ are the hardening moduli of the following form

$$h_{\alpha\beta} = h(\gamma)[q + (1-q)\delta_{\alpha\beta}] \tag{10}$$

where $q$ is a constant ranging from 1 to 1.4 and

$$h(\gamma) = h_0 \mathrm{sech}^2 \left( h_0 \gamma / (\tau_s - \tau_0) \right) \tag{11}$$

where $h_0$ is the initial hardening rate, $\tau_0$ and $\tau_s$ are, respectively, the shear stresses at the onset of yield and the saturation of hardening and $\gamma$ the accumulative shear strain over all the slip systems. The increment of shear $\Delta\gamma^{(\alpha)}$ is calculated as follows:

$$\Delta\gamma^{(\alpha)} = [(1-\eta)^{t}\dot{\gamma}^{(\alpha)} + \eta^{t+\Delta t}\dot{\gamma}^{(\alpha)}]\Delta t \tag{12}$$

where $\eta$ is a control parameter. The substitution of Eq. (12) into Eq. (5) yields

$$\sum_{\alpha=1}^{N} P^{(\alpha)*} \Delta\gamma^{(\alpha)} = (1-\eta)\Delta t \left( \sum_{\alpha=1}^{N} P^{(\alpha)*} {}^{t}\dot{\gamma}^{(\alpha)} \right) + \eta\Delta t \left( \sum_{\alpha=1}^{N} P^{(\alpha)*} {}^{t+\Delta t}\dot{\gamma}^{(\alpha)} \right) \tag{13}$$

By virtue of Eq. (8), we have

$$\sum_{\alpha=1}^{N} P^{(\alpha)*} {}^{t+\Delta t}\dot{\gamma}^{(\alpha)} = \sum_{\alpha=1}^{N} P^{(\alpha)*} \dot{\gamma}_0 \mathrm{sgn}({}^{t+\Delta t}\tau^{(\alpha)}) \left( \frac{\left| P^{(\alpha)*} : {}^{t+\Delta t}\sigma \right|}{{}^{t+\Delta t}g^{(\alpha)}} \right)^{1/m} \tag{14}$$

On combining Eqs. (1, 3, 4, 5, 14), we have the following set of nonlinear equations with Cauchy stresses as unknowns,

$$^{t+\Delta t}\sigma = r - H(^{t+\Delta t}\sigma) \tag{15}$$

where

$$r = \Delta R \cdot {}^{t}\sigma \cdot \Delta R^{T} + \left\{ {}^{4}C_{G} : \left( \Delta \varepsilon - (1-\eta)\Delta t \sum_{\alpha=1}^{N} P^{(\alpha)*}\, {}^{t}\dot{\gamma}^{(\alpha)} \right) \right\} \tag{16}$$

$$H(^{t+\Delta t}\sigma) = \eta \Delta t \dot{\gamma}_0 \left( {}^{4}C_{G} : H_{1}(^{t+\Delta t}\sigma) \right) \tag{17}$$

$$H_{1}(^{t+\Delta t}\sigma) = \sum_{\alpha=1}^{N} P^{(\alpha)*}\, \dot{\gamma}_0 \mathrm{sgn}(^{t+\Delta t}\tau^{(\alpha)}) \left( \frac{|P^{(\alpha)*} : {}^{t+\Delta t}\sigma|}{^{t+\Delta t}g^{(\alpha)}} \right)^{1/m} \tag{18}$$

The predictor-corrector procedure is employed for solving Eq. (15). The predicted values of $^{t+\Delta t}\sigma$ at time $t+\Delta t$ is obtained by substituting $^{t}P^{(\alpha)*}$ and $^{t}g^{(\alpha)}$ into Eqs.(15-18) and using Newton-Raphson iteration scheme. Then the values of $^{t+\Delta t}g^{(\alpha)}$ and $^{t+\Delta t}P^{(\alpha)*}$ are obtained by updating them from the previous values, with the update of the latter essentially depending on the reorientation of single crystal, i.e. the calculation of $T^{crys}$ described in the sequel.

The spin tensor $W$ is additively decomposed as the sum of the plastic spin $W^{p}$ and the lattice spin $W^{*}$, which governs the rate of change of crystallographic lattice axes $c$ and can be expressed as:

$$W^{*} = W - W^{p} = W - \sum_{\alpha=1}^{N} W^{(\alpha)*}\, \dot{\gamma}^{(\alpha)} \tag{19}$$

where

$$W^{(\alpha)*} = (1/2)(m^{(\alpha)*} \otimes s^{(\alpha)*} - s^{(\alpha)*} \otimes m^{(\alpha)*}) \tag{20}$$

We have the following evolution equation for lattice axes $c$

$$c(t+\Delta t) = \exp(W^{*}\Delta t) \cdot c(t) \tag{21}$$

By employing the Cayley-Hamilton theorem and denoting $W^{*}\Delta t$ as $\breve{W}$, the expansion of $\exp(\breve{W})$ is obtained as follows:

$$\exp(\breve{W}) = I + \frac{\sin\omega}{\omega}\breve{W} + \frac{1-\cos\omega}{\omega^2}\breve{W}^{2} \tag{22}$$

where $I$ is the second-order identity tensor and $\omega^2 = -(1/2)\mathrm{tr}(\breve{W}^{2})$. Having updated the lattice spin, we can obtain the reorientation of crystal grain represented by matrix $T^{crys}$.

## 3. Finite Element Implementation and Simulation Results

The crystal plasticity constitutive formulation presented above is implemented into the commercial FE code Marc2007r1 by employing the user material subroutine Hypela2. Following the general FE implementation procedures involving crystal plasticity constitutive integration, we have to choose a fixed sample coordinate system or a local one attached to grain lattice, with respect to which to make the constitutive integration calculation. In the present investigation, the former is preferred. Thus we have to transform all quantities expressed in the local coordinate system into the fixed one, such as grain elastic modulus, Schmid direction tensors. At the beginning of each time increment, the global deformation gradient, rotation tensor, the global strain increment, the magnitude of time step, the Cauchy stress and the solution-dependent state variables are passed into the subroutine Hypela2 from the Marc main program to solve Cauchy stress tensor at the end of the current time step by Eq. (15), followed then by updating the solution-dependent state variables and Cauchy stress including the grain orientation matrix. In simulating machining processes by finite element, the severe plastic deformation ahead of the tool tip will result in the undue distortion of the FE mesh and eventually the breakdown of computation unless the remeshing technique is invoked. On account of such a requirement and also the avoidance of adopting the chip separation criterion, we utilize the capability of global remeshing available within Marc computation platform in preference to Abaqus.

The work-piece used is the copper single crystal with dimension as $10*3*1$ mm$^3$. The crystal anisotropic elastic constants and constitutive parameters are as follows: $C_{11} = 166.1$ GPa, $C_{12} = 111.9$ GPa, $C_{44} = 75.6$ GPa, $\tau_0 = 60.84$ MPa, $\tau_s = 109.51$ MPa, $h_0 = 541.5$ MPa, $\dot{\gamma}_0 = 0.001$ s$^{-1}$, $q=1.2$, $m=0.01$. We use the tetrahedral element to mesh the work-piece.

Shown in Fig.1 are the cutting and thrust forces with cutting performed on different crystallographic planes and in different directions by using cutting tool of round edge and of zero rake and clearance angles. Figure 2 shows the distribution of equivalent Cauchy stress for the case of (001)<100> cutting.

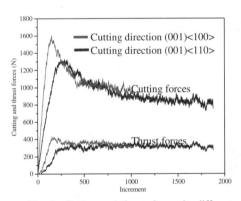

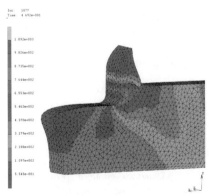

Fig. 1. Cutting and thrust forces in different cutting plane and directions for the cases of zero rake and clearance angles, respectively.

Fig. 2. Distribution of equivalent Cauchy stress for the case of (001)<100> cutting.

Compared with Fig.1, the cutting and thrust forces shown in Fig.3 are obtained by only changing the edge radius of tool, i.e., from a round to an acute edge, and also utilizing an acute rake angle. Figure 4 presents the distribution of Cauchy stress component $\sigma_{11}$ in the case of (001)<100> cutting. An appreciable difference in cutting and thrust forces is observed in Figs.1 and 3 when performing cuttings on the same (001) plane but different crystallographic directions of <110> and <100>, respectively.

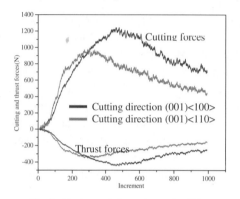

Fig. 3. Cutting and thrust forces in different cutting plane and directions for the cases of acute rake angle, zero clearance angles and acute edge, respectively.

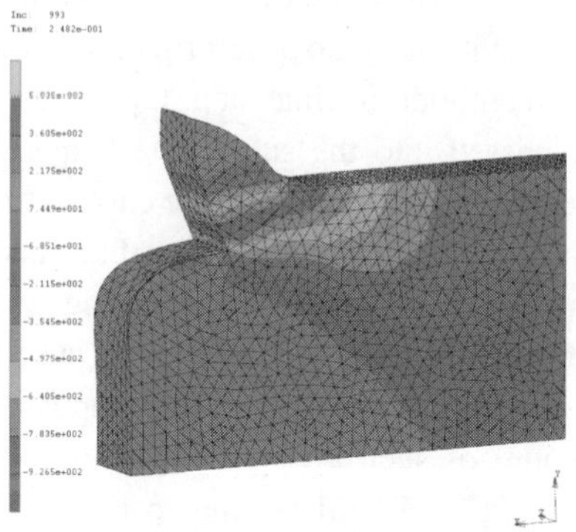

Fig. 4. Distribution of Cauchy stress component 11 in the case of (001)<100> cutting.

## 4. Conlcusion

Based on the simulation results obtained above, conclusions can be drawn as follows: In addition to the extrinsic geometric parameters of tool affecting the cutting forces, the intrinsic feature of grain orientation plays an essential role in determining the cutting forces and eventually the integrity of the machined surface in UPM. The crystal plasticity with its inherent advantages over the phenomenological plasticity model can find a wide application in simulation of the ultra-precision machining processes.

## Acknowledgment

The authors would like to express their sincere thanks to the Research Committee of The Hong Kong SAR of the People's Republic of China for financial support of the research work under the project nos. PolyU 5213/06E and PolyU 5207/07E.

## References

1.    T. Masuzawa,  State of art of micromachining, *Annals CIRP*, 49(2), 2000, 473-488.
2.    J. S. Strenkowski and J. T. Carroll, A finite element model of orthogonal metal cutting, *ASME J. Eng. Ind.* 127(1990), 313-318.
3.    J. M. Huang, and J. T. Black, An evaluation of chip separation criteria for the FEM simulation of machining, *ASME J. Manufact. Sci. Eng.* 118 (1996), 118-545.
4.    J. Mackerle, Finite element analysis and simulation of machining: an addendum: A bibliography (1996–2002), *Int. J. of Machine Tools & Manufacture* 43, 2003, 103–114.
5.    K. S. Zhang, *et al.*, Simulation of microplasticity-induced deformation in uniaxially strained ceramics by 3-D Voronoi polycrystal modelling, *Int. J. of Plas.* 21 (2005), 801-834.

# FINITE ELEMENT ANALYSIS OF THERMAL NANOINDENTATION PROCESS AND ITS EXPERIMENTAL VERIFICATION

HYUN-JUN OH[1], EUN-KYUNG LEE[2]

*National Core Research Center for Hybrid Materials Solution, Pusan National University,*
*Geum-jeong Gu, Busan, 609-735, KOREA,*
*ohjtg1@nate.com, momohime@pusan.ac.kr*

CHUNG-GIL KANG[3†]

*Mechanical engineering department, Pusan National University,*
*Geum-jeong Gu, Busan, 609-735, KOREA,*
*cgkang@pusan.ac.kr*

SANG-MAE LEE[4]

*ERC/NSDM, Pusan National University,*
*Geum-jeong Gu, Busan, 609-735, KOREA,*
*smlee@pnu.edu*

Received 15 June 2008
Revised 23 June 2008

In this paper, deformation behavior of Polymethylmethacrylate (PMMA) during thermal indentation was demonstrated by the finite element method using ABAQUS S/W. Forming conditions to reduce the elastic recovery and pile-up were proposed. Thermal nanoindentation experiments were carried out at the temperature range of 110 ~ 150 °C. The indenter was modeled as a rigid surface. The finite element analysis (FEA) approach is capable of reproducing the loading-unloading behavior for a thermal nanoindentation test and thus comparison between the experimental data and numerical results were demonstrated. The result of the investigation will be applied to the fabrication of the hyper-fine pattern.

*Keywords:* Thermal nanoindentation; Polymethylmethacrylate (PMMA); Hyper-fine pattern; Pile-up; Elastic recovery.

## 1. Introduction

Various nano-probe based lithography technology has been developing to substitute and/or complement optical lithography technology. Nanoindentation process has been widely used because of the advantages that it is simple and low cost when facilitating and operating the initial equipment and that selection of material is flexible.[1-3] On the other hand, the nanoindentation process possesses drawbacks such as wear of the tip, slow

---

†Corresponding Author.

forming velocity, and low efficiency of processing, etc. To improve these drawbacks, high speed patterning,[4] pattering using multi tips,[5] and mold patterned by using a nano probe in application for mass production[6] have been studied.

Oliver et al.[7] have ever conducted the research about material behavior of thermal deformation of PVAc. Li et al. not only measured mechanical properties of silver nanowire but also applied to cutting and manufacturing, by using the nanoindenter. Besides, they reported that nanoindenter can be used in easily cutting and manufacturing nano metallic material. They also have prospected that nanoindented profile can be used in fabricating the drug delivery cell, the slot to be interfaced with a nanodevice, and the data bit for nanoscale storage media.[8] Polymethylmetacrylate (PMMA), which is thermoplastic and acrylic polymer, has been widely used in light waveguide for LCD (liquid crystal display), contact lens, dental resin, DVD disc material, workpiece for nanoimprint, and resist for nanolithography. Polymeric material has excellent biocompatibility and optical characteristics. In order to fabricate high quality polymeric components through nanoimprint lithorgraphy (NIL) process, it is important to measure quantitative properties of high temperature deformation and construct their data base (D/B) as applicable devices go down to micro/nanoscale.

This study demonstrates high temperature nanoindentation process of polymeric material to obtain variation in hardness, elastic modulus, and indentation load through the continuous stiffness method (CSM) using a nanoindenter and hot stage. The effect of temperature variation on mechanical properties of polymer at the maximum indentation depth of 2000 nm is investigated. The finite element analysis (FEA) is performed to simulate nanoindentation process using ABAQUS 6.5-1 software. Using the result for the nanoindentation at elevated temperature obtained from the computer simulation, the stress-strain distribution and occurrence in the pile up, when unloading, are predicted. The result of this investigation will be able to be applied in fabricating the hyper-fine pattern using the high temperature nanoscratch.

## 2.  FEM Modeling and Experiment

### 2.1.  *Experiment*

Nanoindenter XP (MTS, USA) was used in nanoindentation experiment. The Berkovich diamond nanoindenter with more than 40 nm of the radius was used. The polymeric

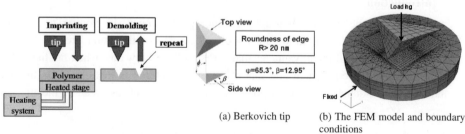

(a) Berkovich tip

(b) The FEM model and boundary conditions

Fig. 1.   Layout of the thermal nano-indentation process on polymer surface for hyperfine pit fabrication.

Fig. 2.   Schematic geometry of Berkovich tip and FEA model.

materials used in this study are 2 mm thick, 15 mm wide, and 15 mm long PMMA plate. Surface roughness of PMMA is 14.4 nm. Figure 1 shows schematic of nanoindentation process at elevated temperature. By repeating molding (imprinting) and demolding through nanoindentation of polymer placed on the hot stage heated by heating system, thermal properties of polymer are obtained while polymer is patterned.

In order to evaluate mechanical properties of PMMA during thermal nanoindentation, nine times nanoindentations were conducted by using CSM at above glass transition temperature, $T_g=105\,°C$, i.e. $110\,°C$ and $130\,°C$. The CSM is the measurement method of mechanical characteristics of material while indenting the material surface up to the indentation depth designated with frequency of 45 Hz. The data were continuously obtained.

## 2.2. Finite element analysis

The three dimensional finite element analysis (FEA) for the nanoindentation process using the Berkovich type indentation tip was carried out.

Figure 2 (a) shows schematic of the Berkovich tip which was used in modeling by using dimension of the nanoindenter tip (MTS, USA).[9] Figure 2 (b) shows FEA mesh generation of the specimen and nanoindenter and their boundary conditions used in computer simulation of nanoindentation. IGES file including surface information of the meshed Berkovich tip was interfaced with ABQUS 6.5/CAE. The specimen used in computer simulation was modeled in axisymmetric structure which is 20 μm in diameter and 6 μm in height. The 8 node-reduced integration elements (C3D8R element type), for which the node number of is 9680, and the 6 node-linear triangular prism elements (C3D6 element type), for which the node number is 5005, were used in FEA modeling. The contact region between the tip and specimen was meshed by using C3D8R element type and the other region without the contact between the tip and specimen used C3D6 element type. Besides, the region being penetrated by the tip was meshed with small elements and the region away from the center of the specimen, at which analysis may be not much affected, was meshed with large elements. The Berkovich tip was modeled with the 4 node rigid quadrilateral element (R3D4 element type) and their minimum mesh size was about 100 nm. In order to simulate nanoindentation process, the contact condition was used, assuming that the tip was master and the specimen was the slave. The three friction coefficient of μ = 0.3, 0.5 and 1.0 was used in computer simulation to investigate

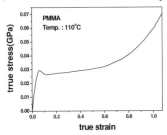

Fig. 3.  True stress – strain curve of PMMA at temperature 110 ℃.

the effect of the friction coefficient on stress and strain distribution and elastic recovery. The elastic-plastic analysis was carried out with the two stages of loading with the maximum penetration of 2000 nm and unloading. Table 1 shows material properties of PMMA at temperature 110℃ used in computer simulation. Elastic modulus and hardness were obtained from thermal nanoindentation conducted in this study to use in computer simulation. Poission ratio and true stress-strain curve were obtained from the literature,[10, 11] as shown in Fig. 3.

Table 1. Material properties of PMMA at temperature 110 ℃.

| Young's modulus (Gpa) | Hardness(Gpa) | Poisson's ratio |
|---|---|---|
| 2.547±0.5 | 0.105±0.05 | 0.40 |

## 3.  Results and Discussion

### 3.1.  *Thermal-nanoindentation test by CSM method*

Figure 4 shows elastic modulus curves obtained by CSM nanoindentation at 110℃ and 130℃. The irregular curves below 300 nm of indention depth show ISE (indentation size effect)[12] resulting from the elastic contact between the end of the tip and the specimen surface because the tip is not ideally keen but rounded. Accordingly, the average value was taken over the penetration depth region of 500 - 1500 nm. Figure 5 shows indentation hardness vs. indentation depth at 110℃ and 130℃ obtained by the CSM. As well as elastic modulus, hardness curves show the ISE phenomenon and thus the average value was taken over the penetration depth region of 500 - 1500 nm. It was shown from the computer simulation that the elastic modulus and hardness decreased as temperature

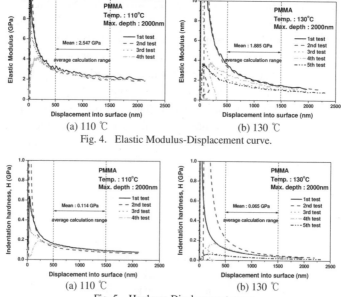

(a) 110 ℃          (b) 130 ℃

Fig. 4.  Elastic Modulus-Displacement curve.

(a) 110 ℃          (b) 130 ℃

Fig. 5.  Hardness-Displacement curve.

increased. It may be due to the fact that polymeric molecules are promoted to be disentangled at elevated temperature and thus molecular movement is free. This causes elastic modulus and harness decreased.

### 3.2. *Comparison with experimental data*

Figure 6 shows Von-mises stress distributions of the specimen when loading and unloading the specimen. As shown in Fig. 6 from stress-strain distribution, elastic recovery is apparent. Figure 7 shows the FEM analysis result of equivalent plastic strain distribution after unloading when indented at the friction coefficient of $\mu = 0.3$ at $110\,°\!C$. Figure 8 shows FEM analysis results for indentation profile of PMMA when indented up to 2000 nm of indentation depth at the friction coefficient of $\mu = 0.3$ at $110\,°\!C$. The elastic recovery and burr (the pile up) were observed at loading-unloading curve. Figure 9 show comparison of load-displacement curves on PMMA with different friction coefficients obtained by the three dimensional FEA and experimental data when indenting up to 1000 nm indentation depth at $110\,°\!C$. The elastic recovery and burr were investigated to predict precision of the nanopatterned structure using load-displacement curves as well as calculating the hardness and elastic modulus. The calculated elastic recovery of about 4% against penetration depth and the experimental elastic recovery of about 33% against penetration depth were obtained. The discrepancy between the calculated and experimental results was caused by assuming that the tip roundness and its surface roughness was zero during computer simulation although the tip was rounded and its surface was rough in reality at the experimental. It also may be due to the fact that work hardening was neglected during simulation. Accordingly, validation of the FEA for

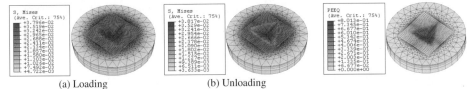

(a) Loading        (b) Unloading

Fig. 6. FEM analysis result of Von-mises stress distribution at $110\,°\!C$ ($\mu = 0.3$).

Fig. 7. FEM analysis result of equivalent plastic strain distribution at $110\,°\!C$ ($\mu = 0.3$).

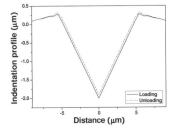

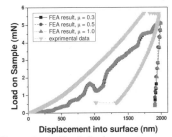

Fig. 8. FEM analysis result for Indentation profile of PMMA (T = $110\,°\!C$, $\mu = 0.3$).

Fig. 9. Comparison of load-displacement curves on PMMA with different friction coefficient obtained by FEA and experimental data.

nanoindentation simulation was verified. Constructing D/B of high temperature material properties through FEA for the thermal nanoindentation process, the elastic recovery results predicted through FEA will be able to be used in nanopattering polymeric structures such as PMMA by using nanoscratch process.

## 4. Conclusion

Demonstrating the deformation behavior of Polymethylmethacrylate (PMMA) during thermal indentation by the finite element method using ABAQUS S/W and thermal nanoindentation experiments at the elevated temperature, the following conclusions were obtained.

(1) It was found from the computer simulation for thermal nanoindentation that the mechanical properties of PMMA such as elastic modulus and hardness tended to decrease as working temperature increased.

(2) Because PMMA is the thermoplastic, mechanical properties decrease as temperature increases. It may be caused by the increase in free volume of polymeric molecules. Therefore, additional experiment is necessary to further investigate thermal behavior according to molecular weight and molecular movement.

Obtaining load-displacement curves through the three dimensional finite element analysis for nanoindentation process, the experimental nanoindentation using the Berkovich tip was validated. For the future study, the FEA considering surface roughness, the initial residual stress, and viscoelasticity of polymeric material is necessary.

## Acknowledgments

This work was supported by the Korean Research Foundation grant (KRF-2006-311-D00309). The authors would like to express their deep gratitude for the financial support of KRF.

## References

1. D. Hardt, B. Ganesan, W. Oi, M. Dirckx, and A. Rzepniewski, "Process Control in Micro-Embossing – A Review", Singapore MIT Alliance Programme (SMA) in IMST, 2004.
2. C. G. Choi, *J. Micromech. Microeng.*, **14**, 945 (2004).
3. W. S. Kim, K. B. Yoon, and B. S. Bae, *J. Mater. Chem.,* **15**, 4535 (2005).
4. K. Ishihara, M. Fujita, I. Matusubara, T. Asano, and S. Noda, *Japanese J. of Applied Physics*, **45**, No. 7, 210 (2006).
5. B. Heidari, I. Maximov, and L. Montelius, *J. Vac. Sci. Technol. B*, **18**, No. 6, 3557-3560 (2000).
6. N. S. Cameron, H. Roberge, T. Veres, S. C. Jakeway and H. J. Crabtree, *Lab Chip*, **6**, 936 (2006).
7. Yoonjoon Choi, Subra Suresh, *Scripta Materialia*, **48**, 249 (2003).
8. X. Li, B. Bhushan, *Surface and Coatings Technology*, **163-164**, 521 (2003).
9. MTS, Nanoindenter XP Manual, 2002.
10. M. Gad-el-Hak, *The MEMS Handbook* (CRC PRESS, 2001)
11. J. Richeton et al., *Int. J. of Solids and Structures*, **44**, 7938, (2007).
12. L. J. Guo, *J. Phys. D.* **37**, 123 (2004).

# Part H
# Phase Transformations

# EFFECT OF TRANSFORMATION VOLUME STRAIN ON THE SPHERICAL INDENTATION OF SHAPE MEMORY ALLOYS

WENYI YAN[1†]

*Department of Mechanical and Aerospace Engineering, Monash University,*
*Clayton, Victoria 3800, AUSTRALIA,*
*wenyi.yan@eng.monash.edu.au*

QINGPING SUN[2]

*Department of Mechanical Engineering, The Hong Kong University of Science and Technology,*
*Hong Kong, P. R. CHINA,*
*meqpsun@ust.hk*

HONG-YUAN LIU[3]

*Centre for Advanced Materials Technology (CAMT), School of Aerospace, Mechanical and Mechatronic*
*Engineering, The University of Sydney, Sydney, NSW 2006, AUSTRALIA,*
*hong-yuan.liu@usyd.edu.au*

Received 15 June 2008
Revised 23 June 2008

The mechanical response of spherical indentation of superelastic shape memory alloys (SMAs) was theoretically studied in this paper. Firstly, the friction effect was examined. It was found that the friction influence is negligibly small. Secondly, the influence of the elasticity of the indenter was investigated. Numerical results indicate that this influence can not be neglected as long as the indentation depth is not very small. After that, this paper focused on the effect of transformation volume contraction. Our results show that the transformation volume contraction due to forward martensitic transformation can reduce the maximum indentation force and the spherical indentation hardness. These research results enhance our understanding of the spherical indentation responses, including the hardness of the smart material SMAs.

*Keywords*: Shape memory alloys; spherical indentation; hardness; friction.

## 1. Introduction

In an indentation test, an indenter is pressed into the surface of a specimen. The curve of the indentation force versus the indentation depth can be recorded, which is the response of the applied load and the specimen's properties. Traditionally, indentation test was applied to evaluate the hardness of materials[1]. With the development of advanced

---

[†]Corresponding Author.

585

instrumented micro- and nano- indentation techniques, improved understanding of indentation mechanisms and advances in analysis methods, indentation test is now being widely applied as a very effective and convenient method to examine the mechanical properties of materials, especially the samples with small size, such as coatings and thin films.[2,3]

Shape memory alloys (SMAs), represented by NiTi, are well known smart materials because of their extraordinary behaviors: shape memory effect and superelastic deformation. Recently, micro-indentation and nano-indentation techniques are playing a growing role in probing mechanical properties of SMAs. Gall et al.[4] studied the instrumented Vickers micro-indentation of single crystal NiTi SMA. Ni et al.[5] reported their indentation experimental results of a NiTi alloy. The dynamic indentation response of NiTi films sputtered on oxidized silicon substrates was examined by Ma and Komvopoulos[6]. Yan et al. have applied dimensional analysis and the finite element method to examine the spherical indentation response of superelastic SMAs[7,8] and proposed a spherical indentation method to measure the transformation stresses.[9]

During the forward austenite to martenstie transformation process, not only the shape of the material changes, which is quantified by the transformation shear strain, but also the volume reduces slightly. Consideration that the volume strain is much smaller than the transformation shear strain components, the transformation volume strain was neglected in these theoretical studies.[7-9] However, experimental research[10] indicated that the transformation volume strain in SMA NiTi might have great influence to hinder the transformation process near a fatigue crack tip with high triaxial stress. Theoretical study[11] also confirmed that the transformation volume contraction strain can increase the effective stress intensity near an advancing crack tip to over ten percent although it is less than a half percent in SMAs. Based on this consideration, the influence of the transformation volume strain on the spherical indentation of superelastic NiTi SMAs was examined in this paper.

## 2. Numerical Model

The commercial finite element package ABAQUS was applied to simulate the spherical indentation of superelastic NiTi SMA. An idealized superelastic SMA model is adopted in the finite element simulation. Under uniaxial tension, this model is illustrated in Fig. 1. The tensile strain due to elastic deformation and forward transformation returns to zero completely after the removal of the tensile stress for a superelastic SMA. Here, the processes of the forward transformation from austenite to martensite and the reverse transformation from martensite to austenite are simplified as perfect, i.e., the forward transformation stress, $\sigma_f$, and the reverse transformation stress, $\sigma_r$, are respectively constant during the forward and reverse transformation processes. As illustrated in Fig. 1, $\varepsilon^{tr}$ is the maximum transformation strain in uniaxial tension or the maximum magnitude of the transformation strain, $E_a$ is the elastic modulus of the austenite, and $E_m$, the elastic modulus of the martensite. Additionally, $v_a$ and $v_m$ state for the elastic Poisson's ratios of the austenite and the martensite, respectively. The values of these material parameters were chosen as $\sigma_f = 600MPa$, $\sigma_r = 400MPa$, $\varepsilon^{tr} = 4\%$, $E_a = E_m = 50GPa$ and $v_a = v_m = 0.3$. During a forward austenite to martensite transformation process, the volume of the material will also change. For example, it equals −0.37% for CuAlNi SMA[12] and about −0.39% for NiTi SMA,[13] which was taken as the transformation volume strain $\varepsilon_V^{tr}$ in the current study.

The indenter tip is normally made of diamond. In most of the published papers on the modeling of indentation, the indenter is considered as a rigid body. In order to examine the effect of the elasticity of the indenter, both of rigid indenter and elastic indenter were considered. The Young's modulus and the Poisson's ratio of the diamond indenter were taken as 1141GPa and 0.07,[2] respectively. As the indenter tip is spherical, an axisymmetric model was established. Fig. 2 shows the finite element mesh with the elastic indenter around the contact area. The radius of the indenter tip is $R = 10$ μm.

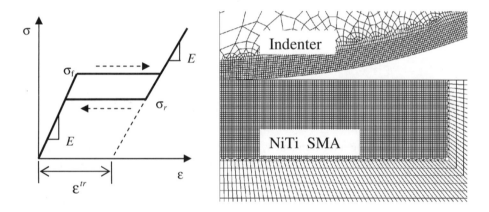

Fig. 1. Illustration of the idealized superelastic SMA model under a uniaxial tension.

Fig. 2. Finite element mesh near the indenter tip.

## 3. Results and Discussion

In the previous simulations,[7,8] the contact friction between the indenter and the specimen was neglected, which caused some concerns. Therefore, the effect of the friction is first examined. Figure 3 shows three indentation curves from the simulations with friction coefficient values of 0, 1.0 and 10.0, respectively. As we can see, there is almost no difference between any of the three curves, which indicates that the influence of the contact friction on the indentation curve can be neglected. Same conclusions can be drawn from the results of the simulations without transformation volume strain and the simulations with rigid indenter.

With regard to the friction effect in a general indentation, previously, people believed that friction between the indenter and the substrate has only a small effect on the hardness and the indentation curve[14,15]. Therefore, the friction effect was simply neglected in many numerical simulations[16-18]. Recent study[19] found that it is only true for an indenter with included angle equal or higher than $60^0$. A spherical indenter has an included angle of $90^0$. Our finding of the negligible friction effect from the spherical indentation on NiTi SMA is consistent with that from the study by Bucaille et al.[19]

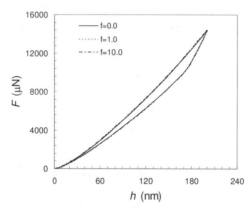

Fig. 3. Simulated spherical indentation curves from an elastic diamond indenter with different friction coefficients. The transformation volume strain is -0.39% in the three calculations.

Instead of assuming a rigid indenter, the elasticity of the indenter is considered in the current study. The effect of the elasticity of the indenter can be evaluated by comparing the indentation curves from the simulation with a rigid indenter and the one with an elastic indenter. The results are shown in Fig. 4(a). As we can see, the difference between the two curves can be neglected if the indentation depth is smaller than 100 nm, i.e., the relative indentation depth, $h/R$, is less than 0.01. However, the influence of the elasticity gradually demonstrates with the increase in the indentation depth, as shown in Fig. 4(a). For example, at the maximum indentation depth of $h/R=0.02$, the maximum indentation force from the rigid indenter simulation is about 7.5% higher than that from the elastic indenter simulation. Therefore, one cannot simply neglect the elasticity of the indenter if the indentation depth is relatively large.

Instead of simulating an elastic indenter, an alternative way to consider the effect of the elasticity of the indenter is to use the equivalent elastic modulus for the specimen. In this way, the indenter is still treated as a rigid body and the equivalent elastic modulus is calculated by

$$\frac{1}{E^*} = \frac{1-v_i^2}{E_i} + \frac{1-v_s^2}{E_s} \tag{1}$$

where $E_i$ and $v_i$ are the elastic constants for the indenter and $E_s$, $v_s$ for the specimen. To verify this equivalent Young's modulus method, we carried out a simulation with a rigid indenter and $E_s = 47.71GPa, v_s = 0.3$ so that this new simulation has the same equivalent Young's modulus of $E^* = 52.43GPa$ as the one with the elastic diamond indenter. The results of the indentation curve from the two simulations are compared in Fig. 4(b). We can clearly see that the new indentation with equivalent Young's modulus cannot predict the same indentation curve if the indentation depth is relatively large. The largest indentation force at the largest depth of $h/R$=0.02 is overestimated by 4.4% although it gives a better prediction than the model which simply neglect the elasticity of the indenter. In fact, the validity of the equivalent Young's modulus method was discussed by Mesarovic and Fleck[20] for elastic-plastic materials. They found that this method is valid in the early stage of elastic-plastic deformation, as shown in Fig. 4(b) that it gives a good prediction when the indentation depth is relatively small. But, with the development of the non-linear plasticity or transformation deformation, the elastic deformation is no longer dominant and therefore, the equivalent Young's modulus method no longer works.

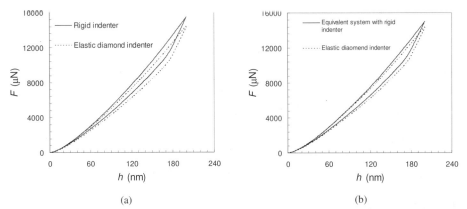

(a)                        (b)

Fig. 4. Spherical indentation curve from an elastic diamond indenter, compared with the one from (a) an idealize rigid indenter and (b) an indentation with equivalent modulus and rigid indenter.

The results from Fig. 3 and Fig. 4 indicate that we can neglect the friction but need to explicitly consider the elasticity of the indenter. Now, we use the frictionless model with an elastic diamond indenter to investigate the influence of the transformation volume strain. Figure 5 shows the simulated indentation curve with transformation volume strain of -0.39%, i.e., the dashed line with ev $= -0.39\%$, compared with the one without transformation volume strain, i.e., the solid line with ev = 0. As we can see, the indentation curve with transformation volume strain is slightly different from the one without transformation volume strain. As this transformation volume strain is negative, which means that the volume contracts during the forward transformation process, the stiffness of the structure is expected to reduce. This is the reason that the indentation force from the curve with transformation volume strain is lower at the same depth after transformation occurs, as shown in Fig. 5. However, because this transformation volume strain $\varepsilon_V^{tr}$ (i.e., ev in Fig. 5) is much smaller than the maximum transformation strain $\varepsilon^{tr}$, (here -0.39% against 4%), the reduction of the structure stiffness is insignificant. That is why the influence of the transformation volume strain is insignificant. For example, the maximum indentation force from the model with transformation volume contraction at the depth of h=200nm is only 2.1% lower.

Furthermore, the reduction of the structure stiffness should enhance with the increase in the transformation zone, i.e., the increase in the indentation depth. Therefore, the influence of the transformation volume strain would enhance with the indentation depth. This prediction is confirmed in Fig. 6. Figure 6 shows the curves of the normalized maximum indentation force $F_m/(R^2 E_a)$ versus the normalized maximum indentation depth $h_m/R$ from the modeling with transformation volume strain of -0.39% and the modeling without transformation volume strain, respectively. Overall, the maximum forces from the modeling with the transformation volume strain slightly lower and this difference is enhanced with the increase in the maximum indentation depth.

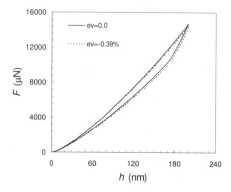

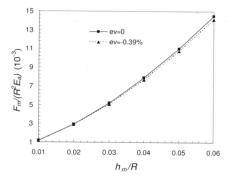

Fig. 5. Simulated indentation curve with transformation volume contraction strain of -0.39%, compared with the one without transformation volume strain.

Fig. 6. Comparison of the variations of the normalized maximum indentation force $F_m/(R^2 E_a)$ with the normalized maximum indentation depth $h_m/R$ from the simulations with transformation volume contraction strain of -0.39% and the simulations without transformation volume strain.

Traditionally, the measurement of the hardness of a material is the major purpose of carrying out indentation tests with different indenter tips. The hardness $H$ is normally defined as the mean pressure the material will support under load, i.e.,

$$H = \frac{F_m}{A_c},$$
(2)

where $F_m$ is the maximum indentation force. $A_c$, named as the contact area, is the projected area of contact under the maximum indentation force $F_m$. The dashed line in Fig. 7 shows the variation of the normalized hardness $H/\sigma_f$ with the normalized maximum indentation depth $h_m/R$ from the simulations with the transformation volume strain of -0.39%. It clearly indicates that the ratio $H/\sigma_f$ evolves from about 2.6 to 5.5 when $h_m/R$ increases from 0.01 to 0.06. These results are also compared with those from the same spherical indentation of an ordinary elastic-perfectly plastic material with a yield strength $\sigma_y = 600MPa$ and Young's modulus $E = 200GPa$. The difference is very obvious for this ordinary metal, and the ratio $H/\sigma_y$ slightly varies from 2.9 to 2.8 when the normalized indentation depth changes from 0.01 to 0.06. These results indicate that the spherical hardness of SMAs due to phase transformation cannot be used as a material constant to give a measure of the forward phase transformation stress. Further more, neglecting the transformation volume strain will overestimate the hardness value, which is demonstrated by the difference between the solid line and the dashed line in Fig. 7. Transformation volume strain is zero for the solid line.

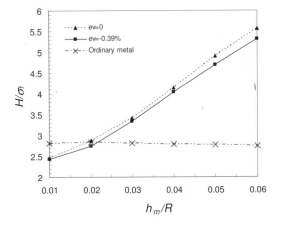

Fig. 7. The variations of the normalized hardness $H/\sigma_f$ with the normalized maximum indentation depth $h_m/R$ from the simulations with the transformation volume strain of $-0.39\%$ and with zero transformation volume strain, compared with an ordinary elastic-perfectly plastic material with $\sigma_y = 600MPa$ and $E = 200$ GPa.

## 4.  Summary

Spherical indentation of NiTi superelastic SMA was re-examined in this paper. Numerical results indicate that the influence of the contact friction on the indentation response is negligibly small. The elasticity of the hard diamond indenter can not be neglected and should be simulated explicitly if the indentation depth is not very small.

These conclusions are consistent with the spherical indentation on elastic-plastic materials. Furthermore, transformation volume contraction during the forward austenite-to-martensite transformation process will reduce the stiffness of the indentation structure and therefore predict a lower indentation force and a lower hardness value. These lower predictions will enhance with the increase in the indentation depth. However, because the transformation volume strain is very small, overall, the influence of the transformation volume strain on the spherical indentation of NiTi SMA response is not significant.

## Acknowledgments

This work was financially supported by the Australian Research Council and the Research Grants Council of Hong Kong. The numerical calculations were carried out at the National Facility of the Australian Partnership for Advanced Computing through an award under the Merit Allocation Scheme to WY.

## References

1. D. Tabor, *Hardness of Metals*, (Clarendon Press, Oxford, 1951).
2. W. C. Oliver and G. P. Pharr, *Journal of Materials Research* **7**, 1564 (1992).
3. Y.-T. Cheng and C.-M. Cheng, *Mater. Sci. Eng. R* **44**, 91 (2004).
4. K. Gall, K. Juntunen, H. J. Maier, H. Sehitoglu and Y. I. Chumlyakov, *Acta Mater.* **49** 3205 (2001).
5. W. Ni, Y.-T. Cheng and D. S. Grummon, *Appl. Phys. Lett.* **82**, 2811 (2003).
6. X.-G. Ma and K. Komvopoulos, *Appl. Phys. Lett.* **84** 4274 (2004).
7. W. Yan, Q. Sun, X.-Q. Feng and L. Qian, *Int. J. Solids Struct.* **44**, 1 (2007).
8. W. Yan, Q. Sun and H.-Y. Liu, *Mater. Sci. Eng. A* **425**, 278 (2006).
9. W. Yan, Q. Sun, X.-Q. Feng and L. Qian, *Appl. Phys. Lett.* **88**, 24192 (2006).
10. A. L. McKelvey A L and R. O. Ritchie, *Metall. Mater. Trans. A* **32**, 731 (2001).
11. W. Yan, C. H. Wang, X. P. Zhang and Y.-W. Mai, *Smart Mater. Struct.* **11**, 947 (2002).
12. D. -N. Fang, W. Lu, W. Yan, T. Inoue and K.-C. Hwang, *Acta Mater.* **47**, 269 (1998).
13. R. L. Holtz, K. Sadananda, M. A. Imam, *Int. J. Fatigue* **21** s137 (1999).
14. A. E. Giannakopoulos, P.-L. Larson, and R. Vestergaard, *Int. J. Solids Struct.* **31**, 2679 (1994).
15. A. E. Giannakopoulos and P.-L. Larson, *Mech. Mater.* **25**, 1 (1997).
16. A. E. Giannakopoulos and S. Suresh, *Scripta Mater.* **40**, 1191 (1999).
17. M. Dao, N. Chollacoop, K. J. Van Vliey, T. A. Venkatesh and S. Suresh, *Acta mater.* **49** 3899 (2001).
18. J. C. Hay, A. Bolshakov and G. M. Pharr, *J. Mater. Res.* **14**, 2296 (1999).
19. J. L. Bucaille, S. Stauss, E. Felder and J. Michler, *Acta Mater.* **51**, 1663 (2003).
20. S. D. Mesarovic and N. A. Fleck, *Proc. Roy. Soc. Lond. A* **455**, 2707 (1999).

# CHARACTERIZATION OF BAINITIC MICROSTRUCUTRES IN LOW CARBON HSLA STEELS

JU SEOK KANG[1]

*Dept. of Materials Science and Engineering, POSTECH, San31 Hyojadong Namgu,*
*Pohang, Gyungsangbukdo, 790-784, KOREA,*
*iloveyou@postech.ac.kr*


CHAN GYUNG PARK[2†]

*Dept. of Materials Science and Engineering, POSTECH, San31 Hyojadong Namgu,*
*Pohang, Gyungsangbukdo, 790-784, KOREA,*
*cgpark@postech.ac.kr*

Received 15 June 2008
Revised 23 June 2008

The austenite phase of low carbon steels can be transformed to various bainitic microstructures such as granular bainite, acicular ferrite and bainitic ferrite during continuous cooling process. In the present study site-specific transmission electron microscope (TEM) specimens were prepared by using focused ion beam (FIB) to identify the bainitic microstructure in low carbon high strength low alloy (HSLA) steels clearly.

Granular bainite was composed of fine subgrains and $2^{nd}$ phase constituents like M/A or pearlite located at grain and/or subgrain boundaries. Acicular ferrite was identified as an aggregate of randomly oriented needle-shaped grains. The high angle relations among acicular ferrite grains were thought to be caused by intra-granular nucleation, which could be occur under the high cooling rate condition. Bainitic ferrite revealed uniform and parallel lath structure within the packet. In some case, however, the parallel lathes showed high angle relations due to packet overlapping during grow of bainitic ferrite, resulting in high toughness properties in bainitic ferrite based steels.

*Keywords*: Granular bainite; acicular ferrite; bainitic ferrite; low carbon steels.

## 1. Introduction

Low carbon high strength low alloy (HSLA) steels are considered as indispensable materials in the constructions of buildings, bridges, linepipes, automobiles and ships. To enhance the safety of such constructions and to reduce the usage of natural resources, high strength and high toughness properties are required. Thus, low carbon bainitic steels rather than ferritic steels have drawn attentions recently[1-3]. The austenite phase of low carbon steels, however, transformed to various bainitic microstructures during continuous

---

†Corresponding Author.

cooling process. Thus, several classification systems on bainitic microstructures in low carbon steels were proposed based on the structure and morphology of bainite e.g. acicular/lath-like ferrite morphology and interlath/intralath carbide precipitations[4-5]. In such studies, the complex bainitic microstructures were mainly identified by using scanning electron microscope (SEM) and/or transmission electron microscope (TEM). However, the conventional SEM and TEM images did not exactly coincide, because of the quite different sample preparation methods in SEM and TEM observations. These conventional analysis methods, thus, were not suitable to characterize complex bainitic microstructures.

The objectives of present study are to identify the low carbon bainitic microstructures and to ascertain the grain boundary characteristics of such microstructures by using advanced analytical method of focused ion beam (FIB) machine.

## 2. Experimental Procedure

The chemical composition of studied steel is given in Table 1. For hot rolling simulations using Gleeble 3500 system, the continuously casted slab was machined into cylindrical samples with 7mmφ × 12mm in dimension. To simulate reheating process, the cylindrical specimens were heated to a specific temperature less than 1150°C, and held for 10 minutes to dissolve Ti and Nb in the steel. And then, the specimens were deformed several times at temperatures between 1100°C and 800°C with a total reduction of 60%. Finally, the deformed specimens cooled down to room temperature with various cooling rates from 0.3°C/s to 50°C/s.

FIB-SEM dual beam machine (SEI 3050) was applied to make site-specific TEM samples. The grain morphologies were observed and misorientation angles between grain and/or sub-grain boundaries were detected by using TEM (JEOL 2100).

Table 1. Chemical composition of studied steel. (wt. %, *:ppm).

| Fe | C | Si | Mn | Ti+Nb | N* | Others |
|------|------|-------|------|--------|------|--------------|
| Bal. | 0.05 | 0.155 | 1.90 | 0.065 | < 50 | Mo, Ni, Cr, B |

## 3. Results

The austenite phase of low carbon HSLA steels could decompose to various bainitic microstructures such as granular bainite, acicular ferrite and bainitic ferrite mainly according to the cooling rate[6-7]. Based on the previous research results[6], the present study will discuss grain boundary characteristics of such microstructures.

At slow cooling rates from 0.5°C/s to 5°C/s, austenite to granular bainite transformation occurred. The prior austenite grain boundaries were clearly preserved and the 2nd phase constituents were observed inside granular bainite grains as shown in Fig. 1(a). Site-specific TEM specimen was made in the boxed region in Fig. 1(a) and the resultant TEM image is exhibited in Fig. 1(b). Dissimilar to scanning ion microscope

(SIM) image, Fig. 1(a), granular bainite was consisted of fine grains in Fig. 1(b). These fine grains were identified as subgrains of granular bainite because misorientation angles between fine grains were mostly less than 5 degree as represented in Table 2. However, in some cases, e.g. misorientation angles between grain 5 and 6, high angle grain boundary was also observed. The high angle grain boundary in Fig. 1(b) is thought to originate from prior austenite grain boundary because the prior austenite grain boundary area was included in site-specific TEM specimen.

Between the subgrains of granular bainite 2nd phase constituents revealing different contrast in TEM image (Fig. 1(c)) were observed. Diffraction pattern of one of these 2nd phase constituent exhibited distorted Kudjumov-Sach relation (Fig. 1(d)) and high resolution high angle annular dark field (HAADF) image (Fig. 1(e)) showed fine twin structures inside that constituent. Thus, the 2nd phase constituent was identified as martensite/austenite (M/A) constituents.

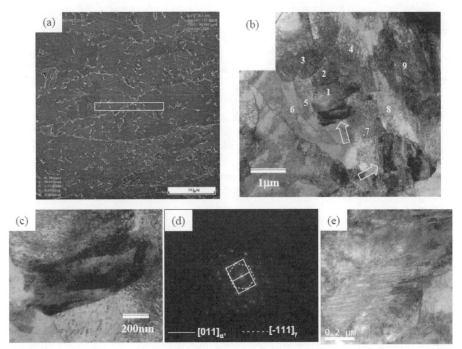

Fig. 1. (a) Scanning ion microscope (SIM) image of granular bainite and (b) Site-specific TEM image of boxed region in (a). The arrows represents 2nd phase constituents. (c) One of 2nd phase constituents in (b). The 2nd phase constituents were identified as M/A by (d) diffraction pattern and (e) fine twins in high angle annular dark field image.

Table 2. Misorientation angle between grains numbered in Fig. 1(b).

| Grain | 1-2 | 2-3 | 2-4 | 1-5 | 5-6 | 7-8 | 8-9 |
|---|---|---|---|---|---|---|---|
| Angle (°) | 3.2 | 1.5 | 4.6 | 3.5 | 12.6 | 8.0 | 4.9 |

From an intermediate cooling rate of 5°C/s to the fastest cooling rate of 50°C/s, needle-like acicular ferrite microstructure was observed. When the acicular ferrite transformation occurred, the prior austenite grain boundary was not preserved. The representative acicular ferrite region in Fig. 2(a) was selected to make site-specific TEM specimen. In the TEM figure, Fig. 2(b), an aggregate of longish grains were clearly observed but the thickness of each grain was randomly distributed from sub-micrometers to a few micrometers. The misorientation angles between adjacent acicular ferrites were summarized in Table 3. Most of acicular ferrite grains had high angle relations larger than 15° among them. It is thought that many embryos of acicular ferrite started to grow simultaneously under high cooling rate conditions. Thus, randomly orientated needle-like acicular ferrite could form.

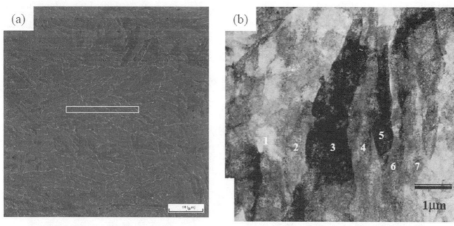

Fig. 2. (a) Scanning ion microscope (SIM) image of acicular ferrite and (b) Site-specific TEM image of boxed region in (a). Acicular ferrite grains reveal needle-like shape.

Table 3. Misorientation angle between grains numbered in Fig. 2(b).

| Grain | 1-2 | 2-3 | 3-4 | 4-5 | 5-6 | 6-7 | 7-8 |
|---|---|---|---|---|---|---|---|
| Angle(°) | 5.1 | 27.6 | 6.9 | 19.3 | 19.0 | 30.0 | 4.9 |

The trace of prior austenite grain boundaries were revealed again with the development of bainitic ferrite at high cooling rate conditions over 20°C/s. The parallel lathes inside the prior austenite grains were the most distinguishable morphological feature of bainitic ferrite. Figure 3(a) represents the morphology of bainitic ferrite very well. When we observed the parallel lathes of bainitic ferrite in a boxed region in Fig. 3(a) closely, however, the lathes actually formed packets. And the lath direction was somewhat different according to packet. The misorientation angles between lathes are summarized in Table 4. As simply analogized from TEM image, the lathes within a packet showed small angle relations, but the packet boundaries revealed high angle relations. In some cases, however, even the lathes looking like parallel showed high angle relations as shown in packet C of Fig. 3(b). High angle relations between parallel-like lathes are thought to originate from the crossing of different packets during growth stage.

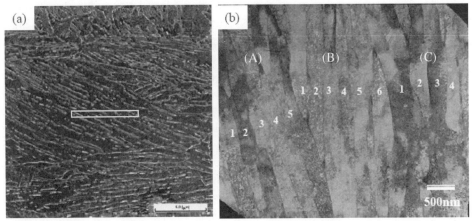

Fig. 3. (a) Scanning ion microscope (SIM) image of bainitic ferrite and (b) Site-specific TEM image of boxed region in (a). The lathes of bainitic ferrite formed packet structure. In this case, three packets, A, B and C, are observed.

Table 4. Misorientation angle between lathes numbered in Fig. 3(b).

| Lath | A1-A2 | A2-A3 | A3-A4 | A4-A5 | A5-B1 | B1-B2 | B2-B3 |
|------|-------|-------|-------|-------|-------|-------|-------|
| Angle(°) | 4.4 | 3.9 | 0.4 | 2.7 | 10.3 | 2.3 | 4.6 |

| | B3-B4 | B4-B5 | B5-B6 | B6-C1 | C1-C2 | C2-C3 | C3-C4 |
|--|-------|-------|-------|-------|-------|-------|-------|
| | 0.3 | 1.6 | 1.5 | 16.7 | 17.4 | 15.6 | 14.1 |

## 4. Discussion

From the continuous cooling transformation (CCT) diagram of the present alloy[7], transformation start temperatures during cooling (Ar3) of granular bainite, acicular ferrite and bainitic ferrite are around 650°C, 560°C and 520°C, respectively. It is reported that the bainitic transformation in low carbon steels occurred with a mixture mode of diffusion and shear and the shear portion gradually increased with increasing cooling rate[8]. Thus, the morphologies of bainitic microstructure changed from globular shape to lath morphology. During the bainitic transformation, the diffusion of carbon from the ferrite/austenite interface to untransformed austenite occurred. And then the transformation temperature of carbon enriched untransformed austenite was more lowered. If a part of those untransformed austenite transformed to martensite during cooling, M/A constituents can form. During granular bainite transformation, carbon diffusion could actively occur due to high transformation temperatures. Thus, M/A constituents can easily form between subgrain boundaries of granular bainite.

In addition to the grain morphologies, the effective grain sizes were also affected by transformation temperature. That is, the nuclei of granular bainite can grow to an extent at high temperatures and then sympathetic nucleation might occur during slow cooling, resulting in low angle subboundaries. Thus, granular bainite shows large effective grain sizes of around 10 to 20 micrometers. The driving force for acicular ferrite transformation is larger than that of granular bainite because it developed under fast cooling conditions. Thus, acicular ferrite could nucleate at austenite grain boundary as well as other acicular ferrite grain boundaries and then it can grow in various directions. Therefore, randomly orientated high angle grain structure could form. As the cooling rate increased further, shear transformation dominantly occurred and bainitic ferrite microstructure form. The packet size of bainitic ferrite is larger than the grain size of acicular ferrite, but some lathes of bainitic ferrite reveal high angle relations. Since fracture crack could deflect at high angle lath boundaries bainitic ferrite based steels could show high toughness properties similar to acicular ferrite based steels with enhanced strength properties.

## 5.  Summary

The grain boundary characteristics of granular bainite, acicular ferrite and bainitic ferrite were studied by the aid of advanced analytical method of focused ion beam. Granular bainite consisted of fine subgrains with $2^{nd}$ phase constituents located between grain and subgrain boundaries. Acicular ferrite was identified as an aggregate of randomly oriented and needle shaped grains. Bainitic ferrite showed parallel lath structure and most of lathes showed similar orientations. In some cases, however, bainitic ferrite's lathes revealed high angle relations, resulting in high toughness properties.

### Acknowledgments

The authors would like to thank POSCO, NCRC, BK21 and NCNT for financial and technical supports.

### References

1.  K. Fujiwara, ISIJ International, **35**, No. 8, pp. 1006 (1995).
2.  J. Y. Koo et al., Proc of 13$^{th}$ int. Offshore and Polar Eng Conf, ISOPE, **4**, pp. 10 (2003).
3.  H. Ohtani et al., Metall. Trans., **21A**, pp.877 (1990).
4.  B. L. Bramfitt and J. G. Speer, Metall. Trans., **21A**, pp. 817 (1990).
5.  G. Krauss and S. W. Thompson, ISIJ international, **35**, No. 8, pp. 937 (1995).
6.  H. J. Jun et al., Mater. Sci. and Eng. **A422**, pp157 (2006).
7.  J. S. Kang et al., Adv. Materials Research, **26-28**, pp. 73 (2007).
8.  Y. Ohmori, **41**, No. 6, pp.554 (2001).

# EBSD STUDY ON THE EVOLUTION OF MICROSTRUCTURES DURING COMPRESSIVE DEFORMATION OF AN AUSTENITIC STAINLESS STEEL

JONG BAE JEON[1], YOUNG WON CHANG[2†]

*Department of Materials Science and Engineering, POSTECH, Pohang, KOREA,*
*blizzard@postech.ac.kr, ywchang@postech.ac.kr*

Received 15 June 2008
Revised 23 June 2008

Microstructural evolution during compressive deformation of a meta-stable austenitic stainless steel has been investigated using an electron-backscattered diffraction (EBSD) technique. A local area tracking method has been adopted in order to observe the successive changes in local microstructures during the compression. The local microstructures could be observed in-situ from this method without using any in-situ stage, in addition to a precisely-controlled real time deformation process. Successive development of grain orientation spread, grain boundary misorientation and deformation-induced phase in the local areas of interest were successfully investigated from this method.

*Keywords*: Austenitic stainless steel; in-situ EBSD observation; martensite nucleation; deformation induced martensitic transformation.

## 1. Introduction

Austenitic stainless steels are now well recognized to have excellent properties in terms of corrosion and oxidation resistance and thus widely used in kitchen utensils, chemical plants and spacecraft, etc[1-2]. Extensive researches have also been performed on the well known transformation induced plasticity (TRIP) phenomena in various austenitic stainless steels since Zackay[3] first reported an enhanced strength-ductility combination in these steels.

The microstructural evolution during plastic deformation of a meta-stable austenitic stainless steel has been investigated in this study in relation to the deformation induced martensitic transformation (DIMT). Several works have already reported some characteristic features of the DIMT[3-9], but few studies have been made so far to clarify the local changes in grain orientation, grain boundary, and phases. This could be achieved by utilizing an in-situ deformation stage, but the cost of an in-situ stage is usually very high together with the limitation in specimen shapes, deformation modes, and the amount of loads. In most of the experiments with an in-situ stage, the load is generally controlled

---

†Corresponding Author.

manually to provide unreliable strain rate during the tests. Due to these several disadvantages of currently-used in-situ stages, a local area-tracking method has been adopted instead in this study, which tracks local area of interest by separating deformation and observation step, viz. (i) deformation is precisely applied to the specimen first using a computer-controlled compressive machine, (ii) microstructure at the definite strain level is observed. These two steps are repeated successively, while the surface of specimen is carefully protected from any physical and chemical damages. This local area-tracking method introduced in this study enabled the specimens to deform in any shape and mode under precisely-controlled strain rate, which can not be achieved in most of the in-situ stages.

The present work has, therefore, been carried out to observe successive changes in local microstructures during several compression steps by using the local-area tracking method introduced above. The method revealed microstructural changes of local grains during the plastic deformation, i.e. evolution of grain misorientation and grain boundaries together with the concurrent phase changes due to DIMT with the help of an electron backscattered diffraction (EBSD) technique, which can provide high resolution crystallographic features of grain, grain boundary and phases.

## 2.  Experiments

An austenitic stainless steel with the composition of Fe-17Cr-7Ni-1Mn-0.5Si-0.3N alloy was first cast utilizing a vacuum-induction melting furnace at POSCO. The ingots were then homogenized at 1200°C for 3 hrs before hot rolled into 3 mm thick sheet at 1050°C. Rectangular-shape compressive specimens with 20 mm thickness, 12mm width and 6.5mm height were machined from the hot rolled sheet. The specimens were then normalized at 1050°C in a sealed quartz tube to remove transformed phases possibly formed during mechanical machining before quenched directly into cold water.

In order to perform local area-tracking, several diamond-shape marks were made on the specimen surface using the Vickers indentor (FM-800, FUTURE-TECH).The local area of interest located near the diamond marks were then first observed with an EBSD (EDAX-TSL, AMETEK) and analyzed with a crystalline orientation analysis software (OIM, AMETEK). Compression tests were then carried out at room temperature under the strain rate of $1.0 \times 10^{-3}$ /s using an electro-mechanical testing machine (INSTRON 5582). After loading to a certain strain level given in Fig. 1, the local area of interest was inspected again with an EBSD in the same manner as before. These steps have identically been repeated until the total compressive strain reached to 0.334, above which crystallographic information is hard to be obtained due to high roughness of deformed surface. Generally, backscattered electron is hard to diffract at grain boundaries, shear bands and highly-refracted surfaces induced by external deformation or chemical etching. In order to preserve the surface from any physical scratch and chemical contamination, the series of steps was carefully conducted and the surface was clearly washed in ethanol prior to EBSD observations after deformation.

## 3. Results and Discussion

Fig. 1 shows true stress-strain curves obtained from a series of compression tests, exhibiting that any unexpected geometrical or microstructural changes did not occur during the tests. In other words, the geometry of rectangular-shape specimens was well kept to be symmetric and also the elastic moduli were found to be same for the whole series of compression tests. In addition, no yield point phenomenon mainly caused by strain aging was observed during the compression tests, implying that the specimen was unloaded and reloaded without any appreciable delay and any undesired adiabatic heating effect. The amounts of total strain, elastic strain, and plastic strain achieved in each compression test are given in Fig. 1. Most of the total strain was recoverable elastic strain

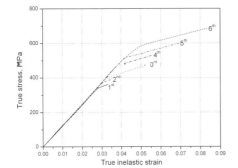

|  | Total strain | Elastic strain | Plastic strain |
|---|---|---|---|
| 1st comp. | 0.0332 | 0.0296 | 0.0036 |
| 2nd comp. | 0.0349 | 0.0319 | 0.0030 |
| 3rd comp. | 0.0578 | 0.0423 | 0.0155 |
| 4th comp. | 0.0557 | 0.0427 | 0.0130 |
| 5th comp. | 0.0692 | 0.0482 | 0.0210 |
| 6th comp. | 0.0831 | 0.0535 | 0.0296 |

Fig. 1. Stress-strain curves at each step of compression.

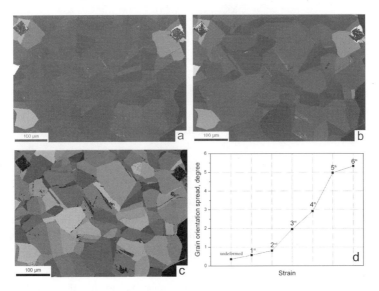

Fig. 2. The grain orientation spread maps of (a) as-received specimen and after (b) the first and (c) the fourth compression step. The angles of grain orientation spread at each step are plotted in Fig. 3 (d).

in these compression tests, since the compression tests were conducted within the low strain ranges in order to avoid high surface roughness expected to generate in high strain range.

Figures 2(a), 2(b), and 2(c) exhibit the grayscale orientation spread maps of grains taken at the same local area of interest before and after the first and fourth compression test, respectively. Diamond-shape marks made for the local area tracking can be found at the left and the right-uppermost part in each figure. The orientation spread of a grain represents the average deviation between the orientation of each point in a grain and the average orientation of the grain. If the orientation of each point within a grain is highly deviated from the average orientation of the grain, then the degree of spread in a grain is high. In the grayscale map of grain orientation spread, the contrast of a grain becomes brighter if the spread is getting higher. For as-received specimen depicted in Fig. 2(a), shear stress was not applied yet, so that the orientation spread is nearly zero. After the first compression step, the orientation of some constituent grain starts to rotate due to the applied shear stress, some grains becoming brighter as can be seen in Fig. 2(b). The color of most grains became close to light-gray after the fourth step as shown in the Fig. 2(c), whose spread angle was about 3° on average. The orientation spread angles of grains at each step are plotted in Fig. 2(d) together with that of as-received specimen. Each compression step is found to cause successively higher orientation spread, which in turn suggests the orientation of grain constituents deviates from the average orientation of the whole grains when the grains are gradually subjected to shear stresses higher than the critical resolved shear stress.

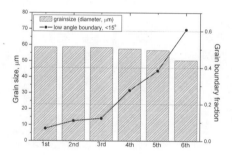

Fig. 3. Grain size variation and the fraction of low angle grain boundaries during the series of compressions.

Fig. 3 shows the grain size variation and the fraction of low angle grain boundaries developed after each step of compression tests. The initial grain size of 58μm decreased slowly finally reducing to 50μm during the series of compression tests. The grain size here is defined as the diameter of grains, whose boundary angles are larger than 2°. This gradual decrease in grain size can be thought due to the increasing fraction of low angle grain boundaries developed during the plastic deformation. In other words, most of low angle grain boundaries were found to evolve into small-sized grains in contrast to the high angle boundaries remaining the same throughout the deformation, so that the increased fraction of low angle grain boundaries simply implies the increased fraction of small-sized grains, which in turn reduces the average grain size. The usual grain refinement process induced by external deformation was not observed during the present compression tests at room temperature. The grain size generally taken in grain refinement process is referred to the grains having high angle boundaries (>15°) and the refinement process is reported to necessarily require severe plastic deformation. Since the total strain applied in this work was less than 0.1 and any appreciable evolution of high angle grain boundaries was not observed, the reduction of grain size is, thus, related to the increase of grains with the boundary angles ranging from 2° to 15°.

It is well known that some of the austenite phases can transform into martensite with the help of external deformation in metastable austenitic steels (DIMT)[5]. The observation of transformed martensites generated by DIMT has usually been made thru diffraction patterns taken by a TEM, which can provide the essential crystallographic information related to the martensite phase. The EBSD technique can also provide reliable crystallographic information to characterize the martensitic transformation.

Fig. 4 shows the successive evolution of shear bands and consequent martensite transformation from austenite during the series of compression tests. No noticeable

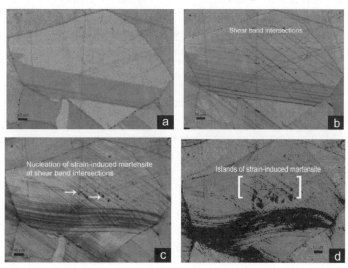

Fig. 4. Phase maps embedded within a quality image map of a local grain: (a) un-deformed, (b) the second step, (c) the fourth, and (d) the sixth step of compression.

microstructural features were found except annealing twins in an un-deformed specimen as can be seen from the Fig. 4(a). After the second step of compressions, numerous shear bands were observed to form, intersecting with each other within the local grain of interest as shown in the Fig. 4(b). Those shear bands were known to be a typical deformation structure in many austenitic stainless steels, while the intersection points of them were reported to become effective nucleation sites of martensite for the DIMT. These nucleated martensites at the shear band intersections can be observed after two further steps of compression as depicted in the Fig. 4(c). The black dots marked with white arrows in the Fig. 4(c) represent the martensite phase nucleated at the shear band intersections. In this type of nucleation, the shear bands were reported to consist of an ε-martensite (HCP) and theoretically the intersection of ε-martensite can nucleate α'-martensite (BCT)[5]. Two additional compression steps made the martensite dots to grow into the arrays of islands as shown in the Fig. 4(d). The black islands within the white bracket in the Fig. 4(d) now represent the grown martensite phase. The large black area below the brackets also observed in the figure could not be confirmed as martensite phase, since crystallographic information could not be obtained due to high surface roughness developed in that area possibly induced by the accumulation of shear bands. These successive images can provide the successive DIMT process within a local grain, like an in-situ observation as have been discussed.

## 4. Conclusion

A local-area tracking method has been introduced in the present study to investigate microstructural evolution within a narrow local region of an austenitic stainless steel, replacing the usual in-situ observation process. The main results obtained thru this local tracking method can be summarized as follows;

• The orientation spread of grains was found to increase during the series of compression steps, which can be explained in relation to the partial rotation of grains caused by an applied shear stress.

• The fraction of low angle boundaries within a local area increased gradually during the successive compression steps to reduce the average grain size, while the high angle boundaries remained the same.

• The DIMT process in a local grain was successive observed to occur, (i) shear band formation, (ii) intersection of them, (iii) nucleation α'-martensite at the shear band intersection, and (v) the growth of α'.

## Acknowledgment

This work was supported by the 2008 National Research Laboratory (NRL) Program of the Korea Ministry of Science and Technology.

## References

1.  H. H. Park  et al., *Met. Mater. -Int.*, **9**, 311 (2003).
2.  W. S. Lee and B. K. Wang, *Met. Mater. Int.*, **12**, 459 (2006).
3.  V. F. Zackay et al., *Trans. ASM*, **60**, 252 (1967).
4.  M.A. Meyers et al., *Acta Mater.*, **51**, 1307 (2003).
5.  G. B. Olson and M. Cohen, *Met. Trans.*, **6A** (1975).
6.  R. Petrov et al., *Mat. Sci. Eng. A.* **447**, 285 (2007).
7.  H. C. Shin et al., *Scrip. Mater.*, **45**, 823 (2001).
8.  H. C. Choi et al., *Scrip. Mater.*, **40**, 1171, (1999).
9.  T. K. Ha and Y. W. Chang, *Acta Mater.*, **46**, 2741, (1998).

# TEXTURES OF EQUAL CHANNEL ANGULAR PRESSED 1050 ALUMINUM ALLOY STRIPS

JAE YEOL PARK[1]

*Department of Automobile Engineering, Doowon Technical College, Ansung-si, Gyonggi-do 456-718, KOREA,*
*jyp@doowon.ac.kr*

SEUNG-HYUN HONG[2]

*Metal Research Team, R&D Division for Hyundai Motor Company and Kia Motors Corporation,*
*772-1 Jangduk-dong, Hwasung-si 445-706, KOREA,*
*hshmsd@hyundai-motor.com*

DONG NYUNG LEE[3†]

*Department of Materials Science and Engineering, Seoul National University, Seoul 151-744, KOREA,*
*and Graduate Institute of Ferrous Technology, Pohang University of Science and Technology, Pohang 790-784,*
*KOREA,*
*dnlee@snu.ac.kr*

Received 15 June 2008
Revised 23 June 2008

In order to improve the deep drawability of aluminum and aluminum alloy sheets, it is desirable to increase the ND//<111> component in their textures. The ND//<111> component is known to develop in shear-deformed fcc alloy sheets. The equal-channel angular pressing (ECAP) is a process in which materials undergo approximately simple shear deformation. The deformation texture of ECAPed strips can change depending on process variables such as the texture of starting strips, the oblique angle, the number of pressing passes, and the shear direction. The deformation and deformation texture have been analyzed.

*Keywords*: 1050 Al alloy; equal channel angular pressing; texture.

## 1.    Introduction

In order to improve the deep drawability of aluminum and aluminum alloy sheet, a lot of studies have been carried out to improve the r-value (the Lankford value). It is well known that the r-value increases with increasing ND//<111> texture in fcc and bcc metals. Therefore, to increase the r-value of fcc metals, it is needed to suppress the {001}<100> texture formation and to develop the ND//<111> texture. Recently, efforts have been made to develop the ND//<111> texture in aluminum alloys by asymmetric rolling.[1-4] The shear texture composed of {001}<110> and {111}<112> orientations in fcc metal sheets is developed by the shear deformation of them. The asymmetric rolling gives rise to the

---

†Corresponding Author.

shear deformation through the sheet thickness, which in turn form the shear texture. The equal-channel angular pressing (ECAP) is a process in which materials undergo simple shear deformation.[5,6] The purpose of this work is to analyze the deformation and deformation textures of ECAPed 1050 Al alloy strips.

## 2.    Experimental Procedure

A commercial 1050 Al alloy sheet of 6.4 mm in thickness was rolled to 2mm in 12 passes under oil-lubrication, with the reduction per pass being about 9.5 %. The rolled sheets were annealed in a salt bath at 350 °C for 1 h. Specimens for ECAP of 20 mm in width, 2 mm in thickness, and 45 mm in length were cut from the annealed sheets with the longitudinal axis parallel to the rolling direction.

The oblique and curvature angles of the ECAP die were 90 and 0°, respectively (Fig 1). The short cross-sectional dimension of the channel was 20 mm × 2 mm. The entrance and exit channels were 50 mm and 20 mm long, respectively. The specimens were subjected to ECAP under lubrication at room temperature with a punch speed of 2 mm/min to minimize an increase in temperature during deformation.

The textures of specimens were measured with an x-ray texture goniometer in the back reflection mode with Fe filtered Co-Kα radiation. The {111}, {200}, and {220} incomplete pole figures were measured at the top surface (s = 0.9), center (s = 0) and bottom surface (s = -0.9) layers of the sheets and were used to calculate orientation distribution functions (ODFs) by the WIMV method.[7] The surfaces of specimens for the x-ray measurements were prepared by removing 0.1 mm in depth with a sand paper, followed by chemical etching in a NaOH solution.

## 3.    Experimental Results

Fig. 2 shows the textures of the rolled and annealed aluminum alloy strips. The texture of the rolled strip can be approximated by the β-fiber. The annealed strip has weak {001}<100> component along with weak deformation-texture remnant.

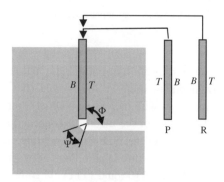

Fig. 1. Oblique angle Φ and curvature angle Ψ in ECAP. In P-pressing, top surface (*T*) and bottom surface (*B*) are same as in previous pressing, and in R-pressing, specimen is pressed after rotation through 180° about pressing direction of previously pressed specimen, by which shear direction becomes opposite to previous one.

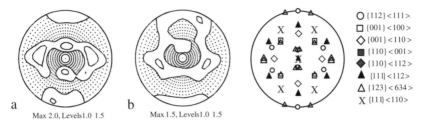

Fig. 2. {111} pole figures of 2 mm-thick (a) rolled and (b) annealed 1050 Al alloy strips.

Fig. 3 shows the measured textures of the top surface, center, and bottom surface layers of the $A$-$P^1$, $A$-$P^2$, $A$-$P^1R^1$, and $R$-$P^1$ specimens. Here $A$ and $R$ mean the annealed and the rolled strips prior to ECAP, and P and R indicate P-pressing and R-pressing in Fig. 1, respectively. The superscript indicates the number of ECAP passes. For example, the $A$-$P^1R^1$ specimen is prepared by P-pressing in the first pass and R-pressing in the second pass of the annealed strip, by which the shear direction in the first pass changes in the second pass.

The top surface layer (s = 0.9) of the $A$-$P^1$ specimen develops a shear texture rotated by 5° about the TD from the ideal shear texture consisting of {001}<110> and

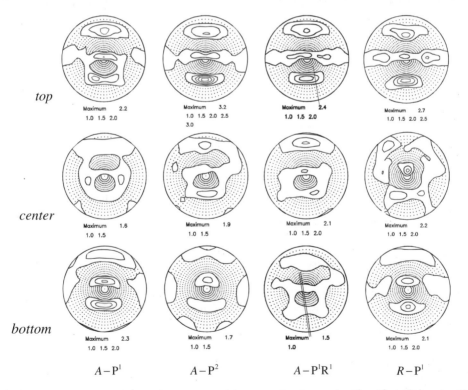

Fig. 3. {111} pole figures of top surface, center, and bottom surface layers of $A$-$P^1$, $A$-$P^2$, $A$-$P^1R^1$, and $R$-$P^1$ specimens.

{111}<112>, while the center layer (s = 0) develops the {110}<772> and {773}<159>, which are rotated by 10° from the ideal shear texture, and the bottom surface layer (s = -9) develops a texture whose major component is approximated by {115}<773> which is rotated by 15° from the ideal shear texture. These results are expected considering the non-uniform strain distribution along the thickness of ECAPed specimens. The texture of the $A$-90P$^2$ specimen is similar to that of the $A$-P$^1$ specimen with the intensity of the former being a little higher than the latter.

The comparison between the textures of the $A$-P$^1$ and $A$-P$^1$R$^1$ specimens indicates that a further rotation of orientation about the transverse axis takes place by shear reversal. This phenomenon was observed in asymmetric rolling[3,8,9] and discussed in detail.[3]

The texture of the $R$-90P$^1$ is similar to that of the $A$-90P$^1$ specimen, but has slightly higher intensity. The higher intensity in the $R$-type specimens compared with corresponding $A$-type specimens can be attributed to the difference in texture between the rolled and annealed specimens (Fig. 2). The initial textures are rotated about the transverse axis after ECAP. Therefore, the initial textures are reflected in ECAPed specimen.

Regardless of the differences in ECAP, the bottom surface layer of most ECAPed specimens has the lowest orientation density. Specimens ECAPed in one pass undergo the least deformation in the bottom layer. These facts imply that the last pass most influences the texture development.

## 4.    Calculation of Deformation

The elasto-plastic FEM analysis of ECAP was carried out using ABAQUS. Two-dimensional analysis was made with no strain in the width direction based on the experimental method. The thicknesses of the inlet and outlet channels were set to be 2 mm and 2.2 mm, respectively, because the specimen of 2 mm in initial thickness was thickened to 2.2 mm after ECAP. The number of elements was $10 \times 50$. Young's modulus and the flow curve of aluminum were approximated by 71 GPa and $\sigma = 179\varepsilon^{0.22}$.[3]

The calculated results are shown in Fig. 4. The strain distributions are not uniform along the thickness. The strains are the lowest in the bottom surface layer due to the corner gap.

## 5.    Calculation of Texture

The non-uniform deformation along the thickness gives rise to non-uniform textures. The textures were simulated using two approaches. One is the crystal-plasticity deformation-gradient (CPDG) method, and the other is the crystal plasticity shear-to-normal strain (CPSNS) method.

In the CPDG method, deformation is calculated by FEM. In the present ECAP deformation, a friction coefficient of 0.1 was used. The displacement gradients obtained from FEM were used to calculate the deformation textures of ECAPed strips based on the

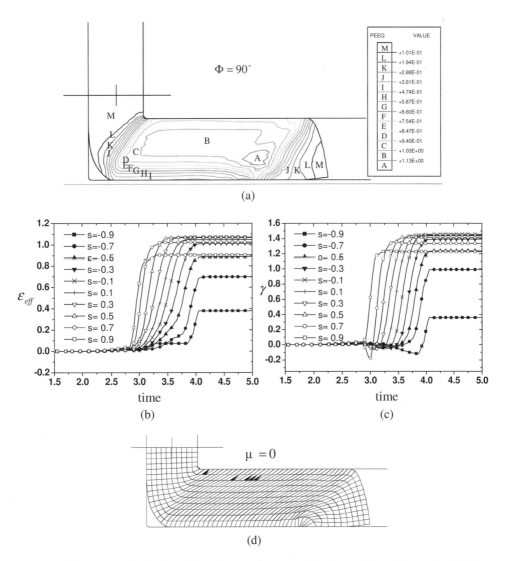

Fig. 4. (a) Effective strain distribution (a,b), engineering shear strain distributions (c), and deformed mesh (d) in specimen during ECAP with $\Phi = 90°$ and friction coefficient $\mu = 0$.

full constraints, rate- sensitive slip of fcc metals.[10] The rate-sensitive slip is expressed as a power law relationship between the rate sensitive shear rate $\dot{\gamma}_s$ and the resolved shear stress $\tau_s$ on each slip system. That is,

$$\tau_s = \tau_0 (\dot{\gamma}_s / \dot{\gamma}_0) |\dot{\gamma}_s / \dot{\gamma}_0|^{m-1} \qquad (1)$$

Here the strain rate sensitivity index $m$, the reference shear stress $\tau_0$, and the reference strain rate $\dot{\gamma}_0$ were assumed to be constant and equal for all slip systems. Isotropic hardening was also assumed. For texture simulations, only the slip distribution is relevant,

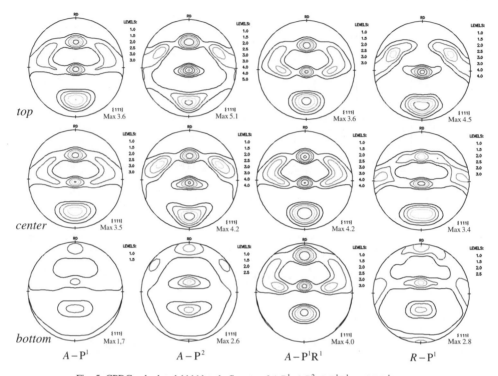

Fig. 5. CPDG calculated {111} pole figures of $A$-$P^1$, $A$-$P^2$, $A$-$P^1R^1$, and $R$-$P^1$ strips.

which is not affected by the magnitude of the stress components under isotropic hardening conditions. For this reason, $\tau_0$ was set to be equal in all grains constituting the samples.

The first step in calculation involves discretizing the initial sheet orientation distribution functions. The initial orientation distribution function of the annealed specimens was represented by 1512 differently oriented grains based on the measured texture (Fig. 2b). Each orientation had a specific volume fraction, which remained unchanged during lattice rotation. A scatter width of 7° was found to reproduce the experimental initial texture. The initial orientation distribution function of the rolled specimen was obtained by plane strain compression of a specimen made of randomly distributed 1006 grains to have a weak β-fiber texture. The strain rate sensitivity index $m$ was set to be 0.02. The strain hardening was not taken into account. The calculated results are shown in Fig. 5. The calculated results are in qualitative agreement with the measured data in Fig. 3, with a little poor agreement in the center layer textures.

The ideal ECAP is expected to undergo simple shear deformation, in which a displacement-gradient component $e_{13}$ has a finite value and other components are zero with the subscripts 1 and 3 indicating the axial direction and the sheet-normal direction, respectively. In real case, the thickness strain $e_{33}$ or the axial strain $e_{11} = -e_{33}$ may not be zero. In this case the ratio of $e_{13}$ to $e_{11}$ is finite. The ratio will be referred to as $\alpha$, Eq. (2). In the CPSNS method, the $\alpha$ values determine displacement gradients.

$$\alpha = e_{13} / e_{11} \qquad (2)$$

The textures were calculated as a function of $\alpha$ with $e_{11} = -0.1$. The calculated results are shown in Fig. 6. It can be seen that the texture is almost not influenced by the sense of shear direction especially at $|\alpha| \geq 15$ because the initial texture effect decreases with increasing deformation. As the $\alpha$ value increases, the texture approaches the ideal shear orientation. The texture intensity changes little at $|\alpha| \geq 5$. The ratio of the intensity of the {001}<110> component to that of the <111>//ND component becomes larger than one at $|\alpha| \geq 30$. In other words, the intensity of the {001}<110> component is higher than that of the <111>//ND component at $|\alpha| \geq 30$. Therefore, to increase the <111>//ND component, a careful control of ECAP is necessary.

The texture of the initial, annealed strip is close to that at $\alpha = 0$ because the strip was assumed to undergo a strain of $e_{11} = -0.1$. The textures of the top surface layers are close to those at $\alpha = -10$ to -15, except that of the $A - P^1$ specimen, which is close to that at $\alpha = -5$. The textures of other specimens can be approximated by that at $\alpha = -5$. This indicates that most ECAPed specimens underwent relatively small shear deformation.

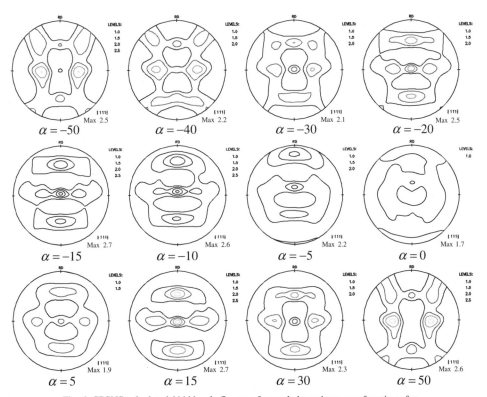

Fig. 6. CPSNS calculated {111} pole figures of annealed specimen as a function of α.

## 6.    Conclusion

ECAPed 1015 Al alloy strips develop, in principle, simple shear textures consisting of the {001}<110> and {111}<112> components. However, the ideal shear deformation cannot be obtained through the strip thickness because of corner gaps. For the oblique angle of 90°, the top surface layers undergo relatively strong shear deformation and the bottom surface layers undergo weak shear deformation. Consequently, textures close to the ideal shear texture develop in the top surface layers and weak shear textures develop in the bottom surface layers, regardless of process variables such as the number of ECAP passes and the shear reversal. Specifically, the top surface layer of the $A$-$P^1$ specimen develops a shear texture rotated by 5° about the TD from the ideal shear texture consisting of {001}<110> and {111}<112>, while the bottom surface layer develops a texture rotated by 15° about the TD from the ideal shear texture. The intensity of the shear textures of ECAPed strips little change at shear-to-normal displacement gradient ratios equal or higher than five. When the ratio increases above 30, the intensity of the {001}<110> component becomes higher than that of the <111>//ND component. Therefore, a careful control of ECAP is necessary to increase the <111>//ND component.

## Acknowledgment

One of the authors (DNL) wishes to express his appreciation for support by POSCO grant funded by 'the supporting project of a research in steel' (2008Z034).

## References

1. C.-H. Choi, K.-H. Kim, S.-Y., Jeong and D. N. Lee, *J. Kor. Inst. Met. Mater.* **35**, 429 (1997).
2. C.-H. Choi, K.-H. Kim and D. N. Lee, in *Synthesis/Processing of Lightweight Metallic Materials – Proc. Synthesis/Processing of Lightweight Metallic Materials II*, ed. C. M. Ward-Close *et al.* (TMS, 1997), pp. 37-48.
3. K.-H. Kim and D. N. Lee, *Acta Mater.* **49**, 2583 (2001).
4. D. N. Lee, *Mater. Sci, Forum* **449-452**, 1 (2004).
5. V. M. Segal. *Mater. Sci. Eng.* **A271**, 322 (1999).
6. J. Y. Park, S.-H. Hong and D. N. Lee, *Mater. Sci, Forum* **408-412**, 1431 (2002).
7. S. Matthies, *Phys. Stat. Sol.* **101**, 111 (1980).
8. S. H. Lee and D. N. Lee, *Int. J. Mech. Sci.* **43**, 1997 (2001).
9. S. K. Lee and D. N. Lee, *Int. J. Mech. Sci.* **50**, 869 (2008).
10. L. S. Toth, P. Gilormini ang J. J. Jonas, *Acta metall.* **36**, 3077 (1988).

# A STUDY ON IMPACT DEFORMATION AND TRANSFORMATION BEHAVIOR OF TRIP STEEL BY FINITE ELEMENT SIMULATION AND EXPERIMENT

TAKESHI IWAMOTO[1†], TOSHIYUKI SAWA[2]

*Department of Mechanical Systems Engineering, Graduate School of Engineering, Hiroshima University,*
*1-4-1 Kagamiyama, Higashi-Hiroshima, Hiroshima, 739-8527 JAPAN,*
*iwamoto@mec.hiroshima-u.ac.jp, sawa@mec.hiroshima-u.ac.jp*

MOHAMMED CHERKAOUI[3]

*The George W. Woodruff School of Mechanical Engineering, Georgia Institute of Technology,*
*801 Ferst Drive N. W. Atlanta, Atlanta, GA, 30332-0405 USA,*
*mcherkaoui@me.gatech.edu*

Received 15 June 2008
Revised 23 June 2008

Due to strain-induced martensitic transformation (SIMT), the strength, ductility and toughness of TRIP steel are enhanced. The impact deformation behavior of TRIP steel is very important because it is investigated to apply it for the shock absorption member in automobile industries. However, its behavior is still unclear since it is quite difficult to capture the transformation behavior inside the materials. There are some opinions that the deformation characteristics are not mainly depending on the martensitic transformation due to heat generation by plastic work. Here, the impact compressive deformation behavior of TRIP steel is experimentally studied by Split Hopkinson Pressure Bar (SHPB) method at room temperature. In order to catch SIMT behavior during impact deformation, volume resistivity is measured and a transient temperature is captured by using a quite thin thermocouple. Then, a finite element simulation with the constitutive model for TRIP steel is performed. The finite element equation can be derived from the rate form of principle of virtual work based on the implicit time integration scheme. Finally, the results between the computation and experiment are compared to confirm the validity of computational model.

*Keywords*: TRIP steel; impact deformation behavior; split Hopkinson pressure bar method; finite element method; transient temperature; thermocouple; volume resistivit.

## 1. Introduction

When steel with metastable austenitic phase undergoes a plastic deformation, the strain-induced martensitic transformation (henceforth SIMT), which a part of the austenitic phase transforms to the martensite by a different mechanism against a usual cooling, occurs[1]. Under a certain specific condition, the steel that accompanies the SIMT can expect the improvement of ductility and toughness as well as remarkably increasing strength without loosing tenacity. The phenomenon of high ductility and high toughness

---

[†]Corresponding Author.

according to the SIMT is called the TRIP (TRansformation-Indueced Plasticity), and steel with TRIP is called TRIP steel[1].

For instance, the low-alloyed high strength steel applying the TRIP is developed in recent years so that the automobile may improve the characteristics such as the body lightening and the high collision safety, and its application to an energy-absorbing member is investigated[2]. Therefore, it is important to study the SIMT behavior under the impact deformation because it can be thought that the crash energy absorption improves by the energy dissipation according to the SIMT.

For the TRIP type-high strength steel, there are a lot of the research works about the effect of the SIMT on formability in deep drawing process, etc[3]. As one of pioneer research works on the dynamic deformation behavior of the TRIP steel, Ishikawa and Tanimura[4] examine the strain rate and the temperature dependence of the shear strength of the TRIP steel by using the split Hopkinson pressure bar (SHPB) technique, and that experimentally investigate the impact deformation behavior of the TRIP steel. After this work, some articles on the dynamic deformation behavior of TRIP steels with the studies on the dual phase steels can be found. (For example, see Ref. 5) However, the influence of SIMT on the impact energy absorption characteristic of the TRIP steel is not clearly examined due to a lack of a measurement on microstructural evolution, and it is effective to reveal the improvement mechanism of the impact energy absorption characteristic by an aid of both experiment and numerical simulation.

In this research work, the impact compression test is first done at the room temperature by using the SHPB technique, and the impact stress-strain curve is measured. During the impact deformation, it is tried to capture the SIMT behavior and the temperature change accurately and in real time by measuring a change in the volume resistivity[6] and using a quite thin thermo-couple[7]. Next, the constitutive model proposed by Iwamoto et al.[8], which the deformation mode dependency of the deformation behavior of the TRIP steel can be expressed, is introduced into the rate form of the principle of virtual work, and a finite element analysis based on the implicit time integration is performed[9]. After the deformation and the transformation behavior under impact compressive deformation are simulated under a variety of test temperatures, the experiment and the computational results are compared and the validity of the computational result is shown.

## 2. Experimental Method

Type 304 austenitic stainless steel, which is a kind of the TRIP steel, is chosen for the test material. The specimen used here has a shape of cylinder with 6 mm in length and 5 mm in the diameter. Then, the solution treatment by 1323K for 30 min is applied for the specimens. Figure 1 shows the schematic diagram of the impact compression testing apparatus based on the SHPB technique used in this experiment. The compressive incident, the reflected and the transmitted waves at the time are recorded by the two strain gauges (KYOWA KSP-2-120-E4) glued on the stress bars, and impact stress-strain curve

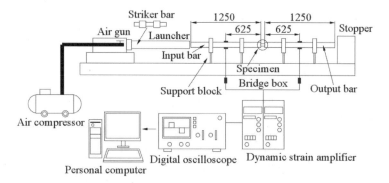

Fig. 1.   Schematic illustration of the impact compressive testing device based on the split Hopkinson pressure bar technique.

of the specimen can be calculated by the theory of stress wave propagation for one-dimensional elastic bar. Here, the material of the stress bar is the bearing steel (SUJ2 in JIS) and the diameter of the stress bars is 16 mm.

In order to catch changes in martensitic phase during the impact test, the volume resistivity[6] of the specimen was measured. In this case, the copper wire of 1 mm in the diameter was fixed to the position corresponding to the parallel part of the test piece by the spot welding, and the direct current of about 1 A was supplied. A quite thin thermo-couple[7] of about 140 $\mu$ m in thickness was fixed directly to the surface of the test piece by the spot welding for the measurement of the transient temperature at the test piece during the impact test. Here, the test temperature and the strain rate are set to 298 K and $1.2 \times 10^3$ /s, respectively.

## 3.  Finite Element Simulation

### 3.1.  *Finite element equation of motion*

Since the inertia term must be considered for a dynamic problem, the rate form of the principle of virtual work based on the Updated Lagrangian method, which involves the inertia term and neglects the body force term, is used[9]. After the multiaxial constitutive equation of the TRIP steel and the kinetics model for SIMT proposed by Iwamoto et al.[8] is introduced into the principle of virtual work, a finite element equation of motion is obtained by obeying the procedure of the conventional finite element method[10] as

$$\mathbf{M}^e \ddot{\mathbf{d}} + \mathbf{K}^e \dot{\mathbf{d}} = \dot{\mathbf{f}}_1 + \dot{\mathbf{f}}_2 + \dot{\mathbf{f}}_3 + \dot{\mathbf{f}}_t , \ \mathbf{M}^e = \int_V \rho \mathbf{N}^T \mathbf{N} dV , \ \dot{\mathbf{f}}_t = \int_{S_t} \mathbf{N}^T \dot{\mathbf{F}} dS , \tag{1}$$

where $\mathbf{d}$ is the nodal velocity vector, $\mathbf{N}$ is the shape function matrix, $\mathbf{M}^e$ is the consistent mass matrix, $\mathbf{K}^e$ is the stiffness matrix, $\dot{\mathbf{f}}_t$ is the nodal force rate vector applied on the boundary. $\dot{\mathbf{f}}_1$, $\dot{\mathbf{f}}_2$ and $\dot{\mathbf{f}}_3$ are the nodal force rate vectors which originate in viscoplasticity, transformation, the temperature change, respectively. The finite difference approximation based on the Houbolt method[10] for $\ddot{\mathbf{d}}$ in the above equation is employed[9].

### 3.2. *Computational conditions*

Figure 2 shows a computational model. The crossed triangular axisymmetric element is used for the discretization of the cylinder of type 304 austenitic stainless steel with the same size as the test piece, and this area is divided into $16\times60$ in each quadrangle of the elements. The nominal strain rate of $-1.2\times10^3$/s was given to the edge side with a shear-free condition and test temperature $T$ is set to 77, 213, 273, 298 K. The displacement rate on the edge is given by using the cosine function for realization of a smooth rise time in strain rate pulse. Moreover, the time step is calculated from a longitudinal sound speed of the elastic wave and the minimum length of the finite element to give the very small time step[10]. The heat transfer condition is only considered on the side surface and it is assumed to do the compressive test in the air.

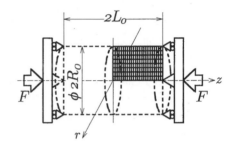

Fig. 2.    Computational model of axisymmetric bar under impact compression.

### 4.   Results and Discussion

### 4.1. *Experimental results*

Figure 3 (a) shows the nominal stress $\sigma_n$ – nominal strain $\varepsilon_n$ curve at the nominal strain rate $\dot{\varepsilon}_n = 1.2\times10^3$ /s. $\sigma_n$ rises as $\varepsilon_n$ increases, and the work hardening is observed. In addition, it is observed that the reproducibility of the experiment is high because four experiments are almost the same. Figure 3 (b) shows the change in volume resistivity –

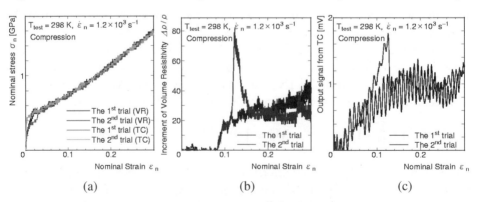

Fig. 3.   (a) Nominal stress - nominal strain, (b) change of volume resistivity - nominal strain and (c) output signal from the thermocouple - nominal strain measured by impact compression test.

$\varepsilon_n$ relationship measured during the impact compression. It can be observed that the change in volume resistivity tends to increase as $\varepsilon_n$ increases from this figure. It can be confirmed that curves obtained by two experiments become the similar, and the reproducibility of the experiment is high. However, the spike is appeared, and it is thought that it is because of noise during a transient response. There is a report that the change in volume resistivity rises along with the generation of SIMT, it undoes according to this report, SIMT is generated in the material, and the stress increases. Figure 3 (c) shows the output signal from the thermo-couple – $\varepsilon_n$ curves during the impact compression. From this figure, it can be observed that the temperature rises during the impact test though the oscillation in the signal is appeared. Moreover, saturation is shown though the peak like the transient response can be observed in the signal.

### 4.2. *Computational results*

Figure 4 shows (a) $\sigma_n$ – $\varepsilon_n$, (b) volume fraction of martensite $f^{\alpha'}$ – $\varepsilon_n$ and (c) average temperature rise $\Delta T$ – $\varepsilon_n$ relationships of TRIP steel obtained by the computation under various test temperature $T$ when the compressive deformation is given at the nominal strain rate $\dot{\varepsilon}_n = 1.2 \times 10^3$ /s. The deformation and transformation behavior as shown in this figure are similarly observed in the case of a low-speed deformation[8]. Therefore, when the basic features of the transformation and the deformation behavior of the TRIP steel can be reproduced by using this model under the impact transformation and it can be understood to obtain the solution stably by using the dynamic implicit solution technique used here.

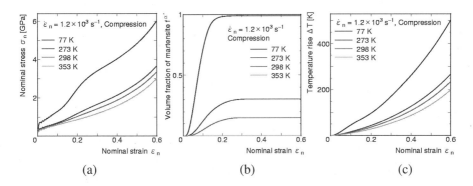

Fig. 4. (a) Nominal stress - nominal strain, (b) volume fraction of martensite - nominal strain and (c) temperature rise - nominal strain under impact compression test obtained by the finite element simulation.

### 4.3. *Comparison of the results obtained by experiment and computation*

Figure 5 shows $\sigma_n$ – $\varepsilon_n$ curves obtained by the experiment and the computation. From this figure, a fairly good agreement can be observed, and it can be understood that the computation is appropriate. Moreover, when Fig. 3 (b) and Fig. 4 (b) are compared, the basic features of the both curves such as the saturation and sigmoidal shape are corresponding, and it is also possible to make the experiment result correspond to the calculation result though a small amount of oscillation is observed in the experiment result. In addition, when Fig. 3 (c) and Fig. 4 (c) are compared, a similar tendency is shown and it can be said that the experimental data is useful though an initial peak might exist in the experiment.

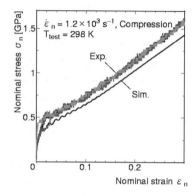

Fig. 5.  Comparison of nominal stress - nominal strain curve obtained by experiment and computation during impact compression.

## 5.  Concluding Remarks

The impact compression test was first done at the room temperature by using the SHPB technique, and the impact stress-strain curve was measured. During the impact testing, the SIMT behavior and the temperature change were captured by measuring a change in the volume resistivity and using a quite thin thermo-couple, respectively. A finite element analysis with the implicit time integration based on the rate form of the principle of virtual work introducing the constitutive model proposed by Iwamoto et al.[8] was performed. The validity of the computation was investigated by comparing with the experiment results.

## Acknowledgments

Financial supports from the Mazda foundation and the Suzuki foundation in Japan are gratefully acknowledged.

## References

1. F. D. Fischer, Q. P. Sun and K. Tanaka, *Appl. Mech. Rev.* **49**, 317 (1996).
2. W. Bleck, *J. Metals* **7**, 26 (1996).
3. K. Sugimoto, S. Hashimoto and Y. Mukai, *J. Mater. Proc. Technol.* **177**, 390 (2006).
4. K. Ishikawa and S. Tanimura, *Int. J. Plasticity* **8**, 947(1992).
5. H. Huh, S. -B. Kim, J. -H. Sung and J. -H. Lim, *Int. J. Mech. Sci.* **50**, 918 (2008).
6. A. W. McReynolds, *J. Appl. Phys.* **17**, 823 (1946).
7. D. Riddle, *Mech. Mater.* **31**, 131(1999).
8. T. Iwamoto, T. Tsuta and Y. Tomita, *Int. J. Mech. Sci.* **40**, 173 (1998).
9. Y. Tomita and T. Fujimoto, in *Dynamic Plasticity and Structural Behaviors — Proc. Plasticity'95*, ed. S. Tanimura *et al.* (Gordon and Breach Publisher, Luxembourg, 1995), pp. 401–404.
10. K. J. Bathe, *Finite Element Procedures* (Prentice Hall, New Jersey, 1996).

# Part I

## Plastic Instability and Strain Localization

# ENERGY ABSORPTION OF EXPANSION TUBE CONSIDERING LOCAL BUCKLING CHARACTERISTICS

KWANG-HYUN AHN[1], JIN-SUNG KIM[2], HOON HUH[3†]

*KAIST, Daejeon, KOREA,*
*ankh-1128@kaist.ac.kr, soitgoes@kaist.ac.kr, hhuh@kaist.ac.kr*

Received 15 June 2008
Revised 23 June 2008

This paper deals with the crash energy absorption and the local buckling characteristics of the expansion tube during the tube expanding processes. In order to improve energy absorption capacity of expansion tubes, local buckling characteristics of an expansion tube must be considered. The local buckling load and the absorbed energy during the expanding process were calculated for various types of tubes and punch shapes with finite element analysis. The energy absorption capacity of the expansion tube is influenced by the tube and the punch shape. The material properties of tubes are also important parameter for energy absorption. During the expanding process, local buckling occurs in some cases, which causes significant decreasing the absorbed energy of the expansion tube. Therefore, it is important to predict the local buckling load accurately to improve the energy absorption capacity of the expansion tube. Local buckling takes place relatively easily at the large punch angle and expansion ratio. Local buckling load is also influenced by both the tube radius and the thickness. In prediction of the local buckling load, modified Plantema equation was used for strain hardening and strain rate hardening. The modified Plantema equation shows a good agreement with the numerical result.

*Keywords*: Expansion tube; energy absorption; local buckling; modified Plantema equation.

## 1. Introduction

An expansion tube is a crash absorbing device with the expanding process by pushing a conical punch into the tube. The expansion tube has some different characteristics from the other types of energy absorbers. Compared to crushing energy absorbers, most commonly used for an energy absorber, an expansion tube has some advantages in energy absorbing. Since the reaction force during the expanding process increases moderately compared to a crushing energy absorber, an expansion tube has the small peak load and it can reduce the crushing deceleration. A smaller peak load can reduce the damage in the main equipment caused by the crushing deceleration. The other advantage of an expansion tube is that the total length of an expansion tube can be used for energy absorption. By this characteristics, an expansion tube can absorb the crash energy as much as crushing energy absorbers, although specific energy absorption of an expansion tube is not efficient compared to the crushing energy absorber.[1] In spite of those advantages, an expansion tube cannot be used for the equipment which needs weight

---

†Corresponding Author.

reduction because of the heavy weight of a punch to expand a tube, but it is being used for heavy equipments such as a train. Further study for space efficiency of an expansion tube should be carried out in order to overcome the disadvantage of an expansion tube.

The tube expanding mechanism can be referred by a tube flaring process. Flaring is understood as a forming process involving expansion of a cylindrical tube which is expanded by a conical punch pushed into the tube. Hill[2] presented a mathematical model of stress flow during tube expansion. It can be a general method for many studies on tube flaring processes. Manabe *et al.*[3-4] conducted a series of experiments related to tube flaring. They investigated effects of the size and the mechanical properties of tubes, lubricants and punch angle. Lu[5] investigated the expansion ratio and punch velocity by using finite element analysis. Daxner *et al.*[6] investigated the effect of instability phenomena − local buckling and necking on the tube end. Almeida *et al.*[7] considered the effect of lubricant on instability phenomena in the flaring process by experiments. Analytical expressions were derived to determine stress and strain fields by Fischer *et al.*[8]

This paper investigates the effect of local buckling on the crash energy absorption of an expansion tube. Local buckling characteristics during a tube expanding process are evaluated since local buckling has a critical effect on energy absorption of an expansion tube. For enhancement of energy absorption, local buckling load is predicted by a modified Plantema equation with respect to the tube shape. Dynamic effect is also considered.

## 2. Analysis Condition

### 2.1. *Finite element model and boundary conditions*

Parametric study is carried out in order to consider the effect of the tube and punch shape on energy absorption of an expansion tube. Energy absorption and the buckling load of an expansion tube are obtained by finite element analysis. An implicit elasto-plastic finite element code, ABAQUS/Standard, is used for the parametric study. An explicit code, ABAQUS/Explicit, is also used for the verification of the rate effect.

The expansion tube shown in Fig. 1(a) is a typical example of an expansion tube used in a train. Fig. 1(b) shows a finite element model and the dimensions of a tube and a

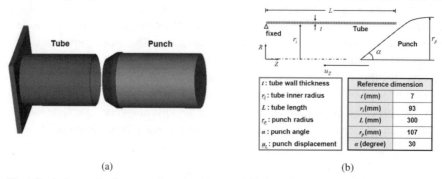

| | Reference dimension | |
|---|---|---|
| $t$ : tube wall thickness | | |
| $r_i$ : tube inner radius | $t$ (mm) | 7 |
| $L$ : tube length | $r_i$(mm) | 93 |
| $r_e$ : punch radius | $L$ (mm) | 300 |
| $\alpha$ : punch angle | $r_p$(mm) | 107 |
| $u_z$ : punch displacement | $\alpha$ (degree) | 30 |

(a)                                                        (b)

Fig. 1. Typical example of an expansion tube: (a) expansion tube and punch; (b) axi-symmetric FE model.

punch. The axi-symmetric condition can be used for efficiency of analysis. Parametric studies were carried out with changing each parameter from the reference dimension in Fig. 1(b).

### 2.2. *Friction and material*

Prior to the parametric study, the friction condition should be determined first since the friction condition is an important factor in an expansion tube due to sliding during the expanding process. In this paper, the friction coefficient is determined by referring to the experimental results.[9] Since analyses results with the friction coefficient of 0.05 give a good coincidence with experimental results,[10] the friction coefficient is selected as 0.05 for following analyses.

In an energy absorber, the material properties are an also important factors. For the crash energy absorption, a high strain hardening material gives the good crashworthiness. In case of an expansion tube, tearing in a tube end can be induced by the tube expanding process. In order to avoid this phenomenon, materials for an expansion tube should have high elongation characteristics as well as high strain hardening. Since TWIP steel has both high strain hardening and high elongation, it is used for a tube material. Tensile tests are carried out to obtain the material properties of TWIP steel for analyses.[10]

### 3.  Energy Absorption of Expansion Tubes

### 3.1. *Parametric study*

In order to design an expansion tube which has high energy absorbing capacity, the effect of the tube and punch shape on energy absorption should be evaluated first. A parametric study is carried out with finite element analysis. Parameters for the parametric study are the tube wall thickness, the tube radius, the punch angle and the expansion ratio. The expansion ratio, $e$, means the ratio of the punch radius to the tube inner radius.

Fig. 2 and Fig. 3 show the effect of the tube and the punch on specific energy absorption of an expansion tube. Specific energy absorption of an expansion tube increases as the tube wall thickness increase as shown in Fig. 2. Specific energy absorption of an expansion tube, however, decreases as the tube radius increases. Hence,

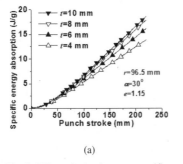

(a)

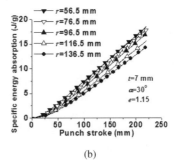

(b)

Fig. 2. Effect of tube shape on specific energy absorption: (a) tube wall thickness; (b) tube radius.

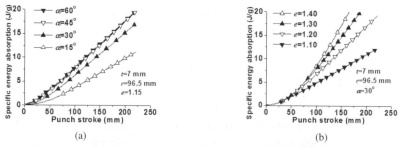

Fig. 3. Effect of punch shape on specific energy absorption: (a) punch angle; (b) expansion ratio.

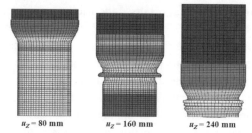

$u_z = 80$ mm          $u_z = 160$ mm          $u_z = 240$ mm

Fig. 4. Deformed shape when buckling occurs ($t$=9 mm, $r_i$=113 mm, $\alpha$ =35 mm, $e$=1.4).

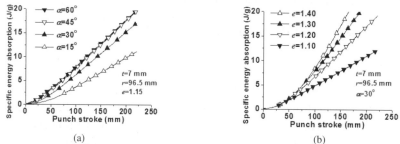

Fig. 5. Effects of buckling: (a) effects of tube radius when tube wall thickness is small; (b) effects of expansion ratio when punch angle is large.

design of an expansion tube with larger $t/r$ can improve specific energy absorption capacity. Fig. 3 shows larger punch angle and expansion ratio can improve specific energy absorption of an expansion tube.

### 3.2. *Effect of local buckling*

Fig. 4 shows deformed shapes when local buckling occurs. Parametric study is carried out in order to find the relation between the buckling characteristics and each parameter. As show in Fig. 5, the buckling easily occur at smaller tube wall thickness, $t$, larger tube radius, $r$, larger punch angle, $\alpha$ and larger expansion ratio, $e$. Fig. 5 also shows that absorbed energy decreases significantly when local buckling occurs. In accordance with the results in Fig. 2 and Fig. 5, the tube with larger $t/r$, larger tube wall thickness and smaller tube radius, can improve not only specific energy absorbing capacity but also local buckling resistance. Therefore, the tube shape with larger t/r can be selected in a

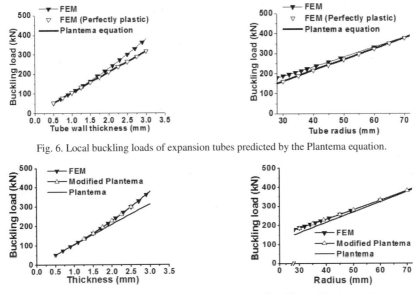

Fig. 6. Local buckling loads of expansion tubes predicted by the Plantema equation.

Fig. 7. Local buckling loads of expansion tubes predicted by modified method.

given design space. Although, the punch with larger $\alpha$ and larger $e$ can improve specific energy absorbing capacity, they have a bad influence on buckling resistance as shown in Fig. 3 and Fig. 5. Therefore, a design guideline is needed to prevent buckling and enhance the energy absorbing.

## 4. Local Buckling Characteristics of Expansion Tubes

### 4.1. *Plantema equation*

The Plantema equation is the most common equation used to predict inelastic local buckling of a tube. For the thin-walled steel tube the Plantema equation is given by $P_{cr}=2\pi rt\sigma_y$. Fig. 6 shows comparison of the buckling load predicted by the Plantema equation and finite element analysis at various tube shapes. The tube dimensions for this analysis are the same as those in the previous chapter. The punch shape is, however, a flat type since the punch angle is not included in the Plantema equation. As shown in Fig. 6, the Plantema equation cannot predict the accurate local buckling load, but predicts the result with perfectly plastic assumption since it did not consider the strain hardening.

### 4.2. *Modified plantema equation*

In order to consider the strain hardening effect with the Plantema equation, averaged compressive stress on fixed ends can be substituted for the yield stress term in the Plantema equation. Fig. 7 shows that modified results show a good agreement with numerical results since the compressive stress includes the strain hardening effect. In order to modify the Plantema equation by using this method, averaged stress on the tube

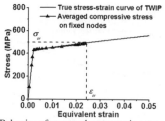

Fig. 8. Behavior of averaged compressive stress on fixed nodes.

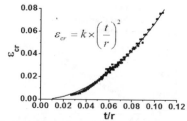

Fig. 9. Buckling strain w.r.t. the ratio of tube wall thickness to radius.

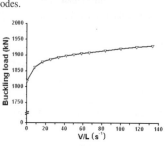

(a)

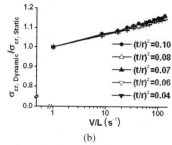

(b)

Fig. 10. Rate effect on local buckling load: (a) Buckling load w.r.t the equivalent strain; (b)Ratio of dynamic to static buckling stress w.r.t. the equivalent strain rate.

ends should be investigated. According to the results in Fig. 8, averaged compressive stress on fixed nodes follows the stress–strain curve of the material. It means the buckling load can be predicted by $P_{cr}=2\pi rt(A+B\varepsilon_{cr}^{\ n})$ where $A$, $B$ and $n$ are material coefficients in Ludwik model. Fig. 9 shows that the analysis result of buckling strain for each shape is proportional to the square value of the ratio of the tube wall thickness to the tube radius. Therefore, the Plantema equation can be modified as follows:

$$P_{cr} = 2\pi rt\left[A+B\left\{k\left(\frac{t}{r}\right)^2\right\}^n\right].$$

(1)

In order to predict the local buckling load at the high crushing speed, strain rate hardening effect should be included in Eq. (1). Fig. 10 shows the local buckling load and stress at various punch velocities. As shown in Fig. 10(b), the local buckling stress increases as the punch velocity increases and it is proportional to the logarithmic value of the equivalent strain rate, $V/L$. Since the ratio of dynamic to static local buckling stress is not influenced by the value of $t/r$, Eq. (1) can be modified as follows:

$$P_{cr} = 2\pi rt\left[A+B\left\{k\left(\frac{t}{r}\right)^2\right\}^n\right]\left[1+a\ln\left(\frac{V/L}{\dot{\varepsilon}_0}\right)\right], \quad \text{where } \dot{\varepsilon}_0 = 1/\sec.$$

(2)

$k$ and $a$ are the fitting parameters determined by the material properties. In case of TWIP steel, $k$ has a value of 6.8 and $a$ has a value of 0.07. By using Eq. (2), local buckling loads for various tube shapes at various punch velocities can be predicted with the fitting parameter $k$ and $a$ of each material. The fitting parameters $k$ and $a$ of each material can be obtained by finite element analysis.

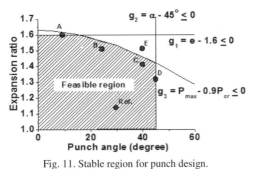

$$g_2 = \alpha - 45° \le 0$$

$$g_1 = e - 1.6 \le 0$$

$$g_3 = P_{max} - 0.9P_{cr} \le 0$$

Fig. 11. Stable region for punch design.

Table 1. Absorbed energy of newly designed model.

| Design point | Shape | | Absorbed energy (kJ) |
|---|---|---|---|
| | $\alpha$ | $e$ | |
| Ref. | 30° | 1.15 | 200.0 |
| A | 10° | 1.6 | 181.4 |
| B | 25° | 1.5 | 343.0 |
| C | 40° | 1.4 | 404.4 |
| D | 45° | 1.3 | 360.1 |
| E | 40° | 1.5 | 295.4 (buckling) |

Fig. 11 shows an example of the punch design using the modified Plantema equation. The tube shape is the same as that in Fig. 1. The first and second constraints, $g_1$ and $g_2$, restrict tearing and outward curling of tubes. The third constraint, $g_3$, is the modified Plantema equation. The buckling load predicted by the modified Plantema equation is evaluated when the punch shape is flat while the punch in an expansion tube has an inclined angle. For that reason, safety factor of 0.9 in the third constraint is introduced to consider the change of punch angle.[10] Table 1 represents the absorbed energy for each design point. Absorbed energy increases by 101.1 % in point C.

## 5. Conclusions

This paper proposes a design guideline to prevent local buckling in an expansion tube by carrying out a parametric study. Local buckling load should be predicted in order to improve energy absorption of the expansion tubes. The Plantema equation, most commonly used for predicting inelastic local buckling load, has been modified to consider the strain hardening and the strain rate hardening effect. The modified Plantema equation gives accurate local buckling loads corresponding to numerical results. The modified Plantema equation proposed in this paper can provide constraints for an expansion tube design to avoid a local buckling.

## References

1. N. Jones, *Structural Impact*. (Cambridge University Press, 1989).
2. R. Hill, *The Mathematical Theory of Plasticity*. (Oxford at the Clarendon Press, 1950).
3. K. Manabe and H. Nishimura, *J. Jpn. Soc. Technol. Plasticity*, **24**, 47 (1983).
4. K. Manabe and H. Nishimura, *J. Jpn. Soc. Technol. Plasticity*, **24**, 276 (1983).
5. Y. H. Lu, *Finite Element Anal. Des.*, **40**, 305 (2004).
6. T. Daxner, F. G. Rammerstorfer and F. D. Fischer, *Comput. Methods Appl. Mech. Engrg.*, **194**, 2591 (2005).
7. B.P.P. Almeida, M.L. Alves, P.A.R. Rosa, A.G. Brito and P.A.F. Martins, *Int. J. Mach. Tools Manuf.*, **46**, 1643 (2006).
8. F. D. Fischer, F. G. Rammerstorfer and T. Daxner, *Int. J. Mech, Sci.*, **48**, 1246 (2006).
9. W. Choi, T. Kwon, S. Kim, H. Huh and H. Jung, *Proc. KSAE spring conference*, (Changwon Korea, 2007), pp.1996-2001.
10. K. Ahn, J. S. Kim and H. Huh, *Numisheet 2008*, (Interlaken Switzerland).

# Part J
# Plasticity in Advanced Materials

# CHANGE IN MICROSCOPIC HARDNESS DURING TENSILE PLASTIC DEFORMATION OF POLYCRYSTALLINE ALUMINUM AND TITANIUM

XIAOQUN WANG [1]

*R & D Center, Panasonic Home Appliances Co. Ltd.*
*Hanghzhou 310018, P. R. CHINA,*


TAKEJI ABE [2][†]

*Tsuyama College of Technology, Tsushima-Fukui 1-8-64-6,*
*Okayama 700-0080, JAPAN,*
*t-abe@po4.oninet.ne.jp*

Received 15 June 2008
Revised 23 June 2008

Microscopic hardness on free surface of polycrystalline metal during plastic deformation is closely related to the inhomogeneous deformation in respective grains. Uniaxial tensile tests were carried out on annealed pure aluminum sheet specimens with different averaged grain size and also on annealed pure titanium sheet specimen. The microscopic hardness was measured with the Vickers type micro-hardness testing machine. The increase in micro-hardness is larger at the grain boundary area than the central area of grains. The increase in the hardness is dependent on the averaged grain size of polycrystalline metals. The experimental results are discussed in relation to Hall-Petch relation concerning the grain size dependence of the yield stress or the flow stress.

*Keywords*: Plasticity; polycrystalline metal; aluminum; titanium; micro-hardness; inhomogeneous deformation.

## 1. Introduction

It is well-known that the hardness of material is related to its mechanical characteristics. The microscopic inhomogeneity of polycrystalline pure aluminum and pure titanium is evaluated by testing the micro-hardness in grains. Namely, the change in micro-hardness associated with plastic deformation is studied, by measuring hardness before and after tensile deformation.

[†]Corresponding Author.

It is recognized that hardness corresponds to the yield stress or the flow stress of metals. Therefore, it is expected that the distribution of flow stress in polycrystalline metals as well as the microscopic plastic deformation behavior are clarified by measuring the microscopic hardness for various area in grains of polycrystalline metals.

It is known that titanium has strong anisotropy of plastic deformation compared with that of aluminum. Hence, the influence of grain orientation on the change in micro-hardness was investigated using pure titanium specimen.

The well known Hall-Petch relation[1, 2] between the yield stress or the flow stress and the grain size is represented as follows

$$\sigma = \sigma_0 + K \, d^{(-1/2)},\tag{1}$$

where $\sigma$ is the yield or the flow stress, d is the grain size, and $\sigma_0$ and K are constants. This is quite unique relation representing the effect of grain size on the flow stress of polycrystalline metals[3, 4]. In the present study, specimens with various grain sizes are prepared for aluminum. Titanium specimen is also used to investigate the effect of grain orientation on the change in micro-hardness.

## 2.   Experimental Method

### 2.1.   *Specimen*

The material used in the present study is polycrystalline pure aluminum and pure titanium for industrial use. The shape of the tensile specimen is shown in Fig. 1. Four specimens with different grain size were prepared for aluminum in order to investigate the effect of grain size on the micro-hardness. The grain size was adjusted by annealing processes. Electro-polishing was used for the preparation of the specimen surface. The resultant averaged grain diameter of the aluminum specimens are about 985 $\mu$m ( Specimen $T_1$ ), 488 $\mu$m ( $T_2$ ), 288 $\mu$m ( $T_3$ ) and 113 $\mu$m ( $T_4$ ), respectively. Meanwhile, the average grain size of the titanium specimen is 180 $\mu$m.  The measuring area is 8×8 mm$^2$ for Specimen $T_1$ and $T_2$, 4×4 mm$^2$ for Specimen $T_3$, and 2×2 mm$^2$ for aluminum Specimen $T_4$ and 2×2 mm$^2$ for titanium specimen.

### 2.2.   *Measurement of micro-hardness and grain orientation*

Tensile tests were conducted on the specimens until the true strain of $\varepsilon = 0.03, 0.06$ and $0.09$ are obtained.

Micro-hardness was measured with Akashi Micro-hardness Measuring Equipment MZT-4. Examples of the load-depth relations are shown in Fig. 2(a) for aluminum and (b) for titanium. For the present study the testing load was chosen as 9.8mN for aluminum and 49mN for titanium, considering the size of indentation pits in relation to the grain size. The basic load of 1/10 of the testing load was initially loaded to avoid the influence of the surface hardened layer. The load was chosen so that the size of the indentation pit is smaller than about 8 $\mu$m. The experimentally measured micro-hardness is denoted as HUT[68] in the following [5].

The orientation of grain in titanium specimen was measured with Link Opal System (Oxford Co.).

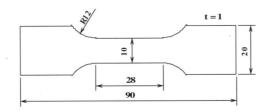

Fig. 1. Shape and dimensions of specimen for tensile test (mm).

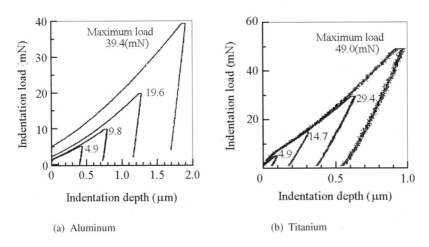

(a) Aluminum

(b) Titanium

Fig. 2. Indentation load-depth curves in micro-hardness measurement.

## 3. Experimental Results and Discussion

### 3.1. *Micro-hardness change with applied strain*

The change in micro-hardness with the applied plastic strain is shown in Fig. 3, for both the central area of grains and the area close to the grain boundaries. The micro-hardness is the averaged value obtained from 30 grains in respective specimens.

It is seen from Fig. 3 that the hardness increases with the strain and that the increase in hardness near grain boundaries is larger than that in the central region of grains. It is considered that this is due to the severe deformation near grain boundaries in order to maintain continuity of strain across the boundaries.

Fig. 4 shows the relation between averaged hardness and grain size for respective specimens. In Fig. 4, micro-hardness is plotted against d $^{(-1/2)}$, and the relation is almost linear, which suggests that a relation similar to Hall-Petch relation for the yield stress or

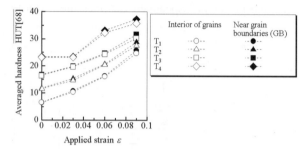

Fig. 3.   Change in averaged hardness with applied strain interior of grains and near grain boundaries of aluminum specimens.

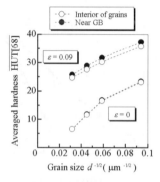

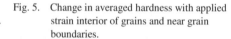

Fig. 4.   Relation between averaged hardness and averaged grain size in aluminum specimens.

Fig. 5.   Change in averaged hardness with applied strain interior of grains and near grain boundaries.

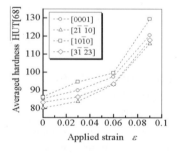

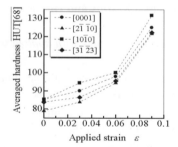

(a)   Interior of grains.

(b)   Near grain boundaries.

Fig. 6.   Change in averaged hardness with applied strain interior of grains and near grain boundaries in respective grains with various grain orientation in titanium specimen.

the flow stress is applicable between micro-hardness and the averaged grain size. The effect of grain size on micro-hardness is smaller after plastic deformation than that before deformation.

The micro-hardness of titanium specimen also increases with the applied plastic strain as shown in Figs. 5 and 6. It is seen that the micro-hardness is large in the order of $[2\bar{1}\bar{1}0]<[3\bar{1}\bar{2}3]<[0001]<[10\bar{1}0]$ of the tensile grain orientations. It is considered that this corresponds to the fact that many multiple slips or twins are observed in the grains or near the grain boundary of $[10\bar{1}0]$ grains.

### 3.2. Micro-hardness distribution in grains

The distribution of hardness in a grain is measured. One example of which is representatively shown in Fig. 7. The measurement was done for five representative grains in respective specimen and the averaged values are shown in Fig. 8.

Figure 8 shows that the micro-hardness increase with the applied strain is larger at the grain boundary area than the central are of grains. The difference is large for the grains with small grain size than that for the grains with large grain size.

Figure 9 shows the similar measurement for the titanium specimen. Again, it is seen that the micro-hardness near grain boundary is larger than that in the central area of grains.

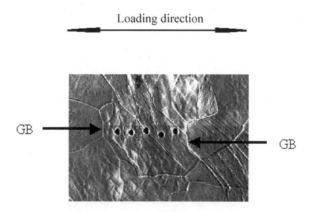

Fig. 7.   An example of position of indenters with various distances from grain boundaries.

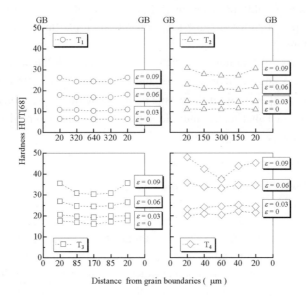

Fig. 8.   Relation between micro-hardness values and distance from grain boundaries in aluminum specimens.

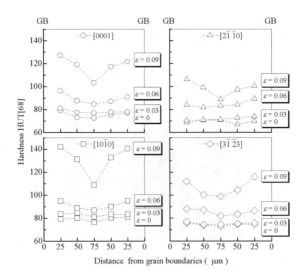

Fig. 9.   Relation between averaged hardness values and distance from grain boundaries in grains with various grain orientations of titanium specimen.

It may be possible to point out that the difference between the values of micro-hardness near the grain boundary and in the center of grains is large for those grains having high averaged value of micro-hardness.

From the experimental data mentioned above, the micro-hardness value H of polycrystalline metals is considered to be represented as follows.

$$H = \text{Hinitial} + \text{Hstrain} + \text{Hposition} + \text{Horientation} \tag{2}$$

Hinitial is the initial micro-harness before plastic deformation and is supposed to be a function of $d^{(-1/2)}$, where d is the grain size. Hstrain is the micro-hardness after plastic deformation and is a function of the strain $\varepsilon$ and $d^{(-1/2)}$. Hposition is the micro-hardness increase depending on the distance from the grain boundary. Horientation is the micro-hardness depending on grain orientation.

Further discussion seems to be necessary for the physical explanation of Eq. (2).

## 4. Conclusions

(1) It is characteristic that micro-hardness is dependent on the grain size before as well as after plastic deformation. Namely, the micro-hardness is large for small grains. Similar relation to Hall-Petch relation for the yield or the flow stress is applicable to the effect of the grain size on micro-hardness.

(2) The initial microscopic hardness is larger at the grain boundary area than in the central area of grains, for both polycrystalline aluminum and titanium. The hardness values increase with plastic strain, though the increase near the boundary area is large for grains with small grain size.

(3) The hardness of polycrystalline titanium is slightly dependent on the orientation of grains. The micro-hardness value is large for the grains having the tensile axis $[10\bar{1}0]$, and then $[0001]$, $[3\bar{1}\bar{2}3]$ and $[2\bar{1}\bar{1}0]$.

## Acknowledgment

The authors are indebted to Professors Naoya Tada and Tashiyuki Torii of Okayama University for their support on the present study.

## References

1. E. O. Hall, *The Deformation and Ageing of Mild Steel: III Discussion of Results, Proc. Phys. Soc.,* **B64**, 747 (1951).
2. J. Petch, *The Cleavage Strength of Polycrystals, J . of Iron and Steel Institute,* **174**, 25 (1953).
3. A. Kelly and R. B. Nicholson (eds.), *Strengthening Methods in Crystals,* (Elsevier, 1971).
4. K. J. Kuzydlowski and B. Ralph, *The Quantitative Description of the Microstructure of Materials,* (CRC Press, 1995), p.282.
5. T. Abe and T. Tsuboi, *Memoir Faculty Engineering., Okayama University,* **35**, 9 (2001).

# FLD OF AZ31 SHEET UNDER WARM STRETCHING AND ITS PREDICTION

TETSUO NAKA[1†], YASUHIDE NAKAYAMA[2]

*Yuge National College of Maritime Technology,*
*1000 Yuge Kamijima Cho Ochi-Gun Ehime, 794-2593, JAPAN,*
*naka@ship.yuge.ac.jp , nakayama@ship.yuge.ac.jp*


TAKESHI UEMORI[3]

*Department of Mechanical Engineering, Kindai University*
*1, Takayaumenobe, Higashi-Hiroshima, 739-2116, JAPAN,*
*uemori@hiro.kindai.ac.jp*


RYUTARO HINO[4], FUSAHITO YOSHIDA[5]

*Department of Mechanical System Engineering, Hiroshima University*
*1-4-1, Kagamiyama, Higashi-Hiroshima, 739-8527, JAPAN,*
*rhino@hiroshima-u.ac.jp, fyoshida@hiroshima-u.ac.jp*


MASAHIDE KOHZU[6], KENJI HIGASHI[7]

*Department of Metallurgy and Materials Science, Osaka Prefecture University*
*1-1, Gakuen-Cho, Sakai, 599-8531, JAPAN,*
*kohzu@mtr.osakafu-u.ac.jp, higashi@mtr.osakafu-u.ac.jp*

Received 15 June 2008
Revised 23 June 2008

Forming Limit Diagrams (FLDs) of a magnesium alloy (AZ31) sheet at various forming speeds (3 to 300 mm·mm$^{-1}$) at several temperatures of 100-250 °C were investigated by performing a punch stretch-forming test. The forming limit strains increased with temperature rise and with decreasing forming speed, where the effect of forming speed was stronger at higher temperatures. To describe such a characteristic of FLD of AZ31, the Marciniak-Kucznski type forming limit analysis was conducted using the Backofen-type constitutive equation ($\sigma = C\varepsilon^n \dot{\varepsilon}^m$). In this analysis, the damage evolution in the necking zone was taken into account based on Oyane's ductile fracture criterion. The numerical results of the FLD show a good agreement with the corresponding experimental observations.

*Keywords*: Forming limit diagram; warm stretching; AZ31.

---

[†]Corresponding Author.

## 1. Introduction

Magnesium alloys are attractive structural metals because of their lightweight characteristic, and in recent years they have been increasingly used for many industrial products such as vehicle components, casings of laptop computers and mobile phones[1, 2]. These magnesium products are manufactured mostly by die-casting and thixomolding[3, 4], but seldom by sheet press-forming since magnesium has low ductility at room temperature due to its hexagonal close-packed structure. To overcome the problem, the best choice of forming technologies would be warm press-forming since the ductility becomes considerably higher when heating-up the sheets[5-9].

For the prediction of formability of sheet metals, it is very popular to use the forming limit diagrams (FLDs), however, only a limited number of papers have been published so far on the warm-FLD for magnesium alloy sheets. As for theoretical predictions of the FLD for rate-sensitive materials, several works have been reported[10-13], where the FLDs were determined analytically based on the Marciniak-Kuczynski (M-K) concept of localized necking. To the authors' best knowledge, such theoretical predictions have seldom been incorporated with discussion of strain and strain-hardening effects on the warm-formability of magnesium alloy sheets.

In the present work, the effects of temperature and forming speed on the FLD for a fine-grain AZ31 magnesium alloy sheet were investigated. The FLDs were determined by performing punch stretch-forming tests at several temperatures from 100 °C to 250 °C at various forming speeds (3 to 300 mm·mm$^{-1}$). The analysis for the FLD determination was also carried out using a rate-dependent constitutive equation of the Backofen-type ($\sigma = C\varepsilon^n \dot{\varepsilon}^m$) together with Oyane's ductile fracture criterion[14].

## 2. Experimental Work

An AZ31 magnesium alloy sheet of 0.8 mm thick and 7 μm grain diameter was used for the experiment. The chemical compositions of the sheet are listed in Table 1. A rectangular sheet specimen was clamped firmly, and stretched by a semi-spherical headed punch of 50 mm diameter. In order to obtain various strain ratios $\beta (= \varepsilon_2/ \varepsilon_1)$, from -0.5 (uniaxial tension) to 1.0 (balanced biaxial stretch), specimens of several aspect ratios (120 mm in the rolling direction × 30 to 120 mm in the transverse direction) were tested. Experiments were performed at various punch speeds of 3, 30 and 300 mm·min$^{-1}$ in an Instron-type (screw-driven) machine at temperatures of 100, 150, 200 and 250 °C. The specimens were heated by an electric heater installed in the punch. In order to register the strain and strain-rate paths and determine the final fracture of the sheet, the deformation of scribed circles (6.3 mm diameter) printed on the surface of the specimens were monitored with a CCD camera and recorded with a video system during the stretch-forming test.

Table 1. Chemical compositions of the specimen. (mass %).

| Al | Zn | Mn | Fe | Si | Cu | Ni | Ca | Pb | Sn | others | Mg |
|----|----|----|----|----|----|----|----|----|----|--------|----|
| 2.9 | 0.82 | 0.67 | 0.0022 | 0.022 | 0.0018 | 0.0008 | 0.001 | 0.001 | <0.001 | 0.3 | bal. |

## 3.  Results and Discussion

### 3.1.  *Experimental results*

Figs 1(a)-(c) show the FLDs for testing temperatures of 150, 200 and 250 °C, respectively, obtained at three different punch speeds of 3, 30 and 300 mm·min$^{-1}$. Although the strain rate was changing during each test, the effective plastic strain rates at the final stage of the tests, for punch speeds of 3, 30 and 300 mm·min$^{-1}$, were of the order $10^{-3}$, $10^{-2}$ and $10^{-1}$ s$^{-1}$, respectively. The effect of punch speed on the FLD becomes much stronger with increase of temperature. This is because the critical resolved shear stress decreases and the number of slip system of the crystals increases with temperature-rise, and the formability improves.

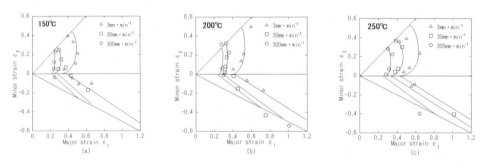

Fig. 1.  FLDs for AZ31 magnesium alloy sheet obtained at various temperatures at several forming speeds.

The forming limit strains at 100 °C were much smaller than those obtained at 150-250 °C (the data are not shown here).

### 3.2.  *Analytical prediction*

The M-K type analysis, which assumes that a thickness imperfection (a groove) develops into localized necking, was performed using the Backofen-type constitutive equation ($\sigma = C\varepsilon^n \dot{\varepsilon}^m$, where $C$, $m$ and $n$ are temperature dependent material parameters). Figs 2(a) and (b) illustrate the idea of the M-K model. Here, areas inside and outside of the groove are denoted by region A and B, respectively. For a given linear strain path for region A (e.g., under $\beta_A = \dot{\varepsilon}_{2A} / \dot{\varepsilon}_{1A} = $ constant), the strain localization in the groove B is calculated based on the force equilibrium in both $x_1$ and $x_2$ directions which corresponds to the major and minor strain ($\varepsilon_1$ and $\varepsilon_2$) directions. Since the thickness in region B is smaller than that in region A, strain develops more rapidly in region B than in region A. The limit strains $\varepsilon_{1A}$ and $\varepsilon_{1B}$ in region A are determined as strains at the final stage of strain localization, when only region B continuously deforms (i.e. $d\bar{\varepsilon}_{1A} / d\bar{\varepsilon}_{1B} \rightarrow 0$). Usually the M-K analysis is conducted for positive strain ratio conditions ($\beta_A = \dot{\varepsilon}_{2A} / \dot{\varepsilon}_{1A} > 0$), however, in the present paper it is also performed for negative strain ratio conditions ($\beta_A \leq 0$). For $\beta_A > 0$ the groove is assumed to be placed in the transverse direction of the major strain (see Fig. 2(a)), while for $\beta_A \leq 0$ it is located in the zero-extension direction (see Fig. 2(b)). For the stress-strain analysis, a yield function proposed by Logan-Hosford was employed, since it can well describes the yield surface of AZ31 sheet at high temperature[15].

In this analysis, the damage evolution in the grove is taken into account using Oyane's fracture criterion where it is assumed that the fracture occurs when the following damage parameter $\phi$ reaches a material's critical value $\phi_c$:

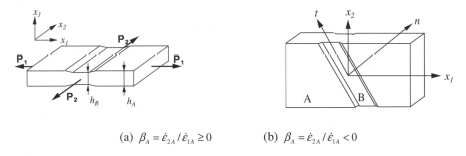

$$(a)\ \beta_A = \dot{\varepsilon}_{2A} / \dot{\varepsilon}_{1A} \geq 0 \qquad (b)\ \beta_A = \dot{\varepsilon}_{2A} / \dot{\varepsilon}_{1A} < 0$$

Fig. 2. Schematic illustration of the M-K analysis.

$$\phi = \frac{1}{b} \int_0^{\bar{\varepsilon}_t} \left( \frac{\sigma_m}{\bar{\sigma}} + a \right) d\bar{\varepsilon} \quad = \phi_c . \tag{1}$$

Here $\sigma_m$ and $\bar{\sigma}$ are the hydrostatic stresses, the effective stresses respectively, and $a$ and $b$ are material constants. As an example, in Fig. 3, thus calculated FLD (indicated by "M-K+Oyane's model"), for a case of tested temperature of 150°C and punch speed of 3 mm·min⁻¹, is compared with the experimental result. In this figure, the result of the conventional M-K analysis not taking account of the damage evolution (indicated by "M-K model") is also shown. M-K+Oyane's model can well predict the FLD, in contrast, the M-K model fails to capture the FLD characteristics especially for a biaxial stretching region. From these results, it is concluded that the damage evolution in the necking zone should be taken into account for accurate prediction of FLDs of the magnesium alloy sheet at high temperatures. Figs 4(a), (b) and (c) show the calculated FLDs for various punch speed conditions at 150, 200 and 250°C, respectively, together with the corresponding experimental data. The predicted FLDs are in good agreement with the experimental results. Comparing these results, Figs 4(a), (b) and (c), we can see that the FLDs are strongly influenced by forming speed at 250 °C, but at lower temperatures they are less sensitive to the forming speed. In the present material model, the exponents of strain hardening and strain-rate hardening (*n*- and *m*-values) were treated as temperature dependent material parameters, and consequently, such effects of temperature and forming speed on FLDs are well captured by the analysis.

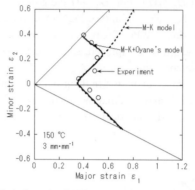

Fig. 3. Comparison of FLD analytical result of AZ31 magnesium alloy sheet (tested temperature 150 °C and punch speed of 3 mm·min⁻¹).

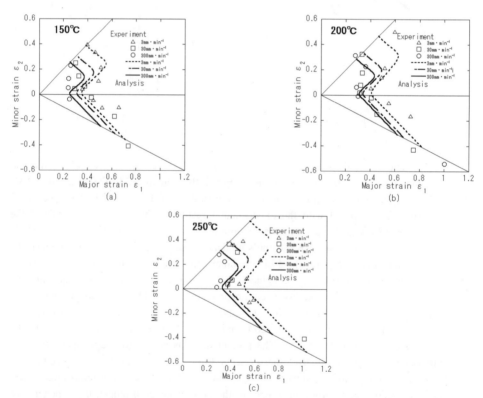

Fig. 4. Experimental measurements and analytical results of forming limit diagrams at various temperatures for AZ31 magnesium alloy sheet.

## 4. Concluding Remarks

The influences of forming temperature and speed on the FLD have been investigated for AZ31 magnesium alloy sheet. The present findings are summarized as follows.

1) The limit strains increased with temperature rise and with decreasing forming speed, where the effect of forming speed was stronger at higher temperatures.

2) The above-mentioned FLD characteristics were well captured by the M-K analysis, where the strain and strain rate exponents (*n*- and *m*-values) were treated as temperature and strain rate. Furthermore, for the accurate prediction of FLDs, the calculation of the damage evolution in the necking zone was essential. In the present analysis, Oyane's fracture criterion was employed.

## References

1. E. Aghion and B. Bronfin, *Mater. Sci. Forum* **350-351**, 19 (2000).
2. H. Friedrich and S. Schumann, *J. Mater. Process. Technol,* **117**, 276 (2001).
3. E. Doege and K. Drőder, *J. Mater. Process. Technol,* **115**, 14 (2001).
4. H. Takuda, T. Morishita, T. Kinoshita and N. Shirakawa, *J. Mater. Process. Technol,* **164-165**, 1258 (2005).
5. J. Kaneko, M. Sugamata, M. Numa, Y. Nishikawa and H. Takada, *J. Jpn. Inst. Met.* **64** 141 (2000).
6. S. Aida, H. Tanabe, H. Sugai, I. Takano, H. Ohnuki and M. Kobayashi, *J. Jpn. Inst. Light Met.* **50** 456 (2000).
7. H. Somekawa, M. Kohzu, S. Tanabe and H. Higashi, *Mater. Sci. Forum.* **350-351** 177 (2000).
8. M. Kohzu, F. Yoshida, H. Somekawa, M. Yoshikawa, S. Tanabe and K. Higashi, *Mater. Trans.* **42** 1273 (2001).
9. M. Kohzu, T. Hironaka, S. Nakatsuka, N. Saito, F. Yoshida, T. Naka, H. Okahara and K. Higashi, *Mater. Trans.* **48** 764 (2007).
10. Kuczynski and T. Pokora, *Int. J. Mech. Sci.* **15** 789 (1973).
11. K. W. Neale and E. Chater, *Int. J. Mech. Sci.* 22 563 (1980).
12. S. Xu and K. J. Weinmann, *CIRP Annals* 47 177 (1998).
13. T. Naka, G. Torikai, R. Hino and F. Yoshida, *J. Mater. Process. Technol,* **113** 648 (2001).
14. M. Oyane, T. Sato, K. Okimoto and S. Shima, *J. Mech. Work. Technol,* **4** 65 (1980).
15. T. Naka, T. Uemori, R. Hino, M. Kohzu, K. Higashi and F. Yoshida, *J. Mater. Process. Technol,* **201** 395 (2008).

# EFFECT OF HIGH TEMPERATURE DEFORMATION ON THE LOW THERMAL EXPANSION BEHAVIOR OF FE-29%NI-17%CO ALLOY

K. A. LEE[1†]

*School of Advanced Materials Engineering, Andong National University*
*Andong, Kyungbuk, 760-749 KOREA,*
*keeahn@andong.ac.kr*


J. NAMKUNG[2], M. C. KIM[3]

*Nonferrous Refining Project Team, RIST*
*Pohang, Kyungbuk, 790-330 KOREA,*
*namkung@rist.re.kr, mckim@rist.re.kr*

Received 15 June 2008
Revised 23 June 2008

The effect of high temperature deformation on the low thermal expansion property of Fe-29Ni-17Co alloy was investigated in the compressive temperature range of 900~1300°C at a strain rate range of 25~0.01 sec.$^{-1}$. The thermal expansion coefficient ($\alpha_{30-400}$) generally increased with increasing compressive temperature. In particular, $\alpha_{30-400}$ increased remarkably as the strain rate decreased at temperatures above 1100°C. Note, however, that $\alpha_{30-400}$ at low compressive temperatures (900°C and 1000°C) increased abnormally at high strain rates. Based on the investigation of various possibilities of change in low thermal expansion behavior, the experimental results indicated that both the appearance of the $\alpha$ phase and evolution of grain size due to hot compression clearly influenced the low thermal expansion behavior of this invar-type alloy. The correlation between the microstructural cause and invar phenomena and theoretical explanation for the low thermal expansion behavior of Fe-29%Ni-17%Co were also suggested.

*Keywords*: Fe-29%Ni-17%Co; hot compression; low thermal expansion; grain size; $\alpha$ phase.

## 1. Introduction

Also known as Kovar,[1] Fe-29%Ni-17%Co alloy is a low expansion alloy utilized in the electronics industry for glass-to-metal sealing. The thermal-expansion behavior of this special alloy during a temperature increase is similar to both glass and ceramic. Fe-29%Ni-17%Co products are usually produced through vacuum induction melting-hot working-cold working-post processing in a variety of forms, i.e., sheet, wire, plate, and rod. The cold working process may be omitted for certain types of vacuum applications. Therefore, the hot working process is very important not only for shaping the product but

---

†Corresponding Author.

also for controlling the property of Fe-29%Ni-17%Co alloy. The near-zero coefficient of thermal expansion, also known as the "invar effect," is related to the spontaneous volume magnetostriction wherein lattice distortion counteracts the normal lattice thermal expansion associated with temperature increase. The compositions of invar-type alloys are pegged at a value wherein saturation magnetization begins to deviate from the Slater-Pauling curve and are also very close to the phase boundary on the FCC side. Invar alloys are known to be magnetically as well as structurally inhomogeneous. The thermal behavior of low thermal expansion alloy is reportedly affected by alloy composition, cold working,[2] and heat treatment.[3] Though several theoretical explanations have been suggested for the invar effect, explaining such experimental results clearly is still somewhat difficult. To date, no research on the effect of high temperature deformation on the invar property of low thermal expansion alloy has been conducted.

This paper sought to investigate the effect of hot compression on the thermal behavior of low thermal expansion alloy, Fe-29%Ni-17%Co alloy, vis-à-vis microstructural evolution. The invar effect of Fe-29%Ni-17%Co alloy was also discussed.

## 2. Experiment

The Kovar alloy used in this work was produced by a process of vacuum induction melting. The ingot was homogenized at 1220°C for 3 hours and hot forged from 4300 cm$^2$ to 350 cm$^2$ square. The chemical composition of this material is shown in Table 1. The compression test at high temperature was performed by using a cylindrical specimen with a diameter of 10 mm and a gauge length of 12 mm. Thereafter, grain size was observed by using both optical microscope and image analyzer. All of the thermal expansion specimens were machined from the above specimens after conducting hot compression. Thermal expansion was measured continuously from 25°C to 600°C to 25°C with heating and cooling rates of 5°C/min. For a detailed understanding of the microstructural evolution during hot compression, X-ray and TEM observations were also conducted.

Table 1. Chemical composition of the Fe-Ni-Co low thermal expansion alloy used in this study.

| Comp. | Fe | Ni | Co | Mn | Si | C | Ti | P | S |
|-------|-----|------|------|-------|-------|-------|------|-------|-------|
| wt.% | Bal. | 27.9 | 17.3 | 0.003 | 0.008 | 0.007 | 0.07 | 0.003 | 0.003 |

## 3. Results and Discussion

After the high temperature compression test, the compressive specimens were re-prepared as specimen for the thermal expansion test. Thermal expansion behavior was then investigated. The typical thermal expansion curve is suggested in Fig. 1. Fig. 2 also shows the variation in the thermal expansion coefficient of the hot-deformed specimen given the strain rate and compressive temperature in Fe-29%Ni-17%Co alloy. In the high temperature range between 1100°C and 1300°C, the thermal expansion coefficient

($\alpha_{30\sim400}$) was found to increase in general with increasing temperature at each strain rate. As the strain rate decreased at each temperature above 1100°C, $\alpha_{30\sim400}$ continuously increased. In the case of 1300°C in particular, the value of thermal expansion coefficient $\alpha_{30\sim400}$ significantly increased by 0.2 as the strain rate changed from 25 sec.$^{-1}$ to 0.01 sec.$^{-1}$. At the low temperature range of 900°C and 1000°C, however, the abovementioned general trend of increase in thermal expansion coefficient with decreasing strain rate changes was not sustained. Moreover, note that $\alpha_{30\sim400}$ increased abnormally particularly at strain rates of 0.5 sec.$^{-1}$ and 25 sec.$^{-1}$ in the low temperature range. From the viewpoint of the fixed strain rate, a high thermal coefficient increase of 0.18 can be detected at a low strain rate of 0.01 sec.$^{-1}$ with increasing temperature from 900°C to 1300°C. At a high strain rate of 25 sec.$^{-1}$, however, a general trend could not be deduced.

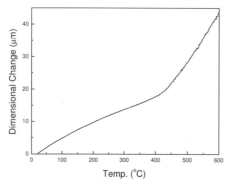

Fig. 1. Typical example of the thermal expansion curve of Fe-29%Ni-17%Co alloy.

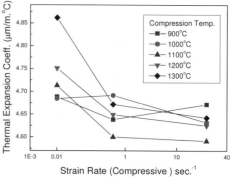

Fig. 2. Variation of thermal expansion coefficient $\alpha$ (30~400°C) based on the strain rate and temperature in hot-deformed sample.

Fig. 3 shows the optical micrograph after hot deformation under different compressive conditions of temperature and strain rate. The area of concentrated deformation -- represented as very small black grains -- was locally observed at a temperature below 1000°C (appeared distinctly at 900°C) especially at a high strain rate. Above 1000°C, however, the shape of the grains turned homogeneous. Grain size apparently increases with increasing compressive temperature and decreasing strain rate. In particular, grain growth proceeded abruptly at high temperatures of 1200°C and 1300°C.

Previous research has shown that the thermal expansion characteristics of low thermal expansion alloy (Fe-Ni alloy) are easily affected by alloy composition, cold working condition,[2] heat treatment,[3] etc. Although the physical origin of Invar characteristics is still unclear, there is no doubt that it is closely related with ferromagnetism. Moreover, $\alpha$-$\gamma$ phase transition, atomic ordering, and $\gamma'$ precipitation are also found to play important roles in the low thermal expansion behavior. The TEM observation in this study did not reveal any significant evidence of $\gamma'$ precipitation in any of the compressed specimens. The mechanism of destroying short-range order also doesn't seem compatible with the present results, considering that in Fig. 2, $\alpha_{30\sim400}$ abruptly increases as the strain rate decreases at the same temperature (1200°C). To confirm the presence of α phase in the γ structure of Kovar alloy, an X-ray analysis of the structure for hot compressed specimens was performed (Fig. 4).

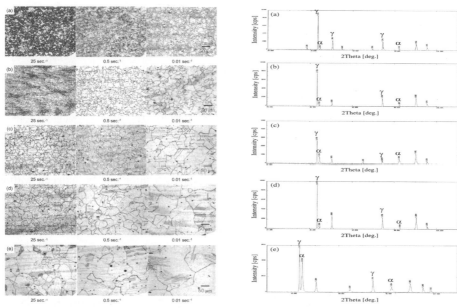

Fig. 3. Optical micrograph according to hot deformation conditions: (a) 900°C, (b) 1000°C, (c) 1100°C, (d) 1200°C, and (e) 1300°C.

Fig. 4. XRD analysis results: (a) 900°C, 0.01 sec.$^{-1}$, (b) 1200°C, 0.01 sec.$^{-1}$, (c) 1300°C, 0.01 sec.$^{-1}$, (d) 1300°C, 25 sec.$^{-1}$, and; (e) 900°C, 25 sec.$^{-1}$.

In the case of low strain rate (0.01 sec.$^{-1}$), the peak of α phase is significantly lower than that of γ phase. The appearance of α phase becomes evident, especially at low temperature and high strain rate conditions, as shown in Fig. 4 (e). The above conditions showing the appearance of α phase (comparatively high strain rate at 900°C and 1000°C) correlated well with that of locally concentrated deformation in the optical micrograph observation. The anomalous increase of $\alpha_{30\sim400}$ at low temperatures in Fig. 2 is attributed to the appearance of α phase.

Nevertheless, the general trend in Fig. 2 that $\alpha_{30\sim400}$ increases with increasing temperature and decreasing strain rate is still unclear. Khasin, et al[4] examined the $\alpha_{30\sim400}$ of Kovar alloy for the samples taken from various zones in billets from different casts. These results indicated that the grain size had an influence on the thermal expansion coefficient. Fig. 5 shows the correlation between grain size and the $\alpha_{30\sim400}$ of hot compressed specimens in this study. It also indicates that grain size has a definite influence on the thermal expansion coefficient and $\alpha_{30\sim400}$ increases with increasing grain size.

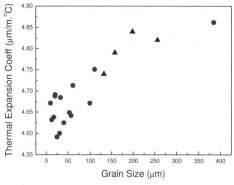

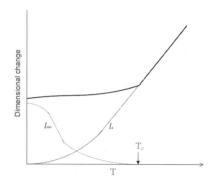

Fig. 5. Correlation between grain size and α (30~400°C)    Fig. 6. Schematic diagram explaining the
of hot-compressed Fe-29%Ni-17%Co alloy.                    thermal expansion behavior of Invar alloys.

The lattice parameter in most materials is known to increase in general with increasing temperature due to the change in the average atomic distance. In addition, in the case of ferromagnetic materials, spontaneous magnetization can be induced without any magnetic field only below the Curie temperature. Thus, some volume expansion can occur due to spontaneous volume magnetostriction. Specific alloys (called invar-type alloys) in ferromagnetic materials are characterized by extremely high positive volume magnetostriction. Spontaneous magnetization in the low thermal expansion invar-type alloy significantly decreases with increasing temperature below the Curie temperature.[3] Low thermal expansion can be said to occur as a consequence of two opposing, thermally activated effects. Fig. 6 shows the schematic diagram explaining the low thermal expansion behavior in invar-type alloys. The alloy expands when heated because of atomic vibration ($l_a$). Such expansion within a specific temperature range is counteracted by a decrease in spontaneous volume magnetostriction ($l_{ms}$). These two appropriate counterbalancing effects may result in low thermal expansion (invar) behavior only below the Curie temperature. Although many theories have been suggested regarding the invar phenomena, the physical origin, cause of particularly large positive magnetostriction, and abrupt decrease in spontaneous magnetization around the Curie temperature have yet to be explained clearly.

Zakharov, et al[5] explained the change in the thermal expansion coefficient by dividing two causes as expressed in equation (1).

$$\Delta\alpha = \Delta\alpha_{in} + \Delta\alpha_{\gamma\to\alpha} \tag{1}$$

In this equation, $\Delta\alpha_{in}$ represents the change in the thermal expansion coefficient (negative value) due to the invar effect; $\Delta\alpha_{\gamma\to\alpha}$ corresponds to that due to the $\gamma\to\alpha$ phase transformation (positive value). Although Zakharov, et al provided this explanation based on cold deformation at and below room temperature, applying the same explanation to the effect of high temperature deformation on the thermal expansion behavior in this study seems to be reasonable.

Equation (1) can be changed to equation (2) considering Colling's assumption,[6] i.e., the change in the thermal expansion coefficient due to the invar effect is attributed to the decrease in spontaneous volume magnetostriction based on temperature. Here, $h_o(T)$ denotes the linear spontaneous magnetostriction and corresponds to 1/3 of the volume of magnetostriction at each temperature.

$$\Delta\alpha = \frac{\partial h_o(T)}{\partial T} + \Delta\alpha_{\gamma\to\alpha} \tag{2}$$

In the magnetostriction study of ferromagnetic materials, E.W. Lee[7] reported based on the experimental results that the volume change for polycrystalline with external magnetic field was different from that of a single crystal. Although there was some difference between the external magnetic induced magnetostriction and spontaneous volume magnetostriction, the grain boundary effect was naturally assumed to be applicable to the spontaneous magnetostriction change in the low thermal expansion alloy as well. As a result, the amount of decrease in spontaneous volume magnetostriction could be changed by the grain size of the compressed specimen as temperature increased. Such change resulted in the difference in the thermal expansion coefficient of the low thermal expansion alloy. Note, however, that the quantitative relationship between the variation in spontaneous volume magnetostriction based on temperature and grain size has yet to be studied. A more detailed study on the low thermal expansion alloy should be conducted to clarify the origin of the invar phenomena and the effect of grain size on the spontaneous volume magnetostriction based on temperature.

## 4. Summary

The thermal expansion coefficient ($\alpha_{30\sim400}$) generally increased with increasing compressive temperature and decreasing strain rate. However, that thermal expansion coefficient $\alpha_{30\sim400}$ at low compressive temperatures (900°C and 1000°C) increased abnormally especially at high strain rates of above 0.5 sec.$^{-1}$, mainly due to the appearance of $\alpha$ phase. It was also apparent that the thermal expansion coefficient ($\alpha_{30\sim400}$) increased as grain size increased with hot working. The origin of the invar phenomena was also discussed based on the microstructure evolution.

## References

1. *Standard Specification for Iron-Nickel-Cobalt Sealing Alloy*, ASTM **10.04**, F15-78 (1993).
2. W. F. Schlosser, *J. Phys. Chem. Solids* **32**, 939 (1971).
3. S. Chikazumi, in *Physics and Applications of Invar Alloys*, ed. H. Saito, Honda Memorial Series on Materials Science, No. 3 (Maruzen Company, LTD, 1978), p. 18.
4. G. A. Khasin et. al, *Stal'* **11**, 1047 (1977).
5. A. I. Zakharov, T. A. Kravchenko and D. S. Barkaya, *Met. Sci. Heat Treat.* **30**, 129 (1988).
6. D. A. Colling, in Thermal Expansion-1971, ed. M. G. Graham and H. E. Hagy (American Institute of Physics, 1972), p. 188.
7. E. W. Lee, *Rep. Progr. Phys.* **18**, 185 (1955).

# THE PLASTICITY OF MONOCRYSTALLINE SILICON UNDER NANOINDENTATION

LI CHANG[1], L.C. ZHANG[2†]

*School of Aerospace, Mechanical and Mechatronic Engineering*
*The University of Sydney, NSW 2006, AUSTRALIA,*
*l.chang@usyd.edu.au, L.Zhang@usyd.edu.au*

Received 15 June 2008
Revised 23 June 2008

This paper focuses on a fundamental understanding of the plastic deformation mechanism in monocrystalline silicon subjected to nanoindentation. It was found that over a wide range of indentation loads from 100 μN to 30 mN and loading/unloading rates from 3.3 μN/s to 10 mN/s, the plasticity of silicon is mainly caused by stress-induced phase transitions. The results indicate that the critical contact pressure for phase transition at unloading is almost constant, independent of the maximum indentation load ($P_{max}$) and loading/unloading rates. However, the shape of the load-displacement curves greatly relies on the loading/unloading conditions. In general, higher $P_{max}$ and lower unloading/loading rates favor an abrupt volume change and thus a discontinuity in the load-displacement curve, commonly referred to as pop-in and/or pop-out events; whereas smaller $P_{max}$ and rapid loading/loading processes tend to generate gradual slope changes of the curves. This study concludes that the difference in the curve shape change does not indicate the mechanism change of plastic deformation in silicon.

*Keywords*: Silicon; plastic deformation; nanoindentation; phase transition.

## 1. Introduction

Monocrystalline silicon is a principal material of semiconductor devices and micro-electro-mechanical systems that require high surface finish and, often, damage-free subsurface. It is therefore important to understand the deformation mechanisms in the material related to micro/nano surface processing. To precisely characterize the local deformation behaviour of silicon, depth-sensing micro/nanoindentation[1-5] has been extensively used to characterize the mechanical properties of silicon, *e.g.*, hardness and Young's modulus, and to understand the phase transformation event, a major mechanism of plasticity in silicon caused by mechanical loading[6]. It has been reported that there is a strong relationship between the shape of the load-displacement curve and silicon's phase transformations during an indentation.[7] For instance, under loading, monocrystalline silicon can transform from its diamond cubic structure (Si-I) to a metallic β-tin phase (Si-II), accompanied by a volume reduction of about 22% and indicated by a distinct displacement discontinuity – "pop-in".[5,8] Upon unloading, the Si-II phase becomes

---

†Corresponding Author.

unstable and undergoes a further phase transformation to amorphous or a mixture of a rhombohedra (Si-XII) and a body-center-cubic (Si-III), leading to a volume expansion signaled by a "pop-out".[2,5,7,8] Moreover, it has been reported that the shape of the load-displacement curves is greatly influenced by the loading conditions, *e.g.*, the magnitude of the maximum indentation load, the rate of loading/unloading and the shape of the indenter used.[9-13] These seem to imply that the phase transformations in silicon are dependent on the contact conditions. Nevertheless, an in-depth understanding of the relationships between the indentation conditions and the plasticity *via* phase transformations is unavailable.

The purpose of this paper is to more deeply investigate the stress related phase transitions in silicon under nanoindentation over a broad range of loads and loading/unloading rates.

## 2. Experimental

The nanoindentation tests were conducted on a nano-TriboIndenter (Hysitron Inc., USA) with a diamond Berkovich indenter in ambient condition. The experiments were performed on precisely polished (1 0 0) surface of monocrystalline silicon with a surface roughness less than 2 nm. The subsurface structure of specimens was examined using cross-sectional transmission electron microscopy (XTEM)[14] to guarantee that the specimens before the indentation test were damage-free. To systematically characterize the deformation behavior of silicon under different loading conditions, the tests were carried out over a wide range of indentation loads from 100 μN to 30 mN and loading/unloading rates from 3.3 μN/s to 10 mN/s. To evaluate the development of the phase transformation, cyclic indentations were also carried out with the maximum indentation load of 30 mN with the loading/unloading rate of 1 mN/s. In each test, five cycles of indentations were carried out. The minimum load of each cycle was 10% of the peak load in order to keep the contact between the indenter and specimens. The holding time at the maximum loads for all the tests was 30 s to minimize the time-dependent thermal shift. For each testing condition, at least 16 tests were repeated.

## 3. Results and Discussion

### 3.1. *Observations*

Fig. 1 shows the deformation behavior of silicon under different loading conditions. When the maximum loads were varied, the duration of the loading and unloading processes was kept the same, 30 s. This means that the tests with different $P_{max}$ had different loading/unloading rates. The hardness and Young's modulus of silicon can by determined based on the load-displacement curves.[15] Fig. 2 shows that the measured Young's modulus and hardness remain almost constants, without an obvious size effect.

As shown in Fig. 1 (b), silicon is perfectly elastic under a very low $P_{max}$, *e.g.* 100 μN. With the further increase of $P_{max}$, pop-in can be observed on the load-displacement curves, associated with the transition from purely elastic to elastic/plastic deformation due to the phase change in silicon from Si-I to Si-II.[16] However, the load-displacement curve is not affected noticeably by the loading rate (Fig. 1). When the maximum indentation loads are relatively low, the unloading displacement curves always show gradual change in slope (elbow). However, with $P_{max}$ = 30 mN (Fig. 1(a)), a clear pop-out appears. In the following section, the influence of loading conditions on the deformation behavior of silicon will be discussed.

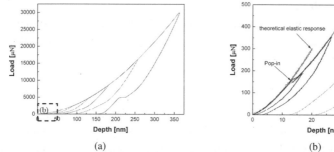

(a)                                                      (b)

Fig. 1. (a) The load-displacement curves measured under different loading conditions and (b) magnified view under ultra-low loads as marked in (a).

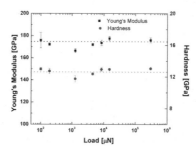

Fig. 2. Hardness and Young's modulus of silicon measured with various peak loads.

### 3.1.1. Influence of loading conditions on the load-displacement curves

Figs.3 and 4 show the variation of contact pressure with the indentation depth, where the contact pressure $P_a$ is defined as[9],

$$P_a = P/A(h_c) \tag{1}$$

in which $P$ is the applied load and $A(h_c)$ is the area of the projected area at contact depth $h_c$. The area function $A$ can be experimentally measured with a standard material, e.g., a fused quartz.[15,17] In the present case, the area function for the indenter tip can be described as[16] $A(h_c) = 24h_c^2 + 1412h_c$. The contact depth can be determined by $h_c = h_f - h_s$, in which $h_f$ is the full indentation depth and $h_s$ is the elastic deflection of the material at the perimeter of the indentation area. According to contact mechanics[18], $h_s = 0.75\sqrt{P_{max}P}/S$ where $S$ is the stiffness of the sample at load $P_{max}$. As shown in the figures, pop-out does not always occur and mostly appears at different contact pressures. This is in agreement with what Juliano *et al* [8] reported. The possibility to obtain pop-out depends very much on unloading rates. Like the effect of loading rate on pop-in[16], a slower unloading process favors a pop-out, indicating that a sudden volume expansion happens in this case due to a phase transformation. With the same unloading rate, pop-out is more likely to appear at a higher fixed $P_{max}$. Yan *et al*[13] reported that there might be a critical pop-out load. However, our result demonstrates that the occurrence of pop-out is uncertain, but is greatly influenced by the unloading rate. For example, pop-out was observed even when $P_{max}$ was very small (*e.g.*, 3mN) when the unloading rate was relatively slow (*e.g.*, 0.1 mN/s), although it occurred only once in the 16 repeated tests.

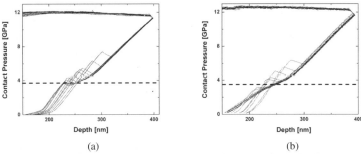

(a)                                         (b)

Fig. 3. Contact pressure versus displacement tested with different loading/unloading rate: (a) 0.5 mN/s, (b) 5 mN/s. $P_{max}$ = 30 mN.

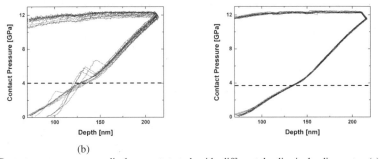

(b)

Fig. 4. Contact pressure versus displacement tested with different loading/unloading rate: (a) 0.5 mN/s, (b) 5 mN/s. $P_{max}$ = 10 mN.

### 3.1.2. *Cyclic indentation*

Fig. 5 shows the evolution of load-displacement curves from 5 cycles of indentations with a peak load of 30 mN. It is clear that phase changes continuously took place during the first few cycles. The microstructure of the material in the deformed zone seems to have been stabilized after a few indentation cycles, because the response of the material becomes entirely elastic in the last two cycles.

The phase transformations do not seem to influence Young's modulus. For example, in the case of Fig. 5, the Young's moduli measured from the five individual cycles are 171 GP, 175 GP, 174 GP, 172 GP, 174 GPa, respectively. This is obvious even by inspecting the curves. The upper parts of the unloading curves (*i.e.*, part ① of the curves) before pop-outs are mostly parallel to each other and show purely an elastic behavior. Similarly, the material's response after the pop-out (*i.e.*, part ② of the curves) is also elastic primarily. Since a pop-out reflects an abrupt phase transition in silicon in the deformation zone underneath the indenter, and because the material behaves purely elastically both before and after the phase transition during unloading, it is reasonable to conclude that plastic deformation in silicon under nanoindentaion is mainly caused by phase transitions.[6, 10, 16] In addition, if we examine the contact pressure at a pop-out, we see that a pop-out always occurs below certain contact pressure (8 GPa in Figs. 3 and 4). This is in agreement with the known conclusion that phase Si-II is stable when the contact pressure in an indentation process is beyond a critical value, 8~12 GPa.

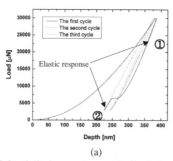 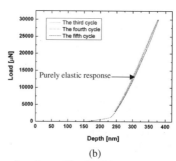

|                    (a)                    |                    (b)                    |

Fig. 5. Load-displacement curves in five indentation cycles when $P_{max}$ = 30 mN with the unloading/unloading rate of 1 mN/s: (a) the first three cycles, and (b) the last three cycles.

### 3.2. Analysis

Comparing Fig. 3 with Fig. 4, we can see that the location of the pop-outs on the curves under nominally identical indentation conditions varies, but the contact pressure for the onset of elbow points is almost constant, independent of loading conditions. The pressure at the elbow is around 4.5 GPa, which is in excellent agreement with the value reported by Juliano *et al.*[9] The elbow pressure keeps constant over a wide range of $P_{max}$ from 300 µN to 30 mN even when the unloading rate also varies from 10 µN/s to 10 mN/s. In other words, $P_{max}$ and unloading rate does not affect the onset of an elbow. Hence we can propose the following mechanism: During unloading, at the point where stress reaches a critical value, phase transition initiates and grows to its surrounding area. On the unloading-displacement curve, this process is reflected by a gradual slope change and governed by the stress field in the indentation zone. As a result, the starting point of an elbow occurs at the same location on the indentation curves. However, if the development of the phase transformation occurs with an abrupt transformation within a considerable volume of material when the average contact pressure reaches a critical value (*e.g.*, less than 8 GPa), a sudden volume increase takes place, leading to a pop-out. Hence, a visible pop-out is the resultant of a sufficient volume of material with critically required stress field for phase transformation. The above mechanism explains the effect of loading conditions on the pop-out event: 1) a higher $P_{max}$ results in a larger deformation zone and thus increases the possibility of a pop-out; and 2) a slower unloading rate tends to generate a pop-out because the microstructure can gradually reconstruct when the stress field varies. The location variation of the pop-outs under nominally identical indentation conditions (Fig. 1) can be interpreted as follows: The process of crystalline nucleation and growth of the phase subjected to transformation is influenced by the concentration of the randomly distributed crystal defects in silicon. The local stress field evolution at indentation is also affected by the randomly distributed surface asperities of the specimens. As a result, the location of a visible pop-out varies, even though the nominal indentation condition is identical.

## 4. Summary

We have investigated the deformation mechanism of silicon subjected to nanoindentation with a Berkovich indenter. It was found that over a wide range of indentation loads from 100 μN to 30 mN and loading/unloading rates from 3.3 μN/s to 10 mN/s, the plasticity of silicon is mainly caused by stress-induced phase transitions. The results indicate that the critical contact pressure for a phase transition is almost constant and is independent of the maximum indentation load and loading/unloading rates. However, the shape of the load-displacement curves greatly relies on the loading/unloading conditions. In general, a higher $P_{max}$ and a lower unloading/loading rate favor an abrupt volume change and thus a discontinuity in the load-displacement curve, commonly referred to as a pop-in and/or pop-out; whereas a smaller $P_{max}$ and a rapid loading tend to generate gradual slope changes on the curves. This study concludes that the difference in the curve shape change does not indicate the mechanism change of plastic deformation in silicon.

### Acknowledgment

The authors appreciate the financial support of the Australian Research Council.

### References

1. G. M. Pharr, W. C. Oliver, D. S. Harding, *J. Mater. Res.* **6,** 1129 (1991).
2. E. R. Weppelmann, J. S. Field, M. V. Swain, *J. Mater. Sci.* **30,** 2455 (1995).
3. A. Kailer, Y. G. Gogotsi, K. G. Nickel, *J. Appl. Phys.* **87,** 3057 (1997).
4. I. Zarudi, L. C. Zhang, *Tribol. Int.* **32,** 701 (1999).
5. J. E. Bradby, J. S. Williams, J. Wong-Leung, M.V. Swain, P. Munroe, *Appl. Phys. Lett.* **77,** 3749 (2000).
6. L. C. Zhang, I. Zarudi, *Int J Mech Sci,* **43** (9), 1985 (2001).
7. V. Domnich, Y. Gogotsi, S. Dub, *Appl. Phys. Lett.* **76,** 2214 (2000).
8. T. Juliano, V. Domnich, Y. Gogotsi, *J. Mater. Res.* **19,** 3099 (2004).
9. T. Juliano, Y. Gogotsi, and V. Domnich, *J. Mater. Res.* **18,** 1192 (2003).
10. L. C. Zhang, H Tanaka, *JSME Int. J. Series* **A 31,** 546 (1999).
11. I. Zarudi, L. C. Zhang, W. C. D. Cheong, T. X. Yu, *Acta. Mater.* **53,** 4795 (2005).
12. I. Zarudi, J. Zou, L. C. Zhang, *Appl. Phys. Lett.* **82,** 874 (2003).

13. J. W. Yan, H. T. Takahashi, X. H. Gai, H. Harada, J. Tamaki, T. Kuriyagawa,et al, *Mater. Sci. Eng.* A **423,** 19 (2006).
14. I. Zarudi, L. C. Zhang, *J. Mater. Process. Tech.* **84,** 149 (1998).
15. W. C. Oliver, G. M. Pharr, *J. Mater. Res.* **7,** 1564 (1992).
16. L. Chang, L. C. Zhang, submitted to *Phil. Mag. Lett.*
17. H. Bei, E. P. George, J. L. Hay, G. M. Pharr, *Phys. Rev. Lett.* **95,** 045501 (2005).
18. I. N. Sneddon, *Int. J. Eng. Sci.* **3,** 7 (1965).

# CHARACTERIZATION OF ELECTRO-RHEOLOGCIAL FLUIDS UNDER HIGH SHEAR RATE IN PARALLEL DUCTS

X. W. ZHANG[1†], C. B. ZHANG[2], T. X. YU[3]

*Department of Mechanical Engineering, Hong Kong University of Science and Technology,*
*Clear Water Bay, Hong Kong, P. R. CHINA,*
*zhangxw@ust.hk, zcxaa@ust.hk, metxyu@ust.hk*

W. J. WEN[4]

*Department of Physics, Hong Kong University of Science and Technology*
*Clear Water Bay, Hong Kong, P. R. CHINA,*
*phwen@ust.hk*

Received 15 June 2008
Revised 23 June 2008

Electro-rheological (ER) fluid is a smart suspension which can be changed promptly from Newtonian to Bingham plastic material when subjected to a high-intensity electric field. This property of ER fluid makes it possible to be applied in adaptive energy absorbers. As the impact velocity encountered in applications could be very large, it is necessary to characterize the ERF under high shear rate. In this study, a capillary rheo-meter with parallel duct was designed and manufactured which is capable of producing a shear rate as high as 5000(1/s). Two giant ER fluids with mass concentration C=51% and 44.5% and a commercial density-matched ER fluid with C= 37.5% were characterized. The experimental results show that when the ER fluids are free of electric field ($E$=0kV/mm), they are Newtonian. However, for the former two ER fluids, the deposition effect is very remarkable and stirring has to be made continuously to keep the suspension stable. With the increase of the electric field intensity, the yield shear stresses of ER fluids increase exponentially but their viscosities do not change much. It is also found that within the parallel duct, the flow of ER fluids exhibits notable fluctuations, whose period increases with the increase of electric field intensity and is independent of the shear rate.

*Keywords*: Electro-rheological; Bingham plastic; high shear rate; parallel duct; fluctuation.

## 1. Introduction

Electro-rheological (ER) fluid is a smart suspension which consists of large mounts of solid particles (e.g., starch, lime, stone, carbon, etc.) dispensed in a dielectric carrier liquid (such as silicon oil, paraffin, kerosene, etc.) [1]. When the ER fluid is subjected to a high-intensity electric field, it can be changed promptly from Newtonian to Bingham-plastic material and this process is usually within tens millisecond [2]. Besides, the yield shear stress and viscosity of the ER fluids can be controlled by the intensity of the applied electric-filed. Thus, the fast response and controllability of ER fluid make it a promising smart material. Since the development of ER fluid by Winslow [3] in 1949, ER fluids have been studied extensively in relation to many applications, such as vibration control [4-5], seismic isolation [6], clutches [7] and so on. Moreover, to increase the force of ER dampers, Gavin [8] and Kuo et al [9] proposed multi-ducts method to generate high damping

---

[†]Corresponding Author.

force. On the other hand, to improve the low yield stress of the previous ER fluids, Wen [10] developed a new material called giant ER fluid whose yield stress is much higher than the conventional ones.

Although numerous studies have been conducted on various applications of ER fluids, most of them aimed at vibration control within the range of low shear rate, while very few works paid attention to the behaviors of ER fluids under impact scenarios. With the application of ER fluids in adaptive energy absorption in mind, the flowing behavior of ER fluids under high shear rate has attracted our attention. However, the conventional rotational rheometer can only measure the ER fluids under low shear rate (<1000/s), because for high shear rate, the centrifugal effect of the particles under electric field becomes quite prominent. As reported in this paper, a capillary rheometer with parallel duct has been designed and manufactured, which could measure the properties of ER fluids under a shear rate as high as 5000/s. Utilizing this capillary rheometer, the flowing behaviors of two giant ER fluids with mass faction 55% and 45%, as well as a density-matched commercial ER Fluid with mass faction 37.5% were experimentally investigated. The experiments show that the flow of the ER fluid under high electric field is not smooth and fluctuation is quite serious. Based on the experimental results, the relation between the nominal shear stress and the shear rate under different electric fields are obtained.

## 2.   Preliminary

In the previous studies, the flowing behavior of ER fluids subjected to high electric field is usually modeled as Bingham plastic with the following constitutive relation

$$\tau = \tau_0 + \mu \cdot \dot{\gamma} \tag{1}$$

where $\tau$ is the shear stress, $\tau_0$ is the yield stress, $\mu$ is the viscosity and $\dot{\gamma}$ is the shear rate.

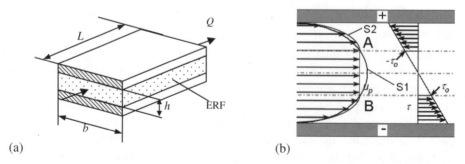

(a)                                      (b)

Fig. 1. The flowing behavior of ER fluid: (a) the parallel duct; (b) the cross-section of the duct.

Consider the flowing of the ER fluid through a parallel duct as illustrated in Fig. 1(a), which has length $L$ and width $b$ while the distance between the electrodes is $h$. When the ER fluid is free of electric field, the contour of the stream lines is a parabola S1 as shown in Fig. 1(b). After applying an electric field to the duct, the stream line will change from

S1 to S2. The flowing in the middle section between A and B is uniform. Then, the relation between the flux of the quasi-steady flow $Q$ and the pressure drop $\Delta P$ through the duct can be obtained from [4],

$$\Delta P^3 - \left(3\tau_y \frac{L}{h} + \frac{12\mu LQ}{bh^3}\right)\Delta P^2 + 4\tau_0^3\left(\frac{L}{h}\right)^3 = 0 \tag{2}$$

If the viscosity of the Bingham plastic fluid is known, for given flux $Q$ and the pressure drop $\Delta P$, the yield stress of the ER fluid could be obtained. However, in the present study, both the yield stress and viscosity of the ER fluids need to be determined, so that Eq. (2) alone is not sufficient to determine the yield stress. However, the shear stress along the boundary can be calculated by

$$\tau = \Delta P \times \frac{h}{2L} \tag{3}$$

On the other hand, if the stream-lines are assumed to be similar to those in a Newtonian flow, then the shear rate can be found as follows,

$$\dot{\gamma} = \pm 6\frac{u_m}{h} = \pm\frac{6Q}{bh^2} \tag{4}$$

where $\dot{\gamma}$ is the nominal shear rate and $u_m$ is the mean flow speed within the duct. Hence, by employing Eqs. (3) and (4), the shear strain-stress curves of the ER fluids under different electric field can be obtained.

## 3. Experimental Setup

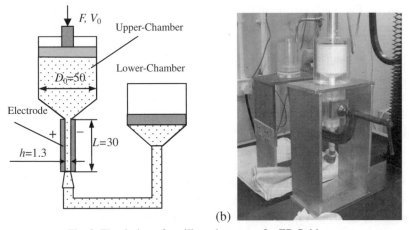

(a)                                    (b)

Fig. 2. The design of capillary rheometer for ER fluids.

As shown in Fig. 2, the capillary rheometer used in this study is composed of three parts, namely, the upper chamber, the lower chamber and the electro-duct. The chambers are cylindrical with an inner diameter $D_0$=50mm, but the duct has a rectangular cross section with $L$=30mm, $b$=10.0mm and $h$=1.3mm. The three parts are connected by means of conic transition sections to minimize the pressure loss. In the experiments, the assembled rheometer was placed on an MTS machine. Initially, the ER fluid was sealed inside the chambers and pipes. The piston of the upper chamber was compressed by the MTS crosshead with a certain velocity, while the piston of the lower chamber was free. The resistant force of the piston, $F$, coming from four mechanisms, is estimated as

$$F = \Delta P_0 \times A_0 = (\Delta P_f + \Delta P_d + \Delta P_c + \Delta P_g) \times A_0 \tag{5}$$

where $\Delta P_f$ is caused by the friction between the pistons and chamber walls, $\Delta P_d$ and $\Delta P_c$ are the pressure drops within the electric duct and the other pipes, respectively, $\Delta P_g$ is resulted from the gravity, and $A_0$ is the cross-sectional area of the chambers.

Since this rheometer is designed to test the flowing behavior of ER fluids within the electric duct, the influences of other sections are undesirable. In order to evaluate the effect of the factors such as pipes and gravity, an FEM simulation was conducted by using the given dimensions and assuming viscosity $\mu$=0.5. It is found from the simulation that the pressure loss due to the pipes system is less than 2% of the total pressure drop. Therefore, in the subsequent analysis, the influences of the pipes and gravity are ignored.

Before the characterization of ER fluids, the rheometer was calibrated using glycerol with mass concentration around C>99.0% under room temperature 20°C. It was found that except the initial stage, the designed rheometer could obtain very smooth loading curves. Using Eqs. (3) and (4), the shear rates and shear stresses are calculated and the measured viscosity of the glycerine is 1.36Pa.s. Compared with the standard viscosity value 1.41Pa.s with C=100%, the accuracy of this rheometer is within 5%.

## 4.  Results and Discussion

In the experiments, two giant ER fluids produced in the smart material lab at HKUST and a commercial ER fluid were tested, which are denoted by ERF No. 1-3. The giant ER fluids are composed of silicon oil and nano-coated particles [10], and their mass concentrations are C=51% and 44.5%, respectively. The commercial ER, which consists of silicone oil and particles of Lithium salt and chloro-fluoro polymer, is a density-matched suspension with C=37.5%. To characterize the ER fluids under high intensity electric field, the loading speed of the MTS machine was taken as $V$=10-450mm/min, and the applied electric field ranged from $E$=0kv/mm to 3.0kV/mm. In trial tests, it was found for the giant ER fluids, the deposit effect of the particles was very serious, which made the properties of the ER fluids rather unstable. Therefore, to ensure the experiments repeatable, the ER fluid was stirred to uniform before every test.

The characteristic curves for ERF-1 under zero electric field are plotted in Fig. 3(a). It is shown that loading was stable, and with the increase of the loading speed, the load $F$ also increased. By using Eqs. (3) and (4), the relations between the shear stress and shear rate of the three ER fluids under zero electric field are obtained and depicted in Fig. 3(b). It is shown that the shear stress has a linear relationship with the shear rate, implying that when free of electric field, the ER fluids behave as Newtonian fluids. It is noted that in Fig. 3(b) the intersections of the fitting lines and the y-axis are not zero, because of the friction. As shown in Fig. 3(b), the viscosities of the three ERFs under zero electric field are $\mu$=0.48 Pas, 0.10Pas and 0.16Pas, respectively.

In the tests under high-intensity electric field, the piston of the upper chamber moved downwards first, then at a certain moment the electric field was applied. Some typical loading curves are shown in Fig. 4. It can be seen that when the electric field was applied, the resistant force increased sharply, since the ERF was changed from Newtonian to Bingham plastic. When subjected to electric field, the loading curves were no longer smooth and exhibited severe fluctuations. It is also found in Fig. 4(a) that for the giant ERF with C=51%, when the loading speed was lower than 200mm/min, the fluctuation was quite serious. When V>200mm/min, the loading becomes stable. Fig. 4(c) shows that the low density giant ERF have more serious fluctuations, and with the same electric field intensity, lower shear rate will cause fluctuations of larger amplitude. Besides, it is shown in Figs. 4(b) and (d) that for higher electric field, the fluctuation period is shorter and the amplitude is larger. A possible explanation of this phenomenon is that for the flowing of ERF in the parallel duct, the particles near the boundary are subjected to larger shear stress than those in the middle, so that the some particles are captured by the electrodes. With the progress of flowing, more and more particles aggregate near the boundary, resulting in an increase of the resistance to the flowing. When the pressure reaches a critical value, the aggregated particles will be pushed away and the pressure drops soon. The repeat of these processes results in the fluctuations as observed.

The characteristic curves for ERF-1 under zero electric field are plotted in Fig. 3(a). It is shown that loading was stable, and with the increase of the loading speed, the load $F$ also increased. By using Eqs. (3) and (4), the relations between the shear stress and shear rate of the three ER fluids under zero electric field are obtained and depicted in Fig. 3(b). It is shown that the shear stress has a linear relationship with the shear rate, implying that when free of electric field, the ER fluids behave as Newtonian fluids. It is noted that in Fig. 3(b) the intersections of the fitting lines and the y-axis are not zero, because of the friction. As shown in Fig. 3(b), the viscosities of the three ERFs under zero electric field are $\mu$=0.48 Pas, 0.10Pas and 0.16Pas, respectively.

In the tests under high-intensity electric field, the piston of the upper chamber moved downwards first, then at a certain moment the electric field was applied. Some typical loading curves are shown in Fig. 4. It can be seen that when the electric field was applied, the resistant force increased sharply, since the ERF was changed from Newtonian to Bingham plastic. When subjected to electric field, the loading curves were no longer smooth and exhibited severe fluctuations. It is also found in Fig. 4(a) that for the giant ERF with C=51%, when the loading speed was lower than 200mm/min, the fluctuation was quite serious. When V>200mm/min, the loading becomes stable. Fig. 4(c) shows that the low density giant ERF have more serious fluctuations, and with the same electric field intensity, lower shear rate will cause fluctuations of larger amplitude. Besides, it is shown in Figs. 4(b) and (d) that for higher electric field, the fluctuation period is shorter and the amplitude is larger. A possible explanation of this phenomenon is that for the flowing of ERF in the parallel duct, the particles near the boundary are subjected to larger shear stress than those in the middle, so that the some particles are captured by the electrodes. With the progress of flowing, more and more particles aggregate near the boundary, resulting in an increase of the resistance to the flowing. When the pressure reaches a critical value, the aggregated particles will be pushed away and the pressure drops soon. The repeat of these processes results in the fluctuations as observed.

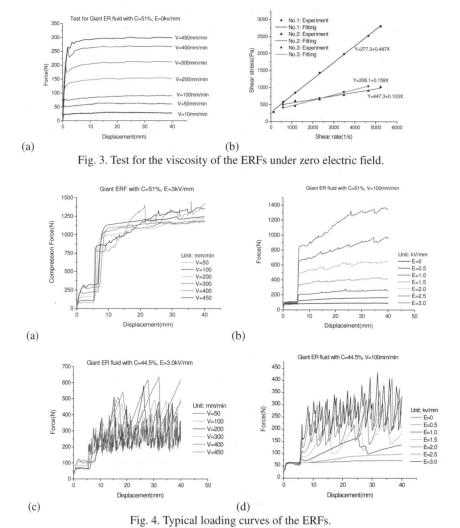

Fig. 3. Test for the viscosity of the ERFs under zero electric field.

Fig. 4. Typical loading curves of the ERFs.

Fig. 5(a) shows the periods of the fluctuations under different electric fields and flow rates. It reveals that this period only depends on the intensity of the electric field. By taking the average of the compressive force as shown in Fig. 4, the shear stress of the ERFs can be obtained and the relations between shear stress and shear rate for Giant ERF C=51% are plotted in Fig. 5(b), in which the intersections of the fitting lines with the $y$-axis give the yield stresses and their slopes indicate the viscosities. The results are shown in Figs. 5(c) and (d). It is found that the yield stress of ERF-1 and ERF-3 increases exponentially with the electric field intensity, while it is linear for ERF-2. In addition, the changes of the ERFs' viscosities are not very large, although high-intensity electric-fields are applied.

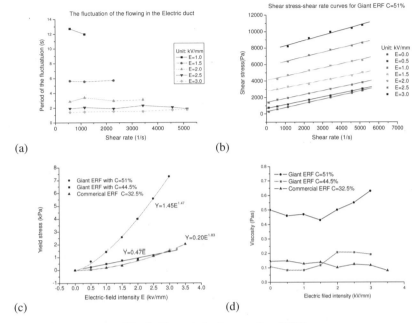

Fig. 5. Characterization results of ERFs.

## 5. Summary

In this study, a capillary rheometer has been designed, manufactured and then used to characterize three ERFs under high shear rate within parallel duct. When free of electric field, the ERFs behave like Newtonian fluids; and when subjected to high electric field they are changed to be Bingham plastic. However, the flowing of the ERFs in parallel duct is not smooth and has severe fluctuations, and the period of the fluctuations only depends on the electric field intensity.

## Acknowledgment

The work reported in this paper is part of a project CERG No.621505 supported by the Hong Kong Research Council. Its financial assistance is gratefully acknowledged.

## References

1. R. Stanway, *Materials Science and Technology*, **20**, 931 (2008).
2. V. M. Zamudio, et al, *Physica A*, **227**, 55 (1996).
3. W. M. Winslow, *Journal of Applied Physics*, **20**, 1137 (1949).
4. R. Stanway, *et al*, *Smart Materials and Structures*. **5**, 464 (1995).
5. G. M. Kamath, *et al*, *Smart Materials and Structures*, **5**, 576 (1996).
6. N Makris, *et al*,*Smart Materials and Structures*, **5**, 551 (1996).
7. K. P. Tan, *et al*, *International Journal of Modern Physics B*, **20**(9), 1049 (2006).

8.   H. P. Gavin, *Smart Materials and structures*, **7**, 664 (1998).
9.   W. H. Kuo, *et al*, *Journal of sound and vibration*, **292**, 694 (2006).
10.  W. J. Wen, et al, *Nature Materials*, **2**, 727 (2003).

# Part K
## Plasticity in Materials
## Processing Technology

# PREDICTION OF THE RESIDUAL STRESS IN THE PROCESS OF STRAIGHTENING WIRE

TAE WON KIM[1], MYOUNG GYU LEE[2]

*Support team, Atthi Inc., 513-14, Jungwon-gu, Seongnam-si, Gyeonggi-do 462-120, KOREA,*
*Eco-Materials Research Center, Korea Institute of Machinery and Materials, Sangnam 66, Changwon,*
*Kyungnam 641-010, KOREA,*
*ccw770@lge. twkim@ahtti.com, mang92@kims.re.kr*

HYUNG-IL MOON[3], HYUNG JONG KIM[4], HEON YOUNG KIM[5†]

*Department of Mechanical & Mechatronics Engineering, Kangwon National University, 192-1 Hyoja-dong,*
*Chuncheon, Gangwon-do 200-701, KOREA,*
*moon@kangwon.ac.kr, khjong@kangwon.ac.kr, khy@kangwon.ac.kr*

Received 15 June 2008
Revised 23 June 2008

Microwire made using a straightening process has high added value and has been adopted in many fields of industry. There is much active research on the straightening process. Very straight microwire can be obtained by removing the residual stress induced during the manufacturing process. Generally, the residual stress is removed or minimized through several drawing steps with heat treatment. This study used finite element analysis to calculate the residual stress during each straightening process and investigated the main reason for a change in stress.

*Keywords*: Wire; Straight Line Treatment; Drawing; Residual Stress; Finite Element Analysis; Stress Relaxation.

## 1. Introduction

With the diverse roles of microwire in industry, the added value of producing truly straight microwire has increased and research into methods to guarantee straight wire is actively underway[1].

Microwire is produced using drawing and annealing processes, which generate tensile and compressive residual stress[2,3]. Compressive residual stress, which is distributed on the surface of the product after drawing, can prevent cracks in the product and inhibit surface wear, while tensile residual stress has an adverse effect on the true straightness and durability of microwire and, consequently, reduces the quality of the product. In particular, tensile stress distributed in the axial and circumferential directions causes cracks to propagate on the wire surface, resulting in fatigue failure[4]. Therefore, it is

---

[†]Corresponding Author.

667

necessary to reduce the tensile residual stress distributed on the surface during the straightening process[5]. So, this study used finite element analysis to simulate the drawing process, which produces the greatest change in the internal stress of wire, and evaluated the effects of the residual stress on the wire initially generated.

## 2.  Finite Element Modeling

A finite element model of wire with a 1.9° bend was constructed to generate an arbitrary initial residual stress, as shown in Fig. 1. This bend is the minimum angle that can generate tensile and compressive residual stress on the surface of the wire. The wire modeled was 0.324-mm-diameter microwire, manufactured by Kwangin Wire. The software used in the analysis was ABAQUS 6.5, and the total length was set to 10mm in order to reduce the analysis time. By imposing symmetry in the longitudinal section, a 1/2 model was adopted. Table 1 lists the mechanical properties of SUS304 wire, and Fig. 2 shows the order and boundary conditions of the finite element analysis used to evaluate the residual stress distribution. In the first stage, the model was subjected to back tensioning, which applies tensile stress in the direction opposite to the direction of drawing (step 2). Then, after the drawing process (step 3), the distribution of residual stress was evaluated. At that time, the drawing die was set as a rigid body, and the non-friction condition between the drawing die and wire was imposed on the assumption that sufficient lubrication was applied. In addition, the section reduction percentage (R.P.) was set to 4.32%, and the half angle of the die was set to 7°. In the reverse bending process, the modeled wire was given a displacement of –0.165mm in direction No. 1, 0.005mm in direction No. 2, and 0.03rad in rotation direction No. 2, while the section on the side of the die was fixed. Consequently, the bent wire was straightened. In the back tensioning process, pressure was applied to the wire section on the opposite side of the wire, which distributed tensile stress on the wire. In the drawing process, the analysis involved passing the wire on the side of the die through the drawing die[6].

## 3.  Residual Stress Evaluated by Stage

### 3.1.  *Reverse bending process*

In the reverse bending process, the initial residual stress was generated by straightening the bent wire. The initial residual stresses generated were a compressive stress of – 32MPa on surface A and a tensile stress of 186MPa on surface B, as shown in Fig. 3. The analysis indicated that after the reverse bending process, the compressive stress was greater than the tensile stress in the wire section.

### 3.2.  *Back tensioning process*

In the back tensioning process, surface pressure is applied in a direction opposite to the drawing direction. In the analysis conducted while arbitrarily changing the amount of surface pressure applied between 50~120MPa, compressive and tensile stresses were distributed on the wire surface, as shown in Fig. 4. Fig. 4 shows that at 80MPa and below, the tensile stress decreases after the process. This may result in a kind of springback that happens as the tensile stress is eased. In order to confirm this, the springback after the reverse bending process was analyzed. In the springback analysis, the restraint on Point A in Direction No. 2 was released after the reverse bending process, and the change in the distribution of the axial stress was examined. Fig. 5 shows the stress distribution after the reverse bending process and the change in the residual stress after the springback analysis. This shows that the springback tends to maintain equilibrium between compressive and tensile stress. Accordingly, the stress reduction phenomenon in Fig. 4 (at below 80MPa) is believed to result from the effect of springback.

Table 1. Mechanical properties of SUS304.

| Modulus of elasticity | 197GPa |
|---|---|
| Yield stress | 415MPa |
| Ultimate strength | 725MPa |
| Poisson's ratio | 0.29 |
| Elongation at break | 50% |

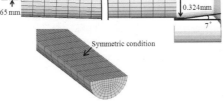

Fig. 1. Finite element model of the wire.

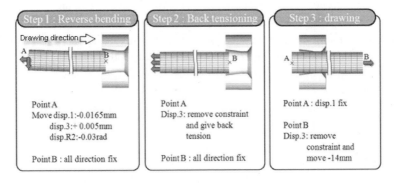

Fig. 2. Analysis procedure of the drawing process.

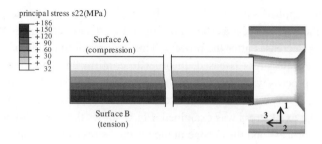

principal stress s22(MPa)

Fig. 3. Axial residual stress after the reverse bending step.

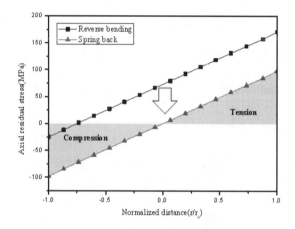

Fig. 4. Variation in the residual stress at surface A.

### 3.3. *Drawing process*

Fig. 6 shows the change in the axial residual stress on the wire surface by stage when the tensile stress was 150MPa. For the surface residual stress generated during the drawing process, the difference in stress between the two surfaces separated by Point A is reversed. This is because the stress is redistributed due to the reduction in diameter. The stresses before passing through the die, namely, a relatively high tensile stress on Surface B and a relatively low tensile stress on Surface A, are both changed into compressive stresses as they pass Point C. At that point, the stress is redistributed internally in the form of tensile stress and, as the wire passes the die (D), an identical strain (strain rate) happens. As a result, a distribution opposite to the initial stress appears. This result is confirmed in Fig. 7, which shows the distribution of stress in the wire during the drawing process.

Fig. 8 compares the axial residual stress between bent and straight wire models when the tensile stress is 150MPa. In the straight wire model, the distribution of the axial residual stress between the two surfaces is symmetrical. By contrast, in the bent wire model, the distribution of stress slants toward one side relative to the center of the wire. This confirms that the initial residual stress is an important variable for predicting the residual stress after the drawing process.

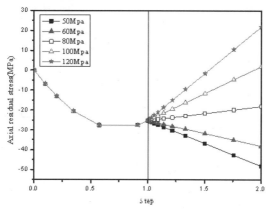

Fig. 5. Variation in the axial residual stress during reverse bending and the springback analysis.

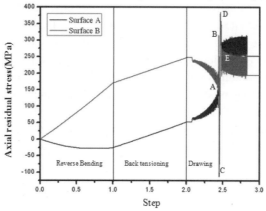

Fig. 6. Variation in the axial surface residual stress during the three steps.

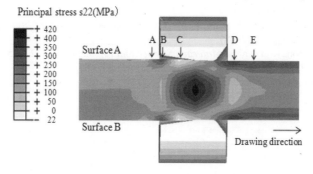

Fig. 7. Distribution of the residual stress during the drawing process.

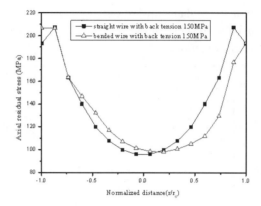

Fig. 8. Axial residual stress distribution of straight and bent wire with a back tension of 150Mpa.

## 4.  Conclusions

This study predicted the residual stress distribution in a microwire during the straightening process using finite element analysis. For this, we applied an initial residual stress to the wire before it passed through the drawing die. The results indicate that

(1) The compressive stress increased with a tensile stress below 80MPa after back tensioning. This is considered to be a kind of springback phenomenon in which the back tensioning condition is eased compared to the reverse bending condition.

(2) During the drawing process, the stress was redistributed and, after the drawing process, the difference in axial residual stress between the two surfaces was reversed.

(3) The effect of the initial residual stress is significant in the analysis of the drawing process. Accordingly, the effect of the initial residual stress should be considered when predicting the residual stress through analysis.

## References

1.  T. Yamashita, K. Yoshida, *Japanl Society of Mechanical Engineering.* **48**, 322 (2005).
2.  W. K. Kim, H. G. Shin, B. H. Kim, H. Y. Kim, *International Journal of Machine Tools & Manufacture.* **47**, 1046 (2007).
3.  B. H. Kim, W. K. Kim, H. Y. Kim, S. M. Yoon, K. H. Na, *5th Japan-Korea Joint Symposium on Micro-Fabrication*, 66 (2005).
4.  S. Norasethasopon, K. Yoshida, *Materials Science and Engineering,* A422, 252 (2006).
5.  D. H. Ko, W. H. Hwang, S. G. Lee, B. M. Kim, *KSPE*, **23**, 162 (2006).
6.  ABAQUS user manual.

# FORMING LIMIT OF AZ31B MAGNESIUM ALLOY SHEET IN THE DEEP DRAWING WITH CROSS-SHAPED DIE

HEON YOUNG KIM[1†], SUN CHUL CHOI[2], HYUNG JONG KIM[3]

*Division of Mechanical Engineering & Mechatronics, Kangwon National University,*
*Chuncheon, 200-701, KOREA,*
*khy@kangwon.ac.kr, cschul@kangwon.ac.kr, khjong@kangwon.ac.kr*

SEOK MOO HONG[4], YONG SEUNG SHIN[5], GEUN HO LEE[6]

*Mechatronics & Manufacturing Technology Center, Samsung Electronics Co., LTD,*
*Suwon, 443-742, KOREA,*
*seokmoo.hong@samsung.com, ysmir.shin@samsung.com, glee@samsung.com*

Received 15 June 2008
Revised 23 June 2008

Magnesium alloy sheets are usually formed at temperatures between 150 and 300°C because of their poor formability at room temperature. In the present study, the formability of AZ31B magnesium alloy sheets was investigated by the analytical and experimental approaches. First, tensile tests and limit dome height tests were carried out at several temperatures between 25 and 300°C to get the mechanical properties and forming limit diagram (FLD). A FLD-based criterion considering the material temperature during deformation was used to predict the forming limit from a finite element analysis (FEA) of the cross-shaped cup deep drawing process. This criterion proved to be very useful in designing the geometrical parameters of the forming tools and determining optimal process conditions such as tool temperatures and blank shape by the comparison between finite element temperature-deformation analyses and physical try-out. The heating and cooling channels were also optimally designed through heat transfer analyses.

*Keywords*: AZ31B magnesium alloy sheet; Forming limit diagram; Cross-shaped cup deep drawing; Heating and cooling channels; Finite element analysis.

## 1. Introduction

Magnesium alloys usually have very poor formability at room temperature because of their hexagonal close-packed (HCP) crystal structure. However, at 200°C or higher, they exhibit considerably improved formability, showing near-superplastic behavior. Conventionally, die casting or squeeze casting have been used for manufacturing magnesium alloy parts, but these methods require the solidification process from the liquid or semisolid state of material, which may result in casting defects and cause safety problems in final products. Recently, warm or hot press-working technology for

---

†Corresponding Author.

magnesium alloy sheets has been recognized as a promising alternative to solve these problems with keeping reasonable productivity. However, there are still difficulties involved in controlling the process parameters such as the forming speed and temperature since magnesium is a very rate-sensitive material especially at elevated temperatures. In addition, thermal softening behavior of magnesium alloys makes it very challenging to predict the forming limit in warm or hot press-working processes using finite elemental analysis.

Research on the material properties and formability of magnesium alloy sheets has been actively conducted for the last decade. Yoshihara et al.[1], Huang et al.[2], and Zhang et al.[3] investigated the formability and the effect of blank holding force in a warm deep drawing process through experiments and finite element analysis. Naka et al.[4] obtained the forming limit diagram (FLD) of an Al-Mg alloy sheet for various temperatures and forming speeds, and Chen et al.[5] evaluated the formability of AZ31 magnesium alloy sheet through conical cup tests and V-bending tests. Recently, the authors[6] have carried out an experimental study to obtain the FLD of AZ31B sheet, and to investigate its springback characteristics at elevated temperatures.

In this study, the stress-strain curves and the FLDs of AZ31B magnesium alloy sheet were obtained through uniaxial tensile tests and limiting dome height (LDH) tests at different temperatures from room temperature to 300°C. The tool set for a cross-shaped cup deep drawing was designed and fabricated on the basis of preliminary heat transfer and forming analyses. Coupled thermal-deformation finite element analyses of the cross-shaped cup deep drawing process for various process conditions were carried out to find the optimum tool temperatures and the minimum tool corner (or fillet) radii which could produce a successful cup without any failure in it. An FLD-based failure criterion, taking the blank temperature into consideration, was adopted to predict the failure initiation in the analyses, and the results of this approach were verified by comparing with the experimental data.

## 2.  Material Tests

### 2.1.  *Tensile test*

Uniaxial tensile tests were conducted at temperatures of 25 (room temperature), 100, 150, 200, 250 and 300°C for the 0.5mm thick AZ31B sheet specimens with the tensile axes of 0°, 45° and 90° orientations from the rolling direction (RD). ASTM E8 sub-size specimens with a gauge length of 25mm and a universal testing machine, INSTRON 5582, with the capacity of 100 kN and a maximum speed of 500 mm/min, were used in the tests. The INSTRON Advanced Video Extensometer® (AVE) enabled us to measure large elongation of the specimens in a hot chamber more easily and more accurately than any other methods available. Tests were performed with a constant crosshead speed of 0.5mm/s, which corresponded to 0.02/s of strain-rate. The stress-strain curves, averaged over the directions, at various temperatures are shown in Figure 1(a). As the temperature

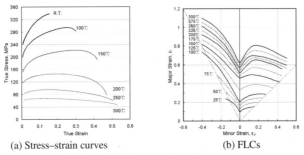

(a) Stress–strain curves            (b) FLCs

Fig. 1. Stress–strain curves and FLCs of AZ31B magnesium alloy sheet at various temperatures.

increases, the flow stress decreases while the total elongation increases, and the material clearly softens at temperatures of 200°C or higher.

### 2.2. *FLD test*

In order to determine the FLD of AZ31B sheet, limiting dome height (LDH) tests were performed. Heat cartridges and thermocouples were installed at proper locations in each tool so that the temperature of die, blank-holder and punch could be controlled separately. Four kinds of specimens were cut into so-called Nakajima specimens with circular cut from the same material sheet as that used in the tensile tests, the narrowest widths of which were 25, 50, 85 and 175mm, with a length of 175mm in the rolling direction. A square grid pattern of 2×2mm was applied on their surfaces using a silk-screening technique instead of the electro-chemical etching method to avoid possible notch effect. Tests were carried out with 0.1 mm/s of punch speed at temperatures of 25, 100, 150, 200, 250 and 300°C. Basically, tests were performed under dry friction conditions without using lubricant, except the additional tests for the 175×175mm specimens, in which water-soluble graphite powder was applied on the blank surface before tests to induce failure in the balanced biaxial mode. Measurement of the principal strains was made using ASIAS developed by the authors[6].

Figure 1(b) shows the FLD of AZ31B magnesium alloy sheet at various temperatures. The FLC$_0$ of AZ31B sheet increased rapidly up to 150°C, and thereafter increased slowly up to 300°C, at which it became more than six times that at room temperature. These FLCs are used to predict the failure initiation in the finite element analysis of forming processes.

### 3. Cross-Shaped Cup Deep Drawing Tests

#### 3.1. *Design of tool set*

To investigate the formability of AZ31B sheet in a practical forming process, a tool set for a cross-shaped cup deep drawing was designed and fabricated by the optimization based on the finite element thermal-deformation analyses. Because lots of variables including geometrical and process parameters, as listed in Table 1, were to be designed, a complicated optimization procedure was needed: first, find the optimum tool temperatures with initial (standard) values of the side lengths, corner/fillet radii and clearance, and with a circular blank; second, find the optimum radii one by one with the optimum tool temperatures obtained at the first step; third, optimize the heating and cooling channels in the corresponding tools; last, find the minimum radii which could produce a successful cups. Both the optimization of the tool temperatures and the minimization of the tool radii were aimed at the lowest maximum thinning over the whole blank, and the optimization of the heating and cooling channels at the most uniform distribution of the tool surface temperatures.

Finite element simulations were carried out using an explicit code PAM-STAMP with the thermal properties of this blank material: the conductivity of 96 W/(m·°C), the specific heat of 1,024 J/(kg·°C), and the interfacial heat transfer coefficient of 4500 N/(s·m·°C) for no clearance (complete contact) and 450 N/(s·m·°C) for 0.1mm of clearance.

Figure 2 shows the change in the maximum thinning at 25mm of punch depth and the maximum drawing depth, which did not cause any failure, depending on the variation of each radius. The boundary between 'safe' and 'failed' was determined by the FLD-based criterion, as would be illustrated in Figure 3.

As the result of optimization, the geometries of the punch and blank were finally determined as shown in Table 1. The location, the number and the diameter of heating channels in the die and blank holder, and cooling channels in the punch and pad were also optimally designed, as described in the previous study by the authors[7].

Table 1. Range of design variables and their optimized values.

| Design variables | Range | Optimized |
|---|---|---|
| **Geometrical parameters** | | |
| Side length_1 ($S_1$), mm | 12, 15, <u>18</u>[†] | 18 |
| Side length_2 ($S_2$), mm | 24, 21, <u>18</u>[†] | 18 |
| Punch corner radius ($R_C$), mm | 1.0, 1.5, <u>2.0</u>[†] | 1.5 |
| Punch fillet radius ($R_F$), mm | 5.0, 10.0, <u>13.0</u>[†] | 10.0 |
| Punch shoulder radius ($R_S$), mm | 1.0, 1.5, <u>2.0</u>[†] | 1.5 |
| Die shoulder radius ($R_D$), mm | 3.0, 5.0, <u>10.0</u>[†] | 5.0 |
| Clearance (C), mm | 0.5, 0.6, <u>0.7</u>[†] | 0.6 |
| **Process Parameter** | | |
| Die Temperature ($T_D$), °C | 200, <u>250</u>[†] | 250 |
| Holder Temperature ($T_H$), °C | 200, <u>250</u>[†] | 250 |
| Punch Temperature ($T_P$), °C | 25, 75, <u>100</u>[†], 150, 200, 250 | 100 |
| Blank shape | <u>Circle</u>[†], Octagon, Square, Rhombus | Octagon |

[†] Standard level of each design variables

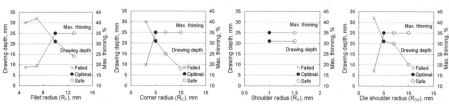

Fig. 2. Maximum thinning and maximum depth depending on tool radii.

### 3.2.  *Optimization of tool temperatures*

To find the optimum tool temperatures which could maximize the forming limit, we tried to do a design of experiments using an orthogonal array $L_8(2^7)$, in which 'smaller-the-better' characteristics of the maximum thinning (negative thickness strain) at 25mm of punch depth was analyzed. The temperature level was 200 and 250°C for the die and holder, 100 and 150°C for the punch, and 150 and 200°C for the pad. The ANOVA table, as in Table 2, shows that the temperatures of the punch and die have greater influence on the maximum thinning than those the blank holder and pad.

Figure 3 illustrates how to apply the maximum thinning-based and FLD-based failure criteria to the forming limit evaluation. When the temperatures of all tools are 200°C, a failure occurred at 8.6mm of punch depth; but when the temperatures of the die, holder, punch and pad were 250, 250, 100 and 150°C, respectively, which could be the optimum temperatures, a successful cup was drawn without any failure up to 25mm of punch depth.

### 3.3.  *Experimental tryout*

A full set of tools was made and installed on the hot formability testing equipment. A water-soluble liquid lubricant for high temperature was applied on the both sides of the blank prior to forming. The average punch speed was set to be around 0.15mm/s during the whole process.

Table 2. ANOVA table for tool temperature optimization.

|            | SS      | DOF | MS     | F-test |
|------------|---------|-----|--------|--------|
| $T_{Die}$    | 24.675  | 1   | 24.675 | 2.638  |
| $T_{Holder}$ | 3.878   | 1   | 3.878  | 0.415  |
| $T_{Punch}$  | 4.337   | 1   | 4.337  | 0.464  |
| $T_{Pad}$    | 49.551  | 1   | 49.551 | 5.298  |
| Error      | 28.056  | 3   | 9.352  | ·      |
| Total      | 110.497 | 7   | ·      | ·      |

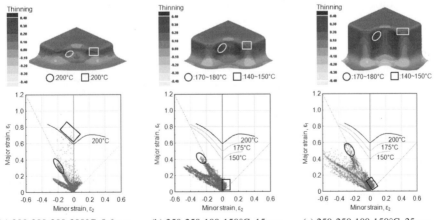

(a) 200-200-200-200°C, 8.6mm     (b) 250-250-100-150°C, 15mm     (c) 250-250-100-150°C, 25mm

Fig. 3. Thinning distribution and FLD-based criterion (temperatures of die-holder-punch-pad, and depth).

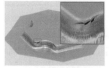 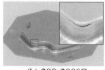  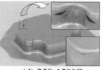

   (a) 200-100°C        (b) 200-200°C        (c) 250-100°C        (d) 250-150°C

Fig. 4. Photographs of the cups drawn in several conditions (temperatures of die/holder-punch/pad).

Figure 4 shows the photographs of the products drawn under several temperature conditions. In these tryouts, the temperatures of the die and holder were set the same, and those of the punch and pad the same for easy control. As predicted above, the blank was successfully formed into a cross-shaped cup of 25mm of depth under almost the same condition (250-100°C) as in the finite element analysis. It is obvious that in the cross-shaped cup deep drawing, the formability of the blank material can be considerably improved by cooling the punch and the pad while heating the die and the blank holder.

## 4. Conclusion

The objective of this study is to investigate the basic formability of the 0.5mm-thick AZ31B magnesium alloy sheet at room and elevated temperatures, and to search for a solution to maximize its forming limit in sheet-forming processes. The results can be summarized as follows:

(1) At temperatures of 200°C or higher, the ductility and formability were greatly improved, while the flow stress decreased to half or lower at room temperature.

(2) The tool set for a cross-shaped cup deep drawing was designed and fabricated through the optimization based on the finite element thermal-deformation analyses. As the result of optimization the corner and fillet radii of the punch, and the blank shapes were determined.

(3) When the temperatures of the die and holder were kept at 250°C by heating, and the punch and pad at 100°C by cooling, the blank was successfully formed up to 25mm (or more)of punch depth without any failure.

(4) The FLD-based criterion proved to be reasonable in predicting possible failure by the finite element analysis. The result showed good coincidence with that of experiment.

## References

1. S. Yoshihara, K. Manabe and H. Nishimura, *J. Mat. Proc. Tech.* **170**, 579 (2005).
2. T. B. Huang, Y. A. Tsai and F. K. Chen, *J. Mat. Proc. Tech.* **177**, 142 (2006).
3. S. H. Zhang, K. Zhang, Y. C. Xu, Z. T. Wang, Y. Xu and Z. G. Wang, *J. Mat. Proc. Tech.* **185**, 147 (2007).
4. T. Naka, G. Torikai, R. Hino and F. Yoshida, *J. Mat. Proc. Tech.* **113**, 648 (2001).
5. F. K. Chen and T. B. Huang, *J. Mat. Proc. Tech.* **142**, 643 (2003).
6. H. J. Kim, S. C. Choi, K. T. Lee and H. Y. Kim, *Materials Trans.* **49**, 1112 (2008).
7. S. C. Choi, D. S. Ko, H. Y. Kim, H. J. Kim, S. M. Hong, S. Y. Ryu and Y. S. Shin, *Proc. Spring Conf. KSTP, 370* (2008).

# MICRO FORMING OF GLASS MICROLENS ARRAY USING AN IMPRINTED AND SINTERED TUNGSTEN CARBIDE MICRO MOLD

JEONGWON HAN[1], BYUNG-KWON MIN[2], SHINILL KANG[3†]

*Department of Mechanical Engineering, Yonsei University,*
*134 Shinchondong, Seodaemungu, Seoul, 120-749, KOREA,*
*snlkang@yonsei.ac.kr*

Received 15 June 2008
Revised 23 June 2008

Glass microlens array with lens diameter under 100 μm was replicated by micro thermal forming process using a tungsten carbide micro mold which was imprinted and sintered from an original master. The original master was fabricated by photolithography and thermal reflow process. The effects of forming process conditions on the form accuracy and the surface roughness of the micro formed glass microlens array were examined and analyzed. Finally, to verify the effectiveness of the proposed process, glass microlens array for the purpose of laser beam focusing was designed and fabricated, and a focused laser beam profile using the fabricated microlens array was measured and analyzed.

*Keywords*: Micro forming; glass microlens array; tungsten carbide micro mold.

## 1. Introduction

Remarkable progress has recently been made in the technology of polymer microlens arrays, which are used in optical data storage, digital display, and optical communication. However, the use of glass in optical components has many advantages over the use of polymer materials. For instance, optical properties of glass materials, such as refractive index and thermal expansion coefficient, are more robust with respect to temperature change. Furthermore, glass materials have excellent acid and moisture resistance. Therefore, demand for glass microlens arrays has been increasing in applications that require high optical performance and environmental stability, such as high density optical data storage and multimedia projector components. In the past, glass microlens arrays were mainly produced with a reactive ion etching (RIE) process[1] or a diamond turning process[2]. However, the mass production at low cost has been found to be rather difficult because of the limitation of etching range of RIE processing parameters and the low production rate of diamond turning process.

The micro thermal forming process is a replication process and is a promising

---

[†]Corresponding Author.

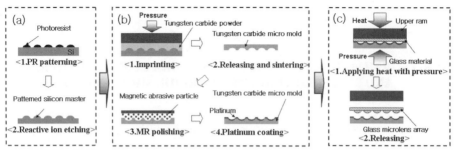

Fig. 1. Schematic diagram of the proposed micro thermal forming process, which consists of (a) original master fabrication, (b) WC micro mold fabrication using imprinting and sintering process, and (c) glass microlens array fabrication by micro thermal forming process.

technique for the mass production of high quality glass microlens arrays at relatively low cost[3]. Tungsten carbide (WC) micro mold has been commonly used for this process[4]. Generally, direct machining process has been used for WC micro mold fabrication[5].

Direct machining process is not appropriate for the fabrication of microlens cavities with diameters of 100 $\mu$m or less. In this study, to overcome the limitation of the conventional WC micro mold fabrication method, the WC micro mold was fabricated by imprinting and sintering process[6]. Then, Glass microlens arrays with lens diameter under 100 $\mu$m were replicated by using a micro thermal forming process. Glass microlens arrays were produced with various micro thermal forming processing conditions. The effects of micro thermal forming process conditions on the form accuracy and the surface roughness of the micro formed glass microlens array were examined and analyzed. Finally, to verify the effectiveness of the proposed process, glass microlens array for the purpose of laser beam focusing was designed and fabricated, and a focused laser beam profile using the fabricated microlens array was measured and analyzed.

## 2.  Fabrication of Tungsten Carbide Micro Mold for Glass Microlens Array

Micro thermal forming process for glass materials is normally operated at a temperature between 300℃ and 1200℃. WC is widely used as a micro mold material in micro thermal processes because of its hardness, creep resistance and thermal stability at high temperature[4]. Figure 1 shows a schematic diagram of the proposed micro thermal forming process: (a) original master fabrication, (b) WC micro mold fabrication using imprinting and sintering process, (c) glass microlens array fabrication by micro thermal forming process. First, original master fabrication process (Fig. 1 (a)) was performed. Photoresist pattern of microlens array was manufactured by using the reflow method[7]. Second, WC micro mold fabrication process (Fig. 1 (b)) was carried out. In this experiment, the WC micro mold composed of micro-cavities was fabricated with imprinting and sintering process that was developed by Kang et al.[6]. Imprinting process (Fig. 1 (b)) replicates the surface shape of the prepared master by applying pressure to mixture of WC powder and bonding agents. The surface of the WC micro mold was polished to refine the coarse texture of the surface generated during the sintering process,

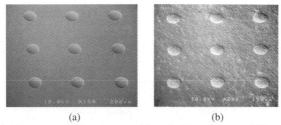

(a)                                              (b)

Fig. 2. SEM image of (a) the patterned silicon master with a microlens diameter of 77 $\mu$m, a sag height of 4.7 $\mu$m, a pitch of 250 $\mu$m and a surface roughness (Ra) of 3.5 nm, and (b) the fabricated WC mold with microlens cavities: a microlens diameter of 58 $\mu$m, a sag height of 3.6 $\mu$m, a pitch of 190 $\mu$m and a surface roughness (Ra) of 53 nm.

as shown in Fig. 1 (b), by using the magneto-rheological (MR) abrasive finishing process[6]. Then, a platinum coating was added to prevent sticking between the glass material and the mold surface during the micro thermal forming process. Scanning electron microscopy (SEM) images of the silicon master and the WC micro mold are shown in Fig. 2.

## 3. Fabrication of Glass Microlens Array by Micro Thermal Forming Process

Figure 3 shows a schematic diagram of the micro thermal forming system used in this study. The machine consists of a system for applying pressure, a WC micro mold, a jig for the WC micro mold, infrared ray lamps for heating, a load cell, and a thermocouple. A glass material was chosen. The glass transition temperature ($T_g$) and yield temperature ($A_t$) of the glass material (K-PSK100, Sumita Optical Glass Inc.) are 390 °C and 415 °C. Diced cube (10 mm x 10 mm x 3 mm) of the glass material was placed on the WC micro mold, as shown in Fig. 1 (c) 1. A two-step pressing method was employed to reduce the shrinkage of the glass material.

The graph in Fig. 4 shows the temperature and the applied compression force during the micro thermal forming process (total processing time: 12 minutes). The two steps are as follows: (1) pressing near the yield temperature with a precisely controlled force, and (2) pressing near the glass transition temperature. The second pressing carried out during cooling from the first pressing temperature to the glass transition temperature.

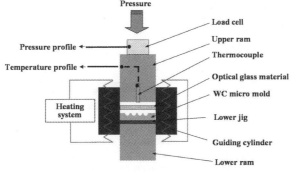

Fig. 3. Schematic diagram of the micro thermal forming machine.

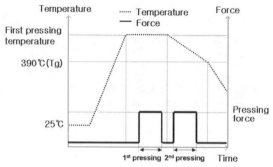

Fig. 4. The variations of the temperature and compression force during the micro thermal forming process.

If the molding process finishes near the yield temperature, the problem of shrinkage of the glass material becomes critical because thermal expansion of the glass occurs significantly above the glass transition temperature. In our approach, the first and second pressing forces were the same.

No significant effect of the compression force on the replication quality was found for compression forces in the range 50 ~ 200 kgf and first pressing temperature of 415 ℃. In our experiment, the compression force was fixed at 50 kgf.

Figure 5 shows three-dimensional surface profiles of glass microlenses fabricated at various first pressing temperatures with a 50 kgf compression force. The glass microlens fabricated using a first pressing temperature of 400 ℃ (Fig. 5 (a)) is defective because insufficient glass material entered the cavity. At the first pressing temperature of 415 ℃ (Fig. 5 (b)), the lowest surface roughness of the glass microlens was achieved. Under this condition, there were no surface defects, i.e., no sticking occurred between the glass material and the WC micro mold surface. At the first pressing temperature

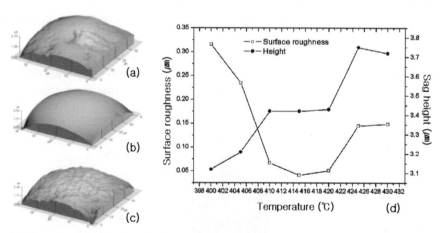

Fig. 5. Effects of varying the first pressing temperature on the sag height and surface roughness of glass microlenses replicated with a compression force of 50 kgf: (a), (b) and (c) show the AFM images of the replicated microlenses' surfaces for various first pressing temperatures, and (d) shows the relationships between the first pressing temperature, the surface roughness (Ra) and sag height of the glass microlenses.

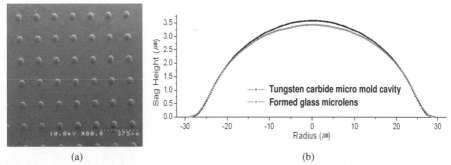

(a)  (b)

Fig. 6. (a) SEM image of the formed glass microlens array with lens diameter of 58 μm and (b) surface profiles of the formed glass microlens (process condition: 415 ℃ first pressing temperature, 50 kgf compression force) and WC micro mold cavity.

of 430 ℃ (Fig. 5 (c)), the surface roughness was excessive: a spire pattern is present on the glass microlens surface, which arises because of sticking between the glass material and the WC micro mold surface. The graph in Fig. 5 (d) shows the variations of the sag height and surface roughness of the formed glass microlenses with respect to the first pressing temperature at a compression force of 50 kgf. The sag height increases as the first pressing temperature increases because the fluidity of the glass material increases. However, from 410 ℃ to 420 ℃ (which is near the yield temperature, 415 ℃) the sag heights are similar. The formed glass microlens array with the best surface roughness was achieved at the yield temperature (415 ℃) of the glass material. Measured values of 10 samples of the formed glass microlens (process condition: 415 ℃ first pressing temperature, 50 kgf compression force) were 3.424 μm average sag height (standard deviation: 0.022 μm), 145 μm average radius (standard deviation: 5.6 μm) of curvature that was measured by surface profiler and 58.04 μm average diameter (standard deviation: 0.19 μm). Figure 6 shows a SEM image of a formed glass microlens array (process condition: 415 ℃ first pressing temperature, 50 kgf compression force) and surface profiles of the formed glass microlens and WC micro mold cavity.

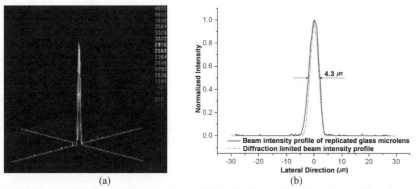

(a)  (b)

Fig. 7. (a) 3-D intensity profile and (b) comparison of diffraction limited beam intensity profile and normalized intensity profile of the focused laser beam (wavelength of 609 nm) using a replicated glass microlens with a diameter of 58 μm, a sag height of 3.42 μm and a surface roughness (Ra) of 44 nm.

## 4. Measurement and Analysis of Optical Properties of Formed Glass Microlens Array

To verify the effectiveness of the proposed micro thermal forming process for glass materials, optical properties of the replicated glass microlens arrays were measured and analyzed. The measured average focal length of replicated glass microlens was 249 $\mu m$ (standard deviation: 6 $\mu m$) using a laser source of 609 nm wavelength at 10 samples. The calculated focal length was 246 $\mu m$ using optical simulation program (Code-V, Optical Research Associates). Measured and calculated focal lengths of glass microlens were very similar. Figure 7 shows the measured light intensity profile of the replicated glass microlens and diffraction limited beam intensity profile at the focal plane. The focused beam spot was measured using a beam profiler and a laser source of 609 nm wavelength. A focused beam spot diameter of 4.3 $\mu m$ at full width half maximum (FWHM) was obtained and diffraction limited beam spot diameter (FWHM) was 3.5 $\mu m$.

## 5. Conclusions

Glass microlens array with lens diameter under 100 $\mu m$ was replicated by a micro thermal forming process using a WC micro mold, which was fabricated by using imprinting and sintering process. To produce a high quality glass microlens array, a two-step pressing method during micro thermal forming process was applied. The geometrical properties of each formed glass microlens were measured, and the effects of various processing conditions on the geometrical properties were analyzed. Finally, to verify the effectiveness of the proposed micro thermal forming process, the focal length and the focused beam spot of formed glass microlens were measured. These experimental results show that the proposed micro thermal forming process can be used effectively in glass microlens array fabrication.

## Acknowledgments

This work was supported by the Korea Science and Engineering Foundation (KOSEF) grant funded by the Korea government (MOST) (No.R0A-2004-000-10368-0).

## References

1. M. Severi and P. Mottier, *Opt. Eng.* 38, 146 (1999).
2. M. Zhou and B. K. A. Ngop, *J. Mater. Process. Tech.* 138, 586 (2003).
3. Y. Aono, M. Negishi, and J. Takano, *Proc. SPIE.* 4231, 16 (2000).
4. H. O. Andren, *Mater. Design.* 22, 491 (2001).
5. Y. Kojima, Optifab 2005, *Technical Digest of SPIE.* TD03, 44 (2005).
6. W. Choi, J. Lee,W. Kim, B. Min, S. Kang, and S. Lee, *J. Micromech. Microeng.* 14, 1519 (2004).
7. S. Moon, N. Lee, and S. Kang, *J. Micromech. Microeng.* 13, 98 (2003).

# FAST SPRINGBACK SIMULATION OF BENDING FORMING
# BASED ON ONE-STEP INVERSE ANALYSIS

YI-DONG BAO[1†,] WEN-LIANG CHEN[2]

*College of Mechanical and Electrical Engineering, Nanjing University of Aeronautics and Astronautics,*
*Nanjing, Jiangsu, 210016, P. R. CHINA,*
*baoyd@nuaa.edu.cn, cwlme@nuaa.edu.cn*

HONG WU[3]
*College of Mechanical Science and Engineering, Jilin University*
*Changchun, Jilin, 130025, P. R. CHINA,*
*wh@jlu.edu.cn*

Received 15 June 2008
Revised 23 June 2008

A simplified one-step inverse analysis of sheet metal forming is a suitable tool to simulate the bending forming since the deformation path of bending forming is an approximately proportion one. A fast spring-back simulation method based on one-step analysis is proposed. First, the one-step inverse analysis is applied to obtain the stress distribution at the final stage of bending. Then, the unloading to get a spring back is simulated by LS-DYNA implicit solver. These processes are applied to the unconstrained cylindrical bending and the truck member rail. The spring-back and member rail widths at the several key sections are compared with experimental ones. It is well demonstrated that the proposed method is an effective way to predict the spring-back by unloading after bending process.

*Keywords*: Bending Forming; one-step inverse analysis; springback.

## 1. Introduction

Spring back prediction is an important issue for the sheet metal forming industry. Spring-back influences the quality and precision of auto-body parts formed by bending. Because spring back causes shape change, which is a crucial trouble in the assembly process, it must be considered at the beginning stage of design for bending. Spring back distortions of sheet metal forming depend on the geometry of tools, the process conditions and the mechanical properties of the sheet metal. The tool tryout stage is a very expensive and time-consuming process. Nowadays, the finite element method provides a technology, which has the potential of moving some of trial and error effort from the expensive physical space to a much less expensive computer simulation technology. Many researchers have been done to analyze spring back phenomenon of sheet metal forming

---

†Corresponding Author.

by numerical method during last decades. Most of them used the dynamic explicit algorithm to simulate the forming process and static implicit algorithm to predict spring back process[1-3]. Some people both used static implicit algorithm to calculate the forming and spring back processes[4-5]. On the other hand, G. Y. Li[6] presented a method which introduce the dynamic explicit algorithm to obtain forming and spring back results.

In this paper, we first review the derivation of one step inverse analysis, including bending effects[7-11]. Then the coupling of the one step inverse analysis and LS-DYNA implicit solver to simulate the spring back of bending forming is presented. The initial stress distribution after bending forming process is predicted by one step inverse analysis quickly. Finally, two numerical results will show efficiency and accuracy of this spring back simulation method.

## 2. Finite Element Formulation of One-Step Inverse Analysis with Bending Effects

In the one-step inverse analysis, only two configurations are considered: the initial flat blank $C^0$ and the final 3D work piece $C$. Using the Kirchhoff assumption, the initial and final position vectors of a material point can be expressed on $C$:

$$\begin{cases} x_q^0 = x_p - u_p + z^0 n^0 \\ x_q = x_p + zn \end{cases} \tag{1}$$

where $u_p$ is the displacement of the point $p$ on the midsurface of the sheet, $n^0$ and $n$ are the normals of the midsurface at $p^0$ and $p$.

The thickness stretch is supposed to be constant through the thickness. So, the inverse Cauchy-Green left tensor $[B]^{-1}$ takes the following form:

$$[B]^{-1} = [F]^{-T} [F]^{-1} \tag{2}$$

The eigen value calculation of $[B]^{-1}$ gives two principal plane stretches $\lambda_1$, $\lambda_2$ and their direction transformation matrix [M]. Then the thickness stretch $\lambda_3$ is calculated by the incompressibility assumption.

Neglecting the transverse shear effects of the thin sheet, the element internal force vector in the global coordinates is obtained:

$$\{F_{int}^e\} = [T]^T \int_{V^e} ([B_m]^T + z[B_b]^T)\{\sigma\} dz dA \tag{3}$$

where $[B_m]$ is membrane matrix and $[B_b]$ is bending matrix. These resultant forces can be obtained by numerical integration through the thickness at the element central using five or seven Gauss integration points.

In the one-step inverse analysis, only the initial $C^0$ and the final configuration $C$ are compared. Therefore, the elastic-plastic deformation is assumed to be independent on the loading path and a total constitutive law is obtained,

$$
\begin{bmatrix} \sigma_{xx} \\ \sigma_{yy} \\ \sigma_{xy} \end{bmatrix} = \frac{2\bar{\sigma}(2+\bar{r})}{3\bar{\varepsilon}(1+2\bar{r})} \begin{bmatrix} 1+\bar{r} & \bar{r} & 0 \\ \bar{r} & 1+\bar{r} & 0 \\ 0 & 0 & 1 \end{bmatrix} \begin{bmatrix} \varepsilon_{xx} \\ \varepsilon_{yy} \\ \varepsilon_{xy} \end{bmatrix}
\tag{4}
$$

where $\bar{\varepsilon}$ and $\bar{\sigma}$ are the total equivalent strain and equivalent stress. The mean planar isotropic coefficient $\bar{r}$ is obtained from the three anisotropic coefficients with the relation, $\bar{r} = (r_0 + 2r_{45} + r_{90})/4$. This operation is performed for each numerical integration point through the thickness.

Since the one-step inverse analysis cannot deal with contact problems depending on the loading history, the tool actions (punch, die and draw-bead) are simply represented by some external forces at the final configuration. The consideration of the tool actions will give the external force vector, leading to the following non-linear equilibrium system.

$$
\{R(U,V)\} = \sum_e (\{F_{ext}^e\} - \{F_{int}^e\}) = 0
\tag{5}
$$

## 3.  Unloading Spring Back Simulation by LS-DYNA Implicit Solver

Although the deformation during spring-back should be predominantly elastic, there may be large deformations, and hence the problem is still fundamentally non-linear. The LS-DYNA implicit solver is a nonlinear finite element method based on elastoplastic large deformation theory. The Belytschko-Lin-Tsay shell element based on Mindlin theory is adopted in subsequent spring back calculation. An incremental-iterative numerical algorithm is implemented in LS-DYNA implicit solver. The advantage of the implicit solver is that the number of load or time steps is typically 100 to 10000 times fewer than would be used in an explicit calculation. So the implicit solver are properly applied to quasi-static problems, such as metal forming, especially, the spring back.

The above one step inverse analysis is used to simulate the bending forming process. The Dynain file which includes the nodes and element connectivity, the stress tensor, and the effective plastic strain for each integration point is output after one step inverse analysis.

## 4.  Numerical Results and Discussion

### 4.1.  *Unconstrained cylindrical bending*

Unconstrained Cylindrical Bending (UCB) is one of Numisheet2002 benchmarks asshown in Fig. 1. The benchmark test of cylindrical bending process is recommended in order to investigate spring back analysis.

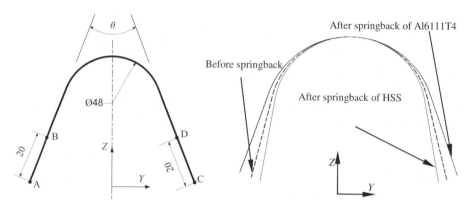

Fig. 1.  Dimension of UCB.                Fig. 2.  Comparison of springback result of UCB.

The blank materials available for unconstrained cylindrical bending are aluminum (Al6111-T4) and high strength steel (HSS). Aluminum samples will be supplied by ALCOA in U.S.A and high strength steels by POSCO in Korea. The flow stress curve is $\sigma = K\varepsilon^n$ *(MPa)* and the material parameters are shown in table1:

Table 1.  Material parameters of Al6111-T4 and HSS.

| Name | E (GPa) | v | YS (MPa) | UTS (MPa) | K (MPa) | n | $r_{avg}$ |
|---|---|---|---|---|---|---|---|
| Al6111T4 | 70.56 | 0.342 | 181.80 | 351.78 | 539.50 | 0.226 | 0.694 |
| HSS | 222.17 | 0.300 | 219.81 | 308.70 | 630.81 | 0.253 | 2.022 |

The initial dimensions of the blank are as follows: Thickness = 1.0mm, length = 120mm, width = 30mm. Angle between line AB and line CD after bending forming is 20 degree (see the Fig. 1).

The bending processes of above materials are simulated by the one step inverse analysis, then the unloading are simulated by LS-DYNA implicit solver based on Dynain files from the one step inverse analysis. The angle between line AB and line CD are experimentally measured before and after spring back. Those are compared with the predictions as shown in table 2. Based on these comparisons, one may say that the spring back tendency of unconstrained cylindrical bending can be predicted by this approach based on one step inverse analysis. The angle $\theta$ between line AB and line CD after spring back becomes larger than before spring back and the spring back angle of Al6111-T4 is greater than one of HSS as shown in Fig. 2. On the other hand, we assume the bottom shape of final configuration used by one step inverse analysis is column surface; however the bottom shape after bending forming should be elliptical column surface. So there are some error between simulation value and experiment one as shown in Table 2.

Table 2. Comparison of spring back angle of two materials.

| Name | Experiment $\theta$ (*Degree*) | Simulation $\theta$ (*Degree*) |
|---|---|---|
| Al6111-T4 | 54 | 50.82 |
| HSS | 35 | 31.66 |

### 4.2. *Bending forming and spring back of truck carling*

The truck carling is one of most important supporting parts. The outer panel of truck carling is usually formed by blanking and bending procedures. The section shape of outer panel of truck carling along axis direction is U shape with variable section. Sometimes there is height difference in the bottom of outer panel. The Fig. 3 shows an outer panel of truck carling made in FAW auto-body. One of the biggest defects is the section shape of sidewall for outer panel will expand or shrink after bending due to spring back. Although the change in part dimension are very small, it will result in difficulties in the following assembling process.

We just measured the deformation results for the right side of truck carling since the left side deformation affects very small for assembling procedure. The real measurement length is 1545 mm. The advantage of this simplified model is we can reduce the simulation and die modification time remarkably. The section dimensions of right side of truck carling are shown in Fig. 4 and the section shape of left side is constant.

The material named 16MnRel is used and the mechanical properties of this material and test conditions are as follow: the initial sheet thickness=7.0mm; the average Lankford value $R_{avg}$=1.05; the friction coefficient=0.15; and the flow stress curve of $\sigma = 849.6(0.00012+\varepsilon)^{0.172}$ (*MPa*).

The die face of truck carling is outer surface from CAD data; however the one step inverse analysis used the central surface, so we should make outer offset 3.5 mm from the initial die face of truck carling. Then this central surface is discretized by finite elements.

Firstly, the bending forming process of truck carling is simulated by one step inverse analysis and the Dynain file is output. The Fig.4 shows the initial blank mesh of truck carling predicted by one step inverse analysis. Secondly, we used LS-DYNA implicit solver to calculate the spring back process based on Dynain file from the one step inverse analysis. Thirdly, we sliced six key sections along trans verse direction and these section

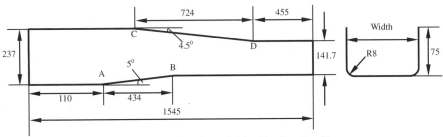

Fig. 3. Section dimensions of right side of truck carling.

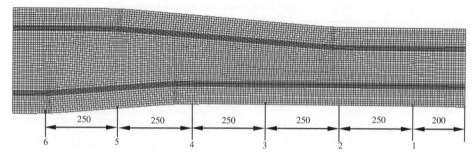

Fig. 4.  Initial mesh and Section positions of right side of truck carling.

positions as shown in Fig.4. Then we can get each section width by section cut function in ETA/Post-Processor. At the same time, the width of corresponding section can be measured by Coordinate Measuring Machine. Finally, we compare these section widths by simulation with the experimental one, the final results are shown in table 3. It is shown that the errors between the predictions and measurements are less than 1.0mm except the section two as shown in Table 3.

Table 3.  Comparison of the section widths by simulation and by experiment.

| Section number | 1 | 2 | 3 | 4 | 5 | 6 |
|---|---|---|---|---|---|---|
| simulation (mm) | 136.74 | 137.01 | 156.03 | 176.01 | 212.53 | 231.89 |
| experiment (mm) | 136.16 | 135.66 | 155.20 | 175.06 | 213.50 | 232.60 |
| error (mm) | 0.58 | 1.35 | 0.83 | 0.95 | 0.97 | 0.79 |

We performed the bending process simulation of truck carling by one step inverse analysis instead of dynamic explicit algorithm. The simulation time of bending was just about 28 seconds whereas the incremental explicit code needed more than 30 times that for the one step inverse one. Thus, the whole simulation time including unloading process for spring back simulation is remarkably reduced. .

## 5.  Conclusion

A fast simulation method based on one step inverse analysis is proposed for bending and unloading for spring back. The one step inverse analysis for bending effects predicted the stress distribution after bending process promptly, which could be used as initial ones for unloading simulation. The unloading for spring back is simulated by the LS-DYNA implicit solver with a Dynain input file from one step inverse analysis. Comparison between the predictions and experiments has proved that the proposed spring back simulation method is efficient and accurate.

## Acknowledgments

This work was funded by the project of high quality personnel from Nanjing University of Aeronautics and Astronautics and this support is gratefully acknowledged.

## References

1. J. S. Shu, H. Hung, *Int. J. Mach. Tools Manufact.*, 1996, 36:423-434.
2. S. W. Lee, D. Y. Yang, *J. of Mats. Proc. Tech.*, 1998, 80-81: 60-67.
3. N. Narasimhan, M. Lovell, *Finite Elements in Analysis and Design*, 1999, 33:29-42.
4. M. Kawka, T. Kakita, A. Makinouchi, *J. of Mats. Proc. Tech.*, 2000, 80-81: 54-59.
5. L. P. Lei, S. M. Hwang, B. S. Kang, *J. of Mats. Proc. Tech.*, 2001, 110: 70-77.
6. G. Y. Li, M. J.Tan, K. M. Liew, *Int. J. Solids Struc.*, 1999, 36:4653-4668.
7. C. H. Lee, H. Huh, *J. of Mats. Proc. Tech.*, 1998, 82: 145-155.
8. Y. Q. Guo, J. L. Batoz, etc., *Computers and Structures*, 2000, 78:133-148.
9. H. Naceur, A.Delaméziere, J. L. Batoz, Y.Q. Guo, *J. of Mats. Proc. Tech.*, 2004, 146:250-262.
10. S. H. Kim, H. Huh, *J. of Mats. Proc. Tech.*, 2002, 130: 482-489.
11. Y. D. Bao, H. Huh, *J. of Mats. Proc. Tech.*, 2007, 187-188:108-102.

# EFFECT OF INITIAL MICROSTRUCTURE ON HOT FORGING OF MG ALLOYS

Y. KWON[1†], Y. LEE[2], S. KIM[3], J. LEE[4]

*Materials Processing Department*
*Korea Institute of Materials Science(KIMS)*
*Changwon 641-831, KOREA,*
*kyn1740@kims.re.kr, lys1668@kims.re.kr, swkim@kims.re.kr, ljh1239@kims.re.kr*

Received 15 June 2008
Revised 23 June 2008

Magnesium alloys still have a lot of technical challenges to be solved for more applications. There have been many research activities to enhance formability of magnesium alloys. One is to design new alloy composition having better formability. Also, low formability of wrought alloys can be improved by optimizing the processing variables. In the present study, effect of process variables such as forging temperature and forging speed were investigated to forgeability of three different magnesium alloys such as AZ31, AZ61 and ZK60. To understand the effect of process variables more specifically, both numerical and experimental works have been carried out on the model which contains both upsetting and extrusion geometries. Forgeability of magnesium alloys was found to depend more on the forging speed rather than temperature. Forged sample showed a significant activity of twinning, which was found to be closely related with flow uniformity.

*Keywords*: Mg alloy; Forging; Texture.

## 1. Introduction

Since magnesium alloys have the highest specific strength among the industrially available alloys, application of magnesium alloys in the field of transportation and consumer electronics industries has increased significantly recently. Because of its light weight property, magnesium alloys are likely to have huge possibilities if process variables are optimized. Especially, forging of Mg alloys can be more beneficial to cope with light weight demand compared with thin sheet forming with a relatively low weight. In the present study, effects of process variables such as forging temperature and forging speed were investigated to understand hot forging behaviors of magnesium alloys like AZ31, AZ61 and ZK60 alloys. All materials used in the present study were cast billets which have quite coarse microstructure. Usually, coarse grained structure might have a very poor formability. On the other hand, extruded billets of Mg alloys surely have a refined grain sizes followed by higher formability. In the present study, however, a

---

[†]Corresponding Author.

feasibility of cast Mg billets was checked for forging stock considering that cast Al billets have been widely used for hot forging stocks.

For this purpose, a model die configuration was suggested in which both upsetting and extrusion could be exerted throughout forming stage. Firstly, a series of compression tests was carried out to get flow behaviors with temperature. Using model die system, a series of forging experiments was done with two billet size samples. To understand the effect of process variables more specifically, numerical simulations have been carried out on the model geometry. Finally, forged microstructures were observed to find out how Mg alloys evolved with strain accumulation.

## 2. Experimental Procedures

All AZ31, AZ61 and ZK60 alloys were received with the cast and homogenized condition as shown in Fig. 1. All the samples carried cast structures and average grain size or primary dendritic arm spacing was measured 380, 580 and 200µm respectively. These cast and homogenized structures seemed to have no preferred orientation initially as shown in Fig. 2.

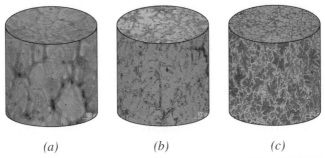

(a)          (b)          (c)

Fig. 1. Microstructure of AZ31 sheet, (a) AZ31, (b) AZ61 and (c) ZK60.

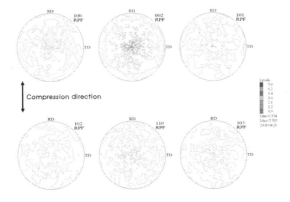

Fig. 2. As-received texture of AZ31 billet showing no preferred orientation.

Forging experiment was designed like a model shown in Fig. 3. Two different billet sizes with a cylindrical shape were prepared as shown in Fig. 3(c) and (d) with diameters of 30mm and 36mm. All the magnesium billets were heated at the target temperature for more than 30mins. To enhance formability die and punch were heated with halogen lamp before forging up to 100 °C. DEFORM-2D program was used to investigate the effect of processing variables, temperature and strain rate. Initial temperature of die and punch was given as 100 °C as controlled in the experiment

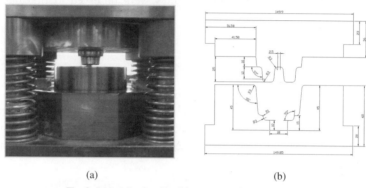

(a)                                                    (b)

Fig. 3. (a) hot forging die, (b) geometry of die and punch.

## 3. Results

Fig. 4 represents example of hot forging experiment taken from AZ31 billet of 36mm at the punch speed of 1mm/min. Formability at certain condition was determined as good, moderate and bad respectively depending on whether external crack opens and how severe crack is. As temperature rises, formability was found to be generally enhanced independent of billet size. Billet size influenced formability as listed in Table 1. Outer surface of smaller diameter billet needs to be flow farther to fill die, which in turn induce more strain and crack initiation. Depending on alloy type, the maximum strain to be applied seems to differ. In the present study, ZK60 alloy seems to have the highest formability. However, it is not true that ZK60 has the highest formability since ZK60 has the finest dendritic arm spacing of 200 µm. So, more careful study needs to be carried out which alloy system has better formability for hot forging process. Table 2 showed the effect of forming speed on hot forging formability. Even though it is not clear in Table 2, overall tendency of forming speed was that higher forming speed leads to a slightly better formability. It is still unclear why higher forming speed leaded to better formability at this stage. However, a limited slip system seemed to activate a large amount of twinning. Then twin might help to accommodate stress concentration and make deformation go on without fracture.

Fig. 5 shows a general failure pattern occurring during hot forging of coarse grained magnesium alloys. It is clear that the first crack initiated from orange peel-like surface relief which was caused by localized slip due to both HCP structure and coarse grain sizes. Fig. 5(b) of room temperature compressed specimen clearly illustrates how surface relief looks like. As shown in Fig. 5, a distribution of surface relief might be dependent on alloy composition. Surface of ZK60 specimen looks relatively clean compared to both AZ31 and AZ61 alloys. Also, location of initial billet gives folding type defect like region 2 type defect in Fig. 5(a). Since grain sizes of three alloys were not same, surface relief pattern might also depend on how slip on basal plane gets influenced with the variation of grain size. ZK60 has the smallest grain size of 200μm while AZ61 has the largest one.

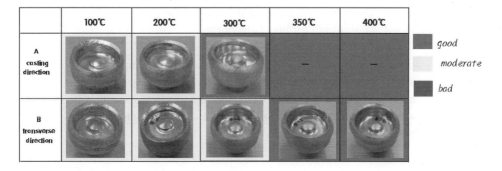

Fig. 4. Hot forging of AZ31 at different temperature with the billet of Φ36.

Table 1. Hot forging in the casting direction with the punch speed of 1mm/min.

|  | AZ31 | | AZ61 | | ZK60 | |
|---|---|---|---|---|---|---|
|  | D30 | D36 | D30 | D36 | D30 | D36 |
| 100℃ | - | △ | - | X | X | △ |
| 200℃ | - | △ | - | △ | O | △ |
| 300℃ | X | O | X | △ | O | O |
| 350℃ | △ | O | △ | △ | O | O |
| 400℃ | △ | - | △ | △ | - | - |

Table 2. Effect of forging speed in the transverse direction with the billet of Φ36mm.

|  | Speed (mm/sec) | 100°C | 200 °C | 300 °C | 350 °C | 400 °C |
|---|---|---|---|---|---|---|
| AZ31 | 1 | △ | △ | △ | O | O |
|  | 5 | X | △ | O | O | O |
| AZ61 | 1 | - | X | △ | △ | △ |
|  | 5 | - | X | △ | △ | △ |
| ZK60 | 1 | △ | △ | O | O | O |
|  | 5 | △ | O | O | O | O |

Fig. 6 shows damage distribution taken from finite element simulation of hot forging of Mg alloys. A fracture area on forged part seems to correspond with the highest damaged area as shown in Fig. 5 and 6. In the present study, Cockcroft and Latham model implemented in DEFORM-2D was used. This model tells that damage level gets higher where higher tensile stress is exerting. These areas might be related with where slip trace gets agglomerated and orange peel –like surface relief prevails.  Actually, magnesium alloys exhibit plastic anisotropy between tensile and compressive loading. Therefore, the finite element simulation done in the present study could not illustrate a real deformation situation. However, a commercial finite element simulation still could predict where crack would open and tell overall forming behavior.

Fig. 7 shows that microstructure of hot forged part differs depending on the accumulated strain. In the area where shear deformation prevails, very refined grains of less than 3μm seem to be recrystallized. On the other hand, less deformed areas still contain coarse grained cast structure. Due to as-received large grain size, twinning was observed to be quite active throughout the forged sample.

(a)                                                            (b)

Fig. 5. (a) Failure occurred during hot forming and (b) surface relief pattern after room temperature compression.

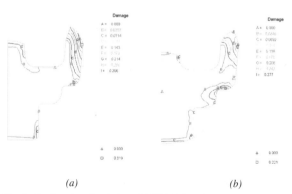

(a)                                        (b)

Fig. 6. Hot forming simulation of AZ31 billet with (a) Φ30 and (b) Φ36.

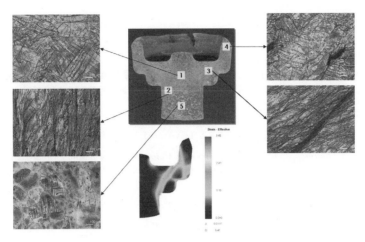

Fig. 7. Hot forged microstructure at various points showing different evolution with strain.

## 4. Conclusion

In the present study, hot forging of three kinds of magnesium alloys with the initial grain size or dendritic arm spacing of over 200μm was carried out. The flow behavior showed a strain softening after peak stress irrespective of strain rate and temperature. Forgeability of magnesium alloys was investigated in terms of forging temperature, speed and billet size. Even in the coarse grain regime, smaller grain size of ZK60 exhibits better formability rather than AZ31 and AZ61 alloys. Crack was found to start from the surface relief where slip localization occurs to insufficient slip system and coarse grain size. This phenomenon could be predicted using Cockcroft and Latham damage model where higher tensile stress appears. In summary, finer grained billet needs for successful forging of magnesium alloys.

## Acknowledgments

This work was supported by grant No. RTI04-01-03 from the Regional Technology Innovation Program of the Ministry of Knowledge Economy (MKE).

## References

1. P. Skubisz, et al : Journal of Materials Processing Technology Vol. 177 (2006), p.210.
2. Byoung Ho Lee, et al : Journal of Materials Processing Technology Vol. 187–188 (2007), p766.
3. Margam Chandrasekaran and Yong Ming Shyan John : Materials Science and Engineering A, Vol. 381 (2004) p.308.

# THE UNLOADING MODULUS OF AKDQ STEEL AFTER UNIAXIAL AND NEAR PLANE-STRAIN PLASTIC DEFORMATION

E.J. PAVLINA[1], B.S. LEVY[2], C.J. VAN TYNE[3†]

*Department of Metallurgical and Materials Engineering, Colorado School of Mines, Golden, CO 80401 USA,*
*epavlina@mines.edu, bslevycons@ameritech.net, cvantyne@mines.edu*

S.O. KWON[4], Y.H. MOON[5]

*School of Mechanical Engineering, Pusan National University, Busan 609-735 KOREA,*
*kwon1558@pusan.ac.kr, yhmoon@pusan.ac.kr*

Received 15 June 2008
Revised 23 June 2008

Springback is a problem in the manufacture of a variety of automotive components. To determine springback, it is necessary to know the strength of the material after plastic deformation and the slope of the unloading curve (i.e. the unloading modulus). Prior investigations have shown that the unloading modulus for steels after plastic deformation has a slope that is lower than the normally accepted value for Young's modulus. Previous studies on the slope of the unloading curve were after uniaxial tensile plastic deformation. In the present study, the unloading modulus for an aluminum killed drawing quality (AKDQ) steel was evaluated after both uniaxial and near plane strain deformation. A tube hydroforming system was used for near plane-strain deformation. The average unloading modulus following uniaxial deformation for the AKDQ steel is approximately 168 GPa. The average unloading modulus for the circumferential stress component after near plane-strain deformation is lower than after uniaxial deformation. For a given amount of overall plastic deformation, the axial component of the unloading modulus is greater than the circumferential component, and with increased plastic strain, the unloading modulus for both components decreases. These results demonstrate that the components of the unloading modulus are dependent on the strain path of the prior plastic deformation.

*Keywords*: Springback; unloading modulus; plane strain deformation; AKDQ steel.

## 1. Introduction

In order to decrease vehicle weight while meeting other performance requirements, it is necessary to use thinner-gauge, higher-strength steels. However, the use of these steels can cause forming problems such as springback after stamping. Uncontrolled springback results in unacceptable distortion and rejection of the stamped part. Therefore, accurate material behavior models are required in order to understand, model, and control springback.

---

[†]Corresponding Author.

Springback occurs during the unloading step after plastic deformation. It depends on the strength of the steel sheet prior to unloading, as well as the "elastic" unloading curve. The slope of the unloading curve is called the unloading modulus. Elastic recovery, some microplastic recovery, and thermal recovery all contribute to the unloading modulus[1]. For advanced high-strength steels (AHSS), the strength prior to unloading is much higher than for the more formable steel grades.

It has been widely reported that the observed modulus on unloading a material following uniaxial tensile plastic deformation is much less than the traditional value of Young's modulus that is used to characterize and quantify the elastic behavior of materials[2-6]. Levy et al.[7] showed that the observed decrease in the unloading modulus stabilizes after prestrains of approximately 3.5%.

Effective modeling for the control of springback requires accurate constitutive equations for the sheet being deformed. Since strain paths in stamped parts are complex, it is important to determine if the strain path of the prior plastic deformation affects the unloading modulus. The objective of the present study is to assess the effect of strain path on the unloading modulus.

## 2. Experimental Procedure

### 2.1. *Uniaxial tensile tests*

In order to evaluate the effects of strain path on the unloading modulus, aluminum killed drawing quality (AKDQ) steel tubes were obtained. The tube dimensions were 76.2 mm in diameter, with a wall thickness of 1.6 mm. The AKDQ tubular material has a yield strength of 236 MPa, a tensile strength of 290 MPa, and a uniform elongation of 0.23 in true strain. Uniaxial tensile specimens with a gauge length of 50 mm were machined from tubes at a location 90° relative to the weld and in the axial direction of the tube. These tensile specimens were plastically deformed in tension to true strains of 0.018 to 0.05 followed by unloading. During the unloading, load was measured with a load cell, and the unloading displacement was measured with an extensometer. These measurements were converted to stress and strain values. The slope of the unloading stress-strain curve was determined from a linear regression through the data after uniaxial plastic deformation for the steel. The unloading modulus was calculated over a stress range of 50 MPa less than the peak strength, down to 20 MPa.

### 2.2. *Tube expansion tests*

Strain gage rosettes were applied to other tubes. The tubes were freely expanded in a hydraulic bulge apparatus to effective strains of 0.018 and 0.026. The deformation path exhibited a strain ratio, $\varepsilon_2/\varepsilon_1$, of 0.06, assuming that the first principal strain component was the circumferential strain and the second principal strain component was the axial strain. The output of the strain gage rosettes was recorded throughout the entire test with particular focus on the depressurization (i.e. unloading) portion of the test. The stress components and strain components were determined from the pressure and strain gage rosette data at a rate of 1 Hz.

The stress components in the tube during unloading were determined by a method developed at Schuler Hydroforming for a free-expansion test for tubes with fixed ends[8]. The Schuler method offers reasonable agreement with similar equations developed by Koç and Altan[9-10]. The circumferential and axial stress components are given by

$$\sigma_\theta = \frac{P_i(2r_0 + 2\Delta r - t_0)^2}{4t_0 r_0} \qquad \sigma_z = \frac{\sigma_\theta}{\dfrac{1}{R} + 1} \qquad (1)$$

where $P_i$ is the internal pressure, $r_0$ is the original tube radius, $\Delta r$ is the change in tube radius, $t_0$ is the original tube wall thickness, and $\overline{R}$ indicates the normal anisotropy of the steel. The stress-strain values in each direction were used to calculate the unloading in that direction.

## 3.  Results

### 3.1.  Stress-strain behavior

Figure 1 shows the effective stress-strain behavior of the AKDQ material deformed in uniaxial tension as a solid line. Stress-strain data points for the plastic hydraulic expansion of the tube along a near plane-strain deformation path are also shown in Fig. 1, in order to compare the two deformation processes. A value greater than one for the normal anisotropy, $\overline{R}$, is needed in order to obtain a reasonable match between the two deformation processes. Unfortunately, the normal anisotropy of the steel was unknown, so the components of the unloading modulus were calculated for a range of $\overline{R}$ values. For the stress-strain behavior, the Schuler method with $\overline{R}$ = 1.75 provides the best consistency with the uniaxial tensile test results. The Schuler method also captures the strain hardening characteristics of the steel.

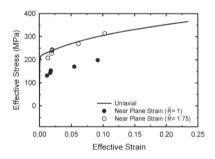

Fig. 1.  Effective stress-strain curves for an AKDQ steel deformed in uniaxial tension shown by the solid line and along a near plane strain deformation path ($\varepsilon_2/\varepsilon_1 = 0.06$) shown by the data points.

### 3.2. *Unloading after uniaxial plastic deformation*

Figure 2 illustrates the effect of prestrain on the unloading modulus after uniaxial tensile deformation. The unloading modulus is essentially constant (168 GPa) over the range of prestrain examined in this study. The value for the unloading modulus appears to stabilize at a constant value for prestrains of 0.015 or less.

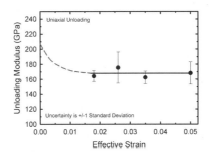

Fig. 2. Average unloading modulus for an AKDQ steel deformed in uniaxial tension.

### 3.3. *Unloading after hydraulic expansion*

Figure 3 shows an example of the unloading behavior for a tube deformed along a near plane-strain deformation path ($\varepsilon_2/\varepsilon_1 = 0.06$) and then followed by a slow release of the pressure. Figure 3(a) shows the strain components as well as the pressure as a function of unloading time. These data and Equation (1) are used to compute the unloading stress-strain curve. The unloading modulus in the axial and circumferential directions is the slope of the stress-strain curve on unloading as illustrated in Figure 3(b).

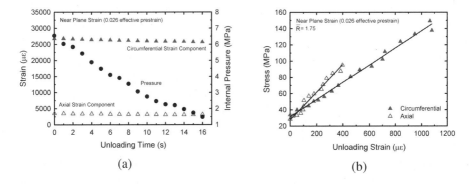

(a)                                                          (b)

Fig. 3. Example unloading curve for the AKDQ steel deformed along a near plane strain deformation path ($\varepsilon_2/\varepsilon_1 = 0.06$) to an effective prestrain of 0.026. (a) Experimental strain components and the internal pressure data as a function of time during the depressurization of the tube (i.e. during unloading). (b) Stress-strain curves for unloading. A normal anisotropy corresponding to $\overline{R} = 1.75$ was used to calculate the stress components in this example.

Figure 4 depicts the axial and circumferential components of the unloading modulus for the AKDQ steel tube. The normal anisotropy was varied from $\overline{R} = 1$ to $\overline{R} = 2$. Regardless of the value of $\overline{R}$, the circumferential component of the unloading modulus is less than the axial unloading modulus. Figure 4 also shows that the circumferential unloading modulus is clearly different from the axial unloading modulus.

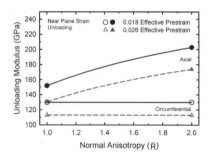

Fig. 4.   Average unloading modulus for an AKDQ steel deformed along a near plane strain deformation path ($\varepsilon_2/\varepsilon_1 = 0.06$).

## 4.  Discussion

The value for the unloading modulus for the AKDQ steel deformed in uniaxial tension as determined in this study is 168 GPa. This value compares reasonably well with other researchers who cite values between 170-180 GPa for comparable amounts of prestrain[5,7].

In this study, the unloading modulus following uniaxial tension stabilizes at a prestrain of 0.015 or less. This value of prestrain is significantly less than the 0.035 stabilization prestrain that was observed by Levy et al.[7]. The uniaxial tensile specimens were taken from tubes that underwent plastic deformation during tube making. The bending strain from tube making is estimated to be 0.02. It would appear that the prior deformation from tube making accounts for the difference in the two values.

For the near plane-strain deformation ($\varepsilon_2/\varepsilon_1 = 0.06$) during tube expansion, the circumferential strain components are much larger than the axial strain components. The difference in the unloading modulus for the two directions may be related to the relative amount of plastic prestrain in each of the component directions. Since the component of prestrain is much less in the axial direction compared to the circumferential direction, the saturation value of the unloading modulus may not have been attained. These results clearly indicate that the various components of the unloading modulus need to be measured, and included in any model for unloading to properly account for the individual stress and unloading components. This consideration is especially important when there are different amounts of plastic strain imparted in each component direction before unloading.

The circumferential component of the unloading modulus following near plane-strain deformation is less than the unloading modulus following uniaxial tensile deformation.

This result clearly indicates that the unloading modulus is dependent on the strain path of the prior plastic deformation.

## 5. Summary

The elastic unloading modulus for an aluminum killed drawing quality (AKDQ) steel was measured after plastic deformation along two different strain paths—uniaxial tension and near plane strain. The unloading modulus after uniaxial tension was found to be 168 GPa, which is consistent with values found in other studies.

For the near plane-strain deformation, the circumferential component of the unloading modulus was found to be less than the axial component. Both unloading modulus components decreased with increased amounts of prior plastic deformation. The circumferential component of the unloading modulus was less than the unloading modulus determined after uniaxial deformation.

The results of this study show that:
- the unloading modulus is dependent on the strain path imposed during the prior plastic deformation,
- for small strains, the unloading modulus decreases with increasing amount of plastic prestrain,
- the magnitude of each plastic strain component should be considered in determining its effect on the unloading modulus in a specific direction, and
- previous deformation, such as tube making, affects the unloading modulus.

## Acknowledgments

Support of this work by the 2007 National Science Foundation (NSF) East Asia and Pacific Summer Institutes for US Graduate Students (Award #0714459) and the Korean Science and Engineering Foundation (KOSEF) is gratefully acknowledged. The grant supported travel and experimental work at Pusan National University. The authors appreciate Dr. Comstock of AK Steel for supplying the tubes used in the study.

## References

1. C. Zener, *Elasticity and Anelasticity of Metals* (University of Chicago Press, Chicago, 1948).
2. M. Blaimschin, K. Radlmayr, A. Pilcher, E. Till, and P. Stiaszny, in Proceedings of the 19th Biennial Congress of the IDDRG, (HDDRG, Egger, Hungary, 1996).
3. R. Cleveland and A.K. Ghosh, in Sheet Metal Forming for the New Millennium, Proceedings of the 21$^{st}$ Biennial Congress of the IDDRG, (NADDRG, Dearborn, MI, 2000), pp. 141-155.
4. R.M Cleveland and A.K Ghosh, *Intl. J. Plasticity* **18**, 769 (2002).
5. L. Luo and A.K. Ghosh, *J. Engr Mat. Tech.* **125**, 237 (2003).
6. S. Thibaud and J.C. Gelin, *Intl. J. of Forming Processes* **5**, 505 (2002).
7. B.S. Levy, C.J. Van Tyne, Y.H. Moon, and C. Mikalsen, SAE Technical Paper 2006-01-0146, (SAE, Warrendale, PA, 2006).
8. E.J. Pavlina, K. Hertel, and C.J. Van Tyne, *J. Mat. Proc. Tech.* **201**, 242 (2008).
9. T. Altan, T. Sokolowski, K. Gerke, and M Ahmetoglu, *J. Mat. Proc. Tech.* **98**, 34 (2000).
10. M. Koç, Y. Aue-u-lan, and T. Altan, *Intl. J. Mach. Tools and Manuf.* **41**, 761 (2001).

# SPRING-BACK CHARACTERISTICS OF GRAIN-REFINED MAGNESIUM ALLOY ZK60 SHEET

SEONG-HOON KANG[1†], YOUNG-SEON LEE[2], JUNG-HWAN LEE[3]

*Applied Plasticity Research Group, Department of Materials Processing, Korea Institute of Materials Science,*
*531 Changwondaero, Changwon, 641-831, KOREA,*
*kangsh@kims.re.kr, lys1668@kims.re.kr, ljh1239@kims.re.kr*

Received 15 June 2008
Revised 23 June 2008

In this work, the effect of grain size on the spring-back characteristic was investigated by carrying out air-bending test using magnesium alloy ZK60 sheet with thickness of 0.5 mm at the various temperatures from room temperature to 300 °C. The angles of the bent specimen before and after unloading were measured in order to quantify spring-back amount. It was found out from the bending tests that when the specimens with grain sizes of 14.66 and 60.71 μm were bent by 90°, the amount of spring-back was relatively small at the testing temperature range and was in the range between -2.5° and 2.5°. On the other hand, the spring-back amount dramatically increased at room temperature and phenomenon of spring-go was observed at high temperature when the specimen with submicro grain size of 0.98 μm was bent by 90°. From this finding, it was confirmed that the different spring-back characteristics according to the grain size takes place and thus the grain size of material is one of the important factors which have an effect on the spring-back.

*Keywords*: Air-bending; Spring-back; Grain size; Magnesium alloy.

## 1. Introduction

It was well known that magnesium alloys have the poor formability and large amount of spring-back after deformation at room temperature due to its hexagonal closed-paced crystal structure and the lower Young's modulus ($\approx$ 45 MPa at room temperature) which are completely different from those of steels and aluminum alloys.

Thus, many researchers[1-3] have made an effort to enhance the formability of magnesium alloys and to reduce the spring-back amount after deformation. One of the effective methods for the improvement of the formability is an equal channel angular pressing (ECAE). This method can give rise to grain-refinement by severe plastic deformation assigning the shear strain to the billet. It was revealed from many research works related with severe plastic deformation that through grain refinement by shear strain accumulated during deformation, the formability of the magnesium alloys can be significantly enhanced at wide range of temperature.

---

[†]Corresponding Author.

On the other hand, most technical researches[4-6] on the spring-back were focused on the effect of various process parameters such as sheet thickness, testing temperature, strain rate (velocity), bending angle, tool geometry, etc. In recent, Paisarn and co-works[7] investigated the effect of process parameters on the spring-back and bending limit during bending of magnesium alloy ZK60 grain-refined by ECAE. Bruni et al.[8] carried out air-bending test of ZK60 magnesium alloy under warm and hot forming conditions and investigated the spring-back amount according to the process parameters. However, there are few researches on the effect of grain size on the spring-back characteristics.

Thus, in this work, the effect of grain size on spring-back amount was investigated by using air-bending tests of magnesium alloy ZK60 sheets. In addition, the effect of testing temperature on spring-back was also investigated.

## 2. Bending Specimens and Experimental Setup

In order to prepare the bending specimens with various grain sizes, the ECAE and annealing processes were adopted. Fig. 1 (a) shows a schematic diagram of ECAE process. The die with the internal angle $\Phi$ of 90.0° and external angle $\Psi$ of 36.9° was used to assign an average effective strain of 0.993 within the ECAE billet at each separate pass.[9]

The ECAE billet was pressed at the temperature of 400 °C for the first four passes and thereafter sequent four pressings were carried out at the temperature of 200 °C using processing route Bc in which the billets are pressed with the rotation angle of 90° along the longitudinal axis in the same direction.

Fig. 2 shows the microstructures and grain size distributions of as-cast and ECAEed billets. It was found out that the coarse grains of average grain size of 60.71 μm is transformed to the fine grains due to accumulated effective strain of 7.944 and dynamic recrystallization during eight-pass ECAE. It was also observed that the most grains are

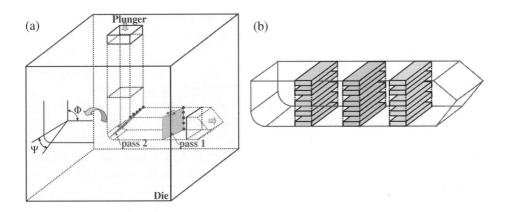

Fig. 1. Schematic diagrams of (a) ECAE with process route Bc and (b) bending specimens machined from ECAEed billet.

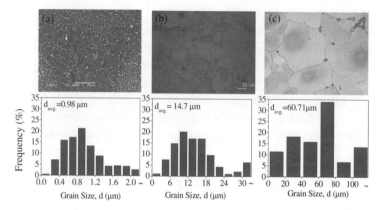

Fig. 2.  Microstructures and grain size distributions of (a) ECAEed-billet, (b) ECAEed-billet annealed at 420°C for 60 minutes and (c) as-cast billet.

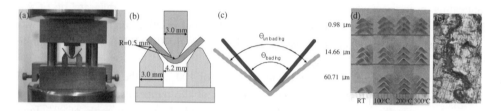

Fig. 3. (a) Air-bending die-set, (b) dimensions of punch/die and die opening width, (c) definition of spring-back, (d) bent specimens according to the grain size and testing temperature and (e) fracture in outside area of specimen with grain size of 60.71 μm after bending at room temperature.

distributed in the range of less than 2.0 μm after eight-pass ECAE and average value is about 0.98 μm.

The microstructurally refined billet with average grain size of 0.98 μm was machined to be bending specimens with the length of 14 mm, width of 8 mm and thicknesses of 0.5 mm along the plane parallel to the extrusion direction as shown in Fig. 1(b). The machined bending specimens were annealed at the temperatures of 420 °C for 60 minutes to obtain the larger grain sizes than grain size of 0.98 μm. The resultant microstructures, grain size distribution and average grain size were shown in Fig. 2 (b) as well.

In order to investigate the effects of grain size and testing temperature on spring-back amount, the small die-set for air-bending test was designed as shown in Fig. 3 (a) and (b). The bending tests were carried out at room temperature, 100, 200 and 300 °C. Punch speed was set to 1.2 mm/min and the stroke was set to 2.77 mm to make the bending angle of 90°. The constant holding time of five seconds after bending of the specimen was maintained for each bending test.

The effect of the various parameters on spring-back amount was analyzed by geometrical measurement of bent specimens. More than three specimens were used in each bending test to obtain the more accurate experimental data. The spring-back

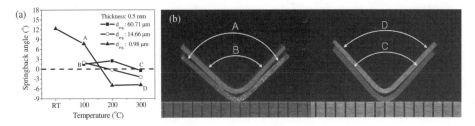

Fig. 4. (a) Measured spring-back amount according to grain size and testing temperature and (b) photos showing the difference in spring-back characteristic.

characteristics of bent specimens were quantified from the angle difference before ($\theta_{loading}$) and after ($\theta_{unloading}$) unloading as shown in Fig. 3 (c).

In addition, the inner and outer radii of bent specimens were also measured by means of image analyzer software to investigate the effects of the grain size and testing temperatures.

## 3. Experimental Results

The photos showing bent specimens and fracture of specimens with grain sizes of 14.66 and 60.71 μm after bending at room temperature were also shown in Fig. 3 (d). The fracture in cases of grain sizes of 14.66 and 60.71 μm is due to its poor formability and large grain sizes.

The measured spring-back amount according to grain size and testing temperature were described in Fig. 4 (a). As depicted, the spring-back amount in cases of grain sizes of 14.66 and 60.71 μm was small in all testing temperatures and its value was in the range between -2.5° and 2.5°. However, the large spring-back amount at the lower temperature (less than 150 °C) and the phenomenon of spring-go at the higher testing temperature (over 150 °C) were observed in case of grain size of 0.98 μm. The photos showing the difference in the spring-back characteristics was shown in Fig. 4(b) to give the better understanding.

The inner and outer radii of the bent specimen with grain size of 0.98 μm appeared to be almost constant regardless of testing temperature in Fig. 5 (a) and (b). On the other hand, these values in cases of grain sizes of 14.66 and 60.71 μm gradually decreased when the testing temperature increased and finally became similar to those in case of grain size of 0.98 μm. It was also observed that the inner and outer radii in case of grain size of 0.98 μm are smaller than those in cases of the larger grain sizes at the temperature of less than 150 °C. For the better understanding, the physical comparison of the differences in inner/outer radii was shown in Fig. 5(c).

Microstructural changes in outside areas of the specimen bent at 300 °C were shown in Fig. 6. As can be seen, even though grain refinement due to dynamic recrystallization in the bending specimen with grain size of 60.71 μm took place during the bending at 300 °C, larger grains are still distributed in the wide area. In case of bending specimen with grain size of 14.66 μm, the grain size after bending at 300 °C was similar to that of initial grain size. On the other hand, the grain growth occurred after bending of the specimen

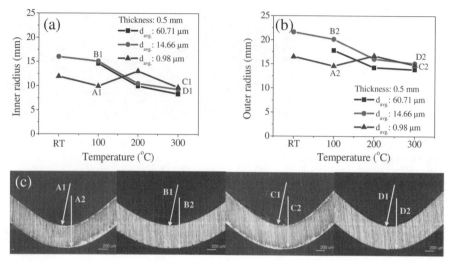

Fig. 5. Measured (a) inner radii, (b) outer radii and (c) photos showing the difference in inner and outer radii.

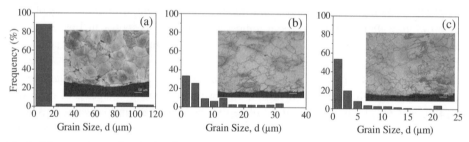

Fig. 6. Microstructures and grain size distributions in the outside areas of the specimens with grain sizes of (a) 60.71, (b) 14.66 and 0.98 μm after bending at 300°C.

with grain size of 0.98 μm. From this, it can be assumed that the phenomenon of spring-go in case of grain size of 0.98 μm might be associated with the grain growth during high temperature bending.

Such grain size effect on spring-back characteristic can be checked from load vs. stroke curves of Fig. 7 measured from bending experiments. As can be seen, the bending load in case of grain size of 0.98 μm is higher than that in cases of grain sizes of 14.66 and 60.71 μm at room temperature. However, opposite result in bending load was obtained at temperature of higher than 200 °C.

## 4. Concluding Remarks

In this work, the effect of grain size on the spring-back characteristic was investigated by air-bending test using ZK60 sheet with thickness of 0.5 mm at the various temperatures. Especially, the large spring-back took place at the lower temperature range and phenomenon of spring-go was observed at the higher temperature range when the grain-

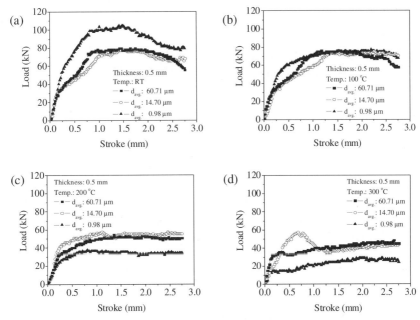

Fig. 7. Measured load vs. stroke curves according to grain size and testing temperature.

refined ZK60 sheet with grain size of 0.98 μm was bent by 90°. It was also found out that the spring-back amounts were relatively small at wide range of temperature when the sheets with relatively larger grain sizes were bent. From this finding, it was confirmed that the grain size of the material is one of important factors which have a significant effect on the spring-back characteristics.

## Acknowledgments

This work was supported by grant No. RTI04-01-03 from the Regional Technology Innovation Program of the Ministry of Commerce, Industry and Energy (MOCIE).

## References

1. S. H. Kang, Y. S. Lee and J. H. Lee, *J. Mater. Process. Tech.* **201**, 436(2008).
2. R. B. Figueiredo and T. G. Langdon, *Mat. Sci. Eng.* **A430**, 151(2006).
3. M. Mabuchi, H. Iwasaki, K. Yanase and K. Higashi, *Scripta Mater.* **36**, 681(1997).
4. T. Sakai, R. Takahashi, T. Furushima and J. I. Koyama, *Keikinzoku/Journal of Japan Institute of Light Metals* **56**, 728(2006).
5. Z. Tekiner, *J. Mater. Process. Tech.* **145**, 109 (2004).
6. J. T. Gau, C. Principe and J. Yu, *J. Mater. Process. Tech.* **191**, 7(1991).
7. R. Paisarn, N. Yugi and N. Koga, *Keikinzoku/Journal of Japan Institute of Light Metals* **55**, 181(2005).
8. C. Bruni, A. Forcellese, F. Gabrielli and M. Simoncini, *J. Mater. Process. Tech.* **177**, 373(2006).
9. Y. Iwahashi, J. Wang, Z. Horita, M. Nemoto and T. G. Langdon, *Scripta Mater.* **35**, 143(1996).

# INCREMENTAL SHEET FORMING WITH LOCAL HEATING FOR LIGHTWEIGHT HARD-TO-FORM MATERIAL

R. HINO[1†], F. YOSHIDA[2]

*Department of Mechanical System Engineering, Hiroshima University,*
*1-4-1, Kagamiyama, Higashi-Hiroshima, Hiroshima, 739-8527, JAPAN,*
*rhino@hiroshima-u.ac.jp, fyoshida@hiroshima-u.ac.jp*

N. NAGAISHI[3]

*Graduate school of Hiroshima University,*
*1-4-1, Kagamiyama, Higashi-Hiroshima, Hiroshima, 739-8527, JAPAN,*
*m070918@hiroshima-u.ac.jp*

T. NAKA[4]

*Department of Maritime Technology, Yuge National College of Maritime Technology,*
*1000, Yuge, Kamijima-cho, Ochi-gun, Ehime, 794-2593, JAPAN,*
*naka@ship.yuge.ac.jp*

Received 15 June 2008
Revised 23 June 2008

A new incremental sheet forming technology with local heating is proposed to form lightweight hard-to-form sheet metals such as aluminum-magnesium alloy (JIS A5083) sheet or magnesium alloy (JIS AZ31) sheet. The newly designed forming tool has a built-in heater to heat the sheet metal locally and increase the material ductility around the tool-contact point. Incremental forming experiments of A5083 and AZ31 sheets are carried out at several tool-heater temperatures ranging from room temperature to 873K using the new forming method. The experimental results show that the formability of A5083 and AZ31 sheets increases remarkably with increasing local-heating temperature. In addition, springback of formed products decreases with increasing local-heating temperature. The developed incremental sheet forming method with local heating has great advantages in not only formability but also shape fixability. It is an effective forming method for lightweight hard-to-form sheet metal for small scale productions.

*Keywords*: Incremental sheet forming; local heating; lightweight sheet metal; forming limit; springback.

## 1. Introduction

Aluminum-magnesium alloy sheets and magnesium alloy sheets are attractive materials for lightweight products. These materials are hard-to-form materials and conventional sheet stamping technology is not capable for forming them since they exhibit limited

---

[†]Corresponding Author.

ductility at room temperature. In order to overcome this problem, warm press forming has been a possible approach[1,2]. However, warm press forming has several drawbacks such as higher cost and, as a result, can be applied only for large scale productions.

On the other hand, new sheet metal forming technology called incremental sheet forming has been investigated actively[3,4]. In this process, a blank is formed by a numerically controlled movement of a simple tool without using expensive dies. Therefore the incremental sheet forming is a cheap and flexible approach which is suitable for small scale productions. Besides, the incremental sheet forming has great advantages in not only flexibility but also formability[4].

Considering the above mentioned characteristics of the lightweight sheet metals and the incremental sheet forming, a combination of heating with incremental forming is a quite attractive choice to form aluminum-magnesium alloy sheets or magnesium alloy sheets for small scale productions. Our aim is to propose a new incremental sheet forming method in conjunction with local heating and understand the process capabilities. For that purpose, a new incremental sheet forming system with local heating is developed. Using the developed forming system, incremental forming experiments of aluminum-magnesium alloy (JIS A5083) sheet and magnesium alloy (JIS AZ31) sheet are carried out at several tool-heater temperatures. Then the effect of the local heating on the formability is discussed based on the experimental results. In addition, the effect of the local heating on residual stress and springback of formed products is also investigated.

## 2. Incremental Sheet Forming with Local Heating

Figure 1(a) illustrates basic ideas of a new incremental sheet forming with local heating. A blank is formed by the relative movement of the moving tool, which is numerically controlled by a computer. The tool is a simple cylindrical punch with hemispherical head of 14 mm diameter. The most important feature of this new incremental forming method is the local heating system. The moving tool has a built-in small heater (200W) and a thermocouple. The moving tool is heated by the heater and the blank is locally heated by heat transfer from the tool surface. Then the ductility of the blank around the tool-contact point becomes higher and the synergistic effect of the incremental forming and the local heating enhances the forming limit of the blank remarkably.

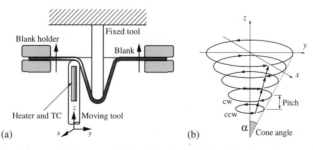

Fig. 1. Schematic illustrations of (a) incremental sheet forming with local heating using moving tool and fixed tool, and (b) tool path for cone-shape forming experiment.

## 3.  Specimens and Experimental Conditions

Experimental conditions of the incremental sheet forming with local heating are summarized in Table 1. Aluminum-magnesium alloy (JIS A5083) sheet of 1 mm thick and magnesium alloy (JIS AZ31) sheet of 0.8 mm thick are used in the present work. A5083 blanks of 250 mm square and AZ31 blanks of 120 mm square are prepared for the experiments. These materials are typical examples of lightweight hard-to-form materials. They exhibit limited ductility at room temperature, but their formability can be improved at elevated temperatures of 400-500K or higher[1,2,5,6].

Simple circular-cone shapes are selected as target shapes of the incremental forming experiments. Figure 1(b) illustrates tool path for the cone-shape forming. The moving tool traces contour of the cone while changing its moving direction clockwise and counter clockwise alternately every one pitch. Tool speed is set to 1 mm/sec and tool pitch is 1 mm. Temperature of the built-in heater of the moving tool is measured by the built-in thermocouple and kept constant during the forming process. Lubricant containing molybdenum disulfide ($MoS_2$) is applied.

Incremental forming experiments are carried out at several tool-heater temperatures ranging from 283 K (i.e. room temperature, RT) to 873 K and at several cone angles from 10 degrees to 80 degrees. Then the effect of the local heating temperature on the formability is investigated. The formability is evaluated based on the forming limit strain measured by means of scribed circles of 6.35 mm diameter. Besides, the formability is also evaluated by the limiting cone angle that is the possible minimum cone angle $\alpha$ for successful forming. In the incremental cone-shape forming, it is well known that the minimum cone angle represents the level of formability[3]. Furthermore, the effect of the local heating temperature on residual stress and springback phenomena is also investigated using some material pieces cut out from the cone-shaped products.

Table 1. Experimental conditions of incremental forming with local heating.

| Material / Blank size [mm] | A5083P-O / 250x250, $t$ = 1 and AZ31-O / 120x120, $t$ = 0.8 |
|---|---|
| Tool diameter [mm] | 14 |
| Tool path | Shown in Figure 1 |
| Product shape | Cone (Cone angle $\alpha$ from 10 to 80 deg.) |
| Tool pitch [mm] | 1 |
| Tool speed [mm/sec] | 1 |
| Tool-heater temperature [K] | From 283 (RT) to 873 |
| Lubricant | SUMIMOLD 201 |

## 4.  Results and Discussion

### 4.1.  *Effect of local heating on formability*

Figure 2 shows samples of the conical shells formed by the incremental forming with local heating at various tool-heater temperatures. In the forming of A5083 sheet, the limiting cone angle can be decreased from 40 degrees to 15 degrees when the tool-heater temperature is raised up from 283 K to 873 K (see Figs. 2(a) and (b)). In the forming of AZ31 sheet, it is impossible to form a blank at the room temperature. However, a cone with 25-degree cone angle is obtained when the tool-heater temperature is 873 K as shown in Fig. 2(c). These results show that the local heating has a significant effect on improvement of formability of A5083 and AZ31 sheets.

(a) Blank: A5083
Tool-heater temp.: 283 K
Cone angle $\alpha$: 40°

(b) Blank: A5083
Tool-heater temp.: 873 K
Cone angle $\alpha$: 15°

(c) Blank: AZ31
Tool-heater temp.: 873 K
Cone angle $\alpha$: 25°

Fig. 2. Samples of incremental cone-shape forming with local heating of A5083 and AZ31 at various tool-heater temperatures and cone angles.

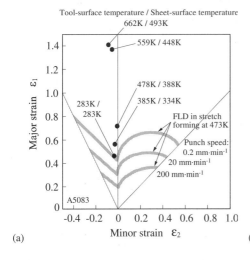 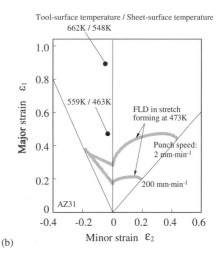

(a) (b)

Fig. 3. Comparison of forming limit strains in incremental forming with local heating with FLDs obtained by warm punch-stretching tests[1,6] for (a) A5083 sheet and (b) AZ31 sheet.

Figure 3 shows the effect of the incremental forming with local heating on formability in comparison with forming limit diagrams (FLDs) obtained by warm punch-stretching tests at 473 K[1,6]. The average major and minor strains of conical shells at their limiting cone angles are treated as forming limit strains in the incremental forming and they are plotted on FLDs as shown by the solid circles in Fig. 3.

This result shows a great advantage of incremental forming with local heating in improvement of forming limit of A5083 or AZ31 sheet. A significant effect of the local-heating temperature on the forming limit is confirmed. In the case of A5083 sheet shown in Fig. 3(a), even the incremental forming at the room temperature exhibits the same forming limit strain with the punch-stretching tests at 473 K. Then the forming limit strain increases with increasing local-heating temperature and a remarkable jump in the formability is observed between the blank-surface temperatures of 388 K and 448 K (i.e. the tool-heater temperatures of 573 K and 723 K). This phenomenon is induced by the high-temperature ductility of the blank which is locally heated around the tool-contact point. The similar tendency is confirmed in the case of AZ31 sheet as shown in Fig. 3(b). The synergistic effect of the incremental forming and the local heating extends the forming limits of the lightweight hard-to-form sheet metals far beyond their FLDs.

### 4.2.  *Effect of local heating on residual stress/springback*

In the incremental forming with local heating, the residual stress and the springback of the products are expected to be reduced due to the temperature dependence of the flow stress. Therefore effects of the local heating on the residual stress and the springback phenomenon are experimentally investigated as follows.

As shown in Fig. 4, a narrow metal strip along the meridian direction or a narrow circular ring is cut out from the cone by using a wire-cut electric discharge machine. When the strip is cut out, the residual stress is released and the strip curls due to springback. Then the curvature change of the strip is as follows:

$$\Delta \kappa_1 = 1/R \tag{1}$$

where $R$ is the radius of the strip curl as shown in Fig. 4(a). Similarly, when the cut-out ring is cut open, the curvature change of the ring is as follows:

$$\Delta \kappa_2 = 1/r_0 - 1/r \tag{2}$$

where $r_0$ and $r$ are the radii of the ring before and after the cut-open as shown in Fig. 4(b). These two parameters, $\Delta \kappa_1$ and $\Delta \kappa_2$, indicate the amounts of residual bending moments due to the residual stresses in the meridian and circumferential directions, respectively.

Figure 5 shows the effect of the local heating on the curvature changes for A5083 cones formed at several tool-heater temperatures from 288 K to 873 K at cone angle of 40 degrees. The springback of the cut-out material reduces remarkably at higher local-heating temperatures. This result indicates that the new incremental sheet forming with local heating provides not only higher formability but also less residual stress/springback at the same time.

### 5.  Conclusions

A new incremental sheet forming technology with local heating for lightweight hard-to-form sheet metals such as aluminum-magnesium alloy sheets or magnesium alloy sheets is proposed. The newly designed forming tool has a built-in heater, and sheet metal is locally heated by heat transfer from the tool surface. As a result, the ductility of the sheet around the tool-contact point becomes higher, and it helps local deformation of the sheet.

Incremental forming experiments of aluminum-magnesium alloy (JIS A5083) sheet and magnesium alloy (JIS AZ31) sheet are carried out at several tool-heater temperatures using the new forming method. It is confirmed that the formability of the sheet metals increases remarkably with increasing local-heating temperature. In addition, springback of formed products decreases with increasing local-heating temperature because the residual stress is reduced. Further study will continue to understand the process mechanics for expanding the process capabilities.

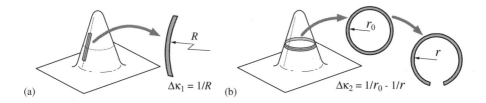

Fig. 4. Experimental methods to examine effects of local heating on residual stress and springback based on curvature changes of (a) strip along meridian direction and (b) circular ring cut out from conical shells.

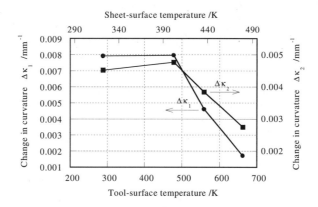

Fig. 5. Relations between local-heating temperature and curvature changes for A5083 formed at several tool-heater temperatures from 288 K to 873 K at cone angle of 40 degrees.

# Reference

1. T. Naka, G. Torikai, R. Hino and F. Yoshida, *J. Mat. Process. Technol.*, **113** (2001), pp. 648-653.
2. F.K. Chen and T.B. Huang, *J. Mat. Process. Technol.*, **142** (2003), pp. 643-647.
3. G. Hussain, L. Gao and N.U. Dar, *J. Mat. Process. Technol.*, **186** (2007), pp. 45-53.
4. M.S. Shim and J.J. Park, *J. Mat. Process. Technol.*, **113** (2001), pp. 654-658.
5. T. Naka, R. Hino and F. Yoshida, *Key Eng. Mat.*, **233-236** (2003), pp. 113-118.
6. K. Fukugi, T. Hironaka, T. Naka, M. Kohzu, K. Azuma and F. Yoshida, *Proc. 54th Japanese Joint Conf. Technol. Plasticity*, (2003), pp. 57-58.

# INFLUENCE OF BILLET SIZE ON THE DEFORMATION INHOMOGENEITY OF MATERIAL PROCESSED BY EQUAL CHANNEL ANGULAR PRESSING

TAO SUO[1†], YULONG LI[2], FENG ZHAO[3]

*School of Aeronautics, Northwestern Polytechnical University, 127 youyi xilu*
*Xi'an 710072, P. R. CHINA,*
*suotao@nwpu.edu.cn, liyulong@nwpu.edu.cn, zhaofeng@nwpu.edu.cn*

Received 15 June 2008
Revised 23 June 2008

Equal channel angular pressing provides a convenient procedure for introducing an ultrafine grained microstructure into materials. In this paper, the deformation distribution of cylindrical billet with different diameters during equal channel angular pressing (ECAP) was simulated using 3D finite element models. The plastic strains in three perpendicular planes of the billet are predicted. And the influence of the friction between billet and channel on the equivalent plastic strain is also determined. The results show that the equivalent plastic strains are inhomogeneous in three directions and the imhomogenuity of the strain distribution inside ECAPed materials is slightly related to their diameters, which means larger scale UFG materials can be achieved via ECAP process.

*Keywords*: Ultrafine-grained materials; Equal Channel Angular Pressing (ECAP); 3D finite element simulation; equivalent plastic strain.

## 1. Introduction

From 90s of last century, equal channel angular pressing (ECAP) has become an important method to produce bulk ultrafine-grained (UFG) materials by severe plastic deformation (SPD).[1] Lots of researches have been performed on investigating the microstructure and mechanical behavior of many metals and alloys.[2-5] The influence of pressing speed, temperature and pressing route on the grain refinement of ECAP processed materials was also analyzed via experimental methods. Furthermore, the finite element method (FEM) was also used to study the influence of factors such as material property, corner angle of ECAP die and friction between billet and dies, on the deformation distribution in billet.[6-8] To meet the requirement of UFG materials in engineering structure applications, scaling to large size of billet is necessary in ECAP. Although ultrafine grained microstructure has been observed for large billets[9], the influence of billet size on deformation homogeneity of ECAPed billet was seldom analyzed.

---

[†]Corresponding Author.

In Ref. 10, 3D finite element method has been used to analyze the deformation heterogeneity of ECAPed billet. The results show the plastic strains are not uniform and the deformation inhomogeneity is varied with the increasing friction coefficient between the billet and channel. In this paper, the 3D finite element method was employed to analyze the influence of billet size on the deformation inhomogeneity of ECAPed billet.

## 2. 3D Finite Element Model

In this paper, the ABAQUS software was used to analyze the deformation distribution of a cylindrical billet during one pass. During present FEM simulation, the top section of the punch was set to move down along the inlet channel at a pressing speed of 5mm/s. For simplicity, the ECAP die was considered to be rigid. As the model is symmetrical about the middle plane of the ECAP die, half of the billet and die was analyzed and the symmetrical boundary conditions were applied. The punch and billet were both cylindrical with the diameter of 8, 12, 16 and 20mm respectively. As slight increase of the billet diameter will lead to remarkable increase of the required computing time, the maximum billet diameter was set to be 20mm in current FEM simulation. The length of the billet was set to be 5d (d is diameter of the billet). The material property of billet was assumed to be elastic-plastic with Young's modulus 60GPa, Poisson's ratio 0.3 and yield stress 98MPa. The rigid ECAP dies have the geometry with the inner corner angle $\Phi=120°$ and outer corner angle $\psi=20°$, both the inlet and outlet channel have the same

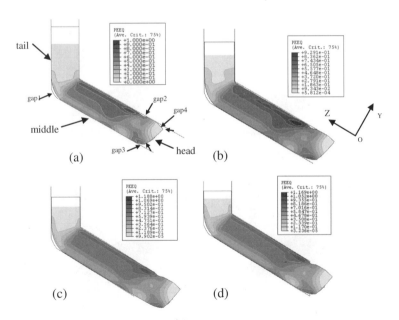

Fig. 1.  Equivalent plastic strain contour in X-plane of billets with different diameter: (a) d=8mm; (b) d=12mm; (c) d=16mm; (d) d=20mm.

diameter as the billet. During simulation, different friction coefficient which was assumed to vary from 0 to 0.15 with an increment of 0.05 was also considered.

## 3.  Finite Element Results and Discussion

Fig. 1(a)-(d) shows the distributions of equivalent plastic strains (EPS) of billets with different diameters in symmetrical plane of the 3D FEM model (called X-plane) under frictionless coefficients. It can be seen the EPS distributions are similar for billets with different diameter. Along the billet axis, there are three distinct deformation regions: tail, middle, and head (see Fig. 1(a)). Along the billet axis, the EPS in middle region is more uniform and larger than that in the tail and head region. Careful inspection of the head region of billets with different diameter shows that the length of the unavoidable insufficient deformation region is larger for billet with larger diameter. This means longer billet is needed to achieve ECAPed billet with longer uniform deformation part.

As bending of the billet during deformation, four gaps (see in Fig. 1(a)) between the billet and channel formed. In the role of the severe shear deformation, the head of billet bends toward the top of outlet channel, which leads to the formation of gap1 and gap3, while the contact between the side surface of head region and die channel leads to the formation of gap2 and gap4. The FEM simulation results of billets with different diameter also show that these gaps vary with friction coefficient (see in Fig. 2). Under friction condition, the bending moment acted on the head of billet may bend the billet

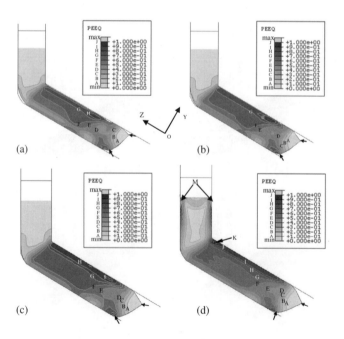

Fig. 2.   EPS contour in X-plane of billet with diameter 8mm: (a) μ=0; (b) μ=0.05; (c) μ=0.10; (d) μ=0.15.

toward the bottom surface of outlet channel and cause the reduction of gap1, gap2 and gap3 with the increase of friction coefficient. For the gap4, its size increases with the increase of μ. As an example, EPS contours of the billet with diameter 8mm in X-plane under different friction coefficients are shown in Fig. 2(a)-(d).

Fig. 3(a)-(d) show the EPS contours in cross-section of billets with different diameter under frictionless condition. It should be noted the cross-sections were cut from the middle region shown in Fig.1, and the right side of each figure corresponds to the surface which contacts with the top surface of the channel during deformation. It can be seen in Fig. 3 that remarkable deformation inhomogeneity exists in the cross-sections of billets. Although the EPS varies slightly along X-direction, large strain gradient along Y-direction is observed. Because the region passing through the inner corner of the channel received larger plastic deformation than that passing through the outer corner, larger plastic strain appears near the right side of the cross-section.

To examine friction effect on deformation homogeneity, distribution of EPS in cross-section of billet with diameter 8mm under different friction condition is shown in Fig.4 (a)-(d). As can be seen, the plastic strain distribution are similar for all friction conditions, except that the location of the maximum strain shifts slightly from left to right with increasing of friction coefficients. It is also interesting to find that a narrow deformation zone appears near the bottom (left) side of the cross-section, in which the plastic strain is smaller than other region. As the friction coefficient increase, this narrow zone shifts to the bottom of the billet. The same phenomena are also observed in billets

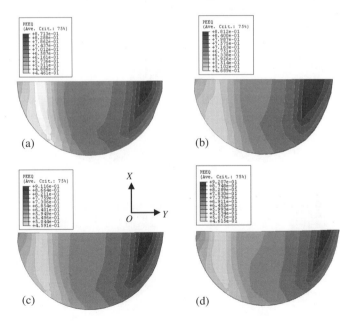

Fig. 3. EPS contour in cross-section of billets with different diameter: (a) d=8mm; (b) d=12mm; (c) d=16mm; (d) d=20mm.

with the diameter of 12, 16 and 20mm, which means that the diameter of billet have only slightly influence on the plastic deformation distribution in the cross-section of ECAPed billets.

To quantify the degree of deformation inhomogeneity, a deformation inhomogeneity index C is defined as follows

$$C = \sqrt{\frac{1}{n}\sum_{i=1}^{n}(\overline{\varepsilon}_i^p - \overline{\varepsilon}_{ave}^p)^2} \Big/ \overline{\varepsilon}_{ave}^p \ . \tag{1}$$

Where $\overline{\varepsilon}_i^p$ is the EPS of node $i$ in cross-section, $n$ is the total number of nodes in cross-section, and $\overline{\varepsilon}_{ave}^p$ is the average of EPS in cross-section. It should be noticed that the numerator of the right side of Eq. (1) is the standard deviation of EPS in cross-section. The average EPS and deformation inhomogeneity index in cross-section of the billets (Z-plane) under different friction conditions are listed in Table. 1.

Following the 2D theoretic model proposed by Iwahashi with the assumption of frictionless die surface [11], the EPS per pass using an ECAP die with 120°and 20° of the inner and outer corner angle respectively can be calculated as 0.635, which is consistent with present FEM simulation of average equivalent strain value in cross-section of billets with different diameter under frictionless condition (see in Table 1).

Comparing the FEM results of billet with the same diameter under different friction conditions, it can be found that the average EPS increase with the rising friction coefficient. The deformation inhomogeneity index first decreases with increasing friction

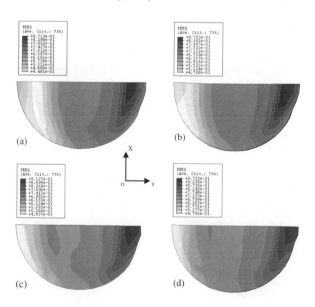

Fig. 4.   Equivalent plastic strain contour in cross-section of billet with diameter 8mm: (a) μ=0; (b) μ=0.05; (c) μ=0.10; (d) μ=0.15.

coefficient till µ=0.10 and then increases slightly, which indicates the remarkable influence of friction condition in the ECAP process.

Table 1. Distribution of EPS in cross-section of billets with different diameter.

| µ | d=8mm | | d=12mm | | d=16mm | | d=20mm | |
|---|---|---|---|---|---|---|---|---|
| | $\overline{\varepsilon}^p_{ave}$ | C | $\overline{\varepsilon}^p_{ave}$ | C | $\overline{\varepsilon}^p_{ave}$ | C | $\overline{\varepsilon}^p_{ave}$ | C |
| 0 | 0.629 | 0.150 | 0.659 | 0.171 | 0.657 | 0.164 | 0.655 | 0.160 |
| 0.05 | 0.644 | 0.143 | 0.676 | 0.155 | 0.671 | 0.150 | 0.666 | 0.151 |
| 0.10 | 0.667 | 0.139 | 0.679 | 0.156 | 0.681 | 0.145 | 0.677 | 0.146 |
| 0.15 | 0.689 | 0.147 | 0.699 | 0.162 | 0.699 | 0.155 | 0.696 | 0.156 |

As can be seen in Table 1, the average EPS varies gently with the increasing billet diameter under the same friction condition. This means to achieve UFG material via ECAP, relative homogeneous deformation may be achieved for larger billet if the friction between billet and ECAP die is not severe.

## 4. Summary and Conclusions

From 3D FE simulations of ECAP process, the conclusions are can be drawn as follows:
  (i) Along the axis of ECAPed billets with different diameter, three regions can be divided according to the plastic deformation distribution in X-plane.
  (ii) For billets with different diameter, the distributions of plastic deformation are similar in both X-plane and cross-section while the effects of friction on the deformation inhomogeneity are remarkable.
  (iii) According to our current simulation results, the influence of billet diameter on deformation inhomogeneity in cross-section of ECAPed billet is gently, which means that larger scale UFG materials can be achieved via ECAP.

## Acknowledgments

The authors would like to thank supports from the 111 project (grant No.B07050) of Northwestern Polytechnical University.

## References

1. V.M. Segal, *Mater. Sci. Eng.* **A197**, 157 (1995).
2. R.Z. Valiev, R.K. Islamgaliev, I.V. Alexandrov, *Prog. Mater. Sci.* **45**, 103 (2000).
3. S. Ferrasse, V.M. Segal, K. T. Hartwig, R. E.Goforth, *J. Mater. Res.* **12**, 1253 (1997).
4. W. J. Kim, J. K. Kim, W. Y. Chao, S. I. Hong, J. D. Lee, *Mater. Letter.* **51**, 177 (2001).
5. I. Kim, J. Kim, D. H. Shin, C. S. Lee, S. K. Hwang, *Mater. Sci. Eng.* **A342**, 302 (2003).
6. H. S. Kim, M. H. Seo, S. I. Hong, *J. Mater Proc Technol.* **130-131**, 497 (2002).
7. S. Li, M.A.M. Bourke, I.J. Beyerlein, D.J. Alexander, *Mater. Sci. Eng.* **A382**, 217 (2004).
8. J. Y. Suh, H. S. Kim, J. W. Park, J. Y. Chang, *Scripta Mater.* **44**, 677 (2001).
9. R. Srinivasan, B. Cherukuri, P.K. Chaudhury, *Mater. Sci. Forum.* **503-504**, 371 (2006).
10. T. Suo, Y. L. Li, Y. Z. Guo, Y. Y. Liu, *Mater. Sci. Eng.* **A432**, 269 (2006).
11. Y. Iwahashi, J. Wang, Z. Horita, M. Nemoto, T. G. Langdon, *Scripta Mater.* **35**, 143 (1996).

# INFLUENCE OF SHOT PEENING ON SURFACE CHARACTERISTICS OF HIGH-SPEED STEELS

YASUNORI HARADA[1†]

*Graduate School of Engineering, University of Hyogo, 2167 Shosya,*
*Himeji, 671-2280, Hyogo, JAPAN,*
*harada@eng.u-hyogo.ac.jp*


KENZO FUKAURA[2]

*Graduate School of Engineering, University of Hyogo, 2167 Shosya,*
*Himeji, 671-2280, Hyogo, JAPAN,*
*fukaura@eng.u-hyogo.ac.jp*

Received 15 June 2008
Revised 23 June 2008

High-speed steels are generally used for the cutting of other hard materials. These are hard materials, and can be used at high temperatures. Therefore, some of them are used for warm metal forming such as forging. However, in the tools used in hot working, an excellent hot hardness and long-life fatigue are strongly required. In the present study, the influence of shot peening on the surface characteristics of high-speed steels was investigated. Shot peening imparts compressive residual stresses on the metal surface, thus improving the fatigue life of the machine parts. In the experiment, the shot peening treatment was performed using an air-type shot peening machine. The shots made of cemented carbide were used. The workpieces were two types, W-type and Mo-type alloys. Surface roughness, compressive residual stress, and hardness of the peened workpieces were measured. It was found that shot peening using the hard shot media was effective in improving the surface characteristics of high-speed steels.

*Keywords*: Shot peening; high-speed steel; residual stress.

## 1. Introduction

Tool steels are used for the shaping of other materials or for the cutting of other materials. The high-speed steel is often used for metal forming tools such as punch and die, though it is suitable for a wide variety of cutting tools. These steels exhibit higher wear resistance and heat resistance than the tool steels. However, several surface treatments have been developed for improving the surface characteristics in the high-speed tool steels. In hot working, excellent hot hardness and impact resistance are strongly required. Much study has been done by researchers on the new material

---

†Corresponding Author.

development. The influences of carbide and inclusion contents on the fatigue properties of high-speed steel[1], the effects of vanadium and carbon on microstructures and abrasive wear resistance of high-speed steel[2], and the grindability of high-speed steel produced by the conventional method and the powder metallurgy method[3] were investigated. On the other hand, in recent years considerable amount of research work has been reported on the coating technologies. The influence of nitrogen compression plasma flow impact on the tribological properties of high-speed steel[4] and the effect of laser surface melting on the microstructure and corrosion behavior[5] were investigated.

In order to enhance wear resistance and fatigue strength in the die tools, the shot peening process is widely utilized. This process is one of the surface treatments. In this process the peening effects are characterized by the fact that the surface layer undergoes large plastic deformation due to the collision of shots. Namely, overlapping dimples develop a uniform layer of residual compressive stress or work hardening. This effect increases the life-time of parts. More recently, in Japan new media have been developed to enhance the peening effect. The peening media called fine-particle or microshot are smaller and harder than the conventional one. These are made of cemented carbide, amorphous alloy, and high-speed steel. The diameter of new media is in the range from 0.02 to 0.15 mm. The use of new media is effective for the hard material such as tool steels. Many studies have been carried out to investigate the effect of new media on the surface characteristics of the tools. However, little is known about the relation between the surface characteristics of high-speed steel and shot peening by using new media[6].

In the present study, using new media, the influence of shot peening on the surface characteristics of high-speed steels were investigated. Vickers hardness, the residual stress, and surface roughness near the surface were measured after shot peening. In addition, the surface conditions and the microstructures were observed by the scanning electron microscope and by the optical microscope.

## 2. Experimental Procedure

### 2.1. *Shot peening*

The apparatus for shot peening was fabricated to examine the effect of shot peening on the surface characteristics in high-speed steels. The workpiece was set on the holder. The distance between an air nozzle and the workpiece is 150mm. The shot velocity and the projection density are controlled in the experiment. The shot made of cemented carbide, which has an average diameter of 0.1 mm, was used for this study. Air pressure is in the range from p=0.4 to 0.8 MPa and the coverage is from C=100 to 800 %. The type of coverage is a standard visual method and coverage is defined as ratio of the dimpled surface to the total surface after shot peening. The experiment was performed at room temperature. The conditions used for the shot peening experiment are summarized in Table 1.

The used tool steels were the commercial high-speed steels, SKH2, SKH4, SKH55, and SKH57. These are of two types, containing Mo or W. The dimensions of the workpieces are 19 mm in diameter, 10 mm in thickness. The workpieces were subjected

Table 1.  Chemical composition of workpieces (mass%).

|   |   | C | Si | Mn | Cr | Mo | W | V | Co |
|---|---|---|---|---|---|---|---|---|---|
| W | SKH2 | 0.73 -0.83 | <0.40 | <0.40 | 3.80 -4.5 | - | 17.0 -19.0 | 0.80 -1.20 | - |
|  | SKH4 | 073 -0.83 | <0.40 | <0.40 | 3.80 -4.5 | - | 17.0 -19.0 | 1.00 -1.50 | 9.0 -11.0 |
| Mo | SKH55 | 0.85 -0.95 | <0.40 | <0.40 | 3.80 -4.50 | 4.60 -5.30 | 5.70 -6.0 | 1.70 -2.20 | 4.50 -5.50 |
|  | SKH57 | 1.0 -1.35 | <0.40 | <0.40 | 3.00 -4.00 | 4.60 -5.30 | 9.0 -11.0 | 3.00 -3.70 | 9.0 -11.0 |

to a standard heat treatment, quenching and triple tempering at 813K. The surface of the workpiece was hand polished on the abrasive papers.

## 2.2.  *Hardness, surface roughness, residual stress*

The distributions of residual stress, Vickers hardness, and surface roughness in the peened workpieces were measured. The workpiece was cut and then the distribution of the hardness at the side surface was measured as a distribution in the thickness direction. The distribution of residual stress in the thickness direction was obtained from the X-ray diffraction method by removing the surface layer of the workpiece using electrochemical polishing.

## 3.  Result and Discussion

### 3.1.  *Vickers hardness*

The shot peening process using the cemented carbide media was performed at room temperature. Vickers hardness of the peened workpieces was examined.

#### 3.1.1.  *Effect of air pressure*
The distribution of the measured hardness for W-type workpiece SKH2 in the thickness direction is given in Fig. 1. The hardness of peened workpiece has peaks near the surface, and the hardness at p=0.8MPa is the highest. However, inside the workpiece, the difference between the hardness at p=0.6 and p=0.8MPa is not large. The plastic deformation induces work hardening to a depth of about 70% of the shot diameter. On the other hand, the distribution of the hardness for the Mo-type workpieces is nearly the same as the W-type workpieces.

#### 3.1.2.  *Effect of coverage*
Shot peening was performed to investigate the effect of the coverage on the hardness. The distribution of the measured hardness for Mo-type workpiece SKH55 in the thickness direction is given in Fig. 2. The air pressure was p=0.6MPa. Vickers hardness

increases as the coverage increases. This is due to the increase of the amount of plastic deformation near the surface. In addition, the distribution of the hardness is nearly the same as the W-type workpieces

### 3.2. *Residual stress*

#### 3.2.1. *Effect of air pressure*

The residual stress of the workpiece peened at room temperature was examined. The distribution of the measured residual stress for W-type workpiece SKH4 in the thickness direction is given in Fig. 3. The coverage was C=200%. The residual stress of the peened workpieces has peaks near the surface, and the maximum appears about 2000MPa near

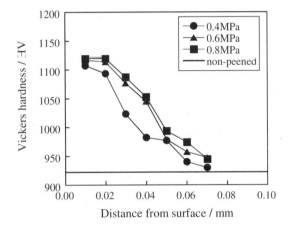

Fig. 1. Distribution of Vickers hardness for W-type workpiece SKH2 in thickness direction.

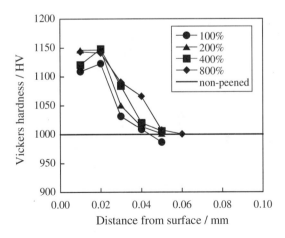

Fig. 2. Distribution of Vickers hardness for Mo-type workpiece SKH55 in thickness direction.

0.01 mm in depth from the surface. The plastic deformation induces compressive stresses to a depth of about 70 % of the shot diameter. A side close to a surface is higher in the mount of plastic deformation than a side far away from the surface. However, there is only a slight difference between the residual stress at p =0.4 and p =0.6 MPa. In addition, the values of residual stress are similar to the Mo-type workpieces. Namely, the maximum appears about 2000 MPa near 0.01 mm in depth from the surface.

### 3.2.2. *Effect of the coverage*

Shot peening was performed to investigate the effect of the coverage on the residual stress. In all the peened workpieces of C=800%, the residual stress of peened workpieces has peaks near the surface. The maximum appears about 2000 MPa near 0.01 mm in depth from the surface. However, the distributions of hardness in the peened workpieces are almost unchanged by varying the coverage. These results are similar to that found in Fig. 3.

## 4.  Surface Roughness

In order to examine the effects of the air pressure and the coverage on surface roughness, shot peening using the cemented carbide media was carried out. The relationship between surface roughness and the air pressure is shown in Fig. 4. As the air pressure increases, there is an immediate sharp increase of surface roughness. However, it is steady over p=0.6MPa. This is due to the small size of media. In addition, the surface roughness values for the peened workpieces are almost unchanged by varying the coverage. As the shot media is very small, the surface condition may not change over time. Shot peening using the cemented carbide media is a very effective means of reducing surface defect.

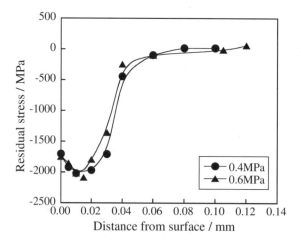

Fig. 3.  Distribution of residual stress for W-type workpiece SKH4 in thickness direction.

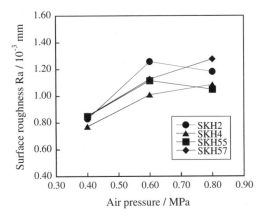

Fig. 4. Variation of surface roughness with air pressure.

## 5. Conclusions

The influence of shot peening on the surface characteristics of high-speed steels were investigated. The hardness and the residual stress of the peened workpiece have peaks near the surface. Especially, the maximum residual stress appears about 2000MPa near surface. On the other hand, the surface roughness values for the peened workpieces are almost unchanged by varying air pressure or coverage. The use of the cemented carbide media was found to cause a significantly enhanced peening effect for the high-speed steels.

## References

1. F. Meurling, A Melander, M. Tidesten, L. Westin, *Int. J. of Fatigue*, **23** (2001) 215.
2. S. Wei, J. Zhu, L. Xu, *Tribology International*, **39** (2006) 641.
3. J. Badger, *Annals of the CIRP*, **56** (2007) 353.
4. N.N. Cherenda, V.V. Uglov, V.M. Anishchik, A.K. S.tamashonak, V.M. Astashinski, A.M. Kuzmickii, A.V. Punko, G. Thorwath, B. Stritzker, *Surface Coatings Technology*, **200** (2006) 5334.
5. C.T. Kwok, F.T. Cheng, H.C. Man, *Surface Coatings Technology*, **202** (2007) 336.
6. Y. Harada, K. Fukaura, S. Haga, *J. Materials Processing Technology*, **177** (2006) 356.

# COLD BUTT JOINING OF LIGHT METAL SHEET BY SHOT PEENING

YASUNORI HARADA[1†]

*Graduate School of Engineering, University of Hyogo,*
*2167 Shosya, Himeji, 671-2280, Hyogo, JAPAN,*
*harada@eng.u-hyogo.ac.jp*

YUJI KOBAYASHI[2]

*Shinto Blastec Company Peening Center, Sintokogio, Ltd.,*
*3-1 Honohara, Toyokawa, 442-8505 Aichi, JAPAN,*
*y-kobayashi@sinto.co.jp*

Received 15 June 2008
Revised 23 June 2008

Aluminum and magnesium materials are very attractive for light weight applications. However, their use is complicated by the fact that dissimilar metals are joined by fusion welding. In the present study, the cold butt joining of light metal sheet with dissimilar material sheet by shot peening was investigated. The shot peening process is widely used to improve the performance of engineering components. In this process the substrate undergoes a large plastic deformation near its surface when hit by many shots. The substrate material close to the surface flows during shot peening. When the dissimilar metal sheets with notched edges are connected without a level difference and then the connection is shot peened, the sheets can be joined by the plastic flow generated by the large plastic deformation during shot peening. In this experiment, an air-type shot peening machine was used. The influences of peening time and shot material on joinability were mainly examined. The joinability was evaluated by tensile test. The joint strength increased with the amount of plastic flow. It was found that the present method can be used to enhance the butt joining of the light metal sheets with the dissimilar material sheets.

*Keywords*: Shot peening; butt joining; plastic deformation.

## 1. Introduction

Because of an intensive demand for the decrease in weight of cars and aircraft, the use of aluminum and magnesium products is increasing. These products are characterized by a large ratio of the strength to the weight. To improve the productivity and the functionality of the light materials, joining methods have been actively investigated[1]. Common liquid-state joining processes are arc and resistance welding. Laser beam joining of dissimilar sheet was investigated[2, 3]. Solid-phase joining processes using plastic deformation, such as cold and friction welding, are also widely used. Especially, the friction stir welding process has a high possibility of making high quality welds for light materials compared

---

[†]Corresponding Author.

to other fusion welding processes[4]. These processes are an important and necessary aspect of manufacturing operations.

In shot peening, the peening effects are characterized by the fact that the surface layer undergoes large plastic deformations due to the collision with the shot. When the machined part with the bore is excessively peened, the accuracy of the shape is lowered as the amount of plastic flow increases. Consequently, plastic flow characterized by shear droop occurs at the edge of the substrate due to shot peening. We have applied the shot peening process to the joining of dissimilar materials[5, 6]. The plastic flow makes the joining of the implant possible. When the dissimilar material is set in a hollow space on the surface of the substrate and then shot peened, it can be joined to the substrate by the overflow material generated by the large plastic deformations that occur during shot peening. This approach has been applied to the butt joining of thin sheets.

In the present study, the butt joining of light metal sheet with a dissimilar material sheet using shot peening was carried out. The influences of peening time and shot material on joinability were examined. Joinability was evaluated by tensile test.

## 2. Experimental Procedure

### 2.1. *Method of butt joining by shot peening*

The butt joining method using shot peening was shown in Fig. 1. The edges of the two sheets are notched. When the connection is shot-peened, the surface layer is deformed by the collision with the shot. The large plastic deformation that occurs on the surface layer generates overflow material in the edge of the sheet that fills the joint cavity between the two sheets. Thus, the sheets are joined without a level difference. In this method, both faces of the connection are shot-peened primarily.

### 2.2. *Conditions of shot peening*

The shot peening treatment was performed using an air-type peening machine (Sintobrator Ltd., MY-30B) with an air-orifice with a diameter of 3 mm and an injection-nozzle with a diameter of 6 mm. Air pressure and peening time were controlled in the experiment. The shots were made of high carbon cast steel and cemented carbide with an average diameter of 0.1 mm. Air pressure was in the range of 0.4-0.8 MPa and the

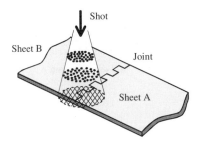

Fig. 1.   Schematic illustration of butt joining by shot peening.

peening time was in the range of 30-240 s. The conditions used for the shot peening experiment are summarized in Table 1.

The sheets were commercial magnesium alloy AZ31B, pure aluminum A1050, aluminum alloy A2017; A5052; A6061, pure titanium TB340, low-carbon steel SPCC, and stainless steel SUS304. A rectangular sheet was used that was 30 mm wide, 60 mm long, and 0.8-1.0 mm thick. The edge was cut into the desired shape of 3 mm wide by 3 mm long. The materials used for the experiment are also summarized in Table 1.

### 2.3. *Evaluation of joint strength*

The joint strength between the joined sheets was evaluated by tensile test using a test machine INSTRON-5582 at a cross head speed of 2 mm/min. The joint strength was defined as the ratio of the maximal load at joint failure to the total contact area; except for the contact area perpendicular to the direction of the tensile axis.

## 3.  Results and Discussion

### 3.1. *Aluminum sheet and dissimilar metal sheets*

The butt joining of aluminum/dissimilar sheets was examined. The appearances of the butt joined sheets by shot peening are shown in Fig. 2. The peening time is 60 s and the shot media is the carbon cast steel. The dissimilar metal sheets are low-carbon steel SPCC, stainless steel SUS304, and pure titanium TB340. In all workpieces, the clearance at a contact zone disappears due to particle bombardment, because shot peening causes large plastic deformation of surfaces. The pure aluminum sheet and the dissimilar metal sheets were successfully joined without a level difference.

Table 1.  Working conditions of shot peening.

| | |
|---|---|
| Shot peening machine | Air peening type |
| Shot media | High carbon cast steel (700HV), |
| | Cemented carbide (1400HV) |
| Shot size | 0.10 mm |
| Air pressure | 0.4, 0.6, 0.8 MPa |
| Peening time | 30 - 300 s |
| Working temperature | Room temperature |
| Sheet | Magnesium alloy AZ31B, |
| t=0.8, 1.0 mm | Pure aluminum A1050, |
| | Pure titanium TB340, |
| | Low-carbon steel SPCC, |
| | Stainless steel SUS304 |
| Atmosphere | Air |

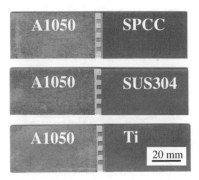

Fig. 2. Appearances of butt joined A1050/dissimilar workpieces after shot peening.

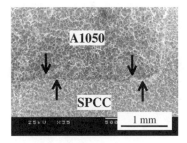

Fig. 3. SEM photomicrograph of the surface of A1050/SPCC workpiece after shot peening.

The surface conditions of the sheet were observed by SEM after shot peening. The upper and lower sides of the workpiece were shot-peened. The peened surface for the A1050/SPCC workpiece is given in Fig. 3. After shot peening for 60 s, the clearance at the joint disappeared due to the collision with the shot. Namely, the convex parts underwent large plastic deformations near the surface due to the collision with the shot.

SEM photomicrograph of the cross-sections observed on the A1050 sheet of the joined A1050/SPCC workpiece (see Fig. 3) after the tensile test is shown in Fig. 4. After the tensile test, no seizure occurred in the fractured surfaces. The scratches generated by the tensile test were observed on sides close to the peened surfaces, since the material near the surface was deformed by the impact of the shot. However, there are no scratches at the midpoint of the cross-section. The scratches increase with the degree of plastic deformation near the surface. It was found that only the joining was formed where the two sheets meet.

Fig. 4.  SEM photomicrograph of the cross-section observed on the A1050 sheet of the joined A1050/SPCC workpiece after tensile test.

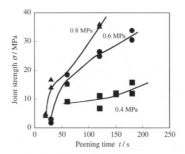

Fig. 5.  Variation of joint strength with peening time for A1050/SPCC workpiece.

The pure aluminum sheet was successfully joined to the dissimilar metal sheets by shot peening, although the contact area of the two sheets was very small. The joint strength of the joined workpieces was evaluated by a tensile test. The joint strength of the A1050/SPCC workpiece was measured, and the effect of the air pressure on joint strength was also examined. Fig. 5 shows variation of the joint strength with peening time for the joined workpiece. The joint strength increases with peening time. On the other hand, the joint strength increased with increasing the air pressure. Especially, when the air pressure is higher, the strength rises rapidly with peening times beyond 60 s. This causes greater plastic deformation of the surfaces.

### 3.2.  *Magnesium sheet and dissimilar metal sheets*

Using shot peening, the magnesium sheet was also successfully joined to the dissimilar metal sheets. Therefore, the joint strength of the AZ31B/dissimilar metal workpieces was measured, and the effect of the shot material on joint strength was also examined. Fig. 6 shows variation of the joint strength with peening time for the AZ31/SUS304 workpiece. The joint strength increases with peening time. Especially, when the workpiece was peened by the cemented carbide media, the strength rises particularly rapidly with

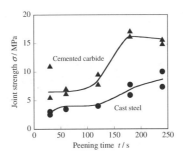

Fig. 6. Variation of joint strength with peening time for AZ31B/SUS304 workpiece.

peening times beyond 120 s due to the existence of space between the dissimilar metal sheets. Compared with the steel media, the cemented carbide media has higher kinetic energy at collision and this causes greater plastic deformation of the surfaces.

In the joined workpieces, the joint strength is lower than the flow stress of base material (see Figs. 5 and 6). This is due to only the material near the surface is deformed by the impact of the shot (see Fig. 4). To enhance joinability, the hard shot peening process or changing the edge shape of the connection will be very efficient.

## 4. Conclusions

The butt joining between the light metal sheets and the dissimilar metal sheets was performed by means of the shot peening process. Tensile strength was measured to examine the influences of peening time and shot media on joinability. The light metal sheets and the dissimilar metal sheets were successfully joined by shot peening. Also, the use of cemented carbide media was very efficient in improving joinability. However, the joint strength is lower than the flow stress of base material. Although further investigations are needed to improve the joinability, we found that the present method can be used for butt joining of the light metal sheets with dissimilar material sheets without melting.

## References

1. C. Connolly, *Ind. Robot*, **34** (2007) 17.
2. F. Vollertsen, M. Grupp, *Steel Research Int.*, **76** (2005) 40.
3. F. Natsumi, K. Ikemoto, H. Sugiura, T. Yanagisawa, K. Azuma, *Int. J. of Materials and Product Technology*, **7** (1992) 193.
4. Y. Sato, S.H.C. Park, M. Michiuchi, H. Kokawa, *Scr. Mater.*, **50** (2004) 1233.
5. Y. Harada, N. Tsuchida, K. Fukaura, *J. Materials Processing Tech.*, **177** (2006) 356.
6. Y. Harada, K. Fukaura, *Key Engineering Materials,* **340-341** (2007) 865.

# EVOLUTION OF STRAIN STATES AND TEXTURES IN AA 5052 SHEET DURING CROSS-ROLL ROLLING

S.H. KIM[1], D.G. KIM[2], H.G. KANG[3], M.Y. HUH[4†]

*Department of Materials Science and Engineering, Korea University, Seoul, 136-701, KOREA,*
*myhuh@korea.ac.kr*

J.S. LEE[5],

*Lightweight Components Team, Korea Institute of Industrial Technology, Gwangju, 500-480, KOREA,*

O. ENGLER[6]
*Hydro Aluminium Deutschland GmbH, R&D Center, P.O. Box 2468, D-53014 Bonn, GERMANY,*

Received 15 June 2008
Revised 23 June 2008

Cross-roll rolling of AA 5052 sheets was carried out using a rolling mill in which the roll axes are tilted by ±7.5° away from the transverse direction of the rolled sample. Besides cross-roll rolling, normal-rolling using a conventional rolling mill was also carried out with the same rolling schedule for clarifying the effect of cross-roll rolling. The evolution of strain states during cross-roll rolling was investigated by texture measurements and by three-dimensional finite element method (FEM) simulation. Cross-roll rolling gives rise to the operation of all three shear components $\dot{\varepsilon}_{12}$, $\dot{\varepsilon}_{13}$ and $\dot{\varepsilon}_{23}$ in the roll gap. This complex shear states during cross-roll rolling strongly reduce the intensities of the deformation texture components.

*Keywords*: Cross-roll rolling; Shear deformation; Shear textures; Finite element method.

## 1. Introduction

In the literature, cross-rolling refers to a rolling procedure in which the rolling direction (RD) is rotated by 90° around the rolling plane normal (e.g. [1-5]). Because of the change in the RD, the evolution of texture and microstructure during cross-rolling differs from that during conventional normal rolling. Commonly, normal rolling of a metallic material gives rise to the formation of a rolling texture and a subsequent recrystallization texture which leads to anisotropic material properties of rolled sheets. However, a proper control of rolling parameters during cross-rolling may weaken the rolling and subsequent annealing textures, thus reducing the anisotropy of rolled sheets [2]

---

[†]Corresponding Author.

Recently, Chino and colleagues have proposed a novel cross-rolling method in which the roll axes are tilted about the normal direction (ND) in the RD-TD plane, as shown in Fig. 1 [6-8]. In this process, a thrust force in the axial direction of the rolls enables imposing a shear strain into the rolled sheet. Since the rolls are crossed in the RD-TD plane, this process is hereafter referred to as "cross-roll" rolling in order to distinguish it from the conventional cross-rolling process in which the RD is rotated after each rolling pass.

The work of Chino *et al.* on cross-roll rolling focused on the improvement of microstructure and formability of magnesium alloy sheets[6-8]. However, detailed studies on the evolution of strain state during cross-roll rolling are still very limited. In the present study, aluminum alloy AA 5052 was cold rolled by using a cross-roll rolling mill. The evolution of strain states during rolling was investigated by texture measurements and by simulation with the finite element method (FEM).

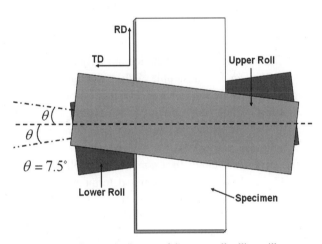

Fig. 1. Schematic diagram of the cross-roll rolling mill.

## 2. Experimental Procedure

The starting material used in this work was the commercial Al-Mg alloy AA 5052 (Al-2.5% Mg-0.3% Cr-0.3% Fe-0.2% Si). The as-received hot band with a thickness of 16 mm was rolled asymmetrically to 6 mm thick sheet and annealed for 1 h at 400 °C in order to provide an initial sample with a fairly random texture and equiaxed grains[9]. Cold rolling was carried out in a dedicated laboratory cross-roll rolling mill in which the roll axes are tilted by ±7.5° away from the transverse direction (TD) in the RD-TD plane (Fig. 1). The process of "cross-roll rolling" is described in detail elsewhere [6-8]. The roll diameter was 160 mm and the roll speed was 2 radian/s. In order to increase the friction between the roll surface and rolled sample, rolling was performed without lubrication.

The sample with a thickness of 6 mm was cold rolled in four passes with maintaining the original RD to a thickness of 1.1 mm, corresponding to a thickness reduction of 80%. Besides cross-roll rolling, normal-rolling using a conventional rolling mill was carried out with the same rolling schedule for clarifying the effect of cross-roll rolling.

In order to interpret the distribution of strain states in the roll gap during cross-roll rolling, three-dimensional finite element method (FEM) simulations were performed with the rigid-plastic FEM code DEFORM[TM]-3D. For FEM calculations of rolling operations the friction between rolls and sheet is of major importance. In the present study, friction was described with the friction parameter $m$, where $m$ is the ratio of the frictional stress to the critical stress and $m=1$ denotes sticking friction. Because of the quite high friction during rolling without lubrication, a friction parameter of $m=0.6$ between the sample surface and rolls was assumed[10]. For texture analysis, pole figure measurements were carried out by means of a conventional X-ray texture goniometer using Cu-Kα radiation [11]. The textures of the surface and center layers were measured separately so as to detect possible through-thickness gradients in strain state upon cross-roll rolling.

## 3.  Results and Discussion

Because the initial texture prior to the deformation strongly affects the rotation of grains during plastic deformation, an initial sample with a random texture is preferable for studying the evolution of deformation textures. For that purpose, the initial sample was prepared by asymmetrical rolling and subsequent annealing, which provided an initial sample having a fairly random texture and equiaxed grains with an average diameter of 25 μm[9].

Cold rolling was carried out either in a conventional rolling mill or in a cross-roll rolling mill (Fig. 1); Fig. 2 shows the pole figures determined from the resulting sheets. The normally rolled sample displayed pronounced texture gradients at the center and surface layers. A typical cold rolling texture of aluminum alloys formed in the center layer, while shear textures developed at the surface layer. As mentioned previously, cold rolling in the present work was performed without lubrication which enhanced the variation of strain states throughout the thickness layers leading to the texture gradient in the normally rolled sheet. In thickness layers close to the sheet center, the deformation is approximated by a plane strain state, where $\dot{\varepsilon}_{11} = -\dot{\varepsilon}_{33}$ and all other components $\dot{\varepsilon}_{22}$, $\dot{\varepsilon}_{12}$, $\dot{\varepsilon}_{13}$ and $\dot{\varepsilon}_{23}$ are assumed to be zero. This leads to the formation of the well-known cold rolling texture of aluminum alloys where the crystal orientations assemble along the so-called β-fiber (Fig. 2(a)) [2, 12-14]. However, factors like roll gap geometry and friction between the roll and the sheet contact surface can cause severe deviations from the plane strain condition in thickness layer close to the surface. Here, the operation of a large shear strain rate $\dot{\varepsilon}_{13}$ gives rise to the formation of shear textures characterized by the rotated cube orientation {001}<110> and {111}//ND orientations as shown in Fig. 2(b).

The evolution of deformation textures during cross-roll rolling completely differs from that during normal rolling, which obviously reflects the different strain states operative upon cross-roll rolling. In the cross-roll rolled sample, the surface layer depicts a very weak, almost random texture, while the pole figure determined from the center layer displays pole intensities ranging from {110}<112> Bs-orientation to {159}<743> which is close to the {123}<634> S-orientation. It should be emphasized that the equivalent orientations are not present in the pole figure, indicating a deviation of the strain state from orthotropic towards monoclinic sample symmetry during cross-roll rolling.

The variation of strain states in the roll gap during normal and cross-roll rolling was studied by three-dimensional FEM simulations using the rigid-plastic FEM code DEFORM$^{TM}$-3D. Fig. 3 shows the variations of the off-diagonal shear strain rates $\dot{\varepsilon}_{12}$, $\dot{\varepsilon}_{13}$ and $\dot{\varepsilon}_{23}$ in the roll gap along a stream line close to the sheet surface during normal and cross-roll rolling. During normal rolling, only the component $\dot{\varepsilon}_{13}$ varies, while the other shear components $\dot{\varepsilon}_{12}$ and $\dot{\varepsilon}_{23}$ are zero. As reported in detail elsewhere[9,10,15], the operation of a large shear component $\dot{\varepsilon}_{13}$ during normal rolling leads to the formation of shear textures as shown in Fig. 2 (b). In contrast, during cross-roll rolling all three shear components $\dot{\varepsilon}_{12}$, $\dot{\varepsilon}_{13}$ and $\dot{\varepsilon}_{23}$ may appear. Accordingly, the formation of a random texture at the surface of the cross-roll rolled sample (Fig. 2(d)) is attributed to the complex strain state during deformation.

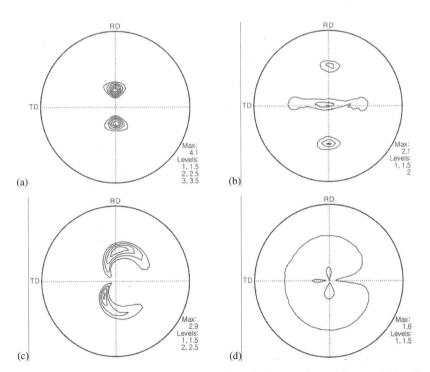

Fig. 2. Evolution of textures after cold rolling. Determined from (a) center layer and (b) surface layer of the normally rolled sample, (c) center layer and (d) surface layer of the cross-roll rolled sample.

The variations of shear strain rates $\dot{\varepsilon}_{12}$, $\dot{\varepsilon}_{13}$ and $\dot{\varepsilon}_{23}$ in the sheet center (Fig. 4) are much simpler than those of the sheet surface (Fig. 3). In the center layer of the normally rolled sheet, all shear strain rates are nearly zero, which is consistent with the assumption of a plane strain condition, leading to the formation of the typical rolling texture as shown in Fig. 2(a). The center layer of the cross-roll rolled sheet displays quite large values of $\dot{\varepsilon}_{23}$, in contrast, while the other shear components $\dot{\varepsilon}_{12}$ and $\dot{\varepsilon}_{13}$ are approximately zero. Thus, the texture as shown in Fig. 2(c) is attributed to the operation of a large $\dot{\varepsilon}_{23}$ shear during cross-roll rolling. It was noted that the evolution of $\dot{\varepsilon}_{23}$ is similar in all thickness layers of the cross-roll rolled sample.

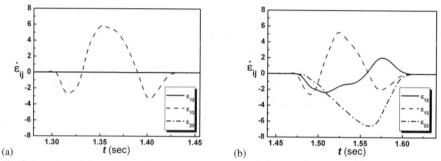

Fig. 3. Variation of shear strain rate components $\dot{\varepsilon}_{12}$, $\dot{\varepsilon}_{13}$ and $\dot{\varepsilon}_{23}$ at surface layer during (a) normal-rolling and (b) cross-roll rolling.

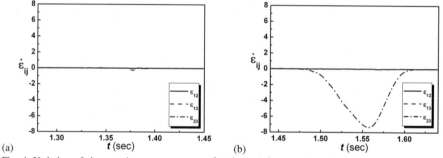

Fig. 4. Variation of shear strain rate components $\dot{\varepsilon}_{12}$, $\dot{\varepsilon}_{13}$ and $\dot{\varepsilon}_{23}$ at center layer during (a) normal-rolling and (b) cross-roll rolling.

## 4. Conclusions

The evolution of strain states during cross-roll rolling was investigated by texture measurements and three-dimensional finite element simulations. The textures of the conventionally rolled sample displayed orthotropic sample symmetry, while those of the cross-roll rolled material depicted monoclinic sample symmetry. The formation of a random texture at the surface of the cross-roll rolled sample is attributed to the evolution of a complex strain state during deformation. The operation of a large $\dot{\varepsilon}_{23}$ shear component during cross-roll rolling leads to the formation of a weak deformation texture in the center of the cross-roll rolled sample.

## References

1. M.Y. Huh, O. Engler, D. Raabe, Textures and Microstruct., 24, 225 (1995).
2. M.Y. Huh, S.Y. Cho, O. Engler, Mater. Sci. Eng. A, 315, 35 (2001).
3. S.H. Hong, D.N. Lee, J. Eng. Mater. Technol.-Trans. ASME, 124, 13 (2002).
4. M.Y. Huh, J.H. Lee, S.H. Park, O. Engler, D. Raabe, Steel Research Int., 76(11), 797 (2005).
5. H.C. Kim, C.G. Kang, M.Y. Huh, O. Engler, Scripta Mater., 57, 325 (2007).
6. Y. Chino, K. Sassa, A. Kamiya, M. Mabuchi, Mater. Sci. Eng. A, 441, 349 (2006).
7. Y. Chino, K. Sassa, A. Kamiya, M. Mabuchi, Mater. Lett., 61, 1504 (2007).
8. Y. Chino, K. Sassa, A. Kamiya, M. Mabuchi, Mater. Sci. Eng. A, 473, 195 (2007).
9. J.J. Nah, H.G. Kang, M.Y. Huh, O. Engler, Scripta Mater., 58, 500 (2008).
10. H.G. Kang, J.K. Kim, M.Y. Huh, O. Engler, Mater. Sci. Eng. A, 452, 347 (2007).
11. V. Randle, O. Engler, Introduction to Texture Analysis: Macrotexture, Microtexture and Orientation Mapping (Gordon and Breach Sci. Publ., Amsterdam, 2000).
12. J. Hirsch, K. Lücke, Acta Metall. 36, 2863 (1988).
13. Cl. Maurice, J.H. Driver, Acta Mater., 45(11), 4627 (1997).
14. W.C. Liu, J.G. Morris, Scripta Mater., 52, 1317 (2005).
15. O. Engler, M.Y. Huh, C.N. Tomé, Metall. Mater. Trans., 31A, 2299 (2000).
16. C.G. Kang, H.G. Kang, H.C. Kim, M.Y. Huh, H.G. Suk, J. Mater. Process. Technol., 187-188, 542 (2007).
17. M.Y. Huh, K.R. Lee, O. Engler, Inter. J. Plast., 20, 1183 (2004).

# MINIMUM WALL THICKNESS OF HOLLOW THREADED PARTS IN THREE-DIE COLD THREAD ROLLING

HUIPING QI[1†], YONGTANG LI[2], JIANHUA FU[3], ZHIQI LIU[4]

*School of Material Science and Engineering, Taiyuan University of Science and Technology*
*Taiyuan, Shanxi 030024, P. R. CHINA,*
*qhp182257@sohu.com, liyongtang@tyust.edu.cn, hua5963@sina.com, alzq_678@yahoo.com.cn*

Received 15 June 2008
Revised 23 June 2008

The cold thread rolling technology was developed rapidly in recent years due to its high efficiency, low cost and perfect mechanical properties of its production. However, researches on the precise thread rolling of the hollow parts were very few. Traditionally, the minimum thickness of the thin-walled threaded parts by thread rolling was mainly determined by the empirical (trial and error) methods. In this study, the forming process of thin-walled thread parts rolled with three thread rolling dies was analyzed. The stress state of the hollow work piece was obtained by solving the statically indeterminate problems. Then, the equations for the minimum wall thickness were derived. Experiments are also performed. The experimental results are generally in good agreement with those by the current theoretical analysis. It could be concluded that the analysis presented in this study can provide a good guidance for the thread rolling of hollow parts.

*Keywords*: Thread; cold rolling; hollow parts; wall thickness.

## 1. Introduction

The cold thread rolling process is a non-chip finish method with high efficiency and low cost. High production rate and good mechanical properties of its product are the greatest advantages. Thereby, the cold thread rolling method has been widely used and drawn great attention. But few studies are carried out on the technological theory and the mechanical analysis of the precise forming process. Especially for hollow parts, since the wall thickness is determined mainly by the empirical (trial and error) method, the production is unscientific and inaccurate, which results in a serious waste of materials. Therefore, technical support and theoretical guidance are necessary for the development and application of this advanced technology. In this paper, the cold thread rolling processes have been studied. The minimum wall thickness of the workpiece is obtained by solving statically indeterminate problems.

---

[†]Corresponding Author.

## 2. Rolling Theory

As shown in Fig.1, the three-die thread rolling is carried out on the three-die thread rolling machine. Three thread rolling dies are installed on the main arbors that are located three symmetrically position. Three thread rolling dies rotate in the same rotational direction and with the same rotational speed while the dies move in radial direction of the work piece. The work piece is pressed and the screw is formed on its surface gradually. The three-die thread rolling is more convenient to operate without the need of supporting the work piece. [1,2]

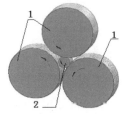

1. the thread rolling dies  2. the workpiece
Fig. 1.  The rolling theory of the three-die thread rolling.

## 3. Failure Forms and Reason

There are several types of failure of the work piece. Firstly, the threaded part is pressed to produce cracks and the longitudinal crack makes the work piece failure. Secondly, the work piece is plastic deformed to elliptic or other irregular shape, which results in the broken of the rolling process. Thirdly, although the rolling can be finished, the product size can not meet the precision requirement.

The common reason for the above three failure forms is that the hollow parts can not bear the rolling force. According to the rolling theory, the radial force and tangential force are applied to the work piece by the thread rolling dies equipped on the thread rolling machine. The radial force makes the work piece deform plastically to produce screw and the tangential force makes the work piece rotate with the rotation of dies to finish rolling. The rolling of the hollow threaded part has a little difference with the solid work piece if the wall thickness of the hollow part is thick enough. But if the wall thickness is very thin, the hollow work piece will be pressed to crack or deformed by the great radial force applied to it, which lead to the failure of the component.

## 4. Minimum Wall Thickness

To solve the minimum wall thickness of the rolled hollow thin-walled threaded part, the stress state of the work piece must be obtained firstly. Ignoring the friction force, the work piece can be simplified as a thin-walled cylinder subjected to three symmetrically positioned forces P (as shown in Fig.2a).

After making a break at sections A and B, the shear force on the section A and B are zero due to the symmetry of the load. Therefore, only the axial force N and the bending moment $M_0$ exist on the two sections. The axial force N can be solved easily according to the equilibrium condition and $M_0$ is the only redundant restrained force, expressed as $X_1$. It's reasonable that only one sixth of the ring be studied because of the symmetry of the configuration.

Because the deflection of the symmetrical section B and C are zero, section C can be regarded as a fixed section, and the zero deflection of section B can be regarded as deformation compatibility condition (Fig.2c).

There is:

$$\delta_{11}X_1 + \Delta_{1p} = 0 \tag{1}$$

Where:

$\Delta_{1p}$ — the deflection of the section B due to the applied axial force $N = \dfrac{\sqrt{3}}{3}P$

$\delta_{11}$ —the deflection of the section B due to applied the bending moment $X_1 = 1$

M and $\overline{M}$ are respectively represent the bending moment of the section where the angle is $\phi$ due to the axial force (Fig.2d) and the bending moment ( Fig.2e).

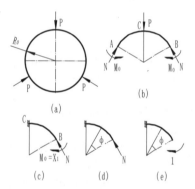

Fig. 2. The stress analytical simple figure of the three-die thread rolling.

There is: $M = -\dfrac{\sqrt{3}}{3}PR_0(1 - \cos\phi)$, $\overline{M} = 1$.

$$\Delta_{1p} = \int_0^{\frac{\pi}{3}} \frac{M\overline{M}R_0 d\phi}{EI} = -\frac{\sqrt{3}PR_0^2}{3EI}\int_0^{\frac{\pi}{3}}(1 - \cos\phi)d\phi = -\frac{\sqrt{3}PR_0^2}{3EI}\left(\frac{\pi}{3} - \frac{\sqrt{3}}{2}\right) \tag{2}$$

$$\delta_{11} = \int_0^{\frac{\pi}{3}} \frac{\overline{M}\,\overline{M}R_0 d\phi}{EI} = \frac{R_0}{EI}\int_0^{\frac{\pi}{3}}1^2 d\phi = \frac{\pi R_0}{3EI} \tag{3}$$

Substituting Eq.(2) and Eq.(3) into Eq.(1) yields

$$X_1 = -\frac{\sqrt{3}}{\pi}PR_0\left(\frac{\pi}{3} - \frac{\sqrt{3}}{2}\right) \tag{4}$$

The total bending moment on the section can be obtained by

$$M(\phi) = M + X_1 = -\left(\frac{3}{2\pi} - \frac{\sqrt{3}}{3}\cos\phi\right)PR_0 \tag{5}$$

According to the Eq.(5) and the symmetry of the ring, the bending moment diagram of the total ring is plotted in Fig.3.

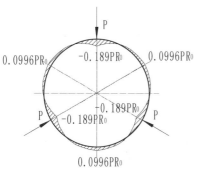

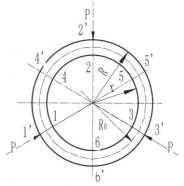

Fig. 3. The bending moment diagram of the workpiece rolled with three dies.

Fig. 4. The forced diagram of the workpiece rolled with three dies.

The internal stress of the one sixth of the ring can be expressed as the function of $\phi$, given by

$$\left.\begin{aligned} M(\phi) &= -\left(\frac{3}{2\pi} - \frac{\sqrt{3}}{3}\cos\phi\right)PR_0 \\ N(\phi) &= \frac{\sqrt{3}}{3}\cos\phi \\ Q(\phi) &= \frac{\sqrt{3}}{3}\sin\phi \end{aligned}\right\} \tag{6}$$

The deformation of the work piece can be analyzed according to the stress state. The plastic deformation occurs at the contact points $1', 2', 3'$ (as shown in Fig.4) firstly. With the increase of the rolling force, the stress of the points 1, 2, 3 reach to the yield limit of the material, where local deformation occurs. For those products with high precision, the stress of points 1.2,3 should be less than yield limit of the materials. But the plastic deformation only occurs at the local round sections $11', 22', 33'$, not the whole annulus. Thus the rolling process can be continued. When the sections $44', 55', 66'$ are in a plastic state, the plastic deformation occurs in the whole annulus and the rolling can't be continued. For the materials with good plasticity, its rollability should be judged by the stresses at points 4, 5 and 6 compared with the yield limit. For the materials with tensile strength close to yield limit such as 40Cr and Cr18Ni9Ti, crack occurs at the sections $11', 22', 33'$ in the work piece before the whole annulus deforms plastically, which form the first failure form. The rollability can be determined by the stresses at points 1, 2 and 3.

According to the analysis above, the stress state of the points 1,2,3 and 4,5,6 could be solved. The distribution of the shear stress on the rectangular section is parabolic curves. The shear stresses on the side of sections $11',22',33',44',55'66'$ are zero. So the shear stresses at the points 1~6 are zero. These points only subject to the stress under the bending moment and stress due to the axial force.

Therefore, the stresses at these points can be given by

$$\sigma = \frac{My}{S\rho} + \frac{N}{F} \tag{7}$$

Where:

y—the distance from the point to the neutral axis

$\rho$—the distance from the point to the center of curvature

S—the static moment of the section about the neutral axis

F—the total area of the cross section

As shown in Fig.2, let outside radius, inner radius, wall thickness and length of the workpiece be separately expressed by R, r, $\delta$, $l$ and $R_0 = (R+r)/2$.

The radius of the neutral layer is $r_0 = \dfrac{\delta}{\ln\left(R/r\right)}$, $S = F \cdot (R_0 - r_0)$

At these points $\rho = r$, $y = r_0 - r$

Substituting corresponding $M$, $y$, $\rho$, $S$, $N$ and $F$ into Eq.(7) yield:

$$\sigma_{1,2,3} = \frac{My}{S\rho} - \frac{N}{F} = \frac{0.189PR_0(r_0 - r)}{l\delta(R_0 - r_0)r} - \frac{\sqrt{3}}{6}\frac{P}{l\delta} \tag{8}$$

$$\sigma_{4,5,6} = \frac{My}{S\rho} + \frac{N}{F} = -\frac{0.0996PR_0(r_0 - r)}{l\delta(R_0 - r_0)r} - \frac{\sqrt{3}}{3}\frac{P}{l\delta} \tag{9}$$

The criterions can be obtained as follows:

$$\sigma_{1,2,3} \le \sigma_s \tag{10}$$

$$\sigma_{4,5,6} \le \sigma_s \tag{11}$$

$$\sigma_{1,2,3} \le \sigma_b \tag{12}$$

According to analysis above, Eq.(10), Eq.(11) and Eq.(12) can be used respectively to judge if the threaded part can be rolled in condition of the product with high precision, the material with high plasticity and the material with low plasticity.

## 5. Experiments

Experiments were carried out on the ZC28-16 thread rolling machine. The system pressure was 8.5 MPa, the rolling force was 10T. Four groups of blanks were chosen for rolling. The sizes of each component are shown in table.2. Substituting corresponding parameters into Eq.(9) and Eq.(11), the minimum wall thickness of the work piece obtained is 7.9mm, and the inner diameter is 28.8mm.

From Fig.5, the product is qualified when the wall thickness is 8.9mm (Fig.5a) and its tooth shape is very close to that of the solid work piece (Fig.5b). The tooth shape is not qualified when the wall thickness is 7.05mm (Fig.5c). The tooth shape is very flat when the wall thickness is 5.45mm (Fig.5d). The rolling can not be finished with an inner diameter is 3.9mm. The results were in a good agreement with the theoretical ones.

Table 2. the size of the blank(mm).

|  | the outer diameter | the inner diameter | the wall thickness | the length |
|---|---|---|---|---|
| group 1 | 44 | 25.2 | 8.9 | 50 |
| group 2 | 44 | 29.9 | 7.05 | 50 |
| group 3 | 44 | 33.1 | 5.45 | 50 |

a. solid workpiece   b. wall thickness is 8.9mm   c. wall thickness is 7.05mm   d. wall thickness is 5. 45mm

Fig. 5. Tooth shape of the production.

| group 4 | 44 | 36.2 | 3.9 | 50 |
|---|---|---|---|---|

## 6. Conclusion

In this study, the minimum wall thickness equations of the work piece in three-die thread rolling are obtained. It is concluded that the failure of the hollow threaded part in rolling is caused by the radial rolling force that exceeding the stress limit of the work piece. The different failure forms of thin-walled work piece in cold thread rolling are caused by the difference of the work piece materials. This study will provide a good guidance for the practice.

In the future, further researches will be carried out to promote the application of this precision and high efficient technology.

## Acknowledgment

This project is supported by National Natural Science Foundation of China (NO.50675145), Key Scientific and Technological Project of Shanxi, China (NO.2006031147).

## References

1. C.H. Cui, Thread rolling, Beijing: mechanical industry publishing house. (1978).
2. X.L. Wang, Processing Technology of Thread Rolling, Beijing: China Railway Publishing House. (1990).

# REPLICATION OF NANO/MICRO QUARTZ MOLD BY HOT EMBOSSING AND ITS APPLICATION TO BOROSILICATE GLASS EMBOSSING

SUNG-WON YOUN[1†], CHIEKO OKUYAMA[2], MASHARU TAKAHASHI[3], RYUTARO MAEDA[4]

*Advanced Manufacturing Research Institute (AMRI), National Institute of Advanced Industrial Science and Technology (AIST), Tsukuba, Ibaraki, JAPAN,*
*youn.sungwon@aist.go.jp, chi-okuyama@aist.go.jp, m.takahashi@aist.go.jp, maeda-ryutaro@aist.go.jp*

Received 15 June 2008
Revised 23 June 2008

Glass hot-embossing is one of essential techniques for the development of high-performance optical, bio, and chemical micro electromechanical system (MEMS) devices. This method is convenient, does not require routine access to clean rooms and photolithographic equipment, and can be used to produce multiple copies of a quartz mold as well as a MEMS component. In this study, quartz molds were prepared by hot-embossing with the glassy carbon (GC) masters, and they were applied to the hot-emboss of borosilicate glasses. The GC masters were prepared by dicing and focused ion beam (FIB) milling techniques. Additionally, the surfaces of the embossed quartz molds were coated with molybdenum barrier layers before embossing borosilicate glasses. As a result, micro-hot-embossed structures could be developed in borosilicate glasses with high fidelity by hot embossing with quartz molds.

*Keywords*: Hot embossing; glass-to-glass embossing; glassy carbon; quartz glass; borosilicate glass; focused ion beam; dicing.

## 1. Introduction

There have been widespread demands for the high-throughout, high-resolution and cost-effective patterning techniques for glasses in the application fields of MEMS such as photonic crystals[1,2] and biochips[3] because glasses have excellent optical properties (e.g., high refractive index, low UV absorption level) and chemical/thermal stability, which are essential for high-performance optical and bio MEMS applications.[1-5] Conventional microstructuring methods of glasses include wet/dry etching, laser machining, powder blasting, and mold replication technique.[1-7] Among these, glass hot-embossing techniques are of interest to fabricate high-precision glass components because they are convenient, do not require routine access to clean rooms and photolithographic equipment, and can be used to produce multiple copies of a quartz mold.[7-10]

As mold materials for glass production, carbide alloys, such as silicon carbide and tungsten carbide, have been widely used in the precision glass molding process (GMP) due to their hardness and dimensional stability at high temperature. In the previous

---

[†]Corresponding Author.

studies, we have proved that a GC is also adequate mold material for high-temperature-embossing of quartz and borosilicate glasses due to its excellent properties (e.g., high operating temperature up to 2000°C, chemical stability, high hardness, and wear resistance).[7,11,12] Excellent chemical stability of GC enables for the ease of demolding,[13] further its amorphous structures allows for the nanoscale processing.[7] Although GC is one of promising mold materials for glass embossing, it is but very expensive.

The motivation of the work is that replicated quartz glasses can be used as mold as well as MEMS components. Due to their excellent UV transmission, quartz molds are generally used for the UV-imprint process which can fabricate nano-/microstructures on permanent use UV curable resist at room temperature under low pressure (<0.5 bar). Because quartz mold is technically more difficult to be prepared and much more expensive as compared to a conventional silicon or nickel molds, a fabrication process for a quartz mold is one of the critical issues that need to be studied for the acceptance of UV-imprint technologies in industrial applications. Quartz molds can also be used for hot-embossing of glasses (those have lower softening temperature than quartz). For the process to be successful, the quartz surface has to be coated with an adequate barrier layer to prevent the glass-to-glass bonding phenomena.

In this study, quartz molds were prepared by hot-embossing with glassy carbon masters and applied to the hot-emboss of borosilicate. Glassy carbon molds for the replication of quartz molds were prepared by two different machining processes, including dicing and FIB milling techniques. Prior to borosilicate embossing with quartz molds, the surfaces of quartz molds were sputtered with molybdenum.

## 2. Experimental Procedure

As a master material for quartz embossing, 3 mm-thick glassy carbon (GS-20, Tokai Carbon, Japan) plate with porosities of 2–3 vol.% was used. The initial surface roughness ($R_a$) for the GC plate measured using an optical interferometer was less than 5 nm. Focused Ga+ ion beam milling tests were performed using computer-controlled FIB system (FIB2000A, Hitachi, Japan), and dicing tests were performed using a dicing saw machine (model DAD 522, Disco) with the 20 µm-wide blade.

Glass embossing and heat treatment tests were performed in vacuum using a hot-embossing equipment. The maximum specification of this equipment is 10 kN of load, 1400°C of heating temperature and 0.07 Pa in vacuum. Alignment accuracy between the upper mold and the lower mold was below 10 µm. A position of the upper mold at vertical axis is controlled with ball screw. A GC sample milled by focused ion beam was heated at 1400°C for 10 min to prevent the surface contamination, which generates by the precipitation of gallium ions during heating in the emboss process. Subsequently, the GC sample was cooled to below 200°C without any coolant in vacuum. Borosilicate glass (IWAKI CODE 7740 Pyrex Glass, Asahi technoglass) and quartz glass (PXST, Asahi technoglass) were used as embossing materials. In hot-emboss process, a master structure on a mold surface was pressed into a substrate at an elevated temperature, and then the applied pressure and the temperature were kept constant for certain time. After completing embossing step, temperature dropped to below 200°C naturally, the glassy carbon mold was removed from the glass substrate. Embossing system provides forced demolding function, and both embossing and demolding was conducted with the upper mold moving speed of 0.3 mm/min. The applied pressure is defined $P_a = L_n/A$, where $L_n$ is the normal load and $A$ is the area of specimen. Residual surface features and the residual surface roughness were determined by interferometric microscopy (ZYGO surface profilometer) and scanning electron microscopy (SEM). For the SEM, sample surfaces were over-coated with a thin Au layer using an auto fine coater (JFC-1300, Jeol).

## 3.   Results and Discussion

### 3.1.   *Preparation glassy carbon negative master by FIB milling and dicing*

Glassy carbon negative masters were prepared by two different types of fabrication techniques including focused ion beam milling and dicing techniques. Figure 1 represents a glassy carbon mold for micro-chamber arrays fabricated by dicing technique under the feed speed of 50 mm/min. The glassy carbon master has array of pyramids with height of 200 µm, bottom width of 400 µm and side wall angle of 45° on its surface. Prior to the hot-emboss tests, the glassy carbon mold was annealed at 1000°C in vacuum. Subsequently, a 100 µm line-and-spacing pattern was fabricated in 1-mm-thick glassy carbon plate by FIB milling. As a machining process, FIB milling offers the high flexibility in the working shapes, the dimensions (a scale ranging from a few tens of nanometers to hundreds of micro-meters), and the material selectivity.[7,11,12,14] These characteristics allow that the mold with nano/microstructures can be milled directly on metal, silicon, glass, carbon substrate, diamond without any pattern transfer or electroplating. The milled glassy carbon surface was then heat-treated at 1400°C for 10 min in vacuum to prevent the Ga contamination[7] that occurs during high temperature imprinting (>250°C) due to the precipitation of the implanted gallium ions.

Fig. 1.   SEM image of the glassy carbon surface machined by dicing saw: array of pyramids with height of 200 µm, bottom width of 400 µm and side wall angle of 45°.

### 3.2. *Fabrication of quartz mold by hot embossing with glassy carbon master*

Quartz positive molds were replicated from the glassy carbon negative master by hot-embossing. Based on the hot-emboss conditions of borosilicate glass given in Ref. 7, complete filling conditions were investigated with different time at the same temperature and pressure, and could be obtained at a press-head temperature of about 1305°C with a press pressure of 0.22 MPa and an embossing time of 400 s (an emboss velocity of 50/min and a heating rate of 0.5°C s$^{-1}$). The SEM images of the replicated quartz patterns are represented in Fig. 2. The obtained results were well in agreement with our previous study.

In Fig. 3, by adapting 35°C higher embossing temperature, the embossing time for complete filling could be reduced to 200 s under the same imprint pressure.

Fig. 2. SEM images of quartz glass surfaces embossed under the embossing temperature of 1305°C and the applied pressure of 0.22 MPa and different embossing time; (a) 100, (b) 300, and (c) 400 s.

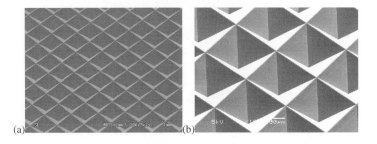

Fig. 3. (a) A SEM image of quartz glass surface embossed under the embossing temperature of 1340°C, the applied pressure of 0.22 MPa, and the embossing time of 200 s. (b) A high magnification SEM image of (a).

### 3.3. *Hot embossing of borosilicate glass with molybdenum sputtered quartz mold*

The final process step is glass-to-glass embossing. For the process to be successful, there are some major concerns that need to be overcome. One of them is that a permanent bonding between the quartz glass and the borosilicate glass may occur because embossing temperature is 100°C higher than annealing temperature of borosilicate. It has been reported that permanent glass bonding occurs above 550°C, which is approximately equal to the glass annealing temperature for borosilicate glass.[15]

A solution to this problem is to coat the surface of quartz mold with a barrier layer. Under the high imprinting temperature, the barrier material should have the following characteristics; (a) good chemical stability to prevent the adhesion to glasses, (b) high hardness and toughness to prevent deformation or breaking, (c) superior resistance to heat shock, and (d) excellent durability to be used repetitively, thereby allowing for the reduction of overall production cost of glass elements.

The barrier material chosen first was sputtered aurum layer with 200 nm in thicknesses. However, the Au layer did not have a role as protecting layer due to its insufficient durability. As shown in Fig. 4, the Au layer was partially delaminated during demolding after the first imprinting test.

Second, molybdenum (having melting point of 2600°C) was chosen as a barrier material. Molybdenum film with 100 nm in thickness deposited by sputtering on glass provides an inexpensive, inert and mechanically durable layer. In Fig. 5, embossing conditions required for complete filling were investigated under different embossing time using a Mo-deposited quartz glass mold. As shown in Fig 5(c), a complete filling could be obtained under the following conditions: the temperature, 650°C, the pressure, 0.22 MPa, and the embossing time, 1200 s. Expected application of the embossed borosilicate sample is a pyramid microlens array for controlling light diffusion. Figure 6 (a) shows the SEM images of quartz mold that was obtained by hot-embossing with the GC milled using a focused ion beam. In Fig. 6(b), the micro patterns were replicated on borosilicate surface with good fidelity under the temperature of 650°C, the pressure of 0.22 MPa and the embossing time of 600 s.

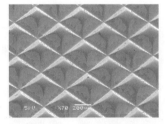

Fig. 4. SEM image of borosilicate glass surface embossed using the Au-coated quartz mold under the temperature of 650°C, the pressure of 0.22 MPa and the embossing time of 300 s.

Fig. 5. SEM images of borosilicate glass surfaces embossed using the molybdenum-coated quartz mold under the embossing temperature of 650°C and the applied pressure of 0.22 MPa and different embossing time; (a) 600 s, (b) 900 s, (c) 1200 s.

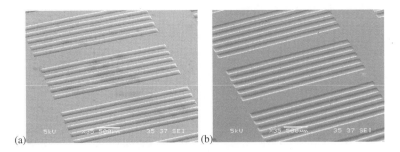

Fig. 6. SEM image of the quartz mold and the replicated borosilicate surfaces; (a) Quartz mold prepared by hot-embossing using the glassy carbon mold milled using focused ion beam and (b) replicated borosilicate by hot-embossing using the molybdenum-coated quartz mold (650°C, 0.22 MPa, and 600 s).

## 4. Summary

In this study, quartz molds were prepared by hot embossing with the glassy carbon masters, and they were applied to the hot-emboss of borosilicate. Glassy carbon masters were machined by dicing and focused ion beam milling techniques. The achieved glassy carbon masters then applied to the hot-emboss process to produce the quartz molds. Finally, micro-hot-embossed structures were developed in borosilicate glasses with high fidelity by hot embossing with the Mo-sputtered quartz molds. This method showed great potential to fabricate multiple quartz molds efficiently and at a very low cost.

## Acknowledgment

The authors would like to thank Mr. Toshihiko Noguchi for technical support.

## References

1. M. Okinaka, S. Inoue, K. Tsukagoshi, and Y. Aoyagi, *J. Vac. Sci. Technol.* **B24**, 271 (2006).
2. S. Ronggui and G. C. Righini, *J. Vac. Sci. Technol.* **A 9 (5)**, 2709 (1991).
3. Y. Utsumi, M. Ozaki, S. Terabe, and T. Hattori, *Jpn. J. Appl. Phys.* **Part 1 42**, 4098 (2003).
4. X. Li, T. Abe, M. Esashi, *Sens. Actuat.* **A 87**, 139 (2001).
5. J. A. Plaza, M. J. Lopez, A. Moreno, M. Duch, and C. Can´e, *Sens. Actuat.* **A105**, 305 (2003).
6. C. Iliescu, J. Jing, F.E.H. Tay, J. Miao, and T. Sun, *Surf. Coat. Technol.* **198**, 314 (2005).
7. M. Takahashi, K. Sugimoto, and R. Maeda, *Jpn. J. Appl. Phys.* **44 (7B)**, 5600 (2005).
8. Y. Hirai, K. Kanakugi, T. Yamaguchi, K. Yao, S. Kitagawa, and Y. Tanaka, *Microelectron. Eng.* **67-68**, 237 (2003).
9. W. Choi, J. Lee, W.-B. Kim, B.-K. Min, S. Kang, and S.-J. Lee, *J. Micromech. Microeng.* **14**, 1519 (2004).
10. M. Okinaka, K. Tsukagoshi, and Y. Aoyagi, *J. Vac. Sci. Technol.* **B 24(3)**, 1402 (2006).
11. S. W. Youn, M. Takahashi, H. Goto, and R. Maeda, *Microelectro. Eng.* **83**, 2482 (2006).
12. S. W. Youn, M. Takahashi, H. Goto, and R. Maeda, *J. Micromech. Microeng.* **16**, 2576 (2006).
13. H. Ito, K. Ito, M. Arai, and K. Sugimoto, T. Matsukura, and T. Kogai, *J. Jpn. Soc. Precision Eng.* **70 (6)**, 807 (2004).
14. A. A. Tseng, *J. Micromech. Microeng.* **14**, R15 (2004).
15. P. Mao and J. Han, *Lab. Chip* **5**, 837 (2005).

# Part L
# Plasticity in Tribology

# COMPARING SLIDING-WEAR CHARACTERISTICS OF THE ELECTRO-PRESSURE SINTERED AND WROUGHT COBALT

J. E. LEE[1], Y. S. KIM[2†]

*Kookmin University, Seoul, KOREA,*
*galelje@hanmail.net, ykim@kookmin.ac.kr*

T. W. KIM[3]

*Ehwa Diamond Ind. Co., Ltd., Osan-si, KOREA,*
*kimtai@ehwadia.co.kr*

Received 15 June 2008
Revised 23 June 2008

Dry sliding wear tests of hot-pressure sintered and wrought cobalt were carried out to compare their wear characteristics. Cobalt powders with average size of 1.5μm were electro-pressure sintered to make sintered-cobalt disk wear specimens. A vacuum-induction melted cobalt ingot was hot-rolled at 800°C to a plate, from which wrought-cobalt disk specimens were machined. The specimens were heat treated at various temperatures to vary grain size and phase fraction. Wear tests of the cobalt specimens were carried out using a pin-on-disk wear tester against a glass (83% $SiO_2$) bead at 100N with the constant sliding speed and distance of 0.36m/s and 600m, respectively. Worn surfaces, their cross sections, and wear debris were examined by an SEM. The wear of the cobalt was found to be strongly influenced by the strain-induced phase transformation of ε-Co (hcp) to α-Co (fcc). The sintered cobalt had smaller uniform grain size and showed higher wear rate than the wrought cobalt. The higher wear rate of the sintered cobalt was explained by the more active deformation-induced phase transformation than in the wrought cobalt with larger irregular grains.

*Keywords*: Sliding wear; wear test; sintered cobalt; wrought cobalt; phase transformation; grain size.

## 1. Introduction

Diamond saws are widely used in construction, stonework, machinery and electronic industries. Cobalt and cobalt-base alloys are regarded as the best bond-metal for a diamond-impregnated segment of the diamond saw, since they have superb diamond-holding capability and exhibit appropriate wear rate to achieve an optimum cutting.[1] However, cobalt has problems of high price and unstable supply, which promoted active research of substitutes of cobalt from 1990's. To develop the substitute, it is essential to understand wear characteristics and wear mechanism of the cobalt with various phase and microstructure.

---

[†]Corresponding Author.

Phase transformation of cobalt is reported to influence sliding wear rate of sintered cobalt.[2] Metallic cobalt commonly possesses a mixture of two phases, ε (hcp) and α (fcc), at room temperature. The ε-Co is known to transform to α-Co allotropically at 722K. It has been reported that the ε phase in a cobalt alloy increases by a transformation of the α-Co during plastic deformation at room temperature through a stress-induced martensitic transformation[3], while Sort and his co-workers reported a reverse transformation that ε phase decreases with the increase of high-pressure-torsion strain.[4] Reported results are contradictory; though Huang et al.[5] had argued earlier that plastic-deformation-induced phase of cobalt depends on the deformation intensity. The phase change of cobalt during deformation is still controversial and so is its effect on its wear.

The present research was performed with the purpose of comparing sliding wear characteristics and wear mechanism of the sintered and wrought cobalt, which were made from cobalt powders and from a hot-rolled cobalt plate, respectively. Grain size and phase fraction of the cobalt were varied through heat treatments, and their effect on the wear was evaluated. Dry sliding wear tests of the sintered and wrought cobalt specimens were carried out at room temperature in the air, and their wear behavior was characterized in conjunction with the phase transformation of cobalt during the wear.

## 2.  Experimental

Sintered-cobalt disk specimens for the wear test were made from cobalt powders with average size of 1.5μm. Diameter and thickness of the disk specimen were 25.4mm and 4mm, respectively. Green compacts of the cobalt powder were electro-pressure sintered with a pressure of 35MPa at 800°C. Wrought-cobalt disk specimens with the same dimension were machined from a hot-rolled cobalt plate. The hot rolling of a cobalt ingot that was induction melted and casted in vacuum was carried out at 800°C. The disks were annealed at temperatures ranging from 150°C to 1200°C for 30min followed by an air cooling. To see the effect of cooling rate, some of the annealed sintered disks (heat treated at 150°C to 750°C) were water quenched, too.

Dry sliding wear tests of the disk specimen were carried out in the air at room temperature against a $SiO_2$ bead using a pin-on-disk wear tester. The adopted sliding speed was 0.38m/sec, and applied wear load and sliding distance were fixed as 100N and 600m, respectively. Wear rate was calculated by dividing the volume loss of a specimen by the sliding distance. Worn surfaces, their cross sections, and wear debris were examined by a scanning electron microscopy (SEM). Phases of the disk specimen and wear debris were identified by an X-ray diffraction (XRD) analysis.

## 3.  Results and Discussion

### 3.1.  *Wear test and specimen analysis results*

Wear rates of the sintered and wrought cobalt specimens heat treated at different temperatures are shown in Fig. 1. The sintered cobalt specimens annealed at different temperatures did not show any significant wear-rate variation until 750°C. Wear rates of the specimen annealed at 900°C and at above gradually decreased with the temperature increase. However, wear rates of all the water-quenched sintered cobalt and wrought cobalt specimens were lower than that of the air-cooled sintered ones.

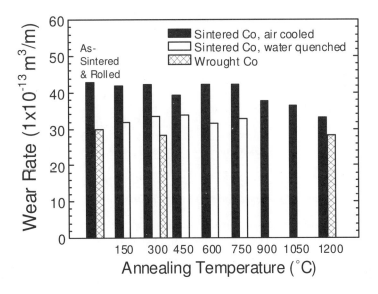

Fig. 1. Wear rates of the sintered and wrought cobalt heat treated at various temperatures.

Hardness, grain size, and phase fraction of the sintered and wrought cobalt specimens are listed in Table 1, together with their heat treatment conditions. The annealing at temperatures below 750°C hardly changed hardness of the sintered and wrought cobalt. The hardness-decrease of the sintered cobalt annealed at higher temperatures (above 900°C) appears to be caused by an increase of grain size. The grain size of the cobalt specimens was measured using a transmission electron microscopy (TEM) and electron backscatter diffraction (EBSD). Like the hardness, the grain size of the sintered cobalt hardly changed with the annealing at temperatures lower than 750°C. The size increased after annealing at temperatures higher than 900°C, but slightly. The phase fraction of the sintered cobalt annealed at various temperatures did not change noticeably. The sintered cobalt presents a mixture of ε-Co (hcp) and α-Co (fcc) after the annealing, of which fractions are approximately 80% and 20%, respectively. The phase fraction looks rather independent of the annealing temperature. Hardness of the wrought cobalt was similar to that of the sintered cobalt. Like the sintered cobalt, the hardness of the wrought did not change appreciably after the heat treatment. However, the grain size of the as-rolled cobalt was much larger than that of the sintered cobalt. The grain size of the wrought specimen also changed significantly after the heat treatment; the size became much larger and the size was not uniform.

Table 1. Hardness, grain size, and phase composition of the sintered and wrought cobalt specimens tested in the present study.

| Specimen | Heat Treatment Condition | Hardness (HRB) | Grain Size (μm) | Phase Composition |
|---|---|---|---|---|
| As-sintered | - | 103.5 | 0.99 | 82% ε-Co, 18% α-Co |
| Sintered and heat treated | 150°C, 30min./A.C.[a] | 105.5 | - | - |
| | 300°C, 30min./A.C. | 104.0 | 1.16 | 78% ε-Co, 22% α-Co |
| | 450°C, 30min./A.C. | 105.0 | - | - |
| | 600°C, 30min./A.C. | 106.0 | 1.11 | - |
| | 750°C, 30min./A.C. | 103.5 | 1.23 | - |
| | 900°C, 30min./A.C. | 97.0 | - | - |
| | 1050°C, 30min./A.C. | 68.0 | - | - |
| | 1200°C, 10min./A.C. | 66.7 | 3.83 | 81% ε-Co, 19% α-Co |
| As-rolled | - | 102.5 | 17.55 | 88% ε-Co, 12% α-Co |
| Rolled and heat treated | 300°C, 30min./A.C. | 103.5 | 20~100 | 78% ε-Co, 22% α-Co |
| | 1200°C, 10min./A.C. | 82.0 | 30~100 | 73% ε-Co, 27% α-Co |

[a]Air Cooled

### 3.2.  *SEM observation of worn surface and wear debris*

Fig. 2 shows SEM micrographs of worn surface and wear debris of the sintered and wrought cobalt specimens. The SEM micrographs of the sintered cobalt display the surface and debris of two different sintered specimens that were annealed at the same temperature (at 750°C), but cooled in a different way: air cooled (Fig. 2 (a), (d)) and water quenched (Fig. 2 (b), (e)). Fig. 2 (c) and (f) are SEM micrographs of the wrought specimen heat treated at 1200°C for 10min. followed by an air cooling.

The worn surface and wear debris of the air-cooled cobalt specimens (both sintered and wrought) look similar. The worn surfaces are shiny and composed of long scratch marks with various widths, though some deformation traces are observed on the wrought-cobalt surface. The surface showed neither major cracks nor any significant trace of debris-detachment, which is a typical worn-surface appearance of most hcp metals. Observation of cross sections of the air-cooled cobalt specimens revealed thin (less than 10μm) detaching surface layers with numerous cracks. The cross-sectional view is unusual, since most metallic materials worn at high loads show a thick deformed surface-layer. Major part of the wear debris collected from the air-cooled cobalt specimens was fine particles (diameter less than 5μm). Cutting-chip like long corrugated debris (a typical pattern of hcp metal wear particles) and plate-like wear particles were also observed.

On the other hand, the worn surface and wear debris of the water-quenched sintered cobalt (Fig. 2 (b) and (e)) look very different. The surface shows trace of large plastic deformation, depressions and debris-detached dents. The surface appearance resembles the worn surface of a metallic material that underwent the delaminating wear with significant surface deformation. Wear debris of the water-quenched sintered cobalt was a mixture, but mostly composed of plate-like flat particles with the thickness ranging from 10μm to 20μm. Most of the wear particles were larger than those from the air-cooled cobalt. Few cutting-chip like corrugated debris was observed.

Fig. 2. SEM micrographs of worn surfaces and wear debris of the sintered and wrought cobalt specimens heat treated at different temperatures: (a), (d) sintered Co annealed at 750°C, 30min., air cooled; (b), (e) sintered Co annealed at 750°C, 30min., water quenched; (c), (f) wrought Co annealed at 1200°C, 10min., air cooled.

### 3.3. *Sliding wear mechanism of the sintered and wrought cobalt*

All cobalt specimens (sintered and wrought) that were air cooled after annealing exhibited typical wear characteristics of an hcp metal: long shallow grooves on worn surfaces, fine wear debris together with long wrinkled cutting-chip like wear particle. Table 1 has shown that the main constituting phase of the cobalt (sintered and wrought) is ε-Co (hcp), which supports the microscopic observation. However, X-ray diffraction (XRD) analysis of wear particles revealed that most of the wear debris from both the sintered and wrought cobalt was α-Co (fcc). The SEM observation of worn surface and the XRD analysis of wear debris strongly indicate that phase transformation of the ε-Co to the α-Co had occurred during the wear of the cobalt. The phase transformation is assumed to promote and accelerate the wear as reported previously.[2, 6]

A correlation is found between the wear-rate variation and the grain-size change, which is given in Fig. 1 and table 1, respectively. Air-cooled sintered cobalt specimens that were annealed at temperatures from 150°C to 750°C had similar grain size of around 1.15μm. Wear rates of the specimens were comparable, which were around $41 \times 10^{-13}$m$^3$/m. When the annealing temperature increased above 750°C, the grain size of the air-cooled cobalt increased, and the wear rate of the cobalt specimen decreased. These findings suggest that the extent of the phase transformation is related with the grain size, since the wear of the cobalt was strongly associated with the phase transformation. The correlation deduced also indicates that the transformation of ε-Co to α-Co is a deformation-induced transformation. Huang et al. had observed such transformation from ball-milled cobalt, and they attributed the hcp (ε) → fcc (α) transformation to the accumulation of structure defects by the severe deformation.[5] Such accumulation of defects would be more operative with small grains. It has been known that stable structure of cobalt is dependent on grain size; as grains become smaller, fcc α-Co becomes more stable structure.[7] Since the allotropic transformation from the ε to the α occurs at 722K, a temperature-rise during the wear might have caused the observed transformation. However, the grain-size dependency of the wear rate supports the deformation-induced transformation.

The wrought cobalt specimens with larger and irregular grain size showed lower wear rate than the air-cooled sintered cobalt. The XRD analysis of the wear debris from the wrought cobalt revealed that the debris is composed of a mixture of α-Co and ε-Co. The significant amount of ε-Co in the wrought-cobalt debris indicates that full transformation of ε-Co to α-Co had not occurred during the wear of the wrought cobalt, because of its larger grain. The wrought cobalt annealed at different temperatures showed a similar wear rate of $29 \times 10^{-13} m^3/m$ in spite of their different grain sizes. This denotes a grain-size boundary which controls the transformation; however, it needs further research.

## 4.  Conclusions

Dry sliding wear tests of hot-pressure sintered and wrought cobalt specimens were carried out to compare their wear characteristics. The cobalt specimens were heat treated at various temperatures to vary their grain size and phase fraction, and the effect of grain-size and phase transformation on the wear was investigated. The phase transformation of the ε-Co (hcp) to the α-Co (fcc) was found to occur during the wear of all cobalt specimens, which accelerated the wear. The phase transformation was reasoned to be a deformation-induced transformation. Wear rate of the wrought cobalt with large and irregular grains was lower than that of the sintered cobalt with smaller uniform grains. The phase transformation of the cobalt showed grain-size dependency; the ε-Co to α-Co transformation seemed to be more effective with smaller grains, which resulted in the higher wear rate.

## Acknowledgments

This research was partially supported by the 2008 research fund of Kookmin University in Korea.

## References

1.  J. Konstanty, *Cobalt as a Matrix in Diamond Impregnated Tools for Stone Sawing Applications* (Wydawnictwa AGH, Krakow, 2002).
2.  Y.-S. Kim, S. H. Kang, and T. W. Kim, *Mater. Sci. Forum* **539-543**, 820 (2007).
3.  P. Huang and H. Lopez, *Materials Letters* **39**, 244 (1999).
4.  J. Sort, A. Zhilyaev, M. Zielinska, J. Nogues, S. Surinach, J. Thibault, and M. D. Baro, *ACTA Mater.* **51**, 6385 (2003).
5.  J. Y. Huang, Y. K. Wu, and H. Q. Ye, *Appl. Phys. Lett.* **66**, 308 (1995).
6.  Y.-S. Kim, J. E. Lee, S. H. Kang, and T.-W. Kim, *Mater. Sci. Forum* **534-536**, 1109 (2007).
7.  E. A. Owen and D. M. Jones, *Proc. Phys. Soc. B* **67**, 456 (1954).

# Part M

Porous, Cellular and
Composite Materials

# A PHENOMENOLOGICAL CONSTITUTIVE MODEL OF ALUMINUM ALLOY FOAMS AT VARIOUS STRAIN RATES

C. LI[1]

*Institute of Applied Mechanics and Biomedical Engineering, Taiyuan University*
*of Technology, Taiyuan, 030024, P.R.CHINA,*
*Engineering College, Shanxi University, Taiyuan, 030013, P.R.CHINA,*
*tydz_lc@126.com*


Z. H. WANG[2†], L.M. ZHAO[3], G.T. YANG[4]

*Institute of Applied Mechanics and Biomedical Engineering, Taiyuan University*
*of Technology, Taiyuan, 030024, P. R. CHINA,*
*wangzh623@yahoo.com.cn, zhaolm@tyut.edu.cn, yanggt@tyut.edu.cn*

Received 15 June 2008
Revised 23 June 2008

A nonlinear elasto-plastic phenomenological constitutive model for aluminum alloy foams subjected to quasi-static and dynamic compression is proposed. The six-parameter model can fully capture the three typical features of stress-strain response, i.e., linearity, plasticity-like stress plateau, and densification phases. Moreover, the parameters of the model can be systematically varied to describe the effect of initial density of foams that may be responsible for changes in yield stress and hardening-like or softening-like behavior at various strain rates. The experimental results at various loading rates are provided to validate the model. It is shown that the proposed model can be used in the selection of the optimal-density and energy absorption foam for a specific application based on certain design criteria..

*Keywords*: Elasto-plastic constitutive model; Aluminum alloy foam; Energy absorption.

## 1. Introduction

Metal foams have a unique combination of properties such as low density, high ratio-stiffness and ratio-strength, and good energy absorption capability[1,2]. These properties offer them the potential uses in automotive, railway and aerospace applications. For example, they had been widely used as internal padding and inside structural elements of the external body as energy absorbers, which may reduce damage from impact.

Generally, impact accidents produce higher loading rates than static or quasi-static cases and which may significantly alter mechanical response of the materials. Therefore, when designing energy absorber in industry using the numerical tools, mathematical description of mechanical properties of foams is needed for strain rates corresponding to impact events. However, the current simulated responses of foam models[3-9] are mostly

---

†Corresponding Author.

derived on the basis of empirically obtained stress-strain curves at a specific strain rate or impact velocity, which could only represent a narrow range of behaviors under specific loading conditions. It is highly desirable to obtain mechanical properties of metal foams from a single constitute equation which is capable of describing the stress-strain behavior at a wide range of strain rates. Such data are essential in realistic numerical simulations for the safety design of structures. Only a few models, as the Gibson model[1], are based on the deformation mechanism and therefore could account for the effects of the characteristic parameters, such as density, cell size etc., However, these models are too complex to be applied in industry since the Gibson density dependency laws or the proposition of other laws must be characterized by analyzing the foam structure.

The present study is an attempt to provide a more comprehensive formulation of phenomenological model which can describe the entire stress-strain response of aluminum alloy foams, including a linear elastic region, a plateau region over a large range of strains and a densification region where the stress rises sharply. The dependency of the model parameters on the density (relative density) and strain rate, which are most important parameter determining the mechanical properties of aluminum alloy foams, are also presented. Such a phenomenological approach can reduce the complexity related to cell morphologies and can facilitate the analysis of engineering structures. Furthermore, it is possible to obtain more precise parameters of the foams in its optimal design, particularly to the optimal density in the specific strain rate.

## 2.    Constitutive Models

In the present paper, a constitutive model which can depict the entire stress-stain response of aluminum alloy foams at different loading rates is presented

$$\sigma = \left( A(\bar{\rho}) \frac{e^{\alpha(\bar{\rho})\varepsilon} - 1}{B + e^{\beta(\bar{\rho})\varepsilon}} + C(\bar{\rho}) \left( \frac{\varepsilon}{1 - \varepsilon} \right)^n \right) \left( 1 + D(\bar{\rho}) \lg \frac{\dot{\varepsilon}}{\dot{\varepsilon}_0} \right) \tag{1}$$

where $\sigma$, $\varepsilon$ are the uniaxial engineering stress and engineering strain, respectively. $\bar{\rho}$ is the relative density (defined as the density of the foam divided by the density of the cell wall material) of the foams. $\dot{\varepsilon}$, $\dot{\varepsilon}_0$ are the average strain rate and quasi-static strain rate, respectively. The first term is used to model the elastic-plateau and densification region, respectively, while the second term is used to depict the dependence of strain rate. The parameters A, B, $\alpha$, $\beta$, C, D are density dependent, while n is density independent. The parameter A describes the yield stress in compression with increasing relative density. The parameter B captures the tensile yield strength. In the current investigation, the parameter B is set equal to one. Parameters $\alpha$ and $\beta$ together capture the features of inelastic response with $\alpha > \beta$ depicting hardening-like response in high-density foams, $\alpha = \beta$ for the ideal plasticity-like response in medium-density foams, and $\alpha < \beta$ for softening-like response in low density foams. The parameters C and n can readily capture the initiation and the intensity of the steep densification phenomenon, respectively, and it is easy to note that the formula have a vertical asymptote corresponding to the physical limit of compression ($\varepsilon = 1$). The last parameter D can depict the dependence of strain rate.

All above these features are illustrated in Fig.1 for various values of parameters among which the parameter A is set as a constant 15. Thus, the model can capture the entirely compressive stress-strain response for aluminum alloy foams subjected to quasi-static and dynamic loadings. It is interesting to note that the first term of the model can provide an equivalent elastic modulus for each density of foam. Its derivative with respect to the strain is

$$\frac{\partial}{\partial \varepsilon}\left( A(\bar{\rho}) \frac{e^{\alpha(\bar{\rho})\varepsilon} - 1}{B + e^{\beta(\bar{\rho})\varepsilon}} + C(\bar{\rho})\left(\frac{\varepsilon}{1 - \varepsilon}\right)^{n} \right) \tag{2}$$

when B is set to unity, the equivalent modulus is given by

$$E = \lim_{\varepsilon \to 0} \frac{\partial \sigma}{\partial \varepsilon} = \lim_{\varepsilon \to 0}\left( \frac{\partial}{\partial \varepsilon}\left( A(\bar{\rho}) \frac{e^{\alpha(\bar{\rho})\varepsilon} - 1}{B + e^{\beta(\bar{\rho})\varepsilon}} + C(\bar{\rho})\left(\frac{\varepsilon}{1 - \varepsilon}\right)^{n} \right) \right) = \frac{A\alpha}{2} \tag{3}$$

Note that near the origin of the stress-strain curve the tangent modulus is equal to the value of $A\alpha/2$, which can be considered the initial elastic modulus of the aluminum alloy foam. Clearly, the modulus varies with parameter A and $\alpha$ but is independent of $\beta$.

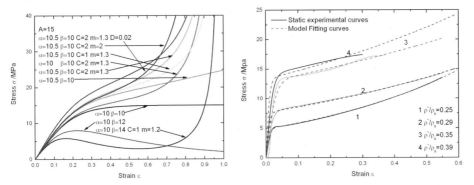

Fig. 1. Illustration of model for various parameter values.   Fig. 2. Comparsion of stress-strain curves from tests and analysis model at the quasi-static strain rate.

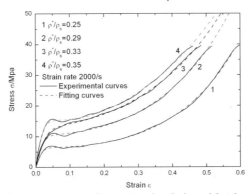

Fig. 3. Comparison of stress-strain curves from tests and analysis model at the strain rate of 2000/s.

## 3.   Experimental Procedure and Results

The open-cell aluminum alloy foams, which were produced by the infiltrating process, were used to validate the proposed phenomenological constitutive model. The composition of the cell wall material is AL-3wt.%Mg-8wt.%Si-1.2wt.%Fe. The relative density of the foams ranges from 0.25 to 0.40 and the average cell sizes are 0.9 mm and 1.6mm, respectively.

Uniaxial compression tests at a quasi-static strain rate of $10^{-3} s^{-1}$ were performed using a servo-hydraulic test machine. All specimens were cylindrical with a diameter of 35mm and a height 30mm. Before tests, each specimen was weighted and measured in order to calculate its effective relative density. For each relative density of foam, at least three repetitions of the compression test were performed.

Dynamic compression test at the strain rates in the range $10^{2} - 10^{4} s^{-1}$ was performed at room temperature using the split Hopkinson pressure bar (SHPB). The cellular material can be treated as a continuum: that is, that the sample dimensions should be as big as possible within the limitations of the Hopkinson pressure bar in order to obtain the actual properties of foam materials, However, to obtain the stress–strain relations with a strain in a large scale, including linear elasticity, collapse plateau and densification, the sample height needs to be minimized. Therefore, the selected specimens are circular cylinders of diameter 35mm and length 10mm. With this choice of specimen dimensions, the specimens have at least 6-10 cells in all directions.

Figures 2 and 3 illustrate the experimentally obtained stress-strain response for these different relative density foams under quasi-static and dynamic compression. It can be seen that the compressive stress–strain curve of aluminum alloy foams exhibits universal three deformation characteristics: an initial linear-elastic region; an extended plateau region where the stress increases slowly as the cells deform plastically; and a final densification as collapsed cells are compacted together. Note that the high relative density foams has a higher the stiffness and yield stress, a smaller strain due to crushing of cells, and an earlier densification region than those of the low relative density foams. It is also clear shown that all the foams in the study exhibit a hardening-like nonlinear response, which has a positive slope that combines crushing and simultaneous densification at a gradually increasing rate, during the crushing stage.

## 4.   Parameters Identification

The identification of the parameter values have been done by using an optimization procedure based on a Gauss-Newton algorithm, in which the parameters are adjusted iteratively to minimize the mean squared error between the experimental data set and the prediction of the nonlinear function from a given set of initial parameters. Only the density dependent parameters are identified for each experimental curve, while the density independent parameters are identified on the whole set of curves together. It should be noted that due to the algorithm is sensitive to the set of chosen initial values, a convergent set of parameters may not be obtained if Eq. 1 is used directly to fit the fully experimental data. The alternative method is, only a section of the data prior to rapid densification phase is used to acquire a convergent parameter set ( A, $\alpha$ and $\beta$ ) by utilizing the portion modeling the elastic-plateau region of Eq.1. Because parameter A can be easily estimated from the yield stress and $\alpha$ and $\beta$ can be estimated from the stress-strain characteristics. The convergent parameter set ( A, $\alpha$ and $\beta$ ) can be obtained in a relatively easy way. And then, the first term of Eq.1, in which the values of the parameter A, $\alpha$ and $\beta$ are known and C and m are unknown, is used to fit the entirely experimental data to obtain the initial parameter values of C and n. In the sequence, all the above predetermined parameters are used as the initial values to obtain a final set of convergent parameters for the function defined in Eq.12 through the overall mean squared error between the experimental data set and the prediction of the nonlinear function.

Figs.2 and 3 show the comparison between model predictions and experimental results of aluminum alloy foams with different relative densities and strain rates under the quasi-static and dynamic compression. The predicted parameter sets for the aluminum alloy foams under different relative densities are given in Tables1 and 2. It can be seen that the model perfectly captures the entirely stress-strain response, and the quality of the fit is remarkable in the elastic-plateau region and densification region. Note that the parameter A clearly reflects the yield stress for different relative density foams. In addition, the hardening-like behavior beyond the yield stress was also fairly captured by the parameters $\alpha$ and $\beta$. The difference between these two parameters reflects the intensity of the slope of plateau region curves. Parameter C and n in Eq.1 modeled effectively the rapid-densification phase. The parameter C reflects the rapid increase in the stress due to densification.

Table 1. Identified parameters for aluminum foams under the quasi-static compression.

| Relative density | Parameter | | | | | |
|---|---|---|---|---|---|---|
| | A (*MPa*) | $\alpha$ | $\beta$ | c | n | d |
| 0.25 | 6.247 | 199.7 | 198.8 | 0.00067 | 1.515 | - |
| 0.29 | 7.652 | 185 | 182.9 | 0.0008 | 1.515 | - |
| 0.35 | 9.856 | 175.7 | 173.7 | 0.001 | 1.515 | - |
| 0.39 | 11.445 | 173.7 | 170.7 | 0.0013 | 1.515 | - |

Table 2. Identified parameters for aluminum foams under the dynamic compression.

| Relative density | Parameter | | | | | |
|---|---|---|---|---|---|---|
| | A (*MPa*) | $\alpha$ | $\beta$ | c | n | d |
| 0.25 | 6.452 | 198.7 | 196.8 | 0.00059 | 1.515 | 0.06455 |
| 0.29 | 7.05 | 193.4 | 190.9 | 0.00073 | 1.515 | 0.06655 |
| 0.33 | 8.512 | 182.6 | 179.7 | 0.00087 | 1.515 | 0.06443 |
| 0.35 | 10.25 | 176.7 | 173.5 | 0.00094 | 1.515 | 0.06343 |

## 5.    Conclusion

A multi-parameter rate-dependent elasto-plastic constitutive model that describes the entire nonlinear stress-strain behavior of open-cell aluminum alloy foams under quasi-static and dynamic compression has been proposed. Its effectiveness was demonstrated by the experimental data obtained from aluminum alloy foams with different initial densities under different loading rates. The constitutive model proposed with six parameters is sufficient to describe the various physical characteristics of the macroscopic stress-strain response effectively. The dependency of the model parameters on the density (relative density) and strain rates responsible for changes in yield stress, hardening-like and densification behavior are also  presented, such that the compressive stress-strain response can be characterized through a single constitutive equation under different loading conditions. The proposed phenomenological constitutive model is more convenient for specific applications in impact safety and crashworthiness analysis in the industry environment. It is also proposed that besides phenomenological approach, further studies on metal foam constitutive model should take into account sub-cellular level deformation mechanism.

### Acknowledgments

The authors sincerely wish to acknowledge the financial support provided by the National Natural Science Foundation of China through grant No. 90716005, 10572100 and 10772130, Natural Science Foundation of Shanxi under grant No. 2004-006, 2007021005 and 2008011007.

### References

1.    L. J. Gibson and Ashby M F. *Cellular solids. Structure and properties.* (Cambridge University Press), Cambridge, 1997.
2.    H. N. G. Wadley. *Phil. Trans. R. Soc. A.* **364**, 31 (2006).
3.    K. C. Rush. *J. Appl. Polym. Sci.* **13**, 297.(1969).
4.    K. C. Rush. *J. Appl. Polym. Sci.* **14**, 1133(1970).
5.    A. Nagy, W. L. Ko, U. S. Lindholm. *J. Cellular plastics,* **10**, 127 (1974).
6.    J. Zhang, N. Kikuchi, V. C. Li, A. F. Yee, et al. *Int. J. Impact Eng.* **21**, 369 (1998).
7.    R. W. Shuttleworth, V. O. Shestopal, P. C. Goss. *J. Appl. Polym. Sci.* **30**, 333 (1985).
8.    Q. L. Liu, G. Subhash, X. L. Gao. *J. Porous Mater.* **12**, 233(2005).
9.    Z. H. Wang, H. W. Ma, L. M. Zhao, G. T. Yang. *Scripta Mater.* **54**, 83(2006).

# EXPERIMENTAL INVESTIGATION ON FRACTURE TOUGHNESS OF INTERFACE CRACK FOR ROCK/CONCRETE

YANG SHUICHENG[1†]

*Department of Civil Engineering, Xi'an Jiaotong University*
*Xi'an, Shaanxi, 710049, P.R.CHINA,*
*yangshuicheng@yahoo.com.cn*

SONG LI[2], LI ZHE[3], HUANG SONGMEI[4]

*Institute of Mechanical and Precision Instrument, Xi'an University of Technology*
*Xi'an, Shaanxi, 710048, P.R.CHINA,*
*songli @xaut.edu.cn*

Received 15 June 2008
Revised 23 June 2008

Fracture toughness is a critical input parameter for fracture-mechanics based fitness-for-service assessments, and it is preferable to determine this by experiment. In the present paper, fracture toughness of rock/concrete bimaterial interface was obtained by the tests which were performed on the universal material tester. A beam specimen with single-edge crack is used to form the different fracture mode mixity. The stress intensity factors of specimen per unit load with different combinations of $K_1$ and $K_2$ were calculated by the mixed hybrid finite element method on the principals of linear elastic interface fracture mechanics. By regressing the critical stress intensity factors of 7 specimen groups, two experiential fracture criterions of mixed crack interface were derived, and the fracture toughness( $K_{1C}^*$ , $K_{2C}^*$ ) of rock/concrete were obtained further.

*Keywords*: Fracture Toughness; Bimaterial Interface; Rock /Concrete; Fracture Criterion.

## 1. Introduction

Most concrete dams experience cracking to some extent along the interface between the concrete dam and the upstream elevated foundation due to the different material properties, the construction defect, the temperature variation, the stress concentration and so on, which affects the safe operation of dams. Of greater concern are those cracks that develop as a result of hydrostatic load application, leading to significant change in the failure resistance of the structure. Due to this growing concern, concrete dams are increasingly coming under the scrutiny of regulatory agencies and other groups that are responsible for dam safety[1].

The fracture toughness is one of key parameters for carrying out linear elastic fracture analysis and safety evaluations, and should be determined by fracture tests and some

---

[†]Corresponding Author.

calculations. Whereas, the fracture toughness of interface cracks is different from the one of homogeneous material, which is related to the mismatch in the material. And the fracture mode on an interface of dissimilar materials is intrinsically mixed due to the asymmetry in the elastic properties across the interface. In the past few decades, several test specimens have been proposed to measure interracial toughness. Charalambides *et al.* [2] proposed a bimaterial notched four-point bending beam specimen and measured about the phase of mode mixities of the stress intensity factors of the stress around the interface crack tips. Then O'Dowd *et al.* [3] analyzed the near tip stress field of the two types of specimens with an interface crack in detail. Yukki *et al.* [4], Xu and Tippur [5], and Toru and Noriyuki [6] measured the fracture toughness of interface cracks covered wide range of mode mixity. They reported strong dependence of fracture toughness on the mode mixity. In addition, reviewed literature indicated that some researches on interface crack for rock/concrete have been made to modify the design method and assess safety of dams based on fracture mechanics theory and experimental methods [7, 8, 9, 10, 11].

However, sufficient experimental works haven't been done to derive fracture criterion on an interface crack for rock/concrete. In this study, fracture tests are performed using single-edge interface crack, and the experiment results are complied into a fracture criterion or fracture toughness using stress intensity factors of an interface crack.

## 2.  Stress Intensity Factors of an Interface Crack and its Calculation

### 2.1.  *Stress intensity factors*

The stress field for a semi-infinite interface crack separating two dissimilar isotropic elastic materials has the form [12].

$$\sigma_{22} + i\sigma_{12} = (\frac{K_1 + iK_2}{\sqrt{2\pi r}})\left(\frac{r}{l_k}\right)^{i\varepsilon} \tag{1}$$

where $r$ is the radial distance from the crack tip. $l_k$ denotes an arbitrary length to normalize the distance $r$. $\varepsilon$ is so-called oscillation index given by

$$\varepsilon = \frac{1}{2\pi} \ln\left(\frac{1-\beta}{1+\beta}\right) \tag{2}$$

where $\beta$ is the second Dundurs's mismatch parameter as shown below

$$\beta = \frac{1}{2} \frac{\mu_1(1-2v_2) - \mu_2(1-2v_1)}{\mu_1(1-v_2) + \mu_2(1-v_1)} \tag{3}$$

and $\mu$ is the shear modulus, and $v$ is Poisson's ratio, and subscripts 1 and 2 refer to the two materials.

It is noted that $K_1$ and $K_2$ for an interfacial crack in dissimilar media are different from those for a crack in a homogeneous material. They are simply the real and imaginary parts of a complex stress intensity factor $K = K_1 + iK_2$ , whose physical meaning can be understood from the complex form as Eq.(1). The effective stress intensity factor $K$ corresponds to the energy release rate and the mode mixity $\psi$ denotes the ratio of mixed-mode can be related to $K_1$ and $K_2$ factors by

$$K = \sqrt{K_1^2 + iK_2^2} \tag{4}$$

$$\psi = \tan^{-1}\left(\frac{K_2}{K_1}\right) \quad K_1 \geq 0 \tag{5}$$

$$\psi = \pi + \tan^{-1}\left(\frac{K_2}{K_1}\right) \quad K_1 < 0 \tag{6}$$

### 2.2. Calculation of stress intensity factors

The hybrid finite element method (FEM) is very powerful for discontinuous problems in mechanics, such as crack growth, interface problems, and so on. In this paper, the stress singularity at the crack tip was taken into account by placing hybrid element at this field, and the displacement continuity between the regular element and hybrid element was ensured on the integral. The hybrid element stiffness matrix of crack tip based on modified potential energy principle with relaxed continuity requirement was derived. And the displacement field, stress field and then the equation of FEM were obtained further. After the displacement vector of node was computed, the stress intensity factors for interface crack could be obtained.

7 groups of specimens with different combination of $K_1$ and $K_2$ were used in the experiment. The geometry, loading and constraint conditions for each group were shown in Fig. 2. Stress intensity factors per unit load and positions of cracking initiation location were listed in Table 1.

Table 1. Calculation result of stress intensity factors per unit load.

| Specimens No. | Figure No. | $\dfrac{K_1 BW^{1/2}}{P}$ | $\dfrac{K_2 BW^{1/2}}{P}$ | $\dfrac{K_2}{K_1}$ | $\psi/(°)$ |
|---|---|---|---|---|---|
| 1 | Fig.2 (i) | 12.010 | - 0.420 | -0.035 | -2 |
| 2 | Fig.2 (ii) a=10, b=6 | 7.360 | - 2.010 | -0.273 | -15 |
| 3 | Fig.2 (ii) a=10, b=4 | 4.607 | - 2.010 | -0.436 | -24 |
| 4 | Fig.2 (ii) a=10, b=2 | 2.120 | - 1.920 | -0.906 | -42 |
| 5 | Fig.2 (ii) a=10, b=1 | 0.989 | -1.896 | -1.917 | -62 |
| 6 | Fig.2 (ii) a=5, b=0.5 | 0.814 | - 5.420 | -6.658 | -81 |
| 7 | Fig.2 (ii) a=5, b=0 | - 0.778 | - 5.531 | 7.109 | -98 |

Note: B is width of specimen and W is height of specimen. a and b are the positions of loading and constraint conditions respectively. Units of size in the table are centimeter.

## 3.  Fracture Experiment

### 3.1.  *Test specimens*

Rock/concrete interface specimen with single-edge crack, as shown in Fig.1, is used to form the different ratio of mixed-mode in the tests. Unit of size in Fig.1 and Fig.2 is centimeter.

Fig.1. Interface specimen with single-edge crack.

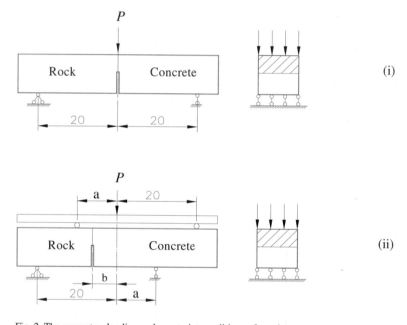

Fig. 2. The geometry, loading and constraint conditions of specimens.

### 3.2.  *Mechanical properties of material and mixture ratio of concrete*

The mechanical properties of material and mixture ratio of concrete were listed in Table 2 and 3.The maximum grain size of aggregate is 20mm.

Table 2. Mechanical properties of material.

| Material | Elastic modulus [MPa] | Unit weight [kN/m³] | Possion Ratio |
|---|---|---|---|
| Concrete | $31.5 \times 10^4$ | 24 | 0.167 |
| Rock | $64.7 \times 10^4$ | 26.8 | 0.210 |

Table 3. Mixture ratio of concrete and its strength.

| Cement [kg/m3] | Sand [kg/m³] | Aggregate [kg/m³] | Water [kg/m³] | Compress Strength [MPa] | Splitting Tensile Strength [MPa] |
|---|---|---|---|---|---|
| 434 | 496 | 1285 | 187 | 36 | 2.84 |

### 3.3. *Specimens preparation*

The cross section of rock specimen was roughed at first, and notches were fabricated in the meantime. After that, the rock specimen was laid in mould and then the concrete was then poured in. The side plates of mould were loosed after 7 days, and specimens were saved in moist room until 28 days for curing.

### 3.4. *Test procedures and experimental results*

The specimen was loaded on a universal material tester. The critical load at fracturing was recorded by an electric recorder. For the interface between rock and concrete is the weakest link, the test procedures showed that the crack propagated along the interface, and the interface fracture situation as shown in Fig.3. Because of the disparity in the concrete properties, six specimens were tested in each of the 7 groups. The finial critical load values $P_f$ were averaged over the maximum loads of six specimens obtained from fracture experiment.

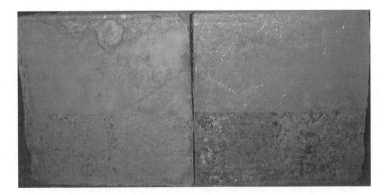

Fig. 3. The interface fracture situation of rock/concrete specimen.

## 4.   Critical Stress Intensity Factors and Fracture Toughness

### 4.1.   *Critical stress intensity factors*

The critical stress intensity factors listed in Table 4 were calculated based on $P_f$ and stress intensity factors per unit load from Table 1. Relationship between $K_{iC}$ and $\psi/(°)$ are shown in Fig.4

Table 4.   Critical stress intensity factors under different ratio of mixed-mode.

| Specimens No. | $P_f$ [kN] | $K_{1C}$ [MPa$\sqrt{m}$] | $K_{2C}$ [MPa$\sqrt{m}$] | $K_{iC} = \sqrt{K_{1C}^2 + K_{2C}^2}$ [MPa$\sqrt{m}$] | $\psi/(°)$ |
|---|---|---|---|---|---|
| 1 | 0.770 | 0.292 | -0.010 | 0.293 | -2 |
| 2 | 1.153 | 0.268 | -0.073 | 0.278 | -15 |
| 3 | 1.740 | 0.253 | -0.111 | 0.277 | -24 |
| 4 | 3.070 | 0.206 | -0.186 | 0.278 | -42 |
| 5 | 4.135 | 0.129 | -0.248 | 0.280 | -62 |
| 6 | 2.033 | 0.052 | -0.349 | 0.352 | -81 |
| 7 | 2.450 | -0.060 | -0.429 | 0.433 | -98 |

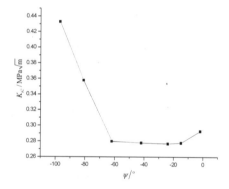

Fig. 4.   Relationship between $K_{iC}$ and $\psi/(°)$ .

Obviously, the critical complex stress intensity factor $K_{iC}$ is not constant, which changes with the mode mixity $\psi$. Therefore, $K_{iC}$ as interface fracture toughness is not appropriate.

### 4.2.   *Fracture criterion and fracture toughness*

It is well known that $K_1$ and $K_2$ for an interfacial crack in dissimilar media are different from those for a crack in a homogeneous material. However, in the meanings of analogy to a crack in a homogeneous material, the $K_1$ and $K_2$ are similar to mode I and II stress intensity factors respectively. Referring to[13, 4-6], parabolic and elliptic experiential fracture criteria of mixed crack were derived respectively. By regressing the critical stress intensity factors of 7 specimen groups, we obtain the following parabolic equation as

$$K_1 K_{1C}^* + 1.329 K_{2C}^* (K_2)^2 = (K_{1C}^*)^2 \tag{7}$$

where $K_{1C}^* = 0.274(\text{MPa}\sqrt{m})$, $K_{2C}^* = 0.384(\text{MPa}\sqrt{m})$.

and elliptic equation as

$$\left(\frac{K_1}{K_{1C}^*}\right)^2 + \left(\frac{K_2}{K_{2C}^*}\right)^2 = 1 \tag{8}$$

in which $K_{1C}^* = 0.270(\text{MPa}\sqrt{m})$, $K_{2C}^* = 0.389(\text{MPa}\sqrt{m})$.

In order to check the relevance of the proposed fracture criteria, the experimental failure data are also reported in Fig.5.

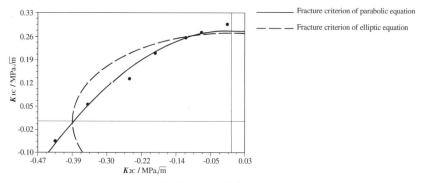

Fig. 5. Criterion in the reduced plan.

It is found that a mixed mode fracture toughness curve can be represented better by a parabolic equation than by an elliptic one. However, the $K_{1C}^*$ and $K_{2C}^*$ obtained from the two fracture criteria are approximate equivalent, and $K_{2C}^* > K_{1C}^*$. Because the $K_1$ and $K_2$ are similar to mode I and II stress intensity factors, $K_{1C}^*$ and $K_{2C}^*$ may be regarded as the effective mode I and II fracture toughness respectively.

## 5. Conclusions

In the present paper, the stress intensity factors of rock/concrete interface with different combinations of $K_1$ and $K_2$ were calculated by the mixed hybrid finite element method, and two experiential fracture criteria were proposed through the experimental investigation, and the fracture toughness( $K_{1C}^*$ , $K_{2C}^*$ ) of rock/concrete were obtained further. The following are conluding remarks of the present paper:

• The hybrid finite element method is effective method for calculating the interface stress intensity factors.

• A mixed mode fracture toughness curve can be represented better by a parabolic equation than by a elliptic one. $K_{1C}^*$ and $K_{2C}^*$ can be regarded as the effective mode I and II fracture toughness respectively, and $K_{2C}^* > K_{1C}^*$ .

• Size effect of interface fracture toughness should be studied further.

## References

1. J.M. Chandra Kishen, Ph.D. Thesis, Univ. of Colorado, Boulder, USA, 1996.
2. P.G. Charalambides, J. Lund, R.M. McMeeking. A test specimen for determining the fracture resistance of biomaterial interfaces, J. Appl. Mech. **56**, 77 (1989).
3. N. P. O'Dowd, C. F. Shih, and M. G. Stout, Test geometries for measuring interfacial fracture toughness, Int. J. Solids and Struct. **29 (5)**, 571 (1992).
4. R. Yuuki, et al, Mixed mode fracture criteria for an interface crack, Eng. and Fract. Mech. **47 (3)**, 367 (1994).
5. L. Xu and H. V. Tippur, Fracture parameters for interfacial cracks: an experimental-finite element study of crack tip fields and crack initiation toughness, Int. J. Fract. **71**, 345 (1995).
6. I. Toru, et al, Mixed mode fracture criterion of interface crack between dissimilar materials, Eng. Fract. Mech. **59(6)**, 725 (1998).
7. Z. Li, et al, Study of fracture criterion of rock/mortar interface crack. J. Dalian University, **37(Suppl.1)**, 57 (1997).
8. S.C. Yang, et al, Experimental investigation on fracture criterion of three-dimensional mixed mode interface crack for rock/concrete, J. Key Eng. Mater. **274-276**, 141 (2004).
9. S.M. Huang, et al, Cracking calculation of arch Dam, J. Xi'an University of Tech. **13(1)**, 18 (1998).
10. S.C. Yang, et al, Application fracture mechanics to arch dam analysis, J. Key Eng. Mater. **261-263**, 57 (2004).
11. J.M. Chandra Kishen and K. D. Singh, Stress intensity factors based fracture criteria for kinking and branching of interface crack: application to dams, Eng. Fract. Mech. 68, 201 (2001).
12. J.R. Rice, Elastic fracture concepts for interfacial cracks, J. Appl. Mech. **55**, 98 (1988).
13. D.Y. Xu, et al, Fracture Criteria of I-II mixed mode of concrete, J. Shuili Xuebao, **6**, 57 (1982).

# INVESTIGATION ON THE EQUIVALENT MATERIAL PROPERTY OF CARBON REINFORCED ALUMINUM LAMINATES

SEUNG-HO SONG[1]

*NCRC, Pusan National University, Busan 609-735, KOREA,*
*mrsong@pusan.ac.kr*

TAE-WAN KU[2], JEONG KIM[3], BEOM-SOO KANG[4†]

*Dept. of Aerospace Engineering, Pusan National University, Busan 609-735, KOREA,*
*longtw@pusan.ac.kr, greatkj@pusan.ac.kr, bskang@pusan.ac.kr*

WOO-JIN SONG[5]

*ILIC, Pusan National University, Busan 609-735, KOREA,*
*woodysong@pusan.ac.kr*

Received 15 June 2008
Revised 23 June 2008

Fiber metal laminates as one of new hybrid materials with the bonded structure of thin metal sheets and fiber/epoxy layers have been developed for the last three decades. These kinds of materials can provide the characteristics of the excellent fatigue, impact and damage tolerance with a relatively low density. Because metal sheets and fiber/epoxy layers are bonded each other, the bonding between two materials is critical. In this study, the bonding strength is investigated experimentally with respect to surface roughness of metal sheets. The equivalent material properties of carbon reinforced aluminum laminates as the input data in the numerical simulation are also investigated and compared with the experimental result. The application of the equivalent material property to the numerical simulation can provide the high degree of efficiency in the build-up of the finite element model and the numerical simulation.

*Keywords*: Fiber metal laminates (FML); Carbon reinforced aluminum laminates (CARAL); Surface roughness; Equivalent material property; Bonding strength.

## 1. Introduction

Weight reduction and improved damage tolerance characteristics were the main drivers to develop new kind of materials for the aerospace/aeronautical industry. Aiming this objective, a new lightweight fiber metal laminates (FML) has been developed. The combination of metal and polymer composite laminates could create a synergistic effect on many properties. The mechanical properties of FML show improvements over the properties of both aluminum alloys and composite materials individually.[1-2] Due to its excellent properties, FML is chosen to serve as skin material in aircraft fuselage and

---

†Corresponding Author.

lower wing skin, and internal parts.[3] Although not commercially available yet, carbon reinforced aluminum laminates (CARAL) consisted of aluminum and carbon/epoxy layers could be expected to be one of next generation of FML.[4] Because high stiffness of carbon fibers allows for efficient crack bridging, crack growth has very low growth rates. At the same time, the impact properties are increased by the existence of metal layer. To extract the maximum performance from CARAL, it was needed to consider many factors which are anisotropic characteristics of carbon/epoxy layers, ply orientations, stacking number and sequences between layers. Equivalent material properties for finite element modeling and its simulation should be needed to obtain high degree of efficiency. Therefore, those of CARAL were calculated by the theoretical, numerical simulation and experiments. Comparison among the results was carried out. From the comparative study, it was found that the numerical approach was in good agreement with the experimental results. Before calculating the equivalent material property, a bonding strength between aluminum and carbon/epoxy layers was experimentally investigated. It is important for reliability of CARAL because metal and composite layers of CARAL are bonded by adhesive film. The bonding strength of adhesive is influenced by the surface roughness of the joining parts.[5-6] Therefore, the bonding strength with respect to the surface roughness of aluminum layer was investigated by bonded joint tests.

## 2.  Bonded Joint Tests

### 2.1.  *Specimen manufacture*

As shown in Fig. 1, a 2mm thick Al 1050, 4 layers of unidirectional carbon/epoxy prepreg (USN-200, SK chemical Inc., Korea), and a layer of adhesive film (BMS5-129, HEATCON Inc., USA) were used for specimens. 6 types of specimen were prepared with respect to surface roughness. For 5 types of the 6 specimens, aluminum surfaces were treated by 220, 320, 400, 600 and 800K mesh abrasive paper and degreased with acetone. For the last type, the aluminum surface was not treated and only degreased with acetone. Fig. 2 shows the micrographs of 6 types of aluminum surface. As the mesh number of abrasive paper increases, the surface roughness decreases. The last micrograph of the surface which was not abraded shows lines along a rolling direction. After the lay-up procedure, the specimens were cured in hot press machine at 120℃ and 7 bar.

### 2.2.  *Bonded joint test method*

This test was carried out in accordance with ASTM D 5868-01 which is a standard test method for lap shear adhesion for fiber reinforced plastic (FRP) and metal bonding.[7] The joint specimens were tested under tensile loading using Instron 8516 at a loading rate of 1.3*mm/min*. The bonded joint specimen and experimental set-up are shown in Fig. 3. SEM photographs of bi-material interface between aluminum and adhesive layer (Fig. 4) had been taken to investigate the bonding status. Circled sections mean the bi-material interface which is well bonded. This shows that the bi-material interface is bonded better as the mesh number of abrasive paper increases.

### 2.3. *Bonded joint test result*

After the tests, significant amount of residual adhesive was found both aluminum and CFRP surfaces. This means that a crack had propagated through the adhesive layer rather than along bi-material interface, and suggests that the adhesive films were successfully bonded to the aluminum and CFRP layers. Fig. 5 shows measured arithmetic surface roughness and joint strength with respect to the mesh number of abrasive paper. It was noticed that the bonded joint strength increased as the surface roughness of aluminum decreased. For the specimen abraded with 220K mesh number of abrasive paper,

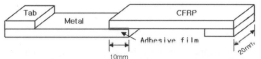
Fig. 1. Geometry of composite bonded single lap joint.

Fig. 2. Microscopic view of abraded aluminum surface structure.

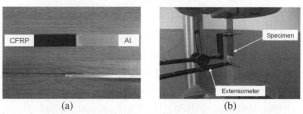
(a)                                        (b)
Fig. 3. Experiment for single-lap bonded joint: (a) single-lap bonded joint specimen: (b) test set-up.

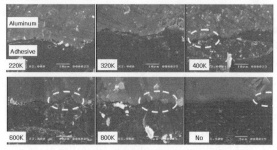
Fig. 4. SEM photographs of bi-material interface between Al and adhesive layer.

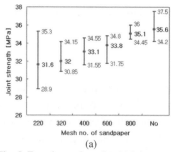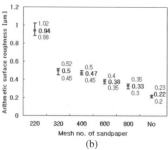

Fig. 5. Experimental results: (a) joint strength: (b) measured Ra (arithmetic surface roughness).

Table 1. Parameters used in mixtures rules.

| Material | Volume fraction | $E_x$ (GPa) | $E_y$ (GPa) | $G_{12}$ (GPa) | $\nu_{12}$ |
|----------|-----------------|-------------|-------------|----------------|------------|
| Al 1050 | 53% | 67.0 | 67.0 | 26.0 | 0.330 |
| CFRP | 47% | 162 | 9.6 | 6.1 | 0.298 |

there was relatively large variation in the joint strength which was undesirable for reliable application of the bonded joint. It is noted that the specimen without pre-treatment derives the highest bonding strength.

## 3. Equivalent Material Properties

### 3.1. *Theoretical, numerical and experimental analyses*

In this study, the equivalent material properties were calculated by the theoretical and numerical analysis. For the theoretical analysis, Eq. (1), the rule of mixtures, was used to calculate elastic properties.

$$E_F = E_C V_C + E_A (1 - V_C)$$ (1)

where $E_F$, $E_C$ and $E_A$ represent Young's modulus of CARAL, CFRP and aluminum, respectively. Also, $V_C$ represents a volume fraction of CFRP. The material properties used for theoretical analysis are shown in Table 1. The equivalent material properties of CARAL were also calculated with finite element analysis by simulating uni-axial tensile tests with commercial software ABAQUS V6.7. 8-node solid elements and 8-node continuum shell elements were used for aluminum and CFRP layers, respectively.[8] Also, only 1/8 of the specimen needed to be modeled due to geometric symmetry of the specimen. The material properties of the aluminum were acquired by a uni-axial tensile test, and those of CFRP were employed as referred reference.[9] Static tensile tests were performed in accordance with ASTM D 3039/D3039M-07 which is a tensile test method for composite materials.[10] Specimens were prepared with the same method mentioned in a bonded joint test section, and tested with the same test machine. Because the aluminum layers without pretreatment leaded the highest bonding strength, the aluminum layers in CARAL were not abraded and only degreased with the acetone. Fig. 6 shows stacking sequence of CARAL and specimens according to fiber directions, 0° and 90°.

## 3.2. *Result and discussions*

The theoretical and numerical elastic constants are shown in Table 2. There is a discrepancy between the theoretical and the experimental results. It was caused by adhesive film and polymer composites since the interface effect and void presence were not considered in the model. Von Mises stress distributions of the tensile test simulation are shown in Fig. 7. Both of two numerical results are in good agreement with the experiment results. Fig. 8 shows engineering stress-strain curves from the tensile tests and FE analysis. From Table 2 and Fig. 8, it is shown that numerical results are in good agreement with experimental results with deviations less than 1.2%. From Fig. 8, it is clear that Al 1050 exhibits a ductile behavior involving small amount of strain, but CFRP exhibits a brittle behavior. Fig. 8(a) also shows that the stress-strain curve for CARAL is predominantly influenced by CFRP because the stress-strain curve of CARAL exhibited similar trend with CFRP's one. On the other hand, the elastic modulus of CARAL, Al 1050 and CFRP were 83.05, 67.0, 162.0 GPa, respectively. Also, the proportional limit of CARAL was about 191.0 MPa, it is related to yield and plastic deformation of Al 1050. Unlike Fig. 8(a), Fig. 8(b) shows that the stress-strain curve of CARAL (90°) is predominantly influenced by Al 1050 because CFRP can sustain almost no load in transverse direction (90°).

Table 2. Equivalent properties of CARAL obtained by analyses.

| Property | Theoretical | Numerical | Experimental |
|----------|-------------|-----------|--------------|
| $E_x$ (GPa) | 111.7 | 83.56 | 83.05 |
| $E_y$ (GPa) | 40.0 | 41.22 | 40.74 |

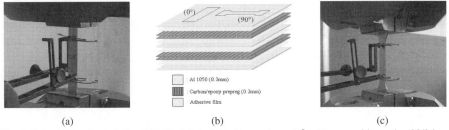

Fig. 6. Uni-axial tension tests for CARAL: (a) straight-side specimen (0°): (b) composition ratio of 3/2 lay-up: (c) dog bone specimen (90°).

Fig. 7. Simulation results: (a) straight-side specimen (0°): (b) dog bone specimen (90°).

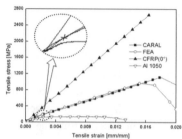 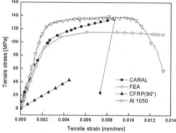

Fig. 8. Engineering stress-strain curves: (a) straight-side specimen (0°), (b) dog bone specimen (90°).

## 4. Conclusion

In this study, the equivalent material properties were calculated by the theoretical and numerical analyses. The calculated equivalent material properties were compared with the experimental results. It was observed that the numerical analysis results were in good agreement with experimental results, but the results of the theoretical analyses were not. The bonded joint strength was also investigated experimentally with respect to surface roughness of the aluminum layer. Because metal and composite layers of CARAL were bonded by adhesive film, it is very important to achieve strong bonding to extract high material performance from CARAL. The result revealed that aluminum layer without pretreatment leaded the highest bonding strength than any other layers. In conclusion, the proposed approach to determine the equivalent material property would play an important role in developing structures with the fiber metal laminates and satisfying increasing practical demands for the aircraft design.

## Acknowledgments

This work has been completed with the support by the National Core Research Center Program from MOST/KOSEF (No. R15-2006-022-02002-0). And the last author would like to acknowledge the partial support of the Korea Science and Engineering Foundation (KOSEF) NRL Program grant funded by the Korea government (MEST) (No. R0A-2008-000-20017-0).

## References

1. A. Vlot and L.B. Vogelsang, *J. Mat. Proc. Tech.*, **103**, 1 (2000).
2. E.C. Botelho, R.S. Almeida, L.C. Pardini, M.C. Rezende, *Int. J. Eng. Sci.*, **45**, 163 (2007).
3. A. Soprano, A. Apicella, L. D'Antonio, F. Schettino, *Int. J. Fatigue*, **18**, 265 (1996).
4. E.C. Botelho, R.A. Silva, L.C. Pardini, M.C. Rezende, *Mat. Res.*, **9**, 247 (2006).
5. D.N. Buragohain and P.K. Ravichandran, *Computers & Structures*, **51**, 289 (1994).
6. K. Uehara and M. Sakurai, *J. Mat. Proc. Tech.*, **127**, 178 (2007).
7. ASTM, D5868-01, 2008.
8. ABAQUS Analysis User's Manual Version 6.7, ABAQUS INC.
9. H.J. Son, J.H. Kwon, J.H. Choi, J.R. Cho, S.R. Cho, in *Structural Engineering and Applications (2) – Proc. KSAS Fall Conference*, ed. I. Lee *et al.* (KSAS, Busan, 2006), pp. 273~277.
10. ASTM, D3039/D3039M-07, 2007.

# MECHANICAL PROPERTIES OF TITANIUM FOAM
# FOR BIOMEDICAL APPLICATIONS

SADAF KASHEF[1], JIANGUO LIN[3], PETER D. HODGSON[4]

*Centre for Material and Fibre Innovation, Deakin University,*
*Waurn Ponds, Victoria, 3217, AUSTRALIA,*
*skas@deakin.edu.au, jianguo.lin@deakin.edu.au, phodgson@deakin.edu.au*


WENYI YAN[2][†]

*Department of Mechanical Engineering, Monash University,*
*Clayton, Victoria, 3800, AUSTRALIA,*
*Wenyi.Yan@eng.monash.edu.au*

Received 15 June 2008
Revised 23 June 2008

Understanding the mechanical behaviour of pure titanium (Ti) foam is crucial for the design and development of Ti foam-based load-bearing implants. In this work, pure titanium foam is fabricated by a powder metallurgical process using the space-holder technique with a spacer size of 500 to 800 μm. Experimental data from static compression testing on the Ti foam are presented. The application of theoretical formulae to predict Young's modulus and yield strength of titanium foams is also discussed. A foam with 63% porosity, 87 ± 5 MPa yield strength, and 6.5 ± 1.3 GPa Young's modulus is found to be appropriate for a number of dental and orthopaedic applications.

*Keywords*: Pure titanium foam; powder metallurgy; compression test; mechanical property.

## 1. Introduction

Solid metal implants with a porous coating have been used broadly in the past to improve the bone-implant interface. These porous coated implants with rough surfaces can prevent encapsulation of the implant and therefore promote long-term interface strength.[1] However, these implants do not completely solve the mechanical mismatch between the implant and bone as they are much stiffer than the bone. This leads to the removal of stress from the bone and causes reduction in the bone mass and resorption of bone; a phenomena known as stress shielding.[2] Therefore, for load bearing applications not only high strength and high toughness are required for the implants, but also a matched stiffness of a biocompatible material to the bone is needed to avoid the stress shielding issue. The stiffness and strength of metal foams can be tailored to be very close to bone by changing the porosity. Hence, metal foams can potentially solve the postoperative problems of fracture and stress shielding of the bone. Titanium (Ti), known as one of the most biocompatible materials,[3] is a suitable material for biomedical applications. Porous Ti foams are flexible, strong, and light weight.[3] Considering these features, open-pore Ti

---

[†]Corresponding Author.

foams could be the most appropriate biocompatible material for implantation. Up to now, a few processes have been emerged for Ti foam production as an implant material. These include plasma-spraying with chemical and thermal treatments,[4] Hot Isostatic Pressing backfilled with argon gas,[5] Mg particles used as spacers[6] and a recently published method of directional freeze-casting by Chino and Dunand.[7] Wen *et al.*[8] and Imwinkelried[9] have used a powder metallurgy method using ammonium hydrogen bicarbonate particles as the spacer to fabricate porous metals; this method has been applied here. Different manufacturing methods can lead to Ti foams with various mechanical properties that can be suitable for one or more orthopaedic/dental applications. This is because the manufacturing parameters in each method can have different effects on the porosity, pore size, and strength of the Ti foam. High porosity foams are achievable with the powder metallurgy method used here. This method is appropriate for materials with a high melting point such as Ti. The pore distribution and pore parameters are also controllable with this simple method.

Hence, to use Ti foam as an implant material, it is important to investigate its tailor-made mechanical properties for specialised implantation at various parts of the body. Simultaneously, the effect of manufacturing parameters (eg. heat treatment) on this class of new porous materials must be considered. For instance, Ti foam has been recently used to manufacture a commercial spinal implant using powder metallurgy method by Synthes GmbH.[9] It is not clear if this method will also be suitable for other biomechanical applications; at present, other than spinal implant,[9] there is no any other commercially available Ti foam implant. In this article, experimental and analytical methods have been used to explore the effect of different parameters and optimize the morphology of this porous material for biomedical applications. The mechanical properties of pure Ti foam with 63% porosity have been examined here by using compression testing and compared against a range of skeletal components. The analytical method is a model based on a cellular network structure by Gibson and Ashby.[10] The results from the analytical method are compared with the experimental data from this work and from literature.

## 2. Experimental Procedures

### 2.1. *Material Fabrication*

Ti powder with the purity of 99.9% with an average particle size of 45μm is used. The Ti powder is first mixed with the ammonium bicarbonate space-holder material, where the particles are 500-800 μm with a purity of 99.0%. The spacer material is chosen based on its chemical properties such that the spacer totally decomposes at a low temperature.[8] The weight ratio of the Ti powder to spacer is chosen to be 60% to 40% to obtain the desired porosity in the foam. The method used here is similar to that applied by Wen *et al.*[1, 8] and Imwinkelried.[9] The steps of the powder metallurgy process are as follows:

  i. Mix Ti powder with spacer particles.

  ii. Cold press at room temperature under about 200 MPa.

  iii. Use heat treatments to remove the spacer and then sinter to obtain the pure Ti foam.

The heating process to make Ti foam is performed in two steps. The first step is to maintain the specimen at 100°C for 10hrs to remove the space-holder and this followed by the sinter stage in argon atmosphere at 1120°C for 7 hrs. The heat treatment cycle affects both the porosity and density.

### 2.2. Static compression test

The compression test of pure Ti foam was performed using cylindrical shape specimens with 15mm in length and 10mm in diameter in an Instron 5567 test frame. The test was carried out on several samples at room temperature with a strain rate of $10^{-3}\,s^{-1}$.

### 3. Analytic Method

An analytical method based on the Gibson and Ashby[10] theory is applied here to investigate the mechanical properties of Ti foam. This method was primarily developed for foams and porous materials based on experimental data of Aluminium (Al) foams. Several assumptions have been set by Gibson and Ashby[10] to derive the theoretical solutions. Gibson and Ashby[10] analysed foam properties in terms of certain parameters such as the Young's modulus ratio, yield strength ratio and porosity. Then the properties are compared with, and calibrated against, experimental data to develop equations appropriate for design.[10]

The Young's modulus of the foam ($E_f$) is related to the Young's modulus of the solid ($E_s$) (Eq. 1) while the yield strength ratio of the metallic foam to solid material in compression is related according to[10] Eq. 2. $\phi$, is the distribution constant and is the volume fraction of solid contained in the cell edges. $\phi=1$ corresponds to fully open cell foams, $\phi=0$ is for fully closed cell foams and $0 < \phi < 1$ represents partially closed cell foams. Eqs. (1) and (2) were derived based on a simple model of open and closed cell foams proposed by Gibson and Ashby.[10] This theoretical model simplifies the complex and irregular microstructure of the metal foams with a hollow cubic cell that has different thickness at the cross section of the edges and faces. These equations are used to calculate $E_f/E_s$ and $\sigma_f/\sigma_s$ for both open and closed cell Ti foams. The calculated Young's modulus ratio and yield strength ratio as a function of material porosity are presented and compared with the experimental results in the next section.

$$\frac{E_f}{E_s} = (\phi\overline{\rho})^2 + (1-\phi)\overline{\rho} \tag{1}$$

$$\frac{\sigma_f}{\sigma_s} = 0.3(\phi\overline{\rho})^{3/2} + (1-\phi)\overline{\rho} \tag{2}$$

### 4. Results and Discussion

The temperature of heat treatment in this work is different from the literature, where the same types of spacer particles are used. Wen *et al.* have removed the space-holder at 200°C after 5 hrs and then sintered at 1200°C for 2 hrs[1, 8] while Imwinkelried removed the space-holder at 95°C after 12 hrs and sintered at 1300°C for 3hrs.[9]  One important issue is the effect of sintering on the foam material  Another distinctive difference in this study is in the size of spacer particles (500-800 μm) compared to 200-500 μm by Wen *et al.*, 425-710 μm by Imwinkelried and 425-600 μm by Esen *et al.* Larger spacer particles give larger pore sizes. In turn, the larger pore size will increase the biological functions of Ti foam but could lower the mechanical properties of the implant material.  The foam density is calculated by measuring the weight and volume of the sample.  The relative density ( $\rho$ ), an important characteristic of the foam, is found from[10] the ratio of $\rho_{foam}$ to $\rho_{solid}$, where the density for Ti[1] is 4.5 g/cm[3].  The porosity, estimated by subtracting relative density from ideal density of 1, is 63% for this fabricated foam.

To investigate the mechanical properties of this Ti foam sample, a static compression test was performed (Fig. 1). The result is compared with the literature data, based on a similar method[1, 8, 9] but with different sintering temperature and spacer particle sizes, [1, 8, 9] and perhaps different sizes of Ti powder.[1, 8, 9] In this study, the Young's Modulus is 6.5 ± 1.3 GPa with a yield strength of 87 ± 5 MPa. The experimental errors were estimated from the results of three test samples.  From the literature, for samples with about 60% porosity, the Young's modulus is reported to be 4.5 GPa[1] or 9 GPa[9] with a Yield strength of 122,[1] and 67.7 MPa.[9]  The mechanical properties of this foam under compression are close to the skull bone where a Young's modulus of 5.6 GPa and yield strength of 96 MPa has been reported.[1]  The Young's modulus of pure Ti foam is also very close to alternate osteon (7.4 ± 1.6 GPa), longitudinal osteon (6.3 ± 1.8 GPa), transversal osteon (9.3 ± 1.6 GPa),[2] cancellous bone of the temporomandibular joint (TMJ) with 7.93 GPa, and femoral condyle (cancellous bone) with 4.9 GPa (Fig. 2).[1]

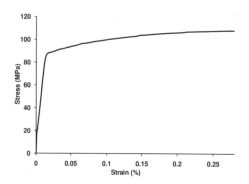

Fig. 1. Compression stress-strain curve for pure Ti foam with 63% porosity.

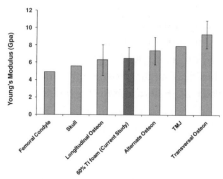

Fig. 2. Young's modulus of different body parts in comparison to 63% Ti foam.

The size and type of the spacer powder, metal powder, the heating temperature, and manufacturing method affect the porosity and mechanical properties of the samples. Fig. 3 demonstrates the data of the Young's modulus ratio of the foam to the solid, which are plotted against porosity. In this figure, the theoretical predictions from Eq. (1) for open-pore foams with $\phi = 1$ and for closed-pore foams with $\phi = 0$ are presented. Experimental data from the literature for a wide range of densities are included and compared with the solid line for open-pore and dash lines for closed or partially closed pore foams from Eq. (1) with values of, $0 \leq \phi \leq 1$. In addition, the Young's modulus ratio from the current study is plotted., In comparison to theoretical values, the Young's modulus ratio of the material in this study is close to that for a distribution constant of 1 (solid line) and therefore could be considered as an open-pore foam. Not all open-pore experimental values are close to the distribution constant of 1. The equation appears to overestimate the actual Young's modulus, possibly due to cell wall folds and broken membranes. This analytical method contains a number of estimates but still gives expressions that are well approximated by the equations. Experimental data supports this result. Therefore, with the known Young's modulus it is possible to use the theoretical model to give an estimate of the pore morphology in Ti foam.

Data for the yield strength of the foams, normalized by the yield strength of solid, are also plotted against porosity in Fig.4. In this figure, the theoretical results for open-pore foams and closed-pore foams are represented by Eq. (2). Experimental data from literature are again included and compared with solid line. The yield strength ratio from current study is also plotted. In comparison to the theoretical values and the literature data, the material in this study is located on the solid line. The data from 11 other studies and this study are all located between distribution constant of 0.4 and 1. This comparison shows that the distribution constant from the theoretical simple model and experiment (i.e. when yield strength ratio and relative density are priori known) are comparable with good approximation. Our foam can again be considered as a fully open pore foam, by analytical predictions presented in Figs. 3 and 4. This is supported by the microstructural observation of the pore morphology as demonstrated. Figs. 5 and 6 shows the SEM microstructure of the pure Ti foam with a relative density of 0.40, irregular cell shape and size, as well as a non-uniform distribution of solid metal in the foamed part. Micro and macro pores enable the body fluid and nutrient transportation and enhance the ingrowth of the new bone tissues through the foam. An open-pore macro-pore also increases the bone ingrowth.[2]

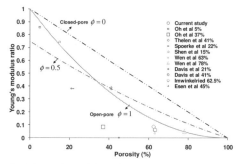

Fig. 3. Young's modulus ratio as a function of porosity.

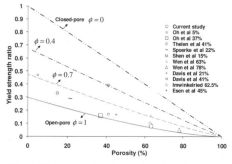

Fig. 4. Yield strength ratio as a function of porosity.

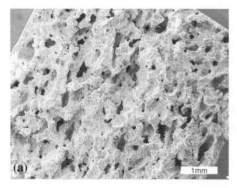

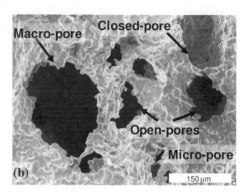

Fig. 5. SEM micrograph of the surface of the Ti foam.          Fig. 6. Close micrograph of some pores.

## 4. Conclusion

It is shown that different manufacturing methods, sintering temperatures, spacer and metal particle size affect the mechanical properties. As different types of bone in the body have different mechanical properties, it is important to choose the best method to tailor the material properties for an implant for a specific body part. The mechanical properties of this foam by compression testing are found to be very close to the skull and therefore, it is a good candidate for such a type of implant. The Young's modulus of the pure Ti foam is also very close to all three types of osteon[2], cancellous bone of the TMJ, and femoral condyle. Therefore, this foam can also be a suitable material for these parts. The analytic predictive approach provided results that are consistent with the experimental findings of this study and literature data. The theoretical model is found to be suitable for estimating the Young's modulus and yield strength of Ti Foams.

## Acknowledgments

This work is financially supported by the Australian Research Council (Project No: DP0770021). The authors thank Dr. Cui'e Wen for her support.

## References

1. C. E. Wen, Y. Yamada, K. Shimojima, et al., *J. Mater. Res.* **17**, 2633 (2002).
2. Y. H. An, in *Mechanical Testing of Bone and the Bone-Implant Interface*, ed. Y. H. An and R. A. Draughn (CRC Press LLC, Florida, USA, 2000), p. 41.
3. G. Vaskiv, G. Gnatyak, and K. Mitina, in *TCSET2002*, Lviv-Slavsko, Ukraine, 2002).
4. M. Takemoto, S. Fujibayashi, M. Neo, et al., *Biomaterials,* **26**, 6014 (2005).
5. N. G. Davis, J. Teisen, C. Schuh, et al., *J. Mater. Res.* **16**, 1508 (2001).
6. Z. Esen and S. Bor, *Scripta Mater.* **56**, 341 (2007).
7. C. Yasumasa and D. C. Dunand, *Acta Mater.* **56**, 105 (2008).
8. C. E. Wen, M. Mabuchi, Y. Yamada, et al., *Scripta Mate.* **45**, 1147 (2001).
9. T. Imwinkelried, in *J. Biomed. Mat. Res. A*, 2007), p. 964.
10. L. J. Gibson and M. F. Ashby, *Cellular Solids: Structure and properties* Cambridge: Cambridge University Press, 1997).

# HOMOGENIZED CREEP BEHAVIOR OF CFRP LAMINATES AT HIGH TEMPERATURE

Y. FUKUTA[1], T. MATSUDA[2†], M. KAWAI[3]

*University of Tsukuba, Tsukuba, JAPAN,*
*y-fukuta@edu.esys.tsukuba.ac.jp, matsuda@kz.tsukuba.ac.jp, mkawai@kz.tsukuba.ac.jp*

Received 15 June 2008
Revised 23 June 2008

In this study, creep behavior of a CFRP laminate subjected to a constant stress is analyzed based on the time-dependent homogenization theory developed by the present authors. The laminate is a unidirectional carbon fiber/epoxy laminate T800H/#3631 manufactured by Toray Industries, Inc. Two kinds of creep analyses are performed. First, 45° off-axis creep deformation of the laminate at high temperature (100°C) is analyzed with three kinds of creep stress levels, respectively. It is shown that the present theory accurately predicts macroscopic creep behavior of the unidirectional CFRP laminate observed in experiments. Then, high temperature creep deformations at a constant creep stress are simulated with seven kinds of off-axis angles, i.e., $\theta = 0°$, 10°, 30°, 45°, 60°, 75°, 90°. It is shown that the laminate has marked in-plane anisotropy with respect to the creep behavior.

*Keywords*: Homogenization; CFRP laminate; creep; high temperature.

## 1. Introduction

CFRP (carbon fiber-reinforced plastics) laminates have high specific stiffness and high specific strength. Thus, the latest aircraft like the Boeing 787, use a large quantity of CFRP laminates in order to save weight. Energy-related industrial products, such as hydrogen vessels for electric vehicles, blades of wind power generators, and so forth, also use CFRP laminates to improve energy efficiency. In these cases, CFRP laminates are expected to encounter considerably severe conditions, i.e., high temperature and high stress, which can cause creep deformation of the laminates. Such creep deformation may result in microscopic failures and macroscopic fractures of the laminates. It is therefore necessary to develop a simulation technique for accurate prediction of the creep behavior of CFRP laminates.

Macroscopic creep behavior of CFRP laminates is microscopically attributable to creep deformation of matrix materials rather than fibers. Thus, a multi-scale simulation which can deal with such microscopic information about constituents of laminates, as well as the macroscopic behavior of the laminates themselves, is considered a useful approach for creep analysis of CFRP laminates. The time-dependent homogenization theory developed by the present authors[1,2] is one of the most promising theories for such multi-scale creep simulation of CFRP laminates. This theory is based on the

---

†Corresponding Author.

mathematical homogenization theory,[3,4] and is capable of analyzing not only the macroscopic time-dependent behavior of composites but also the evolution of microscopic stress and strain fields in the composites. Therefore, the present authors[5,6] have already applied their theory to analysis of the elastic-viscoplastic behavior of CFRP laminates, and showed that the analysis was successful in predicting the elastic-viscoplastic behavior of the laminates, quantitatively. This implies that the theory in question may also provide accurate predictions of the creep behavior of CFRP laminates. However, so far, there has been little research dealing with creep analysis of CFRP laminates using the homogenization theory.

In this study, the time-dependent homogenization theory[1,2] is employed for creep analysis of a CFRP laminate. First, 45° off-axis creep deformation of the laminate at high temperature is analyzed using three kinds of creep stress levels, respectively. Then, high temperature creep deformation at a constant creep stress is simulated with seven kinds of off-axis angles.

## 2. Homogenization Theory

In this section, the time-dependent homogenization theory developed by the present authors[1,2] is briefly explained.

Let us consider a composite that has a periodic internal structure comprised of at least two constituents, and let us define a unit cell $Y$ and Cartesian coordinates $y_i$ $(i = 1, 2, 3)$ as shown in Fig. 1. It is supposed that each constituent in $Y$ obeys an elastic-creep constitutive relation

$$\dot{\sigma}_{ij} = c_{ijkl}(\dot{\varepsilon}_{kl} - \beta_{kl}),$$
(1)

where $\dot{\sigma}_{ij}$ and $\dot{\varepsilon}_{kl}$ indicate microscopic stress and strain rates in $Y$, respectively, $c_{ijkl}$ denotes elastic stiffness, and $\beta_{kl}$ represents creep strain rate and vanishes if elastic. Moreover, the volume average in $Y$ is defined as

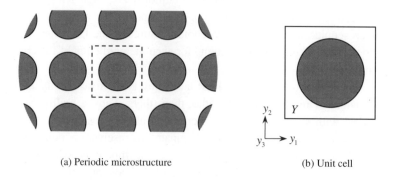

(a) Periodic microstructure                    (b) Unit cell

Fig. 1.   Composite with periodic internal structure and unit cell $Y$.

$$\langle \# \rangle = \mid Y \mid^{-1} \int_Y \# dY \,, \tag{2}$$

where $\mid Y \mid$ indicates the volume of $Y$.

Then, the evolution equation of microscopic stress $\sigma_{ij}$ and the relation between macroscopic stress rate $\dot{\Sigma}_{ij}$ and strain rate $\dot{E}_{kl}$ can be derived as follows:

$$\dot{\sigma}_{ij} = c_{ijpq}(\delta_{pk}\delta_{ql} + \chi_{p,q}^{kl})\dot{E}_{kl} - c_{ijkl}(\beta_{kl} - \varphi_{k,l}) \,, \tag{3}$$

$$\dot{\Sigma}_{ij} = \left\langle c_{ijpq}(\delta_{pk}\delta_{ql} + \chi_{p,q}^{kl}) \right\rangle \dot{E}_{kl} - \left\langle c_{ijkl}(\beta_{kl} - \varphi_{k,l}) \right\rangle \,, \tag{4}$$

where $(\ )_{,i}$ stands for the differentiation with respect to $y_i$, $\delta_{ij}$ signifies Kronecker's delta, and $\chi_i^{kl}$ and $\varphi_i$ denote the characteristic functions determined by solving boundary value problems

$$\int_Y c_{ijpq}\chi_{p,q}^{kl}v_{i,j}dY = -\int_Y c_{ijkl}v_{i,j}dY \,, \tag{5}$$

$$\int_Y c_{ijpq}\varphi_{p,q}v_{i,j}dY = \int_Y c_{ijkl}\beta_{kl}v_{i,j}dY \,. \tag{6}$$

Here, $v_i$ indicates an arbitrary field of perturbed velocity satisfying the $Y$-periodicity with respect to $Y$. The above boundary value problems can be solved using FEM by imposing the $Y$-periodicity of $\chi_i^{kl}$ and $\varphi_i$ on the boundary of $Y$.

## 3. Creep Analysis of Unidirectional CFRP Laminate

Using the homogenization theory described above, creep analysis of a unidirectional carbon fiber/epoxy laminate T800H/#3631 manufactured by Toray Industries, Inc. was performed.

### 3.1. *Analysis conditions*

Two kinds of creep analyses were conducted. First, 45° off-axis creep behavior of the laminate at high temperature (100°C) was analyzed with three kinds of creep stress levels, i.e. 35, 51 and 68MPa, respectively. Then, high temperature creep deformation of the laminate at a constant creep stress (70MPa) were simulated with seven kinds of off-axis angles, i.e., $\theta = 0°$, 10°, 30°, 45°, 60°, 75° and 90°. The off-axis angle means the angle between the fiber axis (the $y_3$-direction in Fig. 2) and a loading direction which is in the $y_1$-$y_3$ plane. Creep time was prescribed to be five hours.

### 3.2. *Fiber arrangement*

The arrangement of carbon fibers, which was unidirectional in the $y_3$-direction, was modeled to be hexagonally periodic on the $y_1$-$y_2$ plane as illustrated in Fig. 2(a).[5,6] Thus, a unit cell Y was defined to be hexagonal as shown in Fig. 2(b). The volume fraction of fibers $V_f$ was taken to be 51.7%. Here, it is noted that $Y$ was taken to be 2D rather than 3D because the laminate is assumed to have no microscopic variation in the fiber direction.[5,6] Since the hexagonal unit cell has point symmetry with respect to the cell center, it is sufficient to consider half of the unit cell as the domain of analysis for solving

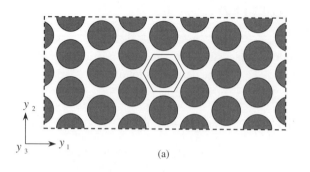
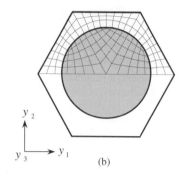

(a)                                                                                          (b)

Fig. 2.   Microstructure of unidirectional CFRP laminate; (a) hexagonal arrangement of fibers, (b) unit cell $Y$ and finite element mesh.

the boundary value problems Eqs. (5) and (6).[5-7] Hence, the upper half of the unit cell was considered and divided into finite elements using four-node isoparametric elements (Fig. 2(b)).

### 3.3. *Microscopic constitutive equations*

The carbon fibers were regarded as transversely isotropic elastic materials. Consequently, the fibers were supposed to have five independent elastic constants, i.e., two Young's moduli $E_{f1}$ and $E_{f3}$, two Poisson's ratios $v_{f12}$ and $v_{f31}$, and one shear rigidity $G_{f31}$, where the subscripts 1, 2 and 3 signify the $y_1$, $y_2$ and $y_3$ directions, respectively. Table 1 shows the five constants of fibers employed in the present work; $E_{f3}$ was provided by the manufacturer, while other constants were obtained by referring to Ref. 8.

The epoxy matrix, on the other hand, was regarded as an elastic-creep material characterized as[5,6]

$$\dot{\varepsilon}_{ij} = \frac{1+v_m}{E_m}\dot{\sigma}_{ij} - \frac{v_m}{E_m}\dot{\sigma}_{kk}\delta_{ij} + \frac{3}{2}\dot{\varepsilon}_0^p \left[\frac{\sigma_e}{g(\bar{\varepsilon}^p)}\right]^n \frac{s_{ij}}{\sigma_e}, \qquad (7)$$

where $E_m$, $v_m$ and $n$ are material constants, $g(\bar{\varepsilon}^p)$ is a material function depending on accumulated viscoplastic strain $\bar{\varepsilon}^p = \int [(2/3)\beta_{ij}^{(\alpha)}\beta_{ij}^{(\alpha)}]^{1/2}dt$, $\dot{\varepsilon}_0^p$ is a reference strain rate, $s_{ij}$ indicates deviatoric part of $\sigma_{ij}$, and $\sigma_e = [(3/2)s_{ij}s_{ij}]^{1/2}$. The material constants and the material function were determined by simulating $45°$ off-axis tensile tests of T800H/#3631 at $\dot{E}_{45°} = 1.0$ and 0.01 %/min. This was because the effect of matrix

Table 1.   Material constants.

| Carbon fiber | $E_{f1} = 1.58 \times 10^4$ | $E_{f3} = 2.94 \times 10^5$ | $G_{f31} = 1.97 \times 10^4$ |
|---|---|---|---|
|  | $v_{f12} = 0.49$ | $v_{f31} = 0.28$ |  |
| Epoxy | $E_m = 3.6 \times 10^3$ | $v_m = 0.35$ | $\dot{\varepsilon}_0^p = 1.67 \times 10^{-4}$ |
|  | $n = 35$ | $g(\bar{\varepsilon}^p) = 115.4(\bar{\varepsilon}^p)^{0.185} + 10$ |  |

MPa (stress), mm/mm (strain), s (time).

viscoplasticity was expected to be significant in such off-axis tests. The material parameters determined are listed in Table 1.

### 3.4. *Results of analysis*

Figure 3 shows 45° off-axis creep curves of the unidirectional CFRP laminate at high temperature (100°C). The stress levels are three kinds, i.e. 35, 51 and 68MPa, which correspond to experiments in Ref 9. In the figure, the solid lines indicate analysis results obtained by the present theory, whereas the markers stand for experimental data.[9] As seen from the experimental data, the creep strain becomes larger as the creep stress level increases, showing the clear stress dependence of in-plane creep behavior of the laminate. Such creep behavior is successfully predicted by the present theory.

Next, high temperature (100°C) creep behavior of the laminate with seven kinds of off-axis angles at a constant stress is shown in Fig. 4. It can be found from the figure that the creep strain hardly occurs at $\theta=0°$ and 10°. By contrast, the creep strain suddenly increases at $\theta-30°$, and can be clearly observed. The creep strain increases as the off-axis angle increases, and reaches the maximum at $\theta=45°$. After that, the creep strain decreases as the off-axis angle increases. These results indicate marked in-plane creep anisotropy of the unidirectional CFRP laminate.

## 4. Conclusions

In this study, the time-dependent homogenization theory developed by the present authors was employed for analyzing creep behavior of a unidirectional carbon fiber/epoxy laminate. It was shown that the theory accurately predicted off-axis creep

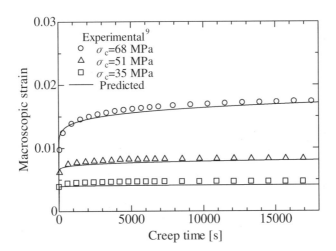

Fig. 3.  Macroscopic creep behavior of unidirectional CFRP laminate at three kinds of creep stress levels (100°C, $\theta = 45°$).

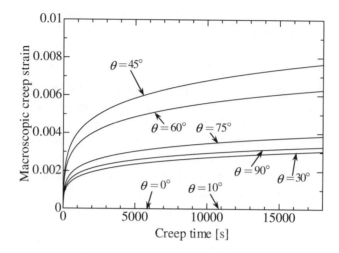

Fig. 4.   Macroscopic creep curves for unidirectional CFRP laminate at constant creep stress with various off-axis angles (100°C).

behavior of the laminate at high temperature. It was also shown that the laminate had marked in-plane creep anisotropy.

## Acknowledgments

The authors acknowledge support, in part, from the Ministry of Education, Culture, Sports, Science and Technology, Japan under a Grant-in-Aid for Young Scientists (B) (No. 20760064).

## References

1.  X. Wu and N. Ohno, *Int. J. Solids Struct.* **36**, 4991 (1999).
2.  N. Ohno, X. Wu and T. Matsuda, *Int. J. Mech. Sci.* **42**, 1519 (2000).
3.  A. Bensoussan, J.L. Lions and G. Papanicolaou, *Asymptotic Analysis for Periodic Structures* (North-Holland Publishing Company, Amsterdam, 1978).
4.  E. Sanchez-Palencia, *Non-Homogeneous Media and Vibration Theory, Lecture Notes in Physics, No. 127* (Springer-Verlag, Berlin, 1980).
5.  T. Matsuda, N. Ohno, H. Tanaka and T. Shimizu, *JSME Int. J.* **A45**, 538 (2002).
6.  T. Matsuda, N. Ohno, H. Tanaka and T. Shimizu, *Int. J. Mech. Sci.* **45**, 1583 (2003).
7.  N. Ohno, T. Matsuda and X. Wu, *Int. J. Solids Struct.* **38**, 2867 (2001).
8.  R.D. Kriz and W.W. Stinchcomb, *Exper. Mech.* **19**, 41 (1979).
9.  M. Kawai and Y. Masuko, *Compos. Sci. Technol.* **64**, 2373 (2004).

# INVESTIGATION ON THERMAL PROPERTIES OF Al/SiC$_p$ METAL MATRIX COMPOSITE BASED ON FEM ANALYSIS

## EUSUN YU[1], JEONG-YUN SUN[2], HEE-SUK CHUNG[3], KYU HWAN OH[4†]

*School of Material Science and Engineering,*

*Seoul National University, Seoul 151-742, KOREA,*

*youes86@snu.ac.kr, scrollin@snu.ac.kr, jguitar1@snu.ac.kr, kyuhwan@snu.ac.kr*

Received 15 June 2008
Revised 23 June 2008

Computational simulations on the thermal analysis of metal matrix composite (MMC) composed of Al and SiC were performed in extended areas of SiC volume fraction. Due to the experimental limitations, only the narrow range of SiC volume fraction has been examined. Through the simulation, which enables current experimental situation to extend, we attempted to explore the dependencies of thermal and mechanical properties on changing the value of volume fraction ($V_f$). To calculate the coefficient of thermal expansion (CTE), variables with temperature and $V_f$ were given in a range from 25℃ to 100℃ and 0 to 100%, respectively. We obtained quantitative results including CTE as a function of $V_f$, which are in a good agreement with previous experimental reports. Furthermore, the stress analysis about thermally expanded MMC was performed. At low volume fraction of SiC, the thermal expansion caused the tensile stress at Al near the interface. However, as the volume fraction of SiC was increased, the stress turned to be compressive, it's because the linked SiC particles contracted the expansion of Al. The MMC of Al matrix face centered cubic site SiC particles has more stress evolutions than the MMC of Al matrix simple cubic site SiC particles at same volume fraction.

*Keywords*: Coefficient of thermal expansion (CTE); Metal matrix composite (MMC); Al/SiC; FEM.

## 1. Introduction

Metal matrix composites (MMC) are active field in industrial applications, such as aerospace structure, automotive and so on[1-3], due to its enhanced thermal and physical properties. By adding strengthening components into metal matrix, it has been known that the enhanced characteristics are resulted from the reinforcement fraction in a given matrix materials, such as Mg/SiCp, Mg/Nip.[4,5] Among the various MMC systems, MMC including aluminum as a matrix has been extensively studied for applying to high strengthening materials system.[6,7] Especially, aluminum is well known as a matrix material with a large coefficient of thermal expansion. Thus SiC particles in Al matrix has been considered as a role of CTE reduction in Al/SiCp system.[8,9] Understanding the thermal properties, for instance, CTE and thermal stress is required to estimate the

---

†Corresponding Author.

thermal stability.[10,11] Finite element method (FEM) is a powerful tool for simulating thermal and mechanical behavior of material. That supplies an institutional analysis taking advantages of graphical and numerical post-processes. Moreover, it helps systematic analysis of material behaviors and properties, including investigation of local stress and strain distribution. Nevertheless, there are few reports of FEM study on the thermal properties of Al/SiCp system compared to that of the experimental researches.

In this work, we investigated the CTE and thermal stress of Al/SiCp MMC system using computational FEM analysis, which enables experimental information to extend over current Al/SiCp MMC system. Two models are prepared – one is SiC inclusions in Al matrix and, the other is Al in SiC matrix. The dispersion conditions are varied following face-centered-cubic (FCC) and simple cubic (SC) to study the contribution of the geometric arrangement. Empirical equations are obtained from the FEM results. They agree quite well with the experimental results from Zhao and Zhang.[8,9]

## 2. Simulation Model

Currently, the control manner of SiCp in Al matrix is not only limited over a large regime due to inter-action between reinforcement material and matrix but also can not show representative nature owing to limited controlled regime over all possible MMC fabrication. In this regard, Computational FEM analysis was carried out for SiCp in Al4032 (a Si-doped aluminum) matrix, which can take into account both thermal and stress enhancers. There have been previous experimental reports, such as the rule of mixture or Kerner's model[12] and Turner's model[13] estimating the behavior CTE as a function of SiC volume change. To extend experimental information, the computational FEM method on a variety of composite material systems allows MMC fabrication to be fruitful with empirical results and computational manners. In this work, we adopted the simulation model based on Kang et al.[14] We assumed a perfect simple cubic system, which is negligible on defects such as pores, with particles arranged in simple cubic structure. Structures of FCC models are also taken into concern. The overall stress distribution was readily confirmed by 3-D structure. FEM calculations are used to investigate displacement due to thermal expansion. Generally, MMC is composed of a metal matrix and reinforcing particles. From the previous reports,[8,9] the arrangement of reinforcement material could play an important role in changing of CTE. In this regard, Three-dimensional simulation was carried out to compute the coefficient of thermal expansion of Al4032/SiCp composite, which was designed to measure line expansion coefficient [Fig.1(a) and (b)]. Fig.1(a) and (b) show that the simulation was carried out for the SiCp with simple cubic(SC and face centered cubic(FCC) arrangement in Al4032, respectively.

For the analysis of metal matrix composite, many researchers suggested the analysis of unit cell of composite.[10,14] Generally, there are computational difficulties to obtain reasonable results based on a small single unit owing to a lack of interaction between reinforcement particle and matrix, on the contrary, the computation with multiple unit cells allows reliable results due to considerable material interaction. In this study, in order to obtain more reliable results, CTE was measured by calculating the change in lattice constant considering close neighbor unit cells.

(a)                                    (b)

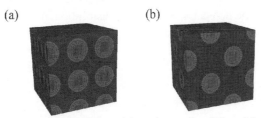

Fig. 1. FEM models for calculation of the coefficient of thermal expansion. SiC particles were located in simple cubic site (a) and face centered cubic site (b).

To successfully predict experiment data, the finite element structure and necessary boundary conditions should be reasonable. SiC particles were put in an array of simple cubic structure, while its size was neglected. The adopted model is designed with Al matrix and SiC reinforcement, however, for the case where SiC volume is larger than that of the metal, MMC model can be reversed, so that the matrix is SiC and metal becomes the particle. In this case, the 3-D model mesh was converted into a finite element mesh with commercially available ABAQUS 6.5.

To find the effect of volume fraction, the Al matrix with SiC reinforcement was used to achieve 9~50% SiC volume fraction and the opposite case was considered to achieve 60~90% SiC volume fraction.

To analyze the CTE, the temperature of finite element mesh was increased from initial temperature of 25 ℃ to 100 ℃ during simulation.

Table 1. Mechanical properties of Al 4032 and SiC [15].

| Material | Temperature (℃) | CTE (ppm/℃) | Young's modulus (Gpa) | Yield Stress (MPa) |
|---|---|---|---|---|
| Al 4032 | 25 | 19.4 | 79 | 315 |
| | 200 | 20.2 | - | 62 |
| | 300 | 21.0 | - | 24 |
| SiC | - | 2.72 | 20.8 | 3440 |

Material properties with pronounced dependence on temperature are used in this simulation. Al4032 and SiC were considered as an isotropic, perfectly elastic material. The material properties are summarized in Table 1.

## 3. Result and Discussion

### 3.1. *Coefficient of thermal expansion analysis*

When SiCp was added in Al matrix with increasing SiCp volume fraction, the CTE value was decreased linearly. The reason that this result is smaller than that of Al, subsequently, SiCp become suppressed state resulting in the difference of their CTE values. From results in Fig. 2 (b), the model conditions between simple cubic and face centered cubic are considered to have a very similar behavior due to few difference of that CTE results. CTE results in this work are evaluated lower 8~ 9 % than experimental data due to ignore of the defects, such as pores, dislocation in MMC [8, 9].

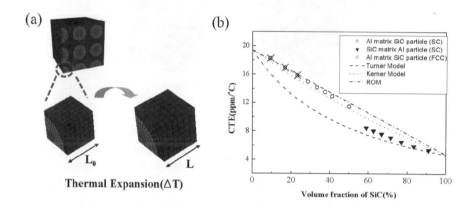

Fig 2. (a) The method to measure the CTE of MMC. To remove the free edge effect, two more shells were calculated and CTEs of MMC were measured in core cell. (b) Calculated CTEs of MMC for simple cubic site SiC, simple cubic site Al, and face centered cubic site SiC. The CTEs of simple cubic site SiC and face centered cubic site SiC MMCs were measured almost same at the volume fraction of 9~ 23% and they showed good matching to ROM. However, at the high volume fraction of SiC, the CTE curve moves like Turner and Kerner models.

As mentioned earlier, the CTE of Al matrix with SiC particles becomes larger than that of the opposite case. Fig. 2(a) reveals that the prototype results varied as a function of temperature. The left image in Fig.2(a) displays the simulated model with non-deformed condition given at 25 ℃ and the right image shows the deformed model with thermal stress up to 100 ℃.

$$CTE = 19.4 - 0.164 \times V_f. \tag{1}$$

In order to obtain qualitative CTE results, we extract the fitted equation (1) from previous results[8,9] dealing with CTE, with unit of ppm/ ℃, and $V_f$ the Volume fraction. The designed model is based on the rule of mixture.

For the opposite case of SiC matrix, when the volume fraction of SiCp is above 50%, the CTE decreased linearly. The obtained tendency on computational manner was close to that predicted from Kernal's model and Turnal's model. The bulk modulus is thought to be a factor to affect this model and the linearity is not observed any more in SiC matrix of Fig. 2(b). However, SiC volume fraction is, indeed, main factor contributing to the CTE of MMC. In this case, the CTE is a function represented as a $2^{nd}$ order polynomial, which is written as

$$CTE = 19.4 - 0.239 \times V_f + 9.310 \times 10^{-4} \times V_f^2. \tag{2}$$

Using this equation, we can predict the CTE's of other regimes as a function of $V_f$.

### 3.2. Stress analysis

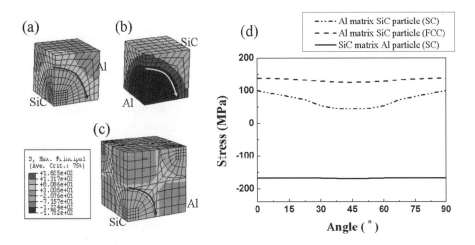

Fig 3. The stress concentration contours of thermally expanded MMC (SiC volume fraction = 30% and Δ T=75 ℃) (a) Al matrix, simple cubic site SiC particles, (b) SiC matrix, simple cubic site Al particles, (c) Al matrix, face centered cubic site SiC particles. (d) The principle stress for Al through the path.

When the MMC was heated, expansion and deformation occurred steadily. Indeed, thermal stress is induced by the difference of lattice constants between the matrix and the particles. Maximum principal stress on each case corresponding to a different cubic arrangement of SiCp is displayed with Al matrix [Fig.3(a)], it shows that the thermal expansion induced the tensile stress at Al near the interface. Fig. 3(b) displays SiC/Alp matrix in simple cubic, where the stress turned to compressive stress since the linked SiC particles contracted the expansion of Al at the volume fraction of increased SiC. For Al/SiCp matrix in FCC [Fig.3(c)], The MMC of Al matrix face centered cubic site SiC particles has more tensile stress evolutions than the MMC of Al matrix simple cubic site SiC particles at same volume fraction. Fig.3(d) shows the computed thermal stress results compared with [Fig.3(a)] and [Fig.3(b)], additionally, the measured difference between [Fig.3(a)] and [Fig.3(c)] is also shown in Fig.3(d). The reason on the difference between [Fig.3(a)] and [Fig.3(c)] can be explained with their arrangement including each different CTE. From this viewpoint of structure arrangement, the more packed structure, such as FCC, allows the stress to increase owing to enhanced stress field induced by surrounding atoms.

Additional consideration on the Fig.3(a) and (b) were described as follows: In case of [Fig.3(a)], the smaller CTE of SiC in heated state renders Al matrix expansion so that plus value of thermal stress were measured. On the other hands, in case of [Fig.3(b)], Al thermal stress in Al/SiC matrix can be obtained with compressed value. (i.e. minus value) because Al particles are isolated  in SiC matrix even though Al has larger CTE rather than SiC matrix.

## 4.  Conclusion

In this work, we have demonstrated the thermal properties of Al/SiCp MMC over all mixing composition via FEM analysis manner. To calculate their coefficient of thermal expansion with FEM analysis, we adopted SC and FCC cubic system composed of Al/SiCp, which have the differences of structure having MMC. The obtained results are in a good agreement with previous experimental reports. Additionally, the calculated CTE values also were confirmed with a qualitative extracted equation. We expect that our approach using computational FEM method facilitates prediction to get promising thermal properties on a variety of MMC.

## Acknowledgments

This work was supported by the Agency for Defense Development and Defense Acquisition Program Administration, and partially supported by the Korea Science and Engineering Foundation (KOSEF) through the SRC/ERC Program of MOST/KOSEF R11-2005-065 (Chung).

## References

1. M. F . Ashby, *Acta Metall. Mater.* **41**, 1313 (1993).
2. N. Chandra, C.R. Ananth, *Compos. Sci. Technol.* **54**, 87 (1995).
3. S. Lemieux, S. Elomari, *J.Mater. Sci.* **33**, 4381 (1998).
4. C.Y.H. Lim, S.C. Lim, M. Gupta, *Wear,* **255**, 629 (2003).
5. S. F. Hassan, M. Gupta, *J. Mater. Sci* **37**, 2467 (2002).
6. Narasimalu Srikanth, Manoj Gupta, *Acta Mater.* **54**, 4553 (2006).
7. A. B. Pandey, B. S. Majumar and D. B. Miracle, *Acta Mater.* **49**, 405 (2001).
8. L.Z. Zhao, M.J. Zhao, X.M. Cao, C. Tian, W.P. Hu, J.S. Zhang, *Compos Sci Technol* **67**, 3404 (2007).
9. Q. Zhang, G. Wu, G. Chen, L. Jiang, B. Luan, *Compos. Part A-Appls.* **34**, 1023 (2003).
10. N. Chawla, V.V. Ganesh, B. Wunsch. *Scripta Mater.* **51**, 161 (2004).
11. N. Chawla, X. Deng, D.R.M. Schnell, *Mat. Sci. Eng.* **426**, 314 (2006).
11. E. H. Kerner, *Proc Phys Soc* **69 B**, 808 (1956).
12. P. S. Turner, *J. Res. Nat. Bureau Stand.* **37**, 239 (1946).
13. C.G. Kang, J.H. Lee, S.W. Youn, J.K. Oh. *J. Mater. Process. Tech.* **166**, 173 (2005).
14. L. C. Davis, C. Andres, J. E. Allison, *Mat. Sci. Eng. A,* **249** 40 (1998).
15. ASM Handbook, Vol 2 (1991), 10th edn. ASM International.

# MULTI-SCALE CREEP ANALYSIS OF PLAIN-WOVEN LAMINATES USING TIME-DEPENDENT HOMOGENIZATION THEORY: EFFECTS OF LAMINATE CONFIGURATION

K. NAKATA[1], T. MATSUDA[2†], M. KAWAI[3]

*University of Tsukuba, Tsukuba, JAPAN,*
*k-nakata@edu.esys.tsukuba.ac.jp, matsuda@kz.tsukuba.ac.jp, mkawai@kz.tsukuba.ac.jp*

Received 15 June 2008
Revised 23 June 2008

In this study, multi-scale creep analysis of plain-woven GFRP laminates is performed using the time-dependent homogenization theory developed by the present authors. First, point-symmetry of internal structures of plain-woven laminates is utilized for a boundary condition of unit cell problems, reducing the domain of analysis to 1/4 and 1/8 for in-phase and out-of-phase laminate configurations, respectively. The time-dependent homogenization theory is then reconstructed for these domains of analysis. Using the present method, in-plane creep behavior of plain-woven glass fiber/epoxy laminates subjected to a constant stress is analyzed. The results are summarized as follows: (1) The in-plane creep behavior of the plain-woven GFRP laminates exhibits marked anisotropy. (2) The laminate configurations considerably affect the creep behavior of the laminates.

*Keywords*: Multi-scale; homogenization; creep; plain-woven laminate; laminate configuration.

## 1. Introduction

In recent years, plain-woven laminates made of plain fabrics and polymeric materials have been used in many high-end industrial products such as components of aircraft and vehicles, and blades of wind power generators. In these situations, plain-woven laminates can be subjected to high stress and high temperature, bringing about creep deformation of the plain-woven laminates. Thus, creep analysis of plain-woven laminates is now an extremely important subject area. Such analysis, however, generally involves difficulties because plain-woven laminates have considerably complex internal structures comprised of fiber bundles and matrix materials (see Fig. 1), and these materials have completely different creep properties, respectively.

To analyze the inelastic behavior of plain-woven laminates with the above-mentioned complex internal structures, a multi-scale simulation technique based on the mathematical homogenization theory[1,2] is highly advantageous, because it enables us to analyze the microscopic stress and strain distributions in laminates as well as the macroscopic behavior of the laminates themselves. In fact, this technique has been already applied to microscopic failure propagation analyses of plain-woven GFRP laminates subjected to an

---

†Corresponding Author.

in-plane on-axis load.[3-5] In addition, the present authors[6] also performed elastic-viscoplastic analysis of plain-woven GFRP laminates using the time-dependent homogenization theory developed by the authors.[7,8] However, so far there is no report in which the creep analysis of plain-woven laminates has been done from a multi-scale point of view.

In this study, multi-scale creep analysis of plain-woven glass fiber/epoxy laminates subjected to a constant stress is performed using the time-dependent homogenization theory.[7,8] Moreover, the effects of laminate configurations of plain fabrics on the creep behavior of the laminates are also analyzed.

## 2.  Homogenization Theory for Plain-Woven Laminates

### 2.1.  *Domain of analysis*

In this study, two patterns of laminate configurations of plain fabrics, i.e. in-phase and out-of-phase, are employed as internal structures of plain-woven laminates.[3-6] The in-phase laminate configuration has no offset of plain fabrics in the $y_1$- and $y_2$-directions (Fig. 1(a)), while the out-of-phase laminate configuration has the phase shift of plain fabrics by $\pi$ in the $y_1$- and $y_2$-directions (Fig. 1(b)). As shown in the previous paper,[6] the plain-woven laminates with the above two laminate configurations have point-symmetric internal structures and the point-symmetry is able to be utilized as a boundary condition for boundary value problems[9], so that we can reduce the domain of analysis.

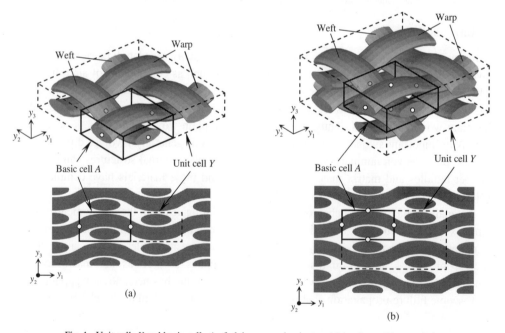

Fig. 1.  Unit cells $Y$ and basic cells $A$ of plain-woven laminates; (a) in-phase, (b) out-of-phase.

This method is briefly explained in the following two paragraphs.

First, for an in-phase laminate configuration, a unit cell $Y$ of the laminate is defined as shown by the dashed lines in Fig. 1(a). In addition, a basic cell $A$ is also defined as indicated by the solid lines in Fig. 1(a). A careful look at the figure reveals that the internal structure of the laminate has point-symmetry with respect to the centers of lateral facets of $A$, which are denoted by the open circles in Fig. 1(a). Perturbed velocity in the laminate, therefore, distributes point-symmetrically with respect to these points. By contrast, the perturbed velocity at the top and bottom facets of $A$ satisfies the $Y$-periodicity because the internal structure is periodic with respect to $A$ in the $y_3$-direction (stacking direction). The use of the point-symmetry and $Y$-periodicity of perturbed velocity as a boundary condition with the previous results[9] enables us to take $A$ as the domain of analysis and to derive the boundary value problems (4) and (5) shown in section 2.2.

Next, with an out-of-phase laminate configuration, a unit cell $Y$ is taken as indicated by the dashed lines in Fig. 1(b). The unit cell has twice the volume of the in-phase laminate configuration, but the same basic cell $A$ is again defined as shown by the solid lines in Fig. 1(b). As indicated in the figure, the internal structure of the laminate has point-symmetry with respect to the centers of the top and bottom facets as well as the lateral facets of $A$, i.e., the centers of all the boundary facets of $A$, which are denoted by the open circles in Fig. 1(b). Thus, the perturbed velocity in the laminate distributes point-symmetrically with respect to these points. Employing the point-symmetric distribution of perturbed velocity as a boundary condition, the same boundary value problems (4) and (5) as for the in-phase laminate configuration are obtained.

## 2.2. Homogenization theory

The basic cell $A$, mentioned above, is then employed as the domain of analysis for the time-dependent homogenization theory.[7,8] Let us consider that a plain-woven laminate is subjected to macroscopically uniform load and exhibits infinitesimal deformation both macroscopically and microscopically. The constituents of the laminate are assumed to have elastic-creep properties and obey the following constitutive equation:

$$\dot{\sigma}_{ij} = c_{ijkl} (\dot{\varepsilon}_{kl} - \beta_{kl}),$$ (1)

where $\dot{\sigma}_{ij}$ and $\dot{\varepsilon}_{kl}$ indicate microscopic stress and strain rates, respectively, $c_{ijkl}$ and $\beta_{kl}$ signify elastic stiffness and creep function, respectively, satisfying $c_{ijkl} = c_{jikl} = c_{ijlk} = c_{klij}$ and $\beta_{kl} = \beta_{lk}$. Then, the evolution equation of microscopic stress $\sigma_{ij}$, and the relationship between macroscopic stress rate $\dot{\Sigma}_{ij}$ and strain rate $\dot{E}_{kl}$ are derived as follows:[6-9]

$$\dot{\sigma}_{ij} = c_{ijpq} \left( \delta_{pk} \delta_{ql} + \chi_{p,q}^{kl} \right) \dot{E}_{kl} - c_{ijkl} \left( \beta_{kl} - \varphi_{k,l} \right),$$ (2)

$$\dot{\Sigma}_{ij} = \left\langle c_{ijpq} \left( \delta_{pk} \delta_{ql} + \chi_{p,q}^{kl} \right) \right\rangle \dot{E}_{kl} - \left\langle c_{ijkl} \left( \beta_{kl} - \varphi_{k,l} \right) \right\rangle,$$ (3)

where $\delta_{ij}$ denotes Kronecker's delta, $( )_{,i}$ indicates the differentiation with respect to Cartesian coordinates $y_i$ $(i = 1, 2, 3)$, and $\langle \ \rangle$ represents the volume average in $A$ as $\langle \# \rangle = | A |^{-1} \int_A \# dA$. Here, $| A |$ stands for the volume of $A$. Moreover, $\chi_i^{kl}$ and $\varphi_i$ are the characteristic functions determined by solving the following boundary value problems:

$$\int_A c_{ijpq} \chi_{p,q}^{kl} v_{i,j} dA = -\int_A c_{ijkl} v_{i,j} dA , \tag{4}$$

$$\int_A c_{ijpq} \varphi_{p,q} v_{i,j} dA = \int_A c_{ijkl} \beta_{kl} v_{i,j} dA , \tag{5}$$

where $v_i$ denotes an arbitrary velocity field satisfying the point-symmetry with respect to the centers of boundary facets of $A$ or the $Y$-periodicity. The above problems are solved using FEM based on the boundary condition described in the previous section. In consequence, compared with the case of using unit cells, the domain of analysis is reduced to 1/4 and 1/8 for the in- and out-of-phase laminate configurations, respectively.

## 3. Creep Analysis of Plain-Woven GFRP Laminates

### 3.1. *Analysis conditions*

Using the present method, in-plane creep deformations of plain-woven glass fiber/epoxy laminates subjected to a constant stress (60MPa) were analyzed under the macroscopic plane stress condition. First, the laminates were elongated at a constant strain rate $(10^{-5} \mathrm{s}^{-1})$ until accomplishing the prescribed creep stress. After that, the stress was held constant for five hours. Defining an off-axis angle as the angle between a loading direction and the warp direction, four kinds of off-axis angles, i.e. $\theta = 0°, 15°, 30°, 45°$ were considered. Two types of laminate configurations stated in the previous section, i.e., the in- and out-of-phase laminate configurations were assumed.

### 3.2. *Basic cell*

The basic cell and its finite element mesh used in the present analysis is illustrated in Fig. 2, which is the same as in the previous paper.[6] The cell was discretized into finite elements using eight-node isoparametric elements (1,624 elements, 1,995 nodes).

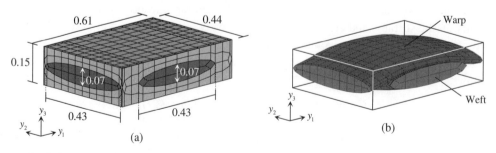

Fig. 2. Basic cell and finite element mesh; (a) full view with dimensions in mm, (b) fiber bundles in basic cell.

### 3.3. *Material properties*

Fiber bundles were regarded as glass fiber/epoxy unidirectional composites and as linear elastic materials. The material properties of fiber bundles were calculated using the homogenization theory[1,2] on the assumption that the fiber volume fraction was 75 % and that the bundles had a hexagonal fiber array. The elastic properties of the glass fibers and epoxy used in the calculation are listed in Table 1. By contrast, the matrix (epoxy) was regarded as an isotropic elastic-creep material and to obey the following constitutive equation:

$$\dot{\varepsilon}_{ij} = \frac{1+v_m}{E_m}\dot{\sigma}_{ij} - \frac{v_m}{E_m}\dot{\sigma}_{kk}\delta_{ij} + \frac{3}{2}\dot{\varepsilon}_0^p \left[ \frac{\sigma_{eq}}{g(\overline{\varepsilon}^p)} \right]^n \frac{s_{ij}}{\sigma_{eq}},$$ (6)

where $E_m$, $v_m$ and $n$ signify material constants, $g(\overline{\varepsilon}^p)$ stands for a hardening function depending on equivalent viscoplastic strain $\overline{\varepsilon}^p$, $\dot{\varepsilon}_0^p$ indicates a reference strain rate, $s_{ij}$ denotes deviatoric part of $\sigma_{ij}$, and $\sigma_{eq} = \left[(3/2)s_{ij}s_{ij}\right]^{1/2}$. These constants are listed in Table 1. Incidentally, material constants in Table 1 are the same as in the previous paper.[6]

### 3.4. *Results of analysis*

Figure 3 shows macroscopic creep curves of the plain-woven GFRP laminates at a constant creep stress, 60MPa, with four kinds of off-axis angles, $\theta = 0°$, 15°, 30°, 45°.

Table 1. Material constants.[6]

| Glass fiber | $E_f = 80\times10^3$ $v_f = 0.30$ | |
|---|---|---|
| Epoxy | $E_m = 5.0\times10^3$ $v_m = 0.35$ $\dot{\varepsilon}_0^p = 10^{-5}$ | |
| | $n = 20$ $g(\overline{\varepsilon}^p) = (\overline{\varepsilon}^p)^{0.50} + 24.5/2.5^{\overline{\varepsilon}^p}$ | |

MPa (stress), mm/mm (strain), s (time).

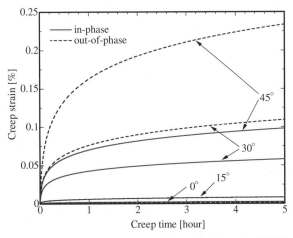

Fig. 3. Creep strain versus creep time relations of pain-woven GFRP laminates.

First, it is seen from the figure that the creep strain is hardly observed at $\theta = 0°$ (on-axis loading). Similarly, considerably small creep strain occurs at $\theta = 15°$. However, with $\theta = 30°$, the creep strain suddenly increases and at $\theta = 45°$, it further increases and reaches the maximum. These results are obtained irrespective of the laminate configurations, suggesting that the plain-woven GFRP laminates have marked in-plane anisotropy with respect to macroscopic creep behavior.

Next, comparing the results of the in-phase laminate configuration with those of the out-of-phase one, there is little difference in the creep strain at $\theta = 0°$ and $15°$. By contrast, the difference between the two laminate configurations becomes clear at $\theta = 30°$ and $45°$. Especially, at $\theta = 45°$, the creep strain of the out-of-phase laminate configuration is more than twice as large as that of the in-phase one. It is thus shown that there is significant dependence of the creep behavior of plain-woven GFRP laminates on the laminate configurations.

## 4.  Conclusions

In this study, creep analysis of GFRP plain-woven laminates subjected to a constant stress was performed using the time-dependent homogenization theory developed by the present authors. Employing a basic cell as the domain of analysis, two types of laminate configurations, i.e., in- and out-of-phase laminate configurations were considered. It was shown that in-plane off-axis creep behavior of the plain-woven GFRP laminates exhibited marked in-plane anisotropy. It was also shown that the laminate configurations of plain fabrics had significant effects on the creep behavior of the plain-woven GFRP laminates.

## Acknowledgments

The authors acknowledge support, in part, from the Ministry of Education, Culture, Sports, Science and Technology, Japan under a Grant-in-Aid for Young Scientists (B) (No. 20760064).

## References

1.  A. Bensoussan, J.L. Lions and G. Papanicolaou, *Asymptotic Analysis for Periodic Structures* (North-Holland Publishing Company, Amsterdam, 1978).
2.  E. Sanchez-Palencia, *Non-Homogeneous Media and Vibration Theory, Lecture Notes in Physics, No. 127* (Springer-Verlag, Berlin, 1980).
3.  N. Takano, M. Zako and S. Sakata, *Trans. Jpn. Soc. Mech. Eng.* **A61**, 1038 (1995), (in Japanese).
4.  N. Takano, Y. Uetsuji, Y. Kashiwagi and M. Zako, *Model. Simul. Mater. Sci. Eng.* **7**, 207 (1999).
5.  V. Carvelli and C. Poggi, *Compos.* **A32**, 1425 (2001).
6.  T. Matsuda, Y. Nimiya, N. Ohno and M. Tokuda, *Compos. Struct.* **79**, 493 (2007).
7.  X. Wu and N. Ohno, *Int. J. Solids Struct.* **36**, 4991 (1999).
8.  N. Ohno, X. Wu and T. Matsuda, *Int. J. Mech. Sci.* **42**, 1519 (2000).
9.  N. Ohno, T. Matsuda and X. Wu, *Int. J. Solids Struct.* **38**, 2867 (2001).

# BENDING BEHAVIOR OF SIMPLY SUPPOTED METALLIC SANDWICH PLATES WITH DIMPLED CORES

DAE-YONG SEONG[1], CHANG GYUN JUNG[2], DONG-YOL YANG[3] †

*Department of Mechanical Engineering, KAIST,*
*Daejeon, 305-701, KOREA,*
*gogoabout@kaist.ac.kr, cgj@kaist.ac.kr, and dyyang@kaist.ac.kr*

DONG GYU AHN[4]
*Department of Mechanical Engineering college engineering, Chosun University,*
*Gwangju ,501-759, KOREA,*
*smart@mail.chosun.ac.kr*

Received 15 June 2008
Revised 23 June 2008

Metallic sandwich plates are lightweight structural materials with load-bearing and multi-functional characteristics. Previous analytic studies have shown that the bendability of these plates increases as the thickness decreases. Due to difficulty in the manufacture of thin sandwich plates, dimpled cores (structures called egg-box cores) are employed as a sandwich core. High-precision dimpled cores are easily fabricated in a sectional forming process. The cores are then bonded with skin sheets by multi-point resistance welding. The bending characteristics of simply supported plates were observed by the defining measure, including the radius ratio of the small dimple, the thickness of a sandwich plate, and the pattern angle (0°/90°, 45°). Experimental results revealed that sandwich plates with a thickness of 2.2 mm and a pattern angle of 0°/90° showed good bendability as the punch stroke under a collapse load was longer than other cases. In addition, the gap between attachment points was found to be an important parameter for the improvement of the bendability. Finally, sandwich plates with dimpled cores were bent with a radius of curvature of 330 mm for the sheet thickness of 2.2 mm using an incremental bending apparatus.

*Keywords*: Sandwich plates, dimpled core, bending behavior.

## 1. Introduction

Metallic sandwich plates are structured plates with low-density cores between two face sheets to reduce their weight and to improve their performance[1, 2]. Various inner structures, including a honeycomb core, truss cores, a corrugated core, perforated sheets, and woven metal have been investigated in reference to these structures[3~13]. The characteristics of these sandwich plates include high strength and good stiffness as a lightweight material. In addition, various studies involving the use of these plates as a heat exchanger, a noise absorption material, a flow channel and a structured catalyst have

---

†Corresponding Author.

been conducted[1, 2]. Though these benefits, the range of the application of metallic sandwich plates is limited due to their low formability compared to commercial sheet metals. As a bendable sandwich plates can expand the application range, dimpled cores (or egg-box) are employed and designed for low-curvature bending.

In Mohr's work, the bendability increased as the thickness of a sandwich plate decreased and the shear strength of the cores increased[14]. High-precision dimpled cores that were easily fabricated in a sectional forming process were employed as a sandwich core due to the difficulty involved in the manufacture of a thin sandwich plate. These cores can then be bonded with skin sheets by multi-point resistance welding, brazing, or an adhesive bonding technique. The three-point bending behavior of simply supported plates was experimentally observed as the defined measure including the radius ratio ($\alpha$) of the dimple, the thickness of the sandwich plate (h), the pattern angle ($\theta$), and the position of a small dimple of a dimpled core.

## 2. Sandwich Plates with Dimpled Cores

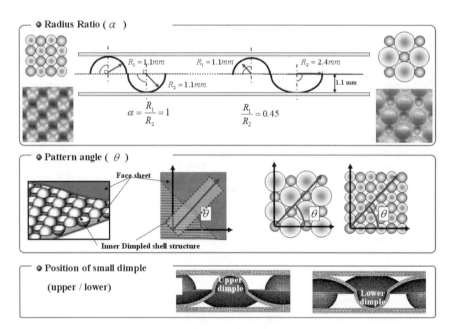

Fig. 1.  Definition of the experimental parameters of a sandwich plate with a dimpled core: radius ratio, pattern angle and position of small dimple.

Schematics that show the experimental parameters of sandwich plates with a dimpled core are shown in Fig. 1. The radius ratio ($\alpha$) is the ratio of the radius of the smaller dimple to the radius of the larger dimple with respect to the reference direction. The pattern angle ($\theta$) is defined as the in-plane angle of the core direction. When $\alpha \neq 1$, the position of the small dimple is an important parameter in bending experiments.

Fabricated sandwich plates with dimpled cores are illustrated in Fig. 2. The effective area of the punch and die used to fabricate the cores was 40 mm × 40 mm. However, the dimensions of the specimens were 20 mm (width) ×160 mm (length) in the three-point bending tests. Piecewise sectional forming techniques were employed to fabricated the desired dimensions of the specimens[15]. The position of the specimen at each step was controlled by gages composed of holes and pins. Dimpled cores can be bonded with face sheets of SUS304 (0.3 mm thickness) by resistance welding[16].

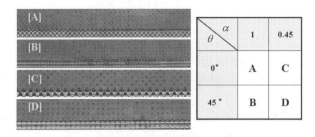

Fig. 2. Fabricated sandwich plates with dimpled cores (resistance welding).

## 3. Definition of the Limit Radius of Curvature

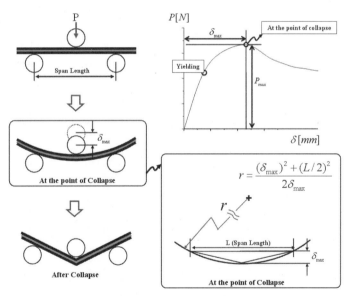

Fig. 3. Definition of limit radius of curvature.

To quantify the bendability of the sandwich plates, the limit radius of curvature (r) was defined from load-stroke curves, as shown in Fig. 3. Loads depending on elastic bending have been shown to increase continuously before the onset of face buckling at the point of collapse, at which a local peak load was found. The punch stroke upon collapse is a quantified value that defines the limit radius of curvature. An arc shape is assumed before a collapse in the calculation of the limit radius of curvature. After a collapse, plastic post-buckling behaviors dominate during the decrease in the load. A triangular shape is assumed due to the plastic hinge of the center region. Thus, the limit radius of curvature can be easily derived by the following equation:

$$r = \frac{(\delta_{max})^2 + (L/2)^2}{2\delta_{max}}$$

(1)

Therefore, the punch strokes at the peak load (or the collapse load) from three-point bending experiments give the limit radius of curvature that is related to the bendability of sandwich plates.

## 4. Static Bending Tests (Simply Supported Bending)

Table 1. Comparison of the limit radius of curvature.

| h | $\alpha$ | $\theta(°)$ | Dimple position | $P_{max}$ (kN) | $\delta_{max}$ (mm) | r |
|---|---|---|---|---|---|---|
| 2.5 | 1 | 0/90 | - | 0.143 | 6.46 | 281.9 |
| | | 45 | - | 0.154 | 4.58 | 395.5 |
| | 0.45 | 0/90 | Upper | 0.144 | 8.09 | 226.4 |
| | | 45 | Upper | 0.120 | 4.26 | 424.3 |
| | | 0/90 | Lower | 0.155 | 5.60 | 324.0 |
| | | 45 | Lower | 0.132 | 3.35 | 539.4 |
| 2.2 | 1 | 0/90 | - | 0.130 | 6.71 | 271.7 |
| | | 45 | - | 0.118 | 4.67 | 388.0 |
| | 0.45 | 0/90 | Upper | 0.126 | 4.99 | 363.2 |
| | | 45 | Upper | 0.112 | 3.38 | 535.0 |
| | | 0/90 | Lower | 0.124 | 4.85 | 373.2 |
| | | 45 | Lower | 0.105 | 3.40 | 531.8 |

The sandwich plates constructed with dimpled cores were measured for their response to simply supported bending. Experiments with respect to pattern angles ($\theta$) of 0°/90° and 45°, radius ratios ($\alpha$) of 1 and 0.45, and the positions of the small dimple of "Upper" and "Lower" dimples were carried out. A standard test equipment (INSTRON) was used under the conditions of a span length of 120 mm and a punch velocity of 5 mm/min[17].

The dimensions of the sandwich specimens, with an average weight of approximately 21g, were 20 mm (width) × 160 mm (length). The load-stroke curves revealed that after a face-yielding load, stable hardening occurred until the punch stroke reached the collapse load caused by face buckling. Thus, the aforementioned limit radii of curvature were measured with respect to the pattern angles ($\theta$), radius ratio ($\alpha$), and the position of the small dimple. Each condition was repeated three times. The average values of the relevant data (the collapse load, punch stroke at the collapse load, and the limit radius of curvature) are described in Table 1.

These results imply that sandwich plates with a thickness of 2.2 mm, a pattern angle of 0°/90°, and a radius ratio of 1 are characterized by good bendability, as the punch stroke at the collapse load was longer than those in other cases. A decrease of the thickness of sandwich plates and an increase in the radius ratio raised the bendability related to the defined limit radius of curvature. In addition, the "Lower" position of the small dimple and a pattern angle of 0°/90° provided good bendability. From these results, it is concluded that the gap between the attachment points is the most important parameter for the improvement of the bendability when face buckling is the main failure mode.

## 5.  Incremental Bending

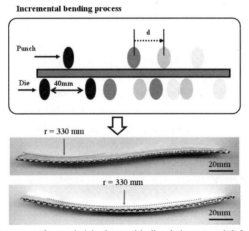

Fig. 4.   Incremental bending process for sandwich plates with dimpled cores and deformed shapes of sandwich plates.

The incremental bending process for a sandwich plate with a thickness of 2.2 mm, a pattern angle of 0°/90°, and a radius ratio of 1 is illustrated in Fig. 4. The apparatus for incremental roll bending[18] that enabled the control of the punch stroke and the span length used for sandwich plate bending had a span length of 40 mm, movement (d) of 10 mm, and a punch stroke of 1.2 mm. For each a rigid structured sandwich plates as dimpled plates can not be simply bent into a curved shape by conventional bending. However, by employing the incremental bending process, the sandwich plates could be successfully fabricated so as to have a radius of curvature of approximately 330 mm.

## 6. Conclusions

In this work, simply supported experiments for metallic sandwich plates with dimpled cores were carried out with reference to the thickness of the sandwich plates (h), the pattern angles ($\theta$), the radius ratio ($\alpha$), and the position of a small dimple. The study concluded that increasing the gap between the attachment points, as the most important parameter for the improvement of bendability, increases the possibility of face buckling, which is the main failure mode. In addition, experimental results have revealed that a sandwich plates with a thickness of 2.2 mm, a pattern angle of 0°/90°, and a radius ratio of 1 showed good bendability related to the defined limit radius of curvature. Sandwich plates with a radius of curvature of approximately 330 mm were fabricated in an incremental bending process.

## Acknowledgments

This work was supported by the Next Generation New Technology Development Program of the Ministry of Commerce and Energy (MOCIE), Grant No. 10028226, and in part by the research project Development of an Ultra-light Metal Sandwich Plate (USP) with a Three-dimensional Inner Structure.

## References

1. H.N.G. Wadley, N.A. Fleck, and A.G. Evans, *Compos. Sci. Technol.* 63, 2331 (2003).
2. J. Banhart, *Prog. in Mater. Sci.* **46,** 559-632 (2001).
3. N. Wick, and J.W. Hutchinson, *Int. J. Solids Struct.* **38,** 5165 (2001).
4. G.W. Kooistra, H.N.G. Wadley, *Mater. Design.* **28,** 507 (2007).
5. L. Valdevit L, J.W. Hutchinson, and A.G. Evans, *Int. J. Solids Struct.* **41,** 5105 (2004).
6. F. Cote, V.S. Deshpande, N.A. Fleck, and A.G. Evans, *Int. J. Solids Struct.* **43,** 6220 (2006).
7. L. Valdevit, Z. Wei, C. Mercer, F.W. Zok, and A.G. Evans, *Int. J. Solids Struct.* **43,** 4888  (2006).
8. S.D. Pan, L.Z. Wu, Y.G. Sun, and Z.G. Zhou, *Mater. Lett.*, **62,** 523 (2008).
9. F.W. Zok, S.A. Waltner, Z. Wei, H.J. Rathbun, R.M. McMeeking, and A.G. Evans, *Int. J. Solids Struct.* **41,** 6249 (2004).
10. H.J. Rathbun, F.W. Zok, S.A. Waltner, C. Mercer, A.G. Evans, D.T. Queheillalt, and H.N.G. Wadley, *Acta Mate.,* **54,** 5509 (2006).
11. D.G. Ahn, S.H. Lee, C.G. Jung, and D.Y. Yang, *J. Mater. Process Tech.* **187,** 521 (2007).
12. D.T. Queheillalt, Y. Murty, and N.H.G. Wadley, *Scripta Mater.* **58,** 76 (2008).
13. S. Chiras, D.R. Mumm, A.G. Evans, N. Wicks, J.W. Hutchinson, K. Dharmasena, H.N.G. Wadley, and S. Fichter, *Int. J. Solids Struct.* **39,** 4093 (2002).
14. D. Mohr, *Int. J. Solids Struct.* **42,** 1491 (2005).
15. M. Z. Li, Z. Y. Cai, Z. Sui, and Q. G. Yan, *J. Mater. Process Tech.* **129,** 333 (2002).
16. J.B. Kim, S.M. Lee, and S.J. Na, *Sci. Technol. Weld. Joi.* **12,** 376 (2006).
17. ASTM, E270 (1995).
18. S. J., Yoon, and D. Y., Yang, *Annals of the CIRP*, **52,** 991 (2003).

# PENETRATION ANALYSIS OF ALUMINUM ALLOY FOAM

NIANMEI ZHANG[1†]

*Institute of Applied Mechanics, Taiyuan University of Technology, Taiyuan, Sanxi, CHINA,*
*College of Physical Sciences, Graduate University of Chinese Academy of Science, Beijing, 100049, CHINA,*
*nianmeizhang@yahoo.com*

GUITONG YANG[2]

*Institute of Applied Mechanics, Taiyuan University of Technology, Taiyuan, Sanxi, CHINA,*
*gtyang@tyut.edu.cn*

Received 15 June 2008
Revised 23 June 2008

Aluminum alloy foam offers a unique combination of good characteristics, for example, low density, high strength and energy absorption. During penetration, the foam materials exhibit significant nonlinear deformation. The penetration of aluminum alloy foam struck transversely by cone-nosed projectiles has been theoretically investigated. The dynamic cavity-expansion model is used to study the penetration resistance of the projectiles, which can be taken as two parts. One is due to the elasto-plastic deformation of the aluminum alloy foam materials. The other is dynamic resistance force coming from the energy of the projectiles. The penetration resistance expression is derived and applied to analyze the penetration depth of cone-nosed projectiles into the aluminum alloy foam target. The effect of initial velocity, the geometry of the projectiles on the penetration depth is investigated.

*Keywords*: Aluminum alloy foam; penetration; elasto-plastics; dynamics.

## 1. Introduction

In the past few years，there has been a strong demand for the use of metal foams, particularly in automobile，railway, building and aerospace industries. Due to the extended plateau stresses of metal foam, they have great potential as energy absorption crash elements in space vehicles[1-3]. The aluminum alloy foam material is also becoming the subject of many investigations because of its outstanding performance against small and medium caliber projectiles, especially when the weight is a designing condition. The main role of the aluminum alloy foam material is the erosion and rupture of the projectile.

T. Kadono[4] studied the depth of cavities and craters caused by hypervelocity impacts in two extreme cases: the penetration of intact projectiles at low impact pressure and the hemispherical excavation at very high impact pressure. They found that the crater produced is

*The project is supported by Natural Science Foundation of China (No. 10772130).
†Corresponding Author.

deep and narrow at low velocities of the projectile penetrating. But at the high velocity, the vestiges become a bowl-shaped crater. Roisman et al. [5] applied velocity–field model to predict oblique penetration of rigid projectile. The velocity–field model has been successfully exploited to study penetration of both rigid and deformable projectiles as well as fragmentation. Based on the dynamic cavity-expansion theory, Forrestal and his colleagues [6] developed the empirical formulae to calculate the penetration resistance of a metallic target. Two terms are included in the Poncelet equation, one is from the static deformation resistance of elasto-plastic target material and the other is the dynamic resistance from velocity effect. And the morphological observations of deformed foam target indicate that the indentation loads of the projectile can be partitioned as the force for plastic crushing the cells of foam material, the force for shearing the cells by the slant face of the projectile [7].

In the present work, based on dynamic cavity expansion theory, the penetration of conical nose projectile to semi-infinite aluminum alloy foam target is analyzed. The formula of penetration depth is recommended for a non-deformable projectile considering the plastic crushing force of target material, dynamic resistance, force of shearing cells and frictional force. The effect of initial velocity of projectile, the mass density of aluminum alloy foam target, the shape of projectile to penetrating depth is analyzed.

## 2.   Penetration process

The conditions of the projectile and material of targets play important roles in both cratering and cavity formation. We consider a conical projectile colliding vertically with a semi-infinite continuous target surface. It is assumed that the impact velocity is not too high so that the damage and deformation of the projectile can be ignored. The penetration process can be divided into three stages, as shown in the Fig. 1.

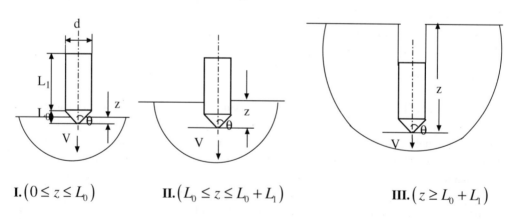

I. $\left(0 \leq z \leq L_0\right)$           II. $\left(L_0 \leq z \leq L_0 + L_1\right)$                    III. $\left(z \geq L_0 + L_1\right)$

Fig. 1. Penetration process.

If the impact is normal to the target, the resistance force of the projectile can be determined by plastic crushing force of target material, dynamic resistance, force of shearing cells and frictional force[6,7,9]. On the first stage ($0 \leq z \leq L_0$), it is

$$F_z = z^2 \left( A + BV^2 \right). \tag{1}$$

where, $A = \pi tg\theta \left[ \left( tg\theta + \mu \right)\sigma_{pl} + \dfrac{\tau}{\cos\theta} \right]$ and $B = \pi tg\theta \left( tg\theta + \mu \right)\sin^2\theta \dfrac{\rho_0}{\varepsilon_D}$.

And $\theta$ is cone angle, $\rho_0$ is the initial mass density of the aluminum alloy foam target.

$\rho_{Al}$ is density of base material of aluminum foam. $\varepsilon_D$ is the locking strain and determined

by the densification ratio[1] $\varepsilon_D = 1 - 1.4\dfrac{\rho_0}{\rho_{Al}}$. $V_0$ is the initial impact velocity of the

projectile. $\sigma_{pl}$ is the plateau stress[1] in rigid-perfectly plastic-locking idealization

model(R-P-P-L) of target material and can be represented as $\sigma_{pl} = 0.3\sigma_{ys}\left(\dfrac{\rho_0}{\rho}\right)^{1.5}$. $\sigma_{ys}$

is the yield strength of aluminum alloy. $\tau$ is the shear strength of aluminum alloy foam. $V$

is the rigid-velocity of a projectile. $d$ is projectile diameter. $\mu$ is the sliding friction

coefficient between the projectile and target. According to the Newton's second law,

$$-F_z = m\frac{dV}{dt} \tag{2}$$

The projectile mass $m$ is modified as

$$m = \rho_p \frac{\pi d^2}{4}\left( \frac{d}{12tg\theta} + L_1 \right) \tag{3}$$

where, $\rho_p$ is mass density of projectile material.

The initial condition for the projectile can be written as

$$At \quad t = 0, \quad \begin{cases} z = 0 \\ V = V_0 \end{cases} \tag{4}$$

On the first stage of penetration $\left( 0 \le z \le L_0 \right)$, the velocity of the projectile varies with the penetration depth increasing

$$V = \sqrt{V_0^2 \exp\left(-\frac{2Bz^3}{3m}\right) - \frac{A}{B}\left(1 - \exp\left(-\frac{2Bz^3}{3m}\right)\right)}. \tag{5}$$

On the second stage $\left(L_0 \leq z \leq L_0 + L_1\right)$, the dynamic resistance yields

$$F_z = L_0^2\left(A + BV^2\right) + C\left(z - L_0\right)$$

where, $C = 2\pi\sigma_{cr}\mu L_0 tg\,\theta$.

The relation between the depth and velocity can be deduced as

$$V^2 = \left(V^2 + \frac{A}{B}\right)\exp\left(\frac{BL_0^2\left(\frac{4}{3}L_0 - 2z\right)}{m}\right) - \frac{Cm}{2B^2L_0^4}e\,xp\left(\frac{2BL_0^2\left(L_0 - z\right)}{m}\right)$$

$$-\left[\frac{AL_0^2 - CL_0}{BL_0^2} - \frac{Cm}{2B^2L_0^4} + \frac{Cz}{BL_0^2}\right] \tag{6}$$

On the third stage $\left(z \geq L_0 + L_1\right)$, the penetrating resistance is $F_z = L_0^2\left(A + BV^2\right) + CL_1$

and the final depth of penetration can be determined as

$$z = \frac{m}{2BL_0^2}\ln\frac{\left(V_0^2 - \frac{A}{B}\right)\exp\left(\frac{4BL_0^3}{m}\right) + \frac{Cm}{2B^2L_0^4}\exp\left(\frac{2BL_0^3}{m}\right)\left[\exp\left(\frac{2BL_0^2L_1}{m} - 1\right)\right]}{\frac{AL_0^2 + CL_1}{BL_0^2}} \tag{7}$$

## 3. Calculation

As an example, the penetration depth is produced from the 7075-T651 aluminum alloy. The final penetration depth from Eq. (7) with $\sigma_{ys} = 448MPa$, $\rho_{Al} = 2710kg/m^3$, $\mu = 0.33$, $\tau = 2.8MPa$ is computed and the results are drawn in Fig. 2~5. Fig. 2 is drawn with varying cone angle $\theta$ and constant $d = 7.6mm$, $L_1 = 8mm$, $V_0 = 800m/s$,

$\rho_0 = 1300 kg / m^3$. The curve shows that the final depth of penetration decreases rapidly

with cone angle increasing. The minimum depth is at $\theta = 90^0$.

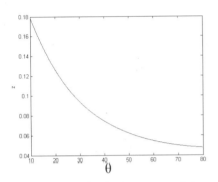

Fig. 2. Penetration depth vs. cone angle θ.

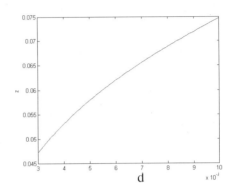

Fig. 3. Final depth vs. projectile diameter d.

The Fig. 3 plots the penetrating depth as a function of projectile diameter d with $45^o$ cone angle. Although the increasing diameter of projectile makes the resistance force increasing, the projectile possesses more kinetic energy to overcome it and reach deeper place.

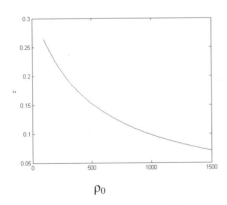

Fig. 4. Dependence of penetration depth to
the density of target material.

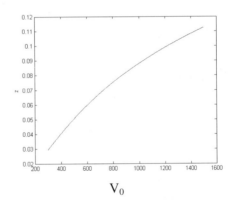

Fig. 5. Variation of the penetration depth to
the initial impact velocity.

The Fig. 4 shows that increasing mass density of aluminum alloy foam causes not only decreasing locking strain, but also increasing of plateau stress. These increase the impacting resistance greatly. The moderate density of foam material will increase the capacity of energy absorbing greatly. The Fig. 5 indicates that more initial kinetic energy of projectile can overcome the greater resistance work. The projectile stops at deeper point.

## 4.   Discussion and Conclusion

Considering the force of shearing cells of aluminum alloy foam and frictional force, the penetration for non-deformable projectile can be calculated with the simple expression (7), which is capable of predicting the penetration depth. The effect of the cone angle, length and diameter of projectile, the mass density of target material and initial impact velocity to the final depth is discussed in detail. The analysis shows that

(i)    The increasing cone-angle will increase the surface area of resistance. More energy of projectile is dissipated in friction.

(ii)   The penetration depth increases greatly if the increment of kinetic energy resulted from the mass of projectile exceeds the increment of work of resistance force.

(iii)  Although the large mass density of target material would decrease the penetration depth, the capacity of energy absorption of foam target would be the optimal at moderate density.

The kinetic energy of projectile plays an important role in the penetrating process. The kinetic energy can be optimized by mass and impact velocity so that to achieve maximum penetration depth. On the other hand, low density aluminum alloy foam has great capacity of energy absorbing. But only moderate density material can be used in the engineering so as to satisfy strength conditions.

## Acknowledgment

The authors wish to acknowledge, with thanks, the financial support from the Natural Science Foundation of China (No. 10772130).

## References

1.    L. J. Gibson and M. F. Ashby, *Cellular Solids : Structure and Properties* (Cambridge University Press, Cam bridge, UK, 1997). J. Callaway, *Phys. Rev.* **B35**, 8723 (1987).
2.    I. W. Hall, M. Guden and G. J. Yu, *Scripta Mater.* **43**, 515 (2000).
3.    V. S. Deshpande, N. A. Fleck, *Inter. J. of Impact Eng.* **24**, 277 (2000).
4.    T. Kadonoa and A. Fujiwara, *Inter. J. of Impact Eng.* **31**, 1309 (2005).
5.    I.V. Roisman, K. Weber, A.L. Yarin, V. Hohler, M.B. Rubin, *Inter. J. of Impact Eng.* **22**, 707 (1999).
6.    M.J. Forrestal, V.K. Luk, *ASME J. Appl. Mech.* **55**, 275 (1988).
7.    U. Ramamurty and M.C .Kumaran, *Acta Mater.* **52**, 181 (2004).
8.    G. Lu, J. Shen, W. Hou, D. Ruan and L.S. Ong, *J. Mech. Sci.* **50**, 932 (2008).
9.    Marvin E Backman and G. Werner, *Int. J. Eng. Sci.* **16**, 1 (1978).

# TWO-SCALE ANALYSIS OF HONEYCOMBS INDENTED BY FLAT PUNCH

TAKASHI ASADA[1†], YUJI TANAKA[2]

*Department of Computational Science and Engineering, Nagoya University,*
*Chikusa-ku, Nagoya 464-8603, JAPAN,*
*asada@mml.mech.nagoya-u.ac.jp, tanaka@mml.mech.nagoya-u.ac.jp*

NOBUTADA OHNO[3]

*Department of Mechanical Science and Engineering, Nagoya University,*
*Chikusa-ku, Nagoya 464-8603, JAPAN,*
*ohno@.mech.nagoya-u.ac.jp*

Received 15 June 2008
Revised 23 June 2008

The fully implicit incremental homogenization scheme developed by Asada and Ohno (2007) for elastoplastic periodic solids is applied to two-scale analysis of honeycomb blocks subjected to flat punch indentation. To this end, the scheme is rebuilt by introducing half unit cells based on the point-symmetric distributions of stress and strain in unit cells, so that analysis domains in unit cells are reduced by half. Then, by assuming the zigzag and armchair types of cell-arrangements, the two-scale analysis of honeycomb blocks is performed. The corresponding full-scale finite element analysis is also performed to reveal the cell-arrangement dependence of cell deformation in the honeycomb blocks. It is shown that the two-scale analysis is macroscopically successful in spite of microscopic limitations.

*Keywords*: Two-scale analysis; homogenization; honeycomb blocks; flat punch indentation.

## 1. Introduction

We consider macro-structures made of heterogeneous solids such as composite materials and cellular solids. Macro-structures of this kind can be analyzed using the same method as that used for structures made of homogeneous solids, if heterogeneous solids are appropriately homogenized. If heterogeneous solids have periodic micro-structures, their homogenized properties can be evaluated using the so-called mathematical homogenization method.[1] If periodic micro-structures are elastic, the mathematical homogenization method is really effective. However, if periodic micro-structures are elastoplastic, the mathematical homogenization method requires that elastoplastic finite element analysis of a unit cell is performed at all integration points in each element in a macro-structure at every increment, resulting in high computational loads. To perform such two-scale elastoplastic analysis, computational schemes have been developed.[2-5]

---

[†]Corresponding Author.

For iteratively determining perturbed displacement increment fields in elastoplastic unit cells, Asada and Ohno[5] investigated the influence of initial fields on convergence. To this end, they developed a fully implicit incremental homogenization scheme that allows any strain increment field to be assumed as an initial field. Then, by applying the scheme to a two-scale analysis of a macro-structure possessing a periodic elastoplastic micro-structure, they showed that the maximum size of incremental steps to attain convergence is strongly dependent on the initial strain increment fields examined. Thus they demonstrated that the computational scheme developed by Asada and Ohno[5] can be effective for reducing computational loads in two-scale elastoplastic analysis.

In this study, the fully implicit incremental homogenization scheme developed by Asada and Ohno[5] is applied to elastoplastic two-scale analysis of hexagonal honeycomb blocks indented by a flat punch. The scheme is rebuilt by introducing half unit cells based on the C-symmetry[6], which is applicable to hexagonal honeycomb cells. The resulting scheme is used to perform the two-scale analysis of hexagonal honeycomb blocks, for which the zigzag and armchair types of cell-arrangements are assumed. The corresponding full-scale finite element analysis is also performed. The two-scale and the full-scale analysis are then compared in the light of an experiment[7], so that validity of the two-scale analysis is discussed.

## 2.  Implicit Homogenization Scheme

Let us consider a macro-structure *B* with an elastoplastic periodic micro-structure, and let *Y* be a unit cell of the micro-structure (Fig. 1). Stress and strain in *Y* are referred to as micro-stress $\boldsymbol{\sigma}$ and micro-strain $\boldsymbol{\varepsilon}$. Then, averaging $\boldsymbol{\sigma}$ and $\boldsymbol{\varepsilon}$ in *Y* gives macro-stress $\boldsymbol{\Sigma}$ and macro-strain $\mathbf{E}$. It is shown that $\boldsymbol{\varepsilon}$ is related with $\mathbf{E}$ as[1]

$$\boldsymbol{\varepsilon} = \mathbf{E} + \tilde{\boldsymbol{\varepsilon}}, \tag{1}$$

where $\tilde{\boldsymbol{\varepsilon}}$ indicates the perturbed strain resulting from *Y*-periodic perturbed displacement $\tilde{\mathbf{u}}$ :

$$\tilde{\boldsymbol{\varepsilon}} = \mathrm{sym}[\nabla_y \tilde{\mathbf{u}}]. \tag{2}$$

Here $\nabla_y$ stands for the gradient with respect to the position vector $\mathbf{y}$ in *Y*.

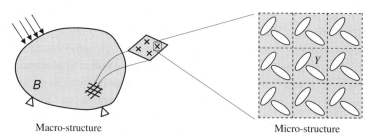

Macro-structure                Micro-structure

Fig. 1. Macro-structure with periodic micro-structure.

Let us suppose that constituents distribute point-symmetrically with respect to the body center of $Y$. Then, as shown by Ohno et al.[6], the field of $\tilde{u}$ satisfies the C-symmetry, i.e., the point-symmetry with respect to all boundary facet centers of $Y$ as well as the body center of $Y$, so that the balance of micro-stress $\sigma$ is expressed in the following weak form:

$$\langle \sigma : \delta\tilde{\varepsilon} \rangle_{Y*} = 0 , \tag{3}$$

where $\delta\tilde{\varepsilon}$ denotes any variation of $\tilde{\varepsilon}$ satisfying Eq. (2) and the C-symmetry, and $\langle\ \rangle_{Y*}$ represents the volume average in a *half unit cell* $Y*$, i.e., $\langle \# \rangle_{Y*} = |Y*|^{-1} \int_{Y*} \# \, dY*$. Here $|Y*|$ indicates the volume of $Y*$.

Incremental elastoplastic constitutive equations in a step from $t_n$ to $t_{n+1}$ can be linearized as[5]

$$\Delta\sigma_{n+1} - \Delta\sigma_{n+1}^{(i)} = \mathbf{D}_{n+1}^{(i)} : (\Delta\varepsilon_{n+1} - \Delta\varepsilon_{n+1}^{(i)}) , \tag{4}$$

where the subscript $n+1$ indicates the values at $t_{n+1}$, the superscript $(i)$ denotes the $i$th iteration, $\Delta$ stands for the increments in the step (e.g., $\Delta\sigma_{n+1}^{(i)} - \sigma_{n+1}^{(i)} - \sigma_n$), and $\mathbf{D}_{n+1}^{(i)}$ represents the consistent tangent modulus at $\Delta\varepsilon_{n+1}^{(i)}$, i.e., $\mathbf{D}_{n+1}^{(i)} = \partial\Delta\sigma_{n+1}^{(i)} / \partial\Delta\varepsilon_{n+1}^{(i)}$.

Since $\sigma_{n+1}$ should satisfy Eq. (3), we can derive the following equitation using Eqs. (1) – (4):

$$\langle \delta\tilde{\varepsilon} : \mathbf{D}_{n+1}^{(i)} : \nabla_y \Delta\tilde{u}_{n+1}^{(i+1)} \rangle_{Y*} = -\langle \delta\tilde{\varepsilon} : \sigma_{n+1}^{(i)} \rangle_{Y*} - \langle \delta\tilde{\varepsilon} : \mathbf{D}_{n+1}^{(i)} : (\Delta E_{n+1} - \Delta\varepsilon_{n+1}^{(i)}) \rangle_{Y*} . \tag{5}$$

The above equation with $Y*$ replaced by $Y$ was derived by Asada and Ohno[5]. Consequently, using the computational algorithm developed by them, Eq. (5) can be iteratively solved at all integration points in a finite element model of $B$. Two-scale analysis can thus be performed as was described in detail by Asada and Ohno[5].

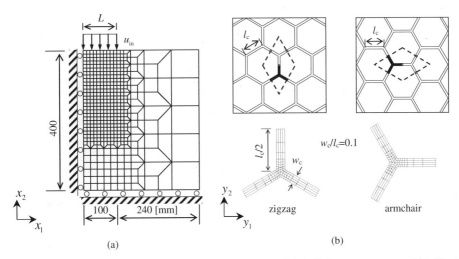

Fig. 2. Honeycomb block subjected to flat punch indentation; (a) right half of a macro-structure, (b) half unit cells $Y*$ for zigzag and armchair types of cell arrangements.

Table 1. Material parameters.

| | |
|---|---|
| Young's modulus, $E$ [GPa] | 69.0 |
| Poisson's ratio, $\nu$ | 0.3 |
| Yield stress, $\sigma_0$ [MPa] | 40.0 |

## 3.  Honeycomb Blocks Indented by a Flat Punch

Figure 2(a) illustrates the right half of the homogenized macro-structure analyzed, which was subjected to flat punch indentation displacement $u_{in}$ on the top surface area of a length $L$ under a plane strain condition of $E_{33} = 0$; $u_{in}$ was varied from zero to $-1.2$ mm, and $L$ was taken to be 100 mm. This macro-structure was divided into finite elements using 2D eight-node reduced integration elements. The zigzag and armchair types of cell-arrangements were assumed to be micro-structures of the macro-structure (Fig. 2(b)), and the half unit cells shown in this figure were considered as $Y^*$ and meshed using 2D four-node non-conforming elements to perform the two-scale analysis. The cell walls were assumed to be elastic-perfectly plastic (Table 1). The ratio of cell wall thickness $w_c$ to cell wall length $l_c$ was taken to be 0.1. The corresponding full-scale finite element analysis was performed in the three cases of $l_c / L = 0.1215$, 0.0491 and 0.0308 to discuss validity of the two-scale analysis.

## 4.  Computational Results and Discussion

The variation in indentation stress $\Sigma_{in}$ computed in the two-scale analysis is shown in Fig. 3(a) when $\Sigma_{in}$ is defined to be the sum of indentation nodal forces divided by $L$. Moreover, the contour of equivalent macro-strain $E_{eq}$ in the macro-structure at $u_{in} = -1.2$ mm in the two-scale analysis is depicted in Fig. 3(b). Here, $E_{eq}$ is empirically

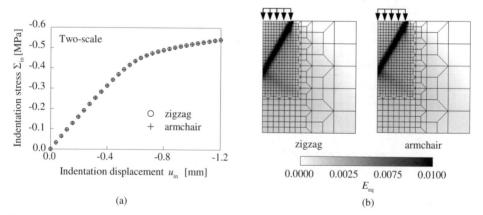

(a)                                                    (b)

Fig. 3. Results of two-scale analysis; (a) relation of indentation stress and displacement, and (b) contour of equivalent macro-strain $E_{eq}$ at $u_{in} = -1.2$ mm.

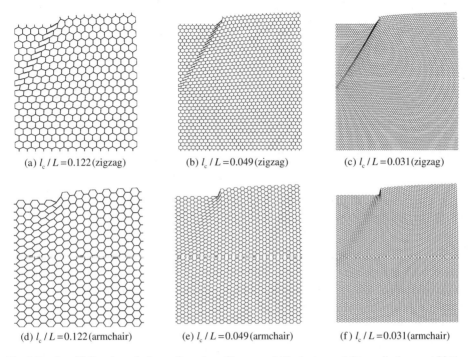

(a) $l_c/L = 0.122$ (zigzag)     (b) $l_c/L = 0.049$ (zigzag)     (c) $l_c/L = 0.031$ (zigzag)

(d) $l_c/L = 0.122$ (armchair)     (e) $l_c/L = 0.049$ (armchair)     (f) $l_c/L = 0.031$ (armchair)

Fig. 4. Results of full-scale analysis; configuration of honeycomb block at $u_{in} = -1.2$ mm; displacement $30\times$.

defined as $E_{eq} = (\frac{2}{3}\mathbf{E}:\mathbf{E})^{1/2}$. It is evident from these figures that the zigzag and armchair types of cell-arrangements gave indistinguishably the same results in the two-scale analysis. This is because the two types of cell arrangements provide almost exactly the same homogenized stress versus strain relations under in-plane loading at small strains even in the presence of plastic deformation.[a] Similar results were obtained in elasto-viscoplastic homogenization analyses of hexagonally fiber-arrayed metal-matrix composites.[10,11] It is further seen from Fig. 3(b) that the macro-structure had a strain localization band making an angle of about 30° with the vertical axis. This angle of macro-strain localization was experimentally observed as the angle of crushing of a cell row in a zigzag cell-arranged honeycomb block indented by a flat punch.[7]

The full-scale finite element analysis confirmed the direction of macro-strain localization in the two-scale analysis mentioned above, when the zigzag type of cell-arrangement was considered (Fig. 4(a)-(c)), or when the armchair cell-arranged honeycomb block of $l_c/L = 0.0308$ was analyzed (Fig. 4(f)). Here it is noted that the two-scale analysis did not predict the following feature revealed in the full-scale analysis:

---

[a] This equivalence of the zigzag and armchair types of cell-arrangements under in-plane loading breaks down at large strains.[8,9]

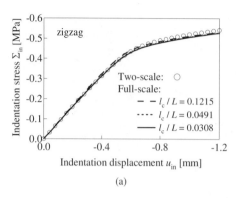 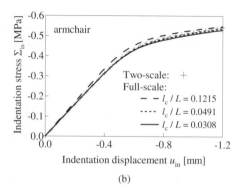

(a)                                    (b)

Fig. 5. Comparison of two-scale and full-scale analyses with respect to the relation of indentation stress $\Sigma_{in}$ and indentation displacement $u_{in}$ ; (a) zigzag and (b) armchair types of cell arrangements.

the zigzag type of cell-arrangement always led to crushing a cell row, while the armchair type of cell-arrangement resulted in a distribution of highly deformed cells near a flat punch edge (Fig. 4(d)-(f)). The full-scale analysis also confirmed the variation in $\Sigma_{in}$ in the two-scale analysis (Fig. 5(a), (b)).

## 5.  Conclusions

In this study, the fully implicit incremental homogenization scheme developed by Asada and Ohno (2007) was rebuilt by introducing half unit cells to reduce computational loads. The resulting scheme was applied to two-scale analysis of hexagonal honeycomb blocks indented by a flat punch. Zigzag and armchair types of cell-arrangements were assumed. The corresponding full-scale finite element analysis was also performed. The two-scale analysis was thus verified with respect to the macroscopic relation of indentation stress and displacement and the direction of a macro-strain localization band, though the two-scale analysis did not predict the cell-arrangement dependence of microscopic deformation revealed in the full-scale analysis.

## References

1.  A. Bensoussan, J.L. Lions and G. Papanicolaou, *Asymptotic Analysis for Periodic Structures* (North Holland, Amsterdam, 1978).
2.  S. Ghosh, K. Lee and S. Moorthy, *Comput. Meth. Appl. Mech. Eng.* **132**, 63 (1996).
3.  K. Terada and N. Kikuchi, *Comput. Meth. Appl. Mech. Eng.* **190**, 5427 (2001).
4.  C. Miehe, *Int. J. Numer. Methods Eng.* **55**, 1285 (2002).
5.  T. Asada and N. Ohno, *Int. J. Solids. Struct.* **44**, 7261 (2007).
6.  N. Ohno, T. Matsuda and X. Wu, *Int. J. Solids. Struct.* **38**, 2867 (2001).
7.  J.W. Klintworth and W.J. Stronge, *Int. J. Mech. Sci.* **31**, 359 (1989).
8.  N. Ohno, D. Okumura and H. Noguchi, *J. Mech. Phys. Solids* **50**, 1125 (2002).
9.  D. Okumura, N. Ohno and H. Noguchi, *Int. J. Solids. Struct.* **39**, 3487 (2002).
10. X. Wu and N. Ohno, *Int. J. Solids. Struct.* **36**, 4991 (1999).
11. N. Ohno, X. Wu and T. Matsuda, *Int. J. Mech. Sci.* **42**, 1519 (2000).

# Part N
# Structural Plasticity

# TUBE HYDROFORMING PROCESS DESIGNE OF TORSION BEAM TYPE REAR SUSPENSION CONSIDERING DURABILITY

KYUNG-TAEK LEE[1], HONG-JUEN BACK[2]

*R&D Center, AUSTEM Co., Ltd, Gajwa 1-dong, Seo-gu, Incheon, 173-299, KOREA,*
*leekt@austem.co.kr, hjbaek@austem.co.kr*

HEE-TAEK LIM[3]

*Metals Root Technology Center, Korea Institute of Materials Science 531 Changwondaero, Changwon,*
*Gyeongnam, 641-831, KOREA,*
*limong@kims.re.kr*

IN-SUK OH[4]

*Hankook ESI, Deungchon 3-dong, Gangseo-gu, Seoul, 157-033, KOREA,*
*limong@kims.re.kr*

HEON-YOUNG KIM[5†]

*Department of Mechanical & Mechatronics, Kangwon National University, 192-1 Hyoja-dong,*
*Chuncheon, Gangwon-do, 200-701 KOREA,*
*khy@kangwon.ac.kr*

Received 15 June 2008
Revised 23 June 2008

Manufacturing processes for automobile suspension components have generally considered the formability but not the durability of the suspension system, even though the latter is very important in the dynamic performance of the vehicle. Suspension systems should be designed with both formability and durability in mind. This paper describes the design of an optimal forming process to control the cross-sectional properties of the torsion beam in rear suspension systems for increased roll durability. Stamping and hydroforming simulations were performed using the finite element method to determine the optimum tube hydroforming process for producing a torsion beam with the best roll durability performance.

*Keywords*: Tube Hydroforming; Tubular Torsion Beam; Roll Durability Performance.

## 1. Introduction

A torsion beam rear suspension system has the advantage of being easier to adjust to the automobile height than any other type of rear suspension. It is also useful for expanding the interior space of the automobile, especially that of the trunk and rear seat. For these reasons, as well as its low production cost, torsion beam rear suspensions are widely used

†Corresponding Author.

in mid-size and compact cars, including 40% of all passenger cars manufactured in Japan[1-3].

Tubular torsion beams can be manufactured by a conventional stamping process or by a more recently developed hydroforming process. Tube hydroforming has many advantages compared to stamping. These include the reduced number and weight of the parts, a simpler manufacturing process, the increased strength and stiffness of the parts, as well as improved dimensional accuracy and material efficiency. Because of these advantages, tube hydroforming is being developed as a method for manufacturing tubular torsion beams that are more economical and effective compared to the currently widespread stamped beams[4,5]. The suspension system must maintain its durability performance and strength despite repeated road impacts. Any deformation and cracks that occur during operation are critical because they affect the driving stability. In particular, the torsion beam in a rear suspension system is repeatedly subjected to torsional loads. Therefore, it must be designed with a sufficient durability margin[6,7].

This paper describes stamping and tube hydroforming processes that can be used to manufacture tube torsion beams for rear suspension systems. A formability evaluation and manufacturing design for the stamping and hydroforming processes were conducted using forming process simulations. In addition, because the rear suspension system of a car must meet certain specifications for load, strength, and stiffness, roll durability tests were performed and the results analyzed to confirm the durability of the systems.

## 2.  Torsion Beam Rear Suspension System

The design goal for the rear suspension system was to maintain the same durability while achieving a weight savings of 20% compared to general pressed parts. Fig. 1 shows the structure of a torsion beam rear suspension system installed in a car. Fig. 2(a) shows the rear suspension system developed in this study. Changing the torsion beam from a plate to a tube required that the thickness of the torsion beam be decreased from 6.0 mm to 2.6 mm. The torsion beam rear suspension system was manufactured by welding the side members, brackets, and tube torsion beam together. Torsion beams were formed using either a stamping process or a tube hyroforming process to produce rectangular sections with rounded edges. The center had a V-shaped section, shown as section B–B in Fig. 2(b). Torsion beams are affected significantly by any forming defects, such as fractures and wrinkles, because of the rapid transition from rectangular to V-shaped cross-sections.

## 3.  Forming Process Simulation

Tube hydroforming and stamping process simulations were performed to determine the best manufacturing method for the tube torsion beam. The tube was made of DP590. Table 1 lists the material properties of DP590.

### 3.1.  *Stamping process simulation*

The formation of a stamped torsion beam was simulated by two separate processes, as shown in Fig. 3. The first stamping process then formed a tube when the upper die was closed after inserting the cores. Inserting the cores in the end holes helped maintain the rectangular cross-section shape during the first stamping process. The second stamping process was used to form the V-shaped section in the middle of the tube. Fig. 3(b) shows how the tube with inserted cores was placed on the blank holder after the first stamping process to be formed by the punch when the upper die descends. Fig. 5 indicates time-displacement relationship between the tools. Fig. 7(a) shows the deformed shape of tube in the stamping process. The final shape of tube has a maximum thinning of 0.06.

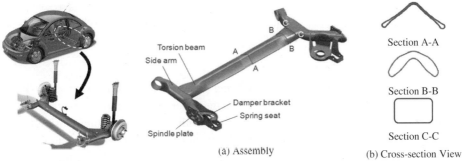

Fig. 1. Torsion beam rear suspension system.

(a) Assembly       (b) Cross-section View

Fig. 2. Tubular torsion beam rear suspension and cross-section.

Table 1. Mechanical properties of DP590 steel.

| Density | $7.8 \times 10^{-6}$ kg/mm$^3$ | Work hardening exponent | 0.1905 |
|---|---|---|---|
| Young's modulus | 210 GPa | Offset strain | 0.001126 |
| Poisson's ratio | 0.3 | Thickness | 2.6 mm |
| Stiffness coefficient | 1085 MPa | Diameter | 101.6 mm |

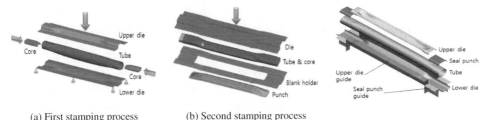

(a) First stamping process      (b) Second stamping process

Fig. 3. Stamping process boundary conditions for the tubular torsion beam.

Fig. 4. Tool sets for the tube hydroforming simulation.

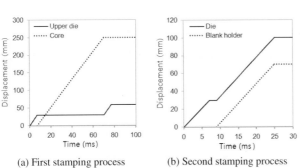

(a) First stamping process      (b) Second stamping process

Fig. 5. Time-displacement relationship between the tools.

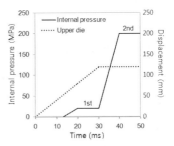

Fig. 6. Loading path for the tube hydroforming process.

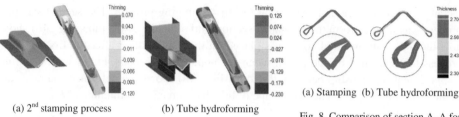

(a) 2<sup>nd</sup> stamping process          (b) Tube hydroforming

Fig. 7. Deformed shapes and thinning distribution for simulation.

(a) Stamping  (b) Tube hydroforming

Fig. 8. Comparison of section A–A for the stamping and tube hydroforming simulations.

### 3.2.  *Tube hydroforming process simulation*

The hydroformed torsion beam was made using pressure sequence hydroforming (PSH). In PSH, hydraulic pressure is maintained in the inner tube as the die is closed. This method makes it possible to form unformed parts with low pressure after closing the die because pre-forming takes place while the die is closing. Fig. 4 shows the tool set and Fig. 6 shows the loading path for the tube hydroforming simulation, while Fig. 7(b) shows the deformed shape of tube in the tube hydroforming process. The upper die contacts the seal punches and punch guides while closing down on both of them. When the seal punches press the seal punch guides to the bottom position, the seal punches and its guides maintain the seal between the ends of the formed tube. As the upper die closes to the bottom position, the hydraulic pressure forms the unformed parts by closing the ends of the tubular blank. The final shape illustrated in Fig. 7(b) has a maximum thinning of 0.13.

### 3.3.  *Comparison of stamping and tube hydroforming simulation results*

Fig. 8 compares section A–A of Fig. 2 for the stamping and tube hydroforming process simulations. Fig. 8(a) shows that the curvature of the edges was not maintained because the forming pressure was not applied during stamping. This is unlike the situation illustrated in Fig. 8(b), where the application of the forming pressure maintained the curvature. The torsion beam manufactured by the tube hydroforming process had a structure that made it possible to distribute alternate torsional loads because of the curvature of the edges.

## 4.  Prototype Durability Testing

Torsion beam prototypes were manufactured using both the stamping and tube hydroforming processes, and then assembled into torsion beam rear suspension systems. The product characteristics produced by the two processes were analyzed though durability tests of the suspension systems.

### 4.1.  *Prototype production*

The prototype torsion beams were produced using stamping and hydroforming processes under the conditions determined by the simulations. The dies are shown in Fig. 9.

### 4.2. *Durability performance test*

The prototypes incorporating the torsion bars produced by the two processes were subjected to static and dynamic tests to evaluate their performance. A roll durability test is particularly important for durability performance testing to estimate the cyclic torsional load. For the purposes of this paper, the durability was tested using 200,000 cycles of torsional stress to compare the prototypes with a plate torsion beam rear suspension system. Fig. 10 shows the roll durability test setup.

For the roll durability tests, the trailing arm was fixed on a pivot and the brake drum was placed on the spindle plate. Two actuators installed on the brake drum alternated pushing and pulling the brake drum to place the torsion bar under torsional stress. If the torsion bar endured the repeated torsional load without fracturing under specified conditions, it satisfied the roll durability performance. Fig. 11 shows a fracture that occurred during the roll durability test of the torsion beam, and Table 2 lists the results of the roll durability tests. The torsion beam used for Case 1 in Table 2 was formed with an upper die first hydraulic pressure of 0.05 GPa, while that for Case 2 was formed with an upper die first hydraulic pressure of 0.01 GPa. The fracture appeared where the cross-section shape changed from a V-shape to a rectangle.

The tests showed that the roll durability performance of the hydroformed prototype was higher than that of the stamped prototype. As shown in Fig. 8, the edge curvature was maintained by hydroforming, but not by the stamping, meaning that the repetitive torsional stress could be effectively distributed in the former but not in the latter. Table 2 shows that a lower first hydraulic pressure produced a part with better roll durability performance. This is because the high pressure creates a defect in the tube, such as the gap shown in Fig. 12. It appears as the side guide is pushed out by the forming inner pressure in the tube when a high first hydraulic pressure is exerted by the closing die.

### 5. Process Adjustment for Increased Durability

As listed in Table 2, the only model with a better durability performance was produced by tube hydroforming with a low first hydraulic pressure. However, accurately forming a torsion beam produced with a low first hydraulic pressure into the desired cross-sectional shape is difficult because wrinkles occur easily. An additional core was used as shown in Fig. 13 to reduce the likelihood of this problem. The core was inserted when the edges of the tube were changed to an elliptical cross-section shape as the tube was pressed by the seal punch. The prototype produced with the additional core had a durability of 317,500 cycles in the roll durability test, an improvement of over 50%.

(a) Stamped  (b) Tube hydroformed

Fig. 9. Die sets for the physical trials of the stamped and tube hydroformed prototypes.

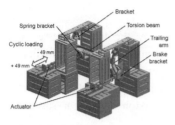

Fig. 10. Roll durability test machine.

Fig. 11. Crack in stamped torsion beam that appeared during roll durability testing.

Table 2. Durability test results of rear suspension prototypes.

| Forming process type | Durability | Forming process type | Durability |
|---|---|---|---|
| Standard (plate) | 215,000 cycles | Tube hydroformed (case 1) | 108,626 cycles |
| Stamped | 63,580 cycles | Tube hydroformed (case 2) | 358,000 cycles |

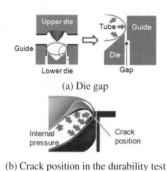

(a) Die gap

(b) Crack position in the durability test

Fig. 12. Defect due to the die gap in tube hydroforming.

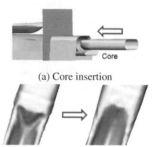

(a) Core insertion

(b) Result of hydroforming with an inserted core

Fig. 13. Tubular torsion beam by formed by tube hydroforming with an inserted core.

## 6. Conclusion

In this paper, a tube hydroforming die and process were developed to produce a tube torsion beam with improved roll durability performance using forming process simulations and roll durability testing. The results are as follows.

(1) The optimal die and process were designed using forming process simulations of stamping and tube hydroforming. The stamping process first formed the ends of the edge after inserting a core, and then formed the middle of the torsion beam into a V-shape. The tube hydroforming process first formed the middle of the torsion beam to a V-shape using the first hydraulic pressure, and then formed the unformed parts using the second hydraulic pressure.

(2) The torsion beam produced by the tube hydroforming process had a higher durability performance than that produced by the stamping process because the forming pressure clearly maintained the curvature of the edges, helping the beam withstand repetitive torsional loads. A high first hydraulic pressure resulted in a low roll durability performance due to surface defects that occurred because of the stress concentration in the gap between dies.

(3) A torsion beam tailored for optimum roll durability performance was manufactured by the tube hydroforming process with a low first hydraulic pressure of 0.01 GPa using inserted cores to prevent wrinkles. The result was a torsion beam with a durability performance of 317,500 cycles, well over the 200,000 cycle standard performance.

## References

1. T.W. Kwon, J.C. Kim, J.H. Jeon, G.W. Jang and W.S. Lee, *KSTP Conf.*, 292 (2006).
2. I.S. Oh, H.Y. Kim, J.M. Ko, D.J. Lee and W.K. Cho, *KSTP Conf.*, 269 (2006).
3. W. Cho, J. Lee, M. Sin and D. Lee, *KSAE Conf.*, 1169 (2006).
4. H.J. Kim, B.H. Jeon, H.Y. Kim and J.J. Kim, *Advanced Tec. of Plasticity, 545* (1993).
5. H.Y. Kim, H.T. Lim, H.J. Kim and D.J. Lee, *Advanced Tec. of Plasticity, Proc. of the 8th ICTP*, 305 (2005).
6. J.S. Kang, *KSAE*, **13**, 164 (2005).
7. D.C. Lee and J.H. Byun, *KSAE*, **7**, 195 (1999).

# THE DEFORMATION OF THE MULTI-LAYERED PANEL OF SHEET METALS UNDER ELEVATED TEMPERATURES

SANG-WOOK LEE[1†]

*Department of Mechanical Engineering, Soonchunhyang University, Asan, Chungnam 336-745, KOREA,*
*swlee@sch.ac.kr*

DONG-UK WOO[2]

*R&D Center, Kukje Machinery Co., Ltd., Okchen, Chungbuk 373-802, KOREA,*
*donguk.woo@dongkuk.com*

Received 15 June 2008
Revised 23 June 2008

A Molten Carbonate Fuel Cell (MCFC) stack consists of several layered unit cells. In each unit cell, the stiff structure of the separator plate contains the softer components, such as electrodes. When surface pressure acts on the stack over an extended period of time at elevated temperatures, the stiffness of the separator plate tends to degrade. Moreover, the demands for large electrode area (to increase the electric capacity of a unit cell) and thinner separator plates (to reduce weight) complicate the design of a separator plate with high stiffness. To evaluate the stiffness of the separator plate at elevated temperatures, we design and test a tiny, multi-layered separator plate specimen using a three-point bending tool. To determine the optimal structure of the separator plate, we investigate three design factors: *angle*, *pitch* and *height*. We adopt the Taguchi method to evaluate the experiments, and use finite element analysis to examine the experimental results. Based on these results, *pitch* is the most effective of these factors. As the *pitch* narrows, the stiffness of the separator plate increases. Therefore, we propose the *pitch* factor as a design criterion for the separator plate of the MCFC stack.

*Keywords*: Multi-layered panel; Three-point bending; Elevated temperature; MCFC; Separator plate.

## 1. Introduction

A Molten Carbonate Fuel Cell (MCFC) stack consists of several layered unit cells. Unit cells have three main components: electrodes, matrices, and separators. Separator plates are composed of two flat, stainless steel skins with a corrugated core between them. These plates protect the other, softer components in the cell, and maintain the passages that fuel gas flow during operation.

The stiffness of the separator plate tends to degrade at elevated temperatures because of the increase in the stack pressure, which is necessary for compacting unit cells.

---

[†]Corresponding Author.

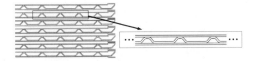

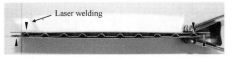

Fig. 1. A multi-layered panel specimen drawn from the original size of the separator.

Fig. 2. A sample of the multi-layered panel specimen, which was completed with laser welding.

Moreover, the demands for larger electrode area to increase electric capacity and thinner separator plates to reduce the weight make the design of separator plates with high stiffness more difficult. Therefore, it is crucial to assess the conventional separator plate design and to find out what design factor has the greatest influence on the stiffness of the separator plate.

Previous research on the bending stiffness of multi-layered panels like the separator plate has been conducted at room temperature[1-3] or at elevated temperatures.[4,5] Also, previous studies have examined springback and the minimization of springback, which is inevitably accompanied with bending.[6-8]

To evaluate the stiffness of the separator plate at elevated temperatures, we design and test a tiny multi-layered panel specimen drawn from the original separator plate using a three-point bending tool. We examine three design factors (*angle*, *pitch*, and *height*) to observe how these factors influence the stiffness of the separator plate. An orthogonal table is adopted as an effective and systematic means of evaluation. We use a multi-stage simulation by the finite element code ABAQUS to analyze the experimental results.

## 2. Experiments

### 2.1. *Preparation of the multi-layered panel specimens*

The separator plate for the MCFC is currently manufactured using sheet metal press forming technology. It is made of SUS 310S, which is a stainless steel material. The plate now under development is too large to handle in the laboratory (around 10,000 cm$^2$). Thus, a multi-layered panel specimen is drawn from the original separator as depicted in Fig. 1.

The multi-layered panel is composed of three pieces of sheet metal. The upper and lower sheets are flat, while the center sheet has corrugations that are formed by press forming. These three sheets are joined by laser welding. The specimen is 71 mm long and 28 mm wide, and includes about ten pitches of corrugations. Fig. 2 shows an example of the multi-layered panel specimen, as well as the four points for laser welding.

### 2.2. *Experimental apparatus for evaluating bending stiffness*

Fig. 3 shows the specifications for a bending stiffness tester, which consists of one depressor and two supports. The distance *l* between the two supports is as follows (refer to Fig. 3):

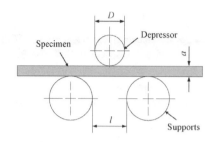

Fig. 3. Schematic drawing of the three-point bending.

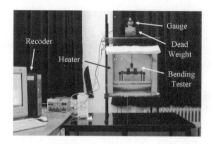

Fig. 4. Experimental setup for measuring the deflection of the specimen.

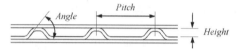

Fig. 5.   The three design factors of the multi-layered panel specimen.

Table 1. Design factors and their levels selected for the Taguchi method.

| Factors | Levels | | |
|---|---|---|---|
| | 1 | 2 | 3 |
| Angle | 50° | 55° | 60° |
| Pitch | 10.2 mm | 7.9 mm | 6.5 mm |
| Height | 0.9 mm | 1.1 mm | 1.3 mm |

$$l = (D + 3a) \pm \frac{a}{2}.  \tag{1}$$

For our experiments, the parameters were set to $D = 16$ mm and $a = 2.5$ mm in average. Thus we get $l = 25$ mm. The diameter of the supports is 20 mm. The stiffness evaluation is based on the deflection at the center of the specimen.

Fig. 4 shows the test apparatus. The bending tester is inserted into the furnace to assess the bending stiffness at elevated temperatures. The dead weight for applying a static load on the specimen is installed over the depressor. The static load is set to 5.6 kgf obtained from several pre-tests. We perform experiments with the following procedures:
 (i) The static load is applied to the specimen through the depressor at room temperature.
 (ii) While the static load is kept in place, the furnace is heated to 750 °C over 6 hours.
(iii) The digital gauge records the deflection of the specimen throughout the process.

### 2.3.  Selection of design factors

We select three kinds of design factors (*angle*, *pitch* and *height*) as key factors that affect the stiffness of the multi-layered panel shown in Fig. 5. Each factor has three levels, as shown in Table 1. We introduce the Taguchi method into the experiments to carry out effective and systematic analysis of the stiffness. We adopt the $L_9(3^3)$ table,[9] which is one of the most frequently referenced orthogonal tables in the Taguchi method.

### 2.4.  Experimental results and discussion

Fig. 6 shows the deflection measured at the center of the specimen at various temperatures. Table 2 presents the arrangement of factors and levels in each run. One can

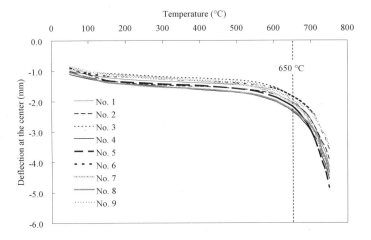

Fig. 6. Experimental results of deflection due to temperature variation measured at the center of the specimen.

Table 2. Layout of Taguchi table $L_9(3^3)$ and the experimental results.

| No. | Factors | | | Deflection |
| | Angle | Pitch | Height | (mm), 650 °C |
|---|---|---|---|---|
| 1 | 50 | 10.2 | 0.9 | -1.94 |
| 2 | 50 | 7.9 | 1.1 | -2.04 |
| 3 | 50 | 6.5 | 1.3 | -1.84 |
| 4 | 55 | 10.2 | 1.1 | -2.33 |
| 5 | 55 | 7.9 | 1.3 | -2.16 |
| 6 | 55 | 6.5 | 0.9 | -1.84 |
| 7 | 60 | 10.2 | 1.3 | -2.06 |
| 8 | 60 | 7.9 | 0.9 | -2.29 |
| 9 | 60 | 6.5 | 1.1 | -1.95 |

Table 3. Analysis of Variance (ANOVA) results for the three factors.

| Source of variation | Sum of squares | DOF | Mean square | $F_0$ | Sig. prob. |
|---|---|---|---|---|---|
| Model | 0.419 | 6 | 0.070 | 1.283 | 0.340 |
| *Angle* | 0.110 | 2 | 0.055 | 1.012 | 0.395 |
| *Pitch* | 0.281 | 2 | 0.141 | 2.584 | 0.120 |
| *Height* | 0.028 | 2 | 0.014 | 0.254 | 0.780 |
| Error | 0.599 | 11 | 0.054 | | |
| Total | 1.018 | 17 | | | |

observe the initial deflection that is caused by the static load on the specimen at room temperature. The change in deflection takes place slowly up to nearly 500 °C. Above 500 °C, the deflection increases very rapidly. This is especially true near 650 °C, at which point the MCFC stack normally works. Thus, the stiffness of the separator plate weakens dramatically near the operational temperature of the stack. Table 2 shows the deflection amount at 650 °C for each run.

We perform Analysis of Variance (ANOVA) for the data in Table 2 to determine the factor that has the greatest influence on the stiffness. Table 3 shows the result of the ANOVA, which determined that *pitch* is the most influential factor. The same conclusion can also be drawn from the mean analysis in Fig. 7. To verify that similar results are obtainable at temperatures besides 650 °C, we describe the significance probability of the three factors along with temperature changes in Fig. 8. *Pitch* remains the dominant factor over all of the tested temperature ranges.

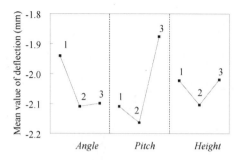

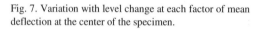

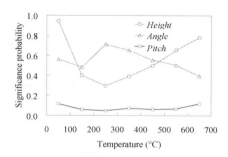

Fig. 7. Variation with level change at each factor of mean deflection at the center of the specimen.

Fig. 8. Significance probability for the three design factors with changes in temperature (experimental result).

Table 4. Material properties according to temperature used in the numerical simulation.

| Temp. (°C) | $E$ (GPa) | $\sigma_y$ (MPa) | $\sigma = K(\varepsilon_0 + \varepsilon_p)^n$ | | |
|---|---|---|---|---|---|
| | | | $K$ (MPa) | $\varepsilon_0$ | $n$ |
| 25 | 174.66 | 207.838 | 1153.975 | 0.024 | 0.459 |
| 550 | 118.137 | 149.342 | 1129.725 | 0.022 | 0.529 |
| 600 | 115.348 | 145.156 | 1052.935 | 0.021 | 0.512 |
| 650 | 105.119 | 142.504 | 734.095 | 0.021 | 0.467 |
| 700 | 84.195 | 133.638 | 501.220 | 0.007 | 0.266 |
| 750 | 55.433 | 123.898 | 321.985 | 0.003 | 0.164 |

Table 5. Comparison of the deflection.

| | Deflection (650 °C) | | |
|---|---|---|---|
| No. | Analysis (mm) | Experiment (mm) | Gap (%) |
| 1 | -1.880 | -1.94 | 3.12 |
| 2 | -1.930 | -2.04 | 5.29 |
| 3 | -1.950 | -1.84 | 5.94 |
| 4 | -2.236 | -2.33 | 3.83 |
| 5 | -2.034 | -2.16 | 5.87 |
| 6 | -1.943 | -1.84 | 5.64 |
| 7 | -1.936 | -2.06 | 5.98 |
| 8 | -1.907 | -2.29 | 16.68 |
| 9 | -1.992 | -1.95 | 2.21 |
| | | Avg. | 6.06 |

## 3.  Finite Element Simulations

We use ABAQUS 6.5[10] to perform finite element simulations that estimate the stiffness of multi-layered panels. The material properties of SUS 310S over the range of temperatures used in the simulations are shown in Table 4 (stress-plastic strain curve, Young's modulus, etc.[5]) and Fig. 9 (coefficient of thermal expansion). Fig. 10 shows the simulation procedure. To ensure the accuracy of the simulated results, we carefully designed the multi-step simulations to reflect the real manufacturing procedure.

Table 5 shows a comparison of the experimental and simulated results. The simulated results agree with the experimental ones, as there is only a 6% average deviation between the two. Therefore, the simulation set-up developed in this work can predict experimental results for specimens that have any other combination of factors and levels.

## 4.  Conclusions

We draw the following conclusions from the results of both experiments and simulations on the multi-layered panel:

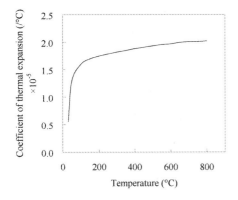

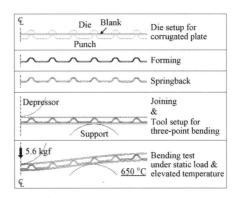

Fig. 9. Variation of the coefficient of thermal expansion with temperature.

Fig. 10. Procedure for the finite element simulation to analyze the three-point bending under elevated temperatures.

(1) The Taguchi method reveals that among the three design factors, *pitch* has the greatest influence on the stiffness of the multi-layered panel at the MCFC stack operational temperature of 650 °C.

(2) The ANOVA results, which are analyzed over wide ranges of temperature from room temperature to 750 °C, also show that *pitch* is the dominant factor.

(3) The simulated results obtained under conditions that closely reflect the real manufacturing procedure for the multi-layered panel are consistent with the experimental results with a deviation of only 6%. Therefore, the simulation procedure used in this study can predict results for other multi-layered panels with different arrangements of sheet metals.

## Acknowledgments

Financial support by the Korean Ministry of Knowledge and Economy through the Electric Power Industry Technology Evaluation & Planning Center is gratefully acknowledged (Project no. R-2004-1-142).

## References

1. J. H. Lim, S. J. Nah, M. H. Koo, K. J. Kang, *J. Kor. Soc. Mech. Eng. A*, **29**(3), 470 (2005).
2. D. Y. Seong, C. G. Jung, S. J. Yoon, D. Y. Yang, D. G. Ahn, *J. Kor. Soc. Prec. Eng.*, **23**(4), 127 (2006).
3. H. J. Rathbun, F. W. Zok, A. G. Evans, *Int. J. Solids Struct.*, **42**, 6643 (2005).
4. C. K. Iu, S. L. Chan, X. X. Zha, *Eng. Struct.*, **27**, 1689 (2005).
5. J. H. Kim, *M.S. Thesis*, Soonchunhyang University, Asan, Korea (2006).
6. S. W. Lee and D. Y. Yang, *J. Mater. Proc. Tech.*, **80-81**, 60 (1998).
7. S. W. Lee and Y. T. Kim, *J. Mater. Proc. Tech.*, **187-188**, 89 (2007).
8. J. Yanagimoto and K. Oyamada, *Annals of the CIRP*, **56**, 265 (2007).
9. G. Taguchi and S. Konishi, *Orthogonal arrays and linear graphs*, ASI (1990).
10. ABAQUS User's Manual (ver. 6.5), ABAQUS Inc., Richmond, USA (2004).

# NUMERICAL ANALYSIS ON MAGNETIC-ELASTO-PLASTIC BUCKLING AND BENDING OF FERROMAGNETIC RECTANGULAR PLATE

YUANWEN GAO[1†]

*Key laboratory of mechanics on western disaster and enviroment,*

*Lanzhou University, Lanzhou, 730000, P.R. CHINA,*

*ywgao@lzu.edu.cn*


HOON HUH[2]

[2]*School of Mechanical, Aerospace and System Engineering, KAIST,*

*Science Town, Daejeon, 307-701 KOREA,*

*hhuh@kaist.ac.kr*

Received 15 June 2008
Revised 23 June 2008

This paper presents an analysis on magnetic-elasto-plastic buckling and bending of a ferromagnetic rectangular plate with simple supports in applied magnetic fields. A numerical code is built to quantitatively simulate the characteristics of the magnetic-elasto-plastic nonlinear coupling problem by using a finite element method. The numerical results indicate that deformation of a ferromagnetic plate with elastic-plastic deformation is larger than that only with the elastic deformation. The incident angle of the applied magnetic field plays an important role in the deformation of the ferromagnetic rectangle plate: not only affects the magnitude of the deformation; but also the mode of the deformation. The configurations and plastic regions of ferromagnetic rectangular plates for different magnitude of the applied magnetic field are simulated and displayed as well.

*Keywords*: Ferromagnetic plate; Elasto-plastic deformation; Magneto-mechanical coupling.

## 1. Introduction

As the extensive applications of electromagnetic materials and structures in modern high technology, the understanding of the magnetic-elasto-plastic behaviors of the ferromagnetic structures in the magnetic field is quite useful in engineering applications, for example, in the safe design and control of fusion reactors, and the electro-magnetic formation processing for ferromagnetic materials.

Moon and Pao[1] investigated the magneto-elastic buckling of a ferromagnetic beamplate for the first time. After that, the magneto-mechanics interaction and coupling problem for ferromagnetic structures attracts many researchers and practitioners in this

---

[†]Corresponding Author.

and neighboring fields[2-6]. However, most of these previous studies are focused on the scope of elastic deformation. As the authors knowledge wide, except Zhou et al.[7], Gao et al.[8], there is few research works conducted for magnetic-elasto-plastic behaviors of ferromagnetic structures materials since the problem is much more complex and difficult to get analysis results for the behaviors. In this paper, we simulated the magnetic-elasto-plastic buckling and bending of ferromagnetic rectangular plate structures with the simple supports. Some numerical results are displayed in the form of characteristic curves of deformation, configuration and residual deformation.

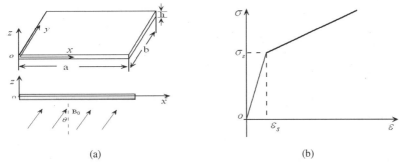

(a)                                                                 (b)

Fig. 1. Schematic drawing of a soft ferromagnetic rectangular plate (a) Ferromagnetic plate in an oblique magnetic field. (b) The constitutive curve of the material employed.

## 2. Fundamental Equations

### 2.1. *Governing equations for ferromagnetic rectangular plate*

Consider a soft ferromagnetic rectangular thin plate with the length, $a$, the width, $b$, and the thickness, $h$, in an applied magnetic field $\mathbf{B}_0$ with the incident angle $\theta$, as shown in Fig. 1(a). When the incident angle equals to zero, we refer to the plate as being in a transverse magnetic field, otherwise, in an oblique magnetic field. According to the bending theory of plate, the governing equation for ferromagnetic plate can be written as follows:

$$\frac{\partial^2 M_x}{\partial x^2} + 2\frac{\partial^2 M_{xy}}{\partial x \partial y} + \frac{\partial^2 M_y}{\partial y^2} + q_z^{em}(x, y) = 0 \tag{1}$$

where $M_x, M_y, M_{xy}$ are bending moments and the twist moment respectively, $q_z^{em}(x, y)$ is an equivalent magnetic force exerted in the middle plane of plate expressed as[6]

$$q_z^{em}(x, y) = \frac{\mu_0 \mu_r \chi}{2} \{ [\mathbf{H}_n^+(x, y, h/2)]^2 - [\mathbf{H}_n^+(x, y, -h/2)]^2 \}$$
$$- \frac{\mu_0 \chi}{2} \{ [\mathbf{H}_\tau^+(x, y, h/2)]^2 - [\mathbf{H}_\tau^+(x, y, -h/2)]^2 \} \tag{2}$$

where $\mu_0$ and $\mu_r$ are the vacuum and relative magnetic permeability of a ferromagnetic medium, respectively; $\chi = \mu_r - 1$ is magnetic susceptibility; $\mathbf{H}_n^+, \mathbf{H}_\tau^+$ are the normal and tangential components of $\mathbf{H}^+$ on the surface of ferromagnetic plate respectively.

## 2.2.  Generalized yield criteria for ferromagnetic rectangular plate

The constitutive curve of the material employed is shown in Fig. 1(b), in which we denote the Young's modulus by $Y$, the hardening parameter by $H'$, which governs the expansion of the yield surface, and the yield stress by $\sigma_s$.

For a linear-strain-hardening material, the Mises yield criterion is[9, 10]

$$F(\{\sigma\}, H') = \bar{\sigma} - \sigma_Y(H') = [\sigma_x^2 - \sigma_x \sigma_y + \sigma_y^2 + 3\sigma_{xy}^2]^{\frac{1}{2}} - \sigma_Y(H') = 0 \tag{3a}$$

$$\sigma_Y(H') = \sigma_s + H'\bar{\varepsilon}_p \tag{3b}$$

where $\bar{\sigma}, \bar{\varepsilon}_p$ are the effective stress and strain respectively.

According to the flow rule of plasticity, the elasto-plastic incremental stress-strain relation can be described as

$$d\{\sigma\} = [D]_{ep} d\{\varepsilon\} \tag{4}$$

with

$$[D]_{ep} = [D]_e - \frac{\{d_D\}\{d_D\}^T}{H' + \{d_D\}\{a\}} \tag{5}$$

where $\{a\} = (\dfrac{\partial F}{\partial \sigma_x}, \dfrac{\partial F}{\partial \sigma_y}, \dfrac{\partial F}{\partial \sigma_{xy}})^T$ is called as the vector of flow, $\{d_D\} = [D]_e\{a\}$.

## 3.  Finite Element Formulation

## 3.1.  The incremental finite element formulation for elasto-plastic deformation

According to the principle of virtual work[10], the incremental equilibrium equations in the $i$-th load step can be expressed as follows

$$[K_T]_i \Delta\{u\}_i = \Delta\{R(\mathbf{H}^+)\}_i \tag{6}$$

here

$$[K_T]_i = \sum_e \int_V [B]^T [D_T^{(i)}]_{ep} [B] dV \tag{7}$$

$\Delta\{R(\mathbf{H}^+)\}$ is the increment of load, which is related to the magnetic field $\mathbf{H}^+$, $[B]$ is the geometric matrix, $[D_T^{(i)}]_{ep}$ is the tangent stiffness, which can be obtained from Eq. (5).

Based on Eq. (6) and (7), the vector of deflection **u**, strain and stress $\varepsilon, \sigma$ of a ferromagnetic plate with elastic-plastic deformation can be easily obtained.

### 3.2. Finite element formula for three dimension magnetic field

For linear electromagnetic materials, the magnetic energy functional of a ferromagnetic rectangular plate can be described as follows[6]:

$$\Pi_{em}(\phi,\mathbf{u}) = \int_{\Omega^+(\mathbf{u})} \mu_0\mu_r(\nabla\phi^+)^2 dv + \int_{\Omega^-(\mathbf{u})} \mu_0(\nabla\phi^-)^2 dv + \int_{\Gamma_0} \mathbf{n_0}\cdot\mathbf{B_0}\phi^- ds \qquad (8)$$

where $\Omega^+(\mathbf{u})$, $\Omega^-(\mathbf{u})$ represent the regions inside and outside of plate respectively. $\Gamma_0$ is the surface of plate, $\phi$ is the magnetic scalar potential of magnetic fields.

Based on the generalized variation principle[6] for a magnetic body and using a twenty-node isoperimetric hexahedron element in a finite element method, one can get the equation for magnetic field as

$$[\mathbf{K}^{em}(\mathbf{u})][\boldsymbol{\Phi}] = [\mathbf{P}] \qquad (9)$$

in which $[\mathbf{K}^{em}(\mathbf{u})]$ is the global magnetic stiffness matrix, $[\boldsymbol{\Phi}]$ is the vector of unknown magnetic scalar potential.

With solving Eq. (9), one can easily obtain $H_x, H_y, H_z$ and then the equivalent magnetic force exerted in the middle plane of ferromagnetic plate of Eq. (2) can be obtained.

## 4. Numerical Results and Discussion

According to the formulation and approaches introduced above, some numerical results are provided in this section. In simulation, the material and geometric parameters of ferromagnetic rectangular plate are set as Young's modulus $Y = 1.2\times10^{11} Pa$, Poisson's ratio $v = 0.3$, Relative magnetic permeability $\mu_r = 10000$, Yield stress $\sigma_s = 40MPa$, Hardening parameter $H' = 1.1\times10^{11}$, the length, width and thickness of the plate are set $a = 2.0m, b = 1.0m, h = 0.01m$ respectively.

Fig. 2 shows the magnetic-elasto-plastic deflection curves of a plate at $(a/4, 4b/5)$ varying with the magnitude of the applied magnetic field $\mathbf{B_0}$ for different incident angle $\theta$. Fig. 2(a) displays the buckling and bending behaviors of a ferromagnetic rectangular plate under the magnetic field with small incident angles $0 \le \theta \le 1^o$, whereas the path of bending deflection varying with the magnitude of applied magnetic field with large incident angles ($5^o \le \theta \le 90^o$) is shown in Fig. 2(b). It can be found only in the transverse magnetic field ($\theta = 0^o$) that the buckling phenomenon of a ferromagnetic plate take place. In an oblique magnetic field, the mode of deformation is only bending. As shown in Fig. 2, the deflection of a ferromagnetic rectangular plate increases with the

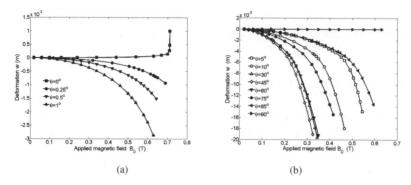

(a)                                    (b)

Fig. 2. Magnetic-elastic-plastic characteristic curves of plate (a) $0 \leq \theta \leq 1^{o}$ ; (b) $5^{o} \leq \theta \leq 90^{o}$ .

increase of the incident angle of applied magnetic field when $0^{o} < \theta < 45^{o}$ , However, the deformation of plate decreases with the increase of the incident angle when $45^{o} < \theta < 90^{o}$ , as special cases, when $\theta = 45^{o}$ , the plate produces the largest deformation and in the case of in-plane magnetic field ( $\theta = 90^{o}$ ), the deformation of a ferromagnetic plate is zero.

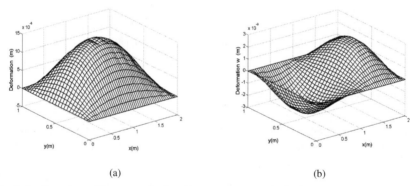

(a)                                    (b)

Fig. 3. Configurations of ferromagnetic plate (a) transverse magnetic field; (b) oblique magnetic field.

Fig. 3 displays the configuration of a ferromagnetic rectangular plate in the transverse magnetic field and the oblique magnetic fields respectively.

Fig. 4 shows the comparison of deflection curves at point $(a/4, 4b/5)$ of in the middle plane of a plate with plastic deformation versus those with elastic deformation (The symbols $e$ and $p$ presents the results for elastic and elasto-plastic, respectively). As shown in Fig. 4, the deflection of ferromagnetic plate with elastic-plastic deformation is obviously larger than those with elastic deformations. The critical value of applied magnetic field for plastic yield of a ferromagnetic rectangular plate, which is corresponding to the separation point of the elastic curves and the elastic-plastic curves for the same incident angle of applied magnetic field, can be observed clearly.

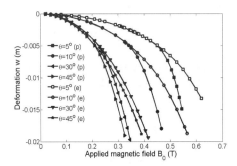

Fig. 4. Comparison of the deflection of the plate with elastic deformation versus the elastic-plastic deformation.

## 5. Conclusion

This paper gives a quantitative simulation on the magnetic-elasto-plastic buckling and bending of a ferromagnetic rectangular plate with simple supports, some characteristic curves of magnetic-elasto-plastic buckling, bending, and configurations are presented. The numerical results indicate that the incident angles of applied magnetic field plays an important role in the deformation of a ferromagnetic rectangular plate: not only in the magnitude of deformation; but in the configuration. When $\theta = 45^o$, the plate generates the largest deformation, while the deformation is zero when $\theta = 90^o$.

## Acknowledgment

This work was supported by the BK21 program of Korea, Program for New Century Talented of China (NCET-06-0896), and Natural Science Fund of China (10672070).

## References

1. F. C. Moon, P. H. Pao, *ASME J. Appl. Mech.* **35**, 53-58 (1968).
2. Y. H. Pao, C. S. Yeh, *Int. J. Engng. Sci.* **11**, 415-436 (1973).
3. A. C. Eringen, *Int. J. Engng. Sci.* **27**, 363-375 (1989).
4. T. Tagaki, J. Tani, Y. M. Matsubara, T. Mogi, *Fusion Eng. Des.* **27**, 481-489 (1995).
5. D. N. Fang, Y. P. Wan, A.K. Soh, *Theor. and Appl. Fract. Mech.* **42(3)**, 317-334 (2004).
6. Y. H. Zhou, X. J. Zheng, *Int. J. Engng. Sci.* **35**, 1405-1417 (1997).
7. Y. H. Zhou, Y. W. Gao, X. J. Zheng, *Int. J. Solids and struct.* **40**, 2875-2887 (2003).
8. Y. W. Gao, Y.H. Zhou, X. J. Zheng, *Key engineering materials.* **276**, 1131-1136 (2004).
9. R. Hill, *The mathematical theory of plasticity* (Oxford Press, London 1956).
10. D. R. J. Owen, E. Hinton, *The finite element in plasticity-theories and practice* (Pineridge Press Limited, Swansea, 1980).

# LOAD-CARRYING CAPACITIES FOR CIRCULAR METAL FOAM CORE SANDWICH PANELS AT LARGE DEFLECTION

W. HOU[1]

*Faculty of Engineering and Industrial Sciences, Swinburne University of Technology,*
*Hawthorn, Vic 3122, AUSTRALIA,*
*whou@groupwise.swin.edu.au*


Z. WANG[2], L. ZHAO[5]

*Faculty of Engineering and Industrial Sciences, Swinburne University of Technology,*
*Hawthorn, Vic 3122, AUSTRALIA,*
*Institute of Applied Mechanics and Biomedical Engineering, Taiyuan University*
*of Technology Taiyuan, 030024, P.R. CHINA,*
*zwang@groupwise.swin.edu.au, zhaolm@tyut.edu.cn*


G. LU[3†], D. SHU[4]
*Faculty of Engineering and Industrial Sciences, Swinburne University of Technology,*
*Hawthorn, Vic 3122, AUSTRALIA,*
*School of Mechanical and Aerospace Engineering, Nanyang Technological*
*University, SINGAPORE,*
*glu@groupwise.swin.edu.au*

Received 15 June 2008
Revised 23 June 2008

This paper is concerned with the load-carrying capacities of a circular sandwich panel with metallic foam core subjected to quasi-static pressure loading. The analysis is performed with a newly developed yield criterion for the sandwich cross section. The large deflection response is estimated by assuming a velocity field, which is defined based on the initial velocity field and the boundary condition. A finite element simulation has been performed to validate the analytical solution for the simply supported cases. Good agreement is found between the theoretical and finite element predictions for the load-deflection response.

*Keywords*: Circular sandwich panel; Large deflection; Metal foam.

## 1. Introduction

Circular panel is a typical structural component and the analytical prediction of its loading response has been extensively studied since 1950s.[1,2] Generally, there are two approaches to calculate the structural response under quasi-static loading: (1) infinitesimal deflection analysis where only bending is considered; and (2) finite deflection analysis where both bending and membrane are taken into account, which is essential when structures undergo large deflection of the order which is of several times

---

[†]Corresponding Author.

the corresponding thickness. The large deflection problems become very complicated because the effect of membrane force induced cannot be neglected[3-8]. Onat and Haythornthwaite observed[3] that simply supported circular plates under a central punch load begins to deform into a shallow cone, and then used an upper-bound technique to calculate the load-carrying capacity of a rigid-plastic circular plate. Onat and coworkers established[4,5] an "exact" limit analysis of shallow conical shells, and then extended the procedure to study large deflection of simply supported and fully clamped circular plates. It has been shown that the previous results overestimate the strengthening effect of the geometry changes. Calladine presented[6] a simplified approach to analyze the large deflection of plates and slabs in the plastic range. Sawczuk attempted[7] to establish a general theory of large deflections of rigid-plastic plates, and it has been found that the results show good coincidence with the experiment. Belenkiy derived[8] upper-bound solutions of problems for perfectly rigid plastic plates using the principles of general form virtual work and stationary of total energy.

Most of the previous works have been focused on the monolithic panels, and in recent years, increasing attention turns to the response of sandwich plates, due to their excellent energy absorbing performances. In this paper, a new yield condition considering the effect of core strength is developed for the sandwich structures. The load-carrying capacities of a simply supported circular sandwich plate with metallic foam core are studied. The influence of the membrane forces on their loading resistant capacities is estimated by assuming a velocity field. Then the theoretical predictions are compared with a numerical simulation, and the results show a reasonable agreement.

## 2. Approximate Solutions of Simply Supported Circular Sandwich Plate at Large Deflection

Consider a simply supported circular sandwich plate of radius $R$ with identical face-sheets of thickness $h_f$ and a core of thickness $H_c$. The core material is a cellular medium with density $\rho_c$ and yield strength $\sigma_c$. Its base material is considered as a rigid-perfectly plastic isotropic solid. The face-sheet material has a density $\rho_f$ and is assumed to be rigid-perfectly plastic with a tensile strength $\sigma_f$. An external load with intensity $p$ is laterally applied to a central circular zone with radius $r = a$. Similar to monolithic solid plate, it is reasonable to assume the initial-velocity distribution would deform the sandwich plate into a circular cone with apex at the center as a first approximation. Therefore, the deflection field can be expressed as

$$w = w_m(1-\rho) \tag{1}$$

where $\rho = r/R$, with $w_m$ being the maximum deflection. Due to nonlinear geometric nature of the problem, the generalized strain components of a circular sandwich panel subjected to a bending moment and a membrane force are expressed as follows:

$$e_r = \varepsilon_r + z\kappa_r, \; e_\theta = \varepsilon_\theta + z\kappa_\theta \tag{2}$$

where $\varepsilon_r = du/dr + 1/2(dw/dr)^2$ and $\varepsilon_\theta = u/r$ are components of tensile strain, and $\kappa_r = -\partial^2 w/\partial r^2$, $\kappa_\theta = -1/r \, \partial w/\partial r$ are principal curvatures of the middle plane. $w$ and $u$ denote the deflection and the radial displacement of the middle plane, respectively. It is noted that, if the plate is freely supported at the outer edge, radial membrane force at the edge would be zero, thus it is reasonable to assume that the strain component of the middle plane is zero at the outer edge. For simplicity, the component is taken as zero throughout the plate, that is $\varepsilon_r = 0$. On the other hand, if $u = 0$ at $\rho = 0$, the following relation is obtained from Eq. (2)

$$u = -\frac{1}{2}\int\left(\frac{dw}{dr}\right)^2 dr = -\frac{w_m^2}{2R}\rho \tag{3}$$

Thus, the displacement field can be defined using only one parameter and the corresponding strain rates can be expressed as

$$\dot{\varepsilon}_r = 0, \dot{\varepsilon}_\theta = -\frac{w_m}{R^2}, \; \dot{\kappa}_r = 0, \; \dot{\kappa}_\theta = \frac{1}{R^2\rho} \tag{4}$$

In the paper, a novel yield criterion of an axisymmetric sandwich element subjected to a circumferential membrane force $N$ and a circumferential bending moment $M$ is developed. Figure 1 shows the typical distribution of the normal stresses on a sandwich cross section, subjected to a bending moment $M$ and a membrane force $N$ simultaneously. Based on the magnitude of $N$, the limited stresses distribution is divided into two regimes, i.e. $0 \le \left|\dfrac{N}{N_0}\right| \le \dfrac{\bar{\sigma}}{\bar{\sigma}+2\bar{h}}$ (Fig.1a) and $\dfrac{\bar{\sigma}}{\bar{\sigma}+2\bar{h}} \le \left|\dfrac{N}{N_0}\right| \le 1$ (Fig. 1b),

where $\bar{\sigma} = \sigma_c/\sigma_f$, $\bar{h} = h_f/H_c$. Then the profile of the stresses can be described by the combination of a symmetric component with respect to the central axis (membrane effect) and an antisymmetric component (bending effect).

According to the stress distribution, circumferential bending moment $M$ and membrane force $N$ in both two cases can be calculated. The corresponding yield locus is expressed as

$$\left|\frac{M}{M_0}\right| + \frac{(\bar{\sigma}+2\bar{h})^2}{4\bar{\sigma}\bar{h}(1+\bar{h})+\bar{\sigma}^2}\left(\frac{N}{N_0}\right)^2 = 1 \qquad 0 \le \left|\frac{N}{N_0}\right| \le \frac{\bar{\sigma}}{\bar{\sigma}+2\bar{h}} \tag{5a}$$

$$\left|\frac{M}{M_0}\right| + \frac{\left[\left(\dfrac{N}{N_0}\right)(\bar{\sigma}+2\bar{h})+(1-\bar{\sigma})\right]^2-(1+2\bar{h})^2}{4\bar{h}(1+\bar{h})+\bar{\sigma}} = 0 \qquad \frac{\bar{\sigma}}{\bar{\sigma}+2\bar{h}} \le \left|\frac{N}{N_0}\right| \le 1 \tag{5b}$$

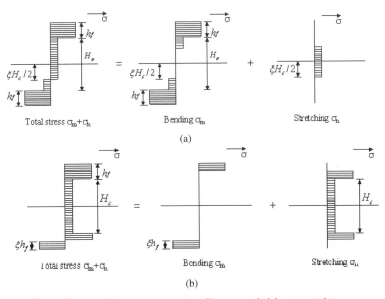

Fig. 1. Sketch of the normal stresses profile on a sandwich cross-section.

Bending moments and membrane forces corresponding to the strain rates defined in Eq. (4) are dependent on the following parameters[4]

$$\beta_1 = -\frac{1}{h}\frac{\dot\varepsilon_r}{\dot\kappa_r} = \frac{0}{0}, \quad \beta_2 = -\frac{1}{h}\frac{\dot\varepsilon_\theta}{\dot\kappa_\theta} = \frac{w_m}{h}\rho, \quad \beta_3 = -\frac{1}{h}\frac{\dot\varepsilon_r + \dot\varepsilon_\theta}{\dot\kappa_r + \dot\kappa_\theta} = \beta_2 \tag{6}$$

where $h = H_c + 2h_f$. Adopting these parameters, the values of membrane forces can be obtained. Then substituting the membrane forces into the yield criterion, we have

$$n_\theta = \frac{-2\overline{w}_m\overline{\sigma}\rho}{\overline{\sigma} + 2\overline{h}}, \quad m_\theta = 1 - \frac{\left(\overline{\sigma} + 2\overline{h}\right)^2}{4\overline{\sigma}\overline{h}\left(1 + \overline{h}\right) + \overline{\sigma}^2}n_\theta^2 \quad \text{when } |n_\theta| \le \frac{\overline{\sigma}}{\overline{\sigma} + 2\overline{h}} \tag{7a}$$

$$n_\theta = -\frac{\left(2\overline{w}_m\rho - (1 - \overline{\sigma})\right)}{\overline{\sigma} + 2\overline{h}}, \quad m_\theta = \frac{\left(1 + 2\overline{h}\right)^2 - \left[|n_\theta|\left(\overline{\sigma} + 2\overline{h}\right) + (1 - \overline{\sigma})\right]^2}{4\overline{h}\left(1 + \overline{h}\right) + \overline{\sigma}}$$

$$\text{when } \frac{\overline{\sigma}}{\overline{\sigma} + 2\overline{h}} \le |n_\theta| \le 1 \tag{7b}$$

where $n_\theta = N_\theta / N_0$, $m_\theta = M_\theta / M_0$, $M_0 = \sigma_f h_f(H_c + h_f) + \sigma_c H_c^2 / 4$

$$N_0 = 2\sigma_f h_f + \sigma_c H_c, \quad \overline{w}_m = w_m / H_c \tag{8}$$

Then dissipation energy can be obtained as

$$D = \int_0^{w_m} \dot{D}\, dw_n = \int_0^{w_m}\int_0^1 2\pi\left(N_\theta\dot\varepsilon_\theta + M_\theta\dot\kappa_\theta\right)R^2\rho\, d\rho\, dw_n \tag{9}$$

The external work by the applied pressure $p$ is

$$E = 2\pi \int_0^a pw_m(1-\frac{r}{R})\,rdr = P\left(1-\frac{2a}{3R}\right)w_m \qquad (10)$$

where $P = \pi a^2 p$. The total energy of the circular plate $\Pi = D - E$, then, from condition $d\Pi/dw_m = 0$, we have

$$\frac{P}{P_L} = \frac{4}{3}\overline{w}_m^2 - 2\overline{w}_m(1-\overline{\sigma}) + (1+2\overline{h})^2 - \frac{(1-\overline{\sigma})}{6\overline{w}_m} \qquad (11a)$$

when $w_m \le \frac{H_c}{2}(1+2\overline{h})$, and

$$\frac{P}{P_L} = \frac{(2\overline{h}+1)^3}{6\overline{w}_m} - \frac{(1-\overline{\sigma})}{6\overline{w}_m} + 2\overline{w}_m(\overline{\sigma}+2\overline{h}) \qquad (11b)$$

when $w_m \ge \frac{H_c}{2}(1+2\overline{h})$, where $P_L = 2\pi\overline{M}/(1-2a/3R)$, $\overline{M} = \sigma_f H_c^2/4$.

It is obvious that Eq. (11) is a general load-displacement relationship, which is valid for various metallic sandwich circular plates with different core strength and geometrical dimensions. If $\overline{\sigma} = \sigma_c/\sigma_f = 1$, then Eq.(11) can reduce to the relationship for the monolithic circular plate[3].

## 3.    Comparison with Simulation Results

FE simulations have been performed using the ABAQUS for two configurations in the simply supported conditions to validate the analytical model. The geometries and material parameters are summarized in Table 1. Only half of the panel was modeled due to symmetry in order to reduce problem size. In the finite element model, the aluminium face sheets with tensile strength 75.8MPa are modeled as a perfect elastic-plasticity material, and the aluminium foam is described using the constitutive relationship by Deshpande and Fleck[9]. A mesh sensitivity study has been performed to ensure an accurate representation of the model.

Comparisons of the load-deflection curves obtained from the numerical and analytical predictions are shown in Fig. 2. It can be seen that the slopes of the curves from simulation are very close to those predicted by Eq. (11). It should be emphasized that an elastic deflection phase is added to the analytical modeling for comparison purpose.    At the early stages of post-yield, the results rising from the present large deflection analysis should be satisfactory for practical applications, because it overestimates deflections, if the shear effect is ignored.

Table 1. Geometries and material parameters of sandwich panels.

| No. | Face sheet thickness(mm) | Core thickness (mm) | Core relative density (%) | Radius of panel (mm) | Core yield stress(MPa) |
|---|---|---|---|---|---|
| Case1 | 1 | 5 | 11 | 160 | 2.01 |
| Case2 | 1 | 5 | 20 | 160 | 4.20 |

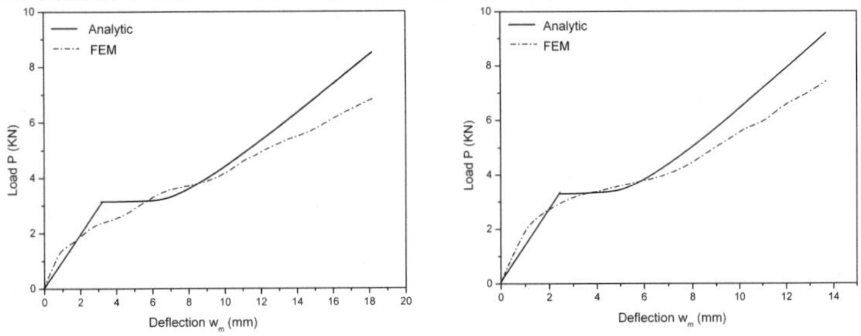

(a) Case 1         (b) Case 2

Fig. 2. Comparison of the load-deflection curves from theoretical calculation and simulation.

## 4. Conclusion

This study has focused on the effect of membrane force on large deflection response of simply supported circular sandwich plates comprising aluminum face sheets and an aluminum foam core. A new yield condition including the effect of core strength is developed for the sandwich structures. The load-carrying capacity at finite deflection is estimated by assuming a velocity field, which is defined based on the initial velocity field and the boundary condition. A finite element simulation has been performed in order to validate the analytical model. The comparison with the simulation results shows a reasonable agreement.

## Acknowledgments

The reported research is financially supported by the Australian Research Council (ARC) through a Discovery Grant and China National Natural Science Funding under the number of 10572100 and 90716005, which are gratefully acknowledged.

## References

1.  G. Lu and T.X. Yu, *Energy absorption of structures and materials* (Woodhead Co. Ltd, Cambridge, 2003).
2.  A. Sawczuk, *Mechanics and Plasticity of Structures* (Ellis Horwood, Chichester, 1989).
3.  E. T. Onat and R. M. Haythornthwaite, *J. Appl. Mech.* **23(1),** 49(1956).
4.  E. T. Onat, and W. Prager, *Proc.R. Netherlands Acad. Sci.* **57,** 336(1954).
5.  R. H. Lance and E. T. Onat, *Journal of Applied Mechanics.* **30,** 199(1963).
6.  C. R. Calladine, *in Engineering plasticity,* ed. J. Heyman and F. A. Lechie (Cambridge University Press, 1968), pp. 93-127.
7.  A. Sawczuk. *in Proc. 11^th Int. Congr. appl. Mech.* (Munich.1964), pp. 224-228.
8.  L. M. Belenkiy. *Journal of Engineering Mechanics.* **133(1),** 98(2007).
9.  V. S. Deshpande, N. A. Fleck. *J. Mech. Phys. Solids.* **48(6),** 1253 (2000).

# EARLY YIELDING OF LIQUID-FILLED CONICAL SHELLS

W. VANLAERE[1†], G. LAGAE[2], R. VAN IMPE[3], J. BELIS[4], D. DELINCÉ[5]

*Ghent University, Ghent, BELGIUM,*
*Wesley.Vanlaere@UGent.be, Guy.Lagae@UGent.be,*
*Rudy.VanImpe@UGent.be, Jan.Belis@UGent.be, Didier.Delince@UGent.be*

Received 15 June 2008
Revised 23 June 2008

In liquid-filled conical shells, the combination of compressive meridional stresses and tensile hoop stresses can lead to instability. The present design rule for these structures is verified in this contribution by means of numerical simulations. Large axisymmetric imperfections are detrimental and may lead to failure by plastic buckling. For these shapes the design rule may be unconservative. The realism of the proposed shape deviations is questioned. However, partially or completely axisymmetric weld depressions can't be excluded from the simulations. This means that an elaborated study is needed and that the design rule must be checked even further.

*Keywords*: Plastic buckling; Thin-walled shell structures; Numerical simulations.

## 1. Introduction

In thin-walled steel shell structures the effect of shape deviations on the buckling load can be very pronounced. At the beginning of the previous century, multiple studies showed that the experimental buckling load may be as low as a fraction of the theoretical buckling load. Elaborated research efforts have led to the conclusion that this discrepancy is mainly due to the unavoidable presence of relatively small geometrical imperfections.

In these studies, the shells collapsed due to elastic buckling. For this type of buckling the strength reduction due to imperfections is most pronounced. Later on new studies have explored the plastic and elastic-plastic buckling of steel shell structures and the sum of these results led to the Eurocode design regulations for shell structures[1]. This part of the Eurocode enables the engineer to design thin-walled steel shell structures, taking into account the effects of plasticity and geometrical imperfections.

In this contribution, the focus is on thin-walled steel liquid-filled conical shells. For these shells, an extensive experimental study was performed by Vandepitte *et al.*[2]. These experiments mostly led to failure caused by elastic instability. Together with a limited numerical study, the experimental results allowed the development of a design rule for the conical shells. The design rule was included in the fourth edition of the ECCS Recommendations of Buckling of Shells[3]. In a modified format – in order to be

---

[†]Corresponding Author.

compatible with the Eurocode – this design rule will also be included in the forthcoming fifth edition of these Recommendations[4].

The aim of our present study is to verify this design rule by means of numerical simulations with the finite element package ABAQUS. Especially the effect of plasticity on the buckling behavior of imperfect conical shells is reported in this contribution.

## 2. Liquid-Filled Conical Shells

### 2.1. *The stresses leading to failure*

A water tower is often constructed in steel. In order to reduce the amount of material that is needed for the structure, the water reservoir is made as a thin-walled conical tank. Although this is an economical solution, the thin shell wall causes the structure to be sensitive to instability. Two buckling phenomena are possible. Elastic buckling is caused by elevated meridional compressive stresses. These stresses appear due to the water load and are most pronounced near the base of the cone, which is therefore the location of the buckles. The presence of the water in the conical tank also leads to tensile hoop stresses. These circumferential stresses have a stabilizing effect and increase the elastic buckling load.

The second buckling phenomenon is plastic buckling. This type of failure appears when the combination of the meridional compressive stresses and the circumferential tensile stresses leads to yielding. The buckles are also located near the base of the cone where the stresses are the highest.

### 2.2. *The design rule*

The design rule in its present format is described in Ref. 4. Due to space limitations, the rule is not repeated here. This new format of the rule is compatible with the Eurocode. Among other things, this means that the predicted buckling load is a function of the three quality classes defined in the Eurocode. The best quality class is Class A, the worst quality class that is still allowed in design is Class C[1].

## 3. Numerical Simulations

### 3.1. *General*

The goal of this study is to verify the present design rule by means of numerical simulations. For this purpose, simulations of a number of cone geometries were performed with the finite element package ABAQUS. The results of these simulations are given in Ref. 4. In this contribution, the results for one of these geometries is discussed in detail and new results are added.

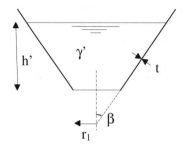

Fig. 1.  The geometry of a liquid-filled conical shell.

The studied cone (Fig. 1) has a base radius $r_1$ of 3 m, a wall thickness $t$ of 10 mm, an apex half angle $\beta$ of 45°, a Young's modulus $E$ of 210 GPa and a yield stress $f_{yk}$ of 240 MPa. The liquid height in the cone $h'_{Rk}$ is equal to 6705 mm. The conical shell is assumed to have shape deviations according to quality class C. For this geometry and these characteristics, the design rule predicts that the conical shell fails if the specific weight of the liquid is 1.62 times the specific weight of water.

### 3.2.  *Results of numerical simulations*

In the numerical model of the conical shell the lower rim is simply supported and the upper rim is free to deform. Different types of numerical analyses are performed for the cone: a linear bifurcation analysis (LBA), a geometrically nonlinear analysis (GNA), a materially nonlinear analysis (MNA), a geometrically and materially nonlinear analysis (GMNA) and a geometrically and materially nonlinear analysis of an imperfect cone (GMNIA). When nonlinear material behavior is included in the simulation, perfect elastic-plastic behavior is assumed. When imperfections are included, the amplitude of the shape deviation $\Delta w_0$ is equal to the maximal value that is allowed in quality class C, i.e. 2.06 times the shell thickness for this geometry[1]. Different types of imperfections are investigated. Two-dimensional cross-sections of the axisymmetric shape deviations are shown in Fig. 2. Initially, the first eigenmode of the cone was included as imperfection shape. This eigenmode is axisymmetric and has a small number of waves in meridional direction. The first half wave can be inward oriented or outward oriented. Both cases are investigated. The second imperfection shape that was studied is an axisymmetric shape with only one half wave in meridional direction. Also for this shape both orientations are investigated. The last imperfection shape that was studied is a weld depression of Type A. This shape was proposed by Rotter *et al.*[5] as a realistic shape deviation. The geometric form is given by

$$\Delta w = \Delta w_0 e^{-\frac{\pi x}{\lambda}} \left( \cos \frac{\pi x}{\lambda} + \sin \frac{\pi x}{\lambda} \right). \tag{1}$$

In this equation, $\Delta w$ is the deviation of the perfect shape perpendicular to the meridian, $\lambda$ is the linear meridional bending half wavelength equal to $2.44\sqrt{rt/\cos\beta}$ and $x$ is the meridional coordinate from the depression centre. Only an inward deviation can appear as a consequence of welding. The results of the simulations are given in Table 1.

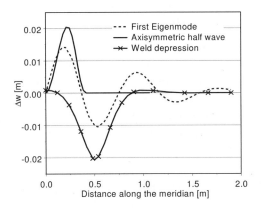

Fig. 2. The investigated axisymmetric imperfection shapes.

In the numerical simulations, the specific weight of the liquid $\gamma'$ is increased until buckling occurred. In Table 1, the ratio of the specific weight of the liquid at failure and the specific weight of water is given. The design rule predicts a ratio of 1.62. The GMNIA analyses lead to different values. Table 1 shows that an outward imperfection near the lower rim is more detrimental than an inward imperfection. Furthermore, even two out of the three imperfection shapes studied with an inward orientation of the imperfection lead to a dramatically lower ratio than the one that was predicted by the design rule. Clearly, this is a dangerous situation.

Table 1. Results of the different numerical simulations and comparison with the design rule.

| | | $\gamma'_{failure}/\gamma'_{water}$ | |
|---|---|---|---|
| Analysis | Imperfection | Inward imperfection | Outward imperfection |
| Design rule | | 1.62 | |
| LBA | | 5.93 | |
| GNA | | 5.14 | |
| MNA | | 3.63 | |
| GMNA | | 2.68 | |
| GMNIA (Class C) | First eigenmode | 1.30 | 1.00 |
| | Axisymmetric half wave | 1.26 | 0.94 |
| | Weld depression | 1.69 | - |

These results can be explained rather easily. The location of the largest stresses is also where the shape deviations are present. If the axisymmetric imperfection is oriented outward, this leads to a locally elevated circumferential stress, entailing early yielding and therefore plastic buckling[4]. Since the design rule is based on an elaborated experimental study – where axisymmetric imperfection shapes were unlikely to appear – it is to be expected that the design rule predicts larger failure loads. The results in Table 1 show that even if the axisymmetric imperfection is oriented inward, this also leads to a significant decrease of the failure load.

### 3.3.  *Partially axisymmetric imperfections*

Apparently the buckling load of an imperfect cone can be below the predicted failure load when the imperfections are axisymmetric. However, questions can be raised about the realistic nature of the imperfections. Imperfections that are completely axisymmetric are unlikely to appear in a real shell and if the imperfection is oriented outward, a large amount of circumferential membrane stretching is required to obtain such a shape. These arguments can be used to conclude that the imperfections with the shape of the first eigenmode or the axisymmetric half wave should be excluded from the simulations. Nevertheless, the weld depression must be taken into account when verifying the design rule. This imperfection appears as a consequence of welding. Even if this imperfection is not completely axisymmetric, its effect can be as detrimental. This was shown by Berry et al.[5] for cylindrical shells. A similar study is presented here with the weld depression Type A for the conical shell. Eq. (1) is modified with a weighting function $f$ given by

$$f = e^{-\frac{\pi\theta}{\rho}}\left(\cos\frac{\pi\theta}{\rho} + \sin\frac{\pi\theta}{\rho}\right). \tag{2}$$

In this equation, $\theta$ is the local circumferential coordinate with its origin in the last point where the imperfection has its complete shape and $\rho$ is the angle over which the imperfection disappears (taken equal to 10° as in Ref. 5). The results of the simulations of the conical shell with the partially axisymmetric weld depression are given in Fig. 3. In the graph, the angular extent of the weld depression is given on the horizontal axis. The ratio of the specific weight of the liquid at failure and the specific weight of water is given on the vertical axis. The graph shows that as soon as the angular extent is 90° or larger, the partially axisymmetric imperfection becomes as detrimental as the completely

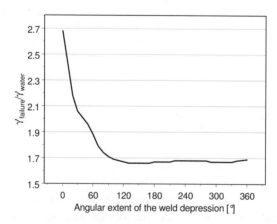

Fig. 3. Effect of the circumferential extent of the imperfection on the buckling strength.

axisymmetric weld depression. It is clear that when verifying the present design rule, the weld depression imperfection shape – completely or partially axisymmetric – can't be excluded from the simulations. The results in Table 1 show that for this geometry, the simulations with the weld depression predict a slightly larger buckling load than the failure load predicted by the design rule. However, for other geometries the design rule may be unconservative. Therefore an elaborated study is needed to verify the safety of the design rule for plastic buckling.

## 4. Conclusions

In this contribution, the design rule for thin-walled steel liquid-filled conical shells is verified by means of numerical simulations. The presence of partially or completely axisymmetric shape deviations may lead to early yielding and thus a plastic buckling load that may be lower than the predicted value for some geometries.

## Acknowledgments

The first author is a Postdoctoral Fellow of the Research Foundation – Flanders (FWO). The authors would like to express their gratitude for the financial support of the FWO.

## References

1. EN 1993-1-6, *Eurocode 3: Design of steel structures – part 1-6: Strength and Stability of Shell Structures* (European Committee for Standardization, Brussels, 2007).
2. D. Vandepitte, J. Rathé, B. Verhegghe, R. Paridaens and C. Verschaeve, in *Buckling of Shells*, ed. E. Ramm (Springer, Berlin, 1982).
3. ECCS, *Buckling of Steel Shells: European Recommendations, 4th Ed.* (ECCS, Brussels, 1988).
4. W. Vanlaere, G. Lagae, W. Guggenberger and R. Van Impe, in *Proc. Eurosteel 2008* (Graz, 2008) in press.
5. P. A. Berry and J.M. Rotter in *Proc. Int. Workshop on Imperfections in Metal Silos* (Lyon, 1996), p. 35.

# Part O
# Superplasticity

# EXPERIMENTAL STUDY ON DEFORMATION MECHANISM OF AZ31 MAGNESIUM ALLOY UNDER VARIOUS TEMPERATURE CONDITIONS

TADASHI INABA[1†], TAKAMASA YOSHIKAWA[2], MASATAKA TOKUDA[3]

*Department of Mechanical Engineering, Mie University,*
*Kurimamachiya 1577, Tsu 514-8507, JAPAN,*
*inaba@mach.mie-u.ac.jp, tkyoshi@mach.mie-u.ac.jp, tokuda@mach.mie-u.ac.jp*

Received 15 June 2008
Revised 23 June 2008

The objective of this study was to experimentally determine the properties of magnesium alloys during plastic deformation under various temperature regimes. Proportional loading tests combining both axial and torsional loads were performed on cylindrical AZ31 Mg alloys in a range from room temperature to 673 K, where superplasticity has a high probability of occurrence, and the crystalline structures of the samples were observed after these tests. The above observations indicate that the deformation mechanisms in AZ31 Mg alloy are dominated by twinning deformation at 300 K and 423 K, and by grain boundary sliding promoted by dynamic recrystallization at 523 K and higher temperatures.

*Keywords*: AZ31 magnesium alloy; deformation mechanism; various temperature; twinning deformation; superplasticity; grain boundary sliding; dynamic recrystallization.

## 1. Introduction

Interest in magnesium alloys has burgeoned in recent years because of its low specific weight and high specific strength, and the promise of energy savings that those properties imply.[1,2] A drawback of these alloys is that their crystal structures are hexagonal close packed, thus offering very few slip planes, and it is therefore very difficult to deform these alloys plastically at room temperature. This has led researchers to examine the phenomenon of superplasticity for deformation of Mg alloys.[3] "Superplasticity" refers to the ability of a material with a fine-grained structure to stretch by several hundred or even thousand percent under tensile loads without necking. This occurs at temperatures exceeding half the absolute melting temperature and at comparatively low strain rates. The most important mechanism controlling this behavior is grain boundary sliding, which does not require slipping within the crystals themselves; this allows the designer to specify large plastic deformations in Mg alloys. Most reports on superplasticity in Mg alloys have examined the behavior at temperatures above 600 K, where superplasticity is easy to obtain. Processing in such high-temperature environments, however, is expensive.

[†]Corresponding Author.

It will be necessary to learn more about plastic deformation in lower temperature zones in order to use Mg in a greater variety of industrial applications.

The objective of this study was to experimentally determine the properties of magnesium alloys during plastic deformation under various temperature regimes. Proportional loading tests combining both axial and torsional loads were performed on cylindrical AZ31 Mg alloys in a range from room temperature to 673 K, where superplasticity has a high probability of occurrence, and the crystalline structures of the samples were observed after the tests. This report describes the results of the above experiment and discusses the mechanisms of plastic deformation in Mg alloys in the different temperature conditions examined.

## 2. Experimental Conditions

The test material in this study was AZ31, a widely used structural magnesium alloy. Its composition was Mg, 3.0wt%Al, 0.8wt%Zn, 0.35wt%Mn, with trace amounts of Fe, Si, Cu, and Ni. The thin-walled tube specimens had an outer diameter of 20 mm and an inner diameter of 17 mm. The gauge lengths were 15 mm apart.

A testing machine capable of placing combined loading on specimens (AG-10TC, Shimadzu) was employed for mechanical testing. This tester contains an electric heater and is capable of conducting tests at temperatures up to 1027 K while applying tensile (or compressive) loads combined with torsional loads on the specimens. In this study, this machine was used to place uniaxial tension, uniaxial compression, simple torsion, and proportional loading combining both axial and torsional components, on AZ31 Mg alloy. Four test temperatures were used: 300 K (room temperature), 423 K, 523 K and 673 K. The equivalent strain rate $d\varepsilon_{eq}/dt$ was defined as shown below, by the von Mises equivalent strain,

$$\frac{d\varepsilon_{eq}}{dt} = \frac{d}{dt}\left[\sqrt{\varepsilon^2 + \left(\frac{\gamma}{\sqrt{3}}\right)^2}\right], \tag{1}$$

where $\varepsilon$ and $\gamma$ represent the axial and shear strains, respectively, $t$ is time and $d\varepsilon_{eq}/dt = 6.0\times10^{-4}$ [1/s] was used as the standard strain rate in all specimens. This is the rate at 673 K at which superplasticity behavior becomes significant. In the proportional loading test, $\theta = \tan^{-1}[(\gamma/\sqrt{3})/\varepsilon] = \pi/6, \pi/4, \pi/3, 2\pi/3, 3\pi/4,$ and $5\pi/6$ were selected. Uniaxial tension, simple torsion and uniaxial compression correspond to $\theta = 0, \pi/2,$ and $\pi$, respectively.

The test specimens were also examined for their crystalline organization after subjection to the mechanical tests. The examination was done with an optical microscope (BHMJ, Olympus). The mean grain diameter was estimated from photographs of the crystal structure.

## 3. Experimental Results

### 3.1. *Mechanical tests*

Figure 1 shows the stress-strain relations in uniaxial tension, uniaxial compression and simple torsion at 300 K, 423 K, 523 K, and 673 K. It can be seen that the yield stresses in all loading modes fell with a rise in temperature. Although the stresses after yield points at 300 K and 423 K indicated work-hardening, at 523 K and 673 K, the stress-strain curves were nearly flat (constant-stress), indicating plastic deformation. Figure 2 shows the equi-strain surface for 4% equivalent strain obtained at the above temperatures under uniaxial tension, uniaxial compression, simple torsion and proportional loading. The vertical and horizontal axes in the figure represent the axial and shear stress components, respectively. The diamond symbols represent actually measured values and the dashed lines represent the von Mises equivalent strain surface, using the tensile stress level as the standard. It can be seen in the figure that the yield stress was lower in uniaxial compression than in tension, approximately half the value. The same asymmetry between tensile and compressive properties was seen at 423 K. This asymmetry was not found at 523 K and 673 K, which are higher than half the absolute value for melting temperature of AZ31 ($T_m$ = 903 K). At the latter temperatures, meanwhile, the shear stress was 20-30% lower than the level that would be expected from von Mises theory, given the tensile stress.

### 3.2. *Observations of crystalline organization*

Figure 3 shows the results of observations of the crystal structure of AZ31 Mg alloy under the unloaded condition. The mean grain diameter was 33 $\mu$m. Figure 4 is a photomicrograph of the same material after uniaxial tensile and uniaxial compressive loading at 300 K. The mean grain diameter after the tensile test was 33 $\mu$m, and after the compressive test, it was 26 $\mu$m, i.e., there was no particular change in the grain size. No extension or compression of grains in the direction of loading was found in either case, but in the figure locations indicated with arrows, there were striations indicating twinning deformation. Additional signs of twinning deformation were seen in the compressive specimen compared with the tensile specimen. Figure 5 is a photomicrograph of the AZ31 after uniaxial tensile and compressive loading at 423 K. As already seen at 300 K, the mean grain diameter in the tensile specimen was 28 $\mu$m and in the compressive specimen, it was 20 $\mu$m. Again, there had been no particular change in grain diameter, although signs of twinning deformation were found.

Figures 6 and 7 are photomicrographs of samples after uniaxial tensile and compressive loads at 523 K and 673 K, respectively. The mean grain size after the tensile load at 523 K was 4 $\mu$m, and after the compressive load, it was 7 $\mu$m. At 673 K, the grain sizes changed to 14 $\mu$m and 10 $\mu$m, respectively; both temperatures caused reductions in grain sizes.

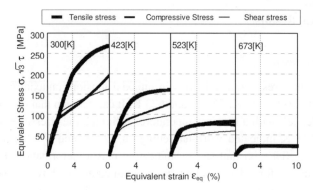

Fig. 1.   Stress-strain relations in uniaxial tension, uniaxial compression and simple torsion at 300 K, 423 K, 523 K, and 673 K.

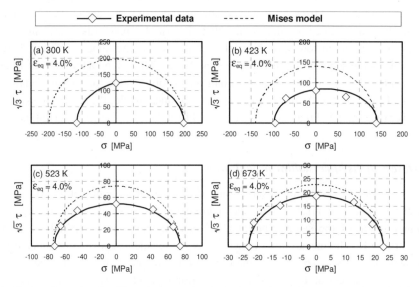

Fig. 2.   Equi-strain surface for 4% equivalent strain obtained at various temperatures under uniaxial tension, uniaxial compression, simple torsion and proportional loading.

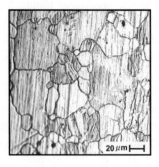

Fig. 3.   Photomicrograph of AZ31 Mg alloy under unloaded condition.

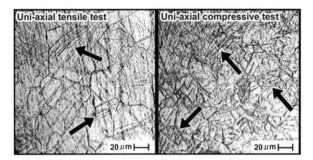

Fig. 4. Photomicrograph of AZ31 Mg alloy after uniaxial tensile and uniaxial compressive loading at 300 K.

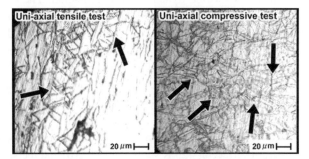

Fig. 5. Photomicrograph of AZ31 Mg alloy after uniaxial tensile and uniaxial compressive loading at 423 K.

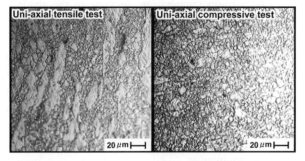

Fig. 6. Photomicrograph of AZ31 Mg alloy after uniaxial tensile and uniaxial compressive loading at 523 K.

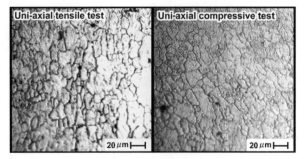

Fig. 7. Photomicrograph of AZ31 Mg alloy after uniaxial tensile and uniaxial compressive loading at 673 K.

## 4. Discussion

Magnesium alloys possess a hexagonal close packed structure. Slip systems are limited to the bottom plane, which has a low critical shear stress. At room temperature, plastic deformation is dominated by twinning deformation, rather than slip within grains. This was confirmed by the present results shown in Fig. 4. Additionally, as shown in Fig. 2(a), the yield strength due to twinning deformation was lower in these specimens under compressive load than under tensile load. This is probably due to the texture formed during extrusion of the original Mg alloy.

In contrast, at 673 K, the grain size was reduced by the deformation at high temperature, i.e., the material showed dynamic recrystallization. The equi-strain surface was symmetric about the results for tensile and compressive stresses, as seen in Fig. 2(d), and the shear stress followed a profile that was lower than predicted by von Mises' theory. The above properties resemble the equi-strain surface observed during the superplasticity displayed by 5083 aluminum alloy[4] and by Pb-Sn alloy[5]. Accordingly, it was deduced that at 673 K, AZ31 deforms plastically under the chief deformation mechanism of grain boundary sliding, assisted by the structural refinement resulting after dynamic recrystallization.

If the results at 423 K and 523 K are considered in light of the above findings, the deformation on equi-strain surface at 423 K shown in Fig. 2(b) and the structure shown in Fig. 5 indicate that plastic deformation is dominated by twinning deformation at this temperature, just as it is at room temperature. In view of the shape of the equi-strain surface in Fig. 2(c) and the observations in Fig. 6, however, the deformation at 523 K seems to be superplastic, as occurred at 673 K.

The above observations indicate that the deformation mechanisms in AZ31 Mg alloy are dominated by twinning deformation at 300 K and 423 K, and by grain boundary sliding promoted by dynamic recrystallization at 523 K and higher. These data suggest that 523 K, or perhaps a slightly lower temperature, is an optimal environmental temperature at which to perform plastic processing of AZ31 Mg alloys in order to take advantage of superplastic properties and their associated mechanisms, and to reduce the costs of the processing environment.

## References

1.  Y. Kawamura, T. Kasahara, S. Izumi, M. Yamasaki, *Scripta Materialia*, **55**, pp.453-456 (2006).
2.  M. Matsuda, S. Ii, Y. Kawamura, Y. Ikuhara, M. Nishida, *Materials Science and Engineering A*, **386**, pp.447-452 (2004).
3.  H. Watanabe, H. Tsutsui, T. Mukai, M. Kohzu, K. Higashi, *Material Science Forum*, **357-359**, pp.147-152 (2001).
4.  M. Tokuda, T. Inaba, H. Ohigashi, A. Kurakake, *International Journal of Mechanical Sciences*, **43**, pp.2035-2046 (2001).
5.  M.K. Khraisheh, H.M. Zbib, C.H. Hamilton, A.E. Bayoumi, *International Journal of Plasticity*, **13**, pp.143-164 (1997).

# Part P
# Time-Dependent Deformation

# THERMAL VISCOELASTIC ANALYSIS OF PLASTIC COMPONENTS CONSIDERING RESIDUAL STRESS

CHEL WOO CHOI[1], KAB SIK JEOUNG[2]

*DA C4 Group, LG Electronics Inc., 391-2 Gaeumjeong-dong,*
*Changwon, Gyeongsangnam-do, 641-711 KOREA,*
*ccw770@lge.com, mccw770@lge.com*

HYUNG-IL MOON[3], HEON YOUNG KIM[4†]

*‡Department of Mechanical & Mechatronics , Kangwon National University, 192-1 Hyoja-dong,*
*Chuncheon, Gangwon-do, 200-701 KOREA,*
*moon@kangwon.ac.kr, khy@kangwon.ac.kr*

Received 15 June 2008
Revised 23 June 2008

Plastics is commonly used in consumer electronics because of it is high strength per unit mass and good productivity, but plastic components may often become distorted after injection molding due to residual stress after the filling, packing, and cooling processes. In addition, plastic deteriorates depending on various temperature conditions and the operating time, which can be characterized by stress relaxation and creep. The viscoelastic behavior of plastic materials in the time domain can be expressed by the Prony series using the ABAQUS commercial software package. This paper suggests a process for predicting post-production deformation under cyclic thermal loading. The process was applied to real plastic panels, and the deformation predicted by the analysis was compared to that measured in actual testing, showing the possibility of using this process for predicting the post-production deformation of plastic products under thermal loading.

*Keywords*: Residual stress; Viscoelasticity; Creep; Stress relaxation; Thermal analysis; Prony series.

## 1. Introduction

Injection processed plastics are widely used in various industrial products such as consumer electronics because of their light weight, and high performance and quality. Moreover, injection molded plastic is frequently used for the outer cases of consumer electronic products due to its advantages of productivity and design flexibility. These cases, however, are subject to deformation and residual stress caused by heterogeneous shrinkage and difference in the cooling temperature of the upper and lower molds. In addition, unlike metallic materials, an intrinsic characteristic of plastics is that they exhibit a large displacement according to weight (load) and Viscoelasticity depending on time and temperature, which can be more pronounced under uncontrolled conditions[1-3].

---

†Corresponding Author.

This report examines the thermal deformation of plastic components starting from the initial condition of the residual stress generated in the molding process. The generally accepted MOLD FLOW and ABAQUS software packages are used in the analysis. Experimental estimates are included for the deformation in a product under specified conditions (e.g., thermal resistance) as well as time-dependent deformation dependence based on linear viscoelastic analysis.

## 2.  Theoretical Background

Post-production deformation in plastics has two causes: changes in physical properties during heating due to residual stress after injection, and material expansion with increasing temperature caused by residual regional deformation after cooling. Computer-aided engineering-based approaches are usually used to predict residual stress during injection-molding and warping after injection with the aim of improving the overall design. However, study of the deformation that occurs after the completion of molding and heat resistance testing have not been put into practical application for product design despite several recent research efforts on these topics[1-3].

### 2.1.  *Injection molding and residual stress*

Plastic injection molding is a manufacturing technique in which high-pressure molten liquefied resin is injected into a mold and then solidified by cooling through two contraction steps. The first step is volume contraction in the liquefied resin due to high pressurization, and the second occurs as it transforms from a liquid to a solid. These types of volumetric contraction largely influence the distortion or twisting in finished parts to cause variations in regional contraction in the part as a form of residual stress and deformation[1].

### 2.2.  *Post-deformation estimation after heating-cooling test*

Plastic has properties of both viscosity and elasticity. The viscoelastic performance is characterized by the stress relaxation and retardation time, which can be simplified as models with serial or parallel connections between a spring and a damper. Therefore, as the production cycle is repeated, the stress relaxation and retardation times also repeat[1, 4]. In addition, effects resulting from the stress relaxation and retardation time cause deformation of the finished product when plastic components undergo repeated heating and cooling[1].

### 2.3.  *Viscoelastic behaviors in materials*

The stress in viscoelastic materials can be represented by the addition of the deviatoric stress caused by shear modulus and the hydrostatic stress caused by bulk modulus. If the strain is very small and the material is not processed into thermoplastics, the regime can be considered linear, and thus the generated stress components can be represented by the

Boltzmann superposition principle[4]. More specifically, the stress at any time $t$ is the sum of the successive accumulated values of the differentiated stress up to the time $t$. The governing equation is as follows

$$\sigma(t) = \int_0^t 2G(\tau - \tau')\dot{e}dt' + \int_0^t 2K(\tau - \tau')\dot{\varphi}dt'$$
(1)

where $e$ is the deviatoric strain, $\varphi$ is the volumetric strain, $G(t)$ is the shear, and $K(t)$ is the bulk relaxation modulus. Equation (1) can be represented in a Prony series as shown in Eqs. (2) and (3),

$$G(t) = G_0\left\{1 - \sum_{i=1}^{N} g_i^p (1 - e)^{\frac{-t}{\tau_i}}\right\}$$
(2)

$$K(t) = K_0\left\{1 - \sum_{i=1}^{N} k_i^p (1 - e)^{\frac{-t}{\tau_i}}\right\}$$
(3)

where $G_0$ (the instantaneous shear) and $K_0$ (the bulk relaxation modulus) are Prony series constants[5].

## 3. Creep and Stress Relaxation Test

Creep and stress relaxation tests were performed to determine the viscoelastic properties for analysis using dumbbell-shaped tensile specimens. The universal testing machine used for the tests was the Instron 5567 (Instron, Norwood, MA, USA) with a 20 kN load cell.

The experimental conditions for the creep and stress relaxation tests were the same as those for uniaxial tension tests. In the creep test, the strain of the specimen was measured in a load under uniaxial tension stress. Then the applied load was modulated over the range of 10 to 408 N, with the temperature at 20°C and 70°C. Figure 1 shows the results of the creep test in which little effect was observed on the viscosity below 20°C since the material was fully solidified. In contrast, a large effect occurred at 70°C.

The test results show that the change in strain depends on the load imposed in the initial stage. This phenomenon can be considered a deviation caused by changing stress over the variable cross-sectional area of the specimen. To illustrate this more clearly, stress relaxation tests were conducted to measure the change in the load while keeping the strain fixed at 0.5%, 1.0%, 1.5%, 2.0%, and 2.5%. The results for each percentage in Fig. 2 show that a similar load change occurred independent of the magnitude of the strain at the initial stage.

As the stress relaxation in normal materials is linear with little change over an extended period of time[4,6], using a linear fit for the analysis is reasonable when the time exceeds 500 s.

## 4.  Analytical Verification

To determine if the viscoelasticity input values were analytically well represented, a verification using one element was carried out. The result in Fig. 3(a) shows a close similarity between the actual test result for stress relaxation and the theoretical values calculated with the Prony series.

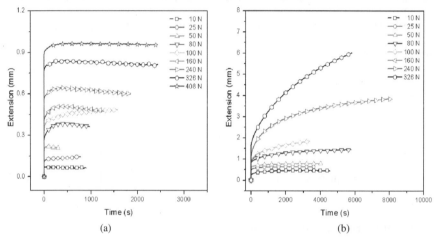

Fig. 1. Creep test results for variation in the initial load. (a) at 20°C and (b) at 70°C.

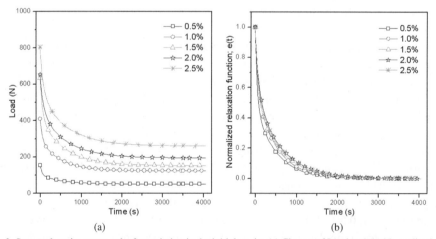

Fig. 2. Stress relaxation test results for variation in the initial strain. (a) Change of Load and (b) Normalized relaxation function $(E(t)=E(\infty)+\{E(0)-E(\infty)\}e(t)$ at $E(t)=\delta\ (t)/\varepsilon)$.

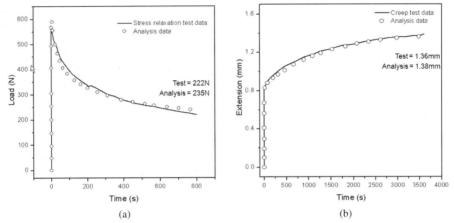

Fig. 3. Comparison of test data with analysis predictions. (a) Stress relaxation using one element, and (b) creep test.

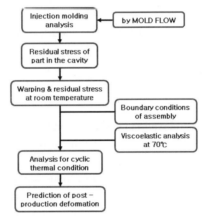

Fig. 4. Analytical process of predicting post-production deformation.

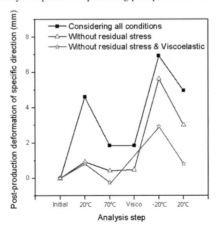

Fig. 5. Prediction of post-production deformation due to cyclic thermal load.

In addition, the analytical reliability was verified in another way by comparing the actual creep test result with theoretical values using its boundary conditions. As shown in Fig. 3(b), this also indicated little difference between the theoretical calculation and the actual tested values.

## 5.  Analysis on Thermal Strain in Control Panel

A comparison was made between the actual test results and the theoretical analysis for heat resistance of an injected plastic control panel. The experimental process is shown in Fig. 4 and the creep effect of temperature was considered only for 70°C. The residual stress calculated with MOLD FLOW was used as the initial condition[7]. The model was based on one-dimensional connection elements[5] and variable boundary conditions to displace the hook position similar to the actual hook movement. Figure 5 shows the result based on these conditions for analyzing the strain of the heat-cooling treatment of the mounted control panel. The overall result also indicates great similarity between the actual test data and the calculated values considering all the assumptions that were made. (Deformation measurement of actual test is about 4mm and simulation result is 4.1~4.3mm).

## 6.  Conclusion

The study can be summarized as follows.
  (1) Creep tests and stress relaxation tests were conducted to analyze viscoelasticity by providing experimental data to check the reliability of theoretical predictions.
  (2) MOLD FLOW was used to extract the calculated residual stress from the results, and those values were used to predict post-production deformation of the product.
  (3) Using one-dimensional connection elements and displaced boundary conditions, this study provided boundary conditions for effectively plotting behaviors around the hook position in analyzing thermal strain.
  (4) This study predicted post-production deformation analytically in heating/cooling conditions for the mounted control panel, and then compared the calculated values to those resulting from the actual tests.

## References

1.  H. Y. Kim, J. J. Kim, *The 1st Japan-Korea Plastics Processing Joint Seminar* (1997).
2.  M. A. Elseifi, S. H. Dessouky, I. L. Al-Qadi, S. Yang, *Transportation Research Board 85th Annual Meeting*, 1 (2006).
3.  C. A. Hieber, S. F. Shen, *Israel Journal of Tech.*, **16**, 248 (1978).
4.  J. D. Ferry, *Viscoelastic properties of polymers* (John Wiley & Sons, New York, 1980).
5.  ABAQUS user manual.
6.  A. Siegmann, A. Buchman, S. Kenig, *Polymer engineering and science*, **22**, 560 (2003).
7.  Mold Flow user manual.

# PREDICTION OF A MODIFIED PTW MODEL FOR VARIOUS TAYLOR IMPACT TESTS OF TANTALUM

JONG-BONG KIM[1†]

*Department of Automotive Engineering, Seoul National University of Technology, 172 Gongneung-2Dong, Nowon-Gu, Seoul, 139-743, KOREA, jbkim@snut.ac.kr*

HYUNHO SHIN[2]

*Department of Ceramic Engineering, Kangnung National University, Kangnung 210-702, KOREA, hshin@kangnung.ac.kr*

Received 15 June 2008
Revised 23 June 2008

The strain hardening part of the Preston-Tonks-Wallace (PTW) model, developed for the description of the plastic constitutive behavior of materials at wide ranges of strain, strain rate, and temperature, has been modified by employing the Voce equation. The prediction capability of the modified PTW (MPTW) has been investigated with reference to Taylor impact test results in the literature, and comparison has been made with the models of Johnson-Cook (JC), Steiberg-Guinan (SG), Zerilli-Armstrong (ZA), and PTW. Of the compared existing models, no model was appropriate for describing the results of various Taylor impact tests. However, the modified PTW is shown to predict fairly accurate results in terms of the length, diameter, and shape of the deformed specimen tested at different temperatures and impact velocity.

*Keywords*: Taylor impact test; Modified PTW model; Tantalum; High strain rate.

## 1. Introduction

Tantalum has been widely used as a liner material for explosively formed projectiles (EFPs) due to its high ductility for easy formation of projectile from liner and high density to ensure an efficient penetration into target. Both the formation and penetration processes take place at wide ranges of strain, strain rate, and temperature. In order to understand such dynamic high-strain-rate events, an experimental approach offers the most accurate results. However, the time and cost constraints hardly permit acquisition of a database with sufficient variation of parameters to construct unambiguous analytical models. A computational approach overcomes these constraints through step-by-step analysis of any event confined to very short period of time, and thus can provide insight which would be difficult to understand solely from the experimental data. Hence,

---

[†]Corresponding Author.

approaches based on modeling and simulations are of interest, especially in the areas of high-strain-rate deformation of materials.

In order to ensure a valid numerical simulation of the high-strain-rate events, employment of a precise constitutive model describing the dynamic mechanical behavior of materials is prerequisite. The parameters in constitutive plasticity models are generally determined from experiment such as the Taylor impact test[1]. This methodology has long been used to estimate the plastic constitutive behavior of metals at high strain rate, high temperature, and large plastic strain. Of the many constitutive models developed for the description of dynamic deformation of materials, the JC model[2], the SG model[3], and the ZA model[4] have been the most widely utilized. Recently, Preston et al.[5] have proposed a new model (PTW: Preston, Tonks, and Wallace) to cover wide ranges of pressure and strain rate. In our separate analysis[6], it was shown that all of the above models reasonably described flow stress and the strain hardening behavior of tantalum only in certain ranges of strain, strain rate, and temperature, for which the models were designed. However, initial yield stress was reasonably predicted by PTW in wide ranges of strain, strain rate, and temperature, except the strain hardening behavior after the initial yielding. This fact indicates that the model can be used in wide ranges provided the strain hardening term in the PTW model is appropriately modified. Here we report that once the strain-hardening term of the PTW model is modified, the modified version predicts the results of Taylor impact tests of tantalum in the literature relatively well as compared to PTW, JC, ZA, and SG over wide ranges of strain rate.

## 2.  The PTW Model and Its Modification

In Preston et al.[5], the plastic constitutive model considering nonlinear dislocation drag effects that are predominant in a strong shock regime is given by,

$$\hat{\tau} = \hat{\tau}_S + \frac{(s_0 - \hat{\tau}_y)}{p} \ln\left[1 - \left[1 - \exp\left(-p\frac{\hat{\tau}_S - \hat{\tau}_y}{s_0 - \hat{\tau}_y}\right)\right] \times \exp\left\{\frac{-p\theta\bar{\varepsilon}}{(s_0 - \hat{\tau}_y)\left[\exp\left(p\frac{\hat{\tau}_S - \hat{\tau}_y}{s_0 - \hat{\tau}_y}\right) - 1\right]}\right\}\right] \tag{1}$$

where, $\hat{\tau}$ is a normalized flow stress defined as

$$\hat{\tau} = \tau / G(\rho,T) = \bar{\sigma} / 2G(\rho,T), \tag{2}$$

and $p$, $\theta$, and $s_0$ are material constants, and shear modulus $G$ is regarded as a function of density ($\rho$) and temperature ($T$). $\hat{\tau}_S$ and $\hat{\tau}$ are the work hardening saturation stress and flow stress, respectively.

The strain hardening term in the PTW model has been modified in a separate work[6] by employing the Voce equation,

$$\hat{\tau} = \hat{\tau}_Y + Ae^{C\log\dot{\bar{\varepsilon}}}(1 - e^{-B\bar{\varepsilon}}) \tag{3}$$

where, $\hat{\tau}_y$ is the initial yield stress[6] and $A$, $B$, and $C$ are the material constants. $A$ is the maximum strain hardening amount when the strain rate is zero, and $B$ controls the saturation speed of the hardening. As $B$ increases, the hardening saturates rapidly. $C$ represents the dependency of the maximum hardening amount on the strain rate. It is assumed that the maximum strain hardening is dependent on the strain rate, and that the effect of the strain rate on the maximum hardening is the same as the effect of the strain rate on the initial yield stress. The constant $C$ is obtained from existing experimental data in the literature[7], which is shown Fig. 1. The logarithmic value of the strain rate and the yield stress has a linear relation, and the slope is presented as follows from Eq. (3).

$$\frac{d(\log_{10}\bar{\sigma}_y)}{d(\log_{10}\dot{\bar{\varepsilon}})} = C\log_{10}e = slope \tag{4}$$

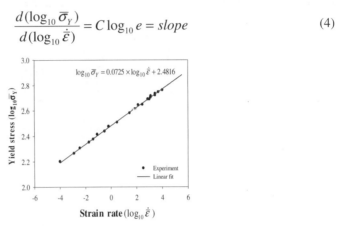

Fig. 1. Change in yield stress of tantalum at 298K as a function of strain rate. Experimental data are adapted from Hoge and Mukherjee[7].

From Eq. (4), the value of the constant $C$ is determined as 0.167 and the material constants for the MPTW model obtained in this manner are listed in Table 1.

Table 1. Material constants for the modified PTW model for tantalum [Eq. (3)].

| $s_0$, $s_\infty$ | $\kappa$ | $\gamma$ | $\beta$ | $G_0$ | $\alpha$ | $A$ | $B$ | $C$ |
|---|---|---|---|---|---|---|---|---|
| 0.009, 0.0016 | 0.55 | 0.00004 | 0.23 | 72.2 GPa | 0.23 | 0.0021 | 1.7 | 0.167 |

## 3. Prediction of Taylor Impact Test

In this section, the modified PTW model's capability in predicting the Taylor impact test[1] is checked through comparison with experimental results shown elsewhere. In the Taylor impact test, a cylindrical specimen is fired to a rigid plate, and then the final length, outer diameter, and the deformed shape are measured. Figure 2 shows examples of the finally deformed shape of the projectile reported by Maudlin et al.[8] and Ting[9]. As seen in Fig. 2, the front part of the specimen undergoes large deformation at a high strain rate, and the rear part undergoes small deformation. As a result, by comparing the deformed shapes obtained by analysis and experiment, the constitutive models used in the analysis can be validated in wide ranges of strain, strain rate, and temperature.

Figure 3 shows the experimentally determined deformed shape at room temperature at an impact velocity of 154 m/s, adapted from Ting[9]. Enclosed in Fig. 3 are the predicted results from JC, SG, ZA, PTW, and MPTW. An in-house hydrocode was used in the analysis of the models, and the reliability of the code was confirmed using a commercial hydrocode, AutoDyn. For all of the simulations in the current work, two-dimensional axi-symmetric analyses were carried out with the assumption of isotropic material properties. Material parameters for SG and PTW models are adapted from Steinberg et al.[3] and Preston et al.[5], respectively. For the ZA model, two different sets of material constants given by Chen and Gray III[10] and by Zerilli and Armstrong[11] are used, which are referred to as ZA1 and ZA2, respectively.

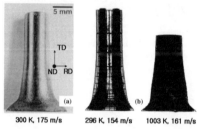

Fig. 2. Examples of deformed tantalum projectiles after the Taylor impact test, adapted from (a) Maudlin et al.[8] and (b) Ting[9].

In Fig. 3(a), JC, ZA2 and PTW more closely describe the length of the specimen than any of the other models, but underestimate. ZA1 and SG overestimate the length. As for the deformed diameter of the frontal part of the specimen, ZA1 gives a similar result, but underestimate. JC, ZA2 and SG overestimate and PTW shows quite large error. In Fig. 3(b), however, the MPTW more accurately predicts the length and frontal diameter of the deformed specimen than any of the other models.

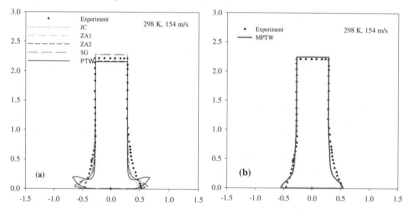

Fig. 3. Deformed shape of the specimen after the Taylor impact test at 298K and 154 m/s, adapted from Ting[9].

Having examined the prediction capability of the MPTW at room temperature, its capability is now checked at high temperature. Figure 4 shows the experimentally obtained[9] deformed shape at 1003 K and impact velocity of 161 m/s. In Fig. 4, JC and

PTW predict the deformed length of the projectile closer to the experiment than any of the other models, but overestimate it. ZA1 overestimates the length, while ZA2 underestimates. SG significantly overestimates the length. As for the diameter of the frontal part, SG is closer to the experiment, ZA1 underestimates, and ZA2 and JC overestimate. PTW significantly overestimates the frontal diameter. These findings indicate that no single model is appropriate for predicting both the length and the frontal diameter of the deformed specimen at high temperature. However, as seen in Fig. 4(b), the MPTW fairly accurately predicts the deformed shape in terms of both the length and diameter of specimen.

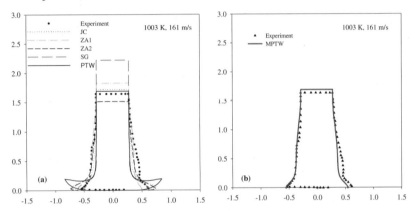

Fig. 4. Deformed shape of the specimen after the Taylor impact test at 1003K and 161 m/s, adapted from Ting[9].

In order to quantify the error in terms of shape, the error in terms of the area of the projectile was used as follows.

$$E_{Shape}(\%) = 100 \frac{A_{Error}}{A_{exp}} = 100 \frac{\int_{r_1}^{r_2} \left| L_{exp} - L_{model} \right| dr + \int_0^L \left| r^+_{exp} - r^+_{model} \right| dz + \int_0^L \left| r^-_{exp} - r^-_{model} \right| dz}{A_{exp}}. \quad (5)$$

The shape error consists of three parts; i.e., the area by length difference in the rear region, the area by radial difference on the right-hand side ($r^+$), and the area by radial difference on the left-hand side ($r^-$). The numerators in Eq. (5) reflect these error sources.

Figure 5 shows the prediction errors for each model from the four experimental data sets, in terms of length, diameter, and shape. From the viewpoint of the length and shape of the deformed projectile for each experimental data set, the MPTW model shows the smallest error among the models. In predicting the diameter of the deformed projectile, the MPTW is the best predictor on an overall basis. In Figs. 3-5, the modified PTW is shown to predict fairly accurate results in terms of the length, diameter, and shape of the deformed specimen tested under the various temperature and impact velocity conditions.

## 4. Conclusion

The Preston-Tonks-Wallace (PTW) model, developed for the description of the plastic constitutive behavior of materials at wide ranges of strain, strain rate, and temperature, has been modified by employing the Voce equation for the strain hardening part. The prediction capability of the modified PTW (MPTW) has been investigated with reference to the various Taylor impact tests in the literature. The models of Johnson-Cook (JC), Steiberg-Guinan (SG), Zerilli-Armstrong (ZA), and PTW were not appropriate for describing the length, diameter, and shape of the deformed specimen at wide range of test condition. However, the modified PTW is shown to predict fairly accurate results in terms of the length, diameter, and shape of the deformed specimen tested under the various temperature and impact velocity. In light of the length and shape of the deformed projectile, for each compared experimental data set, the MPTW model shows the smallest error among the considered models. For the prediction of the diameter of the deformed projectile, the MPTW is the best predictor on an overall basis.

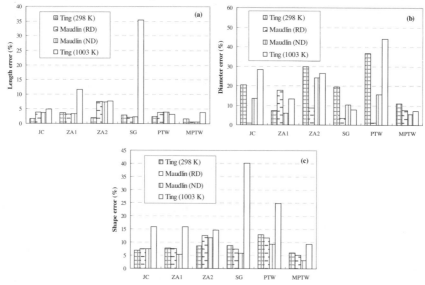

Fig. 5. Prediction errors for each model with the four experimental data sets: in terms of (a) length, (b) diameter, and (c) shape.

## References

1.  G. I. Taylor, *Proc. Roy. Soc. London.* A-194, 289 (1948).
2.  G. R. Johnson and W. H. Cook, in *Proc. of the 7th Int. Symp. on Ballistics.* (Den Hagg, Netherlands, 1983), 541.
3.  D. J. Steinberg, S. G. Cochran, and M. W. Guinan, *J. Appl. Phys.* 51, 1498 (1980).
4.  F. J. Zerilli and R. W. Armstrong, *J. Appl. Phys.* 61, 1816 (1987).
5.  D. L. Preston, D. L. Tonks, and D. C. Wallace, *J. Appl. Phys.* 93, 211 (2003).
6.  H. Shin and J.-B. Kim, submitted to *Int. J. Impact Eng.* (2007).
7.  K. G. Hoge, A. K. Mukherjee, *J. Mater. Sci.* 12, 1666 (1977).
8.  P. J. Maudlin, J. F. Bingert, J. W. House, S. R. Chen, *Int. J. Plasticity.* 15, 139 (1999).
9.  C.-S. Ting, in High Strain Rate Behavior of Refractory Metals and Alloys. Ed. R. Asfahani et al. (1992), 249.
10. S. R. Chen and G. T. Gray III, *Metallurgical and Materials Trans.* 27A, 2994 (1996).
11. F. J. Zerilli and R. W. Armstrong, *J. Appl. Phys.* 68, 1580 (1990).

# EFFECTS OF TEMPERATURE AND FORMING SPEED ON PLASTIC BENDING OF ADHESIVELY BONDED SHEET METALS

MICHIHIRO TAKIGUCHI[1†]

*Department of Maritime Technology, Hiroshima National College of Maritime Technology,*
*4272-1, Higashino, Osakikamijima-cho, Toyota-gun, Hiroshima, 725-0231 JAPAN,*
*taki@hiroshima-cmt.ac.jp*


TETSUYA YOSHIDA[2]

*Department of Electronic Control Engineering, Hiroshima National College of Maritime Technology,*
*4272-1, Higashino, Osakikamijima-cho, Toyota-gun, Hiroshima, 725-0231 JAPAN,*
*yoshida@hiroshima-cmt.ac.jp*


FUSAHITO YOSHIDA[3]

*Department of Mechanical System Engineering, Hiroshima University,*
*1-4-1, Kagamiyama, Higashi-Hiroshima, 739-8527 JAPAN,*
*fyoshida@hiroshima-u.ac.jp*

Received 15 June 2008
Revised 23 June 2008

This paper deals with the temperature and rate-dependent elasto-viscoplasticity behaviour of a highly ductile acrylic adhesive and its effect on plastic bending of adhesively bonded sheet metals. Tensile lap shear tests of aluminium single-lap joints were performed at various temperature of 10-40°C at several tensile speeds. Based on the experimental results, a new constitutive model of temperature and rate-dependent elasto-viscoplasticity of the adhesive is presented. From V-bending experiments and the corresponding numerical simulation, it was found that the *gull-wing bend* is suppressed by high-speed forming at a lower temperature.

*Keywords*: Plastic working; bending; adhesive joints; temperature and rate-dependent shear strength; experimental strain analysis; numerical analysis; constitutive equation.

## 1. Introduction

The present authors have recently proposed a new technique of plastic forming of adhesively bonded sheet metals[1-4], where two flat sheets are adhesively bonded together and then press-formed, which is completely different from the conventional process of adhesive joint. In this technique, suppression of large transverse shear deformation

---

[†]Corresponding Author.

occurring in the adhesive layer during bending is of vital importance, since it would cause the geometrical imperfection of bent sheets (so-called '*gull-wing bend*'[5]).

In this paper, first, tensile lap shear tests of single-lap joints were performed at various temperatures at several tensile speeds. Based on the experimental results, a new constitutive model of temperature and rate-dependent elasto-viscoplasticity of the adhesive is presented. Furthermore, the optimum temperature and forming speed conditions to suppress the geometrical imperfection in the bending process of adhesively bonded sheet metals are investigated through experiments and corresponding numerical simulation.

## 2. Mechanical Properties of Adhesive and Its Constitutive Modeling

The adhesive employed in the present work was an acrylic adhesive *M-372-20* Hardrock (Denki Kagaku Kogyo, Co. Ltd.). This adhesive has high ductility as well as high strength of the same level as conventional epoxy adhesives. In the present research, to investigate the influence on strength and ductility, a tensile lap shear experiment was performed on a single-lap joint at various temperatures of 10-40°C at various speeds ranging from 0.01 to 100 mm/min ( $\dot{\gamma} = 0.00167 \sim 16.7\,\mathrm{sec}^{-1}$ ).

Figure 1 shows the effects of temperature and tensile speed on shear stress at a strain of $\gamma = 1.0$. In this figure, strong temperature and strain rate sensitivity in flow stress is observed.

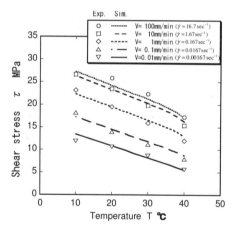

Fig. 1. Effects of temperature and tensile speed on the shear strength (at $\gamma = 1.0$) obtained from single-lap shear experiments.

Based on the above experimental result, as well as the results obtained from previous research[3], a new constitutive model describing both temperature and rate dependency is presented as follows.

The shear strain rate $\dot{\gamma}$ of the adhesive is assumed to consist of an elastic part, $\dot{\gamma}_e$, and a viscoplastic part, $\dot{\gamma}_p$ :

$$\dot{\gamma} = \dot{\gamma}_e + \dot{\gamma}_p. \tag{1}$$

The elastic constitutive equation is given by Hooke's law

$$\dot{\gamma}_e = \dot{\tau}/G, \tag{2}$$

where $\dot{\tau}$ and $G$ stand for the shear stress rate and the shear modulus of elasticity respectively.

The visco-plastic constitutive equation is expressed using overstress theory[6-8], as follows.

$$\dot{\gamma}_p = \left[ -\frac{1}{b} \ln \left\{ 1 - \frac{\tau - (\tau^* - C\gamma_p^{\,n})}{D(1 + k\gamma_p^{\,a})} \right\} \right]^{\frac{1}{m}}. \tag{3}$$

In our previous model, $\tau^*$, $C$, $n$, $D$, $k$, $a$, $b$, and $m$ were treated as material constants. In contrast, in order to take temperature effect into account, here we assume that $G$, $\tau^*$, $C$ and $D$ are temperature dependent material parameters, which are given by the following equations:

$$G = G_{20}\phi_1(T), \quad \tau^* = \tau_{20}\phi_2(T), \quad C = C_{20}\phi_2(T), \quad D = D_{20}\phi_3(T), \tag{4}$$

where $G_{20}$, $\tau_{20}$, $C_{20}$ and $D_{20}$ are the values of these parameters at 20°C, and $\phi_i(T)$, $(i = 1, 2, 3)$ is the following function of temperature $T(°C)$:

$$\phi_i = 1 + w_i \left( 1 - \frac{T}{T_{20}} \right). \tag{5}$$

Here $T_{20} = 20°C$, $w_i (i = 1, 2, 3)$ are material constants, $w_1 = 0.85$, $w_2 = 0.72$, and $w_3 = 0.05$. These material parameters in this model are listed in Table 1.

Table 1. Material parameters of acrylic adhesive resin
(*M-372-20 Hardrock*, Denki-Kagaku Kogyo, Co. Ltd.)

| $G_{20}$ (MPa) | $\tau_{20}$ | $C_{20}$ (MPa) | $n$ | $D_{20}$ (MPa) | $k$ | $a$ | $b$ | $m$ |
|---|---|---|---|---|---|---|---|---|
| 250 | 4.0 | 3.0 | 0.60 | 2.67 | 5.65 | 0.183 | 2.39 | 0.352 |

## 3. Effects of Temperature and Forming Speed on V-bending

### 3.1. *Experimentation*

An aluminium alloy sheet (A5083P-O) of 1.0mm thick was used as adherend. Its stress-strain relation is expressed by the following equation:

$$\sigma = C_{al}\varepsilon^{n_{al}} \quad (C_{al} = 643, \ n_{al} = 0.316). \tag{6}$$

In this paper, V-shaped air-bending[9] was examined. Figure 2 illustrates the experimental set-up for the V-bending experiment. In this figure, $P$, $lp$, $L$, $x$, $\lambda_a$ and $t_a$

represent punch load, lap length, die-span, distance from the bending center, transverse shear deformation of the adhesive layer and thickness of the adhesive layer, respectively.

The shear strain of adhesive layers, $\gamma$, was determined as $\gamma = \lambda_a / t_a$. The thickness of the adhesive layer, $t_a$, was controlled so as to have a value of 0.1mm.

Table 2 shows the experiment conditions for V-bending. To examine the effects of temperature and forming speed on the deformation characteristics of the sheet, particularly in terms of *gull-wing bend*, a span of 20 mm, which is significantly shorter than the lap of 50 mm, was chosen because in this condition the *gull-wing bend* is likely to occur. The punch speed was set at 500 mm/min for all the temperatures. To examine the effect of forming speed, experiments with the punch speed at 1 mm/min and 20 mm/min were also conducted solely at a temperature of 40°C. The punch travel was 8 mm for all the tests.

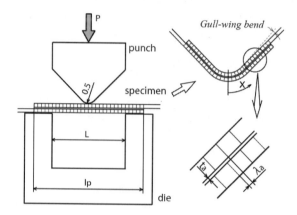

Fig. 2. Experimental set-up for V-bending test.

Table 2. Experimental conditions of V-bending.

| Lap length $lp$(mm) | Die span $L$(mm) | Punch speed $V$(mm) | Temperature $T$(°C) |
|---|---|---|---|
| 50 | 20 | 500 (1,20) | 10 |
| | | | 25 |
| | | | 40 |

### 3.2. *Numerical analysis*

The deformation of the adhesively bonded sheet metals under V-shaped air-bending (three-point bending) was analyzed. In the present paper, the temperature and rate-dependent constitutive model, proposed in Chapter 2, was used for adhesive layer. Using this model, the effect of temperature, as well as the punch speed, on the deformation behavior is clarified analytically. The elasto-visco-plastic simulation code is our original. For details of the rate-dependent deformation analysis, refer to the previous papers[1,4]. In this study, the influence of friction at the punch/specimen and the specimen/die was not taken into account.

### 3.3. *Results and discussion*

Figure 3(a) illustrates the comparison between the experimental and numerical results for the transverse shear strain distributions of the adhesive layer after V-bending conducted at various temperatures. From these results, it is found that the transverse shear strain is affected by tested temperature considerably, i.e., the lower the temperature the smaller the strain. This is because, as the temperature decreases, deformation resistance in adhesive layers increases and the shear deformation becomes difficult. As a result, the shear-induced *gull-wing bend* is reduced markedly with decreasing temperature, as shown Figure 3(b). In addition to the temperature effect, the forming speed also strongly affects the transverse shear of the adhesive layer and *gull-wing bend* as shown in Figures 4(a) and (b). Here, it should be emphasized that all the results of numerical simulation agree well with the corresponding experimental data, and consequently, high accuracy of the present modeling and simulation is confirmed.

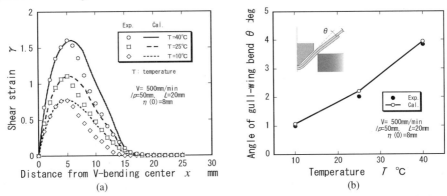

Fig. 3. Effect of the temperature on V-bending  ($lp$=50mm, $L$=20mm, $V$=500mm/min, $\eta(0)$=8mm): (a) shear strain in the adhesive layers; (b) angle of *gull-wing bend*.

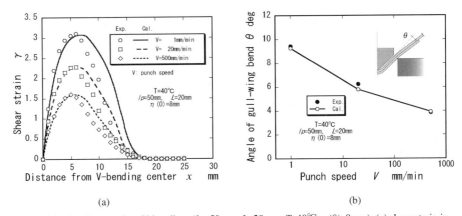

Fig. 4. Effect of the forming speed on V-bending  ($lp$=50mm, $L$=20mm, $T$=40℃, $\eta(0)$=8mm): (a) shear strain in the adhesive layers; (b) angle of *gull-wing bend*.

## 4. Concluding Remarks

This research work was performed as a part of a series of studies on bending process of adhesively bonded sheet metals, and the present paper specifically focused on the effects of temperature and forming speed. The main conclusions obtained are as follows.

(1) The shear strength of the adhesive used in present research is extremely sensitive to the deformation rate and tested temperature. To describe both the effects, an elasto-viscoplasticity constitutive model for adhesive, which takes the rate and temperature effects into account, has been proposed.

(2) The large transverse shear deformation of the adhesive layer and *gull-wing bend*, which are serious defects in press forming of the adhesively bonded sheet metals, are successfully suppressed with high speed forming at a low temperature, since the deformation resistance of the adhesive resin becomes higher under such a forming condition.

(3) V-bending analysis of adhesively bonded sheet metals was performed using the proposed constitutive model. The calculated results are in good agreement with the corresponding experimental data, and consequently, the accuracy of the present modeling and simulation has been confirmed.

The financial support provided by Japan Society for the Promotion of Science (a Grant-in-Aid for Science Research, Fundamental Research 2006–2007: number 18560123) is greatly acknowledged.

## References

1. M. Takiguchi and F. Yoshida, *JSME International Journal, Series A*, **46,** No.1, 68 (2003).
2. M. Takiguchi and F. Yoshida, *Journal of Materials Proccessing Technology*, **113**, 743 (2001).
3. M. Takiguchi, S. Izumi and F. Yoshida, Proc. Inst. Mech. Engrs, **218**, Part C, *J. Mechanical Engineering Science*, 623 (2002).
4. M. Takiguchi and F. Yoshida, *Journal of Materials Proccessing Technology*, **140**, 441 (2003).
5. M. Yoshida, Journal of the Japan Society for Technology of Plasticity, **26**, No.291, 394 (1985).
6. M. C. M. Liu and E. Krempl, *Journal of the Mechanics and Physics of Solids*, **27**, 377 (1979).
7. J. L. Chaboche and G. Rousselier, *Transaction of the ASME, Journal of Pressure Vessel Technology*, **105**, 153 (1983).
8. F. Yoshida, International Journal of Plasticity, **16**, 359 (2000).
9. The Japan Society for Technology of Plasticity ed., "Bending", CORONA PUBLISHING CO. LTD., (1995).

# ANALYSIS OF ELASTO-PLASTIC STRESS WAVES BY A TIME-DISCONTINUOUS VARIATIONAL INTEGRATOR OF HAMILTONIAN

SANG-SOON CHO[1], HOON HUH[2†]

*School of Mechanical, Aerospace & System Engineering, KAIST, 335, Gwahangno, Deadoek Science Town,*
*Daejeon, 305-701 KOREA,*
*ss211003@kaist.ac.kr, hhuh@kaist.ac.kr*

KWANG-CHUN PARK[3]

*Department of Aerospace Engineering Sciences and Center for Aerospace Structures, College of Engineering*
*and Applied Science, University of Colorado, Campus Box 429, Boulder, CO 80309, USA,*
*kcpark@colorado.edu*

Received 15 June 2008
Revised 23 June 2008

This paper proposes a numerical algorithm of a time-discontinuous variational integrator based on the Hamiltonian in order to obtain more accurate results in the analysis of elasto-plastic stress wave. The algorithm proposed adopts both a time-discontinuous variational integrator and space-continuous Hamiltonian so as to capture discontinuities of stress waves. The algorithm also adopts the limited kinetic energy to enhance the stability of the numerical algorithm so as to solve the discontinuities such as elastic unloading and internal reflection in plastic deformation. Finite element analysis of one dimensional elasto-plastic stress waves is carried out in order to demonstrate the accuracy of the algorithm proposed.

*Keywords*: Elasto-plastic stress waves; time-discontinuous variational integrator; Hamiltonian, limited kinetic energy.

## 1. Introduction

A study about the dynamic response of materials under impact or blast load is a crucial research topic in many engineering applications such as crashworthiness of vehicles[1], explosive devices used in aerospace and military industry.[1] When impact or blast load is imposed to a machine, stress waves are propagated through a medium and fracture or damage occurs at a structural weak point such as crack. Therefore, understanding of the stress wave propagation is very important for the reliable and safe design of machines.[2]

Stress waves propagate through a medium rapidly when dynamic forces are applied for very short period of time. The analytical prediction of the propagation of stress waves is very difficult in practical problems since the exact solution of the governing equation

---

†Corresponding Author.

becomes complicated to consider characteristic phenomena such as discontinuities. In spite of such difficulties, many researchers have progressed valuable studies by various methods in recent years.[3] One of remarkable efforts is the numerical simulation using the finite element method, which faces two typical difficulties currently during the numerical procedure. The first difficulty is to control the dispersive and dissipative errors induced at the discontinuous or the singular domain. And the second difficulty is to preserve the shape of stress waves with the small wave length.[3]

This paper is concerned with the analysis of elasto-plastic stress waves by a time-discontinuous variational integrator based on Hamiltonian in order to obtain more accurate computational solution by reducing the dispersive and dissipative errors. The algorithm proposed adopts both time-discontinuous variational integrator and space-continuous Hamiltonian so as to capture discontinuities of stress waves. This algorithm adopts the limited kinetic energy to enhance the stability of the numerical algorithm so as to solve the discontinuities such as elastic unloading and internal reflection in plastic deformation. Finite element simulation of one dimensional elasto-plastic stress waves is carried out in order to demonstrate the accuracy of the algorithm proposed.

## 2.   Formulation for Time-Discontinuous Variational Integrator of Hamiltonian

### 2.1.   *Hamilton's principle*

Hamilton presented in 1834 that the following action integral

$$S = \int_{t_0}^{t_N} L\big(q(t), \dot{q}(t), t\big) dt , \ \ L = (T - V) \tag{1}$$

whose functional variation, $\delta S = 0$, yields the equation of motion, where $q$ denotes the generalized coordinates. In the Eq. (1) $T$ is the kinetic energy, $V$ is the potential energy and $L$ is the Lagrangian. The Hamiltonian is expressed via Legendre's transformations from the Lagrangian as

$$H(q, p; t) = p\dot{q} - L(q, \dot{q}, t) \tag{2}$$

where $p$ is the generalized momentum which can be expressed in Eq. (3) as

$$p = \frac{\partial L}{\partial \dot{q}} \tag{3}$$

From which, one obtains the calculated Hamilton's equation as

$$\dot{q} = \frac{\partial H}{\partial p} \ , \ \ \dot{p} = -\frac{\partial H}{\partial q} \tag{4}$$

A fundamental property of the Hamiltonian, $H$, can be observed from Eq. (5).

$$\dot{H} = \frac{\partial H}{\partial p}\dot{p} + \frac{\partial H}{\partial q}\dot{q} = 0 \tag{5}$$

The equation indicates that the Hamiltonian, $H$, is constant for all t, which must hold for the long-time responses.[5]

The Hamilton's principle shown in Eq. (2) and (4) is an alternative formulation of the differential equations of motion for a physical system as an equivalent integral equation, using the calculus of variations. The principle is also called the principle of stationary action. Although formulated originally for classical mechanics, the Hamilton's principle also applies to classical fields such as the electromagnetic and gravitational fields and has even been extended to quantum mechanics, quantum field theory and criticality theories.

## 2.2. *Finite element formulation of the Hamiltonian*

In order to solve the stress wave propagation in one-dimensional space, the Hamiltonian is discretized with the assumption that the generalized coordinates, $q$, and generalized momentum, $p$, are continuous in the space domain. The displacement and velocity fields can be express as Eq. (6) and (7), respectively.

$$u(x) = \mathbf{Nq}^T = \begin{bmatrix} N_1 & N_2 \end{bmatrix}\begin{bmatrix} q_1 & q_2 \end{bmatrix}^T \tag{6}$$

$$v(x) = \mathbf{NV}^T = \begin{bmatrix} N_1 & N_2 \end{bmatrix}\begin{bmatrix} v_1 & v_2 \end{bmatrix}^T \tag{7}$$

Discretized kinetic and potential energy in the continuous domain, $\Omega$, can be defined as follows:

$$T = \int_\Omega \frac{1}{2}\mathbf{v}^T \rho v d\Omega \tag{8}$$

$$V = U - W = \int_\Omega \frac{1}{2}(\nabla\mathbf{u})^T \mathbf{C}(\nabla\mathbf{u})d\Omega - \left( \int_\Omega \mathbf{u}^T \mathbf{f} d\Omega + \int_\Gamma \mathbf{u}^T \mathbf{t} d\Gamma \right) \tag{9}$$

Substituting Eq. (8) and (9) into Eq. (2), the Hamiltonian can be discretized as:

$$H(\mathbf{q},\mathbf{p};t)=T+V=T+(U-W)$$
$$=\frac{1}{2}\mathbf{p}^T\mathbf{M}^{-1}\mathbf{p}+\frac{1}{2}\mathbf{q}^T\mathbf{K}_\Omega\mathbf{q}-\mathbf{q}^T\left( \int_\Omega \mathbf{N}^T\mathbf{f}d\Omega + \int_{\Gamma_h} \mathbf{N}^T\mathbf{t}d\Gamma \right) \tag{10}$$

where $\mathbf{p}$, $\mathbf{M}$ and $\mathbf{K}_\Omega$ denote the momentum, the mass and the stiffness matrix, respectively. Each matrix is defined as follows:

$$\mathbf{p} = \mathbf{MV}$$
$$\mathbf{M} = \int_\Omega \mathbf{N}^T \rho \mathbf{N} d\Omega \tag{11}$$
$$\mathbf{K}_\Omega = \int_\Omega \nabla\mathbf{N}^T\mathbf{C}\nabla\mathbf{N}d\Omega$$

## 2.3. *Time-Discontinuous Variational Integrator of Hamiltonian*

A variational integrator is applied in this paper based on the Hamiltonian shown in Eq. (10) in order to derive the algebraic finite element equation in the discretized time domain. It is assumed that the displacement and the velocity are discontinuous at the discretized time domain. At first, the action integral of the $k^{th}$ discretized time domain can be defined theoretically from the integration of the Lagrangian as shown in Eq. (12).[4]

$$S_k = \lim_{\varepsilon \to 0} \int_{t_{k-1}+\varepsilon}^{t_{k-1}+\Delta t} \left[ \mathbf{p}^T \dot{\mathbf{q}} - (\frac{1}{2}\mathbf{p}^T \mathbf{M}^{-1}\mathbf{p} + V(\mathbf{q})) \right] dt$$

$$+ \lim_{\varepsilon \to 0} \int_{t_{k-1}-\varepsilon}^{t_{k-1}+\varepsilon} \left[ \mathbf{p}^T \dot{\mathbf{q}} - (\frac{1}{2}\mathbf{p}^T \mathbf{M}^{-1}\mathbf{p} + V(\mathbf{q})) \right] dt \tag{12}$$

$$\approx \mathbf{p}^T \dot{\mathbf{q}}\Delta t - \frac{\Delta t}{2}\mathbf{p}^T \mathbf{M}^{-1}\mathbf{p} - \Delta t V(\mathbf{q}) + \lim_{\varepsilon \to 0} \mathbf{p}^T_{t_{k-1}-\varepsilon} \big\| \mathbf{q} \big\|_{t_{k-1}}$$

The action integral shown in Eq. (12) provides the first order accuracy. For the $k^{th}$ discretized time domain, the action integral can be approximated as

$$\hat{S}_{k+1} \equiv \frac{1}{2}\left( p_{k^+} + p_{k+1^-} \right)^T \left( q_{k+1^-} - q_{k^+} \right) - \Delta t [T + V - \Lambda] + p_{k^-}^T \left( q_{k^+} - q_{k^-} \right) \tag{13}$$

where $T$ and $V$ are the kinetic and potential energy which is expressed as Eq. (14) and Eq. (15), respectively.

$$T = \frac{1}{2}\left( p_{k^+} + p_{k+1^-} \right)^T M^{-1} \left( p_{k^+} + p_{k+1^-} \right) \tag{14}$$

$$V(q_{k+1}) = \frac{1}{2}V\left( \alpha q_{k^+} + (1-\alpha)q_{k+1^-} \right) + \frac{1}{2}V\left( (1-\alpha)q_{k^+} + \alpha q_{k+1^-} \right)$$

$$\alpha = \frac{3 \pm \sqrt{3}}{6} : \text{Gaussian integration point} \tag{15}$$

Internal force at $k^{th}$ discretized time domain can be derived from the potential energy shown in Eq. (15) as follows:

$$F_{q_{k^+}}^{int} = \frac{\partial V}{\partial q_{k^+}} = \frac{1}{3}K(q_{k^+})q_{k^+} + \frac{1}{6}K(q_{k+1^-})q_{k+1^-}$$

$$F_{q_{k+1^-}}^{int} = \frac{\partial V}{\partial q_{k+1^-}} = \frac{1}{6}K(q_{k^+})q_{k^+} + \frac{1}{3}K(q_{k+1^-})q_{k+1^-} \tag{16}$$

A limited kinetic energy, which is denoted as $\Lambda$, is newly introduced in this paper to the action integral as shown in Eq. (12) in order to suppress the abrupt oscillation of the stress waves at discontinuity. The limited kinetic energy is defined as

$$\Lambda \equiv function \left\{ \frac{\phi}{2}\left( p_{k^+} - p_{k+1^-} \right)^T M^{-1}\left( p_{k^+} - p_{k+1^-} \right) \right\}$$

$$\frac{\partial S}{\partial p_{k^+}} = \phi M^{-1}\left( p_{k^+} - p_{k+1^-} \right)$$

$$\frac{\partial S}{\partial p_{k+1^-}} = \phi M^{-1}\left( p_{k^+} - p_{k+1^-} \right) \tag{17}$$

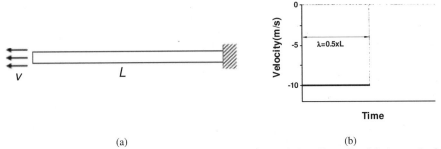

(a)                                                                 (b)

Fig. 1. One-dimensional stress wave propagation: (a) schematic description; (b) profile of the imposed velocity.

Finally, the algebraic finite element equation is obtained in the discretized time domain as expressed in Eq. (18) by taking the variation of the action integral shown in Eq. (13).[4]

$$\delta\left(\sum_{k=0}^{N}\hat{S}_k\right)=0$$

$$F_1^{int}+\frac{1}{2}MV_1+\frac{1}{2}MV_2=MV_0+F_1^{ext}$$

$$F_2^{int}-\frac{1}{2}MV_1+\frac{1}{2}MV_2=F_2^{ext} \tag{18}$$

$$U_1=U_0$$

$$U_2=U_1+\frac{\Delta t}{2}(1-4\phi)V_1+\frac{\Delta t}{2}(1+4\phi)V_2\ ,0\le\phi<\frac{1}{4}$$

## 3. Numerical Analysis of Elasto-Plastic Stress Wave Propagation

A time-discontinuous variational integrator of Hamiltonian is applied to the numerical simulation of the elasto-plastic stress wave propagation. Wave propagation in one-dimensional bar is considered as shown in Fig. 1. The length of the bar is 100 mm and is discretized with 600 elements. The Young's modulus and the density of a bar are assigned as 200 GPa and 8000 kg/$m^3$, respectively. A linear hardening model is used to describe the flow stress of the bar as $\bar{\sigma}=0.2+20\varepsilon^p$[GPa]. The right end side of the bar is entirely fixed and a velocity boundary condition depicted in Fig. 1(b) is imposed on the left end side. A constant velocity of 10 m/s is imposed on the left end side and the velocity boundary condition is suddenly eliminated at the time of 5μsec to describe the loading–unloading condition.[6]

In order to evaluate the efficiency of the method proposed, the results obtained from the proposed time-discontinuous variational integrator of Hamiltonian are compared with those obtained from the conventional finite element method of the continuous Galerkin method with the same finite element discretization. Fig.2 shows the stress waves configurations at two different times: $t_1$=5 and $t_2$=20μs. After an elastic wave followed by a plastic wave is generated, an unloading elastic wave is propagated if the loading is eliminated abruptly. Because elastic wave speed is larger than the plastic wave speed, the unloading elastic wave catches the loading plastic wave. Then two new elastic waves are

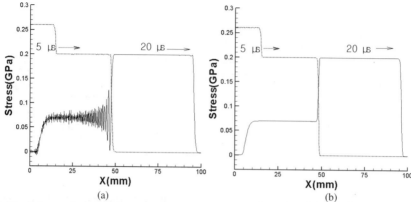

Fig. 2. Simulation result for one-dimensional stress wave propagation: (a) conventional finite element method; (b) time-discontinuous variational integrator of the Hamiltonian.

generated and propagated to the both sides. This is called by the internal reflection.[6] Fig. 2(a) shows the stress profiles obtained from the conventional finite element method. Stress waves are severely oscillating due to the dispersive and dissipative errors induced at the discontinuous region such as the elastic unloading and the internal reflection. The reason is that the conventional finite element method cannot eliminate those two errors arise from the local truncation error of approximation. The oscillation is remarkably reduced as shown in Fig. 2(b) when the time-discontinuous variational integrator of the Hamiltonian is utilized in the numerical simulation. The comparison indicates that the time-discontinuous variational integrator of the Hamiltonian is appropriate in the numerical simulation of the stress wave propagation.

## 4. Conclusion

A time-discontinuous variational integrator of the Hamiltonian is newly proposed in order to obtain more accurate results in the analysis of stress wave propagation. The proposed algorithm adopts both a time-discontinuous variational integrator and a space-continuous Hamiltonian so as to describe discontinuities of stress waves in the numerical simulation with reduced dispersive and dissipative errors. Elasto-plastic numerical simulations for one-dimensional wave propagation are carried out in order to evaluate the efficiency of the proposed algorithm. The results indicate that dispersive and dissipative errors, which appear in the continuous Galerkin method, are remarkably reduced in the proposed algorithm using a time-discontinuous variational integrator of the Hamiltonian.

## References

1. W. J. Kang and H. Huh, *Int. J. Automotive Technology* **1**, 1 (2000) 35.
2. M. A. Meyers, *Dynamic Behavior of Materials* (John Wiley & Sons, Inc., New York, 1994).
3. I. Harari, *Wave Motion* **39** (2004) 279.
4. B. Cockburn, *Z. Angew. Math. Mech.* **83** (2003) 731.
5. J. M. Wendlandt, J. E. Marsden, *Physica D*, **106** (1997) 223.
6. H. Kolsky, *Stress Waves in Solids* (Dover Publications, Inc., New York, 1963).

# VISCO-ELASTIC AND PLASTIC SINGULAR BEHAVIORS OF AN INTERFACE CORNER IN ELECTRONIC PACKAGES

JIN-QUAN XU[1†], HUIMIAO ZHENG[2], HONGLAI ZHU[3]

*Department of engineering Mechanics, Shanghai Jiaotong University*
*800 Dongchuan Road, Minhang, Shanghai, 200240, P.R. CHINA,*
*jqxu@sjtu.edu.cn*

Received 15 June 2008
Revised 23 June 2008

Failure of an electronic package generally initiates from the interface corners, due to the stress and strain singular behaviors near the corners. Since the solder materials are viscous elastic and even maybe viscous plastic, the singular behavior near the solder joint corner is thereby time-dependent, and is quite different from that of an elastic one. It is necessary to clarify such a singular behavior for the strength and life evaluation of an electronic package. This paper firstly has deduced the approximated viscous elastic singular fields theoretically based on the principle of correspondence and the Laplace transformation techniques, and secondly has investigated the viscous plastic singular behavior of an elastic-plastic interface corner by the analogue method; finally the numerical analysis of an electronic package has been carried out to show the details of the singular behaviors.

*Keywords*: Visco-Elasticity; Visco-Plasticity; Interface Corner; Singularity; Electronic Package.

## 1. Introduction

Failure of an electronic package generally initiates from the interface corners of the solder joints. To evaluate the strength and life, the singular behavior near such an interface corner must be analyzed in detail. It is well known that the stress and strain behave singularly near the interface edge of elastic bimaterials[1,2]. Though the solder material usually is a visco-elastic or even visco-plastic one, it is well known that the strain and stress near the interface edge of a solder joint also behave singularly[3-5]. However, unlike that in the elastic and elasto-plastic cases, the singular field in a solder joint is essentially time-dependent. For linear viscous bimaterials, Han *et.al*[6] analyzed the stress field near an interface crack tip by the approximated Laplace transformation techniques, Tang and Xu[7] deduced the stress and displacement field of an interface crack by applying the principle of correspondence directly. For the power law creep materials, Linkens[5] analyzed the asymptotic field near an interface crack tip by the analogue with that of the elasto-plastic interface crack, Kitamura[4] investigated the singular stress distribution near interface edges by numerical calculations. However, for a viscous elastic or plastic interface edge with arbitrary geometry, as it often appears in an electronic

---

[†]Corresponding Author.

package, there is no study reported yet. This paper gives the singular field near the visco-elastic interface edge by applying the principle of correspondence and the Laplace transformation techniques, and investigates the visco-plastic singular behavior by the analogue with that of an elastic-plastic interface edge. Numerical analysis of an electronic package is also carried out to show the details of the singular behavior.

## 2. Theoretical Analysis of Interface Edges in Viscous Bimaterials

### 2.1. *Analysis model*

The arbitrary interface edge model is shown in Fig.1. The most popular interface edges in an electronic package are the cases of $\theta_1 = \theta_2 = 90°$ and $\theta_1 = 90°, \theta_2 = 180°$. The boundary and interfacial continuous conditions can be expressed by Eq.(1).

$$\sigma_{\theta 1} = \tau_{r\theta 1} = 0 \quad \text{at } \theta = \theta_1, \quad \sigma_{\theta 2} = \tau_{r\theta 2} = 0 \quad \text{at } \theta = -\theta_2$$

$$u_{r1} = u_{r2}, v_{\theta 1} = v_{\theta 2}, \sigma_{\theta 1} = \sigma_{\theta 2}, \tau_{r\theta 1} = \tau_{r\theta 2} \quad \text{at } \theta = 0 \tag{1}$$

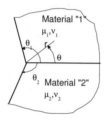

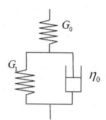

Fig. 1. Analysis model.   Fig. 2. Viscous model.

### 2.2. *The case of linear viscous elastic bimaterials*

The constitutive relationship of linear viscous materials can be expressed in both integral and differential forms. Especially, the differential form can be expressed as

$$P_1(D)s_{ij} = Q_1(D)e_{ij}, \quad P_2(D)\sigma_{kk} = Q_2(D)e_{kk} \tag{2}$$

where $D$ represents the time differential operator $\partial / \partial t$, $P_1, Q_1, P_2, Q_2$ are polynomial functions of $D$, $s_{ij} = \sigma_{ij} - \delta_{ij}\sigma_m, e_{ij} = \varepsilon_{ij} - \delta_{ij}\varepsilon_m$ are deviator stress and strain, and $\sigma_m = \sigma_{ii}/3$, $\varepsilon_m = \varepsilon_{ii}/3$ denote the volume stress and strain. Take the Laplace transformation, one gets

$$P_1(p)\hat{s}_{ij} = Q_1(p)\hat{e}_{ij}(p), \quad P_2(p)\hat{\sigma}_{kk} = Q_2(p)\hat{e}_{kk}(p) \tag{3}$$

here $\hat{f}(p)$ represents $f(t)$ after Laplace transform. The equivalent visco-elastic shear modulus $\tilde{\mu}$ and volume modulus $\tilde{K}$ in the Laplace fields [6] then can be expressed as:

$$2\tilde{\mu} = Q_1(p)/P_1(p), \quad 3\tilde{K} = Q_2(p)/P_2(p) \tag{4}$$

By taking the inversed Laplace transformation of Eq.(4), one can get the time dependent material properties such as $\mu(t), E(t)$, etc.. For example, if the viscous material can be expressed by the standard linear model as shown in Fig.2, the constitutive relationships is:

$$\left(\frac{\eta_0}{G_0 + G_1}\frac{\partial}{\partial t} + 1\right)s_{ij} = \left(\frac{2G_0 G_1}{G_0 + G_1} + \frac{2\eta_0 G_1}{G_0 + G_1}\frac{\partial}{\partial t}\right)e_{ij}, \quad \varepsilon_m = \frac{\sigma_m}{K} \tag{5}$$

where $G_0$ and $G_1$ are the corresponding spring coefficients and $\eta_0$ is the damp coefficient. According to Eqs.(2-4), The time dependent properties can be deduced as:

$$\mu(t) = G_0\left\{1 - G_0[1 - e^{-t/\tau_g}]/(G_0 + G_1)\right\}, \; E(t) = E_0 - (E_0 - E_\infty)[1 - e^{-t/\tau_E}]$$

$$v(t) = v_0 - (v_0 - v_\infty)[1 - e^{-t/\tau_v}] \;, \; \kappa(t) = 3 - 4v(t) \tag{6}$$

Here, $E_i = \dfrac{9K\mu_i}{3K + \mu_i}$, $v_i = \dfrac{3K - 2\mu_i}{2(3K + \mu_i)}, i = 0, \infty$ , $\tau_g = \dfrac{\eta_0}{G_0 + G_1}$ , $\tau_E = \dfrac{E_\infty \, \eta_0}{E_0 \, G_1}$ , $\tau_v = \dfrac{3K(G_0 + G_1) + G_0 G_1}{3K\eta_0 + G_0\eta_0}$

Dundurs' parameters of visco-elastic bimaterials then can be expressed as

$$\alpha(t) = \frac{\mu_1(t)(\kappa_2(t) + 1) - \mu_2(t)(\kappa_1(t) + 1)}{\mu_1(t)(\kappa_2(t) + 1) + \mu_2(t)(\kappa_1(t) + 1)} \;, \beta(t) = \frac{\mu_1(t)(\kappa_2(t) - 1) - \mu_2(t)(\kappa_1(t) - 1)}{\mu_1(t)(\kappa_2(t) + 1) + \mu_2(t)(\kappa_1(t) + 1)} \tag{7}$$

where suffix 1,2 denotes the property of material 1,2, respectively. The stress and displacement fields near the interface edge of elastic bimaterials as shown in Fig.1 can be expressed as follow [2]

$$\sigma_{ij} = \mathbf{K}f_{ij}(\theta)/r^{1-\lambda}, \; u_i = \mathbf{K}r^\lambda F_i(\theta) \tag{8}$$

where $\mathbf{K}$ is the complex stress intensity coefficients of the singular field, $f_{ij}(\theta), F(\theta)$ are complex angular functions only depending on the angle as shown in Fig.1, $\lambda$ is the eigen-value $(1 - \lambda$ is the stress singular order) determined by

$$A\beta^2 + 2B\alpha\beta + C\alpha^2 + 2D\beta + 2E\alpha + F = 0 \tag{9}$$

where

$$K(\theta) = \sin^2(\lambda\theta) - \lambda^2 \sin^2\theta, A = 4K(\theta_1)K(\theta_2), \; B = 2\lambda^2 \sin^2\theta_1 K(\theta_2) + 2\lambda^2 \sin^2\theta_2 K(\theta_1)$$

$$C = 4\lambda^2(\lambda^2 - 1)\sin^2\theta_1 \sin^2\theta_2 + K(\theta_1 - \theta_2), \; D = -2\lambda^2[\sin^2\theta_1 \sin^2(\lambda\theta_2) - \sin^2\theta_2 \sin^2(\lambda\theta_1)]$$

$$E = K(\theta_1) - K(\theta_2) - D, \; F = K(\theta_1 + \theta_2)$$

In the Laplace field, the solutions for elastic and visco-elastic problems have the same form. Let $F(s)$ denote the solution in Laplace field, $f(t)$ the solution in the real time space, $\lambda(s), \alpha(s), \beta(s)$ the eigenvalue and Dundurs' parameters in Laplace fields, then the solution in the Laplace field can be expressed with respect of material properties as $F(s) = F(\lambda(s), \alpha(s), \beta(s))$ . The solution $f(t)$ in real time space can be derived from:

$$f(t) = L^{-1}[F(\lambda(s), \alpha(s), \beta(s))] \tag{10}$$

where $L^{-1}$ is the inversed Laplace transformation operator. It is very difficult to carry out the above inversed transformation exactly, but is relatively easy to obtain $\lambda(t), \alpha(t), \beta(t)$ from $\lambda(s), \alpha(s), \beta(s)$ as shown in Eq.(7)(9). If we neglect the phase effect due to the material property change, then the above inversed transformation can be approximated by

$$f(t) \approx f(\lambda(t), \alpha(t), \beta(t)) \tag{11}$$

This fact means we can just replace the Dundurs' parameters and singular order in the elastic solution by the time-dependent ones to obtain the visco-elastic solutions.

## 2.3. The case of power law creep materials

Power law creep constitutive relationships can be expressed as

$$\dot{\varepsilon} = A\sigma^n e^{-Q/RT} \quad \text{or} \quad \dot{\varepsilon} = A(\sinh(B\sigma))^n e^{-Q/RT} \tag{12}$$

Here $A, B, n$ and $Q$ are material constants, $R$ is the atmosphere constant, $T$ is the absolute temperature. Comparing to the well-known Ramberg-Osgood static plastic constitutive relationship of $\varepsilon = \alpha(\sigma/\sigma_Y)^n$, it can been seen that Eq.(12) gives the strain rate while the static plastic one gives strain. By the analogue analysis of elastic-plastic interface edges [8,9], i.e., replacing the strain by the strain rate, one can get

$$\sigma_{ij} = Ar^{\lambda(t)-1}\tilde{\sigma}_{ij}(\lambda(t),\theta), \quad \dot{\varepsilon}_{ij} = Br^{n(\lambda(t)-1)}\dot{\tilde{\varepsilon}}_{ij}(\lambda(t),\theta). \tag{13}$$

It shall be noted that the eigen-value $\lambda(t)$ is time-dependent and also temperature-dependent, and of course it is geometry dependent. The angular functions $\tilde{\sigma},\tilde{\varepsilon}$ contain the eigen-value, thereby are also time and temperature dependent. Moreover, the singular order of strain is not the same as that of the strain rate since it has to be integrated. By assuming the function $\phi = A(t)r^{\lambda(t)+1}F(\theta)$, $0 < \lambda(t) < 1$ to determine the stresses, theoretically, $\lambda(t)$ can be determined by the dominating equation extended from the strain compatibility as follows

$$\left[\frac{d^2}{d\theta^2} - n(\lambda-1)(n(\lambda-1)+2)\right]\left[\tilde{\sigma}_e^{n-1}[F'' - (\lambda-1)(\lambda+1)F]\right] + 4(1+n(\lambda-1))\frac{d}{d\theta}\left[\tilde{\sigma}_e^{n-1}\lambda F'\right] = 0 \tag{14}$$

where $\tilde{\sigma}_e$ is the angular function of the equivalent stress. Eq.(14) can be only solved by numerical method [8] with the use of Eq.(1), in which the interfacial displacement continuous condition brings the time-dependent behavior to $\lambda(t)$

## 3. Numerical Analysis

To examine the singular behavior of linear viscous interface edge, the model shown in Fig.3 is analyzed with the use of ABAQUS6.5. The bonded materials are considered as epoxy and glass. The material constants of Epoxy as a standard linear visco-elastic model are $G_0 = 1.308GPa$, $G_1 = 0.1453GPa$, $\eta_0 = 1.453GPa \cdot s$. The glass is regarded as elastic with $E_2 = 85GPa$, $v_2 = 0.2$. The analysis is carried out under plane strain condition. The double logarithmic stress distributions at various times are shown in Fig.4, in which the slope of the linear part corresponds to the singular order. It can be seen from Fig.5 that the singular order is time-dependent, and the numerical results agree well with the approximated theoretical results determined by Eq.(9).

To show the singular behavior near the interface corner in an electronic package, the package model shown in Fig.6 is also analyzed. Material constants used in the analysis are shown in Table 1. The thermal conductivity analysis is carried out at first to obtain the temperature distributions. The chip power is 35.7W, the heat transfer coefficient is 40.0 W/(m²K) at the upper surface, and 3500.0 W/(m²K) at the bottom surface. The overall temperature distribution becomes stable after 3 minutes when the chip is switched on, but the local distribution near the edges saturated more rapidly. The constitutive relationships used in the analysis for material 2 and 4 are shown in Table 2. Fig.7 shows the instant stress and strain distribution near the interface edges. It can be seen that there are very sever concentrations at the interface corners, and the interface edges between material 1 and 2, 4 and 5 are the dangerous points. Fig.8 shows the logarithmic stress and strain distributions along the interface between materials 4 and 5. In Fig.8, the normalizing parameter $p$ is the equivalent stress at the middle point of the interface. It can be seen that the distributions are almost linear, and the slope of the distribution represents the singular order. The singular orders obtained numerically are shown in Fig.9. One can found that the stress singularity at the interface edge is very strong at the beginning of switching on, since it can be considered as a thermal shock loading, but it decreases with the time and saturates to a stable value after 90 seconds.

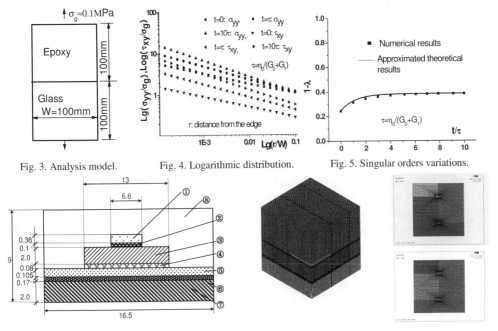

Fig. 3. Analysis model.    Fig. 4. Logarithmic distribution.    Fig. 5. Singular orders variations.

Fig. 6. The power module chip model.

Table 1. Material constants.

| No. | $E$ / GPa | $v$ | $\alpha \times 10^{-6}$ /K | $k$ / W/mK | $\rho$ / kg/m$^3$ | $c$ / J/(kgK) |
|---|---|---|---|---|---|---|
| 1 | 187.0 | 0.25 | 5.05 | 150.0 | 2330 | 678.262 |
| 2 | 16.1 | 0.44 | 29.4 | 35.2 | 11160 | 129.791 |
| 3 | 110.0 | 0.35 | 16.5 | 398.0 | 8960 | 385.186 |
| 4 | 41.6 | 0.36 | 21.7 | 64.2 | 7400 | 234.461 |
| 5 | 110.0 | 0.35 | 16.5 | 398.0 | 8960 | 385.186 |
| 6 | 5.4 | 0.34 | 67.0 | 4.0 | 3100 | 962.964 |
| 7 | 73.0 | 0.33 | 23.6 | 237.0 | 2700 | 900.162 |
| 8 | 0.1 | 0.36 | 50.0 | 0.17 | 890 | 1507.25 |

Table 2. Constants of solder materials (TR=273K).

| Constitutive model | No. | Constants |
|---|---|---|
| If $\sigma > \sigma_V$ <br> $\dot{\varepsilon} = A(\sinh B\sigma)^n e^{-Q/RT}$ | 2 | Q/R=6535K, n=4.2,A=3.16 $s^{-1}$,B=0.18 $MPa^{-1}$ <br> A$_0$ =8.16763-0.05242(T-TR) $MPa^{-1}s^{-1}$ <br> $\sigma_V$=17.34-0.1219 (T-T$_R$)+0.00024565(T-T$_R$)$^2$ |
| If $\sigma < \sigma_V$ <br> $\dot{\varepsilon} = A_0 \sigma e^{-Q/RT}$ | 4 | Q/R=12993 K, n=5.85, B=0.145 $MPa^{-1}$, $A = 2.08 \times 10^6 s^{-1}$ <br> $A_0 = 2.039 \times 10^{-4} e^{8484/T}$ $MPa^{-1}s^{-1}$ <br> $\sigma_V$=17.34-0.1219 (T-T$_R$)+0.00024565(T-T$_R$)$^2$ |

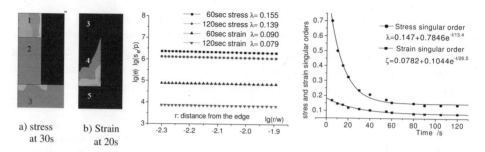

a) stress
at 30s  b) Strain
at 20s

Fig. 7. Equivalent stress/strain.   Fig. 8. Stress and strain on interface 4/5.        Fig. 9. Singular order on interface 4/5.

## 4.   Conclusions

Linear viscous elastic and visco-plastic interface edges have been analyzed theoretically and numerically. The main results can be concluded as:

1) For linear viscous elastic bimaterials, the singular field near the interface edge can be obtained just by replacing the material constants in the elastic solution with the time-dependent ones, which should be obtained by the inversed Laplace transformation.

2) For visco-plastic interface edges, there are also singular field which can be analyzed by the analogue of the power law plastic interface edges. The singular order of strain is different with that of stress and strain rate.

3) Both the singular order and the singular stress distribution near a viscous elastic or plastic interface edge are time dependent, and the temperature change has also strong effects on the distributions.

4) Numerical results of an electronic package show that there are sever stress and strain singularities near the interface edge, and the singular fields will saturate rapidly.

## Acknowledgments

This study is supported by the national natural science foundation of China under Grant No. 10772116.

## References

1.   DB Bogy, *Journal of Applied Mechanics*, **38**, 377(1971).
2.   JQ Xu, *Interface Mechanics* (Science Press, Beijing, 2006).
3.   AQ Xu, HF Nied, *ASME J. Electron. Package*, **122**, 301(2000).
4.   T Kitamura T, K Ngampungpis, H Hirakata, *Engineering Fracture Mechanics*, **74**, 1637(2007).
5.   D Linkens, EP Busso, DW Dean, *Nuclear Engineering and Design*, **158**, 377(1995).
6.   X Han, F Ellyin, Z Xia, *International Journal of Solids and Structure*, **38**, 7981(2001).
7.   LH Tang, JQ Xu, *Chinese Quarterly Mechanics*, **28-1**, 116(2007).
8.   JQ Xu, LD Fu, Y Mutoh, *JSME International Journal*, **45**, 177(2002).
9.   CR Champion, CA Atkinson, *Proc. R. Soc. Lond.*, **A432**, 547(1991).